Tissue Engineering

A volume in the

ACADEMIC PRESS SERIES IN BIOMEDICAL ENGINEERING

JOSEPH BRONZINO, Ph.D., SERIES EDITOR
The Vernon Roosa Professor of Applied Science, Trinity College – Hartford, Connecticut
President of the Biomedical Engineering Alliance and Consortium

Titles in the series

Modelling Methodology for Physiology and Medicine, Carson & Cobelli, ISBN 9780121602451, 2000
Handbook of Medical Imaging, Bankman, ISBN 9780120777907, 2000
Clinical Engineering Handbook, Dyro, ISBN 9780122265709, 2004
Diagnostic Ultrasound Imaging: Inside Out, Szabo, ISBN 9780126801453, 2004
Introduction to Biomedical Engineering, Second Edition, Enderle, Blanchard & Bronzino,
ISBN 9780122386626, 2005
Circuits, Signals, and Systems for Bioengineers, Semmlow, ISBN 9780120884933, 2005
Numerical Methods in Biomedical Engineering, Dunn, Constantinides & Moghe, ISBN 9780121860318,
2005
Hospital Preparation for Bioterror, McIsaac, ISBN 9780120884407, 2006
Introduction to Applied Statistical Signal Analysis, Third Edition, Shiavi, ISBN 9780120885817, 2006
Biomedical Information Technology, Feng, ISBN 9780123735836, 2007
Introduction to Modeling in Physiology and Medicine, Cobelli & Carson, ISBN 9780121602406, 2007
Tissue Engineering, van Blitterswijk et al, ISBN 9780123708694, 2008

of logistics, might provide an 'off the shelf' product. The most frequently used method for facilitating the postponing of implantation is freezing, as discussed in **Cryobiology** (Chapter 13).

Although some applications would involve direct injection of cells into the tissues that one intends to regenerate, most researchers active in the field of tissue engineering, and working with cultured cells, strive for implanting combinations of cells and scaffolds. These so-called hybrid constructs will usually have three dimensions and need to give access to both cells and nutrients or have to allow an out flux of active ingredients and waste products. The chapter **Scaffold Design and Fabrication** (Chapter 14) presents the reader with the different ways to manufacture scaffolds and explains the criteria these scaffolds have to fulfil. The design of scaffolds may also involve incorporating biologically active ingredients that can then be made available to either the cells in the construct or be released to initiate effects in the tissues of the organism into which the device was implanted. The various aspects of such release systems are presented in the chapter on **Controlled release strategies** (Chapter 15). Ideally the researcher working on a hybrid construct and following the chapters, as they have been presented so far, now faces one of the main obstacles of tissue engineering: how to initiate cell growth and extracellular matrix formation in three-dimensional constructs of clinically relevant dimensions (usually centimeters versus millimeters in many *in vitro* and animal studies). These challenges, fundamentals and some solutions are discussed in **Bioreactors for Tissue Engineering** (Chapter 16).

Although not complete or covering all of the subjects, for which one could argue would be essential, the editors feel that these first 16 chapters (as well as this chapter), provide a solid basis for students who wish to acquire an understanding for tissue engineering. Much of the international research effort based on tissue engineering is along the lines presented in these chapters. But it is also fair to say that some of the major challenges are still to be found into making tissue engineering a widespread clinical reality. These clinical aspects are treated in the remaining six chapters.

In making a choice on which clinical applications to discuss, the editors faced a difficult task. Researchers worldwide are currently focusing their attention on the repair or regeneration of many tissues and organs. Almost certainly most, if not all, of them deserve to be presented in this book, but this would lead to an unacceptably thick volume or a lack of sufficient depth per chapter. In selecting the subjects we felt that those tissues that historically moved first into the tissue engineering arena, and have actually found relatively widespread application, certainly deserve to be dealt with here. We feel, therefore, that most would agree that **Tissue Engineering of Skin** (Chapter 17) and the **Tissue Engineering of Cartilage** (Chapter 18) chapters warrant addition to this book on the basis of such historical observations. Another way of selecting which chapters to include involves the shear size of the tissue research field. After all, some of the students using this volume will actually find employment in tissue engineering and chances are that most will end up in the biggest segments. Over the years, it has become clear that both musculoskeletal- and cardiovascular diseases attract many tissue engineers who work towards overcoming the negative affects of these diseases in our society. Characteristically, they are always well-represented at tissue engineering conferences and this explains our choice for **Tissue Engineering of Bone** (Chapter 19). Relevance could be another selection criterion. Neural diseases would be an example of this approach. Not only is this a relevant research area, but the prospect that one day tissue engineering might contribute to relieving the problems of those suffering spinal cord lesions is truly inspiring – hence a chapter on **Tissue Engineering of the Nervous System** (Chapter 20 and Figure I.3). Any scientist, of some reputation, working in the field of tissue engineering will be frequently confronted with a question like: *When will you grow a complete heart or liver?* Many of us may feel that these are not the biggest priority; after all it would be much better to regenerate the damaged part of an organ rather than substituting it fully. Nevertheless, substituting organs is a challenge, if only since it forces us to combine

into the different aspects of this cell and will show that the stem cell is not a single cell type but, in reality, encompasses different categories of cells ranging from the multipotent embryonic stem cell to the apparently less potent adult stem cell. With this knowledge, the reader will subsequently be guided to **Morphogenesis** (Chapter 2) where the formation of tissues in the embryo is discussed. Although most tissue engineers do not venture into the realms of developmental biology and research with a strong focus on developmental biology is all too scarce in tissue engineering, we feel this is an omission and hope that this chapter will urge young scientists to enter into such an essential area. Obviously most tissue engineering constructs will not be implanted into embryos but into human beings after birth. As tissue formation is essentially different in a post-natal tissue when compared to that in the embryo the authors of **Tissue Homeostasis** (Chapter 3) provide insight into how tissues are maintained after birth. Tissues in both the embryo and in an individual after birth usually, although not always, contain multiple cell types. Understanding how these cells interact is recognized as pivotal for the success of tissue engineering. **Cellular Signaling** (Chapter 4) provides such insights. Furthermore, in spite of all media attention that befalls stem cells, in reality cells represent only a small part of the dry weight of living tissue. All, or at least most, cells interact with an extracellular matrix, which, in contrast to the errant opinion of some engineers and even biologists, presents much more than mechanical support and adds substantially to the biological interactions in our body. As tissue engineering typically combines scaffolds with biologically active components as cells or growth factors we felt that a chapter on the biological equivalent: **Extracellular Matrix** (Chapter 5) could not be missed.

At this point in the book a shift is made to the more fundamental engineering aspects. Since not all scaffolds and matrices are completely synthetic, a chapter on **Natural polymers** (Chapter 6) seemed in place as well. Most tissue engineers, actively involved in the design of both synthetic and natural

scaffolds or matrices, prefer to have these degrade after implantation. This is with good cause as the prolonged presence of foreign material in the body may induce a variety of unwanted effects such as implant-associated infection or mutagenesis. Two groups of authors discuss the aspects of implant degradation and this is done in two related chapters: **Degradable polymers** (Chapter 7) and **Degradation of Bioceramics** (Chapter 8). These chapters provide deeper insight into the interactions between materials and a living system; particularly in relation to their function and the release of degradation products. These properties are some of the crucial aspects of influence for the biological performance of a material that interacts with a living system and are discussed with other relevant aspects in the chapter on **Biocompatibility** (Chapter 9), a true interface between biology and engineering.

After considering the above, more fundamental, aspects, the text now gradually moves to knowledge that bears direct practical relevance for the actual process of tissue engineering. In addition, chapters start to increasingly show the multidisciplinary nature of the field. As already explained in the chapter on Stem Cells (Chapter 1), there are different cells that can be used for generating tissues. The chapter on **Cell Sources** (Chapter 10) elaborates further on this subject and also touches on the issue of using autologous versus donor tissue. Having selected a cell source, the researcher now has to obtain these cells in appropriate numbers and sufficient purity and if necessary trigger the cells into the right differentiation, **Cell Culture: Harvest, Selection, Expansion and Differentiation** (Chapter 11) deals with this subject. In order to achieve these goals one will always have to bring sufficient nutrients to the cells. **Cell Nutrition** (Chapter 12) focuses on the various aspects that are pivotal for the understanding of this subject. Theoretically, after having obtained sufficient cells of the right type and state of differentiation, one now has a choice: immediately continue the process towards implantation or first postpone. Choosing to postpone may have several reasons: it would allow a surgeon to set a different date of surgery or, in view

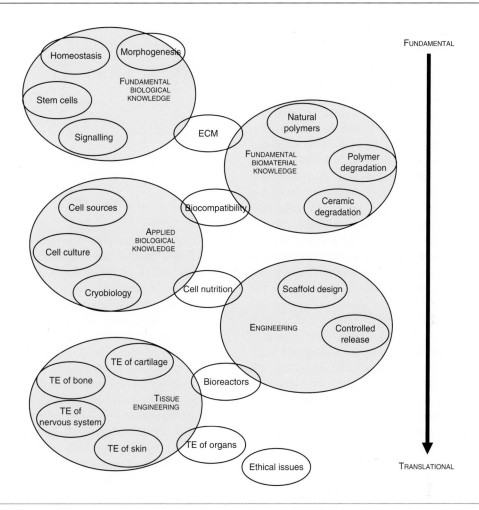

Figure I.2 Diagram sketching the layout of the book (groups biology, technology and evaluation form basic to clinical practice.

both teacher and student will have to make the appropriate selection of what is offered to them. In all instances we have tried to make the text accessible as educational material as opposed to standard scientific publications.

At this stage we were faced with the issue of which subjects to select for publication in this particular book. In essence, in a very basic approach, tissue engineering can be divided into both biological and engineering parts. In principle, most chapters can be placed into one of these categories. However, we

decided to group the chapters in such a way that these related topics are presented while being alternated by clusters from the other disciplines. This avoided one part of the book being completely biological while another part focused only on engineering aspects. In general, the structure of the book moves from fundamental to translational or even clinical (Figure I.2). In this way, the reader will first be confronted with the fundamentals of the cell type that makes tissue engineering truly a part of regenerative medicine: the stem cell. **Stem Cells** (Chapter 1) gives insight

major reasons for this phenomenon. First, bringing a relatively simple medical device from initial idea to a widespread clinical reality frequently takes a minimum of 10 years. As the underlying technology for developing tissue engineering products is less mature, and possibly more complex, it is to be expected that clinical progress in this field will be measured in decades rather than years. Second, tissue engineering is truly a multidisciplinary field where acquired knowledge from individual classical disciplines (e.g. quantum physics, polymer chemistry, molecular biology, anatomy) no longer suffices to make substantial leaps. Individuals active in this field will have to acquire multidisciplinary skills and be willing to look over the borders of their home discipline. The relatively young age of the field does not make it easy to acquire those skills as dedicated textbooks are still scarce and frequently do not address the appropriate audience by either offering a collection of research papers or by only dealing with a selected part of the entire discipline. Without the widespread availability of such textbooks it is to be feared that the rapidly increasing number of graduate courses on tissue engineering may not be as effective as required, which may hamper the development of the field of tissue engineering into a mature scientific discipline.

With this in mind, in 2004 it was decided that it was time to bring together a representative group of internationally active scientists who would be willing to submit chapters for this book, which would address both the multidisciplinary nature of tissue engineering and the underlying base disciplines. During the process, we dedicated ourselves to offering chapters that are not so much complete overviews of the individual sub-disciplines, but rather, per chapter, would offer a small general introduction followed by more in-depth text on issues particularly relevant to tissue engineering. For instance, instead of having a general chapter on cell biology (which could automatically become superficial and partially redundant), we chose to present individual aspects of cell biology in numerous chapters, like those on Stem Cells, Morphogenesis and Cellular Signaling. Furthermore, in a chapter dealing with, for example, Cellular Signaling, we did

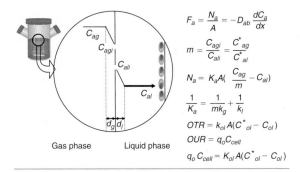

$$F_a = \frac{N_a}{A} = -D_{ab}\frac{dC_a}{dx}$$

$$m = \frac{C_{agi}}{C_{ali}} = \frac{C^{*}_{ag}}{C^{*}_{al}}$$

$$N_a = K_a A\left(\frac{C_{ag}}{m} - C_{al}\right)$$

$$\frac{1}{K_a} = \frac{1}{mk_g} + \frac{1}{k_l}$$

$$OTR = k_{ol} A(C^{*}_{ol} - C_{ol})$$

$$OUR = q_o C_{cell}$$

$$q_o C_{cell} = K_{ol} A(C^{*}_{ol} - C_{ol})$$

Figure I.1 A glimpse at mass transport theory.

not aim for a complete overview of all intracellular pathways. Instead, we concentrated on only a few aspects which are particularly relevant to the field of tissue engineering, and which would simultaneously provide both a general understanding for the working mechanism of most pathways, as well as offer sufficient depth for the few pathways which are discussed in some detail. Ideally, such descriptions would provide fundamental knowledge that can be used for the appropriate understanding of other chapters; like describing BMP signaling for later use in the chapter on Tissue Engineering of Bone.

This still leaves the challenge that the book will be used by students with different backgrounds and in different stages of their training. To accommodate this, we have selected a format where use is made of separate text boxes. Typically, these will address 'classical experiments', 'state of the art experiments' or other dedicated topics which a student or teacher may select at will to provide deeper insight. Sometimes, chapters will offer information that is, for instance, clearly essential for a group of biomedical engineers but might not be fully necessary for the average medical student. A clear example of that is the paragraph on mass transport in the Cell Nutrition chapter (Figure I.1). Understanding all these formulae clearly contributes to understanding the working mechanism of a bioreactor and may even be self evident for a process engineer; whereas such information may be superfluous for a medical student or molecular biologist venturing into the field of tissue engineering. For optimal use of the book,

immersed in a 1% weight/volume solution of poly(lactic acid) (PLA) in methylene chloride for 2 seconds and subsequently shaped into the plaster mold.

- **Biology**: calf chondrocytes (cartilage cells) were harvested from the articular surfaces of a calf, isolated by collagenase digestion, filtered, washed in cell medium, labeled with BrdU to control their viability, and seeded onto the polymeric scaffolds (1.5×10^8 cells in total). The constructs were cultured *in vitro* for 1 week in an incubator under physiological conditions (T = 37°C; p_{co_2} = 5%) and then implanted into a subcutaneous pocket on the back of athymic mice. Three groups of scaffolds were considered: (I) scaffolds seeded with cells; (II) scaffolds reinforced with an external stent and seeded with cells; (III) externally stented scaffolds with no cells.
- **Biochemistry**: after 12 weeks, the constructs were explanted, sectioned, and histologically stained with specific markers for typical cartilage extra cellular matrix components (hematoxylin and eosin for general tissue formation; alcian blue for glycosaminoglycan deposition; Masson's trichrome for collagen formation). Immunohistochemistry was also performed to confirm that present collagen was specific for cartilage (type II).

The results showed extensive cartilage formation in the scaffolds that were seeded with cells, while no cartilage was present in the unseeded scaffolds. Furthermore, scaffolds reinforced with an external stent for the first 4 weeks of implantation maintained the anatomical shape of the ear. In contrast, the other scaffolds lost partially their integrity and appeared of reduced size and distorted shape. From these findings the scientists concluded that cartilage formation is not mature enough in the first 4 weeks to counteract the contraction forces in the healing process.

This experiment was surely a success for those years and definitely contributed to boost the interest in the field. If we consider the state of the art nowadays, however, a number of drawbacks still characterize this study, some of which are mentioned here:

- Skin coverage is missing and is a critical element of any ear reconstruction.
- Bovine immature (young) chondrocytes were used, while clinical application thereof is of course highly unlikely and these cells are now known to be partially unrepresentative for the use of human cartilage cell sources. Such human sources tend to loose differentiation capacity quite fast and are furthermore frequently characterized by necrosis in the center of scaffolds with a clinically relevant size.
- An athymic or immunodeficient mouse model is used here. Obviously, this is nothing more than a useful screening model and large animal models are required to test clinical relevance in the presence of a functional immune system.
- Scaffolds need to provide an adequate mechanical stability to the construct at the time of implantation;
- Implications on the growth rate of the artificial ear compared to the growth rate of a 3-year-old child should be addressed before the final implantation in the patient.

The loss or failure of an organ or tissue is one of the most frequent, devastating, and costly problems in human health care. A new field, tissue engineering, applies the principles of biology and engineering to the development of functional substitutes for damaged tissue. This article discusses the foundations and challenges of this interdisciplinary field and its attempts to provide solutions to tissue creation and repair.

Langer and Vacanti (1993)

If the need is indeed so high, and the field has so extensively grown over the last decade, then why do we still lack frequent clinical successes? There are two

Box 1 The Mouse and the ear: tissue-engineered cartilage in the shape of a human ear

Although there have been several studies since the 1980s where the concepts of tissue engineering were first applied, a formal formulation of the discipline is traced back to the paper of J. Vacanti and R. Langer in *Science* (1993). Since then, the number of studies in tissue engineering has grown rapidly. A landmark study published in 1997 in *Plastic and Reconstructive Surgery* by Y. Cao *et al.* (1997) attracted the interest of a large audience; thanks also to a BBC service on the subject. In this paper it is shown how to successfully regenerate the cartilagineous part of a 3-year-old child's ear. The work is useful also from an educational point of view, as most of the 'ingredients' to perform a classic tissue engineering experiment are present.

To highlight the multidisciplinarity of tissue engineering the study can be divided into three parts:

- **Material Science**: first, a plaster mold of the ear of a 3-year-old child was cast from an alginate impression of the ear. Then, a 100 μm thick non-woven mesh of poly(glycolic acid) (PGA) was

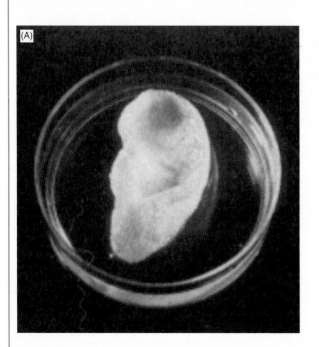
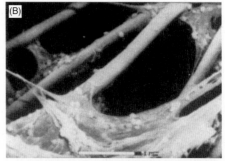

Ear anatomically shaped scaffold. (a) scaffold structure before seeding; (b) SEM micrograph showing cells and extracellular matrix formed in the scaffold; (c) scaffolds implanted subcutaneously on the back of an immunodeficient mouse. Reproduced with permission from Cao, Y., Vacanti, J.P., Paige, K.T., et al., (1997). Transplantation of chondrocytes utilizing a polymer-cell construct to produce tissue-engineered cartilage in the shape of a human ear. Plast Reconstr Surg, 100: 297–302.

these thoughts were all triggered by data obtained through a new technology. Technology continuously shapes our society but sometimes also creates overly optimistic expectations, even among the smartest minds ever. If Darwin in his letter to Fox states that "*any remedy will cure any malady*" (Fox, 1850) this is a statement that could have been done by any ordinary person. The connotation of a 'water cure' gives the statement a comic appearance. But it also shows that even the best among scientists have a hard time to distinguish between real and fake cures. Almost 150 years later we are continuously confronted with similar expectations which are too optimistic where the influence of technology on healthcare is concerned. By the way, it is only fair to Darwin to explain that, further in the same letter, he strongly attacks alternative medicine in the following manner "*You speak about Homœopathy; which is a subject which makes me more wrath, even than does Clairvoyance: clairvoyance so transcends belief, that one's ordinary faculties are put out of question, but in Homœopathy common sense & common observation come into play, and both these must go to the Dogs, if the infinetesimal doses have any effect whatever*". If even the critical Darwin can be triggered into unrealistic expectations, then it is no wonder that healthcare hypes are a more than common phenomenon.

In 1997, media all over the world were aroused by a BBC documentary, *Tomorrow's World*, showing what is now known as the Vacanti mouse (Cao *et al.*, 1997). The term 'Tissue engineering' was no longer seen as an expression familiar only to a limited number of scientists working in the field – it had become well-known to millions of individuals worldwide. Although the Vacanti experiment (see Box 1) is truly exemplary for the discipline of tissue engineering, it is fair to say that the media upheaval was not so much caused by the actual experiment but even more by the spectacular sight of a nude mouse that had apparently grown a human ear on its back. For many, the *Island of Dr Moreau** (Wells, 1896) had become reality and media hype was born. This was not the first media hype on tissue or organ repair and most certainly will not be the last. It would not be difficult to dedicate an entire chapter in this textbook to the promising aspects of our discipline that made it to the media and had a major public impact. It is intriguing to observe that, in contrast, it would be difficult to fill an entire chapter with actual clinical successes. At first sight one would tend to say that the field has over promised and has a history of not delivering on those promises. This statement would be too simple. The eagerness of the media to report on the advances in the field of tissue engineering is not so much caused by publicity eager scientists; the actual cause is the enormous demand in our society for technologies that are able to repair, or even better regenerate, damaged or worn out tissues and/or organs.

On an annual basis many millions of patients undergo surgery for tissue reconstruction. A fair proportion are treated satisfactorily, another portion less effectively, and millions still await treatments that would help them at least in an acceptable way. A now almost classical analysis on the need and commercial opportunity was published in *Science* in 1993 by Langer and Vacanti, the abstract of which is given below.

---

*The *Island of Dr. Moreau* was written by HG. Wells in 1896 and is one of the early examples of a science fiction novel. On his island, Dr. Moreau applies vivisection techniques on animals to create hybrid organisms of animal origin with man-like properties. Interestingly, Wells already realizes that new technologies require new legislation and all of Moreau's monsters have to obey to the following law:

- *Not to go on All-Fours; that is the Law. Are we not men?*
- *Not to suck up Drink; that is the Law. Are we not men?*
- *Not to eat Fish or Flesh; that is the Law. Are we not men?*
- *Not to claw the Bark of Trees; that is the Law. Are we not men?*
- *Not to chase other Men; that is the Law. Are we not men?*

The field of life sciences moves forward at a rapid pace and many of us do not fully realize that this acceleration is a relatively recent phenomenon in the history of mankind. These days even the lesser-educated in our society are aware of the concept 'cell' and the possible benefits that applying isolated, expanded or even manipulated cells may have in healthcare. Nevertheless, the cell as a building block of organisms was unknown until scientists like van Leeuwenhoek in 1675 were able to see the first 'globules' through surprisingly powerful, though primitive, microscopes. Obviously, as they were entering a completely unknown world, where no one had gone before, many of the early assumptions these scientists made were bound to be proven erroneous today. In contrast, one can sometimes be truly impressed by the incredible abilities of these early scientists to grasp the essentials behind entirely new concepts.

It is almost impossible to imagine how van Leeuwenhoek (as quoted above) was able to translate the completely unknown microscopical dimensions of a previously imagined entity like the cell into the very correct example of bladders filled with water. Picture these bags packed tight together on the floor, or filling an empty vessel and the contemporary biologist sees the architecture of the epidermis or a gland. Even the adaptability and variability of the cell shape he already correctly foresaw. We should realize that

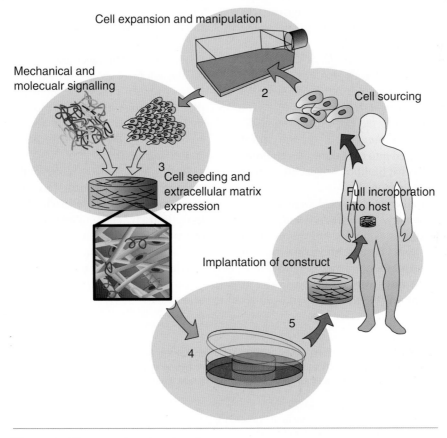

The central tissue engineering paradigm.

"Several times I wrote about bodies consisting of globules, but we must not imagine perfectly round globules, but a number of bladders of animals perfectly round, filled with water, and those bladders lying one next to the other on the earth. . . . and if a large number of these bladders were thrown in an empty barrel or packed tight therein, the round bladders would not maintain their shape but the said bladders would adapt themselves."

Anthoni Leeuwenhoeck, December 20, 1675. Letter addressed to the secretary of the British Royal Society (Leeuwenhoek, 1939)

LEEUWENHOEK.
London published Feb 06 1813 by C.Jones.

"Your aphorism that 'any remedy will cure any malady' contains, I do believe, profound truth—whether applicable or not to the wondrous Water Cure I am not very sure—The Water-Cure, however, keeps in high favour, & I go regularly on with douching &c &c:"

Charles Darwin, September 4. Letter to W.D. Fox. (Fox, 1850) *No.*

Tissue engineering – an introduction

Clemens Antoni van Blitterswijk, Lorenzo Moroni, Jeroen Rouwkema, Ramakrishnaiah Siddappa and Jérôme Sohier

EEN PSALM VOOR VEEL LATER
Systeem, neem aan dat mettertijd
ver achter deze eeuwigheid
geen ster, geen stof, geen licht meer schijnt
en, als Gij wilt, ook Gij verdwijnt,

ook dan nog vraag ik, nu en hier
doodgedrukt op dit papier:
Verschaft ons eeuwig nietmeerzijnd
U dan nog zulk plezier?

Zo niet, laat ons dan wederkeren
als uiterst anderen, want dan
zal ik wellicht nog eens proberen
of alles anders kan

A PSALM FOR MUCH LATER
System, say that as time will flee
far past this eternity
no star, no dust, no light lives on
and, if You wish, You too are gone,

I still can ask right here and now
imprinted on this sheet and dried
so eternally: Why, how,
no, if it still pleases Thee we died?

If not, let us return some day
very different from before
and Who knows, I'll try once more
some other way

Leo Vroman, scientist, poet and artist. Homo Universalis still exists
© Leo Vroman, *Details*, 1999, Querido. Permission acquired through Querido, Amsterdam

Foreword

Tissue engineering is an extremely important area. It generally involves the use of materials and cells with the goal of trying to understand tissue function and some day enabling virtually any tissue or organ on the body to be made *de novo*. To achieve this very important long-range objective requires research in many areas. This book, edited by Professor van Blitterswijk, addresses many of these important topics, and the chapters provide a foundation for the understanding and development of the cell-based systems needed for tissue engineering. One of these areas involves the cells themselves; in this regard, important chapters on stem cell biology and cell sources, and various aspects of cell culture including harvesting, selection, expansion, and differentiation, are provided. Other important themes involving cells, such as cell nutrition and cryobiology, are also examined, as is cellular signaling.

Some of the chapters deal with embryology. Much can be learned from this field as it provides the basic guide as to how tissues can form. The book also discusses morphogenesis and how tissues are generated in the embryo and tissue homeostasis. This is very important for providing vital information as to how tissues can be created. Such chapters may also offer clues for the reader as to how tissue engineering may some day be more successfully accomplished.

There are a variety of very important materials issues in tissue engineering, which are discussed in the book. Cells adhere to the extracellular matrix material in the body; this matrix has an enormous affect on how the cells behave. However, to try to recreate extracellular matrixes, is a difficult task and therefore various materials have been explored to provide substrates for cell growth *in vivo*. The physical and chemical properties of these materials are examined, and important materials that might be used in tissue engineering are discussed. These include natural polymers, degradable polymers, and bioceramics. The book also examines the important issue of tissue compatibility and biomaterials compatibility which are critical if these tissues are to be safe and integrate with the body. In addition, scaffold design and fabrication are discussed so that the reader may have a better understanding of how to develop and manufacture these systems. Ways of using controlled release from materials is examined; controlled release of different factors (e.g. growth factors to promote vascularization) can provide an important means of controlling and improving tissue function.

Once the scaffold system and the cell system are developed, they have to be put together. This is generally done through a bioreactor where the cells and materials are combined with the right type of media and flow conditions inside the reactor. By combining all of these entities, a new tissue may be created. This important issue is also discussed.

The book then turns to important areas of tissue engineering with respect to case studies on individual tissues and organs. There are chapters that specifically discuss skin, cartilage, bone, the nervous system, and various organ systems. Finally, important ethical issues in tissue engineering are discussed.

Overall, this book provides a very useful guide for those who wish to understand important issues such as cell biology, materials science, and bioreactor design with respect to tissue engineering, as well as providing specific examples of how tissue engineering is accomplished.

B. Langer

Oomens, Cees, Eindhoven University of Technology, Eindhoven, The Netherlands

Oudega, Martin, Johns Hopkins University School of Medicine, USA

Pennesi, Giuseppina, Universita' di Genova, Italy

Plant, Giles, The University of Western Australia, Australia

Poelmann, Rob, Leiden University Medical Centre, Leiden, The Netherlands

Post, Jeanine, University of Twente, Enschede, The Netherlands

Price, Richard, Institute of Cell and Molecular Science, Bart's and University of London, UK

Radisic, Milica, University of Toronto, Toronto, Ontario, Canada

Ratcliffe, Anthony, Synthasome Inc, San Diego, CA, USA

Reis, Rui, University of Minho, Guimarães, Portugal

Robey, Pamela Gehron, National Institutes of Health, Bethesda, USA

Roelen, Bernard Antonius Johannes, Utrecht University, Utrecht, the Netherlands

Rouwkema, Jeroen, University of Twente, Bilthoven, The Netherlands

Siddappa, Ramakrishnaiah, University of Twente, Enschede, The Netherlands

Silva, Gabriela, University of Minho, Guimarães, Portugal

Silva, Simone, University of Minho, Guimarães, Portugal

Sohier, Jérôme, University of Nantes, Nantes, France

Sousa, Rui, University of Minho, Guimarães, Portugal

Svalander, Peter, Swedish Biotechnology AB, Göteborg, Sweden

Tallheden, Tommi, Göteborg University, Göteborg, Sweden

Timmins, Nicholas, University Hospital Basel, Switzerland

Vunjak-Novakovic, Gordana, Columbia University, New York, NY, USA

Welin, Stellan, Professor Biotechnology Culture Society, Department of Health and Society, Linköping University, Sweden

Wendt, David, University Hospital Basel, Switzerland

Williams, David, University of Liverpool, Liverpool, UK

Woodfield, Tim, University of Twente, Enschede

Yaszemski, Michael, Department of Orthopedic Surgery, Mayo Clinic College of Medicine, Rochester (MN), USA.

Zhang, Zheng, University of Twente, Enschede, The Netherlands

Gilbert,Thomas, University of Pittsburgh, Pittsburgh, USA

Gomes, Manuela, University of Minho, Guimarães, Portugal

Grijpma, Dirk, University of Twente, Enschede, The Netherlands

de Groot, Klaas, University of Twente, Enschede, The Netherlands

Habibovic, Pamela, University of Twente, Enschede, The Netherlands

Harvey, Alan, The University of Western Australia, Australia

Hierck, Beereend, Leiden University Medical Centre, Leiden, the Netherlands

Hirschi, Karen, Baylor College of Medicine, Boston, USA

Hodges, Steve, Wake Forest University Health Sciences, Winston-Salem, USA

Hubbell, Jeffrey Alan, Ecole Polytechnique Fédérale de Lausanne, Lausanne, Switzerland

Hutmacher, Dietmar Werner, National University of Singapore, Singapore

Jansen, John, Department of Periodontology & Biomaterials, Radboud University Nijmegen Medical Center, The Netherlands

Janssen, Frank, University of Twente, Enschede, The Netherlands

Jukes, Jojanneke, University of Twente, Enschede, The Netherlands

Karperien, Marcel, University of Twente, Enschede, The Netherlands

Kruyt, Moyo, Department of Orthopaedics, G05.228, University Medical Center Utrecht, P.O. Box 85500, 3508 GA Utrecht, The Netherlands

Kuleshova, Lilia, National University of Singapore, Singapore

Levenberg, Shulamit, Israel Institute of Technology, Haifa, Israel

Lewis, Jennifer, University of Illinois at Urbana-Champaign, Urbana, Illinois, USA

Lindahl, Anders, Sahlgrenska University Hospital, Göteborg, Sweden

Malafaya, Patrícia, University of Minho, Guimarães, Portugal

Malda, Jos, Queensland University of Technology, Australia

Mano, João, University of Minho, Guimarães, Portugal

Martin, Ivan, University Hospital Basel, Switzerland

Meijer, Gert, Department of Oral Maxillofacial Surgery, University Medical Center Utrecht, The Netherlands

Mikos, Antonios, Department of Bioengineering, Rice University, Houston (TX) U.S.A.

Mistry, Amit, Department of Bioengineering, Rice University, Houston (TX) U.S.A.

Moroni, Lorenzo, University of Twente, Enschede, The Netherlands

Myers, Simon, Institute of Cell and Molecular Science, Bart's and University of London, London, UK

Myers-Irvin, J., University of Pittsburgh, Pittsburgh, USA

Navsaria, Harshad, Institute of Cell and Molecular Science, Bart's and University of London, UK

Ni, Ming, School of Engineering and Applied Science, University of Pennsylvania, Philadelphia, USA

Oliveira, Joaquim, University of Minho, Guimarães, Portugal

Olivo, Christina, Department of Immunology, University Medical Center Utrecht, The Netherlands

List of contributors

Anthony, Edwin, Queen Mary's University of London, UK

van Apeldoorn, Aart, University of Twente, Enschede, The Netherlands

Atala, Anthony, Wake Forest University of Medicine, Winston-Salem, USA

Azevedo, Helena, University of Minho, Guimarães, Portugal

Baaijens, Frank, Eindhoven University of Technology, Eindhoven, The Netherlands

Badylak, Stephen, McGowan Institute for Regenerative Medicine, Pittsburgh, USA

Barrère, Florence, University of Twente, Enschede, The Netherlands

van den Beucken, Jeroen, Department of Periodontology & Biomaterials, Radboud University Nijmegen Medical Center, The Netherlands

van Bezooijen, Rutger, Leiden University Medical Centre, Leiden, The Netherlands

Bianco, Paolo, Universita' la Sapienza, Roma, Italy

van Blitterswijk, Clemens Antoni, University of Twente, Enschede, The Netherlands

de Boer, Jan, University of Twente, Enschede, The Netherlands

Both, Sanne, University of Twente, Enschede, The Netherlands

Brittberg, Mats, University of Göteborg, Göteborg, Sweden

Cancedda, Ranieri, Universita' di Genova & Istituto Nazionale per la Ricerca sul Cancro, Genova, Italy

Claase, Menno, University of Twente, Enschede, The Netherlands

D'Amore, Patricia, Schepens Eye Research Institute, Boston, USA

Dalton, Paul, University of Southampton, UK

DeRuiter, Marco, Leiden University Medical Centre, Leiden, The Netherlands

Deschamps, Audrey, University of Twente, Enschede, The Netherlands

Dhert, Wouter, Department of Orthopaedics, G05.228, University Medical Center Utrecht, P.O. Box 85500, 3508 GA Utrecht, The Netherlands

van Dijkhuizen-Radersma, Riemke, Progentix BV, Bilthoven, The Netherlands

Ducheyne, Paul, Center for Bioactive Materials and Tissue Engineering, University of Pennsylvania, Philadelphia, USA

Feijen, Jan, University of Twente, Enschede, The Netherlands

Frey, Peter, Centre Hospitalier Universitaire Vaudois, Switzerland

van Gaalen, Steven, Department of Orthopaedics, G05.228, University Medical Center Utrecht, P.O. Box 85500, 3508 GA Utrecht, The Netherlands

el Ghalbzouri, Abdoelwaheb, Leiden University Medical Centre, Leiden, The Netherlands.

Gibbs, Susan, VU Medical Centre, Amsterdam, The Netherlands

Contents

Academic Press is an imprint of Elsevier
84 Theobald's Road, London WC1X 8RR, UK
30 Corporate Drive, Suite 400, Burlington, MA 01803, USA
525 B Street, Suite 1900, San Diego, CA 92101-4495, USA

First edition 2008

Notice
No responsibility is assumed by the publisher for any injury and/or damage to persons or property as a matter
of products liability, negligence or otherwise, or from any use or operation of any methods, products, instructions
or ideas contained in the material herein. Because of rapid advances in the medical sciences, in particular, independent
verification of diagnoses and drug dosages should be made

British Library Cataloguing in Publication Data
A catalogue record for this book is available from the British Library

Library of Congress Cataloguing in Publication Data
A catalogue record for this book is available from the Library of Congress

ISBN: 978-0-12-370869-4

For information on all Academic Press publications visit our
web site at http://books.elsevier.com

Typeset by Charon Tec Ltd (A Macmillan Company), Chennai, India.
www.charontec.com

Printed in China

10 11 12   10 9 8 7 6 5 4 3

Tissue Engineering

Senior Editor: Clemens van Blitterswijk

Editors: Peter Thomsen, Anders Lindahl, Jeffrey Hubbell, David Williams, Ranieri Cancedda, Joost de Bruijn, Jérôme Sohier

AMSTERDAM • BOSTON • HEIDELBERG • LONDON • NEW YORK • OXFORD
PARIS • SAN DIEGO • SAN FRANCISCO • SINGAPORE • SYDNEY • TOKYO
Academic Press is an imprint of Elsevier

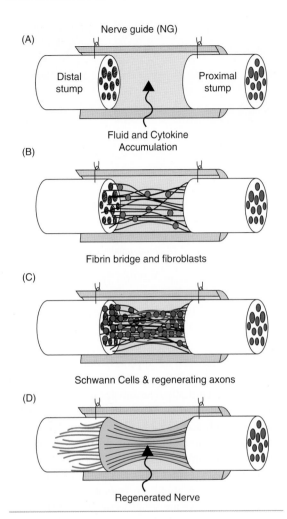

(A)

Nerve guide (NG)

Distal stump

Proximal stump

Fluid and Cytokine Accumulation

(B)

Fibrin bridge and fibroblasts

(C)

Schwann Cells & regenerating axons

(D)

Regenerated Nerve

Figure I.3 Guided nerve regeneration.

multiple tissues into one system. In a sense one might say that tissue engineering can be compared to say, a house where the elements provided (i.e. windows, doors, etc.) need to fit together to create a home. This convinced us to contribute a chapter on **Tissue Engineering of Organ Systems** (Chapter 21).

There are some areas that may not directly involve actually generating a tissue or organ substitute, but are still pivotal to the overall development and acceptance of the field. Therefore, the editors decided to add one more chapter. Ultimately, tissue engineering

constructs have to be placed into a human recipient. This puts major responsibility on the researchers and clinicians that are actively involved in this process. Naturally, a minimum requirement would be that the construct is safe and may not harm the patient; preferably it would be more effective than existing treatments. Apart from this, an important issue for any new technology, which is frequently underestimated by scientists, concerns the ethical aspects of such technologies. It requires little imagination to see that tissue engineering would certainly rank among those disciplines most prone to ethical scrutiny. Not only should we be sincerely interested in the ethical aspects of our field but we should thoroughly realize that public acceptance, and the ethical debate related to that, will be an important determinant for the widespread application of the technology we develop. In view of this, **Ethical Issues on Tissue Engineering** (Chapter 22) has been included. A careful reader may have noticed that this chapter does not start by giving a definition on tissue engineering, but a schematic overview. This was a deliberate choice. Very often students are confronted with definitions which they are expected to accept as the ultimate truth. In reality, any definition has both supporters and opponents, and this is certainly true in the field of tissue engineering. Therefore, instead of expecting everybody to agree to the same set of definitions, we have endeavored to encourage the reader to use the definitions in which he or she believes.

It is the hope of the editors that the chapters that have been included both in this book, and online, will provide a solid basis in tissue engineering. Apart from explaining what the purpose of this textbook is, and how we tried to achieve this goal, this introduction chapter might further be useful in that it should exceed the purpose of the other individual chapters. While each of the remaining chapters focuses on a particular element of tissue engineering, they are not specifically dedicated to giving an overall view of the field. More particularly, all those areas where developments take place are both exciting and crucial for the further development of our discipline.

In treating some of these aspects it may become clear where the field is moving towards. Not only will it reveal that indeed in the years to come tissue engineering may likely contribute to the well-being of millions, which gives justification to the field being there at all, but it will also provide something else: *Tissue Engineering is fun!* This is a subject frequently forgotten but not justifiably so. It is relevant that researchers not only find justification in their efforts leading to the benefits of society but have fun on the way as well. After all, a subject like tissue engineering is quite adventurous and usually involves combining knowledge of different disciplines into one research project. The complexity of such combinations makes the field intellectually challenging and the possibilities are such that there is ample space for individual efforts still to be recognized. The fact that a successful experiment may bring the technology closer to clinical reality subsequently provides further rewards. That this opinion is also shared by those not professionally active in our field was illustrated by a *Time* Magazine article, around the turn of the century, ranking the 'Tissue Engineer' among the hottest future professions of the twenty-first century (Time Magazine online, http://www.time.com/time/magazine/article/0,9171,997028,00.html).

Let us look back at some of the early statements at the beginning of this chapter. Many people active in the field feel we are progressing too slowly and wonder if we have not been over-promising. We should realize that comments on the slow progress of the field can only concern the apparent lack of clinical success. Science is progressing and at a rapid rate. In this perspective, it might be useful to look again at the Vacanti mouse (Cao *et al.*, 1997, Box 1). At the time this was a state of the art experiment. A cast of a human ear formed the fundament for a porous nonwoven poly(glycolic acid) (PGA)/poly(lactic acid) (PLA) scaffold which was subsequently filled with cultured cartilage cells. Then the hybrid construct was placed under the skin of an immuno-deficient mouse. Irrespective of the outcome of the specific Vacanti experiment, we

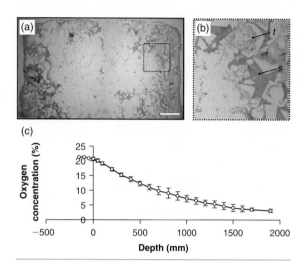

Figure I.4 Insufficient nutrition transport is one of the main problems that Tissue Engineers have to face (scale bar = 1 mm).

now know that such an approach would, in general, present major problems. One reason being that implants of a clinically relevant size in general do not allow optimal nutrient transport to the cells in the center of such scaffolds, as illustrated in Figure I.4. After implantation this lack of nutrients is directly related to poor and late ingrowth of blood vessels into the scaffold. At the time, this was a fundamental obstacle and even today it still presents a significant challenge. However, we are progressing. To illustrate this we should introduce the Levenberg mouse (Levenberg *et al.*, 2005) (Box 2) – less well-known than that of Vacanti, the Levenberg mouse is perhaps even more interesting. The concept of this study was that a carefully generated hybrid construct of which the pores were filled with cultured tissue still faces a major risk after implantation. Where, during culture nutrients can be forced through the scaffold by processes such as agitation or perfusion, this usually is no longer the case after implantation. In most cases the scaffold will be dependent on nutrients coming from blood vessels. In an average natural tissue, cells will typically

Box 2 Cells nutrition: engineering vessels

One of the major limitations of tissue engineering is the inability to provide a sufficient blood supply in the initial phase after implantation. As long as a proper vascularization has not been established, the implant has to rely on diffusion for the supply of nutrients and the removal of waste. This can lead to nutrient limitations, which can result in improper integration or even death of the implant. Since tissue engineering has primarily been performed on implants of small dimensions for research purposes, the impact of this problem has so far been limited. However, in order to grow tissues on a larger, clinically relevant, scale, this problem has to be dealt with.

Several strategies to enhance early vascularization, for instance the targeted delivery of angiogenic growth factors, have been studied. However, these strategies still rely on the ingrowth of host endothelial cells and therefore vascularization will still take a considerable time. This paper describes a novel approach to enhance vascularization after implantation. Endothelial cells (HUVEC or endothelial cells from embryonic stem cells) were added to engineered muscle tissue, which lead to the formation of a vascular network inside the muscle tissue *in vitro*. The prevascular network was stabilized by adding smooth muscle precursor cells (mouse embryonic fibroblasts, MEF).

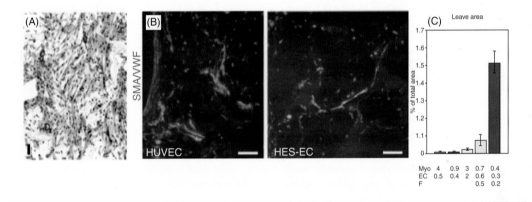

In vitro results after 10 days. A, Endothelial cells (brown) organize into vascular structures containing lumen; B, MEF differentiate into smooth muscle cells (red) and colocalize with endothelial cells (green); C, effect of different seeding ratios on the formation of vascular structures. Myo = myoblasts, EC = UVEC and F = MEF. Number of cells in millions seeded per scaffold. Note the positive effect of the addition of MEF. C, Reproduced with permission from Levenberg, S., Rouwkema, J., et al. (2005). Engineering vascularized skeletal muscle tissue. Nat Biotechnol, 23: 879–884.

After implantation, the muscle tissue integrated properly with the surrounding tissue, indicating that the added vascular structure did not have a negative effect on the differentiation of the muscle tissue. Moreover, the prevascular network fused with the blood vessels of the host and became functional vessels transporting blood. This leads to enhanced vascularization and better survival of the implant.

This study shows that *in vitro* prevascularization can be a successful strategy to improve vascularization after implantation. Which means that this technique could be a way to solve the problem of nutrient limitations in large tissue constructs after implantation. It is likely that this strategy does not only work for muscle tissue engineering, but could be used as a more general technique to successfully engineer diverse tissues with clinically relevant sizes.

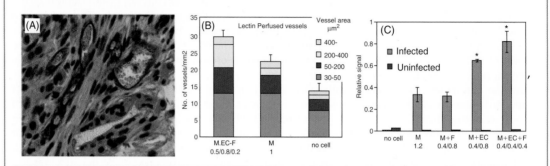

In vivo results after 2 weeks. A, Human vascular structures (brown, human-specific CD31 staining) have connected to the vasculature of the host and carry blood; B, quantification of the amount of perfused vessels after 2 weeks of implantation. Note that prevascularized samples (M + EC + F) contain more perfused vessel; C, survival of cells in the implant determined with a luciferase assay. Note that prevascularized samples (M + EC and M + EC + F) display a better survival (confirmed with TUNEL assay). B and C, Reproduced with permission from Levenberg, S., Rouwkema, J., et al. (2005). Engineering vascularized skeletal muscle tissue. Nat Biotechnol, 23: 879–884.

find such a source of nutrients (a capillary) within roughly 200 microns. As an implant of clinically relevant size may measure centimeters, it becomes clear that most cells in such a scaffold may quickly starve to death by lack of vascular supply. After all, it may take a week before blood vessels have sufficiently penetrated a hybrid construct after implantation. A week is really a long time, as surgeons who have to operate on patients whose bowel was constricted for only a day can tell us, as much of the tissue may have become necrotic by then. Levenberg had an appealing idea: rather than waiting for spontaneous vascular ingrowth after implantation, she would provide a cultured vascular network prior to implantation. If Vacanti, with his mouse, provided the walls, taps and sanitary equipment in a house, Levenberg would now provide the plumbing to make it all work *and* connect it to the water supply and sewer system. The Levenberg mouse is discussed later in more detail,

but for now we will describe the experimental design briefly.

The goal was to grow a muscle substitute, supplemented with a vascular network. The cells of a muscle cell line were grown in co-culture with endothelial cells (the cells which line our blood vessels) from the human umbilical cord. As it is known that in natural capillaries endothelial cells are lined by supporting cells, in part of the experiment murine smooth muscle precursor cells were added to perform this function. The cells were grown in a non-woven PLA/PGA scaffold and after culture placed in immunodeficient mice. The authors investigated the functionality of the graft in several ways and the overall outcome was truly interesting. Not only had the endothelial cells formed structures that resembled a capillary network, but even better, mouse red blood cells were now found, within the construct, in capillaries made from the human cells. The graft had connected to the host vascular network and the network was functional, as proven by superior cell survival for those hybrid constructs that contained the human endothelial network. In an editorial this was described as a landmark paper (Jain *et al.*, 2005).

Critics may still state that this is an interesting concept that nevertheless bears little clinical relevance as nobody (in the foreseeable future) is likely to implant a mouse cell line into a human being. Furthermore, only for very few individuals, would clinicians have access to, preferably autologous, umbilical cord endothelial cells. Frankly, even the most optimistic researcher would find these arguments compelling. Therefore, quite a few parameters will have to be changed and optimized for this technology to become a clinical reality. This all fits in with the earlier remark that it easily takes a decade to get from early concept to clinical reality. The next example will show that it may take even longer than a single decade before a concept is translated into a clinical and commercial success.

In the 1960s and 1970s, Urist pioneered in the field of bone morphogenetic proteins. His classical paper of 1965 (Urist, 1965) is described in the Chapter 5 (Box 1). Urist had discovered that the demineralized fraction of bone contains a protein fraction that is able to induce bone formation in a non-species specific way, even in non-bony sites. He first hypothesized that the bone formation was induced by a non-collagenous protein acting as bone morphogen. Obviously, this was a finding of great importance and researchers worldwide started to further investigate the phenomenon of bone induction. It was not until 15 years later that the bone morphogenetic proteins (BMPs) were isolated and their existence definitely proven by Sampath and Reddi (1981, 1983). Nevertheless, their role in osteoinduction mechanisms has been debated – and is still today – although it has been clearly demonstrated over the last 25 years that BMPs alone could induce bone formation in non-bony sites of many different mammals. The lack of consensus around the BMPs action further delayed the development of BMPs as a potential tool for bone regeneration and it was not until the beginning of the 1990s that industries

and researchers really saw the interest of the protein and went on the way to develop and commercialize it for clinical use in fracture, non-union and spinal fusion treatment. During the 10 years necessary for the protein to be approved by the regulatory organisms, it became the most extensively studied orthobiologic product in history. Researchers would focus on how to translate their findings into clinical practice by optimizing, standardizing and industrializing the processes. In parallel, quality and regulatory protocols had to be developed as well. Finally, in 2001, BMP-7 was approved for spinal fusions and in 2002, BMP-2 followed. The protein is mixed with collagen sponges that allow its delivery to the site of surgery, from which it will induce bone formation and fusion of two vertebras. The treatment of tibial fractures by BMP-2 was further approved in 2004. Since then, the number of patients successfully treated with BMPs has not ceased to increase, making it one of the most successful commercial and healthcare successes in the field of orthopedics and in the field of tissue engineering. Overall, it took 30 years from the concept of BMPs expressed by Urist to the successful commercialization of a beneficial product. Interestingly, the availability and subsequent sales of the product now reveals that the market size for bone replacement might well be larger than ever expected, stimulating researchers in both industry and academia to try to do a better and, hopefully, faster job, than those who went before them.

The example above clearly illustrates the difficulty that tissue engineers have to face. As much fundamental knowledge is still lacking, they have to conjointly investigate and unravel the basic mechanisms taking place during tissue homeostasis and regeneration, and at the same time develop the tools to exploit and enhance these natural mechanisms. This dichotomy becomes evident when one considers the range of matrices used for tissue engineering applications.

One of the first tissue engineering paradigms was the use of supportive matrices that provide associated cells with a substrate appropriate for attachment, multiplication, and differentiation towards the desired tissue. Further than the necessary degradability of such a substrate and eventual remodeling, the exact requirements of the matrix were not known with regard to cell interaction and response. Therefore, as a first step, many different matrices were evaluated to determine the most prone to replace the cell environment, from natural or synthetic origin. Collagen sponges made from purified bovine collagen fibers and poly(α-hydroxy) acids (such as poly(lactic-co-glycolic) acids, PLGA) scaffolds appeared suitable at first. The former is still widely in use and provides the cells with a nature-derived matrix familiar to them. However, the non-optimal control and reproducibility of the properties of this material combined with the potential transmission of diseases raises concerns for clinical applications. Polymeric scaffolds such as PLGA on the contrary can be produced in a variety of reproducible shapes and properties such as degradation and

mechanical properties while being free of any pathogenic substances. However, the scaffolds produced with such polymers are foreign to the cells, thus creating important drawbacks. The lack of interaction with the polymer results in uncontrolled response of the cells, among which insufficient cell attachment (which cause tremendous difficulties during the seeding of the cells), or undesired changes of cell morphology. These non-physiological responses are the result of the material non-compliance with the necessary cell requirements. But what are these requirements and what do the cells find in their natural surroundings that they do not in a polymer scaffold? This knowledge is still not completely available and it is the task of tissue engineers to obtain and use it. Following this track, other materials have been introduced that try to more closely reproduce some key aspects of the extracellular matrix (ECM) (Box 3).

Box 3 Mimicking the extracellular matrix

One of the latest and most important realizations in the design of supportive structures for tissue engineering is the importance of mimicking the natural cell environment. Native tissue exist within a three-dimensional (3-D) viscoelastic milieu, the extracellular matrix (ECM), with which cells interact constantly and which guides their development or homeostasis. Accordingly, there is an increasing agreement that 3-D matrices such as hydrogels provide valuable systems that are closer to physiologic situations than most conventional biomaterials.

In addition to the imitation by hydrogels of the biochemical composition, structure or mechanical properties of the ECM, the creation of synthetic matrices that can as well mimic the natural interactions occurring between cells and ECM provides the tools to better understand, control, and guide the tissue regeneration. With this in mind, a novel line of research has been conducted by Hubbell and collaborators ((1995) and Lutolf and Hubbell (2005) over the past two decades, focusing on the incorporation within hydrogels of key functions of the ECM). This was attained by different strategies, based on the functionalization and modular design of synthetic hydrogels with bioactive domains of natural ECM components recognized as interacting with the cells. For instance, the biological recognition between the cells and their 3-D milieu can be achieved by incorporating the RGD tripeptide of fibronectin in the hydrogels design during synthesis. The local remodeling by the cell of their immediate surrounding, or in other words the cell-controlled degradation of the hydrogels and the cell migration, is achieved by the inclusion within the hydrogel of sites that are sensitive for proteases usually excreted at the cell surface during tissue repair (such as hyaluronidase, plasmin and matrix metalloproteinases-MMP). Finally, the presentation of growth factors to the cells is elegantly obtained by covalently attaching to the matrix material recombinant growth factors containing a protease cleavage site. Doing so, the cells induce a controlled and localized release by proteases active at their surface.

A good example of this approach can be found in the publication of Lutolf *et al.* (2003) in which proteolytically-sensitive networks bearing adhesion peptides and entrapping or bounding growth factors are presented. Hydrogels containing cells were prepared using vynil sulfone-functionalized

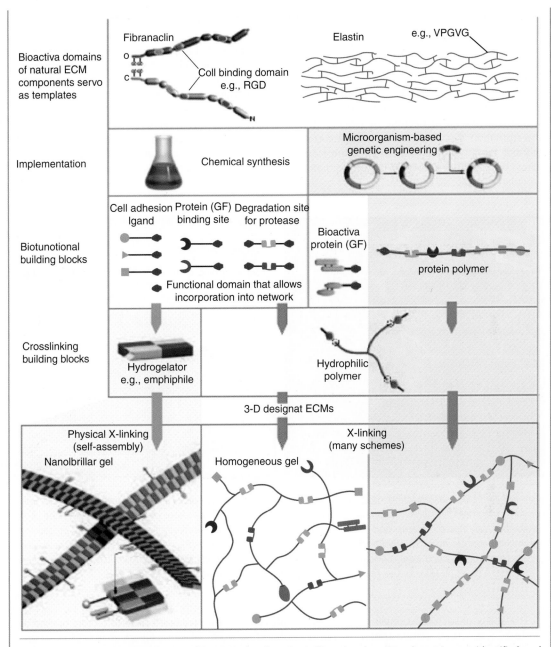

Design strategies of hydrogels mimicking ECM key functions. Bioactive domains of proteins are identified and synthesized by chemical strategies or by recombinant technologies. They include cell-adhesive ligands, growth factor binding sites or modified growth factors with cleavage sites and domains for protease degradation. Hydrogels can then be obtained by physical cross-linking of the selected components, either by physical or chemical mechanisms. Reproduced with permission from Lutolf, M.P. and Hubbell, J.A. (2005). Synthetic biomaterials as instructive extracellular microenvironments for morphogenesis in tissue engineering. Nat Biotechnol, 23: 47–55.

poly(ethylene glycol) macromers that were conjuguated by a Michael-type addition to cysteine-containing peptides at physiological temperature and pH. The peptides selected contained a cell adhesion motif based on the RGD peptide and a MMP sensitive sequence. The network supported adhesiveness of the cells (fibroblast) which spread and migrated within the gel. The migration was controlled and regulated by the MMP-sensitive sites repartition and by the level of secretion of MMP by the cells, which is increased in necrotic or healing tissues. The migration and cellular multiplication allowed the formation of complex morphologies *in vitro* and *in vivo*, in which the material was remodeled without loosing its integrity. *In vivo*, the incorporation of vascular endothelial growth factor (VEGF) allowed the complete remodeling of the material by infiltration of connective tissue and extensive new vascularization.

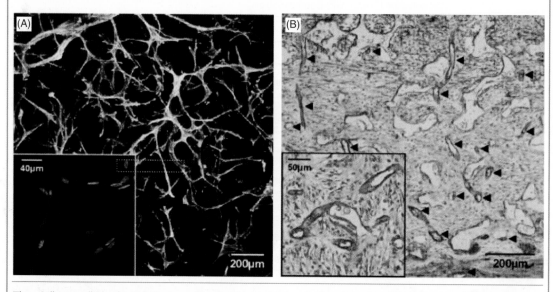

The cell-controlled degradation and migration of the hydrogels allowed the formation of complex network morphologies while retaining the material integrity and mechanical properties. (A), The inset shows the cell nuclei (green) and actin cytoskeleton (red). When implanted subcutaneously in the rat, a complete remodeling of the hydrogel into native tissue was observed; (B), with the formation of new blood vessels (arrowheads) and connective tissue infiltration. Reproduced with permission from Lutolf, M.P., Raeber, G.P., Zisch, A.H., et al. (2003). Cell-responsive synthetic hydrogels. Adv Mater, 15: 888–892.

The ability of such functionalized hydrogels to provide cell with localized growth factors release by protease was further demonstrated in another publication from Sakiyama-Elbert *et al* (2001). There, a recombinant β-nerve growth factor (β-NGF) variant that expresses a domain substrate for factor XIIIa, which cross-links the growth factor into fibrin has been produced. A protease cleavage site was included between the growth factor and the fibrin-coupling site to enable a localized release by cell surface proteases. The use of this functionalized gel induced an increase of dorsal root ganglia nerve ingrowth up to 100% compared to non-functionalized gels or gels containing unbound growth factor, demonstrating the advantage of an on-demand delivery of growth factor to cells.

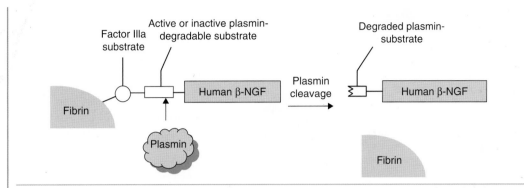

β-NGF variant incorporated to fibrin hydrogels and allowing a cell-controlled release of the growth factor under protease action. Reproduced with permission from Sakiyama-Elbert, S.E., Panitch, A. and Hubbell, J.A. (2001). Development of growth factor fusion proteins for cell-triggered drug delivery. FASEB J, 15: 1300–1302.

The development of synthetic matrices that mimic the natural functions and structure of the ECM, although not completely and perfectly, gives the tools to tissue engineers to selectively and modularly refined their materials in view of providing an environment that dictates and guides cells to tissue regeneration.

The first improvement is to provide the cells with a three-dimensional watery environment (hydrogels) instead of a flat surface on which to attach (at the cell scale, a polymeric scaffold is a flat surface). Furthermore, these hydrogels can be functionalized with moieties that interact in a defined way with cells. For instance, the attachment and spreading of the cells within such matrices can be achieved by the incorporation of specific peptide sequences known to play a role in cell adhesion (as the RGD sequence for instance). The ECM being constantly remodeled by the cell *in vivo*, the hydrogels can also include protease sensitive sites that will allow the cell-controlled degradation of the matrix during cell migration. Finally, as ECM is a natural reservoir for signaling molecules and growth factors (present in minute quantities but of high potency) the hydrogels should provide the signals controlling cell fate.

Although these key elements might not be sufficient to reach the final goal of fully mimicking the ECM, they do represent a significant first step that should allow researchers to further identify crucial needs of the cell.

Aside from the importance of understanding and creating the most optimal environment for the cells, another open question of possibly higher importance concerns the cells themselves: what are the most suitable cells to use for any given tissue? Pivotal knowledge regarding this fundamental question is also lacking. One could logically think that the most suitable cell source for cartilage or bone, for instance, is cartilage and bone, respectively. This is logical indeed and was the first approach to be followed. However, many hurdles render this strategy not as straightforward as expected. The first difficulty was seen in the number of cells necessary to provide a basis for tissue growth. The extraction of cells from native tissue is generally done in humans by biopsy and usually results in an insufficient cell number (Figure I.5). The cells so collected therefore have to be multiplied. This step is generally done by culturing them on polystyrene culture flasks that allow cell attachment. Although almost any cell types can be expanded, provided the right and specific conditions are supplied, their multiplication immutably induces the loss of their particular phenotype; in other words, they

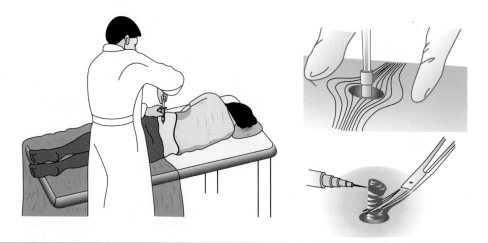

Figure I.5 Although human biopsies are the main cell source, they do not provide a sufficient number of cells.

dedifferentiate. For instance, a cartilage cell of round morphology will turn to a fibroblastic-like cell (of stretched and elongated shape) after some divisions (von der Mark *et al.*, 1977). As a result, once a sufficient number of cells have been reached, they have to be re-differentiated to the desired phenotype, which is not easily attainable.

Another approach, which has gained a constant increase in interest, consists of using undifferentiated cells as starting material. After all, a complete organism consisting of billions of highly specialized cells originates only from a single undifferentiated one. Even in adults, there are pools of undifferentiated cells which allow self-renewal of the organism over a life-span; in other words, cells that form the stem of all the others. The first of these stem cells to be identified after the Second World War were the hematopoietic cells that allow the renewal of blood (Till and McCulloch, 1961). Since then, other stem cells have been discovered in the bone marrow, peripheral blood, brain, spinal cord, dental pulp, blood vessels, skeletal muscle, heart, epidermis, mucosa of the digestive system, cornea, liver, and pancreas. These cells, once isolated, can be differentiated in highly specialized tissues (multipotency) if provided the right conditions, which are not entirely established. This is one of the issues that tissue engineering has to

face in order to collect the fruits promised by the use of stem cells.

Even though the apparent slow progress of the field can be understood, it should not hide the fact that successful applications already exist. With current techniques, which may be qualified as crude by some, the effectiveness of tissue engineering for some applications can already be demonstrated clinically, as was done recently by Atala *et al.* (2006) in a milestone article (see Box 4). This research group focused on the application of tissue engineering to urological diseases and, more specifically, bladder-related diseases. In this particular study, young patients suffering from an invalidating congenital malformation that induces high pressure in the bladder and, as a result poor and low compliancy, were treated with a tissue engineering approach. Native bladder cells were isolated and expanded *in vitro* prior to seeding on a composite scaffold of collagen and polyglycolic acid. The scaffold was designed to replace or augment the bladder size to improve its compliancy and patient continence. After a short period of *in vitro* culture – to allow cell attachment to the scaffold – the engineered constructs were implanted for a period up to 61 months. The main outcome of this study was the definitive improvement of the bladder compliance and capacity while restoring physiological function. In addition, no side effects caused by the tissue engineered construct were found.

Although this study treats a particular organ that can already be considered as simpler than, for instance, bone or skin (which are still by far surpassed in complexity by a kidney or our brain), it still shows the potency of the tissue engineering strategy. It should serve to convince those who do not see tissue engineering evolving fast enough that, although its progress is slower than all of us would wish, it has already reached the stage of treating numerous patients and adding valuable quality of life.

Box 4 Tissue Engineered Bladder: A seminal clinical study

Generating an organ is the holy grail of tissue engineering. In contrast to the classical grail ours will actually be found. As a matter of fact, there will be many grails and a small but interesting one has been described by Atala *et al.* (2006) which has made a seminal attempt in applying the basic research in bladder tissue engineering into clinical application.

Traditionally patients with end-stage bladder disease are treated with cystoplasty using gastrointestinal segments. However, such segments results in many complications such as metabolic disturbances, urolithiasis, increased mucous production, and malignant disease.

The authors successfully used an alternative approach using autologous engineered bladder tissues for reconstruction. Their earlier animal model experiments, using autologous cells in combination with biodegradable matrix from normal and diseased bladders, demonstrated similar functional properties encouraging the authors to engineer human bladder tissues by seeding autologous cells on different matrices in patients with end-stage bladder diseases requiring cystoplasty. The authors, based on their initial preclinical studies, decided to use collagen matrix derived from decellularized bladder submucosa. Their additional animal experiments with a collagen and polyglycolic acid (PLG) with omental coverage improved tissue vascularization and performed better in long-term.

The study included seven patients with patients being implanted with collagen scaffold without omental wrap and four patients with collagen-PLG with omental wrap. A bladder biopsy sample (1–2 cm^2) was obtained through a small suprapubic incision. The initial size of the bladder mould ranged

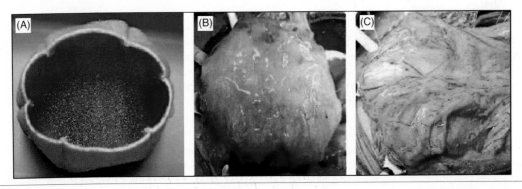

Construction of engineered bladder *Scaffold seeded with cells (A) and engineered bladder anastamozed to native bladder with running 4–0 polyglycolic sutures (B). Implant covered with fibrin glue and omentum (C). Reproduced with permission from Atala, S., Bauer, S.B., et al. (2006). Tissue-engineered autologous blassers for patients needed cystoplasty.* The Lancet, 367: 1241–1246.

from 70–150 cm^2 with a thickness of around 2 mm. The exterior surface of the scaffold was seeded with smooth muscle cells at a concentration of $50 \times 10^6/cm^3$. After 48 hours, urothelial cells were seeded by coating the inside of the scaffold at a concentration of $50 \times 10^6/cm^3$ and maintained at 37°C until implantation. The patients were followed up to five years postoperatively.

All of the patients urodynamic studies demonstrated that the mean leak point pressure decreased postoperatively by 13% and 29% in collagen and collagen-PLG scaffold with omental wrap, respectively. Further, the bladder capacity was found to have decreased by 30% in collagen scaffold. However, the collagen-PLG scaffold with omental wrap showed 1.58-fold increase in the bladder capacity. The postoperative compliance found to be increased by 15% and 67% in collagen and collagen-PLG with omental wrap respectively. The irregular bladder pressure found preoperatively was substantially improved postoperatively.

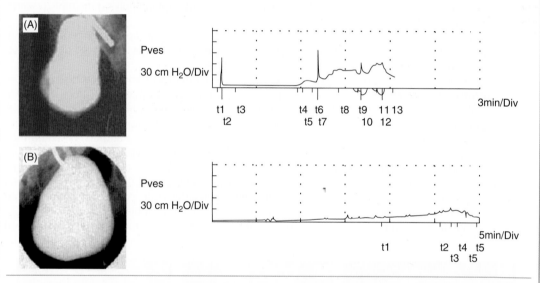

Preoperative (A) and 10-month postoperative (B) cystograms and urodynamic findings in patient with a collagen-PGA scaffold engineered bladder *Note irregular bladder on cystogram, abnormal bladder pressures on urodynamic study preoperatively, and improved findings postoperatively. Reproduced with permission from Atala, S., Bauer, S.B., et al. (2006). Tissue-engineered autologous blassers for patients needed cystoplasty. The Lancet, 367: 1241–1246.*

Further, all of the patients had normal serum sodium, potassium, chloride, phosphorus the and arterial blood gases postoperatively with a normal mucus production. Morphological analysis of the implanted engineered bladders demonstrated a tri-layered structure consisting of a urothelial cell-lined luman surrounded by submucosa and muscle indicating the implanted bladder was anostmosed to host tissue. This study not only demonstrates that reconstructed engineered bladders showed improved functional parameters that were durable over a period of years but also stands as milestone towards successful tissue engineering in other disciplines.

Morphological analysis of implanted engineered bladders. (A, B, C): Cystoscopic biopsies of implanted engineered bladders 31 months after augmentation shows extent of regeneration. Engineered bladder tissue showed tri-layered structure, consisting of lumen lined with urothelial cells (U) surrounded by submucosa (S) and muscle (M). Haemotoxylin and eosin. A, immunocytochemical analysis with anti-pancytokeratin AE1/ AE3 antibodies; B, and anti-α smooth muscle actin antibodies; C showed presence phenotypically normal urothelium and smooth muscle; (D, E, F): native bladder tissue. Magnification 100×. Reproduced with permission from Atala, S., Bauer, S.B., et al. (2006). Tissue-engineered autologous blassers for patients needed cystoplasty. The Lancet, 367: 1241–1246.

These final considerations will undeniably persuade the reader of the long way that Tissue engineers have already progressed and of the long way that still remains. A journey where unexpected hurdle will certainly appear, always making room for renewed excitement and fun. Indeed, Tissue engineers will have to become as multiple as the tissues they wish to regenerate and as pluripotent as the cells they use. We hope this textbook will contribute to this and to a faster progress of the field towards a clinical reality by enlightening young researchers to the state of the art of tissue engineering and to the challenges still lying ahead.

References

Atala, A., Bauer, S.B., Soker, S., Yoo, J.J. and Retik, A.B. (2006). Tissue-engineered autologous bladders for patients needing cystoplasty. *Lancet*, 367(9518): 1241–1246.

Cao, Y., Vacanti, J.P., Paige, K.T., Upton, J. and Vacanti, C.A. (1997). Transplantation of chondrocytes utilizing a polymer-cell construct to produce tissue-engineered cartilage in the shape of a human ear. *Plast Reconstr Surg*, 100(2): 297–302. discussion 303–304.

Hubbell, J.A. (1995). Biomaterials in tissue engineering. *Biotechnology (NY)*, 13(6): 565–576.

Jain, R.K., Au, P., Tam, J., Duda, D.G. and Fukumura, D. (2005). Engineering vascularized tissue. *Nat Biotechnol*, 23(7): 821–823.

Langer, R. and Vacanti, J.P. (1993). Tissue engineering. *Science*, 260(5110): 920–926.

Leeuwenhoek, A.v. (1939) *Alle de brieven van Antoni van Leeuwenhoek – The collected letters*. Deel 1: Br. 1[1]–21[14], 1673–1676, ed. C.G. Heringa, Amsterdam, p. 454.

Levenberg, S., Rouwkema, J., Macdonald, M., Garfein, E.S., Kohane, D.S., Darland, D.C., Marini, R., van Blitterswijk, C.A., Mulligan, R.C., D'Amore, P.A. and Langer, R. (2005). Engineering vascularized skeletal muscle tissue. *Nat Biotechnol*, 23(7): 879–884.

Lutolf, M.P. and Hubbell, J.A. (2005). Synthetic biomaterials as instructive extracellular microenvironments for morphogenesis in tissue engineering. *Nat Biotechnol*, 23(1): 47–55.

Lutolf, M.P., Raeber, G.P., Zisch, A.H., Tirelli, N. and Hubbell, J.A. (2003). Cell-responsive synthetic hydrogels. *Advanced materials*, 15(11): 888–892.

Sakiyama-Elbert, S.E., Panitch, A. and Hubbell, J.A. (2001). Development of growth factor fusion proteins for cell-triggered drug delivery. *FASEB J*, 15(7): 1300–1302.

Sampath, T.K. and Reddi, A.H. (1981). Dissociative extraction and reconstitution of extracellular matrix components involved in local bone differentiation. *Proc Natl Acad Sci USA*, 78(12): 7599–7603.

Sampath, T.K. and Reddi, A.H. (1983). Homology of bone-inductive proteins from human, monkey, bovine, and rat extracellular matrix. *Proc Natl Acad Sci USA*, 80(21): 6591–6595.

Till, J.E. and McCulloch, C.E. (2006). A direct measurement of the radiation sensitivity of normal mouse bone marrow cells. *Radiat Res*, 14: 213–222.

Urist, M.R. (1965). Bone: formation by autoinduction. *Science*, 150(698): 893–899.

von der Mark, K., Gauss, V., von der Mark, H. and Muller, P. (1977). Relationship between cell shape and type of collagen synthesised as chondrocytes lose their cartilage phenotype in culture. *Nature*, 267(5611): 531–532.

Wells, H.G. (1896). *The Island of Doctor Moreau*. London: William Heinemann. 3p. [v]–x[1]–219[1] p. 18.

Chapter 1
Stem cells

Jojanneke Jukes, Sanne Both, Janine Post, Clemens van Blitterswijk, Marcel Karperien and Jan de Boer

Chapter objectives:

- To recognize the defining properties of stem cells
- To identify the major differences between embryonic and adult stem cells
- To understand that the mechanisms that regulate self-renewal are complex
- To understand how stem cells can differentiate into a more specialized cell
- To learn how researchers can isolate and characterize embryonic stem cells
- To know where adult stem cells can be found in the body
- To understand the challenges for tissue engineers when using stem cells

"The essence of knowledge is, having it, to apply it; not having it, to confess your ignorance"

Confucius

1.1 What defines a stem cell?

Stem cells can be defined by two properties: the ability to make identical copies of themselves (self-renewal) and the ability to form other cell types of the body (differentiation) (Figure 1.1). These properties are also referred to as 'stemness'. Stem cells may potentially provide an unlimited supply of cells that can form any of the hundreds of specialized cells in the body. It is because of these properties that stem cells are an interesting cell source for tissue engineers.

Stem cells can be divided into two main groups: embryonic and adult or somatic stem cells. Embryonic stem cells are responsible for embryonic and fetal development and growth. In the human body, adult stem cells are responsible for growth, tissue maintenance and regeneration and repair of diseased or damaged tissue.

1.1.1 Stem cell self-renewal

During a stem cell division, one or both daughter cells maintain the stem cell phenotype. This process is called self-renewal. Stem cells can divide symmetrically or asymmetrically. It is the balance between symmetrical and asymmetrical divisions that determines the appropriate numbers of stem cells and differentiated daughters.

During a symmetric cell division, both daughter cells acquire the same fate; either undifferentiated (new stem cells) or differentiated.

During an asymmetric cell division, one daughter cell becomes a new stem cell; the other differentiates into a more specialized cell type (see Figures 1.1 and 1.2). Asymmetric cell divisions are controlled by intrinsic

and extrinsic mechanisms. Intrinsic mechanisms rely on the asymmetric partitioning of cell components, such as cell polarity factors or cell fate determinants. In the extrinsic mechanism, the two daughter cells are positioned asymmetrically in their environment and receive different external signals (Morrison and Kimble, 2006).

The past 25 years of research have given some insight into the mechanism by which a cell maintains its undifferentiated fate. Since self-renewal involves both proliferation and the maintenance of an undifferentiated phenotype, multiple pathways are involved. Stem cells from different tissues or at different stages of developmental potential (pluripotent or multipotent) use different mechanisms to regulate self-renewal. The pathways regulating self-renewal are depending on the context. Factors that might stimulate differentiation of one cell type, might be involved in the maintenance of self-renewal of another stem cell. Some mechanisms and interactions are still unknown, some are debatable, and others are well described. Self-renewal of embryonic and adult stem cells is described in section 1.2.3 and 1.3.3.

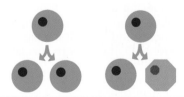

Figure 1.1 Stem cell characteristics. Upon cell division, a stem cell (green circle) can produce a new stem cell (self-renewal), and a differentiated daughter cell (orange hexagon). On the left, a symmetrical is shown and on the right an asymmetrical cell division.

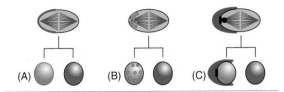

Figure 1.2 Controls of asymmetric stem cell division. Three simple mechanisms are shown. Stem cells are orange, differentiated cells are green. A, Asymmetric localization of cell polarity regulators (red) initiates the asymmetric division; B, Cell fate determinants (red) can be segregated to the cytoplasm of one daughter cell, as shown here, or they can be associated with the membrane, centrosome or another cellular constituent that is differentially distributed to the daughters; C, Regulated orientation of the mitotic spindle retains only one daughter in the stem-cell niche (red), such that only that daughter cell has access to extrinsic signals necessary for maintaining stem-cell identity. This mechanism achieves an asymmetric outcome, even though the division itself is intrinsically symmetric. In an alternative but similar model, the daughter cell placed away from the niche is exposed to signals that induce differentiation. Reproduced with permission, Morrison, S.J. and Kimble, J. (2006). Asymmetric and symmetric stem-cell divisions in development and cancer. *Nature*, 441: 1068–1074.

1.1.2 Differentiation

1.1.2.1 Can a stem cell become everything it wants to be?

The second defining property of a stem cell is its ability to differentiate into a more specialized cell. The number of cells types a stem cell can differentiate into is determined by its potency:

- Totipotent stem cells have the ability to form an entire organism. The fertilized oocyte and the cells after the first cleavage divisions are considered totipotent.
- Pluripotent stem cells are able to form all three germ layers including germ cells, but not the extra-embryonic tissue as placenta and umbilical cord. Cells of the inner cell mass of the blastocyst are pluripotent. When these cells are brought into culture, they are called embryonic stem cells.
- Multipotency means the ability to form multiple cell types. Mesenchymal stem cells can differentiate into cells that form bone, cartilage and fat.
- Oligopotent stem cells can differentiate into two or more lineages, for example neural stem cells that can form a subset of neurons in the brain.
- Unipotency is the ability to form cells from a single lineage, for example spermatogonial stem cells.

The term omnipotence is not used for stem cells, but is used in religions as one of God's characteristics.

1.1.2.2 Stem cells, precursor cells and differentiated cells

Once a stem cell leaves its niche (see section 1.3.4) and is no longer under control of intrinsic and extrinsic factors that maintain the undifferentiated phenotype, they will start to differentiate. This cell will become a progenitor or precursor cell, or a transit amplifying cell ('transit', because they are in transit from a stem cell to a differentiated cell; 'amplifying' because the continuing cell divisions amplify the number of differentiated progeny). The committed cell can differentiate further along a specific lineage, until it is terminally differentiated into the mature phenotype (Figure 1.3). These cells are presumably irreversibly blocked in their ability to proliferate, but they can perform specialized functions for a long period of time before they die.

The progenitor cells can divide many times, ultimately giving rise to thousands of fully differentiated cells that have originated from one stem cell division. This explains why the number of stem cells is so small and that stem cell division rate is low. For example, in bone marrow only an estimated 1 in 10,000 to 15,000 cells is considered to be a stem cell. Nevertheless, billions of new blood cells are formed every day.

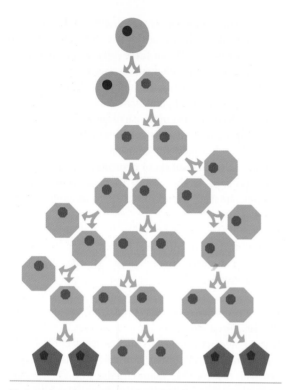

Figure 1.3 One stem cell division results in many differentiated cells via transit amplifying cells. The stem cells are green circles, the progenitor, transit amplifying cells are orange hexagons, and the differentiated cell are red pentagons.

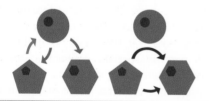

Figure 1.4 Dedifferentiation, redifferentiation and transdifferentiation. On the left, a differentiated cell dedifferentiates (pink arrow), and redifferentiates (red arrows) into the same phenotype (pentagon) or a different phenotype (hexagon). On the right, a differentiated cells transdifferentiates (blue arrow) into another differentiated phenotype, sometimes via an intermediate cell type (circle).

1.1.2.3 Dedifferentiation, redifferentiation and transdifferentiation

Differentiation might not entirely be a one-way street. Some differentiated cells can dedifferentiate into a less mature phenotype (Figure 1.4). Chondrocytes for example, when removed form their extracellular matrix and cultured *in vitro* on tissue culture plastic, will loose their cartilage phenotype. They stop expressing the cartilage-specific marker collagen type II and change morphology from a rounded chondrocyte to a stretched fibroblast-like cell (von der Mark *et al.*, 1977). When growth factors, for example TGFβ (transforming growth factor), are added to the culture medium, they will redifferentiate into chondrocytes and start expressing collagen type II again.

Transdifferentiation is a switch of a differentiated cell into another differentiated cell, either within the same, or into a completely different tissue (Figure 1.4). Transdifferentiation does not necessarily involve dedifferentiation and redifferentiation. When the switch of gene expression happens quickly, there will be coexistence of markers from both cell types for a short time. Transdifferentiation can be induced by modifying the gene expression of cells. An example of induced transdifferentiation in mammals is the conversion of pancreatic cells to hepatocytes (reviewed by Slack and Tosh, 2001). Whether or not transdifferentiation occurs *in vivo* is still controversial.

1.1.2.4 Plasticity of stem cells

For many years, researchers thought that adult stem cells could only generate cells of the tissue in which they reside. However, experiments in the past 10 years have shown that adult stem cells may be capable of differentiating across tissue lineage boundaries, sometimes even across germ layers (Figures 1.5 and 1.6A). This is called plasticity: the ability of adult stem cells from one tissue to generate the specialized cell type of another tissue. For example, hematopoietic stem cells might contribute not only to the formation of blood cells, but also to the formation of for example skin, liver, brain and heart. Some studies claim the contribution of brain and muscle derived stem cells to the formation of blood cells.

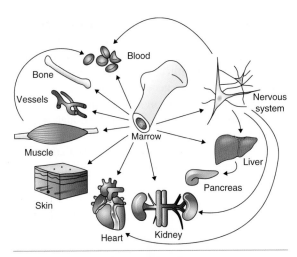

Figure 1.5 Adult stem cell plasticity: Too good to be true? Studies in mice yielded evidence, now being reassessed, that stem cells from a variety of tissues can produce progeny in different organs. Bone marrow, which has several types of stem cells, seems particularly versatile. Reproduced with permission from Holden C, Vogel G. (2002). Stem cells. Plasticity: time for a reappraisal? *Science*, 296: 2126–2129.

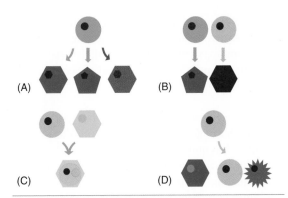

Figure 1.6 Plasticity of adult stem cells (A) and possible explanations for apparent plasticity (B-D). A, An adult stem cell from one tissue (light green), cannot only form differentiated progeny of its own tissue (dark green), but also of another tissue (red, on the right); B, A heterogeneous population of (stem) cells (light green and blue) results in diverse differentiated cells (dark green and blue); C, Fusion of a stem cell (blue) with a differentiated cell (yellow) results in a hybrid cell displaying a differentiated phenotype; D, True plasticity would result in differentiation outside the stem cells own tissue, as indicated by GFP signal of the differentiated cell on the left (green nucleus). The actual signal might be background staining of autofluorescent neighboring cells on the right.

Recent literature on stem cell plasticity demonstrates a substantial amount of papers dealing with the controversy that surrounds plasticity. There are many possible explanations for the apparent plasticity of adult stem cells (reviewed in Wagers and Weissman, 2004). The stem cell population used for experiments might not be homogeneous (Figure 1.6B). In experiments performed with such heterogeneous cell populations, distinct cell types could contribute to the observed outcome. Many experiments were performed with bone marrow-derived cells, a population known to be very heterogeneous, even after some purification steps. Ideally, the experiments should be performed with clonally derived stem cells. The stem cell populations, although isolated from one tissue, might also contain circulating stem cells. Hematopoietic stem cells, for example, can circulate in the blood, thereby contaminating many non-hematopoietic tissues.

Another explanation of plasticity might be cell fusion. The resulting cells are tetraploid hybrid cells (Figure 1.6C). Spontaneously fused bone marrow cells can subsequently adopt the phenotype of the recipient cells, which, without detailed genetic analysis, might be interpreted as plasticity (Terada *et al.*, 2002; Ying *et al.*, 2002). Others claim that in some tissues, such as liver and muscle, cell fusion is a natural process. Whether this fusion process results in functional tissue cells is, however, still unclear.

Technical problems might also account for some of the plasticity claims. Many of these experiments have been performed with Green Fluorescent Protein-labeled (GFP) stem cells. However, skeletal muscle fibers for example exhibit autofluorescence, resembling the GFP signal. Consequently, in experiments where GFP-labeled stem cells were analysed for their plasticity, the fluorescent signal might not have come from apparently transdifferentiated stem cells, but from the autofluorescing muscle fibers (Jackson *et al.*, 2004) (Figure 1.6D).

1.1.2.5 *Differentiation of stem cells* in vitro

It is one challenge to keep stem cells undifferentiated in culture; it is quite another challenge to differentiate the cells into the desired tissue. A major challenge is differentiating the pluripotent or multipotent stem cells into a homogeneous population of cells. Directed differentiation of stem cells *in vitro* typically involves changing the culture medium. Addition of growth factors, cytokines, or other proteins to the culture medium can induce differentiation into a specific lineage. This involves cell signaling and transcriptional responses, as described in Chapter 5. Another option is changing the culture environment of a cell, for example culturing in 3D pellets instead of 2D adherent cultures. Stem cells can also be co-cultured with cells of the differentiated phenotype, either in direct contact, in a trans-well system through which only medium components and no cells can diffuse, or by adding conditioned medium of those differentiated cells. Further examples of differentiation *in vitro* are given in Chapter 11, Cell Culture.

1.1.2.6 *Epigenetics and differentiation*

All differentiated cells originate from the same fertilized egg, and thus contain the same genetic material. However, the cell morphology and function changes dramatically during differentiation. Growth factors that can stimulate stem cells to differentiate into the neuronal lineage, cannot stimulate heart cells to become neurons. The diversity in cell types is caused by differential gene expression patterns. During differentiation, stem cell self-renewal genes have to be silenced, and only the tissue-specific genes have to be transcribed. Upon cell division, this expression pattern has to be passed onto the daughter cells. This stable change in gene expression is coordinated by epigenetic mechanisms. Epigenetics, an emerging field of research, may be defined as the stable alterations

in gene expression potential that arise during development (differentiation) and cell proliferation, without altering the DNA sequence. This is regulated through the chromatin structure by different mechanisms. Two major epigenetic mechanisms are DNA methylation and histone modifications.

DNA can be covalently modified through methylation, primarily on cytosines of the dinucleotide sequence CpG. Regions of the genome that have a high density of CpGs are called CpG islands. Many

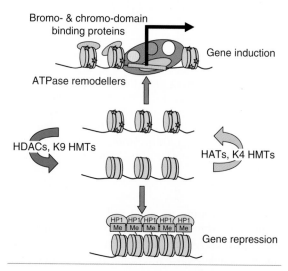

Figure 1.7 The histone switch. Targeted modifications under the control of histone methylases (HMTs), histone acetyltransferases (HATs) and histone deacetylases (HDACs) alter the histone code at gene regulatory regions. This establishes a structure that contains bromo- and chromo-domains that permits recruitment of ATP-dependent chromatin remodelling factors to open promoters and allow further recruitment of the basal transcription machinery. Deacetylation, frequently followed by histone methylation, establishes a base for highly repressive structures, such as heterochromatin. Acetylated histone tails are shown as yellow stars. Methylation (Me) is shown to recruit heterochromatin protein 1 (HP-1). Reproduced with permission from Adcock, I.A., Ford, P., Ito, K. and Barnes, P.J. (2006). *Respiratory Research* 7: 21.

tissue-specific promoter regions contain CpG islands, and the hypermethylation of promoter-associated CpG islands suppresses gene expression. The methylation patterns are passed on to daughter cells during cell division, by the action of DNA methyltransferases.

Chromatin is the structure of genomic DNA of eukaryotic cells that is compacted on nucleosomes. Nucleosomes consist of a histone octamer containing two of each of the histones H2A, H2B, H3 and H4 around which the DNA is wound. Post-translational modification of histone proteins at their N-terminal 'tail' include methylation, acetylation, ubiquitylation, sumoylation, phosphorylation and addition of ADP-ribosyl groups (Figure 1.7).

These modifications can influence chromatin compaction and accessibility for transcriptional complexes. More condensed chromatin, marked by histone methylation, is less accessible for gene transcription. In stem cells, chromatin is less compacted, marked by acetylated histones, than in differentiated cells. Differentiation is accompanied by a successive restriction in the repertoire of genes that can be expressed. This implies a close relationship between differentiation potential and chromatin remodeling.

Epigenetics also plays a role in the maintenance of pluripotency. Recent research suggests that cell fate can be reset by epigenetic reprogramming. By demethylating DNA, for example demethylation of the *oct4* promoter region, cells can regain a pluripotent phenotype. The chromatin can be remodeled to be more accessible for gene transcription. There is growing evidence that epigenetic modifications are the core machinery required for nuclear reprogramming and cell-fate conversion. These remarkable findings suggest that epigenetics provide an important new research field for improving regenerative medicine. An example of epigenetic reprogramming is given in the State of the Art Experiment.

State of the Art Experiment

Reprogramming of adult cells into a pluripotent ES-cell like state

Human embryonic stem cells are a promising cell source for treatment of diseases and injuries, because they provide an unlimited supply of cell that can differentiate into cells of all three germ layers (Thomson *et al.*, 1998). However, there are still technical (immune rejection of transplanted ES cells) and ethical problems (use of human embryos) before ES cells can be used in patients. The problem of rejection can be overcome by a technique called somatic cell nuclear transfer. In this technique, the nucleus from an egg is replaced by the nucleus of a somatic cell (Wilmut *et al.*, 1997). In another experiment, an ES cell was fused with a somatic cell (Cowan *et al.*, 2005; Tada *et al.*, 2001). The ES cell reprograms the somatic cell chromosomes to an embryonic state. Both experiments are technically challenging and do not solve the ethical problem. However, from these experiments it has become clear that somatic cells can be reprogrammed into an ES cell-like state.

In a major breakthrough article, Takahasi *et al.* (2006) hypothesized that the factors that play an important role in the maintenance of the ES cell identity may be able to induce pluripotency in somatic cells. They selected 24 genes as candidate factors and after several experiments they narrowed this down to 4 transcription factors: Oct-4 and Sox2, which function in maintaining pluripotency in early embryos and ES cells, and c-Myc and Klf-4, which contribute to the maintenance ES cell phenotype and their rapid proliferation. By retroviral introduction of the four transcription factors in mouse fibroblasts, these differentiated somatic cells were reprogrammed into a pluripotent state. These cells were called induced pluripotent stem (iPS) cells. The iPS cells displayed a similar morphology and growth phenotype as ES cells, expressed some ES cell markers. After subcutaneous implantation into immunodeficient

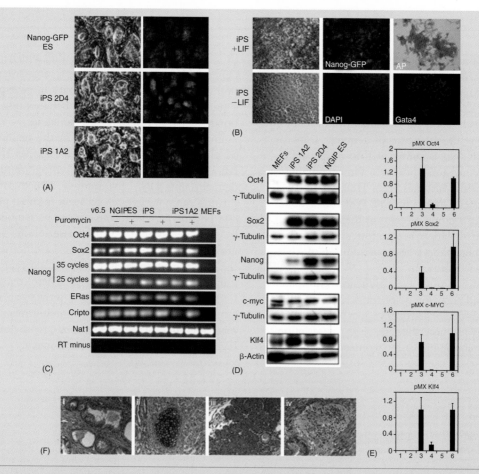

ES Cell-like Properties of Nanog-Selected iPS Cells. (A) Morphology and Nanog promoter-driven GFP expression in ES cells (Nanog-GFP ES) and two iPS cell lines grown on feeders in the absence of puromycin selection; (B) effect of LIF withdrawal on iPS cells. Cells were grown for three passages without feeders. In the presence of LIF, 2D4 iPS cells maintain an ES-like morphology, express endogenous Nanog as indicated by GFP expression, and are alkaline phosphatase (AP) positive. Upon LIF withdrawal, iPS cells upregulate the primitive endoderm marker Gata4 as detected by immunostaining. A phase contrast image and counterstaining of the same cells with DAPI is shown; (C) RT-PCR analysis of ES cell marker gene expression in Nanog-GFP (NGiP) ES cells, and two iPS cell lines grown with and without continued puromycin selection, as well as in wild-type ES cells (V6.5) and MEFs as additional reference points. Primers for Oct4 and Sox2 are specific for transcripts from the respective endogenous locus. Nat1 was used as a loading control; (D) Western blot analysis for expression of Nanog, Oct4, Sox2, c-myc, and Klf4 in iPS cell lines, MEFs, and Nanog-GFP (NGiP) ES cells. Anti-tubulin and anti-actin antibodies were used to control for loading; (E) Quantitative PCR analysis of pMX retroviral transcription in (1) wild-type MEFs, (2) wild-type ES cells, (3) cells from the heterogeneous iPS line 1A2 before sorting and subcloning, (4) 1D4 iPS, (5) 2D4 iPS, and (6) MEFs infected with the respective pMX virus. Transcript levels were normalized to β-actin. It should be noted that the retroviruses in the 2D4 iPS line appear completely silenced, whereas the heterogeneous 1A2 line still shows abundant expression of the exogenous factors. Error bars represent the standard deviation of triplicate reactions; (F) Teratoma derived from iPS line 1A2 showing differentiation into cell types from all three germ layers: epithelial structures (i), cartilage with surrounding muscle (ii), glandular structures (iii), and neural tissue (iv). Reproduced with permission from Maherali, N., Sridharan, R., Xie, W. et al. (2007). Directly reprogrammed fibroblasts show global epigenetic remodeling and widespread tissue contribution. Cell Stem Cell, 1: 55–70.

mice, they formed teratomas containing tissues originating from of all three germ-layers. However, iPS cells were not identical to ES cells. The gene expression pattern and epigenetic state was different, and the iPS cells failed to produce chimaeras.

In three recent publications, a second generation of iPS cells was presented (Maherali *et al.*, 2007; Wernig *et al.*, 2007; Okita *et al.*, 2007). All groups used Nanog for the selection of reprogrammed cells, and this proved to be a better approach. The iPS cells obtained with this selection strategy did generate viable adult chimaeras (see Figure 1.12), contributed to the germ line, and had an epigenetic state that was similar to that of ES cells.

If it is possible to reprogram human somatic cells as well, this technique may solve the ethical problems surrounding the use of human ES cells. Theoretically, it will enable the generation of patient specific iPS, thereby circumventing immuno-rejection. However, there are still numerous problems to be solved. The efficiency of reprogramming should be increased. So far, the experiments were only performed with mouse cells. The method has to be applied to human cells, which most likely will require other factors. For therapeutic applications, the use of retroviruses to introduce the factors should be avoided.

1.2 Embryonic stem cells

Embryonic stem cells do not exist in the body. When cells are isolated from the inner cell mass of the blastocyst, they can be massively expanded in the laboratory, while maintaining their pluripotency (self-renewal). These *in vitro* propagated cells are called embryonic stem cells. Mouse ES cells were the first to be isolated (Evans and Kaufman, 1981; Martin, 1981). The next major breakthrough was in 1998, when Thomson *et al.* (1998) isolated ES cells from human embryos.

1.2.1 Isolation of embryonic stem cells

Mouse embryonic stem cells can be isolated from super-ovulated or naturally mated females. After 3.5 days, the pregnant mice are sacrificed and the blastocyst stage embryos are flushed from the uterine horn. The blastocyst contains the trophectoderm (the outer layer of cells), a fluid filled cavity called the blastocoel, and an inner cell mass. The embryos are transferred to a culture dish, and after attachment, the inner cell mass can be isolated from the rest of the embryo by aspirating it into a pipette. The ICM is then transferred to a new dish, and examined for undifferentiated morphology (Figure 1.8).

The cell colonies that grow from these cells have to be dissociated every few days, to prevent differentiation of the embryonic stem cells. This is usually done by the addition of trypsin, an enzyme that dissociates cells from each other and from the plastic of the culture dish. The single cells will form new colonies, and some of these colonies will remain undifferentiated.

To keep mouse embryonic stem cells undifferentiated, they have to be grown in optimal conditions. The mouse ES cells attach and grow on a feeder layer prepared from mouse embryonic fibroblasts, which are mitotically inactivated, either by irradiation, or treatment with the toxic antibiotic mitomycin-C. The feeders cells do not replicate, but they do produce mostly unknown factors that keeps the embryonic stem cells undifferentiated. The discovery of Leukemia Inhibitory Factor (LIF) in 1988 (Williams *et al.*, 1988), allowed researcher to grow mouse ES cell in the absence of a feeder layer.

Flushing the ovary ducts of a pregnant woman is not an option for the isolation of human embryonic stem cells. Therefore, surplus embryos of IVF treatment that are donated after informed consent of the parents are used. First an oocyte is fertilized *in vitro* by a sperm cell. The zygote, the fertilized egg, is grown *in vitro* until it reaches the blastocyst stage. Instead of being transferred to the uterus, these 5-day old blastocysts are used to isolate human embryonic stem cells. The blastocysts contain approximately 200–250 cells, of which 30–34 cells form the inner cell mass. The trophectoderm can be removed by mechanical surgery (cutting with a small scalpel) or immunosurgery (antibodies break down the trophectoderm).

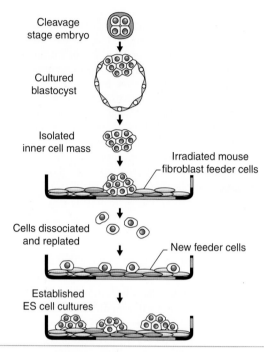

Cleavage
stage embryo

Cultured
blastocyst

Isolated
inner cell mass

Irradiated mouse
fibroblast feeder cells

Cells dissociated
and replated

New feeder cells

Established
ES cell cultures

Figure 1.8 Derivation of human ES cell lines. Human blastocysts were grown from cleavage-stage embryos produced by in vitro fertilization. ICM cells were separated from trophectoderm by immunosurgery, plated onto a fibroblast feeder substratum in medium containing fetal calf serum. Reproduced with permission from Odorico, J.S., Kaufman, D.S. and Thomson, J.A. (2001). Multilineage Differentiation from Human Embryonic Stem Cell Lines, *Stem Cells*, 19: 193–204.

The inner cell mass is cultured on a feeder layer, similar to the isolation of mouse embryonic stem cells. However, LIF cannot keep human ES cells undifferentiated. Human ES cells have to be cultured on a feeder layer in the presence of serum or serum-replacement in combination with basic fibroblast growth factor (bFGF).

Several human ES cell lines cannot be dissociated by the use of trypsin. Therefore, these colonies are mechanically dissected by cutting them in pieces with a knife made of a glass capillary (Figure 1.9). The colony pieces are then transferred to a new dish with feeder cells.

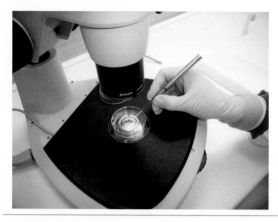

Figure 1.9 Human ES cell colony transfer. Colonies are cut into pieces with a cutting pipette made from a glass capillary and transferred to a dish with new feeder cells.

Embryonic stem cells are very sensitive to temperature and pH change, and when colonies overgrow, they also tend to differentiate. Therefore, ES cells have to be cared for every day, also in the weekend and during holidays.

1.2.2 Characterization of embryonic stem cells

The derived embryonic stem cell lines will be cultured for months, to ensure their self-renewal capacity. Human embryonic stem cells have been reported to proliferate for years and go through hundreds of population doublings (Hoffman and Carpenter, 2005).

There are some markers that can be used to determine the undifferentiated state of embryonic stem cells. The best characterized is Oct4. Undifferentiated cell express the *Pou5f1* gene, which encodes for the transcription factor Oct4. Loss of pluripotency of ES cells is often accompanied by a down regulation of Oct4 expression. Other markers are the enzyme alkaline phosphatase (ALP), stage-specific embryonic antigen (SSEA)-1 for mouse ES cells, and SSEA-3 and SSEA-4 for human ES cells, and tumor rejection antigen

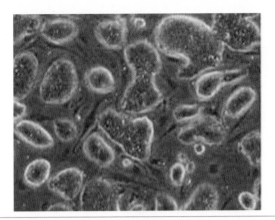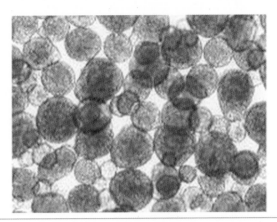

Figure 1.10 The onset of ES cell differentiation. Left: Mouse embryonic stem cells in colonies attached to culture plastic. Right: Embryoid bodies floating in the culture medium.

TRA1-60 and TRA1-81 for human ES cells. ES cells also express high levels of telomerase.

The pluripotency of the ES cells can be identified both *in vitro* and *in vivo. In vitro* differentiation generally starts with the formation of embryoid bodies (EBs): free floating aggregates of randomly differentiating cells (Figure 1.10). When ES cells are placed in a non-adherent bacterial dish or in small droplets hanging from a bacterial lid (hanging drop method), they will spontaneously form cell aggregates in which cells start differentiating in a fashion that resembles early post-implantation embryos. Cell types of all the three germ layers (ectoderm, mesoderm and endoderm) are formed. Once the EBs are allowed to attach to a culture dish, differentiated cells will grow out of the aggregates. These can be identified by morphology (for example spontaneously contracting cardiac muscle cells), or immunostaining for specific cell types.

An *in vivo* method for determining the pluripotency of embryonic stem cells is the injection of embryonic stem cells under the skin or in the kidney or testis of an immunodeficient mouse. A benign tumor, called teratoma, will form and advanced tissue types of all three germ layers can be identified (Figure 1.11), for example gut epithelium (endodermal), cartilage and bone (mesodermal) and neural tissue (ectodermal).

The ultimate proof of pluripotency of mouse ES cells is the formation of chimeric mice, in which the cells have contributed to the formation of all tissues, including germ cells. This has only been achieved for mouse embryonic stem cells. First the researcher has to test whether the number of chromosomes is normal and whether the chromosomes are not damaged. This can be done by karyotyping. Next, an ES cell can be injected into the cavity of a blastocyst, and transferred to the uterus of a pseudo-pregnant mouse. The offspring are chimeric mice, of which all tissues are composed partly of host cells and partly of the donor embryonic stem cells (Figure 1.12).

1.2.3 Self-renewal of embryonic stem cells

1.2.3.1 *Self-renewal of mouse ES cells*
Cytokines or growth factors have to be added to the culture medium to keep mouse ES cells undifferentiated. One signaling pathway involved in the self-renewal of mouse ES cells is the LIF-STAT3 pathway. Leukemia Inhibitory Factor (LIF) binds to a receptor complex of the LIF and gp130 receptor, which triggers the activation of the transcription factor STAT3 (Signal Transducer and Activators of Transcription). LIF cannot support self-renewal in the absence of serum,

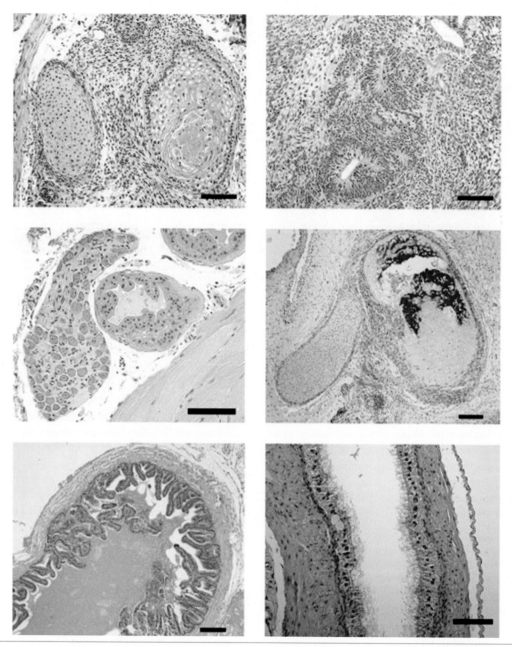

Figure 1.11 Histology of differentiated elements found in teratomas formed in the testis of immunedeficient SCID mice following inoculation of two human ES cell colonies (HES-1 and HES-2). A, Cartilage and squamous epithelium, HES-2; B, Neural rosettes, HES-2; C, Ganglion, gland, and striated muscle, HES-1; D, Bone and cartilage, HES-1; E, Glandular epithelium, HES-1; F, Ciliated columnar epithelium, HES-1. Scale bars: (A–E) 100 μm; (F) 50 μm. Reproduced with permission from Reubinoff, B.E., Pera, M.F., Fong, C.Y. *et al.* (2000). Embryonic stem cell lines from human blastocysts: somatic differentiation *in vitro*. *Nat Biotechnol*, 18: 399–404.

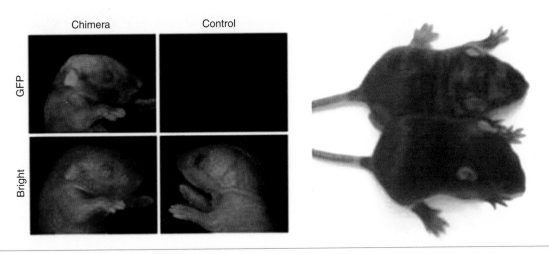

Figure 1.12 Chimeric mice from ES-like cells. Left: Cells from iPS line 2D4 that carried a randomly integrated GFP transgene were injected into blastocysts. Surrogate mothers gave birth to GFP-positive pups. A nonchimeric pup not expressing GFP is shown; Right: Ten-day-old chimeric mouse derived from blastocyst-injected 2D4 iPS cells, shown next to a wild-type littermate. iPS-derived cells are responsible for the agouti coat color (iPS cells: induced pluripotent stem cells as described in the State of the Art Experiment). Reproduced with permission from Maherali, N., Sridharan, R., Xie, W., *et al.* (2007). Directly reprogrammed fibroblasts show global epigenetic remodeling and widespread tissue contribution. *Cell Stem Cell*, 1: 55–70.

indicating that its activity is dependant on one or more factors that are present in the serum. A possible candidate is BMP4 (bone morphogenetic protein 4), which induces the expression of transcription factors of the Id (inhibitors of differentiation)-family (Ying, 2003). These transcription factors block differentiation.

One of the master genes of mouse ES cell pluripotency is the transcription factor Nanog (Chambers *et al.*, 2003; Mitsui *et al.*, 2003). Nanog is named after Tir Na Nog, Land of Ever Young or Land of Eternal Youth in Irish mythology. *In vivo*, Nanog is expressed in the morula. In the blastocyst stage, expression is limited to the inner cell mass (ICM). *In vitro*, Nanog is enriched in undifferentiated ES cells, but is downregulated in differentiating ES cells and in adult tissues. In mouse ES cells, high levels of Nanog can maintain pluripotency in the absence of LIF, and Nanog enables human ES cells to grow undifferentiated in the absence of feeders.

Overexpression of Nanog does not seem to affect phosphorylated STAT3 levels, nor does elevated STAT3 signaling result in changed Nanog expression. Despite all this, a functional STAT3 binding site is present in the Nanog promoter region. How Nanog and STAT3 signaling cooperate in the maintenance of pluripotency is still largely unclear.

In contrast, a direct interaction between Nanog and BMP4 signaling has been described. BMP4 signaling is mediated via the activation of SMAD1. Nanog can interact with SMAD1, and thereby inhibits the actions of BMP. Since BMPs are also involved in mesodermal differentiation, the inactivation of the BMPs by Nanog can help in maintaining the undifferentiated state of ES cells.

The expression of Nanog is regulated by a number of transcription factors (Figure 1.13). The most important are Oct4 and Sox2, which can bind to an Oct4/Sox2 motif in the Nanog promoter, thereby activating Nanog transcription. However, in *oct4*-deficient mice, Nanog is still expressed. Therefore, other factors must also be involved in regulating Nanog expression. FoxD3, a transcription factor of the

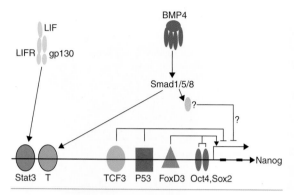

Figure 1.13 Regulation of Nanog expression. FoxD3 and Oct4/Sox2 bind to the proximal region of Nanog promoter and support its expression. TCF3 and p53 also bind to the promoter and negatively regulate.

forkhead family, is highly expressed in ES cells and can bind to an enhancer in the Nanog promoter, thereby activating gene transcription. Interestingly, Nanog can also positively autoregulate its own expression.

To allow differentiation during embryonic development, Nanog must also be negatively regulated. The tumor suppressor p53 and Tcf3, a transcription factor in the Wnt pathway, are considered to be negative regulators of Nanog expression in ES cells.

As for Nanog, other key transcription factors involved in maintaining pluripotency, i.e. Oct4, Sox2 and FoxD3, can positively autoregulate their expression. In addition, they also bind to each others promoter region, thus forming a negative feedback loop. Nanog, Oct4 and Sox2 cooperatively regulate the expression of many target genes. In conclusion, several key transcription factors involved in maintenance of pluripotency activate or inhibit each other's expression and they simultaneously regulate a set of target genes. These cooperative actions result in the formation of a regulatory network, that balances the maintenance of self-renewal and the ability of stem cell differentiation (reviewed in Pan and Thomson, 2007).

1.2.3.2 Self-renewal of human ES cells

One might expect that mouse and human embryonic stem cells self-renewal are regulated by similar mechanisms. However, it appears to be regulated by many different pathways. The LIF/STAT pathway fails to maintain self-renewal of human embryonic stem cells (Daheron, 2004) and BMP4 seems to stimulate differentiation of human ES cells. Apparently, the feeder cells produce other unknown factors to support the self-renewal of human ES cells. TGFβ family signaling, and especially the TGFβ/Activin/Nodal pathway, plays a role. Activin A may be one of the critical factors produced by the feeder cells. Prolonged culture in the absence of feeders is possible in the presence of TGFβ1, bFGF and LIF (Amit *et al.*, 2004). It is well known that bFGF can maintain the pluripotency of human ES cells (Figure 1.14) but the detailed mechanism is still unclear. Many other pathways have been associated with the self-renewal of human ES cells, but there appear to be differences between individual ES cell lines and culture conditions (feeder or feeder-free cultures on extracellular matrix products such as laminin or Matrigel).

The transcriptional regulatory network involved in self-renewal is being elucidated piece by piece. Oct4 is expressed by mouse and human ES cells, but in itself is insufficient to maintain self-renewal. Nanog expression is high in human ES cells, and is reduced in differentiating cells. Some similarities between the mechanism through which Nanog regulates the maintenance of pluripotency in mouse and human ES cells, have been identified, but the exact regulatory network is not clear yet.

1.2.4 Differentiation of human embryonic stem cells

Human embryonic stem cells might be an ideal cell source for tissue engineering and regenerative medicine, because of their indefinite proliferation capacity and pluripotency. Embryonic stem cells are often mentioned as a promise for the cure of Parkinson's disease, diabetes and cardiovascular diseases. This inevitably means that human ES cells have to be differentiated into respectively neurons, insulin-producing cell and cardiomyocytes. Indeed, many articles are published

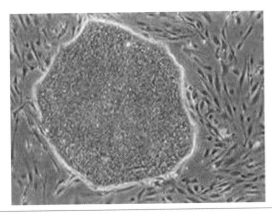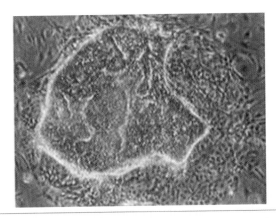

Figure 1.14 Human ES cell colonies on feeder cells. The colony on the left displays successful self-renewal. The colony on the right contains many differentiated cells, as can be seen from the heterogenic pattern of the colony and the outgrowth of cells.

in which the *in vitro* and *in vivo* differentiation of human ES cells is described. Useful cell types such as neurons (Carpenter *et al.*, 2001; Reubinoff *et al.*, 2001; Zhang *et al.*, 2001), cardiomyocytes (He *et al.*, 2003; Mummery *et al.*, 2003), hepatocytes (Lavon *et al.*, 2004; Rambhatla *et al.*, 2003), pancreatic beta cells (Assady *et al.*, 2001), endothelial cells (Levenberg *et al.*, 2002), blood cells (Chadwick *et al.*, 2003; Kaufman *et al.*, 2001) and chondrocytes (Vats *et al.*, 2006) have all been successfully derived in the laboratory. Levenberg and co-workers (2003) demonstrated the differentiation of human ES cells on polymeric scaffolds into 3D structures with characteristics of developing neural tissues, cartilage, liver, or blood vessels.

1.2.5 Applications of embryonic stem cells

1.2.5.1 *Application of mouse ES cells*
It is obvious that mouse ES cells will never find clinical applications. However, the knowledge of mouse ES cells was used for the isolation, growth and differentiation and thus possible future application of human ES cells. Furthermore, mouse ES cells are used as a model system to study early embryonic development and differentiation.

Another valuable feature of mouse ES cells is the ability to create genetically modified mice. When genes are introduced into an ES cell, and these ES cells contribute to the germ cells (eggs or sperm) of the chimeric progeny, it is possible to breed a line of genetically changed mice. In a knock-out mouse, the function of a gene is disturbed, and the biological function of the gene can be studied. When a mutation is introduced that is known to be the cause of a human genetic disease, the mice may serve as a model for this human disease. Much of the knowledge of stem cell self-renewal is also based on the use of these knock-out, knock-in and gene over-expression models.

1.2.5.2 *Application of human ES cells*
The first step that has to be taken before it is feasible to use human ES cells in therapeutic application is the optimization of the culture method. The current method of colony growth on feeder cells is time-consuming and labor-intensive, and large-scale propagation of these cells is not possible. Feeder-free growth of enzymatically passaged human ES cells would be a huge step forward.

Another optimization of the culture protocol would be the derivation and growth under animal product free conditions. After the discovery of

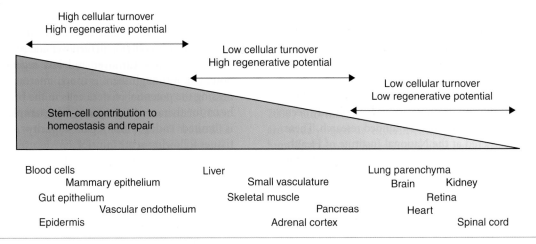

High cellular turnover
High regenerative potential

Low cellular turnover
High regenerative potential

Low cellular turnover
Low regenerative potential

Stem-cell contribution to
homeostasis and repair

Blood cells	Liver	Lung parenchyma
Mammary epithelium	Small vasculature	Brain	Kidney
Gut epithelium	Skeletal muscle	Retina
Vascular endothelium	Pancreas	Heart
Epidermis	Adrenal cortex	Spinal cord

Figure 1.15 Stem-cell contribution to homeostasis and repair. Reproduced with permission from Rando, T.A. (2006). Stem cells, ageing and the quest for immortality. *Nature*, 441: 1080–1086.

performed with stem cells from multiple patients, in order to validate the obtained results.

1.3.2 Characterization of adult stem cells

Adult stem cells are rare in the human body. The isolation of stem cells generally results in a heterogeneous population of cells. Many cell surface markers have been identified for the various stem cells. Combinations of several markers, or the absence of other markers allows researchers to enrich or purify the population. Many of these markers are CD molecules (Cluster of Differentiation).

Human haematopoietic stem cells are not as well defined as mouse haematopoietic stem cells yet. CD34 was the first differentiation marker to be recognized and is still the most commonly used marker to obtain enriched populations of human HSCs. This population still contains many other cell types. Other markers include the absence of CD38, the presence of CD43, CD45RO, CD45RA, CD59, CD90, CD109, CD117, CD133, CD166 and Lin. A combination of CD34 with one or more of these markers results in a highly purified population.

Mesenchymal stem cells can be characterized by the presence of STRO-1 (from bone marrow STROmal cells) and the absence of CD34. More markers have been identified, but the isolation of a homogeneous mesenchymal cell population has not been achieved (see also the discussion in Classical Experiment).

For tissue engineers, functional characterization is more important. The potency of a stem cell population to differentiate into various tissues can be analyzed *in vitro*, but the ultimate proof is formation of functional tissue *in vivo*.

For haematopoietic stem cells, the best described *in vivo* assay is the long-term repopulation assay. First, a mouse receives a dose of irradiation sufficient to kill its blood producing cells. Next, haematopoietic stem cells are injected into this lethally irradiated mouse. When the mouse recovers, and the injected cells have repopulated the entire haematopoietic system, these cells can be retransplanted into the next lethally irradiated mouse. When this mouse recovers as well, these cells are considered long-term stem cells capable of self-renewal (see Figure 2.1, Chapter 10).

For human mesenchymal stem cells, the *in vitro* formation of adipose tissue, cartilage, and mineralization

Multilineage potential of adult human mesenchymal stem cells

The paper by Pittenger and co-workers in the *Science* issue of April 1999 (Pittenger *et al.*, 1999) marked the broad introduction of the term 'mesenchymal stem cells' for a population of cells originally identified by Friedenstein in the 1960s as plastic-adherent colony-forming units fibroblast (CFU-F)

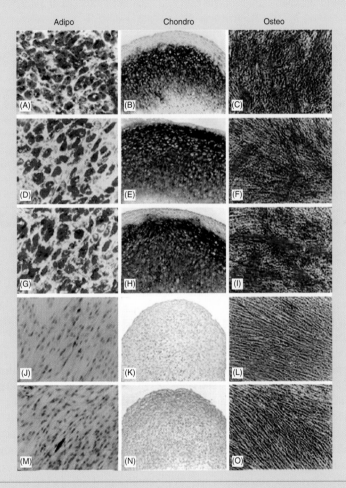

Isolated marrow-derived stem cells differentiate to mesenchymal lineages. Cultured cells from donors were tested for the ability to differentiate in vitro to multiple lineages. Donors (A through C) 158, (D through F) 177, and (G through I) 260 were each shown to differentiate appropriately to the adipogenic (Adipo), chondrogenic (Chondro), and osteogenic (Osteo) lineages. Adipogenesis was indicated by the accumulation of neutral lipid vacuoles that stain with oil red O (A, D, and G), and such changes were not evident (J) with Hs27 newborn skin fibroblasts or (M) with 1087Sk adult mammary tissue fibroblasts. Chondrogenesis was shown by staining with the C4F6 monoclonal antibody to type II collagen and by morphological changes (B, E, and H), which were not seen by similarly culturing (K) Hs27 or (N) 1087Sk cells. Osteogenesis was indicated by the increase in alkaline phosphatase (C, F, and I) and calcium deposition, which was not seen in the (L) Hs27 or (O) 1087Sk cells. Reproduced with permission from Pittenger, M.J., Mackay, A.M., Back, S.C. et al. (1999). Multilineage potential of adult human mesenchymal stem cells. Science, 284: 143–147.

isolated from bone marrow (Friedenstein *et al.*, 1970). Pittenger and co-workers obtained fifty bone marrow aspirates from 19–57-year-old donors and selected the mononuclear fraction from the aspirate using a density gradient. Most of the cells from this fraction belong to the haematopoietic lineage and will not adhere when brought into culture. A small percentage of the cells (0.001–0.01%) did adhere and they developed into symmetric colonies of cells. As such, the only difference with Friedenstein's isolation method is the purification of the mononuclear fraction. Flow cytometry analysis revealed that all cells were consistently positive for markers like, SH2, SH3, CD29, CD44, CD71, CD90 and CD106, but negative for the haematopoietic marker CD34. Despite this apparent homogeneity, MSCs isolated through this method still display large heterogeneity with respect to growth rate, phenotypic plasticity and colony morphology. Efficient expansion was achieved, with 50-375 million hMSCs within two cell passages. hMSCs, however, do not express telomerase and are subject to proliferative senescence. The researchers further demonstrated efficient differentiation of hMSCs into the adipogenic, osteogenic and chondrogenic lineage by varying culture conditions.

This landmark paper, which has been cited 2500 times within 8 years after publication, strongly advocates human mesenchymal stem cells as a model system for questions of cell biological nature, such as cell fate decision, plasticity and senescence as well as a readily available source of cells for tissue engineering purposes. However, the name "mesenchymal stem cell" raised an ongoing debate on the nature of and nomenclature for mesenchymal cells with multi-potentiality isolated from various parts of the body including bone marrow and fat. The senescent phenotype of hMSCs in culture argues against their stemness. Therefore, many researchers rather use terms like marrow stromal cells, mesenchymal progenitor cells or skeletal progenitor cells to describe a population of cells, which cannot be discriminated from hMSCs on basis of marker gene expression or differentiation potential.

is described in Classical Experiment. However, mineralization *in vitro* is not a proof of functional bone formation. Therefore, mesenchymal stem cells will have to be implanted and analyzed for bone formation (de Bruijn *et al.*, 1999; Haynesworth *et al.*, 1992). *In vivo* experiments with human mesenchymal stem cells can only be performed in immunodeficient animals (mice and rats) and are mostly performed ectopically (not in bone). Orthotopic implantations of mesenchymal stem cells in large bone defects cannot be performed in humans. Therefore, researchers use large animal models, like the goat, to analyse the bone forming capacity of goat mesenchymal stem cells (see Chapter 20, Tissue Engineering of Bone). Similarly, when bone marrow stromal cells are induced into the myogenic lineage, the cells do not only show characteristics of muscle cells *in vitro*. Upon transplantation, they also differentiate into muscle fibers (Dezawa *et al.*, 2005).

1.3.3 Self-renewal of adult stem cells

Developmental signaling pathways such as Notch, Wnt and Hedgehog signaling are involved in the self-renewal of many adult stem cells, all in a context-dependent manner. Besides these extrinsic factors, self-renewal is also regulated intrinsically.

Stem cell self-renewal is regulated through the chromatin structure. Polycomb group proteins (PcG) can repress transcription of genes linked to differentiation by regulating chromatin structure. In particular Bmi-1, Mel-18 and Rae-28 (members of the Polycomb repression complex 1) are involved in the maintenance of self-renewal in haematopoietic stem cells (HSCs). Overexpression of Bmi-1 promotes HSC self-renewal, by enhancing symmetrical cell divisions. *Bmi-1*-deficient cells have increased expression of the cell cycle inhibitor p16^{Ink4a}, which results in senescence and increased expression of p19^{Arf}, which is linked to apoptosis. Thus, the mechanism by which *Bmi-1* modulates HSC self-renewal seems to be through stimulation of symmetric cell division and the induction of survival genes and simultaneously the repression of anti-proliferative genes (Park *et al.*, 2003).

Other genes that are required for self-renewal are thought to negatively regulate differentiation. For example, *tlx* and *N-Cor* promote self-renewal of neural stem cells by inhibiting the differentiation towards astrocytes. *Sox* genes are involved in both self-renewal and differentiation of neural stem cells in a time and context dependant manner.

Furthermore, control of cell cycle and proliferation machinery is required for self-renewal regulation. Deficiency of the early G1 phase regulator $p16^{Ink4c}$ leads to increased HSC self-renewal. The late G1 phase regulator $p21^{cip1}$ deficiency leads to increased proliferation and thereby stem cell exhaustion. These mechanisms are summarized in Figure 1.16 and reviewed in Molofsky *et al.* (2004). In conclusion, somatic stem cells have tissue-dependent mechanisms for self-renewal.

1.3.4 Stem cell niche

The idea that stem cells are located in specific anatomical locations in adult tissue called 'niches' was introduced in 1978 by Schofield (1978). The niche is the stem cell microenvironment that provides a sheltering environment for the stem cells, in which the balance between stem cell quiescence and activity is maintained. Stem cells reside in the niche for an indefinite period of time while self-renewing and producing differentiated progeny. The balanced interaction between the stem cell, the niche cells, the extracellular matrices and secreted factors ensures the maintenance of the stem cell phenotype, and guides a differentiating daughter away from the niche (Ohlstein *et al.*, 2004).

Many recently characterized niches appear to be simple in structure and operate using common mechanisms. In a simple niche, stem cells are locked to niche cells by adherens junctions and to the extracellular matrix through, amongst others, integrins. The niche positions the stem cells to receive intercellular signals to control growth and inhibit differentiation (Figure 1.17a) In a more complex niche, different stem cells might be localized in the same niche, or more cell types contribute to the niche (Figure 1.17b). In a different type of niche, the storage niche, quiescent stem cells reside. These reserve stem cells are activated in case of wounding and subsequently divide and migrate to contribute to repair injured tissue (Figure 1.17c).

As a tissue engineer, it is important to realize that a stem cell, once isolated, purified and cultured, is devoid of its niche, and as a result, is not likely to behave as it would in the body. It might be crucial to understand the signals a cell receives when it is located in the niche, to be able to keep cells undifferentiated when placed in culture. Examples of stem cells and their niches are given in another chapter.

1.3.5 Replicative senescence and immortality

Embryonic stem cells can divide indefinitely. However, when a cell becomes differentiated, it has a restricted proliferation capacity. Mesenchymal stem cells also have a limited life span when grown *in vitro*. After 50–70 cell divisions (Harley *et al.*, 1990), the cell cannot divide anymore, goes into replicative senescence (referred to as the 'Hayflick phenomenon', in honour of Dr. Leonard Hayflick who was the first to publish it in 1965) and dies.

A phenomenon associated with senescence is telomere shortening. The ends of chromosomes are protected from degradation by a special chromatin structure known as the telomere. Telomeres consist of tandem repeats of TTAGGG, and during each cell division, repeats are lost as a result of incomplete replication. This successive shortening of telomeres eventually leads to loss of genetic material and results

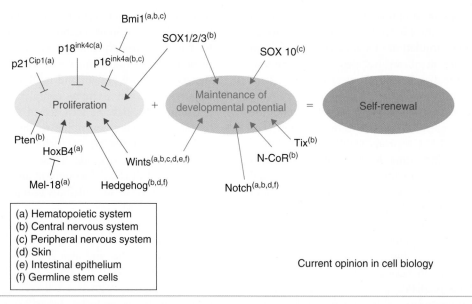

Figure 1.16 Regulators of somatic stem cell self-renewal can affect the ability of stem cells to proliferate, retain their developmental potential, or both. The maintenance of developmental potential includes establishing the competence to express each potential fate as well as inhibiting the act of lineage commitment and/or differentiation. It is possible that distinct mechanisms are employed to regulate competence as opposed to the actual decision to commit and/or differentiate. However, the precise mechanisms by which the depicted gene products regulate developmental potential is uncertain. This figure represents recent work in progress as additional regulatory proteins will exist, and the mechanisms by which the depicted proteins regulate self-renewal will continue to be elucidated. Reproduced with permission from Molofsky, A.V., Pardal, R. and Morrison, S.J. (2004). Diverse mechanisms regulate stem cell self-renewal. *Current Opinion in Cell Biology*, 16: 700–707.

in cell death. Embryonic stem cells express high levels of telomerase. The enzyme telomerase (Greider *et al.*, 1985) adds telomeric repeats to the chromosome ends, thus protecting the shortening of the chromosomes. In most other cells, telomerase activity is low or undetectable. Telomerase has been proposed as the key to cellular immortality, turning off the clock, which counts off the number of cell division before senescence.

For research purposes, the unlimited availability of cells is highly desired. So besides embryonic stem cells, immortal or immortalized cell lines can be used. This can either be tumor cell lines, such as the HeLa cell line, which was isolated from a tumor biopsy of a patient called Henrietta Lacks (hence the name HeLa) in 1951, and has been in culture ever since. Alternatively, cells can be immortalized

for example by introducing SV40 large TAg (Simian Vacuolating Virus 40 large T antigen), which is a powerful immortalizing gene. Cells that express SV40 large T antigen escape senescence but continue to lose telomeric repeats during their extended life span. After extended population doublings, they will eventually cease to proliferate as a result of chromosomal instability.

Cells can also be immortalized by restoring telomerase activity. When hTERT (human telomerase reverse transcriptase), the catalytic subunit of telomerase, is retrovirally introduced into human mesenchymal stem cells, the lifespan in these cells is extended. The cells maintain the ability to proliferate and differentiate over 3 years in culture. Thus, telomerization of human MSC by hTERT

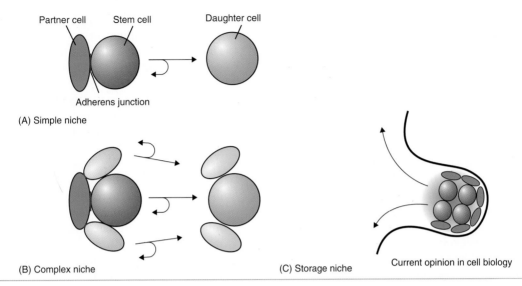

Figure 1.17 Proposed niche types: (a) Simple niche. A stem cell (red) is associated with a permanent partner cell (green) via an adherens junction (blue). The stem cells divides asymmetrically to give rise to another stem cell and a differentiating daughter cell (orange); (b) Complex niche. Two (or more) different stem cells (red and pink) are supported by one or more partner cells (green). Their activity is coordinately regulated to generate multiple product cells (orange and yellow) by niche regulatory signals; (c) Storage niche. Quiescent stem cells are maintained in a niche until activated by external signals to divide and migrate (arrows). Reproduced with permission from Ohlstein, B., Kai, T., deCotto, E. and Spradling, A. (2004). The stem cell niches: theme and variations. *Current Opinion in Cell Biology*, 16: 693–699.

overexpression may maintain the stem cell phenotype (Simonsen *et al.*, 2002).

1.4 Future perspective

Are stem cells a body self-repair kit? Can everybody get their own ES cells? Is it a hype, hope, or reality? Is it science, or science fiction? Is it too good to be true, too fast, too much, too soon, or too early to tell? The expectations are high, but can scientists meet them?

A lot of attention is being paid to stem cells, because their defining properties make them an ideal candidate to cure diseases. But a simple definition does not repair damaged tissue or replace failing organs. Much more knowledge is necessary, both fundamental and applied. If we look at the field of stem cell biology, the fundamental knowledge on the stem cell niche, self-renewal, early developmental processes, and epigenetics is growing every day. The next challenge is to apply this knowledge in the field of tissue engineering. Improved stem cell harvesting, stem cell culture, differentiation protocols and reprogramming of differentiated cells into a stem-cell like state will improve the chances of successful clinical applications. To achieve this, stem cell biologists and tissue engineers have to work together to bridge the gap between fundamental and applied science.

Human embryonic stem cell research is still in its infancy: they were discovered just 10 years ago. As already mentioned in section 1.2.5.2, embryonic stem cell researchers in tissue engineering still face many technical hurdles. A recent international collaborative study from the International Stem Cell Initiative (ISCI) compared 59 human ES cell lines from 17 laboratories from 11 countries (Adewumi *et al.*, 2007). The results from collaborative effort can be used to improve isolation, culture and differentiation protocols to make large-scale, clinical applicable culture available.

Adult stem cells have not been taken from the lab into the clinic at large scale. Bone marrow transplantation is by far the most successful therapy so far. In these transplants, a diseased of absent stem cell population is replaced by a fresh stem cell population. The body then instructs these cells to differentiate into the appropriate tissue. If we would like to replace damaged organs, we have to be able to differentiate the cells into a complex three dimensional tissue *in vitro* and/or *in vivo*. Therefore, we have to understand the complex cell-cell interactions that contribute to the function of the tissues. We might be able to differentiate a mesenchymal stem cell into the osteogenic lineage, but that does not result in a piece of vascularized bone as seen in the body. Stem cells differentiated into cardiomyocytes will have to be coupled to the patient's heart cells to be functional. So it is not just (initiation of) differentiation that has to be investigated, in the end only a functional tissue will contribute to tissue repair. Many studies have been performed with mouse stem cells, or human stem cells implanted into an immunodeficient mouse or rat. Because mice and men are not the same, successful results in mice will not automatically lead to (fast) successful results in humans.

The list of hurdles seems endless, but progress is being made. Progress towards the ultimate goal: an ES-like cell for every patient (see State of the Art Experiment)? It will take a lot of small steps, some breakthroughs, and unfortunately also unsuccessful experiments. During our efforts in the lab, Coldplay sings to us: "*Nobody said it was easy*". But realize that the combining factor between *fun*damental stem cell biology and engineered *funct*ional tissue is FUN!

1.5 Snapshot summary

1. Two defining properties of stem cells are their ability to self-renew and their ability to differentiate.
2. Self-renewal is orchestrated by a complex network of intrinsic and extrinsic factors, which are species and tissue specific.
3. Embryonic stem cells are pluripotent. Most adult stem cells are multipotent.
4. Differentiation of cells is not always a one-way street. Cell fates can be reset by epigenetic reprogramming.
5. Adult stem cells might display plasticity.
6. Embryonic stem cells are isolated from the inner cell mass of a blastocyst and exist only *in vitro*. Adult stem cells can be isolated from various tissues.
7. Embryonic stem cells have to be characterized *in vitro* and *in vivo*, to confirm their self-renewal capacity and pluripotency.
8. Adult stem cells are rare and stem cell division rate is low in the body.
9. The stem cell niche is the micro-environment where the stem cells reside *in vivo*.
10. Mesenchymal stem cells go into replicative senescence when cultured *in vitro*.

References

Adewumi, O., Aflatoonian, B., Ahrlund-Richter, L., Amit, M., Andrews, P.W., *et al.* (2007). Characterization of human embryonic stem cell lines by the International Stem Cell Initiative. *Nature Tiotechnology*, 25(7): 803–816.

Amit, M., Shariki, C., Margulets, V. and Itskovitz-Eldor, J. (2004). Feeder layer and serum-free culture of human embryonic stem cells. *Biology of Reproduction*, 70(3): 837–845.

Assady, S., Maor, G., Amit, M., Itskovitz-Eldor, J., Skorecki, K.L. and Tzukerman, M. (2001). Insulin production by human embryonic stem cells. *Diabetes*, 50(8): 1691–97.

Becker, A.J., McCulloch, C.E. and Till, J.E. (1963). Cytological demonstration of the clonal nature of spleen colonies derived from transplanted mouse marrow cells. *Nature*, 197: 452–454.

Carpenter, M.K., Inokuma, M.S., Denham, J., Mujtaba, T., Chiu, C.P. and Rao, M.S. (2001). Enrichment of neurons and neural precursors from human embryonic stem cells. *Experimental Neurology*, 172(2): 383–397.

Chadwick, K., Wang, L., Li, L., Menendez, P., Murdoch, B., *et al.* (2003). Cytokines and BMP-4 promote hematopoietic differentiation of human embryonic stem cells. *Blood*, 102(3): 906–915.

Chambers, I., Colby, D., Robertson, M., Nichols, J., Lee, S., *et al.* (2003). Functional expression cloning of Nanog, a pluripotency sustaining factor in embryonic stem cells. *Cell*, 113(5): 643–655.

Cowan, C.A., Atienza, J., Melton, D.A. and Eggan, K. (2005). Nuclear reprogramming of somatic cells after fusion with human embryonic stem cells. *Science* (New York, NY), 309(5739): 1369–1373.

Daheron, L., Opitz, S.L., Zaehres, H., Lensch, W.M., Andrews, P.W., *et al.* (2004). LIF/STAT3 signaling fails to maintain self-renewal of human embryonic stem cells. *Stem Cells* (Dayton, Ohio), 22(5): 770–778.

de Bruijn, J.D., van den Brink, I., Mendes, S., Dekker, R., Bovell, Y.P. and van Blitterswijk, C.A. (1999). Bone induction by implants coated with cultured osteogenic bone marrow cells. *Advances in Dental Research*, 13: 74–81.

Dezawa, M., Ishikawa, H., Itokazu, Y., Yoshihara, T., Hoshino, M., *et al.* (2005). Bone marrow stromal cells generate muscle cells and repair muscle degeneration. *Science* (New York, NY), 309(5732): 314–317.

Evans, M.J. and Kaufman, M.H. (1981). Establishment in culture of pluripotential cells from mouse embryos. *Nature*, 292(5819): 154–156.

Friedenstein, A.J., Chailakhjan, R.K. and Lalykina, K.S. (1970). The development of fibroblast colonies in monolayer cultures of guinea-pig bone marrow and spleen cells. *Cell and Tissue Kinetics*, 3(4): 393–403.

Greider, C.W. and Blackburn, E.H. (1985). Identification of a specific telomere terminal transferase activity in Tetrahymena extracts. *Cell*, 43(2 Pt 1): 405–413.

Harley, C.B., Futcher, A.B. and Greider, C.W. (1990). Telomeres shorten during ageing of human fibroblasts. *Nature*, 345(6274): 458–460.

Hayflick, L. (1965). The Limited in Vitro Lifetime of Human Diploid Cell Strains. *Experimental Cell Research*, 37: 614–636.

Haynesworth, S.E., Goshima, J., Goldberg, V.M. and Caplan, A.I. (1992). Characterization of cells with osteogenic potential from human marrow. *Bone*, 13(1): 81–88.

He, J.Q., Ma, Y., Lee, Y., Thomson, J.A. and Kamp, T.J. (2003). Human embryonic stem cells develop into multiple types of cardiac myocytes: action potential characterization. *Circulation Research*, 93(1): 32–39.

Hoffman, L.M. and Carpenter, M.K. (2005). Human embryonic stem cell stability. *Stem cell Reviews*, 1(2): 139–144.

Jackson, K.A., Snyder, D.S. and Goodell, M.A. (2004). Skeletal muscle fiber-specific green autofluorescence: potential for stem cell engraftment artifacts. *Stem cells*, (Dayton, Ohio) 22(2): 180–187.

Kaufman, D.S., Hanson, E.T., Lewis, R.L., Auerbach, R. and Thomson, J.A. (2001). Hematopoietic colony-forming cells derived from human embryonic stem cells. *Proceedings of the National Academy of Sciences of the United States of America*, 98(19): 10716–10721.

Lavon, N., Yanuka, O. and Benvenisty, N. (2004). Differentiation and isolation of hepatic-like cells from human embryonic stem cells. *Differentiation; Research in Biological Diversity*, 72(5): 230–238.

Levenberg, S., Golub, J.S., Amit, M., Itskovitz-Eldor, J. and Langer, R. (2002). Endothelial cells derived from human embryonic stem cells. *Proceedings of the National Academy of Sciences of the United States of America*, 99(7): 4391–4396.

Levenberg, S., Huang, N.F., Lavik, E., Rogers, A.B., Itskovitz-Eldor, J. and Langer, R. (2003). Differentiation of human embryonic stem cells on three-dimensional polymer scaffolds. *Proceedings of the National Academy of Sciences of the United States of America*, 100(22): 12741–12746.

Maherali, N., Sridharan, R., Xie, W., Utikal, J., Eminli, S., *et al.* (2007). Directly Reprogrammed Fibroblasts Show Global Epigenetic Remodeling and Widespread Tissue Contribution. *Cell Stem Cell*, 1: 55–70.

Martin, G.R. (1981). Isolation of a pluripotent cell line from early mouse embryos cultured in medium conditioned by teratocarcinoma stem cells. *Proceedings of the National Academy of Sciences of the United States of America*, 78(12): 7634–7638.

Martin, M.J., Muotri, A., Gage, F. and Varki, A. (2005). Human embryonic stem cells express an immunogenic nonhuman sialic acid. *Nature Medicine*, 11(2): 228–232.

Mitsui, K., Tokuzawa, Y., Itoh, H., Segawa, K., Murakami, M., *et al.* (2003). The homeoprotein Nanog is required for maintenance of pluripotency in mouse epiblast and ES cells. *Cell*, 113(5): 631–642.

Molofsky, A.V., Pardal, R. and Morrison, S.J. (2004). Diverse mechanisms regulate stem cell self-renewal. *Current Opinion in Cell Biology*, 16(6): 700–707.

Morrison, S.J. and Kimble, J. (2006). Asymmetric and symmetric stem-cell divisions in development and cancer. *Nature*, 441(7097): 1068–1074.

Mummery, C., Ward-van Oostwaard, D., Doevendans, P., Spijker, R., van den Brink, S., *et al.* (2003). Differentiation of human embryonic stem cells to cardiomyocytes: role of coculture with visceral endoderm-like cells. *Circulation*, 107(21): 2733–2740.

Ohlstein, B., Kai, T., Decotto, E. and Spradling, A. (2004). The stem cell niche: theme and variations. *Current Opinion in Cell Biology*, 16(6): 693–699.

Okita, K., Ichisaka, T. and Yamanaka, S. (2007). Generation of germline-competent induced pluripotent stem cells. *Nature*, 448: 313–317.

Pan, G. and Thomson, J.A. (2007). Nanog and transcriptional networks in embryonic stem cell pluripotency. *Cell Research.*, 17(1): 42–49.

Park, I.K., Qian, D., Kiel, M., Becker, M.W., Pihalja, M., *et al.* (2003). Bmi-1 is required for maintenance of adult self-renewing haematopoietic stem cells. *Nature*, 423(6937): 302–305.

Pittenger, M.F., Mackay, A.M., Beck, S.C., Jaiswal, R.K., Douglas, R., *et al.* (1999). Multilineage potential of adult human mesenchymal stem cells. *Science* (New York, NY), 284(5411): 143–7.

Rambhatla, L., Chiu, C.P., Kundu, P., Peng, Y. and Carpenter, M.K. (2003). Generation of hepatocyte-like cells from human embryonic stem cells. *Cell Transplantation*, 12(1): 1–11.

Reubinoff, B.E., Itsykson, P., Turetsky, T., Pera, M.F., Reinhartz, E., *et al.* (2001). Neural progenitors from human embryonic stem cells. *Nature Biotechnology*, 19(12): 1134–1140.

Schofield, R. (1978). The relationship between the spleen colony-forming cell and the haemopoietic stem cell. *Blood Cells*, 4(1–2): 7–25.

Simonsen, J.L., Rosada, C., Serakinci, N., Justesen, J., Stenderup, K., *et al.* (2002). Telomerase expression extends the proliferative life-span and maintains the osteogenic potential of human bone marrow stromal cells. *Nature Biotechnology*, 20(6): 592–596.

Slack, J.M. and Tosh, D. (2001). Transdifferentiation and metaplasia – switching cell types. *Current Opinion in Genetics and Development*, 11(5): 581–586.

Suemori, H., Tada, T., Torii, R., Hosoi, Y., Kobayashi, K., *et al.* (2001). Establishment of embryonic stem cell lines from cynomolgus monkey blastocysts produced by IVF or ICSI. *Dev Dyn*, 222(2): 273–279.

Tada, M., Takahama, Y., Abe, K., Nakatsuji, N. and Tada, T. (2001). Nuclear reprogramming of somatic cells by in vitro hybridization with ES cells. *Curr Biol*, 11(19): 1553–1558.

Takahashi, K. and Yamanaka, S. (2006). Induction of pluripotent stem cells from mouse embryonic and adult fibroblast cultures by defined factors. *Cell*, 126(4): 663–676.

Terada, N., Hamazaki, T., Oka, M., Hoki, M., Mastalerz, D.M., *et al.* Bone marrow cells adopt the phenotype of other cells by spontaneous cell fusion. *Nature*, 416(6880): 542–545.

Thomson, J.A., Itskovitz-Eldor, J., Shapiro, S.S., Waknitz, M.A., Swiergiel, J.J., *et al.* (1998). Embryonic stem cell lines derived from human blastocysts. *Science* (New York, NY), 282(5391): 1145–1147.

Thomson, J.A., Kalishman, J., Golos, T.G., Durning, M., Harris, C.P., *et al.* (1995). Isolation of a primate embryonic stem cell line. *Proceedings of the National Academy of Sciences of the United States of America*, 92(17): 7844–7848.

Thomson, J.A., Kalishman, J., Golos, T.G., Durning, M., Harris, C.P. and Hearn, J.P. (1996). Pluripotent cell lines derived from common marmoset (Callithrix jacchus) blastocysts. *Biology of Reproduction*, 55(2): 254–259.

Till, J.E. and McCulloch, C.E. (1961). A direct measurement of the radiation sensitivity of normal mouse bone marrow cells. *Radiation Research*, 14: 213–222.

Vats, A., Bielby, R.C., Tolley, N., Dickinson, S.C., Boccaccini, A.R., *et al.* (2006). Chondrogenic differentiation of human embryonic stem cells: the effect of the microenvironment. *Tissue Engineering*, 12(6): 1687–1697.

von der Mark, K., Gauss, V., von der Mark, H. and Muller, P. (1977). Relationship between cell shape and type of collagen synthesised as chondrocytes lose their cartilage phenotype in culture. *Nature*, 267(5611): 531–532.

Wagers, A.J. and Weissman, I.L. (2004). Plasticity of adult stem cells. *Cell*, 116(5): 639–648.

Wernig, M., Meissner, A., Foreman, R., Brambrink, T., Ku, M., *et al.* (2007). *In vitro* reprogramming of fibroblasts into a pluripotent ES-cell-like state. *Nature*, 448: 318–324.

Williams, R.L., Hilton, D.J., Pease, S., Willson, T.A., Stewart, C.L., *et al.* (1988). Myeloid leukaemia inhibitory factor maintains the developmental potential of embryonic stem cells. *Nature*, 336(6200): 684–687.

Wilmut, I., Schnieke, A.E., McWhir, J., Kind, A.J. and Campbell, K.H. (1997). Viable offspring derived from fetal and adult mammalian cells. *Nature*, 385(6619): 810–813.

Ying, Q.L., Nichols, J., Chambers, I. and Smith, A. (2003). BMP induction of Id proteins suppresses differentiation and sustains embryonic stem cell self-renewal in collaboration with STAT3. *Cell*, 115(3): 281–292.

Ying, Q.L., Nichols, J., Evans, E.P. and Smith, A.G. (2002). Changing potency by spontaneous fusion. *Nature*, 416(6880): 545–548.

Zhang, S.C., Wernig, M., Duncan, I.D., Brustle, O. and Thomson, J.A. (2001). *In vitro* differentiation of transplantable neural precursors from human embryonic stem cells. *Nature Biotechnology*, 19(12): 1129–1133.

Chapter 2
Morphogenesis, generation of tissue in the embryo

Marcel Karperien, Bernard Roelen, Rob Poelmann, Adriana Gittenberger-de Groot, Beerend Hierck, Marco DeRuiter, Dies Meijer, and Sue Gibbs

Chapter contents

Chapter objectives:

- To understand how the three germlayers and the neural crest arise during embryogenesis
- To recognize the cellular origin of the heart and the sequential steps in heart development
- To appreciate the importance of smooth muscle cell-endothelial cell interactions in blood vessel development
- To know the cellular origin of the Schwann cells, the difference between myelinating and non-myelinating cells and their importance in peripheral nerve development
- To understand the cellular origin of the skin and the sequential steps that can be recognized in skin formation
- To identify the differences between intramembranous and endochondral bone formation and the role of the osteoblasts, chondrocytes and osteoclasts in both processes
- To know the basic principles by which cellular differentiation and specification is induced during organogenesis
- To understand the importance of the mutual signaling between keratinocytes and dermal fibroblasts for proper skin development, and the most important signaling cascades that are involved in the formation of the heart, blood vessels, peripheral nerves, skin and skeleton

"Recapitulation of inductive processes used in organogenesis by the embryo is a prerequisite for successful tissue engineering."

Marcel Karperien May 2006

"How superior to our mind is Nature's own experiment."

Leo Vroman 1992

2.1 Introduction

Tissues in the human body are the result of millions-of-years of evolution. This process has resulted in the natural selection of a tissue structure that is optimally adapted to its function in the body. Tissue engineering aims at repairing damaged tissue that is insufficient or cannot be repaired by the intrinsic repair mechanisms present in almost all organs of an individual. In most instances, these intrinsic repair mechanisms recapitulate many of the processes involved in the formation of an organ during embryogenesis. Incorporation of these mechanisms in tissue engineering procedures will significantly contribute to the success of the construct in the body. Thus important lessons for tissue engineering can be learned from the formation of organs during embryogenesis. For example, on (i) the origin of cells that contribute to the formation of a particular organ, (ii) the growth factors and their interrelationship in the formation of an organ, (iii) the mechanisms by which undifferentiated precursor cells are induced to specialize into an organ specific cell type, (iv) the subsequent steps in organ formation and (v) the interaction between cells and their environment consisting of both the extracellular matrix and neighboring cells. Ideally, inclusion of all these aspects in the tissue engineering procedure will provide the best change of success.

This chapter aims at providing a background on the formation of various organs during embryogenesis that are subject of tissue engineering, like the heart, blood vessels, peripheral nerves, skin, bone and cartilage. It starts with providing an overview of the formation of pluri- and multipotent cell populations and their positioning in the overall body plan. Subsequently, some of the basic mechanisms by which these cell populations are recruited in organogenesis are discussed in greater detail.

2.1.1 The formation of pluripotent cells

All tissues in the adult mammal originate from the fertilized oocyte, called zygote. In most species, ovulation takes place when the oocyte has been arrested at the metaphase II stage of meiosis. During the two steps of meiosis the cell's chromosome number is reduced from the diploid to the haploid condition. After ovulation the oocyte is transported to the oviduct where it is fertilized by the sperm cell. Meiosis of the oocyte is completed and a female and male pronucleus form that will ultimately fuse to form the diploid zygotic nucleus. The first divisions of the fertilized egg are not accompanied by cellular growth and are therefore referred to as cleavage divisions and arising cells are called blastomeres. At the early cleavage stages each blastomere is still totipotent and thus is capable of forming an entire organism (Tsunoda and McLaren, 1983). In mouse embryos, however, some specification has already occurred. The blastomere with the sperm entry point divides first and usually develops into extra-embryonic tissue but this process is still reversible (Piotrowska and Zernicka-Goetz, 2001).

The first differentiation occurs at the morula stage after the 3rd series of cleavage divisions (8 cell stage). At this stage, the embryo undergoes compaction, a process involving flattening of the blastomeres and an increase in contact of the blastomeres. The cells that are on the outside of the morula differentiate into trophectoderm, a cell layer mediating implantation and formation of extraembryonic structures such as the placenta on which most mammalian embryos are largely dependent for their nutrition (Tarkowski and Wroblewska, 1967). The polarized trophectoderm cells create a fluid-filled cavity in the embryo by active transportation of water to the inside of the embryo. The embryo is now called a blastocyst and contains besides the trophectoderm and a fluid filled cavity or blastocoel also an inner group of cells known as the inner cell mass. The inner cell mass consists of pluripotent cells that will ultimately form the entire embryo. These cells can be isolated from the inner cell mass and under the right conditions kept into culture and are then called embryonic stem (ES) cells. ES cells have the capacity to self renew and are truly pluripotent as they can differentiate into every cell type of the embryonic and adult organism, making them excellent cells for tissue engineering purposes. In mouse genetics, ES cells are an important tool for the generation of genetically modified animals.

The first stages of embryonic development occur when the embryo travels through the oviduct prior

Classical Experiments

Generation of the first embryonic stem cell lines

Already in the 1950s it was recognized that the 129Sv strain of mice exhibited a high incidence in the formation of spontaneous testicular tumors, so-called teratocarcinomas. Teratocarcinomas can also be induced in mice by the grafting of early embryos into extrauterine sites (Stevens, 1967). Individual cells from these tumors could be transplanted to other mice where they formed new tumors, demonstrating the existence of cells with self-renewal capacity or stem cells in the tumors (Kleinsmith and Pierce, 1964). Teratocarcinomas contain a wide variety of differentiated cells that were derivatives of the stem cells. The exact potency of these stem cells remained unclear until chimaeric mice were created by injecting few of the tumor cells into blastocyst stage embryos. The tumor cells started to participate in the development of many cells and tissues, including germ cells. This demonstrated that the teratocarcinoma cells were pluripotent (Mintz and Illmensee, 1975; Papaioannou et al., 1975). The cell lines that were derived from teratocarcinomas were called embryonal carcinoma cell lines. The pluripotency of the embryonal carcinoma cells raised many questions. These cells were after all derived from embryos. It was hypothesized that the early embryo contained cells that divided and remained pluripotent unless they received signals for differentiation, as normally occurred during embryogenesis. It seemed logical to assume that these cells could be directly isolated from the embryo and kept in culture.

Martin Evans and Matthew Kaufman, then from the University of Cambridge, were the first to directly generate progressively growing cultures from preimplantation mouse embryos. By delaying the implantation of blastocyst stage embryos, they obtained embryos with a bigger-than-normal inner cell mass. Using their knowledge of culture media obtained from working with embryonal carcinoma cells, they were able to culture cells from these large blastocysts. The cells strongly resembled embryonal carcinoma cells; they too formed teratomas when injected into mice and differentiated in vitro. Evans and Kaufman published their seminal paper on the derivation of pluripotent embryonic cells in 1981 (Evans and Kaufman, 1981). In the same year Gail Martin from the University of California independently established pluripotent cell lines from mouse blastocysts (Martin, 1981). Instead of using delayed implantation embryos, Martin was able to successfully culture the embryonic cells by using a feeder layer of fibroblasts and conditioned medium from embryonal carcinoma cells. Martin coined the term embryonic stem (ES) cells for these embryo-derived cell lines. The pluripotency of the ES cells was convincingly demonstrated by the formation of healthy germ-line chimaeras (Bradley et al., 1985).

The aim of the underlying studies was now at reach, namely to use mutation-carrying cells for the generation of mice bearing these mutations to assess their functional consequence. Gene-targeting by homologous recombination in ES cells was developed (Smithies et al., 1985), which proved very fruitful. Indeed, the production of many knock-out mice and the knowledge of gene function by studying these mice would not have been possible without ES cells.

to implantation in the uterus. Under the appropriate culture conditions, this process can be mimicked *in vitro*. Thus isolated oocytes (human or from other mammals) can be fertilized *in vitro* and cultured up until the early blastocyst stage. From these blastocysts it is possible to generate ES cells that could be applied in regenerative medicine and tissue engineering. Further development of the embryo, however, requires implantation in the uterus and the formation of the germ layers by gastrulation.

2.1.2 Preimplantation development and preparation for gastrulation

When the embryo has reached the uterus, at around 3 to 4 days after fertilization in mice and 5 to 7 days in humans, it frees itself from the zona pellucida, a thick layer of glycoproteins that has until this stage surrounded and protected the embryo. The embryo is now ready to implant in the uterine tissue. At this time point human and mouse embryos have reached the blastocyst stage. Indeed, mouse and human embryos immediately implant in the uterus but implantation in other mammalian species can occur much later when gastrulation has already started or, as in the case of the egg-laying monotremes, never occurs. Upon implantation, uterine trophoblast cells differentiate into decidual cells. Particulary in mouse embryos, these cells form a specific structure, the decidua, which enwraps the whole embryo. Subsequent steps in embryonic development aim at establishing the proper structure in which specification of the 3 germ layers, the ectoderm, endoderm and mesoderm can occur by a process called gastrulation. As a first step a single cell layer of primitive endoderm is formed from the inner cell mass cells at the site of the blastocoelic cavity. This primitive endoderm will not contribute to any part of the newborn organism, but instead gives rise to the endoderm in the visceral and parietal yolk sacs. The visceral endoderm is later replaced by definitive endoderm during gastrulation (Lawson and Pedersen, 1987).

The cells from the inner cell mass are organized into an epithelial-like structure. Here, a clear difference in mouse and human embryology can be recognized (Figure 2.1). In the mouse, the inner cell mass forms a cup-shaped structure called the egg cylinder, As the cup elongates, a proamniotic cavity is formed inside the cup by the process of programmed cell death or apoptosis (Coucouvanis and Martin, 1999) and the

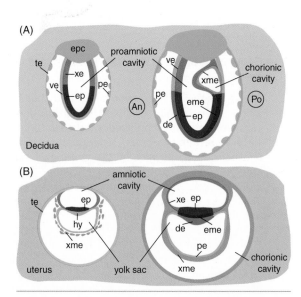

Figure 2.1 Embryonic development around implantation. Schematic representation of early mouse (A) and human (B) postimplantation development shortly before (left) and after (right) the onset of gastrulation. Gastrulation is characterized by the formation of mesoderm and definitive endoderm from the embryonic ectoderm (epiblast). In human embryos, extraembryonic mesoderm is already formed before gastrulation has started. Only the epiblast (ectoderm), embryonic mesoderm and definitive endoderm will form the future fetus. Abbreviations: An, anterior; de, definitive endoderm; eme, embryonic mesoderm; ep, epiblast; hy, hypoblast; epc, ectoplacental cone; pe, parietal endoderm; Po, posterior; te, trophectoderm; ve, visceral endoderm; xe, extraembryonic ectoderm; xme, extraembryonic mesoderm.

embryo assumes the shape of a hollow cylinder with an inner layer of ectoderm and an outer layer of visceral endoderm (Figure 2.1A). The upper part of the cylinder near the ectoplacental cone (the future placenta) contains extraembryonic ectoderm. The lower part of the cylinder contains the embryonic ectoderm or epiblast. The epiblast contains pluripotent cells that will form the future embryo (Figure 2.1A). In contrast, the human embryo adopts a planar morphology and produces an amniotic cavity between the former inner cell mass, which has been transformed into the embryonic ectoderm or epiblast and the overlying trophectoderm. The primitive endoderm or hypo-blast proliferates and forms another hollow structure, the yolk sac. Before gastrulation starts, extraembryonic mesodermal cells delaminate from the epiblast and line the amniotic cavity and the yolk sac from the outside (Figure 2.1B). The amnion and yolk sac now consist of two sheets of tissue, extraembryonic ectoderm/extraembryonic mesoderm and primitive endoderm/extraembryonic mesoderm, respectively. Neither structure will contribute to the fetus. The embryo has now adapted a structure in which the specification of the embryonic ectoderm or epiblast into the germ layers can occur.

2.1.3 Gastrulation and the establishment of the germ layers

In the mouse, at about 6.5 days after fertilization and in the human at about 13 to 15 days, the multilayered body plan is established through gastrulation. During gastrulation, which commences with the formation of a so-called primitive streak, the three definitive germ layers that construct the adult organism are established: the outer ectoderm, the inner endoderm and the interstitial mesoderm. The ectoderm forms the outer part of the skin, brain cells, nerve cell, parts of the eye like the lens, epithelial structures of the mouth and anus, the pituitary gland, parts of the adrenal glands and pigment cells. The endoderm forms the lining of the gastrointestinal and respiratory tracts, plus the liver, pancreas, thyroid gland, thymus and the

lining of the bladder. The mesoderm gives rise to skeletal muscle, heart and blood vessels, connective tisssue, kidneys, urethra, gonads, bone marrow, blood, bone, cartilage and fat. During gastrulation extensive cell movements take place through which cells acquire new positions and new neighbors with which to interact.

The primitive streak is first visible as a thickening of the epiblast at the midline of the embryo, caused by a delamination of loosely attached mesodermal cells from the epiblast (Figure 2.2). The primitive streak commences at the posterior part of the embryo and progresses to anterior whereby the anterior-most tip of the primitive streak, coined 'node', functions as the organizing center. The mesodermal cells spread

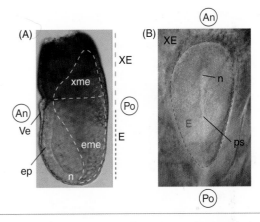

Figure 2.2 Embryos at the onset of gastrulation. Photographs of a gastrulating 7-days-old mouse, side view (A) and 15-days-old horse, dorsal view (B) embryo. The mouse embryo has adopted a cylindrical shape whereas embryos of many other mammalian species, like the horse and including the human, have a planar morphology. The white dotted line in A indicates the newly formed mesoderm, part of which will become extraembryonic, the black dotted line in B indicates the embryonic region. Abbreviations: An, anterior; E, embryonic region; eme, embryonic mesoderm; ep, epiblast; n, node; ps, primitive streak; ve, visceral endoderm; XE, extraembryonic region; xme, extraembryonic mesoderm. Micrograph A courtesy of S. Chuva de Sousa Lopes.

distally between the epiblast and the endoderm as a new layer of cells. In addition, the visceral endoderm (or hypoblast in human embryos) is displaced by other cells from the epiblast which form the definitive endoderm (Figure 2.3). These cell movements are accompanied by an increase in cell number and a decrease in cell doubling time so that in the mouse, the cell number increases from about 660 at the start of gastrulation to 15000 some 24 hours later (Power and Tam, 1993; Snow and Bennett, 1978). Before gastrulation, the mouse embryo is a symmetrical cup-shaped structure. The first morphological sign of asymmetry is the formation of the primitive streak in the future posterior region of the epiblast at the junction with

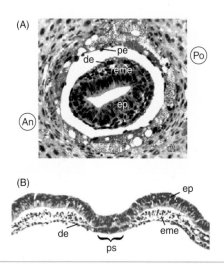

Figure 2.3 Histology of gastrulating embryos. Transversal sections through the embryonic regions of gastrulating 7-days-old mouse (A) and 14-days-old horse (B) embryo. The epiblast, newly formed mesoderm and endoderm are already visible. Note the inverted organization of the mouse embryo, with the ectoderm present in the epiblast on the inside and the endoderm on the outside. Abbreviations used, An, anterior; Po, posterior; de, definitive endoderm; dt, decidual tissue; eme; embryonic mesoderm; ep, epiblast; pe, parietal endoderm; ps, primitive streak. Micrographs courtesy of S. Chuva de Sousa Lopes (A) and B. Rambags (B).

the extra-embryonic ectoderm. The establishment of an anterior-posterior axis in the embryo is largely dependent on the surrounding visceral endoderm, that itself is principally an extra-embryonic tissue. The *T* gene, deleted in mouse *Brachyury* mutants, is important for the formation of the primitive streak and before gastrulation commences is expressed on one side of the epiblast that marks the future posterior side. During gastrulation *T* is expressed in the newly formed mesoderm (Wilkinson *et al.*, 1990). The importance of the extra-embryonic visceral endoderm in defining the anterior-posterior axis is demonstrated by mutant embryos that lack expression of the endoderm-specific genes *HNF3β* and *Lim1*, in which *T* expression is no longer restricted to the posterior side of the embryo and an enlarged primitive streak and absence of anterior-posterior polarity is observed (Perea-Gomez *et al.*, 1999). Although much is still unknown, it is most likely that indeed the visceral endoderm provides the epiblast cells with information about their spatial identity. Also the extraembryonic ectoderm is important in providing instructions to the epiblast. The extraembryonic ectoderm expresses Bone morphogenetic protein (BMP)4, a member of the transforming growth factor (TGF)-β family of secreted proteins, at the junction between embryonic and extraembryonic regions. BMP4 signaling, directly and supplemented with signaling via the visceral endoderm, regulates the formation of the allantois (extraembryonic mesoderm) and the precursors of the primordial germ cells from epiblast cells (de Sousa Lopes *et al.*, 2004; Lawson *et al.*, 1999). Also factors like BMP2, which is expressed in the visceral endoderm and BMP8b which is expressed in the extraembryonic ectoderm are important but not essential for primordial germ cell formation (Ying and Zhao, 2000, 2001). Thus, as a very early event in embryonic development, germ cells of the future individual are specified.

2.1.4 Establishment of the body plan by morphogen signaling

During gastrulation not only the germ layers are specified but cells will also receive instructions on their future position and role in the developing fetus by morphogen signaling. An important factor in this process is Nodal, which is like BMP4 a TGF-β family member that is recognized as a central molecule in gastrulation. Nodal is first expressed in the proximal part of the epiblast shortly before gastrulation and then becomes localized to the node, hence its name. Nodal signaling is essential for the formation of mesoderm and definitive endoderm (Zhou *et al.*, 1993). The highest levels of Nodal are found in the node and with increasing distance to the node, the concentration of Nodal will gradually decrease establishing a so-called morphogen gradient. Cells along this gradient are exposed to different levels of Nodal, providing the cell with important instructions on its future position and role in the fetus. Morphogen gradients are important and recurrent mechanisms by which undifferentiated cells receive instructions on their future position and role in the body.

During gastrulation, pattern formation along the anterior-posterior axis is further specified by the products of the *Hox* genes, a class of transcription factors that are essential to establish the body plan. The mammalian *Hox* genes are located on 4 chromosomal clusters and they are expressed in an ordered sequence that is reflected by their position in the clusters, with 3′ gene being expressed earlier and with a more anterior expression boundary than 5′ genes (so-called spatio-temporal colinearity). Expression starts posterior in the mesoderm and ectoderm of the primitive streak and the anterior-most expression domains shift towards anterior as the embryo develops. The regulation of *Hox* gene expression is still largely unknown. It is proposed that one or more morphogen signaling gradients are involved in the sequential activation of *Hox* gene expression. These gradients hierarchically control a number of genetic mechanisms including transcriptional enhancers and repressors of gene transcription.

Because of its inaccessibility, many questions about gastrulation in the human embryo remain unanswered. Most of the knowledge comes from hysterectomies and abortions and molecular data on these samples are lacking. Since before and during gastrulation the human embryo has adopted a planar-like morphology opposed to the cylindrical shape of the mouse embryo, extrapolation of the molecular data of the mouse is not without pitfalls (Bianchi *et al.*, 1993).

2.1.5 Neural crest cells

Besides the 3 germ layers that are established during gastrulation, a fourth pluripotent cell population is required for organogenesis. This population consists of neural crest cells. Although neural crest cells are derived from the ectoderm, they are sometimes called the fourth germ layer because of their importance. Neural crest cells arise relatively late in embryonic development after establishment of the general body plan and their formation is tightly linked with the development of the central nervous system. The differentiation of the central nervous system starts with formation of the neural tube from the neural plate. The neural plate is a layer of ectodermal cells in the dorsal midline of the embryo. Signals from the underlying notochord (the future nucleus pulposa) instruct cells in the neural plate to proliferate and adapt a tube-like shape that will eventually generate the entire central nervous system. Neural crest cells are generated at the interface of the neurectoderm with the surface ectoderm, the so-called neural plate border (Figure 2.4).

From their source of origin, neural crest cells migrate extensively to specific places in the embryo where they generate a diverse group of differentiated cells that can be divided in four distinct subgroups, cranial, cardiac, vagal and trunk cells that together give rise to sympathetic and parasympathetic neurons, glia cells, fat, cardiac mesenchyme, melanocytes, skin, connective tissue of salivary, thymus, adrenal, thyroid and pituitary glands, smooth muscle cells of arteries, tooth and bone and cartilage particularly of the face and cranium. The neural crest cells therefore have an extreme plasticity, much of which is not understood (Bronner-Fraser and Fraser, 1988). Due to this extreme plasticity, the pluripotent neural crest cells would provide an excellent source for tissue engineering and regenerative medicine particularly for restoration of cranial and facial birth defects. However, neural crest cells are extremely difficult

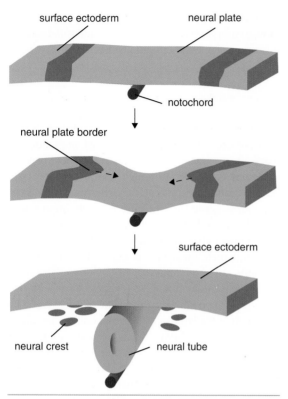

Figure 2.4 Formation of the neural crest. Midline ectoderm differentiates into neural cells to form a neural plate (yellow) by signals from the underlying notochord (red). Induction of neural crest (blue) cells takes place at the border between surface ectoderm (green) and the neural plate by signals from the mesoderm and the surface ectoderm. The sides of the neural plate fold together and close to form a neural tube. The neural crest cells delaminate from the ectoderm and migrate to their destinations.

to isolate due to their high mobility and to the fact that they are derived from a transient structure. The neural crest disappears soon after the neural tube is closed.

It is thought that the surface ectoderm, together with the underlying mesoderm, induces the neural plate to form neural crest cells by signals that include BMPs, in particular BMP4 and BMP7, Fibroblast Growth

Factors (FGFs) and Wnts. As they arise, neural crest cells undergo an epithelial to mesenchymal transition characterized by expression of members of the Snail family of zinc finger transcription factors (Nieto, 2001). Migration is probably initiated by a decrease in the amount of cell surface adhesion molecule N-cadherin. Epithelial to mesenchymal transition is an important mechanism by which cells can dramatically change their fate. It plays an essential role in organogenesis.

2.2 Cardiac development

Rebuilding the injured heart after infarction is a major challenge for tissue engineering. Presently, various cell-based approaches are used. Initially, this work started with committed myoblasts, but now a variety of undifferentiated cells is used, such as endothelial progenitors, mesenchymal stem cells, resident cardiac stem cells and ES cells. Various cell types from different origins are involved in heart formation. Important lessons can be learned from the complex interplay between these cell types that eventually give rise to the four-chambered cyclically beating heart during embryonic development. The heart, being the first functionally active organ, must continuously adapt to the growing needs of the embryo, whereas in postnatal life it has to support the ever-changing activity levels of the body. Figure 2.5 provides a time table of the sequential formation of different parts of the heart during embryogenesis.

Here we will provide a framework for understanding the multiple origins of the cardiac cell populations, including the cardiomyocytes, fibroblasts, smooth muscle and nerve cells. Furthermore, the increasing complexity of the heart's architecture will be explained, in particular chamber formation, cardiac septa and valves, the coronary circulation and the conduction system. Finally, from the myriad of transcription and growth factors, important during the building of the heart, a brief selection will be provided.

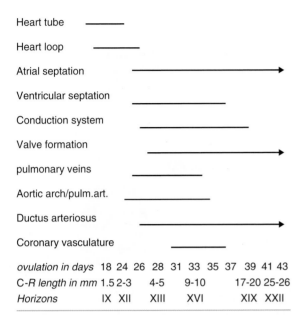

	ovulation in days	18	24	26	28	31	33	35	37	39	41	43
	C-R length in mm	1.5	2-3		4-5		9-10			17-20	25-26	
	Horizons		IX	XII		XIII		XVI			XIX	XXII

Figure 2.5 Time table of human heart development during embryogenesis. Note the different developmental windows for the various parts of the heart. Days since ovulation as a measure of developmental stage is compared with Crown-Rump length (CR) and with the human stages of development indicated by Streeter's Horizons.

2.2.1 Cell interactions

In the earliest phase of cardiac development the bilateral anterior tips of the unsegmented splanchnic mesoderm, the cardiogenic plates, are characterized mainly by the interaction of the transcription factors GATA 4/5/6 (Laverriere *et al.*, 1994) and NKX2.5 (Jiang *et al.*, 1999). The splanchnic mesoderm refers to cells of the inner lining of the body cavity. The transcriptional machinery for defining the cardiomyocyte is far from clear until now and seems more complicated than for instance for skeletal muscle, in which the transcription factor MyoD is elemental and sufficient for muscle cell differentiation. Evidence exists that the transcription factor myocardin and the

BMP pathway govern cardiomyocyte differentiation, although myocardin is also present at the onset of smooth muscle cell differentiation (Callis *et al.*, 2005; Wang *et al.*, 2001). The cardiogenic plates fuse in the midline (Figure 2.6A) and the precursor cells differentiate into cardiomyocytes and endocardial cells (see Gittenberger-de Groot *et al.*, 2005, for a schematic overview). The cardiomyocytes become the contracting part of the heart. The space between myocytes and endocardial cells, which have adapted an epithelial phenotype, becomes filled by cardiac jelly produced by the myocytes. The endocardium gives rise to mesenchymal cushion cells (Person *et al.*, 2005; Ramsdell and Markwald, 1997) by a process known as epithelial to mesenchymal transition (EMT) (Hay, 2005) and, as a consequence, the cardiac jelly will transform into the thick endocardial cushions containing extracellular matrix and mesenchymal cells. Parts of these cells are also recruited from the neural crest (see below). This is a complex process requiring various growth factors and many transcription factors. The atrioventricular cushions reside in the transition zone of the primitive atrium and ventricle, the so-called Atrio Ventricular (AV) canal, whereas the outflow tract (OFT) cushions are found in the conotruncal transition between ventricle and aortic sac (Figure 2.6B). Both cushion systems are in close proximity in the inner curvature of the heart. The OFT cushions are important in separation of the OFT into the aorta and the pulmonary trunk, while the AV cushions will play an important role in AV septation (Wenink *et al.*, 1986) by heterologous fusion with the interventricular or primary fold and with the intra-atrial spina vestibuli (see Figure 2.6B and Gittenberger-de Groot *et al.*, 2005).

2.2.2 Extracardiac cell populations

Two extracardiac cell populations are very important in heart formation, the cardiac neural crest (NCC) and the pro-epicardial organ (PEO). Both primordia have in common that their original epithelium (being neurectoderm and splanchnic epithelium, respectively) transforms by epithelial-mesenchymal transition

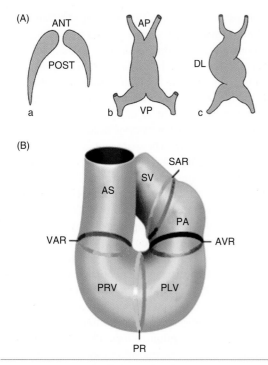

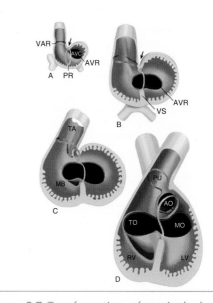

Figure 2.6 The first stages in heart development. (A) Schematic representation of (a) the bilateral formation of the mesoderm-derived cardiogenic plates, which are not quite symmetric. (b) After fusion, these form a straight heart tube wit an arterial (AP) and a venous (VP) pole. (c) Thereafter, the tube starts its rightward (dextral) looping (DL); (B) schematic drawing of the looped heart tube with the cardiac chambers and the transitional zones. Following the blood flow from venous to arterial, we can distinguish the sinus venosus (SV), the sinoatrial ring (SAR), the primitive atrium (PA), the artrioventricular ring (AVR), encircling the atrioventricular canal, the primitive left ventricle (PLV), the primary fold or ring (RR), the primitive right ventricle (PRV), the outflow tract ending at the ventriarticular ring (VAR) and the aortic sac (AS). (Adapted from Gittenberger-de Groot, A.C., Bartelings, M.M., DeRuiter, M.C. and Poelmann, R.E. Basics of cardiac development for the understanding of congenital heart malformations. *Pediatr Res*, 57: 169–176, 2005.)

Figure 2.7 Transformation of a single heart tube into a 4-chambered pumping heart. Schematic representation of the remodeling of the cardiac chambers and the transitional zones at the ventricular level: (A) internal view (compare with Figure 2.6B) of the looped heart tube. The transitional zones are, going from the venous to the arterial pole, the atrioventricular ring (AVR, dark blue), the primary ring (PR, yellow), and, at the distal end of the myocardial outflow tract, the ventriculoarterial ring (VAR, bright blue); (B) during looping, with tightening of the inner curvature (arrow), the right part of the AVR moves to the right of the ventricular septum (VS); (C) start of formation of the inflow tract of the right ventricle by excavation of the PR. The lower border is formed by the moderator band (MB); (D) completion of the process with formation of a tricuspid orifice (TO) above the right ventricle (RV) and the aortic orifice (Ao) and the mitral orifice (MO) above the left ventricle (LV). It is easily appreciated that there is aortic-mitral continuity, whereas the distance between the TO and the pulmonary orifice (Pu) is marked. (Adapted from Gittenberger-de Groot, A.C., Bartelings, M.M., DeRuiter, M.C. and Poelmann, R.E. Basics of cardiac development for the understanding of congenital heart malformations. *Pediatr Res*, 57: 169–176, 2005.)

(EMT) into migratory mesenchyme. Subpopulations of NCC migrate into the pharyngeal arches (to form e.g. the arterial vessel wall, adventitial fibroblasts and smooth muscle cells) and into the secondary heart field, a part of the dorsal body wall adjacent to the heart tube (vide infra), subsequently migrating into and contributing to the OFT and the AV-cushions. The PEO starts as a small grape-like structure extending from the inner lining of the body cavity, the splanchnic mesoderm. It expands and migrates as an epithelial sheet over the outer surface of the myocardium to form the epicardium (Virágh *et al.*, 1993; Vrancken Peeters *et al.*, 1995). After complete coverage of the heart tube, EMT will give rise to subepicardial mesenchyme. A large number of these so-called epicardial derived cells (EPDC) migrate into the heart and differentiate into cardiac fibroblasts, smooth muscle cells and adventitial fibroblasts of the coronary vessels (Gittenberger-de Groot *et al.*, 1998). In bulk number the fibroblasts comprise the largest cell population, providing for the extracellular matrix evolving into the fibrous heart skeleton. This system mechanically supports the cardiac valves and the tendinous apparatus during the various phases of life. After cardiac infarction the EPDC derived fibroblasts provide for the scar tissue giving integrity to the cardiac wall, although it is of course not contractile. However, in early embryonic development it is responsible for the induction of the architecture of the cardiac wall. Experimental ablation of the PEO results in an aberrant coronary vasculature feeding a paper thin myocardium almost devoid of trabeculations. This resembles a failing heart, eventually leading to embryonic death (Gittenberger-de Groot *et al.*, 2000). Therefore, the importance of these epicardium-derived cells must not be underestimated.

2.2.3 Secondary heart field

Activity of the cranial and caudal splanchnic mesoderm, also called the secondary heart field results in a continued addition of cardiomyocytes (Abu-Issa *et al.*, 2002; Kelly, 2005; Mjaatvedt *et al.*, 2001) and probably also endocardial cells (Noden, 1990) to the arterial and

most likely also to the venous pole. Important gene expression patterns involved in secondary heart field formation are reviewed by Kelly (2005). It is conceivable that by the activity of the secondary heart field also neural crest cells become incorporated into the OFT septation complex and even in the atrioventricular endocardial cushions.

2.2.4 Looping and chamber formation

Initially, the heart is formed as a single almost strait tube. By a series of geometrical changes this tube finally transforms into a double pump, separately serving lungs and body. Various mechanisms involve looping of the primary tube, wedging of the outflow tract between left and right atrium, expansion of the chambers, atrial and ventricular septum formation, valve differentiation and formation of the pharyngeal arterial system. Recent descriptions of the morphological changes important for the understanding of congenital anomalies have been provided elsewhere (Gittenberger-de Groot *et al.*, 2005), while part of the genomic coding underlying chamber formation has been reviewed (Moorman and Christoffels, 2003) (Figure 2.7). It is evident that one simple transcriptional code for all cardiomyocytes does not exist, as left and right ventricular and atrial and outflow tract myocytes present with a variety of different gene expression patterns in successive time-windows (Moorman and Christoffels, 2003; Srivastava and Olson, 2000; VandenHoff *et al.*, 2004).

The early heart tube is originally connected to the dorsal embryonic body wall by the mesocardium. The mesocardium disrupts centrally and allows contact of the left and right parts of the pericardial coelomic cavity. The heart tube will keep its contact with the dorsal body wall at the arterial and venous poles, the only part that is completely surrounded by pericardial cavity being the primitive ventricle. The cardiac tube, never being completely straight, will loop as a complex three-dimensional skewed U-shaped curve with the ventricle at the bottom of the U, allowing for longitudinal expansion of the tube and subsequent expansion of the cardiac chambers. Furthermore, additional longitudinal growth is established by activity of the secondary heart field (Kelly, 2005) recruiting cells to both the outflow tract region and the sinus venosus part. The inner curvature of the heart, where OFT and AV cushions meet, is usually very tight.

2.2.5 Septation and valve formation

Septation of the outflow tract requires the coordinated interactions of the OFT cushions, containing also neural crest cells, with the surrounding cardiomyocytes. AV canal septation requires fusion of the AV cushions, harboring also neural crest cells and Epicardial derived cells (EPDC). As a final step in septum formation cardiomyocytes penetrate parts of the fused cushions. Differentiation of the OFT semilunar valves and AV mitral and tricuspid valves and their tendinous apparatus, involves coordinated interactions of the endocardial cushion cells with migrated NCC and probably also invaded EPDC in conjunction with the surrounding cardiomyocytes (Gittenberger-de Groot *et al.*, 2000).

2.2.6 Conduction

Contractions of the heart tube starts as a peristaltic movement (de Jong *et al.*, 1992) with a pacemaker functioning at the venous pole, but this will soon change in a cyclic base-to-apex-contraction, that in time will be followed by an apex-to-base constriction (Chuck *et al.*, 1997; Rentschler *et al.*, 2001; Sedmera *et al.*, 2004) in a time-specific manner. In yet to establish time windows these alterations in contraction mode are paralleled by differentiation of the myocardial entity acquiring fast and slow conduction properties (Christoffels *et al.*, 2000). The formation of the conduction system probably requires the involvement of NCC (for the induction or insulation of the central conduction system, Poelmann *et al.*, 1999) and EPDC for the peripheral Purkinje fibers (Gittenberger-de Groot *et al.*, 1998) as well as important signaling molecules such as platelet derived growth factors (Van Den Akker *et al.*, 2005) and endothelin (Hall

et al., 2004), the expression of the latter is shear stress dependent (Groenendijk *et al.*, 2005).

2.2.7 Concluding remarks

The formation of the heart is a complex and time-consuming process, which is not finished before birth. The multiple transcriptional interactions and molecular and cellular contributions framed in exact developmental windows provide an overwhelming myriad of possibilities, of which only a few combinations are adequate enough to result in a normally structured and functioning heart. Many combinations prove to be embryolethal or result in dramatic congenital malformations. Reconstructing and engineering the heart for major diseases must be based on thorough knowledge of the multiple interactions at all levels of development, function and maintenance of the heart. The different origin of various cell populations involved in heart formation provides a challenge for tissue engineering.

2.3 Blood vessel development

A clear need for a suitable arterial replacement has prompted researchers to look beyond autologous and synthetic tissue replacements toward the engineering of vessels using mesenchymal, hematopoietic and embryonic stem cells. In embryonic development the formation of the vasculature is tightly linked with heart development. Thus, when the heart starts to beat (about 4 weeks after conception) a network of endothelial-lined vessels through which the blood will be transported to and from the yolk sac, has developed in the meantime. This almost two-dimensional extra-embryonic plexus will rapidly grow out and gives rise to organ-specific vasculatures within the fast growing embryo. Except for the capillaries each vessel will be enveloped by one or more layers of specialized mural cells largely depending on their position and function within the vasculature.

Here the embryonic origin and development of the multiple cell populations involved in vessel formation, such as endothelial cells (ECs), smooth muscle cells (SMCs), fibroblasts and pericytes, is related to their contribution to the various types of vessels (muscular and elastic arteries, capillaries and veins). Recent studies lift a tip of the veil clouding the molecular pathways involved in embryonic endothelial and smooth muscle cell differentiation and the role of environmental factors like haemodynamic forces in the morphogenesis of the vessel wall. Although the combination of these factors will determine the ultimate phenotype of the vascular cells, one has to keep in mind that a single endothelial or a single smooth muscle cell phenotype does not exist. In fact, there is a large large heterogeneity in endothelial and smooth muscle phenotypes reflecting their multiple origins and their site and tissue-specific role in controlling homeostasis of the body.

2.3.1 Origin of the endothelial cells

The embryoblast organizes into a three-layered embryo consisting of an outer ectoderm, a medial mesodermal

DEVELOPMENTAL FATE MAP OF THE CARDIOVASCULAR SYSTEM

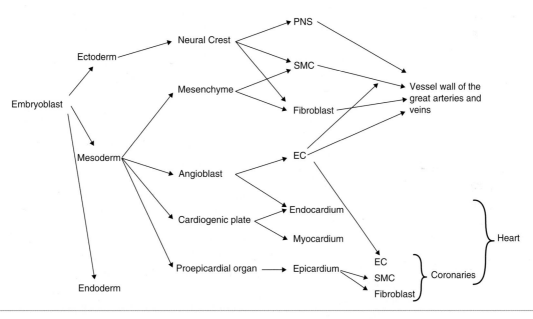

Figure 2.8 Developmental origin of cells contributing to the cardiovascular system. The mesodermal compartment of the three-layered embryo gives rise to the major part of the mural cells within the embryo. Angioblasts are the first vascular cells that differentiate by vasculogenesis within the mesoderm. Then the crescent shaped cardiogenic plate arises giving rise to the endocardium and myocardium of the heart. The remaining mesenchyme will later on differentiate into SMC and fibroblasts of most of the organs. The vascular SMC and fibroblasts in the head and trunk, however, are derived from the mesectodermal neural crest population. Although the endothelial lining of the coronaries are derived by angiogenesis within the septum transversum the SMC and fibrobasts differentiate from a specialized region within the splanchnic mesothelial lining of the coelomic cavity. Abbreviations: Endothelial cell (EC); Peripheral nerve system (PNS); Smooth muscle cell (SMC).

layer and an inner endoderm. Like the endocardium the endothelial cells (EC) of the blood vessels differentiate in situ within the so-called splanchnic mesoderm that faces the endoderm (Figure 2.8). This process is called vasculogenesis. The endothelial precursors line up, connect, lumenize and form a plexus that gives rise to both the embryonic and extra-embryonic vessels on the yolk sac and later on the placenta. VEGF, TGFβ and FGF signalling pathways are seen as key players in the initial establishment of this endothelial network. This endothelial plexus in the splanchnic mesoderm expands into the somatic mesoderm by the formation of new sprouts. This process is also termed angiogenesis. Recent studies demonstrate that even before the first heart beat an imprinting of arterial and venous identity exists within this capillary plexus. Arterial identity is characterized by the expression of genes like ephrin-B2, neuropilin-1, Notch3, DLL4 and gridlock (Herzog *et al.*, 2001; Lawson *et al.*, 2001; Moyon *et al.*, 2001a, 2001b; Villa *et al.*, 2001). EphB4 and Neuropilin-2 indicate a venous identity (Lawson *et al.*, 2002; Torres-Vazquez *et al.*, 2003; Wang *et al.*, 1998). Blood flow is necessary for ongoing differentiation and shaping of the vasculature. Experimentally changing the blood flow can induce a phenotypic shift of the initial identity demonstrating the plasticity of the endothelial cells in the early stages of vasculogenesis (le Noble *et al.*, 2005).

2.3.2 Origin of the smooth muscle cells

The next step in vessel wall differentiation is the recruitment of smooth muscle cells (SMC) (arteries and veins) or pericytes (precapillaries) to the preformed endothelial scaffolding. The location of the vessel within the embryo determines the origin of the SMCs (Figure 2.8). In the greater part of the body the SMCs have a mesodermal origin. In the head and the thoracic region the arteries and veins recruit their SMCs from the ectodermal cardiac neural crest cells (Bergwerff *et al.*, 1998; Kirby and Waldo, 1995). The SMCs around the coronary vessels are derived from the proepicardial organ, a proliferative population

of cells within the coelomic mesothelium (Vrancken Peeters *et al.*, 1997). Moreover, even differentiated ECs of the dorsal aorta can contribute to SMCs by transdifferentiation (DeRuiter *et al.*, 1997), a potency that can still be found in adult ECs. Sharp borderlines demarcate the segments of the vasculature that are composed of SMCs from different origins. Cell tracing studies have demonstrated that within elastic arteries these borderlines coincide with a change in morphology and functionality (Bergwerff *et al.*, 1998; Molin *et al.*, 2004). Origin-related SMC differences are e.g. reflected in the differences in glycosylation of the TGFβ type II receptor resulting in a different response to TGFβ1 in the aortic arch and descending aorta (Topouzis and Majesky, 1996). Vascular abnormalities like Marfan (disrupted fibrillin-1), Williams' syndrome (disrupted elastin), Loeys-Dietz syndrome (TGFβ receptor 1 and receptor 2 mutations) and Char syndrome (mutations in the transcription factor AP-2) have preferential sites of occurrence that probably reflect the embryonic origin of the SMCs.

2.3.3 Stabilization of the vessel wall

With the recruitment of SMCs from the surrounding mesoderm or neural crest the vessels stabilize. In short the growth factor PDGF-B expressed by the angiogenic ECs initiates the recruitment of the PDGF receptors expressing mural cells (pericytes and SMCs) towards the newly formed vessel (Hellström *et al.*, 1999). In turn these cells produce Angiopoietin 1, that after binding the Tie2 receptor on the ECs down regulates the angiogenic activities of the ECs. This is an example of the intensive communication between the endothelial cells and underlying pericytes and SMCs and vice versa. In the meantime, endothelium-derived TGFβ signalling is important for the differentiation of these peri-endothelial cells.

2.3.4 Pharyngeal arch arteries

The formation of arteries and veins in the embryo is a tightly regulated process (Figure 2.9). The combined

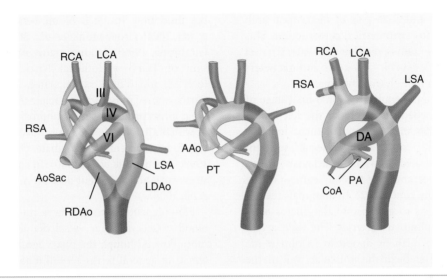

Figure 2.9 Remodeling of the pharyngeal arch artery system. During embryonic development the pharyngeal arch artery system remodels from a symmetrical into an asymmetric system. In the left picture the first two pairs of pharyngeal arch arteries have already disappeared. The fifth pair does not fully develop in mammals. As a consequence of the disappearance of the right dorsal aorta (RDAo) and the right sixth pharyngeal arch artery (VI) the system becomes asymmetrical resulting in an aortic arch connecting the ascending aorta (AAo) from the heart with the descending aorta at the left side of the trachea and oesophagus. The ductus arteriosus, a shunt between the pulmonary and systemic circulation, will close after birth. Each vessel segment indicated with a different colour has its own embryonic origin and characteristic phenotype. Abbreviations: Aortic sac (AoSac); Coronary arteries (CoA); Ductus arteriosus (DA); Left carotid artery (LCA); Left dorsal aorta (LDAo); Left subclavian artery (LSA); Pharyngeal arch arteries (III, IV and VI); Pulmonary arteries (PA); Pulmonary trunk (PT); right carotid artery (RCA); right subclavian artery (RSA). (Adapted from Molin, D.G.M., DeRuiter, M.C., Wisse, L.J. *et al.* (2002). Altered apoptosis pattern during pharyngeal arch artery remodelling is associated with aortic arch malformations in Tgf beta 2 knock-out mice. *Cardiovasc Res*, 56: 312–322.)

action of signalling, migration, proliferation, apoptosis and differentiation is often demonstrated in the development and extensive remodelling of the pharyngeal arch system. Cardiac output is fed into the aortic sac and the attached pharyngeal arch arteries that drain into the dorsal aorta. A number of arterial segments appear and disappear in a programmed manner involving e.g. neural crest cells (Bergwerff *et al.*, 1998; Le Lièvre and Le Douarin, 1975). In an early stage of development only one pair of pharyngeal arteries connects the heart to the vasculature of the embryo. Subsequently, the second and third pair

of arteries appear. During development of the fourth pair, the first pair starts already to disintegrate. Apoptosis or programmed cell death is an important mechanism in the breakdown of specific segments. The fourth pair will form part of the aortic arch and the subclavian artery. Its unique morphology is associated with common anomalies such as the type B interruption and arteria lusoria (Bergwerff *et al.*, 1999). The fifth pair is not developing fully in mammals and birds (DeRuiter *et al.*, 1993). During development of the sixth pair, the second pair starts to disintegrate. Consequently, mammals end up with the third, fourth and sixth pair of pharyngeal arch arteries in an almost symmetrical construction. This vascular system contains a number of collateral vessel segments that have specific curvatures and diameters with impelling complex haemodynamics.

Before reaching the adult situation a final stage of remodelling is needed. In mammals the left fourth artery and in birds the right fourth turns into the aortic arch. The contralateral arterial segment becomes incorporated into the subclavian artery, due to the disappearance of a specific segment of the dorsal aorta (Molin *et al.*, 2002). In mammals the left sixth artery persists to give rise to the ductus arteriosus and the pulmonary arteries. The right artery disappears before birth by apoptosis (Molin *et al.*, 2002). Shortly after birth the muscular wall of the ductus arteriosus, part of the sixth artery, contracts and the vessel closes allowing expansion of the pulmonary circulation (DeReeder *et al.*, 1988; DeReeder, 1989; Slomp *et al.*, 1997). The composition of the extracellular matrix of the various arterial segments in particular as well as their response to various members of the TGFβ growth factor family differs greatly (Bergwerff *et al.*, 1996; Molin *et al.*, 2003). Anomalies during this complex remodeling occur frequently (Molin *et al.*, 2002). The normal high incidence of apoptosis in the fourth arch arteries of the wild type mouse is decreased to back ground levels in the TGFβ2 mutant mice. Growth factor-driven apoptosis, therefore, is included in the remodeling mechanism of the arterial tree. This is also applies for Vascular Endothelial Growth Factor (VEGF) (Stalmans *et al.*, 2003).

2.3.5 Haemodynamics and vessel formation

From the onset of circulation the blood exerts mechanical forces on the ECs and the surrounding undifferentiated mesenchyme which appear to be essential epigenetic factors in vascular differentiation. It is known that the expression of many genes is regulated by shear stress that is a component of the fluid flow inside a blood vessel (Groenendijk *et al.*, 2004; Groenendijk *et al.*, 2005; Topper and Gimbrone, 1999). Among many others, the expression of TGFβ, endothelial Nitric Oxid Synthase (eNOS) and VEGF is regulated by shear stress. Furthermore, pulsatile and relatively high levels of shear stress protect against apoptosis (Li *et al.*, 2005) in which nitric oxide (NO) produced by the enzyme eNOS is a potent intermediate (Dimmeler *et al.*, 1999; Groenendijk *et al.*, 2005). In contrast, low shear stress areas and turbulent flow promote apoptosis (Li *et al.*, 2005).

Flow depends among others on the diameter of blood vessels, a larger vessel diameter allowing for more flow. Although the pharyngeal arch artery system is in general terms a symmetrical construction, the diameter of the various vascular segments, more specifically the right and left-sided counterparts, varies to some degree. The vessels destined to disappear by a high frequency of apoptosis always show a slightly smaller diameter (Molin *et al.*, 2002).

2.4 Development of the peripheral nerve tissue

Driven by an enormous clinical need, peripheral nerve regeneration has become a prime focus within the field of tissue engineering. The recent research, contributing to our current understanding of the development of a functioning peripheral nerve during embryogenesis, is described here. The various tissue components that make up the peripheral nerve originate from different germ layers. The neurons and glial cells, Schwann cells, are derived from the either the neuroectoderm or neural crest cells, while the nerve sheath, which protects the nerves, is derived from the mesoderm (Le Douarin and Kalcheim, 1999; Bunge *et al.*, 1989). Thus, the development and organization of the peripheral nerves with their myelinated and non-myelinated axonal fibres and their protective epithelial layer involves complex cellular interactions and molecular mechanisms.

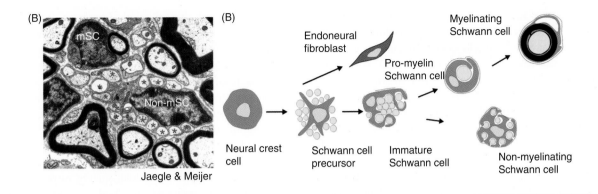

Figure 2.10 Development of the Schwann cell. The mature Schwann cell phenotypes that can be distinguished in the nerve differentiate through a number of intermediate stages from the neural crest: (A) This electronmicrograph of a transverse section of a typical mixed nerve (containing both sensory and motor fibers) shows myelinating Schwann cells associated with a single larger axon and a non-myelinating Schwann cell that ensheaths multiple small axons (asterisks) accommodating them in cytoplasmic cuffs; (B) the schematic drawing illustrates the different intermediate stages that can be distinguished in the development of a Schwann cell from the neural crest. Lineage tracing studies in mice have indicated that also the endoneurial fibroblasts are derived from the neural crest cells that populate the embryonic nerves. The basal lamina that surrounds the axon Schwann cell units is produced by the Schwann cell. Immature Schwann cells have not produced a continuous basal lamina yet. For further details see text.

2.4.1 Development of the Schwann cell lineage

The peripheral nerve contains two types of Schwann cells, myelin forming Schwann cells that ensheath large caliber axons (diameter larger than 1 μm) and non-myelinating Schwann cells that accommodate multiple lower caliber axons in cytoplasmic cuffs (Figure 2.10). Both cell types are derived from an immature Schwann cell, which itself is the product of a common precursor cell, the Schwann cell precursor (Jessen and Mirsky, 2005). From studies in the mouse and rat, it was found that Schwann cell precursors are formed early during embryonic nerve development, around embryonic day 12-13 (E12-E13 mouse). At this time, neural crest stem cells have populated the outgrowing axon bundles (Jessen and Mirsky, 2005). The cellular interactions and cues that divert the fate of the neural crest stem cell to the Schwann cell lineage at the eventual exclusion of

other fate options is still largely unknown. However, it is known that the survival and ensheathment of axons by Schwann cell precursors critically depends on the axonally derived neuregulin1 protein (Esper *et al.*, 2006; Jessen and Mirsky, 2005; Taveggia *et al.*, 2005). Neuregulin1 proteins come in at least thirteen different forms and are generated from a single neuregulin1 gene (Nrg1) through alternative splicing and promoter usage (Falls, 2003). The major protein form expressed by sensory and motor neurons is the membrane bound, type III isoform of neuregulin1. Migrating neural crest cells express the neuregulin1 receptor ErbB3/ErbB2 and maintenance of ErbB3 expression depends on the Sry box transcription factor Sox10 (Britsch *et al.*, 2001). Both Sox10 and ErbB3/ErbB2 are essential for survival and proliferation of Schwann cell precursors. Mice homozygous for null alleles of Sox10, ErbB3, ErbB2 and Nrg1 have severely reduced numbers or no Schwann cell precursors (Britsch *et al.*,

2001; Meyer and Birchmeier, 1995; Riethmacher *et al.*, 1997; Woldeyesus *et al.*, 1999). Interestingly, lack of embryonic Schwann cells results in massive sensory and motor neuron cell death in these embryos, demonstrating that neurons not only rely on target-derived trophic support but also critically depend on Schwann cell support (Riethmacher *et al.*, 1997). Currently, the molecular nature of this support is not known.

Schwann cell precursors migrate, proliferate and ensheath axons and differentiate into immature Schwann cells. Immature Schwann cells populate the fetal nerves and differ from Schwann cell precursors in several ways. While the survival of Schwann cell precursors depends strictly upon axonal contact, immature Schwann cells have acquired the ability to survive in the absence of axonal contact by establishing an autocrine survival pathway (Meier *et al.*, 1999). Insulin-like growth factor (IGF1), neurotrophin-3 (NT-3) and platelet-derived growth factor (PDGFβ) are important components of this pathway. The survival of immature and mature Schwann cells in the absence of axonal contact is of physiologic importance as successful regeneration of injured peripheral nerves greatly depends on axonal contact with living Schwann cells in the denervated nerve stump. Indeed, these so-called reactive Schwann cells secrete a range of neurotrophic factors and cytokines, such as nerve growth factor (NGF), ciliary neurotrophic factor (CNTF), leukemia inhibitory factor (LIF) and interleukin-6 (Il-6), which stimulate axonal growth (Scherer and Salzer, 2001).

2.4.2 Radial sorting and myelination of peripheral nerves

From mouse midgestation, the migration, proliferation and ensheathment of axonal fibres continues until around birth, when the number of Schwann cells and axons are eventually matched. The process of radial sorting and ensheathment of axons relies on interactions between the Schwann cell and its basal lamina (Colognato *et al.*, 2005). In particular laminin, which is a major component of the basal lamina, is of importance for proper axonal sorting by Schwann cells. It

has been found that although immature Schwann cells have not elaborated a continuous basal lamina yet, they do express laminin and laminin receptors (integrins and dystroglycan) and that in the absence of these factors, many axon bundles remain clustered (Colognato *et al.*, 2005). This suggests that Schwann cell basal lamina interactions are required for the elaboration of the cellular processes that interdigitate and segregate the axon bundles. The molecular mechanism by which these interactions result in changes in cell shape and radial sorting must merge with signaling pathways involved in cytoskeletal rearrangements. In line with this, it was found that activation of ErbB2/ErbB3 by Nrg1 results in the rapid association of the focal adhesion kinase (FAK) protein suggesting a link between Nrg1 signaling and axonal ensheathment (Vartanian *et al.*, 2000). Indeed, axons in sensory neuron cultures of Nrg1 typeIII mutant animals are poorly ensheathed (Taveggia *et al.*, 2005).

During the process of radial sorting, individual Schwann cells select larger caliber axons in a 1:1 relationship. These Schwann cells will exit the cell cycle and initiate myelin formation. In contrast, groups of lower caliber axons (mainly the fibres of nociceptive and most autonomic neurons) remain associated with a single Schwann cell. These so-called non-myelin forming Schwann cells accommodate and ensheath the individual axons in cytoplasmic intentions or cuffs. These groups of axons-glia are called Remak fibers (see Figure 2.10A).

The fate decision to become either a myelin-forming or a non-myelin-forming cell is governed by axonal cues. The exact nature of these cues has long remained elusive and is still not understood in full detail. However, again Nrg1 signaling appears to play a major role in the initiation and extent of myelination. The extent of myelination, that is, the number of layers of compact myelin and the length of the internode, the distance between two successive nodes of Ranvier, which play an essential role in action potential conduction, correlates with the diameter of the axon. Thus, thicker axons have thicker myelin and longer internodes. It has been demonstrated that myelination is graded to the strength of the Nrg1 signal. For

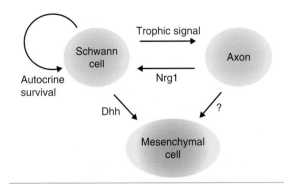

Figure 2.11 Signaling between cell populations in the peripheral nerve. The figure summarizes the cell types and the cellular and molecular interactions that shape the peripheral nerve tissue and that are discussed in the text.

example, forced expression of Nrg1 typeIII on the axonal surface of sympathetic neurons, which normally express very low levels of Nrg1 and are not myelinated, induce Schwann cells to form a myelin sheath (Figure 2.11) (Taveggia *et al.*, 2005). Furthermore, overexpression of typeIII Nrg1 in peripheral neurons in transgenic mice results in the formation of thicker myelin sheaths. Additionally, the decreased strength of the Nrg1 signal in ErbB2+/−, Nrg1+/− mice results in reduced myelin thickness (Michailov *et al.*, 2004).

The diameter of an axon appears to be determined by signals from its specific target. Thus, axons of sensory neurons that innervate muscle spindle organs have a much larger radius than those of sensory neurons that innervate the skin. Different targets express different neurotrophic factors that aid in guiding and supporting the innervating neuron. Both qualitative and quantitative aspects of neurotrophic support appear to regulate axonal diameter. For example, it was found that the experimental increase of the target field of sympathetic axons that innervate the submandibular gland, which expresses NGF, results in an increase in diameter of the innervating axons and myelination of these otherwise unmyelinated axons (Voyvodic *et al.*, 1989). It is therefore anticipated that a mechanism exists that couples axon diameter with

the level of Nrg1 typeIII expression on the axonal surface (Taveggia *et al.*, 2005).

2.4.3 Initiation of myelination

Nrg1 signaling activates an intracellular signaling pathway in Schwann cells that includes the PI3 and cAkt/PKB Kinases. Activation of this pathway is required for initiation of myelination (Ogata *et al.*, 2004; Maurel and Salzer, 2000). This and other unknown signaling pathways converge on the nucleus to activate a transcriptional program of myelination, involving the coordinate regulation of a large number of genes. The onset and progression of the myelination program is controlled by a set of transcription factors that include the POU homeodomain proteins Oct6 and Brn2, the Sry box proteins Sox2 and Sox10 and the zinc finger protein Krox20 (Bermingham *et al.*, 1996; Britsch *et al.*, 2001; Le *et al.*, 2005). Sox2 is expressed in early stages of the Schwann cell lineage and its downregulation is required for myelination to commence (Le *et al.*, 2005). The progression of Schwann cell differetiation from a promyelinating Schwann cell, a cell that has a acquired a 1:1 relationship with its axon and ceases to proliferate, to a myelinating cell requires the action of Oct6. Upregulation of Oct6 is mediated through a remote enhancer element called the Oct6 Schwann cell enhancer (Ghazvini *et al.*, 2002). It is on this element that many of the signals discussed above converge to activate myelin formation. A second POU protein, Brn2, closely related to Oct6, plays a partially redundant role as in the absence of both Oct6 and Brn2 the initiation of myelination is severely inhibited. A major target of Oct6 and Brn2 regulation is Krox20. Once Krox20 is activated, Oct6 and Brn2 expression is gradually extinguished. Activation of myelin genes requires Krox20, as in the absence of Krox20 myelination is blocked. Additionally, Sox10 cooperates with Krox20 in regulating expression of the major myelin gene mpz and mbp. Thus, a genetic hierarchy of a limited number of transcription factors, including Oct6, Brn2 and Krox20, initiate, together with Sox10, the myelination program.

2.4.4 Structure of the peripheral nerve sheath

While nerve tracts in the central nervous system are protected by the rigid bony structure of the skull and vertebral column, a compact connective tissue matrix protects the peripheral nerves. Three layers can be distinguished and these have been termed epineurium, perineurium and endoneurium. The epineurial and perineurial cells of the nerve sheath are derived from the mesoderm (Bunge *et al.*, 1989).

While the epineurium is a condensation of connective tissue surrounding the peripheral nerve, the perineurium is a multilayered cellular sheath surrounding individual fascicles of the nerve. The layers consist of concentric sleeves of flattened epithelial cells that interdigitate and are connected by tight junctions at their margins.

The endoneurium mainly consists of a thick layer of collagen fibrils running parallel with the axon-Schwann cell units. Axons, Schwann cells and fibroblasts are the major cellular component of the intrafascicular space. The majority of endoneurial fibroblasts are derived, like the Schwann cells, from the neural crest (Joseph *et al.*, 2004).

2.4.5 Development of the peripheral nerve sheath

The peripheral nerve sheath develops from mesenchymal cells surrounding the embryonic nerves. From E15 onwards, mesenchymal cells assemble concentrically around the developing nerve and around birth a recognizable perineurium is present. At this stage the perineurium is still leaky, but becomes impermeable a few weeks after birth (Sharghi-Namini *et al.*, 2006).

Genetic experiments have demonstrated that the formation of a structurally and functionally normal perineurium depends on the Schwann cell-derived signaling molecule desert hedgehog (Dhh). Dhh is one of three members of the hedgehog family of intercellular signaling molecules that play key roles in embryonic pattern formation and organogenesis (Hammerschmidt *et al.*, 1997). Hedgehog molecules signal through the patched and smoothened receptor molecules expressed on target cells. Activation of the hedgehog receptor results in the activation and translocation to the nucleus of one of the Gli family of transcription factors (Gli1, 2 and 3). This results in activation of a set of target genes among which is the hedgehog receptor patched. Dhh is expressed in Schwann cell precursors of the embryonic nerve at around E12. Its receptor, patched is first detectable in presumptive perineurial cells around E15 in the mouse. In dhh mutant animals patched expression in presumptive perineurial cells is strongly reduced and the development of all three layers of the nerve sheath is severely affected (Parmantier *et al.*, 1999). The number of perineurial cells is reduced and only a few layers are formed. Additionally these abnormal perineurial sheaths extend into the endoneurium to form mini-fascicles. The structurally abnormal perineurium of dhh animals proved permeable for large proteins and neutrophils could penetrate the nerve from the surrounding tissue (Parmantier *et al.*, 1999). It was also found that the nerve blood barrier was compromized. Thus, it is likely that the compromized function of the perineurium and breakdown of the blood-nerve barrier contributes to the structural myelin abnormalities and increased axonal degeneration observed in Dhh-/- mice (Sharghi-Namini *et al.*, 2006).

The fact that a thin layer of perineurial cells is formed in the absence of dhh signaling, suggests that dhh is not required for the initial recruitment of mesenchymal cells. The nature of this mesenchymal recruitment signal is still unknown, but the observation of a perineurial like sheath in genetic mutants that lack Schwann cells suggest that the initial signal might be derived from the axon.

2.4.6 Summary and perspective

The development of the peripheral nerve tissue is orchestrated through a number of cellular and molecular interactions that are now understood in some detail. Clearly, the Nrg1-ErbB2 signaling axis plays a pivotal role in Schwann cell survival, migration and differentiation. It is intuitively obvious that additional signals must exist that act on this crucial pathway to drive successive stages of Schwann cell development. Some of these signals are provided through Schwann cell basal lamina interactions but many more remain to be discovered. In addition to providing essential support for neurons, Schwann cells also orchestrate the development of the protective perineurial sheath through secretion of Dhh. Importantly; Schwann cells provide trophic and structural support for regenerating axons (Scherer and Salzer, 2001). Ideally, artificial nerve grafts that aim to guide and support the regenerating nerve into a distal, denervated nerve stump should include autologous Schwann cells. However, to obtain autologous Schwann cells healthy nerves need to be sacrificed. Exciting new developments

suggest that the stem cell population of the hair follicle bulge might provide an alternative source of Schwann cells (Amoh *et al.*, 2005). The successful application of these cells in a clinical setting will require a detailed understanding of the biology of these stem cells and the development of procedures to derive and expand Schwann cells.

2.5 Embryonic skin development

Tissue-engineered living skin substitutes have found their way into clinical practice for treatment of severe skin defects. While the application of these *in vitro* generated skin substitutes have greatly improved treatment outcome of severe burn defects, many issues remain to be resolved to end up with an aesthetic and functional equivalent that matches the natural skin. Here the

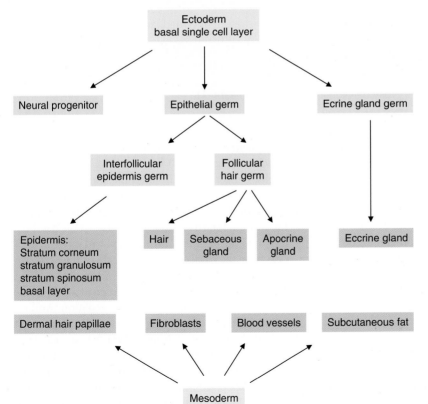

Figure 2.12 Schematic diagram of the cellular origin of the skin.

Table 2.1 Timelines for embryonic skin development

Gestation (weeks)	Fetal length (cm)	Developing structure
5	0.14	ectoderm: basal single cell layer and intermittent periderm
7		Langerhans cell precursors from hematopoietic origin
8		mesoderm: loosely arranged mesenchymal cells embedded in ground substance; Merkel cell differentiation from ectoderm
10	5.0	epithelial germ: basal cell layer, stratum intermedium, periderm
12		epithelial germ: melanocyte precursors in intermediate layer; early dermis: argyrophilic reticulum fibers (collagen III), abundant fibroblasts
13		Eccrine glands start to develop
17		Hair and sebaceous gland development
19	17	epidermis: basal layer, 2–3 intermediate layers, flattened periderm
20		apocrine glands; fat cells mainly in brown fat (very little white fat)
22		dermis: elastic fibers
23	19	epidermis: keratinisation in stratum intermedium (keratohyalin granules), premature stratum corneum
24		Formation of vernix caseosa
32		papillary and reticular dermis
36	40–50	fully developed immunocompetent skin

sequential steps and factors involved in embryonic skin development are summarized and correlated with the formation of skin substitutes *in vitro*.

Embryonic skin develops from the ectoderm and mesoderm via a complex process of cell proliferation, differentiation, migration and apoptosis (Figure 2.12 and Table 2.1). The ectoderm is the progenitor of the follicular and interfollicular epidermis and is also the neural progenitor (Moreau and Leclerc, 2004). The mesoderm is the progenitor of the dermis and subcutaneous fat.

2.5.1 Interfollicular epidermis

In 5 week gestational skin, the epidermis consists of a single layer of ectodermal cells with intermittent

areas containing a second suprabasal layer – the periderm (Figures 2.13A and 2.14E). Already after 7 weeks, Langerhans cells (immune surveelance cells of the skin), derived from bone marrow cells, begin to appear in the developing epidermis. By the 10th week, the stratum intermedium forms between the basal cell layer and the periderm through upward movement of cells from the basal cell layer. The cells of the periderm become large and protrude into the amniotic cavity. Merkel cells (nerve/touch cells) differentiate from the ectoderm between weeks 8–12 of gestation. Around 12 weeks, melanoblasts migrate from the neural crest into the ectoderm (Hirobe, 2005). These cells later develop into melanocytes – the pigment forming cells in the epidermis. The appearance of melanoblasts takes place in a craniocaudal direction following the 'Lines of Blaschko' (Figure 2.15) (Bolognia

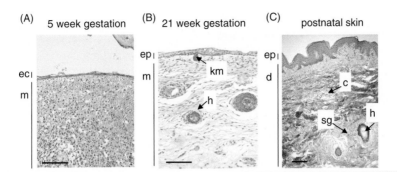

Figure 2.13 Developing embryonic skin: (A) Embryonic skin at 5 weeks gestation. Note the single layered ectoderm (ec) containing sporadic periderm cells, the cell dense mesenchyme (m) and the absence of capillaries; (B) embryonic skin at 21 weeks gestation. Note the multilayered developing epidermis (ep) lacking a stratum corneum. The epidermal keratinocyte mass (km) that develops into the hair follicle and sebocytes can be seen budding into the dermis. Mesenchymal cells are becoming less dense in the dermal matrix and precursor hair follicles are visible (h); (C) fully developed new born skin. Note the stratified epidermis (ep), endothelial capillaries (c) and low density of fibroblasts (single cells) in the dermis (d). Fully developed hair follicles (h) and sebaceous glands (sg) are visible deep in the dermis. Bar = 100 μm.

et al., 1994). After 21 weeks of gestation, fetal skin becomes thickened due to increasing numbers of stratum intermedium cells and the periderm becomes flattened (Figures 2.13B and14F). At 23 weeks keratinization takes place in the upper cell layers and small keratohyalin granules form. The cells of the periderm are shed into the amniotic fluid, leaving only fragments of degenerated periderm cells above the keratinized cells of the newly formed stratum corneum. The epidermis continues to form by upward movement of differentiating keratinocytes from the proliferating basal layer until a fully differentiated, stratified epidermis is formed (Figures 2.13C and 2.14G) (Mack *et al.*, 2005). During the last trimester of gestation (>24 weeks), the newly formed stratum corneum of the fetus is protected from the amniotic fluid by a white, greasy biofilm called vernix caseosa. Vernix caseosa consists of water-containing corneocytes embedded in a lipid matrix and the basic structure shows certain similarities with the stratum corneum (Rissmann *et al.*, 2006). After birth, the vernix caseosa is absorbed and the epidermis forms a competent barrier to the environment preventing infection and dehydration of the infant (Haubrich, 2003).

2.5.2 Follicular epidermis

The ectoderm basal cell layer, proliferates and differentiates into the keratinizing interfollicular epidermis as described above. Additionally, the ectoderm forms the eccrine glands (sweat) and hair germs which then further differentiate into hair, subbaceous glands and apocrine glands (Figure 2.12) (Fu *et al.*, 2005; Schmidt-Ullrich and Paus, 2005). In early gestation (13 weeks), when fetal skin is composed of two or three epidermal layers, eccrine glands begin to form. Eccrine gland islands composed of epidermal cells gradually migrate down into the dermis to form the juvenile sweat glands at week 18–20. Hair follicles begin to form at 17 weeks. The formation of these follicles represents a proteotypic interaction between the neuroectoderm and mesoderm that is provided by three different stem cell sources, epidermal, neural crest and mesenchyme. The epidermal keratinocyte mass that differentiates into hair follicles and sebocytes buds into the deeper layers of the dermis Figures 2.13B and 2.14F). Beneath each bud lies a group of mesenchymal cells (fibroblasts) from which the dermal hair papillae and connective tissue sheath is later

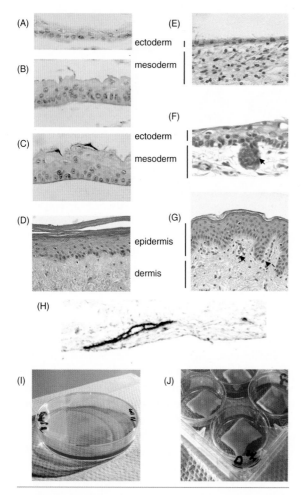

Figure 2.14 *In vitro* versus in vivo developing skin: (A) Cross section of a keratinocyte sheet cultured submerged under the culture medium on a feeder layer of irradiated 3T3 mouse fibroblasts. Fibroblasts are washed away before harvesting and therefore are not visible. Note the similarity with 5 week gestational ectoderm (E); (B) keratinocyte sheet cultured as described in (A) but with addition of retinoic acid to the culture medium which stimulates keratinocyte proliferation. Note the resemblance to 21 week gestation skin (F), only the follicular bud is absent (arrow); (C) keratinocyte sheet as described in (A) but with the addition of vitamin D as well as retinoic acid to the culture medium. Retinoic acid stimulates proliferation and stratification whilst vitamin D

Figure 2.14 (Continued) stimulates differentiation. Note the keratohyalin-like granules in the upper layers making this culture resemble developing epidermis at 23 weeks gestation; (D) reconstructed epidermis cultured on fibroblast populated human dermis. As the cultures are grown at the air-liquid interface, with nutrients from the medium diffusing in through the dermis, complete epidermal differentiation occurs. Note the fibroblasts populate the dermis at a similar frequency to that observed in (G) but significantly less than in (E). Also, no endothelial capillaries are present; (H) fibroblast–endothelial cell co-culture stimulates endothelial capillary formation (blue shows haematoxylin staining of fibroblast nuclei; red shows EN4 immunostaining of endothelial cells) and deposition of extracellular matrix from fibroblasts; (I) Petri dish showing experimental set-up for culturing under submerged conditions as described in (A), (B), (C) and (H); (J) experimental set-up for culturing at the air–liquid interface as in (D).

formed. Neural crest cells give rise to the melanocytes of the hair follicle pigmentary unit. After 21 weeks of gestation, the structures of skin appendages (eccrine glands, hair follicles and sebaceous glands) can be detected (Figure 2.13B). After 28 weeks of gestation, as fetal skin becomes even thicker, the number of eccrine glands and hair follicles are increased and the structure of the appendages become mature. At birth, the sebaceous glands are considerably larger than they are in infancy and secrete the vernix caseosa.

2.5.3 Dermis

In 5 week gestational skin, the dermis consists of loosely arranged mesenchymal cells that are embedded in a ground substance (Figure 2.13A). At 12 weeks, argyrophilic reticulum fibres appear. As these fibres increase in number and thickness, they arrange themselves in bundles. Simultaneously, mesenchymal cells develop into fibroblasts and endothelial cells (Figure 2.12). The fetal dermis shows many more fibroblasts than adult dermis (compare Figures 2.13A and B with

2.12C). Also, the dermis of the foetus contains a large amount of collagen type III, in contrast to adult skin which contains a large amount of collagen type 1. Elastic fibres appear in the dermis at 22 weeks. As gestation continues, elastic fibres increase in number until at 32 weeks, a well developed network is formed in the reticular and papillary dermis which is indistinguishable from that found in new born infants.

Fat cells begin to develop in the subcutaneous tissue after 20 weeks. Three types of cell can be found, (a) spindle shaped, lipid free mesenchymal precursor cells, (b) young type fat cells containing two or more small lipid droplets and (c) mature fat cells possessing one large lipid droplet.

2.5.4 *In vitro* models for embryonic and new born skin

The various stages of skin development can be mimicked *in vitro* by adjusting and finely tuning the culture conditions under which the cells are grown. One should not just consider the tissue itself (in this case skin) but also the environment around the tissue. For example, embryonic ectoderm is in a wet environment engulfed by amniotic fluid on one side and the developing mesoderm rich in fibroblasts on the other side. This provides an extremely wet and nutrient rich environment. Upon birth, the environment changes significantly. The skin is now exposed to the air. Nutrients reach the epidermis by diffusion from blood vessels in the underlying dermis. Therefore, conditions switch from a wet to a dry environment and also from a nutrient rich to a nutrient poor environment. This transition is thought to trigger the final stage in the formation of fully developed skin thus forming a competent barrier to the environment. Therefore when trying to generate different stages of skin development, for example the ectoderm transition to epidermis, one should try to copy these environmental changes. Cultures which resemble early ectoderm can be generated from newborn or adult keratinocytes by culturing the keratinocytes on a dense feeder layer of lethally irradiated fibroblasts completely submerged

in culture medium (Figure 2.14A, I) (Gibbs *et al.*, 1996; Green *et al.*, 1979). The fibroblasts provide a contact point for the initially seeded keratinocytes to nestle up to, but importantly the fibroblasts are metabolically active resulting in the secretion of growth factors and cytokines essential for keratinocyte viability and proliferation. The result is a keratinocyte sheet forming which consists of approximately two layers of unkeratinized epidermis showing no similarity to new born skin. The keratinocyte sheet cultured in this way visible resembles early ectoderm at 5 weeks gestation (compare Figure 2.14A with 2.14E).

In order to progress to the next stage of embryonic epidermal development *in vitro*, additional nutrients are required. Addition of, for example retinoic acid (a factor known to be involved in healthy skin development) to the primitive ectoderm like culture stimulates cell division and upward migration of keratinocytes. This increases the number of cell layers and results in cultures which visibly resemble embryonic skin at 21 weeks gestation (compare Figures 2.14 B and 2.14F). Addition of vitamin D in combination with retinoic acid further increases the number of layers, epidermal differentiation, keratinization and the appearance of small keratohyalin-like granules in the upper layers (Figure 2.14C) (Gibbs *et al.*, 1996). These cultures visibly resemble embryonic skin at approximately 23 weeks gestation.

In order to mimic the final stage in epidermal development, the culture environment has to be changed significantly – copying the environment change at birth. The most profound change is exposure of the skin to the air in place of the warm and wet environment of the amniotic fluid. *In vitro* this can be mimicked by raising the keratinocyte cultures to the air–liquid interface (Figure 2.14J) (Gibbs *et al.*, 1997). The transition from submerged culture conditions to air exposed culture conditions stimulates the final step in skin development – the formation of the stratum corneum.

The developing dermis can be copied *in vitro* by co-culturing fibroblasts and endothelial cells (Hudon *et al.*, 2003; Ponec *et al.*, 2004). Cell–cell interactions and optimal culture conditions stimulate individual

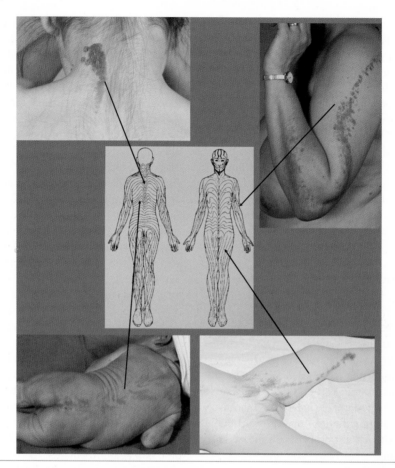

Figure 2.15 Lines of Blaschko. More than a century ago, the lines of Blaschko were described by a dermatologist named Alfred Blaschko (Happle and Assim, 2001; Rott, 1999; Traupe, 1999; Bolognia *et al.*, 1994). The lines were based on the observation of patterned skin lesions that were linear on the extremities, S-shaped on the anterior trunk and V-shaped on the back. The lines (or mosaicism) result when a postzygotic mutation occurs leading to a subject with two or more genetically different populations of cells that originate from the genetically homozygous zygote. In most cases, the linear patterns of abnormal skin are surrounded by normal skin. Blaschko's lines are thought to trace pathways of ectodermal development. Several studies have demonstrated that the genotypes of keratinocytes and underlying fibroblasts do not correlate in mosaic skin conditions, confirming the different developmental patterns of ectoderm and mesoderm. The epidermis and its appendageal structures, the melanocytes, the vascular system and the subcutaneous fat, separately or in combination may be involved in the morphological manifestations which follow Blaschko's lines. The central figure illustrates the Lines of Blaschko originating from the neural crest. Upper left/right and lower right show subjects with epidermal nevi; lower left shows a baby with 'linear and whorled nevoid hypermelanosis'. Photographs are supplied by Dr. M. Wintzen, dermatologist, VU university medical center, NL.

endothelial cells to migrate towards each other to form capillary-like tubes (Figure 2.14H). Fibroblasts synthesize extracellular matrix components thus forming a human collagen–elastin rich dermal matrix similar to *in vivo*.

Whereas the development of the interfollicular epidermis and dermis can be mimicked *in vitro*, development of the folicullar epidermis is still an unsolved challenge to scientists.

2.5.5 Cell–cell interactions and growth factors

The skin development and thereafter the maintenance of skin integrity (homeostasis) is dependent on a complex interplay between cell types within the ectoderm (epidermis) and mesoderm (dermis). Cell homing to the correct location, proliferation and differentiation is regulated by growth factors, cytokines and chemokines that act in autocrine and paracrine loops. The result is the formation of fully developed skin with constant ratios of keratinocytes:melanocytes (36, 1), keratinocytes:Langerhans Cells (53, 1) and keratinocytes:fibroblasts (Hoath and Leahy, 2003). The formation of cell units can be mimicked *in vitro* and give an insight into embryonic skin development. Melanocytes co-cultured with keratinocytes under submerged conditions (ectoderm-like culture) maintain a constant ratio for up to three passages (3 weeks) and when co-seeded onto acellular dermis and cultured at the air-liquid interface form a differentiated epidermis interdispersed with melanocytes in the basal layer (Gibbs *et al.*, 2000a; Staiano-Coico *et al.*, 1990). Melanosomes formed in the melanocytes are transferred to keratinocytes and cap the nuclei just as *in vivo* in order to protect the dividing keratinocyte population from harmful ultraviolet irradiation after birth. It has been reported that fetal keratinocytes possess a greater stimulatory effect on proliferation of melanocytes than neonatal keratinocytes as fetal keratinocytes produce and release more mitogens than newborn keratinocytes. Similar to melanocytes, Langerhans cell precursors (CD34+ cord blood derived) can be introduced into submerged keratinocyte cultures which when air-exposed, develop into fully immuno-competent Langerhans cells in the epidermis in similar ratios found in fully developed skin (Facy *et al.*, 2005). Co-cultures of keratinocytes and fibroblasts (irradiated feeder layer) maintain the keratinocyte stem cell population *in vitro* and enables large amounts of keratinocytes, including stem cells, to be amplified for transplantation as sheets onto large burns wounds (Green *et al.*, 1979). Upon transplantation, the ectoderm-like culture develops into a differentiated adult epidermis as the environmental conditions change (submerged, nutrient rich environment changes to air-exposed, nutrient poor environment). Soluble factors secreted by keratinocyte–fibroblast interactions are involved in formation of the basement membrane (El Ghalbzouri and Ponec, 2004). The basement membrane is essential for the attachment of the epidermis to the dermis and inheritory defects can result in blistering (bullous disease). Soluble factors secreted by keratinocytes result in homing of fibroblasts into the dermis (Gibbs *et al.*, 2006), whereas soluble factors secreted by fibroblasts stimulate formation of the epidermal layers (basal layer, spinous layer, granular layer, stratum corneum) (El Ghalbzouri *et al.*, 2002). One of these factors is keratinocyte growth factor (FGF7). Supplementation of keratinocyte growth factor to the culture medium induces keratinocyte proliferation and differentiation and can be used to replace living fibroblasts in the dermal matrix (Figure 2.16) (Gibbs *et al.*, 2000b). Table 2.2 summarizes the properties of some keratinocyte and fibroblast derived growth factors involved in skin development (Botchkarev *et al.*, 1999; Botchkarev and Sharov, 2004; Gibbs *et al.*, 2000b; Heng *et al.*, 2005; Hirobe, 2005).

2.5.6 Summary

The development of embryonic skin involves environmental, matrix and cell–cell interactions, Very importantly, it involves a finely tuned differential and sequential secretion of soluble factors which regulate

 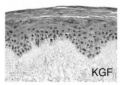 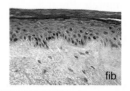

Figure 2.16 Keratinocyte growth factor (KGF) can be used to replace fibroblasts (fib) in stimulating epidermal morphogenesis. In the absence of KGF or fibroblasts (fib) the stratified layers of the epidermis are not properly formed. KGF is a growth factor made and secreted by fibroblasts which stimulates keratinocyte proliferation and differentiation and is used as a substitute for fibroblasts (left panel) in co-culture experiments with human keratinocytes.

Table 2.2 Growth factors secreted by keratinocytes and fibroblasts which are involved in skin development

Growth factor	Secreted by	Target cell	Property
Bone morphogenic protein	?	Ectoderm / KC	Differentiation of ectoderm and epidermis
Noggin	mesoderm	ectoderm	Neural differentiation, hair follicle formation suppression of ectoderm,
Epiregulin	KC	KC	Proliferation
Epidermal growth factor	Hair follicle?	KC / Fib	Proliferation, migration
Keratinocyte growth factor	Fib	KC	Proliferation, differentiation
Tumor growth factor-α	KC / Fib	KC / Fib	differentiation, migration, extracellular matrix synthesis
Vascular endothelial growth factor	KC	Endo	Angiogenesis
Tumor necrosis factor-α	KC / Fib	KC / Fib / LCp	KC / Fib proliferation, LCp differentiation to LC
Tumor growth factor-β	KC	Fib / LCp	Fibroblast differentiation to myofibroblast, LCp differentiation to LC
Granulocyte macrophage-colony stimulating factor	KC	LCp / MC	LCp differentiation to LC, MC proliferation, melanogenesis, dendritogenesis
Basic fibroblast growth factor	KC	MC / Endo / Fib	Proliferation, angiogenesis
Hepatocyte growth factor	KC / Fib	MC	Proliferation
α-MSH	KC	MC	Melanogenesis, dendritogenesis
Nerve growth factor	KC	MC	Melanogenesis, dendritogenesis
Endothelin 1	KC	MC	Proliferation, melanogenesis, dendritogenesis
Stem cell factor	KC / Fib	MC	Proliferation, melanogenesis, dendritogenesis

KC: keratinocyte; Fib: fibroblast; LCp: Langerhans Cell precursor; LC: Langerhans Cell; MC: melanocyte; Endo: endothelial cell.

cell growth and differentiation. *In vitro*, a number of phases of embryonic skin development can be mimicked such as development of the interfollicular epidermis including keratinocytes, melanocytes and Langerhans cells. Also a capillary network in a fibroblast populated dermal matrix can be generated from a mesenchymal cell mix. However, the formation of appendages such as hair follicles and eccrine glands, which involve extensive dermal and epidermal interactions, are still an unsolved challenge for tissue engineering.

2.6 Skeletal formation

The skeleton is a dynamic living tissue consisting of an abundant mineralized extracellular matrix combined with bone forming and bone resorbing cells. The extracellular matrix can be divided into two types, bone and cartilage, which differ in their matrix composition and physical properties in line with their respective functions. Three cell types are involved in skeletal formation, the cartilage producing chondrocyte, the bone forming osteoblast and the cartilage and bone resorbing osteoclast. Major challenges for tissue engineering of the skeleton are, (i) the repair of damaged articular cartilage covering the distal ends of the bone in the joints, (ii) the repair of non-union fractures and (iii) the improvement of the union of synthetic protheses with native bone. The purpose of this section is to provide a general overview of skeletogenesis during fetal development including the origin of the bone forming and degrading cells and the basic principles underlying the maintenance of skeletal integrity during adulthood.

The adult human skeleton contains 213 bones. Depending on their location in the body, these bones have one or more specific functions. They provide structural support and integrity to the body and protect vital internal organs such as the brain, heart and lungs. Bones play an essential role in body movement by providing attachment sites for muscles. They are also involved in the regulation of mineral homeostasis.

Furthermore, they provide the appropriate environment for the development of the hematopoietic system inside the bone marrow cavity. Bones are formed either by endochondral ossification or by intramembranous ossification.

Three cell lineages are responsible for the formation of the skeleton during embryonic development. The paraxial and lateral plate mesoderm and the neuroectoderm-derived cranial neural crest. The lateral plate mesoderm will form the long bones in the limbs. The paraxial mesoderm will form the axial skeleton (vertebral column and rib cage) and together with the neural crest will shape the craniofacial bones. The formation of bones starts during the fourth week of human development and begins with the formation of the bones at the base of the skull. New bones are formed in an anteroposterior direction. With exception of the bones in the jaws, the formation of the craniofacial bones begins later. The shape of each bone is specifically adapted to its function, which depends on its place in the skeleton. Even before the onset of skeletogenesis, cells at the site of the future bone already contain information on the shape and structure of the bone they will form. This information is provided, amongst others, by the homeodomain containing transcription factors of the four different Hox-gene clusters (see also section 2.1.4). Each cell that will contribute to a distinct bone expresses a specific combination of hox-genes and this combination provides detailed instructions on the shape of the future bone (Van Den Akker *et al.*, 2001; Deschamps and van Nes, 2005).

2.6.1 Skeletal precursor cells

Osteoblasts and chondrocytes are both derived from the multipotent mesenchymal stem cell (MSC). This cell also gives rise to a variety of related mesodermal cell lineages such as fibroblasts, muscle cells, tendons and adipocytes. The MSC is derived from either the mesoderm or the cranial neural crest. In fetal development MSCs can be isolated from virtually any tissues in relatively high quantities. In adulthood, most tissues possess only a limited number of MSCs, except for the bone marrow in which relatively high amounts of MSCs reside throughout life. These cells play an important role in the continuous cycle of bone formation and degradation by providing a source for new osteoblasts. Signaling by the Wnt-family of morphogens plays an important role in the maintenance of the undifferentiated stem cell population (Niemann, 2006; Hartmann, 2006). Wnts stimulate stem cell proliferation and renewal and simultaneously block the initiation of differentiation by increasing the nuclear localization and the transcription potential of β-catenin, an essential and critical

intracellular mediator of the wnt-signal transduction pathway (see also Chapter 4.4.3). The Wnt-family consists of 19 different members which use at least 10 different receptors. Their activity is tightly controlled by a variety of extra- and intracellular antagonists. It is currently largely unknown which combination of these factors is involved in maintenance of the MSC population (Figure 2.17) (Gordon and Nusse, 2006).

At sites of future bones, the MSC will give rise to a skeletal precursor cell which can differentiate into either an osteoblast or a chondrocyte. The direction of differentiation depends on the activity of two transcription factors, Sox-9 and RunX2. Knock-out

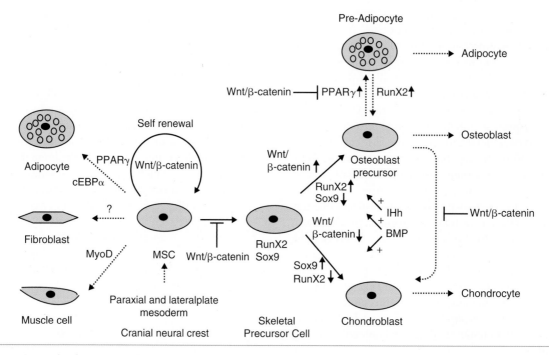

Figure 2.17 The formation of skeletal precursor cells. Skeletal precursor cells are derived from the Mesenchymal Stem Cell that also give rise to adipocytes, tendon and fibroblasts. Wnt/β-catenin is responsible for stem cell proliferation and renewal. The MSC differentiates into a skeletal precursor cell that expresses the master transcription facors for osteoblast differentiation (RunX2) and chondrocyte differentiation (Sox9). Wnt/β-catenin signaling directs the precursor into the osteoblast lineage and prevents transdifferentiation into chondrocytes by augmenting RunX2 activity and repressing Sox9. Also Indian hedgehog (IHh) and BMP-signaling are involved in this process. The osteoblast precursor is still able to differentiate into adipocytes. The balance between the transcription factors PPARγ (adipocyte specific) and RunX2 (osteoblast specific) determines the fate of the cell. See text for further details.

State of the Art Experiment

Wnt is a family of growth factors that belong to one of the major morphogenic pathways. Abberations in this pathway underlie a variety of developmental disorders and cancer. One of the intracellular effectors of Wnt-signaling is β-catenin, a key component of the so-called canonical Wnt-signaling pathway. In the absence of Wnt-signaling, cytoplasmic β-catenin is phosphorylated by the enzyme Glycogen-Synthase Kinase 3β (GSK3β). Phosphorylation targets β-catenin to the proteasome where it is inactivated by degradation. In the presence of Wnt, GSK3β is inhibited and unphosphorylated β-catenin can accumulate in the cytoplasm and eventually translocate to the nucleus. Here it can act as an activator of gene transcription, resulting in the expression of a wide variety of target genes. The importance of this signaling pathway in embryonal development is underscored by classical gene knock out experiments. Loss of β-catenin, results in early embryonic lethality shortly after gastrulation. At this time point organogenesis has not been started, thus this conventional knock out approach does not allow to examine the role of Wnt/β-catenin in, for example, bone formation. To circumvent this problem developmental biologists use a conditional gene targeting approach. In such an approach, the gene of interest, in this case β-catenin, is genetically manipulated in ES cells. The manipulation aims at the introduction of a short DNA sequence, a so-called lox-P sites, at both ends of the gene resulting in a 'floxed' gene. The lox-P sites do not interfere with the normal function of β-catenin. They are, however, a target for the enzyme CRE-recombinase. This enzyme specifically deletes all the DNA sequences that are present between two lox-P sites and is normally not present in a mouse. This enzyme can be introduced into a new transgenic mouse line. In these mouse lines, CRE-recombinase expression is often controlled by a tissue-specific promoter, resulting in the expression of the CRE enzyme in the cell type of interest. When the cell type of interest is formed during embryogenesis, the promoter is turned on and the CRE enzyme becomes expressed. In itself, the CRE-enzyme has no effect on the development of the embryo. However, in case the CRE-enzyme expressing mouse line is crossed with a transgenic mouse line with a floxed gene, the expression of the CRE enzyme results in the efficient deletion of the DNA-fragments between the loxP sites. This recombination only occurs in cells of the embryo in which the expression of the CRE-recombinase is turned on, resulting in tissue or cell type specific deletion of the gene. Day et al. (2005) and Hill et al. (2005) used such an approach to manipulate β-catenin in skeletal precuros cells. Hill et al used the Prx1 promoter to drive the expression of the CRE-enzyme in paraxial and lateral plate mesoderm. Day et al. used the Dermo1 promoter that is also expressed in skeletal progenitors in the skull. Inactivation of β-catenin in skeletal precursor cells present in these cell populations caused ectopic formation of chondrocytes at the expense of osteoblast differentiation during both intramembranous and endochondral ossification. They both used the conditional gene targeting strategy to express a stabilized form of β-catenin in the same cell populations with opposite results. The stabilized form of β-catenin is resistant to GSK3β phosphorylation and cannot be degraded by the proteasome. Thus even in the absence of Wnt, the Wnt-signaling pathway is activated. Day et al. showed that ectopic activation of canonical Wnt-signaling leads to enhanced osteoblastic mediated bone formation and the suppression of chondrogenesis. Using the Prx1 promoter, Hill et al. showed that ectopic expression of a stabilized β-catenin completely blocks the differentiation of mesenchymal cells into skeletal precursor cells. The authors conclude that canonical Wnt/β-catenin signaling is essential for skeletal lineage differentiation from mesenchymal precursors. It prevents the transdifferentiation of osteoblastic cells into chondrocytes, which are the default cell types arising from mesenchymal precursors in the embryo. It shows that the Wnt-signaling pathway is a key mechanism by which the specification of intramebranous and endochondral ossification is controlled during embryogenesis. The results of these studies are fundamental to the current model for skeletal precursor cell differentiation as depicted in Figure 2.17.

The results of these and other studies have important implications for tissue engineering of bone and cartilage. For example, a scaffold that has intrinsic capacity to activate Wnt-signaling or β-catenin will be beneficial for bone regeneration but will be very inefficient in repair of cartilage. Conversely, an ideal scaffold for cartilage tissue engineering will be capable of actively repressesing Wnt-signaling in the stem cells or primary chondrocytes that are seeded on the material.

studies in mice have shown that Sox-9 is indispensable for chondrocyte differentiation, while RunX2 is indispensable for osteoblast differentiation. The skeletal precursor cell expresses both transcription factors. Intracellular levels of β-catenin controlled by wnt-signaling determine the direction of differentiation. High levels of β-catenin inhibit Sox-9 and potentiate RunX2 activity resulting in osteoblast differentiation. In contrast, low levels of β-catenin result in unopposed Sox-9 activity and thus in the formation of chondrocytes (Figure 2.17) (Hill *et al.*, 2005a; Day *et al.*, 2005b; Glass and Karsenty, 2006). In embryogenesis, chondrocyte differentiation appears to be the default differentiation route of the MSC.

2.6.2 Endochondral ossification

The axial and appendicular skeleton is formed by endochondral ossification, i.e. the skeletal elements are preshaped in a cartilaginous mold which is subsequently replaced by bone (Kronenberg, 2003). This process starts with the condensation of loosely connected mesenchymal cells at sites of the future bones (Figure 2.18). This is accompanied with a rearrangement of the vasculature resulting in an avascular condensed mesenchyme surrounded by blood vessels. The transcription factor Sox-9 is responsible for the condensation of the skeletal precursor cells. The combination of increased cell-cell contacts resulting in signaling of amongst others cadherins, low oxygen conditions due to the absence of vessels and low levels of wnt-signaling, is believed to trigger the initiation of chondrocyte differentiation. It is characterized by the production of an abundant extracellular matrix consisting predominantly of collagen 2 and glycosamine glycans which are typical cartilage markers. In the middle of the cartilage anlage, chondrocytes start to arrange in typical columns and display high proliferative activity. Somewhat later these chondrocytes stop proliferating and differentiate further into hypertrophic chondrocytes. This is characterized by a dramatic increase of the cell's volume and a reshuffling of the extracellular matrix, which now contains collagen 10. The hypertrophic chondrocytes start to mineralize their matrix forming the primary ossification center and subsequently die. At the same time, the chondrocytes signal to the surrounding perichondrial cells in which osteoblast precursors reside by producing Indian hedgehog (IHh). IHh initiates osteoblast differentiation and the formation of the bone collar. It is the most important coupling factor between chondrocyte and osteoblast differentiation during endochondral ossification (Kronenberg, 2003a). The hypoxic conditions in the center of the bone anlage, induces the expression of the transcription factor Hif1α, an essential transcription factor for the expression of Vascular Endothelial Growth Factor (VEGF). VEGF recruits the in growth of endothelial cells from the surrounding lateral perichondrium (Schipani, 2006). With these in growing vessels both osteoclasts and osteoblasts enter the bone resulting in the resorption of cartilage, its replacement by newly formed bone and the formation of the bone marrow cavity. In this process a cartilaginous matrix containing collagen-2, collagen-10 and glycosamin glycans is replaced by a bone matrix predominantly consisting of collagen 1. Somewhat later, chondrocytes start to hypertrophy in the center of the cartilaginous heads at the distal ends of the bones and mineralize their extracellular matrix resulting in the

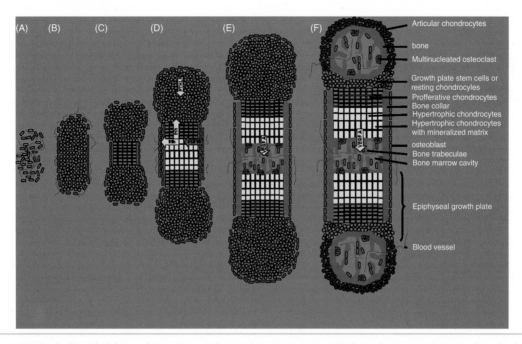

Figure 2.18 Endochondral bone formation. Schematic presentation of the subsequent stages of endochondral ossification. It starts with loosely connected mesenchyme invaded with blood vessels (red) (A). The mesenchymal cells start to condense and the vasculature is rearrangement. The condensed cells differentiate into chondrocytes that are surrounded by a perichondrium (green) (B). Cells in the middle of the bone anlage arrange into columns and start to proliferate (C) and subsequently start to hypertrophy (yellow) (D). Chondrocytes in the transition zone secrete Indian Hedgehog (IHh) that stimulates the formation of a bone collar (blue), stimulates chondrocyte proliferation and stimulates the expression of PTHLH in the periarticular region of the long bone. Osteoblasts secrete FGF18 that inhibits chondrocyte proliferation. Terminally hypertrophic chondrocytes start to mineralize their matrix (black lines). At this time point mononuclear osteoclast precursors (red dots) reside in the perichondrium. These cells differentiate into multinucleated osteoclasts that together with bloodvessels and osteoblasts (bleu) invade the hypertrophic cartilage and start to resorb the mineralized cartilage matrix. VEGFA, expressed due to the hypoxic conditions in the cartilage, plays an important role in the attraction of blood vessels. This results in the formation of a bone marrow cavity. The osteoblasts replace the chondrocyte matrix with a bone matrix (blue) (E). Somewhat later, this process is repeated in the secondary ossification centers in the cartilaginous heads of the long bones. The formation of the secondary ossification centers divides the cartilage into articular chondrocytes (dark blue) covering the bone surface in the joints and in epiphyseal growth plate cartilage that is responsible for bone elongation (F).

formation of the secondary ossification center. This is followed by in growth of vessels, osteoclasts and osteoblasts. The secondary ossification center demarcates the separation of two types of chondrocytes, articular chondrocytes that cover the distal ends of the bones having a relatively low metabolic activity and epiphyseal growth plate chondrocytes that have a high metabolic activity and are solely responsible for bone elongation. The signaling mechanisms involved in this process are largely unknown.

2.6.3 Chondrocyte differentiation

The transcription factor Sox-9 is indispensable for the early stages of chondrocyte differentiation by regulating cell condensations and the production of early chondrogenic markers like collagen 2. Further differentiation of chondrocytes requires the expression of two closely related transcription factors, Sox-5 and Sox-6. To enable hypertrophic differentiation, the expression of Sox-9 must be downregulated in late proliferative chondrocytes. The transcription factor RunX2 is involved in hypertrophic differentiation. The subsequent stages of chondrocyte differentiation are characterized by the expression of typical marker genes (Figure 2.19).

Chondrocyte proliferation and hypertrophic differentiation is regulated by a variety of growth factors. Most notably is the Indian Hedgehog/Parathyroid hormone related peptide (PTHrP) negative feedback loop in which PTHrP inhibits hypertrophic differentiation by keeping the chondrocytes in a proliferation competent stage (Kronenberg, 2006). Besides its role in the regulation of the onset of chondrocyte hypertrophy, IHh is a potent regulator of chondrocyte proliferation (Lai and Mitchell, 2005). TGFβ is another important growth factor involved in chondrogenesis by stimulating the

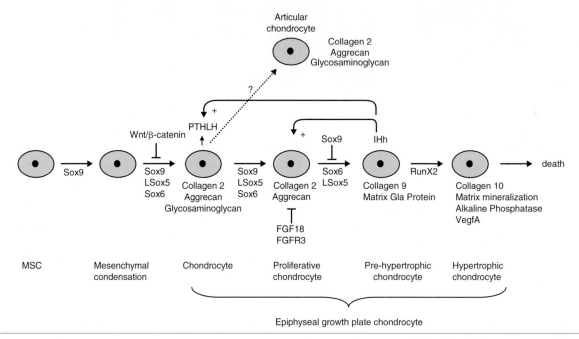

Figure 2.19 Chondrocyte differentiation. The differentiation of chondrocytes from Mesenchymal Stem Cells (MSCs) can be divided into various stages that are characterized by the expression of typical markers. The transcription factor Sox9 is a positive regulator and Wnt/β-catenin is a potent negative regulator of chondrogenesis. To enable hypertrophic differentiation, Sox9 needs to be downregulated. Other important transcription factors are Lsox5 and sox6 that are particularly important in early differentiation of chondrocytes while RunX2 is involved in hypertrophic differentiation. Indian Hedgehog (Ihh) is expressed by pre hypertrophic chondrocytes and stimulates chondrocyte proliferation. It also induces PTHLH expression in early stage chondrocytes. Osteoblast derived FGF18 is a potent negative regulator of chondrocyte proliferation. The molecular mechanism that results in the formation of articular chondrocytes instead of epiphyseal growth plate chondrocytes is presently unknown.

production of a chondrogenic extracellular matrix such as glycosamin glycans (Grimaud *et al.*, 2002). Chondrogenic differentiation is also stimulated by BMPs (Tsumaki and Yoshikawa, 2005). In the absence of both BMP-receptors in double BMP-receptor 1A and 1B knock out mice, chondrocyte differentiation and bone formation does not occur. FGF, particularly FGF18, is an important class of negative regulators of chondrocyte proliferation. The effect of FGF18 is in part mediated by down regulation of IHh expression in late proliferative and early hypertrophic chondrocytes (Ohbayashi *et al.*, 2002; Ornitz, 2005).

2.6.4 Intramembranous ossification and osteoblast differentiation

The flat bones of the skull and the flat part of the clavicle are formed by intramembranous ossification. In this process cells directly deposit a mineralized bone matrix without a cartilage intermediate. This process also starts with condensation of mesenchymal cells. In contrast to endochondral bone, these condensations occur in vascularized regions of mesenchyme. The condensed cells directly differentiate into osteoblasts which start to produce a mineralized bone matrix. The sequential activation of two transcription factors is essential for the formation of osteoblasts from skeletal precursor cells. This differentiation route is initiated by RunX2 and requires relatively high levels of Wnt-signaling. The subsequent activation of a second transcription factor, Osterix (Osx), is needed for further maturation of the cells and the production of a mineralized bone matrix. Indeed, knock out mice lacking either RunX2 or Osx lack bone formation due to the absence of mature osteoblasts. In Osx knock out mice osteoblasts are arrested in a somewhat later differentiation stage in comparison to the RunX2 knock out mice (Day *et al.*, 2005; Hill *et al.*, 2005; Komori *et al.*, 1997; Nakashima *et al.*, 2002; Otto *et al.*, 1997). Other transcription factors like Msx2, ATF4, deltaFosB and Fra1 and -2 further facilitate and cooperate with Runx2 and Osx in osteoblast differentiation and production

of a mineralized bone matrix (Karsenty and Wagner, 2002; Nakashima and de Crombrugghe, 2003).

The life-cycle of the osteoblast can be divided into various phases. In the first phase, the number of osteoprogenitors is increased by rapid proliferation. Subsequently, the cells begin to secrete large quantities of extracellular matrix predominantly consisting of collagen 1 followed by a maturation phase in which the matrix is prepared for matrix mineralization. Each of these phases is characterized by the expression of typical markers (Figure 2.20). At the end of the life span, the osteoblast has three choices, it can die by apoptosis, it can become fully embedded in the extracellular bone matrix and differentiate further into an osteocyte or it can be become a quiescent bone lining cell that covers the bone surface. Osteocytes are single cells fully surrounded by a mineralized bone matrix. They are in close contact with each other and the bone surface with a network of cell extensions. These cell protrusions are localized in small channels, the caniculae. The lining cells are quiescent cells no longer involved in bone formation. Upon the appropriate signals these cells can, however, resume their activity and start participating again in the formation of a bone matrix (Manolagas, 2000).

A large number of growth factors are involved in the regulation of osteoblast differentiation. Members of the TGFβ-Bone Morphogenetic Protein (BMP) superfamily, particularly the BMPs, IHh and Wnt family members can mediate initiation of osteoblast differentiation from uncommitted precursors. These factors act amongst others by inducing the expression of RunX2. Other factors like Insulin-like Growth factor I (IGF-I), TGFβ, Fibroblast growth factors (FGFs) to name a few play a role in osteoblast proliferation, matrix production and mineralization. Of particular interest are members of the Wnt-family, which are involved in successive stages of osteoblast differentiation such as in the initiation of osteoblast differentiation, matrix production and osteoblastic cell death. Furthermore, they are involved in the regulation of the coupling between bone formation and bone resorption. Besides control by locally produced growth factors, bone formation by osteoblasts

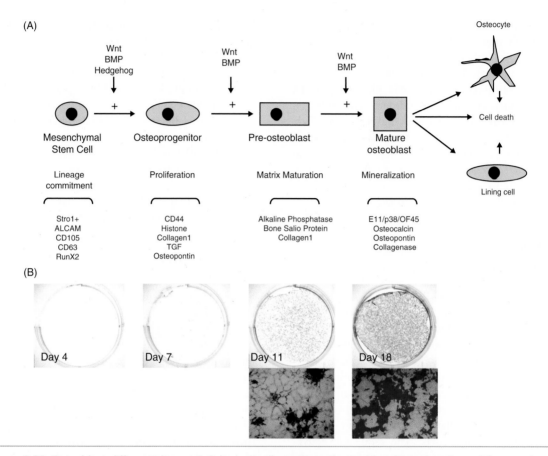

Figure 2.20 Osteoblast differentiation: (A) Schematic illustration of osteoblast lineage cells and frequently used markers of the stages of maturation. Wnt and BMPs have positive effects on osteoblast differentiation in all stages of development. The actions of Hedgehog are limited to the earliest phases of differentiation from mesenchymal stem cells; (B) mesenchymal stem cells can be used to mimic the subsequent stages of osteoblast differentiation *in vitro*. In the initial phases of the differentiation process at day 4 and 7, the undifferentiated MSCs become committed and start to proliferate. At day 11 cells start to mature the extracellular matrix and the first bone nodules appear (grey spots). This can be visualized by staining for Alkaline Phosphatase activity resulting in blue staining of positive cells. Seven days later, mature osteoblasts are present that have deposited a mineralized matrix. This can be visualized by staining with Alizarin red which colors mineralized bone matrix red.

is controlled by systemic factors, such as sex-steroids, Growth Hormone and Parathyroid Hormone and by the hypothalamus via the sympathic nervous system (Karsenty and Wagner, 2002; Manolagas, 2000; Ducy *et al.*, 2000).

2.6.5 Osteoclast differentiation

The osteoclast is a multinucleated highly specialized cell specifically equipped for the resorption of a mineralized matrix (Figure 2.21). Osteoclasts are

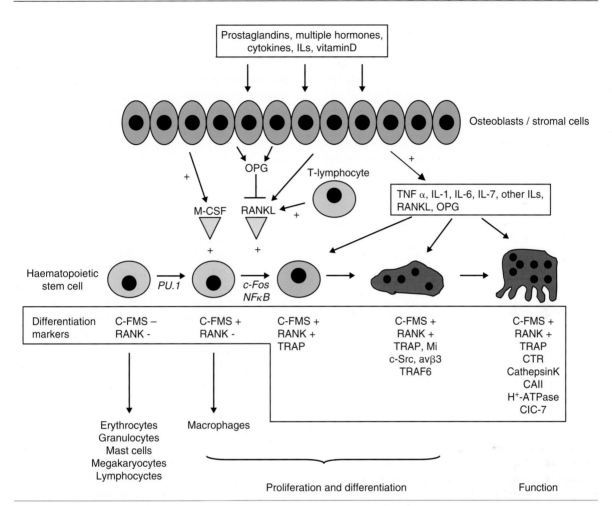

Figure 2.21 Osteoclast differentiation. Osteoclasts are derived from the haemotopoietic stem cell. These cells become committed to the myeloid lineage under the influence of the transcription factor PU-1 and and start to express c-FMS and RANK, the receptors for m-CSF and RANKL, respectively. In subsequent differentiation steps the transcription factors c-Fos and NFκB are of critical importance. Osteoclastogenesis can be divided into various stages that are characterized by the expression of typical marker genes. The formation of osteoclasts is tightly controlled by osteoblasts and stromal cells that express m-CSF and RANKL. Furthermore they express a negative regulator of RANKL, osteoprogerin (OPG). The RANKL/OPG ratio is a critical component in the regulation of osteoclast formation and activity. Besides RANKL and OPG, osteoblasts and stromal cells express a variety of other growth factors and cytokines that are either pro- and anti osteoclastogenic. The expression of all of these factors is tightly controlled by prostaglandins, a variety of hormones like vitamin D and estrogens, Interleukins (ILs) and cytokines. Besides osteoblasts and stromal cells, T-lymphocytes also express RANKL and can thus influence osteoclastogenesis and bone resorption.

derived from the heamatopoeitic stem cell. This cell gives rise to erythrocytes, granulocytes, mast cells, megakaryocytes, lymphocytes and macrophages. The transcription factor PU-1 and the growth factor Macrophage-Colony Stimulating Factor (M-CSF) are required for the establishment of the macrophage lineage. The derivation of osteoclasts from this lineage requires a direct interaction between the osteoclast precursor and stromal cells or osteoblasts involving the expression of Receptor Activator of Nuclear Factor Kappa κ (RANK) by the osteoclast precursor and the membrane associated ligand (RANKL) by the osteoblast. The interaction between RANK and RANKL is required for all subsequent stages of osteoclastogenesis including the fusion of mononucleated precursor cells into a multinucleated functional osteoclast. The subsequent stages of osteoclast development are characterized by specific marker genes (Figure 2.21) (Glass and Karsenty, 2006; Karsenty and Wagner, 2002; Roodman, 2006).

The interaction between RANK and RANKL can be antagonized by the decoy receptor osteoprotegrin (OPG). The major source of RANKL and OPG are cells from the osteoblast lineage although also T-cells appear to be involved. A large variety of stimuli like cytokines (e.g. IL-1, IL-6, TNFα, hormones (Vitamin D, Parathyroid Hormone, estrogens) are involved in the regulation of the osteoblastic expression of RANKL and OPG. The balance between these two proteins determines whether osteoblasts can support osteoclastogenesis or not. The formation of osteoclasts is tightly linked with osteoblast differentiation, since the expression of RANKL is controlled by RunX2. Furthermore, the expression of OPG is repressed by Wnt-signaling which simultaneously increases osteoblast activity (Glass and Karsenty, 2006).

2.6.6 Bone remodeling

While the majority of organs in the human body are static, i.e. do not undergo major changes once the formation of the adult organ is completed, bone is continuously renewed by a process called remodeling. It starts as soon as the first bone is formed. In this process, old fatigued bone is replaced by new mechanically sound bone. It is an essential mechanism in the preservation of bone strength and in skeletal adaptation to changing environmental conditions such as in load bearing of the skeleton. Remodeling is largely orchestrated by the osteocyte and osteoblast and is controlled by environmental, hormonal and hypothalamic functions. For example, weight bearing of the skeleton induces a fluid shear stress in the caniculae which is sensed by the osteocytes. The osteocytes signal to the osteoblasts at the bone surface. Amongst others, the secreted Wnt-antagonist SOST is an osteocyte expressed gene that appears to play an important role in this process (van Bezooijen *et al.*, 2004). Depending on the signal, osteoblastic bone formation is either activated or repressed. In addition, these signals may influence the balance between OPG and RANKL expression thereby modulating osteoclast formation and bone resorption. Disturbances in the balance between bone formation and resorption are at the basis of all skeletal disorders.

2.7 Future developments

In the past decade our understanding of the molecular processes that are involved in tissue formation during embryonal development has increased exponentially. A large part of this success can be attributed to earlier work that has resulted in the isolation and culture of Embryonal Stem (ES) cells. This has enabled the development of relatively easy and straightforward strategies for genetic manipulation of gene function in ES cells and the subsequent use of these genetically engineered cells for the generation of transgenic animal models. This allowed, for the first time, the experimental evaluation of gene function in its natural environment; the developing embryo. Based on the analysis of these models, many key players in tissue formation have been identified. These include the elucidation of key morphogenic signals, their signal transduction pathways and their targets in the nucleus; the master transcription factors that induce lineage commitment of undifferentiated stem cells. Furthermore, many of the underlying mechanisms involved in inter-cellular communication and cell-matrix interactions and their importance in organ development have been revealed. Nevertheless, there are many aspects that are still unclear, particularly in stem cell biology. Examples are the molecular mechanisms that underly stem cell renewal and maintenance of pluri- and multipotency. This will be an area of active research in the coming years and insights in these pathways will be of eminent importance for tissue engineering.

Another important milestone that has contributed significantly to the progress in the field of developmental biology is the completion of the human genome project and as a spin off the genomes of important experimental models like mouse and rat. Combination of genome information and recent advancements in genetic manipulation has resulted in the formation of various commercial and non-commercial consortia that aim at the systematic knock out of all genes in the mouse genome. These approaches will result in an overwhelming amount of new information on gene function in tissue formation in embryonal development and in health and disease. Technological advancements now enable the analysis of genome wide gene transcription profiling by microarray analysis. The interpretation of these enormous data sets require novel bioinformatics approaches and this will be a field of large growth in the near future. It can be expected that this type of analysis eventually results in the elucidation of hierarchical and non-hierarchical genetic networks in formation of tissue from stem cells. These networks will reveal the relationship between an array of morphogens and growth factors and their signaling pathways. While major attention has been focused on gene identification and gene expression, the effectors of gene function, the proteins, have been relatively neglected. This will be an area of very active research that will result in the identification of the so-called proteome; a detailed overview of all expressed proteins in a particular cell type in combination with post-translational modifications.

A major challenge for tissue engineering in the coming years is to translate the recent and coming advancements in the field of developmental biology into functional applications in regenerative medicine. For example, novel research in the field of bone formation will undoubtedly result in the identification of the key signaling molecules that are responsible for the differentiation of osteoblasts or chondrocytes from mesenchymal stem cells. It is the task of tissue engineering to incorporate these findings in a novel generation of biomaterials either by incorporating the respective biologicals into the scaffold or by designing instructive biomaterials specifically activating these pathways in the cells that are seeded on the scaffold. Critically analyzing the progress in developmental biology, particularly in the area of stem cell biology, will show the route for successful tissue engineering in the future.

2.8 Summary

The inner cell mass in the blastocyst contains pluripotent cells that give rise to all tissues in the adult body. When brought into culture these cells are called Embryonic Stem cells that retain pluripotency and self renewal *in vitro*.

In gastrulation, the 3 germ layers ectoderm, mesoderm and endoderm are formed by a complex process involving cell-cell interactions, cell movements and gradients of morphogens by which the body plan is established.

Pluripotent neural crest cells originate from the neural plate border. They migrate via clearly defined patterns to all tissues in the body where they actively participate in tissue formation. They give rise to cell types of ectodermal, mesodermal and endodermal origin.

The heart is formed by mesoderm and ectoderm derived cell types. The mesodermal cells are present in the cardiogenic plates and will give rise to cardiomyocytes and endocardial cells. The cardiac neural crest, which will contribute to formation of e.g. the arterial vessel wall and the pro-epicardial organ, which will form the epicardium, originate from the ectoderm by epithelial-mesenchym transition.

The formation of the heart is a complex and time-consuming process. Initially the heart is formed as a single beating tube that by a series of geometrical changes transforms into a four chamber septated beating heart.

In blood vessel formation, endothelial cells are derived from the splanchnic mesoderm. The smooth muscle cells and pericytes are derived from the mesoderm or the neural crest.

Blood flow and communication between endothelial cells and surrounding smooth muscle cells or pericytes will shape and define the identity of the vessel. In the absence of blood flow, vessels will not develop properly or even detoriate.

In peripheral nerve tissue, the neurons and glial cells (also called Schwann cells) are derived from either the neuroectoderm or neural crest, while the nerve sheath is derived from the mesoderm.

Neurons can only survive in close contact with Schwann cells. Vice versa, mature Schwann cells can only survive in close contact with neurons. Only immature Schwann cells can survive without axonal contact.

The Nrg1-ErbB2 signaling axis plays a pivotal role in Schwann cell survival, migration and differentiation.

In adult skin, the follicular and interfollicular epidermis is derived from the ectoderm. The mesoderm is the progenitor of the dermis and subcutaneous fat.

The development of embryonic skin involves environmental, matrix and cell-cell interactions. These processes can be mimicked to some extent *in vitro* by changing the culture conditions from a wet to air-exposed environment and by co-culture of dermal fibroblasts and keratinocytes.

The major bone forming cells, the osteoblast and chondrocyte, are derived from the paraxial and lateral plate mesoderm and the neuroectoderm-derived cranial neural crest. The bone-degrading osteoclasts are derived from the mesenchymal hematopoeitic cell lineage.

Bone is an active tissue that is continuously remodeled. It is formed either by intramembranous ossification or by endochondral ossification. In intramembranous ossification, mesenchymal stem cells directly differentiate into a bone matrix depositing osteoblast. In endochondral bone formation, skeletal elements are preformed in a cartilaginous mold that is replaced by bone.

Osteoblasts and chondrocytes have a common precursor, the multipotent MSC. The differentiation into either cell lineage is governed by a complex array of signaling molecules including BMPs and Wnts of which the intracellular wnt-target β-catenin appears to play central role.

References

Abu-Issa, R., Smyth, G., Smoak, I., *et al.* (2002). Fgf8 is required for pharyngeal arch and cardiovascular development in the mouse. *Development*, 129: 4613–4625.

Amoh, Y., Li, L., Campillo, R., *et al.* (2005). Implanted hair follicle stem cells form Schwann cells that support repair of severed peripheral nerves. *Proc Natl Acad Sci USA*, 102: 17734–17738.

Bergwerff, M., DeRuiter, M.C., Poelmann, R.E., *et al.* (1996). Onset of elastogenesis and downregulation of smooth muscle actin as distinguishing phenomena in artery differentiation in the chick embryo. *Anat Embryol*, 194: 545–557.

Bergwerff, M., Verberne, M.E., DeRuiter, M.C., *et al.* (1998). *Neural crest cell contribution to the developing circulatory system. Implications for vascular morphology?* Circ Res, 82: 221–231.

Bergwerff, M., DeRuiter, M.C., Hall, S., *et al.* (1999). Unique vascular morphology of the fourth aortic arches, possible implications for pathogenesis of type-B aortic arch interruption and anomalous right subclavian artery. *Cardiovasc Res*, 44: 185–196.

Bermingham, J.R., Scherer, S.S., O'Connell, S., *et al.* (1996). Tst-1/Oct-6/SCIP regulates a unique step in peripheral myelination and is required for normal respiration. *Gene Develop*, 10: 1751–1762.

Bianchi, D.W., Wilkins-Haug, L.E., Enders, A.C. and Hay, E.D. (1993). Origin of extraembryonic mesoderm in experimental animals, relevance to chorionic mosaicism in humans. *Am J Med Genet*, 46: 542–550.

Bolognia, J.L., Orlow, S.J. and Glick, S.A. (1994). Lines of Blaschko. *J Am Acad Dermatol*, 31: 157–190.

Botchkarev, V.A., Botchkareva, N.V., Roth, W., *et al.* (1999). Noggin is a mesenchymally derived stimulator of hair-follicle induction. *NatCell Biol*, 1: 158–164.

Botchkarev, V.A. and Sharov, A.A. (2004). BMP signaling in the control of skin development and hair follicle growth. *Differentiation*, 72: 512–526.

Bradley, A., Evans, M., Kaufman, M.H. and Robertson, E. (1985). Formation of germ-line chimaeras from embryo-derived teratocarcinoma cell lines. *Nature*, 309: 255–256.

Braude, P., Bolton, V. and Moore, S. (1988). Human gene expression first occurs between the four- and eight-cell stages of preimplantation development. *Nature*, 332: 459–461.

Britsch, S., Goerich, D.E., Riethmacher, D., *et al.* (2001). The transcription factor Sox10 is a key regulator of peripheral glial development. *Genes Dev*, 15: 66–78.

Bronner-Fraser, M. and Fraser, S.E. (1988). Cell lineage analysis reveals multipotency of some avian neural crest cells. *Nature*, 335: 161–164.

Bunge, M.B., Wood, P.M., Tynan, L.B., *et al.* (1989). Perineurium originates from fibroblasts, demonstration *in vitro* with a retroviral marker. *Science*, 243: 229–231.

Callis, T.E., Dongsun, C. and Wang, D. (2005). Bone morphogenetic protein signaling modulates myocardin transactivation of cardiac genes. *Circ Res*, 97: 992–1000.

Christoffels, V.M., Habets, P.E.M.H., Franco, D., *et al.* (2000). Chamber formation and morphogenesis in the developing mammalian heart. *Dev Biol*, 223: 266–278.

Chuck, E.T., Freeman, D.M., and Watanabe. M., and Rosenbaum, D.S. (1997). Changing activation sequence in the embryonic chick heart. *Implications for the development of the His-Purkinje system. Circ Res*, 81: 470–476.

Colognato, H., ffrench-Constant, C. and Feltri, M.L. (2005). Human diseases reveal novel roles for neural laminins. *Trends Neurosci*, 28: 480–486.

Coucouvanis, E. and Martin, G.R. (1999). BMP signaling plays a role in visceral endoderm differentiation and cavitation in the early mouse embryo. *Development*, 126: 535–546.

Day, T.F., Guo, X., Garrett-Beal, L. and Yang, Y. (2005). Wnt/β-catenin signaling in mesenchymal progenitors controls osteoblast and chondrocyte differentiation during vertebrate skeletogenesis. *Dev Cell*, 8: 739–750.

de Jong, F., Opthof, T., Wilde, A.A., *et al.* (1992). Persisting zones of slow impulse conduction in developing chicken hearts. *Circ Res*, 71: 240–250.

De La Fuente, R., Viveiros, M.M., Burns, K.H., *et al.* (2004). Major chromatin remodeling in the germinal vesicle (GV) of mammalian oocytes is dispensable for global transcriptional silencing but required for centromeric heterochromatin function. *Dev Biol*, 275: 447–458.

de Sousa Lopes, S.M., Roelen, B.A., Monteiro, R.M., *et al.* (2004). BMP signaling mediated by ALK2 in the visceral endoderm is necessary for the generation of primordial germ cells in the mouse embryo. *Genes Dev*, 18: 1838–1849.

DeReeder, E.G., Girard, N., Poelmann, R.E., *et al.* (1988). Hyaluronic acid accumulation and endothelial cell detachment in intimal thickening of the vessel wall. *The normal and genetically defective ductus arteriosus. Am J Pathol*, 132: 574–585.

DeReeder, E. G. (1989). Maturation of the ductus arteriosus; a model for intimal thickening. 1–103. 1989. Leiden.

DeRuiter, M.C., Gittenberger-de Groot, A.C., Poelmann, R.E., *et al.* (1993). Development of the pharyngeal arch system related to the pulmonary and bronchial vessels in the avian embryo. *Circulation*, 87: 1306–1319.

DeRuiter, M.C., Poelmann, R.E., VanMunsteren, J.C., *et al.* (1997). Embryonic endothelial cells transdifferentiate into mesenchymal cells expressing smooth muscle actins *in vivo* and *in vitro. Circ Res*, 80: 444–451.

Deschamps, J. and van Nes, J. (2005). Developmental regulation of the *Hox genes* during axial morphogenesis in the mouse. *Development*, 132: 2931–2942.

Dimmeler, S., Hermann, C., Galle, J. and Zeiher, A.M. (1999). Upregulation of superoxide dismutase and nitric oxide synthase mediates the apoptosis-suppressive effects of shear stress on endothelial cells. *Arterioscler Thromb Vasc Biol*, 19: 656–664.

Ducy, P., Schinke, T. and Karsenty, G. (2000). The osteoblast, a sophisticated fibroblast under central surveillance. *Science*, 289: 1501–1504.

El Ghalbzouri, A., Gibbs, S. and Lamme, E. *et al.* (2002). Effect of fibroblasts on epidermal regeneration. *Br J Dermatol*, 147: 230–243.

El Ghalbzouri, A. and Ponec, M. (2004). Diffusible factors released by fibroblasts support epidermal morphogenesis and deposition of basement membrane components. *Wound Repair Regen*, 12: 359–367.

Esper, R.M., Pankonin, M.S. and Loeb, J.A. (2006). Neuregulins, Versatile growth and differentiation factors in nervous system development and human disease. *Brain Res Brain Res Rev*, 51: 161–175.

Evans, M.J. and Kaufman, M.H. (1981). Establishment on culture of pluripotential cells from mouse embryos. *Nature*, 292: 154–156.

Facy, V., Flouret, V., Regnier, M. and Schmidt, R. (2005). Reactivity of Langerhans cells in human reconstructed epidermis to known allergens and UV radiation. *Toxicol In vitro*, 19: 787–795.

Falls, D.L. (2003). Neuregulins, functions, forms and signaling strategies. *Exp Cell Res*, 284: 14–30.

Flach, G., Johnson, M.H., Braude, P.R., *et al.* (1982). The transition from maternal to embryonic control in the 2-cell mouse embryo. *Embo J*, 1: 681–686.

Fu, X., Li, J., Sun, X., *et al.* (2005). Epidermal stem cells are the source of sweat glands in human fetal skin, evidence of synergetic development of stem cells, sweat glands, growth factors and matrix metalloproteinases. *Wound Repair Regen*, 13: 102–108.

Ghazvini, M., Mandemakers, W., Jaegle, M., *et al.* (2002). A cell type-specific allele of the POU gene Oct-6 reveals Schwann cell autonomous function in nerve development and regeneration. *Embo J*, 21: 4612–4620.

Gibbs, S., Backendorf, C. and Ponec, M. (1996). Regulation of keratinocyte proliferation and differentiation by all-trans-retinoic acid, 9-cis-retinoic acid and 1,25-dihydroxy vitamin D3. *Arch Dermatol Res*, 288: 729–738.

Gibbs, S., Vicanova, J., Bouwstra, J., *et al.* (1997). Culture of reconstructed epidermis in a defined medium at 33 degrees C shows a delayed epidermal maturation, prolonged lifespan and improved stratum corneum. *Arch Dermatol Res*, 289: 585–595.

Gibbs, S., Murli, S., De Boer, G., *et al.* (2000a). Melanosome capping of keratinocytes in pigmented reconstructed epidermis – effect of ultraviolet radiation and 3-isobutyl-1-methyl-xanthine on melanogenesis. *Pigment Cell Res*, 13: 458–466.

Gibbs, S., Silva Pinto, A.N., Murli, S., Huber, M., Hohl, D. and Ponec, M. (2000b). Epidermal growth factor and keratinocyte growth factor differentially regulate epidermal migration, growth and differentiation. *Wound Repair Regen*, 8: 192–203.

Gibbs, S., Hoogenband van den, H.M., Kirschig, G., *et al.* (2006). Autologous full-thickness skin substitute for healing chronic wounds. *Br J Dermatol*, 155(2): 267–274.

Gittenberger-de Groot, A.C., Vrancken Peeters, M.-P.F.M., Bergwerff, M., *et al.* (2000). Epicardial outgrowth inhibition leads to compensatory mesothelial outflow tract cuff and abnormal septation and coronary formation. *Circ Res*, 260: 373–377.

Gittenberger-de Groot, A.C., Bartelings, M.M., DeRuiter, M.C. and Poelmann, R E. (2005). Basics of cardiac development for the understanding of congenital heart malformations. *Pediatr Res*, 57: 169–176.

Gittenberger-de Groot, A.C., Vrancken Peeters, M.-P.F.M., Mentink, M.M.T., *et al.* (1998). Epicardial derived cells, EPDCs, contribute a novel population to the myocardial wall and the atrioventricular cushions. *Circ Res*, 82: 1043–1052.

Glass, D.A. and Karsenty, G. (2006). Canonical Wnt signaling in osteoblasts is required for osteoclast differentiation. *Ann NY Acad Sci*, 1068: 117–130.

Gordon, M.D. and Nusse, R. (2006). Wnt signaling, multiple pathways, multiple receptors and multiple transcription factors. *J Biol Chem*, 281: 22429–22433.

Green, H., Kehinde, O. and Thomas, J. (1979). Growth of cultured human epidermal cells into multiple epithelia suitable for grafting. *Proc Natl Acad Sci USA*, 76: 5665–5668.

Grimaud, E., Heymann, D. and Redini, F. (2002). Recent advances in TGF-β effects on chondrocyte metabolism. *Potential therapeutic roles of TGF-β in cartilage disorders. Cytokine Growth Factor Rev*, 13: 241–257.

Groenendijk, B.C.W., Hierck, B.P. and Gittenberger-de Groot, A.C. and Poelmann, R.E. (2004). Development-related changes in the expression of shear stress responsive genes KLF-2, ET-1 and NOS-3 in the developing cardiovascular system of chicken embryos. *Dev Dyn*, 230: 57–68.

Groenendijk, B.C.W., Hierck, B.P., Vrolijk, J., *et al.* (2005). Changes in shear stress-related gene expression after experimentally altered venous return in the chicken embryo. *Circ Res*, 96: 1291–1298.

Hall, C.E., Hurtado, R., Hewett, K.W., *et al.* (2004). Hemodynamic-dependent patterning of endothelin converting enzyme 1 expression and differentiation of impulse-conducting Purkinje fibers in the embryonic heart. *Development*, 131: 581–592.

Hammerschmidt, M., Brook, A. and McMahon, A.P. (1997). The world according to hedgehog. *Trends Genet*, 13: 14–21.

Happle, R. and Assim, A. (2001). The lines of Blaschko on the head and neck. *J Am Acad Dermatol*, 44: 612–615.

Hartmann, C. (2006). A Wnt canon orchestrating osteoblastogenesis. *Trends Cell Biol*, 16: 151–158.

Haubrich, K.A. (2003). Role of Vernix caseosa in the neonate, potential application in the adult population. *AACN Clin Issues*, 14: 457–464.

Hay, E.D. (2005). The mesenchymal cell, its role in the embryo and the remarkable signaling mechanisms that create it. *Dev Dyn*, 233: 706–720.

Hellström, M., Kalén, M., Lindahl, P., *et al.* (1999). Role of PDGF-B and PDGFR-? in recruitment of vascular smooth muscle cells and pericytes during embryonic blood vessel formation in the mouse. *Development*, 126: 3047–3055.

Heng, B.C., Cao, T., Liu, H. and Phan, T.T. (2005). Directing stem cells into the keratinocyte lineage *in vitro*. *Exp Dermatol*, 14: 1–16.

Herzog, Y., Kalcheim, C., Kahane, N., *et al.* (2001). Differential expression of neuropilin-1 and neuropilin-2 in arteries and veins. *Mech Dev*, 109: 115–119.

Hill, T.P., Spater, D., Taketo, M.M., *et al.* (2005). Canonical Wnt/β-catenin signaling prevents osteoblasts from differentiating into chondrocytes. *Dev Cell*, 8: 727–738.

Hirobe, T. (2005). Role of keratinocyte-derived factors involved in regulating the proliferation and differentiation of mammalian epidermal melanocytes. *Pigment Cell Res*, 18: 2–12.

Hoath, S.B. and Leahy, D.G. (2003). The organization of human epidermis, functional epidermal units and phi proportionality. *J Invest Dermatol*, 121: 1440–1446.

Hudon, V., Berthod, F., Black, A.F., *et al.* (2003). A tissue-engineered endothelialized dermis to study the modulation of angiogenic and angiostatic molecules on capillary-like tube formation *in vitro*. *Br J Dermatol*, 148: 1094–1104.

Jessen, K.R. and Mirsky, R. (2005). The origin and development of glial cells in peripheral nerves. *Nat Rev Neurosci*, 6: 671–682.

Jiang, Y., Drysdale, T.A. and Evans, T. (1999). A role for GATA-4/5/6 in the regulation of Nkx 2.5 expression with implications for patterning of the precardiac field. *Dev Biol*, 216: 57–71.

Joseph, N.M., Mukouyama, Y.S., Mosher, J.T., *et al.* (2004). Neural crest stem cells undergo multilineage differentiation in developing peripheral nerves to generate endoneurial fibroblasts in addition to Schwann cells. *Development*, 131: 5599–5612.

Karsenty, G. and Wagner, E.F. (2002). Reaching a genetic and molecular understanding of skeletal development. *Dev Cell*, 2: 389–406.

Kelly, R.G. (2005). Molecular inroads into the anterior heart field. *Trends Cardiovasc Med*, 15: 51–56.

Kirby, M.L. and Waldo, K.L. (1995). Neural crest and cardiovascular patterning. *Circ Res 77*, 211–215.

Kleinsmith, L.J. and Pierce, G.B. (1964). Multipotentiality of single embryonal carcinoma cells. *Cancer Res*, 24: 1544–1551.

Komori, T., Yagi, H., Nomura, S., *et al.* (1997). Targeted disruption of Cbfa1 results in a complete lack of bone formation owing to maturational arrest of osteoblasts. *Cell*, 89: 755–764.

Kronenberg, H.M. (2003). Developmental regulation of the growth plate. *Nature*, 423: 332–336.

Kronenberg, H.M. (2006). PTHrP and skeletal development. *Ann NY Acad Sci*, 1068: 1–13.

Lai, L.P. and Mitchell, J. (2005). Indian hedgehog, its roles and regulation in endochondral bone development. *J Cell Biochem*, 96: 1163–1173.

Laverriere, A.C., Macniell, C., Mueller, C., *et al.* (1994). GATA-4/5/6: a subfamily of three transcription factors transcribed in developing heart and gut. *J Biol Chem*, 269: 23177–23184.

Lawson, K.A. and Pedersen, R.A. (1987). Cell fate, morphogenetic movement and population kinetics of embryonic endoderm at the time of germ layer formation in the mouse. *Development*, 101: 627–652.

Lawson, K.A., Dunn, N.R., Roelen, B.A., *et al.* (1999). Bmp4 is required for the generation of primordial germ cells in the mouse embryo. *Genes Dev*, 13: 424–436.

Lawson, N.D., Scheer, N., Pham, V.N., *et al.* (2001). Notch signaling is required for arterial-venous differentiation during embryonic vascular development. *Development*, 128: 3675–3683.

Lawson, N.D., Vogel, A.M. and Weinstein, B.M. (2002). Sonic hedgehog and vascular endothelial growth factor act upstream of the Notch pathway during arterial endothelial differentiation. *Dev Cell*, 3: 127–136.

Le, N., Nagarajan, R., Wang, J.Y., *et al.* (2005). Analysis of congenital hypomyelinating Egr2Lo/Lo nerves identifies Sox2 as an inhibitor of Schwann cell differentiation and myelination. *Proc Natl Acad Sci USA*, 102: 2596–2601.

Le Douarin, N. and Kalcheim, C. (1999). *The Neural Crest*. Cambridge: Cambridge University Press.

Le Lièvre, Cs. and Le Douarin, N.M. (1975). Mesenchymal derivatives of the neural crest, analysis of chimaeric quail and chick embryos. *J Embryol Exp Morphol*, 34: 125–154.

le Noble, F., Fleury, V., Pries, A., *et al.* (2005). Control of arterial branching morphogenesis in embryogenesis, go with the flow. *Cardiovasc Res*, 65: 619–628.

Li, Y.S., Haga, J.H. and Chien, S. (2005). Molecular basis of the effects of shear stress on vascular endothelial cells. *J Biomech*, 38: 1949–1971.

Mack, J.A., Anand, S. and Maytin, E.V. (2005). Proliferation and cornification during development of the mammalian epidermis. *Birth Defects Res C Embryo Today*, 75: 314–329.

Manolagas, S.C. (2000). Birth and death of bone cells, basic regulatory mechanisms and implications for the pathogenesis and treatment of osteoporosis. *Endocr Rev*, 21: 115–137.

Martin, G.R. (1981). Isolation in culture of a pluripotent cell line from early mouse ebmryos cultured in medium conditioned by teratocarcinoma stem cells. *Proc Natl Acad Sci USA*, 78: 7634–7638.

Maurel, P. and Salzer, J.L. (2000). Axonal regulation of Schwann cell proliferation and survival and the initial events of myelination requires PI 3-kinase activity. *J Neurosci*, 20: 4635–4645.

Meier, C., Parmantier, E., Brennan, A., *et al.* (1999). Developing Schwann cells acquire the ability to survive without axons by establishing an autocrine circuit involving insulin-like growth factor, neurotrophin-3 and platelet-derived growth factor-BB. *J Neurosci*, 19: 3847–3859.

Meyer, D. and Birchmeier, C. (1995). Multiple essential functions of neuregulin in development. *Nature*, 378: 386–390.

Mintz, B. and Illmensee, K. (1975). Normal genetically mosaic mice produced from malignant teratocarcinoma cells. *Proc Natl Acad Sci USA*, 72: 3585–3589.

Michailov, G.V., Sereda, M.W., Brinkmann, B.G., *et al.* (2004). Axonal neuregulin-1 regulates myelin sheath thickness. *Science*, 304: 700–703.

Mjaatvedt, C.H., Nakaoka, T., Moreno-Rodriguez, R., *et al.* (2001). The outflow tract of the heart is recruited from a novel heart-forming field. *Dev Biol*, 238: 97–109.

Molin, D.G.M., DeRuiter, M.C., Wisse, L.J., *et al.* (2002). Altered apoptosis pattern during pharyngeal arch artery remodelling is associated with aortic arch malformations in Tgf β 2 knock-out mice. *Cardiovasc Res*, 56: 312–322.

Molin, D.G.M., Bartram, U., Van der Heiden, K., *et al.* (2003). Expression patterns of Tgfβ1-3 associate with myocardialization of the outflow tract and the development of the epicardium and the fibrous heart skeleton. *Dev Dyn*, 227: 431–444.

Molin, D.G., Poelmann, R.E., DeRuiter, M.C., *et al.* (2004). Transforming growth factor β-SMAD2 signaling regulates aortic arch innervation and development. *Circ Res*, 95: 1109–1117.

Moorman, A.F.M. and Christoffels, V.M. (2003). Cardiac chamber formation, Development, genes and evolution. *Physiol Rev*, 83: 1223–1267.

Moreau, M. and Leclerc, C. (2004). The choice between epidermal and neural fate, a matter of calcium. *Int J Dev Biol*, 48: 75–84.

Moyon, D., Pardanaud, L., Yuan, L., *et al.* (2001a). Plasticity of endothelial cells during arterial-venous differentiation in the avian embryo. *Development*, 128: 3359–3370.

Moyon, D., Pardanaud, L., Yuan, L., *et al.* (2001b). Selective expression of angiopoietin 1 and 2 in mesenchymal cells surrounding veins and arteries of the avian embryo. *Mech Dev*, 106: 133–136.

Nakashima, K., Zhou, X., Kunkel, G., *et al.* (2002). The novel zinc finger-containing transcription factor osterix is required for osteoblast differentiation and bone formation. *Cell*, 108: 17–29.

Nakashima, K. and de Crombrugghe, B. (2003). Transcriptional mechanisms in osteoblast differentiation and bone formation. *Trends Genet*, 19: 458–466.

Niemann, C. (2006). Controlling the stem cell niche, right time, right place, right strength. *Bioessays*, 28: 1–5.

Nieto, M.A. (2001). The early steps of neural crest development. *Mech Dev*, 105: 27–35.

Noden, D.M. (1990). Origins and patterning of avian outflow tract endocardial tissues and cushion mesenchyme. *Anat Rec*, 226: 72A–73A.

Ogata, T., Iijima, S., Hoshikawa, S., *et al.* (2004). Opposing extracellular signal-regulated kinase and Akt pathways control Schwann cell myelination. *J Neurosci*, 24: 6724–6732.

Ohbayashi, N., Shibayama, M., Kurotaki, Y., *et al.* (2002). FGF18 is required for normal cell proliferation and differentiation during osteogenesis and chondrogenesis. *Genes Dev*, 16: 870–879.

Ornitz, D.M. (2005). FGF signaling in the developing endochondral skeleton. *Cytokine Growth Factor Rev*, 16: 205–213.

Otto, F., Thornell, A.P., Crompton, T., *et al.* (1997). Cbfa1, a candidate gene for cleidocranial dysplasia syndrome, is essential for osteoblast differentiation and bone development. *Cell*, 89: 765–771.

Papaioannou, V.E., McBurney, M., Gardner, R.L. and Evans, M.J. (1975). The fate of teratocarcinoma cells injected into early mouse embryos. *Nature*, 258: 70–73.

Parmantier, E., Lynn, B., Lawson, D., *et al.* (1999). Schwann cell-derived Desert hedgehog controls the development of peripheral nerve sheaths [see comments]. *Neuron*, 23, 713–724.

Perea-Gomez, A., Shawlot, W., Sasaki, H., *et al.* (1999). HNF3β and Lim1 interact in the visceral endoderm to regulate primitive streak formation and anterior-posterior polarity in the mouse embryo. *Development*, 126: 4499–4511.

Person, A.D., Klewer, S.E. and Runyan, R.B. (2005). Cell biology of cardiac cushion development. *Int Rev Cytol*, 243: 287–335.

Piotrowska, K. and Zernicka-Goetz, M. (2001). Role for sperm in spatial patterning of the early mouse embryo. *Nature*, 409: 517–521.

Poelmann, R.E. and Gittenberger-deGroot, A.C. (1999). A subpopulation of apoptosis-prone cardiac neural crest cells targets to the venous pole, multiple functions in heart development. *Dev Biol*, 207(2): 271–286.

Ponec, M., El Ghalbzouri, A., Dijkman, R., *et al.* (2004). Endothelial network formed with human dermal microvascular endothelial cells in autologous multicellular skin substitutes. *Angiogenesis*, 7: 295–305.

Power, M.A. and Tam, P.P. (1993). Onset of gastrulation, morphogenesis and somitogenesis in mouse embryos displaying compensatory growth. *Anat Embryol (Berl)*, 187: 493–504.

Ramsdell, A.F. and Markwald, R.R. (1997). Induction of endocardial cushion tissue in the avian heart is regulated, in part, by TGFβ-3-mediated autocrine signaling. *Dev Biol*, 188: 64–74.

Rentschler, S., Vaidya, D.M., Tamaddon, H., *et al.* (2001). Visualization and functional characterization of the developing murine cardiac conduction system. *Development*, 128: 1785–1792.

Riethmacher, D., Sonnenberg-Riethmacher, E., Brinkmann, V., *et al.* (1997). Severe neuropathies in mice with targeted mutations in the ErbB3 receptor. *Nature*, 389: 725–730.

Rissmann, R., Groenink, H.W., Weerheim, A.M., *et al.* (2006). New Insights into Ultrastructure, Lipid Composition and Organization of Vernix Caseosa. *J Invest Dermatol*, 126, 1823–1833.

Roodman, G.D. (2006). Regulation of osteoclast differentiation. *Ann NY Acad Sci*, 1068: 100–109.

Rott, H.D. (1999). Extracutaneous analogies of Blaschko lines. *Am J Med Genet*, 85: 338–341.

Scherer, S.S. and Salzer, J.L. (2001). Axon-Schwann cell interactions during peripheral nerve degeneration and regeneration. In *Glial Cell Development* (Jessen, K.R. and Richardson, W.D., eds). Oxford: Bios Scientific Publishers Ltd, pp. 299–300.

Schipani, E. (2006). Hypoxia and HIF-1alpha in chondrogenesis. *Ann NY Acad Sci*, 1068: 66–73.

Schmidt-Ullrich, R. and Paus, R. (2005). Molecular principles of hair follicle induction and morphogenesis. *Bioessays*, 27: 247–261.

Sedmera, D., Reckova, M., Bigelow, M.R., *et al.* (2004). Developmental transitions in electrical activation patterns in chick embryonic heart. *Anat Rec A Discov Mol Cell Evol Biol*, 280: 1001–1009.

Sharghi-Namini, S., Turmaine, M., Meier, C., *et al.* (2006). The structural and functional integrity of peripheral nerves depends on the glial-derived signal desert hedgehog. *J Neurosci*, 26: 6364–6376.

Slomp, J., Gittenberger-de Groot, A.C., Glukhova, M.A., *et al.* (1997). Differentiation, dedifferentiation and apoptosis of smooth muscle cells during the development of the human ductus arteriosus. *Arterioscler Thromb Vasc Biol*, 17: 1003–1009.

Smithies, O., Gregg, R.G., Boggs, S.S., *et al.* (1985). Insertion of DNA sequences into the human chromosomal b-globin locus by homologous recombination. *Nature*, 317: 230–234.

Snow, M.H. and Bennett, D. (1978). Gastrulation in the mouse, assessment of cell populations in the epiblast of tw18/tw18 embryos. *J Embryol Exp Morphol*, 47: 39–52.

Srivastava, D. and Olson, E.N. (2000). A genetic blueprint for cardiac development. *Nature*, 407(6801): 221–216.

Staiano-Coico, L., Hefton, J.M., Amadeo, C., *et al.* (1990). Growth of melanocytes in human epidermal cell cultures. *J Trauma*, 30: 1037–1042.

Stalmans, I., Lambrechts, D., Desmet, F., *et al.* (2003). *VEGF, a modifier of the del22q11 (DiGeorge) syndrome?* *Nat Med*, 9: 173–182.

Stevens, L.C. (1967). The biology of teratomas. *Adv Morphog*, 6: 1–31.

Tarkowski, A.K. and Wroblewska, J. (1967). Development of blastomeres of mouse eggs isolated at the 4- and 8-cell stage. *J Embryol Exp Morphol*, 18: 155–180.

Taveggia, C., Zanazzi, G., Petrylak, A., *et al.* (2005). Neuregulin-1 type III determines the ensheathment fate of axons. *Neuron*, 47: 681–694.

Topouzis, S. and Majesky, M.W. (1996). Smooth muscle lineage diversity in the chick embryo. *Two types of aortic smooth muscle cell differ in growth and receptor-mediated transcriptional responses to transforming growth factor-β. Dev Biol*, 178: 430–445.

Topper, J.N. and Gimbrone, M.A. Jr. (1999). Blood flow and vascular gene expression, fluid shear stress as a modulator of endothelial phenotype. *Mol Med Today*, 5: 40–46.

Torres-Vazquez, J., Kamei, M. and Weinstein, B.M. (2003). Molecular distinction between arteries and veins. *Cell Tissue Res*, 314: 43–59.

Traupe, H. (1999). Functional X-chromosomal mosaicism of the skin, Rudolf Happle and the lines of Alfred Blaschko. *Am J Med Genet*, 85: 324–329.

Tsumaki, N. and Yoshikawa, H. (2005). The role of bone morphogenetic proteins in endochondral bone formation. *Cytokine Growth Factor Rev*, 16: 279–285.

Tsunoda, Y. and McLaren, A. (1983). Effect of various procedures on the viability of mouse embryos containing half the normal number of blastomeres. *J Reprod Fertil*, 69: 315–322.

van Bezooijen, R.L., Roelen, B.A., Visser, A., *et al.* (2004). Sclerostin is an osteocyte-expressed negative regulator of bone formation, but not a classical BMP antagonist. *J Exp Med. 199*,, 805–814.

Van Den Akker, A.E., Fromental-Ramain, C., de Graaff, W., *et al.* (2001). Axial skeletal patterning in mice lacking all paralogous group 8 Hox genes. *Development*, 128: 1911–1921.

Van Den Akker, N.M., Lie-Venema, H., Maas, S., *et al.* (2005). Platelet-derived growth factors in the developing avian heart and maturating coronary vasculature. *Dev Dyn*, 233(4): 1579–1588.

Van Den Hoff, M.J., Kruithof, B.P. and Moorman, A.F. (2004). Making more heart muscle. *Bioessays*, 26(3): 248–261.

Vartanian, T., Goodearl, A., Lefebvre, S., *et al.* (2000). Neuregulin induces the rapid association of focal adhesion kinase with the erbB2-erbB3 receptor complex in schwann cells. *Biochem Biophys Res Commun*, 271: 414–417.

Villa, N., Walker, L., Lindsell, C.E., *et al.* (2001). Vascular expression of Notch pathway receptors and ligands is restricted to arterial vessels. *Mech Dev*, 108: 161–164.

Virágh, Sz., Gittenberger-de Groot, A.C., Poelmann, R.E. and Kálmán, F. (1993). Early development of quail heart epicardium and associated vascular and glandular structures. *Anat Embryol*, 188: 381–393.

Voyvodic, J.T. (1989). Target size regulates calibre and myelination of sympathetic axons. *Nature*, 342: 430–433.

Vrancken Peeters, M.-P.F.M., Mentink, M.M.T., Poelmann, R.E., *et al.* (1995). Cytokeratins as a marker for epicardial formation in the quail embryo. *Anat Embryol*, 191: 503–508.

Vrancken Peeters, M.-P.F.M., Gittenberger-de Groot, A.C., Mentink, M.M.T., *et al.* (1997). Differences in development of coronary arteries and veins. *Cardiovasc Res*, 36: 101–110.

Wang, H.U., Chen, Z.-F. and Anderson, D.J. (1998). Molecular distinction and angiogenic interaction between embryonic arteries and veins revealed by ephrin-B2 and its receptor Eph-B4. *Cell*, 93: 741–753.

Wang, D., Chang, P.S., Wang, Z., *et al.* (2001). Activation of cardiac gene expression by myocardin, a transcriptional cofactor for serum response factor. *Cell*, 105: 851–862.

Wenink, A.C.G., Gittenberger-de Groot, A.C., Oppenheimer-Dekker, A., *et al.* (1986). Septation and valve formation, similar processes dictated by segmentation. *Congenital Heart Disease, Causes and Processes*, pp. 513–529.

Wilkinson, D.G., Bhatt, S. and Herrmann, B.G. (1990). Expression pattern of the mouse T gene and its role in mesoderm formation. *Nature*, 343: 657–659.

Woldeyesus, M.T., Britsch, S., Riethmacher, D., *et al.* (1999). Peripheral nervous system defects in erbB2 mutants following genetic rescue of heart development. *Genes Dev*, 13: 2538–2548.

Ying, Y. and Zhao, G.Q. (2000). Detection of multiple bone morphogenetic protein messenger ribonucleic acids and their signal transducer, Smad1, during mouse decidualization. *Biol Reprod*, 63: 1781–1786.

Ying, Y. and Zhao, G.Q. (2001). Cooperation of endoderm-derived BMP2 and extraembryonic ectoderm-derived BMP4 in primordial germ cell generation in the mouse. *Dev Biol*, 232: 484–492.

Zhou, X., Sasaki, H., Lowe, L., *et al.* (1993). Nodal is a novel TGF-β-like gene expressed in the mouse node during gastrulation. *Nature*, 361: 543–547.

Chapter 3
Tissue homeostasis

Anders Lindahl

Chapter objectives:

- To know the definition of the terms regeneration and homeostasis
- To recognize that different tissues have different regeneration capacity
- To acknowledge that the brain and heart are regenerating organs
- To understand how tissue regeneration can be analyzed by labeling experiments

- To recognize a stem cell niche
- To understand that stem cell niches share common regulatory systems
- To understand that tissue regeneration has consequences and offer opportunities in the development of tissue engineering products

"I'd give my right arm to know the secret of regeneration"

Oscar E Schotte, quoted in Goss (1991).

3.1 Introduction

The ability to regenerate larger parts of an organism is connected to the complexity of that organism. The lower developed the animal, the better the regeneration ability. In most vertebrates, the regeneration potential is limited to the musculoskeletal system and liver. In the hydras (a 0.5 cm long fresh-water cnidarian) the regeneration is made through morphollaxis, a process that does not require any cell division. If a part of the organism is lost through traumatic injury, other cells adapt to the new situation and the result is a smaller but fully functional hydra.

In higher organisms like Salamanders, belonging to the urodele amphibians (a family of adult vertebrates that can regenerate their limbs after amputation), the regeneration is initiated by epimorphosis, a process characterized by dedifferentiation and high proliferation of the local cells. The epimorphosis process starts immediately after the amputation of the limb when a wound epidermis is formed through the migration and proliferation of epithelial cells. In a zone underlying the epidermis, mesenchymal cells (muscle, cartilage and bone) lose their phenotype and start to dedifferentiate into blastemal cells which are the progenitor cells of the regenerating limb. Within the blastema, the cells undergo proliferation and redifferentiation into the various cell types needed to regenerate the limb. During this process the initial muscle cells show plasticity by being able to rediferentiate into muscle, cartilage and bone (Figure 3.1).

This ability of limb regeneration has obviously been lost in humans although we have the capacity for regeneration after injuries in a few tissues such as the liver, where we are able to regenerate a substantial part if it is resected surgically, bone and connective tissue. Furthermore, young children and

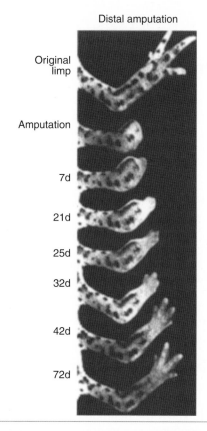

Distal amputation

Original limp

Amputation

7d

21d

25d

32d

42d

72d

Figure 3.1 Tissue regeneration in salamander. (From Goss, R.J. *Principles of Regeneration*, NY: Academic Press, 1969.)

even adults have the ability to regenerate finger tips. Regeneration of these tissues mirrors embryonic control mechanisms of differentiation from stem cells. In the skin, brain, bone and muscle of mammalians, a subset of the cell population are selected during embryonic and fetal life to be used during childhood growth and for tissue regeneration throughout life.

Box 1 Regeneration potential of tissues during lifetime

- 3 tons of blood
- 500 kg intestinal epithelium corresponding to 40 km of intestine

---

This concept has also been demonstrated in the heart tissue (Laflamme *et al.*, 2002).

Embryonic development results in a determined cellular structure where each cell has its function in a specific place. During subsequent growth, where the organism becomes larger cells proliferate but within their determined fate. Some animals like the fish, continue to grow throughout life although mammals stop growing in size; although the cell proliferation continues resulting in a constant renewal of cells in the body.

All organs are developed by the end of the first trimester, and the process continues during the rest of the gestation of the fetus growth, and after birth until the growth spurt in adolescence is over. Most people consider this to be the end stage of human growth but nothing could be more wrong. We stop growing, and throughout life cells and tissues in the human body are exposed to internal and external stress-factors which lead to injury and loss by apoptosis or necrosis. However, to sustain the function of the tissues, the affected cells need to be replaced through a process called regeneration. During our lifetime, we are growing at a constant rate with a cell turnover of about 1% of the body weight each day (Box 1).

In some sense the organs of the body can be characterized as a cellular homeostasis where cells are constantly produced in order to balance the continuous cell death (Figure 3.2). The balance between cell production and cell death must be properly maintained since even small differences will give dramatic effects; if the liver cells produced exceeds the cell death by 1%, the liver weight would be equal to the initial body weight within a time period of 6–8 years.

Tissue homeostasis

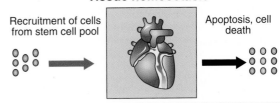

Recruitment of cells from stem cell pool

Apoptosis, cell death

Figure 3.2 In the tissue regeneration process there is a balance between cell proliferation and cell death.

The tissue maintenance also has to control the supporting tissues for a specific organ: organs need mechanical strength which is provided by the extracellular matrix produced by the connective tissue cells (fibroblast, chondrocytes, bone cells). Most tissue also need a blood supply provided by the capillaries and lining endothelial cells as well as innervation provided by the nerve cells. Cellular debris cleaning and tissue defense against bacteria and viruses are provided by macrophages and lymphocytes. All of these cells are necessary to support the specialized cells of each organ; the cleaning process of the liver cells, the muscle cell contraction in the heart and the specialized endocrine cells.

The internal structure of each organ must maintain the intricate mixture of cells although the specialized cells are renewed. This is orchestrated by the internal determination memory of each specialized cell and its stem cells that pass the information on to its progeny. However, the specialized cell is highly adaptive to the surrounding environment and can change its phenotype. Articular cartilage chondrocytes can be isolated and propagated in monolayer cultures where the typical type II collagen production is changed to a type I connective tissue type. However, when reimplanted into a cartilage environment, these cells turn on their type II collagen synthesis again. A second example of adaptation properties are the epidermal skin cells where mechanical load gives a different thickness of the skin. Just compare the back of your hand with the front of your palm or your foot soil. Bone tissue is also constantly adapting structurally to the changing load of the tissue. Not all organs regenerate in the same way – instead different organs have different rates of regeneration.

3.2 Tissues with no potential of regeneration

Although our body has the capacity to regenerate to a large extent, there are tissues with no or limited capacity of regeneration. For example, the central part of our lens consists of lens lamellas that are embryonic fossils that will not have changed since they were developed during embryonal life. The cell populations of the photo receptor of the retina and the auditory organ of Corti are not regenerated. In mammals, all cells in the auditory organ of Corti becomes terminally mitotic by embryonic day 14 and a similar phenomenon occurs in avian basal papilla by embryonic day 9 and thus one would expect that neither avian nor mammals would be able to regenerate auditory cells (Rubel *et al.*, 1995). However, in contrast to mammals, cells in the avian cochlear epithelium re-enter the cell cycle, and divide and differentiate into new hair cells after experimental loss of hair cells. A challenge for the future is to decipher the differences between the species and hopefully in the future auditory cells can be induced to regenerate also in humans.

3.3 Tissues with slow regeneration time

Our bones are constantly being renewed by an active process which involves the breakdown and rebuilding of new bone matrix by the osteoblasts. This constant renewal, governed by the continuous optimization of the load-bearing role of the bone by a

functionally adaptive remodeling activity (which is more active in growing bone), is dominated by high-magnitude, high-rate strains presented in an unusual distribution. Adaptation occurs at an organ level, involving changes in the entire bone architecture and bone mass. This process is continuously ongoing and results in a total turnover time of 3 years for the whole bone structure.

Cartilage has a similar turnover rate over time, although the individual matrix components are renewed at various rates. The two major extracellular components in articular cartilage, collagen type II and aggrecan, are relatively longlived in the tissue and as a consequence undergoes non-enzymatic modifications by reducing sugars, thus ending in accumulation of advanced glycation endproducts. These accumulated endproducts reflect the half life of the components that for collagen is estimated to be 100 years and for aggrecan to be 3,5 years. The cellular turnover in cartilage is probably limited and the localization of stem cells in articular cartilage is so far undetermined. However, the common dogma of cartilage as homogenous tissue with only one cell type, the chondrocyte, producing the extracellular matrix consisting of mainly collagen type II fibres, and the high molecular weight aggregating proteoglycan aggrecan is changing. The tissue is instead heterogeneous with distinct cellular characteristics in different zones from the surface zone with flattened discoid cells secreting surface proteoglycans, through the middle zone. Rounded cells produce not only collagen type II and aggrecan but also cartilage intermediate layer protein and the deep and calcified zones with larger cells produce type X collagen and alkaline phosphatase. Furthermore, cells isolated from different compartments of the articular cartilage demonstrate a phenotypic difference between the cells located at the surface and cells in the deeper layer when subjected to agarose suspension cultures (Archer *et al.*, 1990; Aydelotte *et al.*, 1988). Within the mesenchymal tissue hyaline cartilage chondrocytes are usually considered tissue restricted and without the broader differentiation potential seen in bone marrow derived mesenchymal stem cells (Caplan, 1991).

This concept has been challenged by the demonstration that isolated articular chondrocytes are able to take on several phenotypic identities within the mesenchymal lineage; cartilage, adipose cells, osteoblast-like cells and muscle cells (Barbero *et al.*, 2003). In contrast to mesenchymal stem cells, chondrocytes form only cartilage, and not bone, in an *in vivo* osteochondrogenic ceramic implantation assay (Tallheden *et al.*, 2003) indicating a different *in vivo* default pathway.

Within articular cartilage there are subgroups of clonogenic cells that have heterogenous the capacity for hyaline cartilage formation (Barbero *et al.*, 2003) which could explain the plasticity seen in primary isolated articular chondrocytes. However, for articular chondrocytes, multipotency can be demonstrated even on the clonal level (Brittberg *et al.*, 2005).

Recent studies have demonstrated that the surface layer is involved in regulation of joint development and growth as well as responsible for the post-foetal and adult appositional growth of the joint (Archer *et al.*, 2003). The surface cells harbor a progenitor cell population with high growth potential that participate actively in the repair of cartilage injury by migration into the defect.

3.4 Tissues with a high capacity of regeneration

If the liver is resected, cells undergo limited dedifferentiation allowing them to reenter the cell cycle while maintaining all critical differentiated functions. Cells subsequently start to divide and the liver is regenerated in a short time – usually within weeks. This regeneration works through a homogenous process of simultaneous proliferation of the hepatocytes. The liver does not overgrow its original size; instead it adapts itself to the organism. This can be demonstrated by the fact that adult resected liver lobes implanted into children adjust to the smaller host. This process is potentially controlled by a circulating homeostatic factor – the hepatocyte growth factor.

Human regeneration in intestine and epidermis of the skin is characterized by a high turnover rate where cells constantly proliferate and differentiate; a process that is finalized by a discarding of the cells into the intestine, or off the body. The regeneration is maintained by a special resident cell type, the stem cell situated at the basal lamina. These cells persist throughout life their function is to maintain homeostasis, and effect tissue regeneration and repair (Figure 3.3).

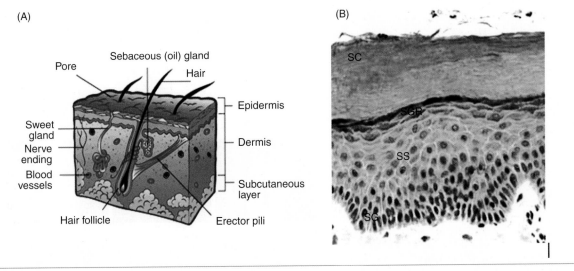

Figure 3.3 Schematic drawing of skin with hair follicle (left) and microscopic picture of epidermis (right) Abbreviations: SC = stratum corneum, SGR = stratum granulosum, SS = stratum spinosum, SG = stratum germinative.

Classical Experiment

Clonogenic keratinocytes are long-term multipotent stem cells (Claudinot *et al.*, 2005)
In order to demonstrate that cells in the hair root sheet are multipotent stem cells, single keratinocytes were isolated from the upper or lower regions of whisker follicles of adult rats. Clones were labeled by using a defective retrovirus bearing a β-gal gene and subcloned, and labeled cells (passages 11–12) were injected into newborn mouse skin. Transplanted clonogenic keratinocytes, whether from the upper or lower regions, contributed to all epithelial lineages of hairy skin, including sebaceous glands, and participated in the hair cycle of pelage follicles for months. Note the variable degree of chimerism and that the epidermis is not labeled in long-term grafts (Lower in A and C). Insets show duration of transplantation in days in A and C; (B). Schematic representation of the hair cycle phases of a pelage follicle.

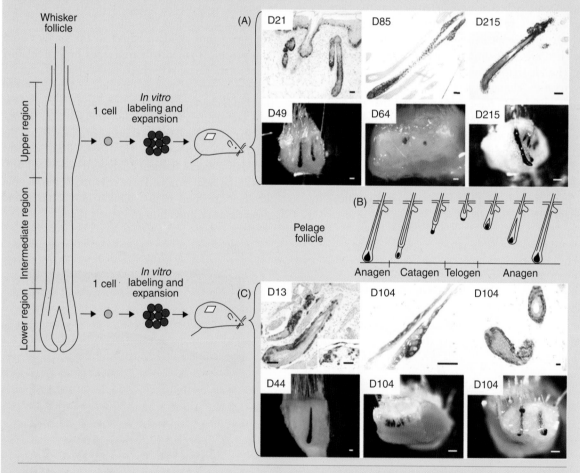

Multipotency of clonogenic keratinocyte. Scale bars: biopsies stained for β-gal in toto, 100 μm; microscopic sections, 50 μm. (From Claudinot, S., Nicolas, M., Oshima, H. et al. (2005). Long-term renewal of hair follicies from clonogenic multipotent stem cells. Proc Natl Acad Sci USA, *102(41): 14677–14682.)*

In the mid-1980s, researchers developed a method for growing a type of human skin cells called keratinocytes (which populates the skin's upper, or epidermal, layer) outside the body (Rheinwald and Green, 1975). The secret behind the culture method was that the keratinocyte stem cells could be propagated and that the contaminating fibroblasts of the dermis were growth inhibited. The technique is based on a culture system of mouse-derived fibroblast cells that inhibit contaminating fibroblast growth as well as the differentiation of the keratinocytes. The medium was further enriched with a growth factor Epidermal Growth Factor (EGF) that stimulates keratinocyte migration and proliferation. After several days in such an environment, few starting keratinocytes grew into a sheet of epidermal-like tissue.

Further investigation of the potential of the skin has revealed that the hair follicle contains stem cells with a pluripotency to produce skin, sebaceous glands and hair follicles (see Classical Experiment).

3.5 Tissues where regeneration was not considered – the paradigm shift in tissue regeneration

3.5.1 The brain as a regenerative organ

Until recently, the common understanding has been that nerve cells of the human brain no longer have the capacity for renewal. The only period where cells could be renewed was through the embryogenesis or during the prenatal period. In the adult brain plasticity existed, but only through increased complexity of new synapses, dendrites and neuritis. This belief could seemingly be unfounded with the knowledge of the existence of stem cells in most tissues of the body. However, the inability of the brain to self-repair is clinically obvious and the lack of appropriate techniques has hindered the discovery of adult brain regeneration. With the emerging radioactive techniques in the 1960s, researchers used ^{3}H-thymidine that could be incorporated into the DNA of dividing cells and visualized by autoradiography to demonstrate nerve cell renewal in rats, guinea pigs and cats. The scientific attitude towards these experiments was skeptic and little attention was attributed to the studies. The data was revisited in the mid 1970s, and neurogenesis was demonstrated in the dentate gyrus and olfactory bulbs of adult rats by alternative methods using electron microscopy. The work was followed by other researchers demonstrating regeneration in the adult brain of mammals and birds. In 1998 researchers were able to demonstrate cell division in human post mortem tissues using the thymidine marker 5-bromo2′deoxy-uridine (BrdU) (Box 2) (Eriksson et al., 1998), discoveries that were later confirmed by others (see State of the Art Experiment).

3.5.2 The heart as a regenerative organ

Scientific data published in recent years have radically changed the view of the regenerative potential of the heart, and thus opened the possibility of cell therapy as well as new pharmacological concepts for treatment of cardiac insufficiency. According to the dominating dogma, the heart tissue is regarded as postmitotic without regenerative capacity and it has from birth a finite number of cells that is gradually

---

Box 2

To measure DNA synthesis or cell proliferation, 5-bromo2′-deoxy-uridine (BrdU) can be incorporated into DNA in place of thymindine. Cells, which have incorporated BrdU into DNA, can be quickly detected using a monoclonal antibody against BrdU. The binding of the antibody is achieved by denaturation of the DNA. This is usually obtained by exposing the cells to acid, or heat.

State of the Art Experiment

Newly generated cells can be detected in the adult human brain in patients previously treated with BrdU (Eriksson *et al.*, 1998).

In order to investigate the fundamental question whether neurogenesis occurs in the adult human brain tissue specimens were obtained

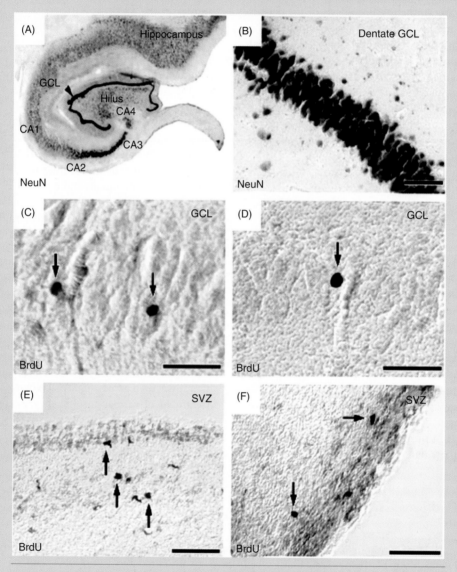

Detection of newly formed cells in the human brain using various techniques (see text). All scale bars represent 50 μm. (From Eriksson, P.S., Perfilieva, E., Bjork-Eriksson, et al. (1998). Neurogenesis in the adult human hippocampus. Nat Med, 4(11): 1313–1317.)

postmortem from patients who had been treated with the thymidine analog, bromodeoxyuridine (BrdU), that labels DNA during the S phase. Areas of the human brain previously identified as neurogenic in adult rodents and monkeys was used for the experiment. With immunofluorescent labeling for BrdU in combination with the neuronal markers, NeuN, calbindin or neuron specific enolase (NSE) demonstration of new neurons generated from dividing progenitor cells in the dentate gyrus of adult humans was possible. a, localization for the neuronal marker NeuN; b, the hippocampal dentate gyrus granule cell layer (GCL) visualized with immunoperoxidase staining for NeuN; c, differential interference contrast photomicrograph showing BrdU-labeled nuclei (arrows) in the dentate granule cell layer (GCL); d, differential interference contrast photomicrograph showing a BrdU-labeled nucleus (arrow) in the human dentate GCL. BrdU-positive nuclei have a rounded appearance and resemble the chromatin structure of mature granule cells and are found within the granule cell layer; e, differential interference contrast photomicrograph showing BrdU-positive cells (arrows) adjacent to the ependymal lining in the subventricular zone of the human caudate nucleus. Cells with elongated nuclei resembling migrating cells are in the rat subventricular zone (SVZ); f, differential interference contrast photomicrograph showing BrdU-positive cells (arrows) with round to elongated nuclei in the subventricular zone of the human caudate nucleus.

reduced over time. The only compensating mechanism for loss of heart tissue is thus hypertrophy and not through proliferation of individual cardiomyocytes. This view is in contrast to the biological reality in newt and zebra fish where regeneration of heart tissue is seen after injury (McDonnell and Oberpriller, 1984) (Poss *et al.*, 2001) (Figure 3.4).

The dogma of non-existing regeneration potential of cardiomyocytes has been challenged by individual researchers (McDonnell and Oberpriller, 1983; Oberpriller and Oberpriller, 1974) but the finding of y chromosome-containing cardiomyocytes, smooth muscle cells and endothelial cells in female-donated hearts in male recipient patients, gave new support to the view of the human heart as a regenerating organ (Quaini *et al.*, 2002). The demonstration of dividing myocytes in the normal and pathologic heart tissue has given further support to the hypothesis that the heart harbors a pluripotent stem cell niche supporting a normal slow regeneration (Beltrami *et al.*, 2001).

Cardiomyocyte progenitors with a specific phenotype (Lin(−) c-kit(+)) have been identified in rats. The cells are able to self renew and have multipotency with ability to form myocytes, smooth muscle cells and endothelium. When cells are injected into injured cardiac tissue the cells are able to induce regeneration of myocardium, including new vessels and endothelium (Beltrami *et al.*, 2003). Recently it was demonstrated that stem cells with the Sca-1(+) marker in mice have a homing potential to injured myocardium and the effect was due to both cell fusion and regeneration of new cells (Oh *et al.*, 2003). These results clearly demonstrate that heart tissues in mammalians that mainly consist of terminally differentiating cells also has the potential of a classic regeneration organ (Beltrami *et al.*, 2003). Interestingly, a stem cell niche has been demonstrated in the mouse heart (Figure 3.5). In these niches containing cardiac stem cells and lineage committed cells the supporting fibroblasts and myocytes are connected to the stem cells by connexins and cadherins (Urbanek *et al.*, 2006).

3.6 Consequence of regeneration potential for the tissue engineering concept

The regeneration capacity of a single organ requires careful considering when approaching a therapeutic

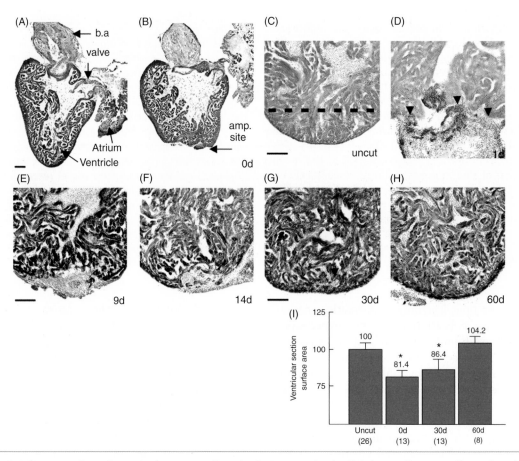

Figure 3.4 Regeneration of ventricular myocardium in the resected zebrafish heart. Hematoxylin and eosin stain of the intact zebrafish heart (A) before, and (B) after about 20% ventricular resection, b.a., bulbous arteriosus; (C) An intact ventricular apex at higher magnification, indicating the approximate amputation plane (dashed line). All images in this and subsequent figures display longitudinal ventricular sections of the amputation plane; (D) 1 dpa. The large clot is filled with nucleated erythrocytes (arrowheads); (E) 9 dpa. The heart section is stained for the presence of myosin heavy chain to identify cardiac muscle (brown) and with aniline blue to identify fibrin (blue) (accessed from http://www.sciencemag.org/cgi/content/full/298/5601/2188#R5#R5). The apex is sealed with a large amount of mature fibrin; (F) 14 dpa. The fibrin has diminished, and the heart muscle has reconstituted; (G) 30 dpa. A new cardiac wall has been created, and only a small amount of internal fibrin remains (arrowhead). (H) 60 dpa. This ventricle shows no sign of injury; (I) Quantification of healing at 0, 30, and 60 dpa. Values represent the size of the largest ventricular section (mean $\pm$ SEM; *$P < 0.05$); parentheses indicate the number of hearts examined (accessed from http://www.sciencemag.org/cgi/content/full/298/5601/2188#R5#R5). Scale bars, 100 μm. (From Poss, K.D., Wilson, L.G. and Keating, M.T. (2002). Heart regeneration in zebrafish. *Science*, 298(5601): 2188–2190.)

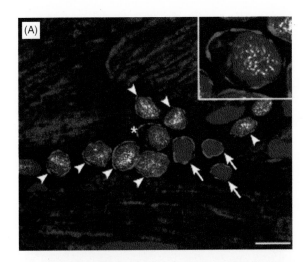

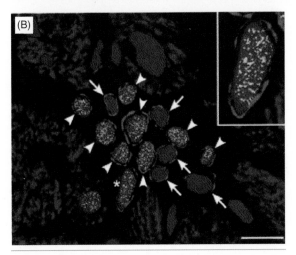

Figure 3.5 Clusters of Primitive and Early Committed Cells in the Heart, A, Cluster of 11 c-kit^POS cells (green) with three expressing c-kit only (arrows), seven expressing Nkx2.5 (white dots; arrowheads) in nuclei (blue, propidium iodide, PI), and 1 Nkx2.5 and α-sarcomeric actin in the cytoplasm (red; asterisk, see inset); B, Cluster of 15 c-kit^POS cells with five c-kit^POS cells only (arrows), eight expressing MEF2C (yellow dots; arrowheads), and one expressing MEF2C and α-sarcomeric actin (asterisk, see inset). Bars, 10 μm. (From Beltrami, A.P., Barlucchi, L., Torella, D. *et al.* (2003). Adult cardiac stem cells are multipotent and support myocardial regeneration. *Cell*, 114(6): 763–776.)

concept. If each organ is schematically described as an equilibrium between cells produced and cells discarded, the newly implanted tissue construct needs to consider the fact that this equilibrium has to be maintained by adding progenitor cells, or by not stopping or interfering with the normal process of regeneration.

The stem cell niche concept is central to most of the approaches we are discussing in tissue engineering. Stem cell niches are composed of a cellular microenvironment that supports the stem cells and enables them to maintain tissue homeostasis (Moore and Lemischka, 2006). A cellular interaction between the stem cells and the niche cells exists, with the goal to fulfill the lifetime support of differentiated cells. The niche cells shelter the stem cells from differentiation stimuli, apoptotic signaling and other stimuli that would challenge the reserve of stem cells. The stem cell niche also protects the stem cell pool from overactivity, e.g. unnecessary hypertrophy and must be activated in a timely sense to produce progenitor and transiently amplifying cells. Thus, the overruling control is to balance between cell quiescence and activity.

Today there are three identified and well-defined stem cell niches that harbor stem cells responsible for the maintenance of a continuous turnover of cells during its life time. These are the interstitial stem cell niche (ISCN) of the intestine, the hair follicle epidermal stem cell niche (HFSCN) and the hematopoietic stem cell niche in the bone marrow (HSCN).

The stem cell niches share common properties and requirements. All harbor stem cells that give rise to several different cell lineages. The transient amplifying cells that are produced must migrate into their proper location in order to fulfill the correct function. They are dependent on the surrounding cells consisting of mesenchyme and for the HSCN on the osteoblasts. The intricate signaling from the mesenchyme is one of the control mechanisms regulating the stem cell niche.

Interestingly, the three stem cell niches share interesting entities. An anatomical organization coordinates stem cell control and fate, both stimulating and repressor signaling systems are integrated and intercellular signaling pathways are shared. Bone morphogenetic protein (BMP) functions as a negative regulator on stem cell proliferation by suppressing nuclear β-catenin accumulation. This is counteracted by Wnt signaling (Figure 3.6).

Numerous regulatory signaling pathways have been revealed by global gene expression profiles of quiescent and activated HSCN as well as more committed progenitor populations (Eckfeldt *et al.*, 2005). Furthermore, a comprehensive genome analysis of an HSC supportive cell population has been described which further enlightens these interesting structures. The future of tissue engineering and regenerative medicine lies in increased knowledge of anatomical organization and location of stem cell niches in different organs. Future efforts will be focused on constructing proper microenvironments *in vitro* to accurately mimic the *in vivo* function of the niches, including understanding of the migration behavior of the transient amplifying (TA) cell population.

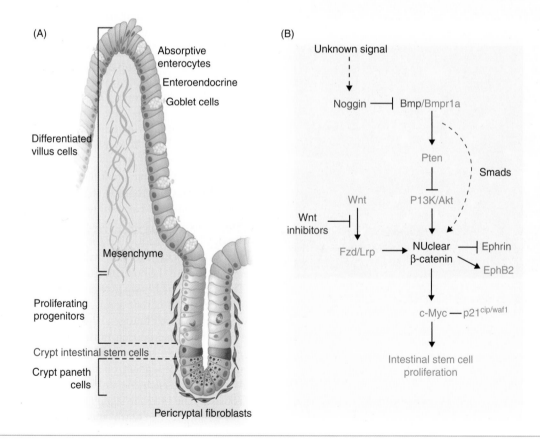

Figure 3.6 Stem cells within their niche in the small intestine. A, Schematic diagram of the major types and spatial orientations of cells found within the crypt niche and the villus; B, Interactive signaling pathways that mediate ISC proliferation. Colors represent the cell types sending and receiving the signals as displayed in (A). (From Moore, K.A. and Lemischka, I.R. (2006). Stem cells and their iches. *Science*, 311(5769): 1880–1885.)

3.7 Cell migration of TA cells

One central issue regarding stem cell niches and the maintenance of an organ is how a TA cell is moved from the stem cell niche to the proper location. Cell migration must be properly maintained in any potential scaffold produced. This is either provided by the laid down extra cellular matrix or by, e.g. nanofibers, giving the cells a proper attachment and migration potential. Examples of how nature has solved these issues are given in the hair follicles which show how cell trafficking is controlled for transient amplifying cells. Hair shafts contain a multipotent stem cell region – the bulge region – that contains three-potent epidermal cells that are able to form the epidermis (the skin), the sebaceous glands and the hair (Oshima *et al.*, 2001) (Figure 3.7).

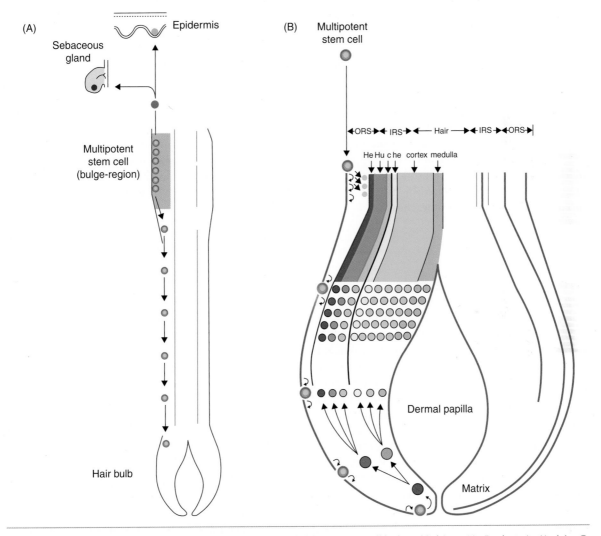

Figure 3.7 Migration behavior of TA hair root cell generated from stem cell bulge. (Oshima, H., Rochat, A., Kedzia, C. *et al.* (2001). Morphogenesis and renewal of hair folliclies from adult multipotent stem cells. *Cell*, 104(2): 233–245.)

With time, the stem cell moves by trafficking along the hair to the root. Similar trafficking can be found after implantation of mouse embryonic stem cells in the mouse brain using labeled cells traceable by magnetic resonance imaging.

3.8 Future developments

The emerging understanding of tissue regeneration will open new frontiers for tissue engineering researchers. The tissue engineering concept of engineering whole functional organs will probably change towards manipulating the enormous potential of autologous tissue regeneration. One could envision local implantation of chemical substances able to enhance stem cell recruitment. Perhaps even nanoconstructed small stem cell niches could be implanted that locally will enhance tissue regeneration. Furthermore, large fibers unable to support cell migration of TA cells and transduction of cellular biomechanics might be replaced by nanofibers produced by electrospinning or other means.

3.9 Summary

- Tissue regeneration after injury is limited to a few tissues in vertebrates including humans with the liver, blood and bone as examples of tissues with full regeneration capacity after injury.
- Tissues in the body undergo a constant renewal over time with approximately 1% of the body renewed each day. Tissues with high regeneration capacity are the intestine skin and blood. Others tissue like bone have a slow cellular turnover.
- Skin transplantation to burn victims was introduced in the 1980s. By the development of a stem cell culture system for human keratinocytes based on fibroblast feeder layers and addition of EGF a $1 \, cm^2$, skin bisopsy is able to regenerate $2 \, m^2$ of skin.
- Each tissue is maintained in a steady-state equilibrium where the cell loss due to apoptosis, shredding of cells or for other reasons is balanced by cell renewal from stem cells.
- The brain and heart were considered nonregenerating but recent knowledge gained from labeling experiments of human tissue revealed cell regeneration capacity in the human brain and heart.
- Most organs contain a defined stem cell niche where transient amplifying cells are constantly released. Well-defined stem cell niches are identified in the skin, intestine and bone marrow.
- Stem cell niches share common cellular signaling pathways usually well-conserved over species borders.
- When discussing the concept of tissue regeneration by artificial means, i.e. by implantation of scaffolds with or without cells, one has to acknowledge the normal regeneration mechanisms in the tissue. The stem cell niche releases cells into the organ and cells migrate to the location of function. Such migration behavior should be considered in the tissue engineering approach by the introduction of a nanoscale 3D web able to support migration.

References

Archer, C.W., Dowthwaite, G.P., *et al.* (2003). Development of synovial joints. *Birth Defects Res Part C Embryo Today*, 69(2): 144–155.

Archer, C.W., McDowell, J., *et al.* (1990). Phenotypic modulation in sub-populations of human articular chondrocytes *in vitro. J Cell Sci*, 97(Pt 2): 361–371.

Aydelotte, M.B., Greenhill, R.R., *et al.* (1988). Differences between sub-populations of cultured bovine articular chondrocytes. II. Proteoglycan metabolism. *Connect Tissue Res*, 18(3): 223–234.

Barbero, A., Ploegert, S., *et al.* (2003). Plasticity of clonal populations of dedifferentiated adult human articular chondrocytes. *Arthritis Rheum*, 48(5): 1315–1325.

Beltrami, A.P., Barlucchi, L., *et al.* (2003). Adult cardiac stem cells are multipotent and support myocardial regeneration. *Cell*, 114(6): 763–776.

Beltrami, A.P., Urbanek, K., *et al.* (2001). Evidence that human cardiac myocytes divide after myocardial infarction. *N Engl J Med*, 344(23): 1750–1757.

Brittberg, M., Sjogren-Jansson, E., *et al.* (2005). Clonal growth of human articular cartilage and the functional role of the periosteum in chondrogenesis. *Osteoarthritis Cartilage*, 13(2): 146–153.

Caplan, A.I. (1991). Mesenchymal stem cells. *J Orthop Res*, 9(5): 641–650.

Claudinot, S., Nicolas, M., *et al.* (2005). Long-term renewal of hair follicles from clonogenic multipotent stem cells. *Proc Natl Acad Sci USA*, 102(41): 14677–14682.

Eckfeldt, C.E., Mendenhall, E.M., *et al.* (2005). Functional analysis of human hematopoietic stem cell gene expression using zebrafish. *PLoS Biol*, 3(8): e254.

Eriksson, P.S., Perfilieva, E., *et al.* (1998). Neurogenesis in the adult human hippocampus. *Nat Med*, 4(11): 1313–1317.

Goss, R.J. (1969). *Principles of Regeneration*. NY: Academic Press.

Goss, R.J. (1991). The natural history (and mystery) of regeneration. In *A History of Regeneration Research. Milestons in the Evolution of a Science* (Dinsmore, C.E., ed.). Cambridge: Cambridge University Press, pp. 7–23.

Laflamme, M.A., Myerson, D., *et al.* (2002). Evidence for cardiomyocyte repopulation by extracardiac progenitors in transplanted human hearts. *Circ Res*, 90(6): 634–640.

McDonnell, T.J. and Oberpriller, J.O. (1983). The atrial proliferative response following partial ventricular amputation in the heart of the adult newt. A light and electron microscopic autoradiographic study. *Tissue Cell*, 15(3): 351–363.

McDonnell, T.J. and Oberpriller, J.O. (1984). The response of the atrium to direct mechanical wounding in the adult heart of the newt, Notophthalmus viridescens. An electron-microscopic and autoradiographic study. *Cell Tissue Res*, 235(3): 583–592.

Moore, K.A. and Lemischka, I.R. (2006). Stem cells and their niches. *Science*, 311(5769): 1880–1885.

Oberpriller, J.O. and Oberpriller, J.C. (1974). Response of the adult newt ventricle to injury. *J Exp Zool*, 187(2): 249–253.

Oh, H., Bradfute, S.B., *et al.* (2003). Cardiac progenitor cells from adult myocardium: homing, differentiation, and fusion after infarction. *Proc Natl Acad Sci USA*, 100(21): 12313–12318.

Oshima, H., Rochat, A., *et al.* (2001). Morphogenesis and renewal of hair follicles from adult multipotent stem cells. *Cell*, 104(2): 233–245.

Poss, K.D., Wilson, L.G., *et al.* (2002). Heart regeneration in zebrafish. *Science*, 298(5601): 2188–2190.

Quaini, F., Urbanek, K., *et al.* (2002). Chimerism of the transplanted heart. *N Engl J Med*, 346(1): 5–15.

Rheinwald, J.G. and Green, H. (1975). Serial cultivation of strains of human epidermal keratinocytes: the formation of keratinizing colonies from single cells. *Cell*, 6(3): 331–343.

Rubel, E.W., Dew, L.A., *et al.* (1995). Mammalian vestibular hair cell regeneration. *Science*, 267(5198): 701–707.

Tallheden, T., Dennis, J.E., *et al.* (2003). Phenotypic plasticity of human articular chondrocytes. *J Bone Joint Surg Am*, 85-A(Suppl 2): 93–100.

Urbanek, K., Cesselli, D., *et al.* (2006). From the Cover: Stem cell niches in the adult mouse heart. *Proc Natl Acad Sci USA*, 103(24): 9226–9231.

Chapter 4
Cellular signaling

Jan de Boer, Abdoelwaheb el Ghalbzouri, Patricia d'Amore, Karen Hirschi, Jeroen Rouwkema, Rutger van Bezooijen and Marcel Karperien

Chapter objectives:

- To understand the paradigm of cell signaling and to be able to outline it for a given signaling pathway
- To be able to use the nomenclature to annotate the classes of molecules involved in cell signaling
- To be able to list signaling pathways and explain their role in tissue homeostasis and tissue engineering
- To explain methods which may be applied to manipulate cell signaling in a given cell type
- To appreciate the complexity of signaling events to which a cell is exposed both *in vitro* and *in vivo*
- To value the amount of effort and time in the discovery of signal transduction pathways and the implementation of signaling in tissue engineering

"Tissue engineering requires, amongst others, fundamental insights in the regulatory mechanisms that are involved in normal formation of organs. Knowledge of the signal transduction pathways will allow rational manipulation of these mechanisms ultimately improving quality of the implants. "

Marcel Karperien (2005)

"Does matrix produce a specific diffusible chemical agent that induces the cells of the host to differentiate into osteoblasts? The answer is no. The system is more complex than a simple chemical stimulus and direct cell response....
Michael Urist in the discussion section of his seminal paper on osteoinductivity by decalcified bone matrix."

(Urist, 1965)

4.1 General introduction

4.1.1 Paradigm of cellular signaling

Tissue engineering, in analogy with development and wound healing, is a dynamic process in which the right cell type should be at the right place at the right time in order to establish a normally functioning tissue. Communication between cells and the rest of the body plays a crucial role in the coordination of cell number, position and function. The way cells function in a tissue environment will be discussed in other chapters of this book. In this chapter we will discuss the molecular mechanism by which cells communicate, referred to as cellular signaling. Generally speaking, cellular signaling is initiated by generation of a ligand; i.e. a molecular entity generated by a sending cell to bring about a change in the physiology of a responding cell (Figure 4.1). Putative responses include altered proliferation, differentiation, migration, cytokine production or extracellular matrix production. An illustrative example of cell signaling is the contraction of muscle cells brought about by the local delivery of acetylcholine from motor neurons to muscle cells.

The cell's hardware, i.e. the repertoire of proteins expressed by the cell, determines whether and how it responds to a certain signal. Since the gene expression profile is different for every cell type, cells will respond differently to certain signals. A striking example is the hormone glucocorticoid, which triggers cell death in lymphocytes but stimulates osteogenic differentiation of mesenchymal stem cells. Moreover, cells respond differently to ligands at different concentrations. To add to this complexity, cells in the body are usually

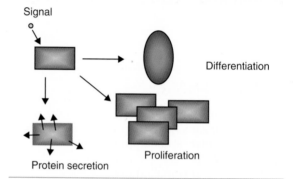

Signal

Differentiation

Proliferation

Protein secretion

Figure 4.1 Cell signaling leads to changes in the responding cell. Reception of the signal by the responding cell leads to changes in its physiology. Three potential responses are depicted.

not exposed to a single signal at a single concentration but to a cocktail of hormones, cytokines and growth factors. Therefore, for a full recognition of the role of a ligand in a biological process, its presence should be considered in a dose- and context-dependent manner.

A central paradigm can be recognized in most events of cellular signaling, which consists of three distinct steps (see Figure 4.2):

1. **Signal initiation**: an extracellular ligand binds to a receptor on the surface of the cell. Ligand binding changes activity of the receptor, thus generating the signal.
2. **Signal transduction**: The activated receptor triggers a signal transduction cascade in which intracellular proteins are activated, ultimately leading to activation of a so-called transcription factor in the cell's nucleus.

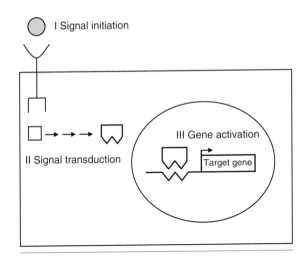

Figure 4.2 The paradigm of cell signaling. Although many variations on this theme excist, many signaling events follow the paradigm of signal intitiation by an extracellular ligand, followed by signal transduction into the nucleus where the change in gene expression underlies physiological adaptation of the cell as depicted in Figure 4.1.

3. **Gene activation**: The transcription factor binds to regulatory sequences in target genes, resulting in gene activation, protein synthesis and changed cellular physiology.

Below we discuss the main types of signals, receptors, signaling molecules and the machinery for gene activation.

4.1.2 Signal initiation

The cellular signal is initiated by the interaction of a ligand with its receptor. A wide variety of molecules in the body serve as ligands. They can be categorized into:

1. Diffusible molecules.
2. Extracellular matrix proteins.
3. Membrane-bound ligands.

In the first group, a sending cell produces a protein (e.g. vascular endothelial growth factor) or organic substance (e.g. steroid hormone) and secretes it into

its environment. The action range of diffusible ligands varies. Long range or endocrine action is seen for instance for parathyroid hormone, which is produced in the thyroid glands but affects bone remodeling throughout the body. Short range or paracrine activity is seen, e.g. for interleukins, which are produced in the dermis of the skin but control epidermal biology at a distance of a few layers of cells. In addition to ligands produced for the purpose of signaling, some intermediates of cell metabolism, e.g. calcium, glucose and oxygen, can also trigger a physiological response by acting as a ligand. A profound example of this is the serum level of calcium, which is sensed by receptors on the membrane of pituitary cells. As a response to low calcium levels, the cells will secrete parathyroid hormone, which affects the bone remodeling process throughout the body.

Equally important as responding to distantly generated signals, cells have to be aware of and adapt to their direct environment. Many extracellular matrix proteins such as collagens and proteoglycans are ligands for specific receptors. Moreover, cells communicate by direct contact of proteins expressed on the membrane of both cells. Typically, one protein acts as ligand and another protein acts as receptor.

Irrespective of their identity, ligands share the characteristic that they physically interact with a receptor. This induces a change in the receptor leading to its activation. Receptors are transmembrane proteins, which interact with the signal molecule in the extracellular space and transduce the signal into the cell. Examples of these types of receptor families include the receptor tyrosine kinases, receptor serine/threonine kinases, the G-Protein Coupled Receptor family and integrins (see below for more detailed descriptions of the signaling pathways). An alternative mechanism is used by hydrophobic compounds such as steroid hormones, which can pass the membrane and bind to their receptor in the cytoplasm.

4.1.3 Signal transduction

The signal, initiated by the ligand/receptor interaction has to be transduced from the membrane to the nucleus. Every ligand/receptor combination triggers

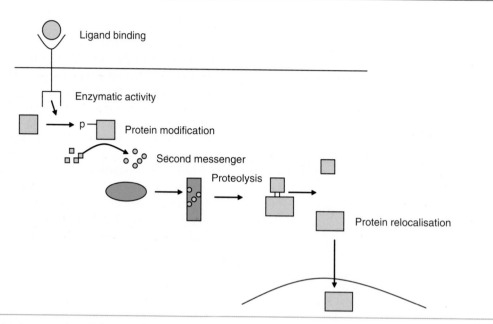

Figure 4.3 Molecular relay of the signal. Hypothetical flowchart of a signal into the nucleus depicting some of the most relevant molecular mechanisms involved in signal transduction.

a unique cascade of molecular interactions in the cytoplasm, resulting in the relay of the signal into the nucleus. One might consider a given signal transduction pathway as a flow chart, in which upstream and downstream events can be recognized (see Figure 4.3). In many cases the signal is conveyed by direct protein/protein interaction and subsequent modification of target proteins, leading to their activation. For example, binding of BMP2 to its receptor leads to activation of the receptor's serine/threonine kinase activity and phosphorylation of target proteins (discussed in more detail in section 4.4.2). Some of the most prevalent protein modifications are summarized in Table 4.1. Alternatively, activated proteins drive the production or relocalization of so-called second messengers, which are small molecules such as cyclic AMP, phospho-inositol-2-phosphate (PIP2), calcium and nitric oxide, which act as ligands for target proteins, hence the name second messenger.

Protein activity can be controlled in several ways. The enzymatic activity of the protein can be switched on, as discussed before for BMP receptor activation. Furthermore, proteins can be delocalized from an inactive to an active site of the cell as seen for instance with the transcription factor NFκB, which is actively retained from the nucleus by binding to the inhibitory protein IκB (discussed in section 4.2.2). Another way to control protein activity is its abundance. Cell signaling can in some cases inhibit the constitutive proteolytic degradation of a certain protein, which then accumulates and exerts its function (see paragraphs 4.4.3 and 4.5.3 on Wnt and Hedgehog signaling respectively).

An important consequence of the signal transduction cascade is that the signal becomes amplified: one ligand/receptor interaction leads to activation of many downstream proteins.

4.1.4 Gene activation

The last step in the signal transduction cascade is gene activation (see Figure 4.4). Gene activity is mainly

Table 4.1 Enzymatic activities involved in signal transduction

Enzyme	Activity	Putative effect
Kinase	Protein phosphorylation	Changed enzymatic property or protein stability
Phosphatase	Protein dephosphorylation	Changed enzymatic property or protein stability
Ubiquitin ligase	Protein ubiquitination	Protein degradation
Proteases	Protein site specific cleavage	Degradation, activation
Fatty acid esterase	Palmitoyl/myristoyl addition to protein	Relocalization to the plasma membrane
SUMO ligase	Protein SUMO modification	Protein degradation
Lipase	Cleavage of phospholipids	Generation of second messengers, such as PIP2 and IP3
Cyclase		Generation of second messenger cAMP and cGMP

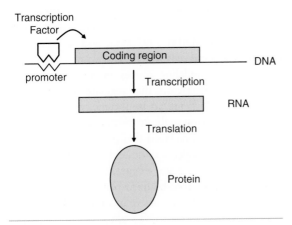

Figure 4.4 Activation of gene transcription. The activation of gene transcription by transcription factors is the final stage of the cell signaling event, resulting in the anticipated physiological response.

controlled by the rate of transcription of the gene, the process in which a messenger RNA is produced by the RNA polymerase holocomplex. Signal transduction impinges on the initiation of the transcription process by activating so-called transcription factors, which are sequence-specific DNA-binding proteins. The transcription factor is either retained in the cytoplasm and is translocated upon activation or the transcription factor is already present in an inactive form in the nucleus where it becomes activated. Once actively present in the nucleus, they bind to unique DNA sequences in the promoter region of target genes. The promoter is the DNA sequence adjacent to the protein-encoding DNA sequence, and contains the regulatory elements that control gene transcription. Every gene has a unique promoter sequence and therefore recruits one or more different transcription factors. For instance, the Runx2 gene, which is a master regulator of osteogenic differentiation, contains binding sites for the SMAD transcription factors, which are activated by the BMP signal transduction cascade. Once bound to the promoter, the transcription factor activates the RNA polymerase complex, resulting in transcription of the gene into a messenger RNA, which will be translated in the cytoplasm into a protein (Figure 4.4).

4.1.5 Variations on a theme

The paradigm of cellular signaling described above is conceptually a good starting point but it should be noted that many signal transduction pathways use variations on this theme or use only a part of the route to convey a signal from outside the cells to evoke a biological response. First of all, some ligands directly bind and activate a transcription factor, such as vitamin D3 (see section 4.4.4). Moreover, the path length of the cascade differs strikingly from a single step as for instance in BMP signaling (see section 4.4.2) to an elaborate cascade seen with integrin signaling (see section 4.2.3). Some signal transduction pathways exert their effect without activation of gene transcription. A noticeable example of this is the family of Rho kinases, which is downstream of some of the main signal transduction pathways and control cytoskeletal organization (see section 4.3.4).

Another important phenomenon in cellular signaling is crosstalk, in which the signals of two different ligands converge on the same effector molecule. Crosstalk may occur at several levels in the signaling cascade. In some cases, co-activation of a signal transduction molecule by two signaling pathways is required. For instance, CREBP activation requires phosphorylation at serine-133 and serine-129, which are downstream events of G-protein coupled receptor signaling and insulin signaling respectively. In other cases, the transcription factors downstream of two or even more signal transduction pathways have to bind simultaneously to a promoter in order to activate gene transcription. Finally, many signaling molecules are not uniquely activated by a single signal transduction pathway. For instance, mitogen-activated protein kinase activation can be downstream of receptor tyrosin kinases but also of integrins.

Below we discuss a number of signal transduction pathways in detail to illustrate their complexity but also to point out how detailed knowledge on signaling can help in the rational design of tissue engineering approaches.

4.2 Cellular signaling in skin biology

4.2.1 Skin signaling overview

Skin wound healing is characterized by an orderly sequence of events that begins immediately after injury. The healing process can be broadly classified into four phases: haemostasis, inflammation, proliferation and remodeling. These phases proceed in a systematic fashion, in which specific signaling routes are responsible for proper, organized and controlled healing (Figure 4.5). Clarification of the signaling pathways involved in these processes will be advantageous for skin tissue engineering. This will result in new strategies to reintegrate these events in the development of new skin matrices that will cover epidermal and dermal wounds.

In the haemostasis and inflammation phase, macrophages, lymphocytes, neutrophils and platelets have to kill bacteria and clear the wound bed. Here transforming growth factor (TGF)-β facilitates polymorphonuclear neutrophil migration from the surrounding blood vessels where they extrude themselves from these vessels. These cells cleanse the wound and clear it of debris. After this process, these cells will release some factors that will act to attract various cell types to the site of the scene (so called chemotactic agents are released (see Table 4.2) (Martin, 1997; Witte and Barbul, 1997). In addition, these cells initiate and sustain the proliferative and remodeling phases by synthesizing and releasing growth factors that regulate wound repair. Dermal fibroblasts are an important cell type during this process serving multiple purposes. One of them is the production of collagen that provides strength and integrity to the healed wound and a path to guide the keratinocytes. While the extracellular matrix (ECM) proteins are produced by fibro-blasts, other cells such as the keratinocytes are active in producing cytokines (tumor necrosis factor (TNF)-α, various interleukins), growth factors (TGF-α, TGF-β, VEGF, IGF) and signaling polypeptides to restore the epithelium. Recent studies, in particular those utilizing genetically manipulated animal models, have highlighted the impact of TGF-β on various aspects of wound healing, and surprisingly, not all of its effects are conducive to optimal healing. Intriguingly, TGF-β1 mutations, or mutations in the cell signaling intermediate Smad3, lead to normal or even accelerated cutaneous wound healing responses (Ashcroft *et al.*, 1999). Endothelial cells in their turn produce granulation tissue to repair the vascular integrity, and like keratinocytes, secrete enzymes that degrade basal lamina components so that the epithelialization process can take place.

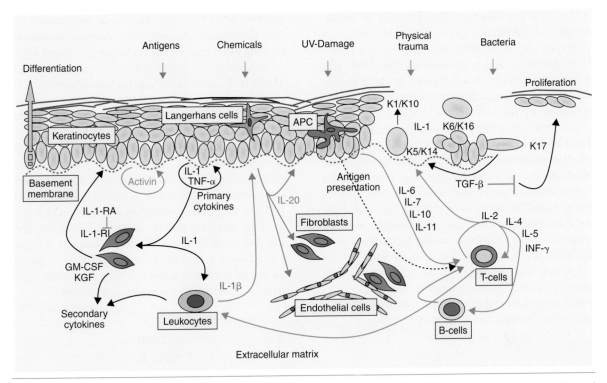

Figure 4.5 Schematic overview of signaling pathways in the regenerating skin. When keratinocytes are activated by a specific signal (i.e. chemical insults, bacteria, UV-irradiation, physical trauma) they will express a number of specific keratins, integrins, and interleukins. The homeostasis of the skin is regulated by an exquisitely finely tuned exchange of signaling factors among the dermal and epidermal cells. Abbreviations: APC, Antigen Presenting cells; IL, interleukins; K, keratins; TNF-α, Tumor Necrosis Factor-α; GM-CSF, Granulocyte Monocyte-Colony Stimulating Factor; KGF, Keratinocyte Growth Factor; TGF-β, Transforming Growth Factor; INF-γ, Interferon-γ.

Table 4.2 Signals involved in recruitment of cells to the woundbed

Cell type	Signals
Keratinocytes	EGF, PDGF, IL-1, IL-6, IL-GF1.
Fibroblasts	PDGF, platelet factor 4 (PF4), TGF-β, IL-4, hypoxia, fragments of fibronectin and types I and III collagen.
Endothelial cells	bFGF, TNF-α, IL-8, hypoxia, heparin, fibronectin.
Monocytes/ Macrophages	Monocyte chemoattractant protein-1 (MCP-1), macrophage inflammatory protein-1a (MCP-1a), platelet-derived endothelial cell growth factor (PD-ECGF), thrombin, TGF-β, fibronectin, and elastin.
Neutrophils	Fibrinopeptides, fibrin lysis products, C5a, formyl methionyl peptides (from bacterial proteins), platelet activating factor (PAF), TNF-α, PDGF, PF4.

The most common cytokine involved in keratinocyte and fibroblast *proliferation* is IL-1α/β. IL-1 triggers release of keratinocyte growth factor (KGF) from fibroblasts, and in this way keratinocytes indirectly promote their own proliferation. IL-1α is constitutively expressed in human keratinocytes while in other cells it is released under stimulation.

The final *remodeling phase* is characterized by synthesis and degradation of collagen and depends on a balance between factors that endorse synthesis of ECM components and enzymes that degrade these components – e.g. matrix metalloproteases (MMPs). For instance, MMP-11 is involved in remodeling of the extracellular matrix and may function through release of growth factors from the ECM. Stat3 plays a crucial role in transducing a signal required for migration and is essential for skin remodeling, mainly in hair cycle and wound healing (Sano *et al.*, 1999). In cytoskeletal modulation, integrins signal to the nucleus via Rho family or MAPK pathways.

4.2.2 Interleukin signaling

Interleukins (IL) are a class of proteins (cytokines and lymphokines) that are released by cells of the immune system and act as intercellular mediators in developmental regulation, tissue repair, hemopoiesis, inflammation, and specific and nonspecific immune responses. For example, IL-1 promotes keratinocyte and fibroblast proliferation and induces the expression of intercellular adhesion molecules in endothelial cells and fibroblasts. The signal transduction events initiated by IL-1 are described below (see Figure 4.6). First, binding of IL-1 to its receptor, IL-1RI, induces recruitment and interaction of IL-1RacP to IL-1RI, which results in a high affinity complex for IL-1. IL-1RI and IL-1RAcP recruit MyD88, a member of the so-called IRAK (IL-1 Receptor Associated Protein Kinases) family and TRAF-6. Together, they activate the IκK complex, consisting of the inhibitory protein IκK and the transcription factor NF-κB. The TRAF-6 signal will result in phosphorylation of the inhibitory IκB subunit, thereby releasing the NF-κB

dimmer from the complex. NF-κB will translocate into the nucleus and activate gene transcription by binding to κB binding sites in target genes such as IL-6 (Shimizu *et al.*, 1990) and IL-8 (Kunsch and Rosen, 1993). IL-1 signaling can also lead to the activation of c-Jun N-terminal kinase (JNK) and other mitogen-activated protein kinases (MAPK) that results in the phosphorylation and activation of AP1.

The relevance of interleukin signaling is highlighted in hyperactivated skin, (caused by e.g. chemical insults) or in psoriasis, where the IL-balance is shifted towards overproduction of IL-1α or IL-1β. The misbalance in IL-1 expression has recently been used as an indicator for skin irritation, and the cosmetic industry is testing thousands of chemicals and substances for IL-1 expression. For this, tissue-engineered skin is used as an animal substitute using IL-1 as a marker discriminating non-irritant from irritant compounds.

4.2.3 Integrin signaling

Integrins are heterodimeric transmembrane proteins consisting of α and β subunits that serve both as adhesive and signaling receptors in various cell types. They transmit information in both directions across the plasma membrane (bidirectional signaling) to integrate the intracellular and extracellular environments. Depending on the α and β composition, the extracellular domains can bind to a number of different ligands, thus establishing physical contact between the cell and its extracellular matrix (see Table 4.3 with integrin-ligand combinations).

In normal undamaged epidermis integrin expression is confined to the basal cell layers and the outer root sheath of the hair follicles. They play a prominent role during skin re-epithelialization. While migrating over the collagenous matrix, keratinocytes are constantly breaking and making new integrin contacts. During this process the integrin surface repertoire changes and provides a means of altering the strength and ligand preferences of cell adhesion. By using the integrin α2β3 as a model, investigators have shown that in the resting phase, integrins normally bind their

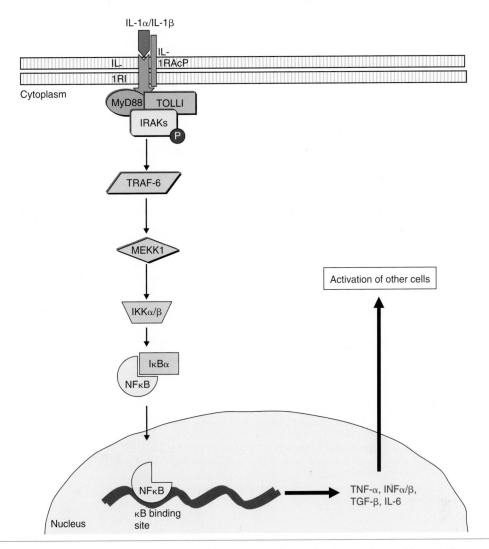

Figure 4.6 Interleukin signaling in keratinocytes. The IL-1 signal transduction pathway. After binding, recruitment of MyD88, TOLLIP, and IRAKs will follow. Thereafter TRAF-6 will interact, which causes activation of several protein kinases such as MEKK1. This finally results in the activation of several transcription factors, such as NFκB. IRAK, IL-1 receptor associated protein kinases; TRAF6, tumor necrosis factor receptor-associated factor-6; IkB, inhibitory kB; IKK, IkB kinase, MEKK1; MAP kinase kinase-1; TOLLIP, Toll-interacting protein.

ligands with low affinity. After activation, a cellular signal induces a conformational change in the integrin cytoplasmic domain which results in low-to a high affinity ligand binding state. Thus, this 'inside-out'

signal transduction appears to be mediated through the integrin cytoplasmic domains (inside-out signaling, Figure 4.7). For example, sequential activation of PKG, p38, and ERK kinases regulate the activation

Table 4.3 Integrin α-β subunit combinations and ligands.

	β1	β2	β3	β4	β5	β6	β7	β8
α1	Coll							
α2	Coll							
α3	Lam							
α4	VCAM, Fn						MAdCAM-1, VCAM-1, Fn	
αE							E-cadh	
α5	Fn							
α6	Lam			Lam				
α7	Lam							
α8	Fn,Vn,Tn							
α9	Tn							
α10	Coll							
α11	Coll							
αV	Fn,Vn,Op,Fn		Fn,Vn,Op,Tn,vWF		Vn	Fn,Tn		Fn,Coll,Lam,Vn
αIIb			Fb,Vn,Fn,vWF					
αL		ICAM-1,2,3						
αM		iC3b,Fx,Fb, ICAM-1						
αX		iC3b,Fb,ICAM-1						
αD		ICAM-3						

Abbreviations: Coll, collagen; Fb, fibrinogen; Fn, fibronectin; ICAM-1,2, 3, intracellular adhesion molecule-1,2,3; lam, laminin; MAdCAM-1, rnucosal atldrrssin cell adhesion rnolerulc-1; Op, osteopontin; Tn, tenascin; VCAM-1, vascular. Cell adhesion molecule-1; Vn, vitronectin; vWF, von Willebrand's factor.Empty boxes indicate α-β subunit combinations that have not been observed. Many additional ligands have been proposed, but often these are not as well established or interact more weakly compared to one listed. Adapted from (Hemler, 1999).

of integrin$\alpha_{\text{IIb}}\beta_3$ on platelets (Li *et al.*, 2006). Inside out signaling also protects the host from excessive integrin-mediated cell adhesion, which could have profound pathological consequences (Qin *et al.*, 2004). Illumination of the mechanisms of integrin affinity modulation should help explain rapid changes in cell adhesion that occur during e.g. skin re-epithelialization.

Conversely, binding of the intercellular cytoplasmic ligands initiates downstream signaling, referred to as outside-in signaling. This signaling path include enzymes (e.g. focal adhesion kinase/c-Src complex, Ras and Rho GTPases) and adapters (e.g. Cas/Crk, paxillin) that assemble within dynamic adhesion structures, referred to as focal adhesions (Qin *et al.*, 2004). Depending on the ECM composition, integrins are able to activate various signaling components such as cytoskeletal proteins (talin, α-actinin, WASP), adapter molecules (JAB, paxillin, RACK1, p130CAS, afflixin), Ca^{2+} binding proteins (CIB and calreticulin),

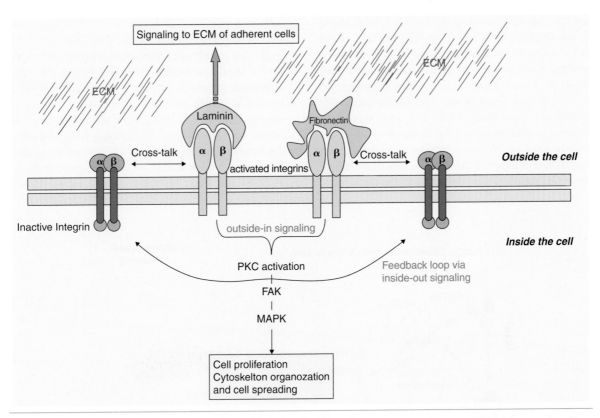

Figure 4.7 Schematic overview of integrin-mediated activation leading to inside-out and inside-out signaling. Integrin engagement leads to PKC activation and autophosphorylation of focal adhesion kinases (FAK), these outside-in signals will among others, result in the activation of the MAPK pathway, which finally leads to cell proliferation and cell spreading. Inside-out signaling occurs when changes within the cell lead to changes in the affinity of the integrin pair for its extracellular target. PKC, Protein kinase-c; FAK, focal adhesion kinases; MAPK, mitogen-activated protein kinase.

protein tyrosine kinases, including FAK and the Src-family kinases (Figure 4.7) (Parsons and Parsons, 1997) . These signals, in concert with signals derived from other growth factors, regulate cell behavior in a complex tissue microenvironment. For example, α6β4 promotes migration through activation of phosphatidylinositol 3-kinase (PI3-K) (Mercurio *et al.*, 2001) but signaling via the Rac, Rho and the MAPK pathway is also documented to play an important role in this process (Mainiero *et al.*, 1997; O'Connor *et al.*, 2000).

4.3 Cellular signaling in vascular biology

4.3.1 Vascular signaling overview

The development of blood vessels is an important area for tissue engineering. Apart from engineering blood vessels for bypass surgeries, blood vessel formation becomes an important issue for other engineered tissues after implantation. In the initial phase after implantation, engineered tissues have to rely on diffusion for the delivery of oxygen and nutrients and the removal of waste products. Over time, blood vessels from the host will grow into the engineered tissue, allowing for a better transport of these components. If the process of vascularization is too slow, however, it can lead to non-optimal integration or even partial death of the engineered tissue due to nutrient limitations. There has been significant progress in understanding the molecules and mechanisms involved in the assembly, remodeling and maturation of blood vessels. The factors that mediate these various phases are delivered to the endothelial

cells in a paracrine or juxtacrine (cell communication in which ligand and receptors are both anchored in the cell membrane) manner, making vessel formation a locally controlled event (see Figure 4.8). The formation of vessels can occur via vasculogenesis, a process in which endothelial precursors coalesce in situ to form vessels *de novo*, or via angiogenesis in which new vessels arise by budding from pre-existing vessels. Signaling pathways stimulated by growth factors such as vascular endothelial growth factor (VEGF) and basic fibroblast growth factor (bFGF)

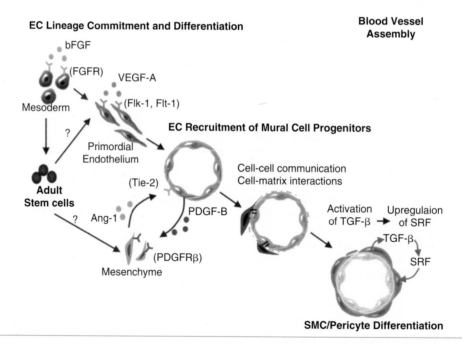

Figure 4.8 Blood vessel assembly. Blood vessel assembly can be divided into different stages. First, endothelial progenitor cells are attracted to the site and stimulated to differentiate into endothelial cells. VEGF is the key factor governing these processes. When endothelial cells have organized into a vascular structure, they secrete PDGF-B to recruit mural cell progenitors. Communication between the endothelial cells and the mural cell progenitors is important for the differentiation of the mural progenitors into pericytes or smooth muscle cells (depending on the vessel type) and the proper remodeling of the vessel. TGF-β is amongst the factors that are important for this process. EC, endothelial cell; SMC, smooth muscle cell; bFGF, basic fibroblast growth factor; FGFR, fibroblast growth factor receptor; VEGF-A, vascular endothelial growth factor A; Flk-1, vascular endothelial growth factor receptor 2; Flt-1, endothelial growth factor receptor 1; Ang-1, angiopoietin 1; Tie-2, angiopoietin 1 receptor; PDGF-B, platelet derived growth factor B; PDGFRβ, platelet derived growth factor receptor β; TGF-β, transforming growth factor β; SRF, serum response factor.

are known to play critical roles in the initial stages of vascular development, namely in the establishment of a primitive vascular plexus from multipotent mesodermal progenitors.

Once a capillary plexus is formed, it is remodeled via series of steps collectively referred to as remodeling. We know from various knockout and mutagenesis studies in multiple developmental model systems that a number of signaling pathways, including platelet derived growth factor B (PDGF B), transforming growth factor β (TGFβ) as well as angiopoietin-1 and -2 (ang-1 and ang-2) are involved (Dickson *et al.*, 1995; Lindahl *et al.*, 1997). These factors signal through their cognate receptors to yield a stable two cell capillary. Signaling between the endothelial cells, as well as between the endothelial cells and pericytes (or smooth muscle cells), also occurs through gap junctions. The major steps in blood vessel assembly and the signaling events involved are depicted in Figure 4.8. In this chapter, we will focus on some signaling pathways that regulate different stages of blood vessel assembly.

4.3.2 Receptor tyrosine kinase signaling

Several tyrosine kinase receptors act during vascular morphogenesis, including the VEGF receptors, VEGFR1 (Flt-1), VEGFR2 (Flk-1); PDGF receptor β (PDGFRβ) and the angiopoietin receptor, Tie-2. The VEGF receptors are primarily involved in the early stages of blood vessel assembly. The activation of these receptors leads to the migration, proliferation and eventually differentiation of endothelial cells to form a new, lumen-containing vessel (Cross and Claesson-Welsh, 2001).

As can be seen in Figure 4.9, the binding of VEGF to VEGFR2 can lead to the activation of several signaling pathways involved in angiogenesis. In this section, we will discuss the MAPK signaling pathway in more detail. The pathway begins with the activation of VEGFR2 by binding of the growth factor VEGF. This provides the binding site for the adapter protein Grb2 that in turn localizes Sos to the plasma membrane. Sos activates Ras by exchange of GTP for GDP (Tarnawski, 2000). The Ras-GTP binds directly

to a serine-threonine kinase, Raf, forming a transient membrane-anchoring signal. Active Raf kinase phosphorylates a dual specificity kinase, mitogen-activated kinase MEK, and activates it (Kolch, 2005). The activated MEK phosphorylates ERK1/ERK2, which subsequently translocates to the nucleus, where it can phosphorylate the transcription factor Ets-1. Ets-1 then regulates the expression of multiple angiogenic genes like MMP-1, MMP-9, integrin β3 and VE cadherin (Watanabe *et al.*, 2004).

Endothelial cells within the primitive capillary network govern the subsequent acquisition of mural cells (pericytes or smooth muscle cells) that will make up the surrounding vessel wall. Proliferating endothelial cells recruit mural cell precursors via secretion of PDGF-B that acts as a chemoattractant and mitogen for multipotent progenitors (Cross and Claesson-Welsh, 2001; Kolch, 2005; Tarnawski, 2000), derived from the mesenchyme surrounding the endothelial tubes (Lindahl *et al.*, 1997). Conversely, the mesenchymal progenitors secrete soluble factors, Ang-1 and Ang-2 (Hirschi *et al.*, 1998), which both bind the receptor Tie-2 that is expressed by endothelial cells. Ang-1 and Ang-2 have opposing roles in vessel assembly. Unlike Ang-2, which activates Tie-2 in some cells, and blocks its function in others, Ang-1 consistently activates Tie-2 and promotes vessel formation (Folkman and D'Amore, 1996). Mice lacking either Ang-1 or Tie-2 exhibit similar phenotypes, in which the embryos die in utero mid-gestation with endothelial tubes, but no associated mural cells. Defective mural cell recruitment is proposed as the cellular defect (Folkman and D'Amore, 1996); thus, a major biological role of the Tie-2 signaling pathway is mediating endothelial cell recruitment of mural cell precursors during vascular development.

The addition of growth factors like VEGF to engineered tissues could initiate faster vessel formation after implantation. However, in order to get stable, functional vascular structures, other factors are necessary as well. Therefore, a thorough understanding of receptor tyrosine kinase signaling pathways, involved both in initial vessel formation and vessel maturation, is important to be able to enhance the vascularization of engineered tissues after implantation.

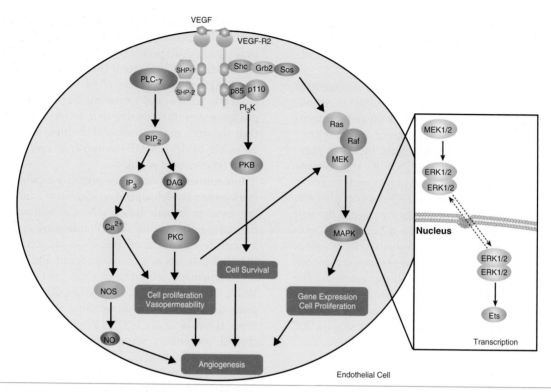

Figure 4.9 Receptor tyrosin kinase signaling. Schematic presentation of a receptor tyrosin kinase signaling pathway involved in angiogenesis. Binding of VEGF to its receptor leads to the activation of several signaling pathways that are involved in angiogenesis. Activation of these pathways leads to endothelial cell proliferation, differentiation but also migration. One of the pathways involved is the MAPK pathway. The MAPK pathway is activated through a Ras/Raf complex, but it should be noted that an alternative route via PKC also plays an important role in endothelial cells. VEGF, vascular endothelial growth factor; VEGF-R2, vascular endothelial growth factor receptor 2; SHP, SH phosphatase; Grb2, growth factor receptor-bound protein 2; PI$_3$K, Phosphoinositide-3 kinase; PLC-γ, phospholipase C γ; PIP$_2$, phosphatidylinositol bisphosphate; IP$_3$, phosphoinotide-3,4,5-triphosphate; DAG, Diacylglycerol; PKC, protein kinase C; NOS, nitric oxide synthase; NO, nitric oxide; PKB, protein kinase B; MEK, MAPK/ERK kinase; MAPK, mitogen-activated protein kinase; ERK, extracellular signal-regulated kinase.

4.3.3 Rho kinase signaling

The coordination of multiple signaling pathways, as discussed above, is required to achieve the well-controlled endothelial cell proliferation and migration needed for blood vessel formation and vascular remodeling. Directed cell migration requires dynamic changes in cell shape that are regulated, in part, by changes in the arrangement of the actin-cytoskeleton and involve the formation and turnover of cell-extracellular matrix adhesions. Extracellular factors, such as diffusible growth factors and their receptors, signals on adjacent cells, and signals from the extracellular matrix, can initiate cell migration. There are various cellular components involved in and needed for endothelial cell migration.

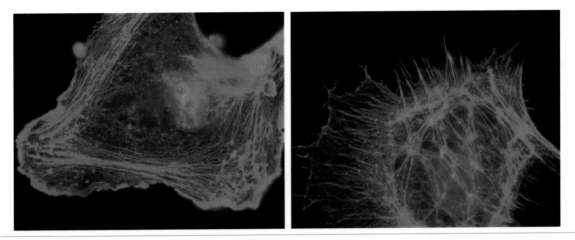

Figure 4.10 Cytoskeletal reorganization induced by Rac signaling. The effect of Rac on cytoskeletal organization in human umbilical vein endothelial cells. In these cells, actin is stained red with fluorescently labelled phalloidin and the nucleus is stained blue. The left cell is a normal cell which only received an empty vector, the right cell received a vector containing dominant-negative Rac1 (N17Rac), meaning that Rac is not active in this cell. As can be seen, the lack of Rac activity prevents cell polarization and the formation of actin-rich lamellipodia (the actin-rich regions at the edge of the cell).

Rho GTPases are low molecular weight GTP-binding proteins known to play key roles in mediating cell shape and migration in response to extracellular growth factors. Rho proteins are active when bound to GTP and inactive when GDP-bound. RhoA, Rac1, and Cdc42 are among the Rho-GTPases known to regulate the actin cytoskeleton and cell adhesion. RhoA regulates the formation of contractile actin-myosin filaments to form stress fibers (Ridley and Hall, 1992), Rac1 mediates lamellipodia formation and cell motility, and Cdc42 mediates filopodia formation and maintains cell polarization (Etienne-Manneville and Hall, 2002; Kozma *et al.*, 1995; Nobes and Hall, 1995; 1999; Ridley *et al.*, 1992).

Forward cell movement requires Rac activation and subsequent induction of protrusions at the cell's leading edge (Nobes and Hall, 1999). Rac plays a crucial role in cell migration and spreading through induction of lamellipodia that form the advancing cell edges in both processes (see Figure 4.10) (Hall, 1998). These lamellipodia are actin-rich adhesive structures that concentrate high-affinity integrins at a location that facilitates their binding to extracellular matrix proteins (Kiosses *et al.*, 2001). Other than focal adhesions, which function as an anchor for actin contraction, lamellipodia seem to actively advance the leading edge of the cell (Ballestrem *et al.*, 2000). RhoA effects the actin cytoskeleton organization through the activation of Rho-associated kinases (ROCKs). ROCKs phosphorylate various downstream target proteins, including LIM-kinases, ERM proteins and Adducin (Noma *et al.*, 2006). The overall physiological effect of ROCKs in endothelial cells is to prevent actin depolymerization. The phosphorylation of LIM kinase for instance, inhibits this enzyme which phosphorylates the cofilin/actin-depolymerizing factor complex involved in the depolymerization and severing of actin filaments (Maekawa *et al.*, 1999).

Since Rho signaling is involved in endothelial cell migration, a deeper understanding of this signaling pathway is necessary for blood vessel tissue engineering or for the prevascularization of other engineered tissues *in vitro*. By interfering with this pathway,

success rates. In subsequent experiments, this was found to be due to characteristics of the extracellular bone matrix and not due to the bone inductive proteins (BMPs) that were later identified and act non-species specific (Sampath and Reddi, 1983).

A typical decalcified implant was enveloped in loose, highly vascular, inflammatory, and fibrous connective tissue 3 weeks after implantation. Trabecular interstices and old vascular channels were infiltrated with wandering histiocytes/macrophages, body giant cells (multinucleated cells larger than osteoclasts), large and small lymphocytes, and fibroblasts that were related to the process of resorption of the dead decalcified bone matrix. The earliest deposits of new bone appeared at 4 to 6 weeks after implantation. Whenever bone induction occurred there was a pool of osteoprogenitor cells and small capillaries. New bone was formed by palisades of deeply basophilic plump osteoblasts and always occurred within the old bone matrix and never

extended outside the implants. New bone contained newly incorporated osteocytes, calcified rapidly, and was always easy to distinguish from the old dead bone by a thin line of cement substance. The whole process of bone formation with a modularly cavity was completed within a few months.

In an attempt to identify the molecular nature of the bone inducing compound in demineralized bone matrix, bone samples were treated with chemical compounds and tested for their osteoinductive capacity. Decalcification with equimolar solutions of different acids showed that different acids could be used. Decalcification, however, had to be complete and should not affect deanimate tissue proteins, since this induced widespread inflammation and total disintegration of the matrix that never resulted in new bone formation. Heating of the bone samples to 70°C, sufficient to induce shrinkage of collagen fibers, impeded but did not prevent osteogenesis, showing that collagen was not the inducer for new bone formation. Furthermore, blockage of sulfhydryl and other reactive groups dehydrated and shrank the decalcified matrix and this appeared to prevent osteogenic induction by retarding the cellular ingrowth and excavation of matrix. Together, these experiments showed that the properties of the bone matrix determined the formation of new bone. The existence of a diffusible chemical agent released from the decalcified bone matrix that induced cells of the host to differentiate into osteoblasts, however, was firmly rejected. Instead a more complex mechanism based upon the Speman's theory of induction was suggested. In the eighties, it was discovered that the inductors for osteogenesis were, in contrast to Urist's hypothesis, matrix bound BMPs (Sampath and Reddi, 1981; Wang *et al.*, 1988; Wozney *et al.*, 1988). The discovery of BMP is a classical example of how basic research on cellular signaling can be translated into clinical applications. BMP-2 and BMP-7 (also known as OP-1) were recently approved for clinical use as chemical adjuncts in fracture, non-union, and spinal fusion treatment.

Rabbit Muscle Pouch-Homogenous hci-Decalcified Bone-8 Weeks Postoperative

None	Toluidine blue
70°C	Fta Acid
Lyophilized	Fluorodinitro-Phenol
70% Alcohol	β-Propriolactone
100°C	Nitrous Acid

Osteo-induction by chemically modified demineralized bone. Radiographs showing new ossicles formed from implants of homogenous decalcified bone in rabbits, treated in various ways and implanted in the anterior abdominal wall. Fluorodinitrobenzene, b-propriolactone, or nitrous acid prevented formation of ossicles in implants of decalcified bone matrix. (From Urist, M.R. Bone: formation by autoinduction. Science, 150(698): 893–899, 1965.)

4.4.2 TGF-β superfamily

The TGF-β superfamily consists of various sub-families, i.e. TGF-βs, BMPs, activins, inhibin, and anti-Mullerian hormone. Here we focus on the major group of the TGF-β superfamily involved in bone biology, the BMPs, which are secreted growth factors that were originally identified by their ability to induce ectopic bone and cartilage formation (see Classical Experiment) (Urist, 1965). The BMP subfamily consists of more than 20 members (van Bezooijen *et al.*, 2006). BMPs exert their effects through distinct combinations of two different types of serine/threonine kinase receptors, i.e. type I receptors and type II receptors. Receptor activation involves ligand-induced hetero-oligomerization of two sequentially acting kinases, with the type I receptor acting as a substrate for the type II receptor kinase (Figure 4.12). The activated, phosphorylated type I receptor then propagates the signal through phosphorylation of receptor Smads (R-Smads1, 5, and 8) that assemble into heteromeric complexes with Co-Smad4 and translocate to the nucleus, where they regulate transcription of target genes. In addition, BMPs stimulate other pathways that are distinct from the Smad pathway, i.e. activation of JNK and p38 MAP kinase pathways.

Negative regulation of BMP activity occurs at nearly every step in the BMP/Smad pathway. Intracellularly, inhibitory Smads (I-Smad6 and 7) antagonize R-Smads by competing with them for interaction with activated receptors or heteromeric complex formation with Co-Smad4 (Figure 4.12). I-Smads also prevent activation of R-Smads by recruitment of ubiquitin ligases, like Smad ubiquitination regulatory factor 1 (Smurf-1) and Smurf-2, which add a ubiquitin moiety to Smad proteins, earmarking them for destruction by the proteasome. Extracellularly, soluble pseudoreceptors and antagonists inhibit BMP activity. Extracellular antagonists include noggin, chordin, DAN, cerberus, and gremlin (Canalis *et al.*, 2003). Of these antagonists, chordin and noggin have been shown to antagonize BMP signaling by blocking binding of BMPs to their receptor. The others are thought to act in a

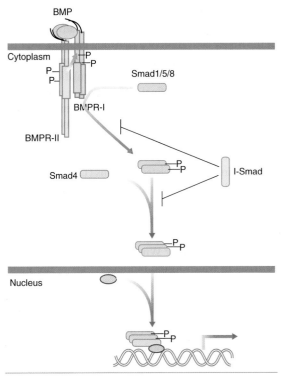

Figure 4.12 Model for activation of bone morphogenetic protein (BMP) receptors. BMP-mediated heteromeric complex formation of two type I and two type II BMP receptors (BMPRs) induces activation of the BMPR-I by the BMPR-II kinase. The activated BMPR-I propagates the signal downstream through phosphorylation of R-Smads1/5/8. These phosphorylated Smads assemble into heteromeric complexes with Co-Smad4 and translocate into the nucleus where they regulate gene transcription.

similar way. In tissue engineering, strategies can be designed to control BMP signaling activity at each of the regulatory steps, both outside and inside the cell.

4.4.3 Wnt signaling

Like BMPs, Wnts are secreted growth factors with pivotal roles in a variety of cellular activities, including cell

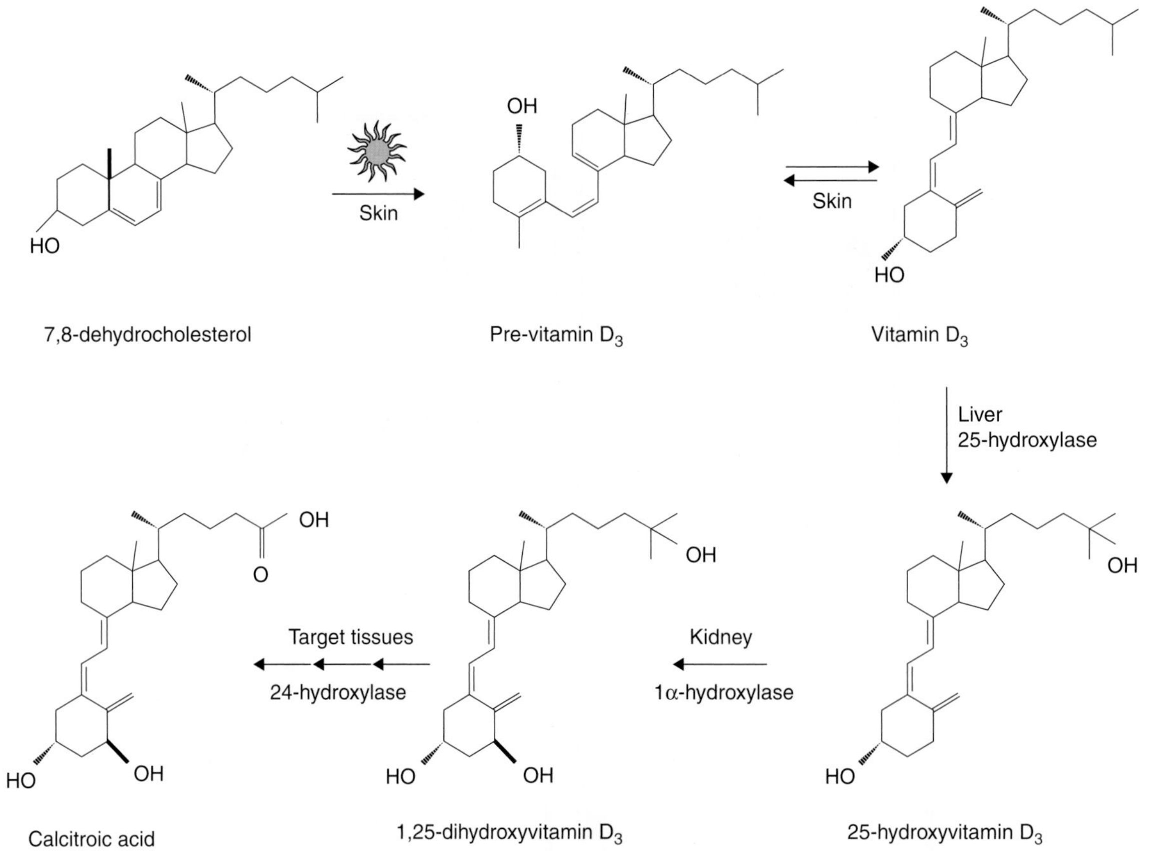

Figure 4.14 Vitamin D_3 synthesis, activation, and catabolism. Vitamin D_3 is produced in the skin by the photolytic cleavage of 7-dehydrocholesterol followed by thermal isomerization. Vitamin D is transported to the liver via the serum vitamin binding protein, where it is converted to 25-hydroxyvitamin D_3, the major circulating metabolite of vitamin D_3. The final step, 1α-hydroxylation, occurs primarily, but not exclusively, in the kidney, forming 1,25-dihydroxyvitamin D_3, the hormonal form of the vitamin. Catabolism inactivation is carried out by 24-hydroxylase, which catalyzes a series of oxidation steps resulting in side chain cleavage.

which cover the distal ends of the long bones and vertebrae (Kronenberg, 2003). In tissue engineering, the major aim is to regenerate damaged articular cartilage.

There are important functional and structural differences between articular and growth plate chondrocytes. The latter are characterized by high activity and are the driving force of longitudinal bone growth. This process ceases at the end of puberty due to growth plate fusion

resulting in the disappearance of growth plate chondrocytes by apoptosis and osteoclastic resorption (van der Eerden *et al.*, 2003). In contrast, articular chondrocytes are relatively inert and play no role in bone elongation. Furthermore, in normal circumstances they are protected from apoptosis and resorption concomitant with their role in joint mobility. In pathophysiology, destruction of articular cartilage is, amongst others, induced by a wide array of cytokines expressed by

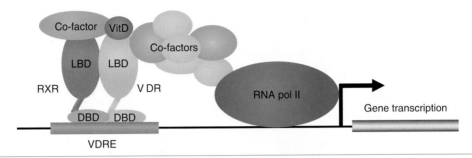

Figure 4.15 Vitamin D receptor-mediated gene activation. 1,25-Dihydroxyvitamin D$_3$ regulated gene transcription. A liganded VDR-RXR heterodimer recruits co-factors and binds to VDRE using the DNA binding domains. Complex formation with basal transcription machinery and histone modifiers enables activation of gene transcription. DBD, DNA binding domain containing the two zinc fingers; LBD, ligand binding domain; VDR, vitamin D receptor; VDRE, vitamin D responsive element; VitD, 1,25-dihydroxyvitamin D$_3$.

mononuclear cells and chondrocytes themselves. These cytokines induce the expression of matrix metalloproteases and activate osteoclasts, both of which stimulate matrix decay (Martel-Pelletier, 2004).

Elucidation of the signaling pathways involved in the formation of articular cartilage during endochondral ossification will be advantageous for cartilage tissue engineering. It will allow rational strategies to recapitulate these events in the implant.

The past years our understanding of the signaling pathways involved in endochondral bone formation has increased rapidly. Multiple growth factors and hormones activating a wide diversity of signal transduction pathways play a key role in this process (see Figure 4.16). Examples are members of the BMP/TGFβ-superfamily (Tsumaki and Yoshikawa, 2005), members of the Wnt-family (Day *et al.*, 2005; Hill *et al.*, 2005), Indian Hedgehog (IHh) (Lai and Mitchell, 2005), Parathyroid Hormone related Protein (PTHLH) (Karaplis and Deckelbaum, 1998) and fibroblast growth factors (FGF) (Ornitz, 2005). It has become clear that these factors are part of a large regulatory network with multiple checks and balances. An important aspect in this network is paracrine communication between cells of different lineages, exemplified by signaling between the FGF-receptor 3 and its ligand FGF18. In mouse fetuses, FGF18 is expressed in osteoblasts precursors in the perichondrium, which surrounds

the cartilage anlage, while FGFR3 is highly expressed in proliferative chondrocytes (Liu *et al.*, 2002).

For proper bone formation it is important that chondrocytes progress through the distinct stages of endochondral bone formation in an orderly, synchronous manner. PTHLH and IHh are two growth factors that play a central role in this process. They are linked in a negative feedback loop involving paracrine signaling between growth plate stem cells and differentiating, early hypertrophic chondrocytes (see Figure 4.16). In this feedback loop IHh is expressed by early hypertrophic chondrocytes and controls the expression of PTHLH by growth plate stem cells. PTHLH, in turn, acts on proliferative chondrocytes and inhibits their differentiation into hypertrophic cells, thereby decreasing IHh expression. This feedback loop critically controls the pace and synchrony of chondrocyte differentiation during endochondral ossification. Whether PTHLH and IHh are also involved in establishing distinct zones in articular cartilage is presently unknown. Like PTHLH, the growth factor TGFβ which is invariantly used in chondrogenic culture media inhibits hypertrophic chondrocyte differentiation and simultaneously stimulates the production of a chondrogenic extracellular matrix. Another growth factor frequently used in chondrogenic culture media is basic FGF (FGF2). This growth factor is also a potent inhibitor of chondrocyte proliferation. TGFβ

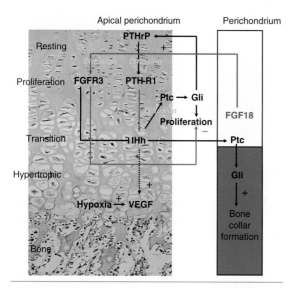

Figure 4.16 Regulation of chondrocyte proliferation and differentiation by PTHLH, IHh and FGF18. Chondrocyte proliferation and differentiation is controlled by a large regulatory network of signaling molecules. Some of the signaling molecules involved have a restricted expression pattern: i.e. PTHLH is specifically expressed in the resting zone of the growth plate in which the stem cells reside. IHh is expressed in early hypertrophic chondrocytes in the transition zone of the growth plate and FGF18 is expressed by perichondrial cells. They reach their target cells by diffusion resulting in paracrine communication not only in the cartilage itself but also between cartilage and perichondrial cells. The receptor for PTHLH is PTHR1. This receptor is expressed in proliferating chondroctyes. Activation of this receptor prevents differentiation of chondrocytes into hypertrophic chondrocytes and reduces IHh expression (−). IHh binds to its receptor Ptc. This results in activation of the transcription factor Gli. IHh has three distinct effects in the growth plate: (1) IHh stimulates PTHLH expression in the resting zone of the growth plate (+). This constitutes a negative feedback loop in which IHh stimulates PTHLH expression and PTHLH represses IHh expression. This feedback loop tightly controls the pace of chondrocyte differentiation; (2) IHh stimulates chondrocyte proliferation; and (3) IHh induces via Ptc and Gli osteoblast formation in the

Figure 4.16 (Continued) perichondrium resulting in the formation of the bone collar. IHh, in combination with hypoxia, plays an indirect role in the stimulation of Vascular Endothelial Growth Factor (VEGF) expression in hypertrophic chondrocytes. This factor plays a role in terminal chondrocyte differentiation and the replacement of hypertrophic chondrocytes by bone by attracting bloodvessels from the underlying bone. FGF18 diffuses to FGFR3-expressing chondrocytes in the proliferation zone of the growth plate. Activation of the FGFR3 results in IHh dependent and independent inhibition of chondrocyte proliferation. Diffusion of growth factors is indicated by green arrows. Negative interactions are indicated by red arrows. Positive interactions are indicated by black arrows.

and FGF inhibit chondrocyte proliferation independently of the actions of PTHLH (Minina *et al.*, 2002).

4.5.2 G-protein coupled receptor signaling

PTHLH is a small peptide hormone which binds with high affinity to the type 1 Parathyroid Hormone Receptor (PTHR1, see Figure 4.17). PTHR1 is highly expressed in proliferative chondrocytes. As discussed above, activation of the PTHR1 by PTHLH inhibits chondrocyte differentiation. The PTHR1 is a member of the G-protein coupled receptor (GPCR) superfamily. This family contains many different receptors that play an important role in physiology, including the adrenergic receptor, light and odor receptors and receptors for small peptide hormones, like Vasopressin, Glucagon and Serotonin. All GPCRs have a common structure: They consist of a single amino acid chain containing an N-terminal extracellular domain, a transmembrane domain in which the amino acid chain traverses seven times through the cell membrane, and a cytoplasmic C-terminus. The N-terminus and the extracellular amino acids of the transmembrane domain

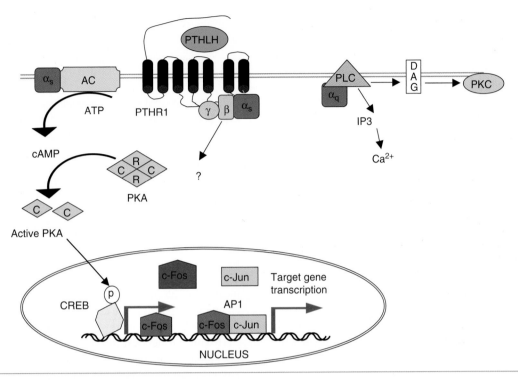

Figure 4.17 G-protein coupled receptor signaling. Binding of the PTHR1 by PTHLH results in the activation of two classes of G protein α subunits. Activation of Gα$_s$ results in the conversion of ATP into cAMP by the membrane bound enzyme adenylate cyclase (AC). cAMP binds to the regulatory subunits of Protein Kinase A (PKA). This releases the catalytic subunits (C) which phosphorylate cAMP Response Element Binding Protein (CREB) resulting in the activation of transcription of target genes. One of the target genes is c-Fos. Together with c-Jun they are part of the AP1 family of transcription factors.

are involved in binding of the ligand. The cytoplasmic amino acids of the transmembrane domain and the C-terminus are involved in signal transduction (Gensure et al., 2005). All G-protein coupled receptors transduce their signal via heterotrimeric G-proteins, consisting of an α, β, and γ-subunit. In the absence of a ligand, the heterotrimeric G-protein is bound to the receptor and is inactive. Binding of a ligand induces a conformational change in the receptor. This results in dissociation of the α-subunit from the β- and γ-subunits. Although the heterodimeric β and γ-subunit have some signaling activity and are involved in turning the activated receptor off, the most important signaling

properties reside in the α-subunit. Three distinct classes of α-subunits can be recognized, namely Gα$_s$, Gα$_q$ and Gα$_I$ (Lefkowitz and Shenoy, 2005; McCudden et al., 2005; Pierce et al., 2002).

Most GPCRs transduce their signal by using 1 class of α-subunits. Only a few receptors, like the PTHR1, can activate 2 classes of α-subunits. Gα$_s$–subunits are involved in the activation of cyclases, which are anchored in the cell membrane. Cyclases convert Nucleotide TriPhosphates (NTP) into cyclic monophosphate nucleotides (cNMP) which act as second messengers. Activation of the PTHR1 results in the release of the Gα$_s$-subunit and the activation of adenylate cyclase, which converts ATP into cAMP. Subsequently, cAMP binds to the regulatory subunit of Protein Kinase A (PKA), thereby relieving its inhibitory actions on the catalytic subunit of PKA. The catalytic form of PKA phosphorylates proteins on serine residues. One of the main targets of PKA is the transcription factor cAMP Response Element Binding Protein (CREBP). This transcription factor is bound in an inactive form to specific DNA sequences present in promoters of a wide variety of genes. PKA activates CREBP by phosphorylation of serine 133, resulting in activation of gene transcription. One of the target genes of CREBP is the transcription factor c-Fos which belongs together with c-Jun to the AP1-family of transcription factors (Swarthout *et al.*, 2002). The cAMP/PKA pathway can be activated by various chemical substances like forskolin, which is an activator of adenylate cyclase, and Isobuthyl Methyl Xanthine (IBMX), which is an inhibitor of cAMP degradation. Both agents are frequently used in tissue engineering, particularly in the formation of adipocytes from mesenchymal progenitor cells.

Gα_q-subunits are responsible for activation of Phospholipase Cβ, leading to the formation of the second messengers diacylglycerol (DAG) and 1,4,5-inositol triphosphate (IP3) by lipolysis of triglycerides present in the cell membrane. DAG in turn activates Protein Kinase C, which like PKA, phosphorylates

proteins on serine residues thereby modulating their activity (Pierce *et al.*, 2002) (McCudden *et al.*, 2005).

4.5.3 Hedgehog signaling

IHh plays a central role in the regulation of endochondral bone formation. It is expressed by early hypertrophic chondrocytes and signals to 3 distinct cell populations: (i) growth plate stem cells where it controls PTHLH expression, (ii) proliferative chondrocytes where it controls chondrocyte proliferation independ from PTHLH expression. (iii) mesenchymal precursors in the perichondrium where it induces osteoblast differentiation (Lai and Mitchell, 2005). Hedgehog proteins or agonists may have potential application in tissue engineering of bone and cartilage (Edwards *et al.*, 2005). A complex pattern of posttranslational modifications results in the formation of an active hedgehog (Ingham, 2001) molecule capable of diffusing through the growth plate (Figure 4.18). These modifications start with the removal of the N-terminal signal peptide, required for protein secretion. Subsequently, the Hh precursor undergoes autocatalytic internal cleavage catalyzed by the C-terminus of the protein. The N-terminal domain of hedgehog (Hh-N) has been shown to account for all known signaling activity, while the role for the C-terminal part (Hh-C) is less clear. During the autocatalytic cleavage, a cholesterol moiety is attached

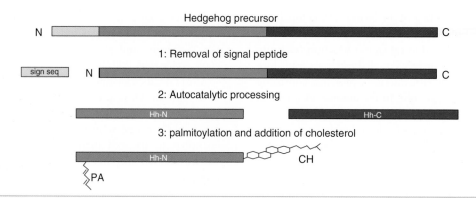

Figure 4.18 Protein processing required for Hedgehog activation. Three sequential steps are needed for the formation of an biological active hedgehog molecule from an inactive precursor. Abbreviations: N, N-terminus; C, C-terminus, sign seq, signal peptide; PA, palmitoleic acid; CH, cholesterol.

to the C-terminal part of Hh-N, which is important for regulation of the spatial distribution of the Hedgehog signal by sequestering the protein to the cell membrane. Thereafter, the protein is palmitoylated, resulting in 30-fold higher biological activity (Ingham, 2001; Nusse, 2003).

Two transmembrane proteins are required to transduce the Hh-signals: the 12-transmembrane protein Patched (Ptc) and the 7-transmembrane receptor Smoothened (Smo, see Figure 4.19). Ptc is required for Hh binding and Smo is required for downstream signaling. Binding of Hh to Ptc alleviates the inhibitory effect of Ptc on Smo. This process is poorly understood and may involve a cytoplasmic inhibitor that is transported over the membrane by Ptc (Nusse, 2003).

Despite the structural similarity with GPCRs, signaling events downstream of Smo do not involve heterotrimeric G-proteins. In the absence of Hh, a protein complex inactivates Gli transcription factors by proteolytic cleavage. This results in the physical separation of the DNA binding domain from the transcription activation domain. Subsequently, the N-terminal fragment which contains the DNA-binding domain but lacks transactivating functions translocates to the nucleus, where it suppresses the expression of Hh target genes. In the presence of Hh, the proteolytic cleavage of Gli is inhibited and a full length Gli protein translocates to the nucleus. This protein contains both a DNA-binding and a transcription-activation domain and thus can activate gene transcription (Nusse, 2003).

4.6 Future developments: Understanding and implementing principles of cellular signaling in tissue engineering

The molecular processes underlying cellular signaling show a high degree of complexity, as demonstrated

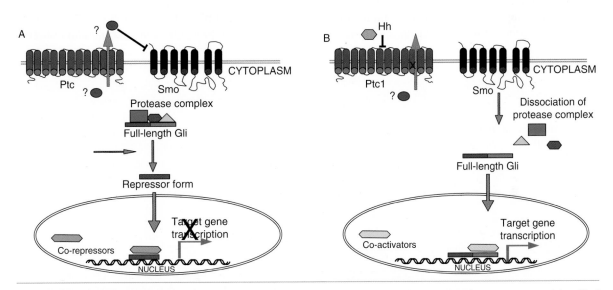

Figure 4.19 The hedgehog signaling pathway: (A) In the absence of hedgehog molecules, an as yet unidentified molecule is transported via Ptc across the membrane. This molecule inhibits Smo from signaling. A protease complex in the cytoplasm cleaves the transcription factor Gli resulting in the physical separation of the DNA binding domain from the transactivation domain. The DNA binding domain translocates to the nucleus where it acts as a repressor of gene transcription; (B) in the presence of hedgehog molecules, the transport of the inhibitor across the membrane by Ptc is prevented and its inhibitory actions on Smo are alleviated. Due to a poorly understand mechanism the protease complex is dissociated. Gli is not cleaved anymore and translocates as an intact transcription factor containing both a DNA and a transactivation domain to the nucleus where it activates gene transcription.

by the content of this chapter. To fully understand the molecular ins and outs of a biological process, countless numbers of proteins have to be considered as candidate players. High throughput analytical methods are increasingly used to simultaneously analyse thousands of molecules, i.e. micro-array technology for gene expression profiling and mass spectrometry for protein analysis. Critical to this development is improved software to interpret the data. Considering the complexity of signal transduction, network modeling of signal transduction will play an increasingly important role in the future. Another challenge for the future lies in the extrapolation of the molecular knowledge to tissue engineering. Many signaling pathways have been delineated using model cell lines, but cells are exposed to a dynamic 3-dimensional environment in tissue engineering. We anticipate that principles of wound repair and developmental biology will be increasingly incorporated. Because the signaling events that play a role in tissue engineering should ideally be studied *in situ*, molecular imaging techniques such as bioluminescent imaging, magnetic resonance imaging and positron

emission topography will be increasingly used to non-invasively image gene expression and enzymatic activity of proteins. These technological developments will increase our knowledge of the molecular basis of tissue engineering. Once critical cellular signaling events have been identified in a certain tissue engineering problem, the next challenge lies in implementation of the knowledge for which a whole toolbox is currently being developed. For instance, ligands involved in tissue regeneration can be incorporated into biomaterials. Extracellular matrix proteins or peptides mimicking their action can be used to coat biomaterials whereas growth factors, cytokines, hormones and small molecules can be released from scaffolds. The same compounds can also be used to control the culture medium of expanding and differentiating stem cells. Furthermore, materials can be designed in such a way that they elicit the desired response. Both the chemical composition and micro-architecture have an influence on the biological behavior of the cells growing on them. An example is discussed in Box 1. Finally, tissue formation can be regulated by systemic administration

---

State of the Art Experiment
Cell signaling control through biomaterials

Some biomaterials are endowed with properties to manipulate surrounding cells through impinging upon their signaling cascades. These bio-active materials are extremely interesting for tissue engineering

A

1024µm²   10000µm²

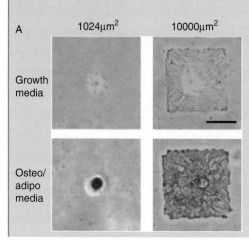

Growth media

Osteo/ adipo media

Size dependent differentiation of hMSCs. Cells grown in adipo/osteogenic medium on a small adhesive patch cannot fully spread and consequently differentiate into adipocytes. In contrast, spreading hMSCs grown on large adhesive patches differentiate into osteoblasts (From McBeath, R., Pirone, D.M., Nelson, C.M., et al. Cell shape, cytoskeletal tension, and RhoA regulate stem cell lineage commitment. Dev Cell, 6(4): 483–495, 2004.)

because they allow the control of tissue formation *in situ*. An instructive example of this is in the paper by McBeath (McBeath *et al.*, 2004) in which human mesenchymal stem cell (hMSC) differentiation is orchestrated through Rho signaling indirectly by controlling the shape of the cells. In the body, several mesenchymal cell types are known, each with a different function and a concomittant shape to support this function. For instance, adipocytes have a rounded morphology for optimal storage of fat droplets whereas muscle cells are elongated for optimal contractile properties. Using different cocktails of growth factors, MSCs can be differentiated into different lineages *in vitro* and it was noted that osteogenic differentiation was favored when cells were seeded at low density, whereas adipogenic differentiation is more efficient at high seeding density. It was hypothesized that cell density influences differentiation through cell shape and to investigate this, MSCs were seeded onto patches of either 1024 or 10000 µm^2 of fibronectin which were printed onto a further non-adhesive surface. In medium allowing both adipogenic and osteogenic differentiation, cells on the small patches were rounded and differentiated into adipocytes, whereas MSCs seeded onto the large patches were spread and differentiated into osteoblasts.

This suggests that not only function dictates shape but also that the shape of the cells dictates its differentiation potential. To verify this, the authors used pharmaceuticals that interfere with the cytoskeletal architecture and thus the shape of the cells. Indeed, when cytoskeletal tension was reduced, adipogenic differentiation was favored. Because the Rho signaling cascade is known to control cytoskeletal

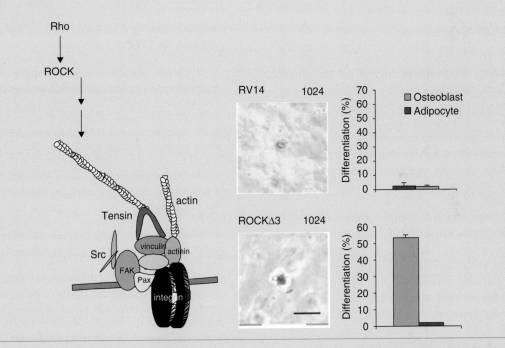

Rho/ROCK signaling influences cell fate determination; Rho and ROCK regulate the cytoskeletal architecture; hMSCs grown on a small patch but transfected with dominant active Rho (RV14) do not differentiate into adipocytes. In contrast, hMSCs transfected with dominant negative ROCK differentiate into osteoblasts.

organization, the authors used a genetic approach to either activate or inhibit RhoA and its downstream target ROCK. By activating Rho/ROCK signaling, osteogenesis was enhanced, whereas inhibition of the pathway stimulated adipogenesis. Overexpression of constitutively active ROCK was able to drive MSC into osteogenesis even when seeded onto small finbronectin patches. The paper demonstrates that new signaling pathways controlling differentiation are still being uncovered. More importantly, it is possible to interfere with these patways by intelligent design of biomaterials. Based on the expanding knowledge on cellular signaling and development of new technologies in biomaterials research, these fields can be expected to further intertwine in the future.

of pharmacological agents. A special promise for the future are so-called small interfering RNAs (siRNAs), which are molecules that can be designed to target the messenger RNAs of any desired gene. The efficacy of systemically administered siRNA has recently been demonstrated (Soutschek *et al.*, 2004) and might also be applied to control the growth and differentiation of tissue engineered grafts.

4.7 Summary

1. Most cell signaling events follow the three-step paradigm of signal initiation, signal transduction and gene activation.
2. Manipulation of signal transduction for tissue engineering purposes should be considered in its *in vivo* context with phenomena such as cell type and crosstalk with other signaling mechanisms.
3. Intricate signaling crosstalk between different cell types exists in skin, bone, cartilage and vascular biology.
4. Signaling pathways are regulated at many different levels, i.e. extracellularly, in the cytoplasm and in the nucleus. This provides engineers multiple means to manipulate the cell appropriately.
5. The intricate details of many signaling pathways have yet to be discovered and may provide new tools for tissue engineering in the future.
6. Control of cell differentiation can be achieved through modifying the cell shape.
7. Basic research on the BMP signal transduction pathway lay the foundation for the discovery of BMPs and their subsequent successful introduction in the clinic.

8. High-throughput phenotypical screening plays an important role in the application of cell biology in pharmaceutical sciences and is likely to become similarly important in tissue engineering.

References

Ashcroft, G.S., Yang, X., Glick, A.B., Weinstein, M., Letterio, J.L., *et al.* (1999). Mice lacking Smad3 show accelerated wound healing and an impaired local inflammatory response. *Nat Cell Biol*, 1: 260–266.

Ballestrem, C., Wehrle-Haller, B., Hinz, B. and Imhof, B.A. (2000). Actin-dependent lamellipodia formation and microtubule-dependent tail retraction control-directed cell migration. *Mol Biol Cell*, 11: 2999–3012.

Canalis, E., Economides, A.N. and Gazzerro, E. (2003). Bone morphogenetic proteins, their antagonists, and the skeleton. *Endocr Rev*, 24: 218–235.

Cross, M.J. and Claesson-Welsh, L. (2001). FGF and VEGF function in angiogenesis: signaling pathways, biological responses and therapeutic inhibition. *Trends Pharmacol Sci*, 22: 201–207.

Day, T.F., Guo, X., Garrett-Beal, L. and Yang, Y. (2005). Wnt/beta-catenin signaling in mesenchymal progenitors controls osteoblast and chondrocyte differentiation during vertebrate skeletogenesis. *Dev Cell*, 8: 739–750.

Dickson, M.C., Martin, J.S., Cousins, F.M., Kulkarni, A.B., Karlsson, S. and Akhurst, R.J. (1995). Defective haematopoiesis and vasculogenesis in transforming growth factor-beta 1 knock out mice. *Development*, 121: 1845–1854.

Dusso, A.S., Brown, A.J. and Slatopolsky, E. (2005). Vitamin D. *Am J Physiol Renal Physiol*, 289: F8–28.

Edwards, P.C., Ruggiero, S., Fantasia, J., Burakoff, R., Moorji, S.M., *et al.* (2005). Sonic hedgehog gene-enhanced tissue engineering for bone regeneration. *Gene Ther*, 12: 75–86.

Etienne-Manneville, S. and Hall, A. (2002). Rho GTPases in cell biology. *Nature*, 420: 629–635.

Folkman, J. and D'Amore, P.A. (1996). Blood vessel formation: what is its molecular basis? *Cell*, 87: 1153–1155.

Gensure, R.C., Gardella, T.J. and Juppner, H. (2005). Parathyroid hormone and parathyroid hormone-related peptide and their receptors. *Biochem Biophys Res Commun*, 328: 666–678.

Hall, A. (1998). Rho GTPases and the actin cytoskeleton. *Science*, 279: 509–514.

Hemler, M.E. (1999). Integrins. In *Guidebook to the extracellular matrix, anchor and adhesion proteins* (Kreis,, T. and Vale, R., eds). New York: Oxford University Press, pp. 196–212.

Hill, T.P., Spater, D., Taketo, M.M., Birchmeier, W. and Hartmann, C. (2005). Canonical Wnt/beta-catenin signaling prevents osteoblasts from differentiating into chondrocytes. *Dev Cell*, 8: 727–738.

Hirschi, K.K., Rohovsky, S.A. and D'Amore, P.A. (1998). PDGF, TGF-beta and heterotypic cell-cell interactions mediate endothelial cell-induced recruitment of 10T1/2 cells and their differentiation to a smooth muscle fate. *J Cell Biol*, 141: 805–814.

Ingham, P.W. (2001). Hedgehog signaling: a tale of two lipids. *Science*, 294: 1879–1881.

Karaplis, A.C. and Deckelbaum, R.A. (1998). Role of PTHrP and PTH-1 receptor in endochondral bone development. *Front Biosci*, 3: 795–803.

Kawano, Y. and Kypta, R. (2003). Secreted antagonists of the Wnt signaling pathway. *J Cell Sci*, 116: 2627–2634.

Kiosses, W.B., Shattil, S.J., Pampori, N. and Schwartz, M.A. (2001). Rac recruits high-affinity integrin alphavbeta3 to lamellipodia in endothelial cell migration. *Nat Cell Biol*, 3: 316–320.

Kolch, W. (2005). Coordinating ERK/MAPK signaling through scaffolds and inhibitors. *Nat Rev Mol Cell Biol*, 6: 827–837.

Kozma, R., Ahmed, S., Best, A. and Lim, L. (1995). The Ras-related protein Cdc42Hs and bradykinin promote formation of peripheral actin microspikes and filopodia in Swiss 3T3 fibroblasts. *Mol Cell Biol*, 15: 1942–1952.

Kronenberg, H.M. (2003). Developmental regulation of the growth plate. *Nature*, 423: 332–336.

Kunsch, C. and Rosen, C.A. (1993). NF-kappa B subunit-specific regulation of the interleukin-8 promoter. *Mol Cell Biol*, 13: 6137–6146.

Lai, L.P. and Mitchell, J. (2005). Indian hedgehog: its roles and regulation in endochondral bone development. *J Cell Biochem*, 96: 1163–1173.

Lefkowitz, R.J. and Shenoy, S.K. (2005). Transduction of receptor signals by beta-arrestins. *Science*, 308: 512–517.

Li, Z., Zhang, G., Feil, R., Han, J. and Du, X. (2006). Sequential activation of p38 and ERK pathways by cGMP-dependent protein kinase leading to activation of the platelet integrin alphaIIb beta3. *Blood*, 107: 965–972.

Lin, R. and White, J.H. (2004). The pleiotropic actions of vitamin D. *Bioessays*, 26: 21–28.

Lindahl, P., Johansson, B.R., Leveen, P. and Betsholtz, C. (1997). Pericyte loss and microaneurysm formation in PDGF-B-deficient mice. *Science*, 277: 242–245.

Liu, Z., Xu, J., Colvin, J.S. and Ornitz, D.M. (2002). Coordination of chondrogenesis and osteogenesis by fibroblast growth factor 18. *Genes Dev*, 16: 859–869.

Maekawa, M., Ishizaki, T., Boku, S., Watanabe, N., Fujita, A., *et al.* (1999). Signaling from Rho to the actin cytoskeleton through protein kinases ROCK and LIM-kinase. *Science*, 285: 895–898.

Mainiero, F., Murgia, C., Wary, K.K., Curatola, A.M., Pepe, A., *et al.* (1997). The coupling of alpha6beta4 integrin to Ras-MAP kinase pathways mediated by Shc controls keratinocyte proliferation. *Embo J*, 16: 2365–2375.

Martel-Pelletier, J. (2004). Pathophysiology of osteoarthritis. *Osteoarthritis Cartilage*, 12(Suppl A): S31–S33.

Martin, P. (1997). Wound healing - aiming for perfect skin regeneration. *Science*, 276: 75–81.

McBeath, R., Pirone, D.M., Nelson, C.M., Bhadriraju, K. and Chen, C.S. (2004). Cell shape, cytoskeletal tension and RhoA regulate stem cell lineage commitment. *Dev Cell*, 6: 483–495.

McCudden, C.R., Hains, M.D., Kimple, R.J., Siderovski, D.P. and Willard, F.S. (2005). G-protein signaling: back to the future. *Cell Mol Life Sci*, 62: 551–577.

Mercurio, A.M., Rabinovitz, I. and Shaw, L.M. (2001). The alpha 6 beta 4 integrin and epithelial cell migration. *Curr Opin Cell Biol*, 13: 541–545.

Miller, J.R. (2002). The Wnts. *Genome Biol*, 3. Reviews3001

Minina, E., Kreschel, C., Naski, M.C., Ornitz, D.M. and Vortkamp, A. (2002). Interaction of FGF, Ihh/Pthlh and BMP signaling integrates chondrocyte proliferation and hypertrophic differentiation. *Dev Cell*, 3: 439–449.

Nobes, C.D. and Hall, A. (1995). Rho, rac, and cdc42 GTPases regulate the assembly of multimolecular focal complexes associated with actin stress fibers, lamellipodia, and filopodia. *Cell*, 81: 53–62.

Nobes, C.D. and Hall, A. (1999). Rho GTPases control polarity, protrusion, and adhesion during cell movement. *J Cell Biol*, 144: 1235–1244.

Noma, K., Oyama, N. and Liao, J.K. (2006). Physiological role of ROCKs in the cardiovascular system. *Am J Physiol Cell Physiol*, 290: C661–668.

Nusse, R. (2003). Wnts and Hedgehogs: lipid-modified proteins and similarities in signaling mechanisms at the cell surface. *Development*, 130: 5297–5305.

O'Connor, K.L., Nguyen, B.K. and Mercurio, A.M. (2000). RhoA function in lamellae formation and migration is regulated by the alpha6beta4 integrin and cAMP metabolism. *J Cell Biol*, 148: 253–258.

Ornitz, D.M. (2005). FGF signaling in the developing endochondral skeleton. *Cytokine Growth Factor Rev*, 16: 205–213.

Parsons, J.T. and Parsons, S.J. (1997). Src family protein tyrosine kinases: cooperating with growth factor and adhesion signaling pathways. *Curr Opin Cell Biol*, 9: 187–192.

Pierce, K.L., Premont, R.T. and Lefkowitz, R.J. (2002). Seven-transmembrane receptors. *Nat Rev Mol Cell Biol*, 3: 639–650.

Qin, J., Vinogradova, O. and Plow, E.F. (2004). Integrin bidirectional signaling: a molecular view. *PLoS Biol*, 2: e169.

Ridley, A.J. and Hall, A. (1992). The small GTP-binding protein rho regulates the assembly of focal adhesions and actin stress fibers in response to growth factors. *Cell*, 70: 389–399.

Ridley, A.J., Paterson, H.F., Johnston, C.L., Diekmann, D. and Hall, A. (1992). The small GTP-binding protein rac regulates growth factor-induced membrane ruffling. *Cell*, 70: 401–410.

Sampath, T.K. and Reddi, A.H. (1981). Dissociative extraction and reconstitution of extracellular matrix components involved in local bone differentiation. *Proc Natl Acad Sci U S A*, 78: 7599–7603.

Sampath, T.K. and Reddi, A.H. (1983). Homology of bone-inductive proteins from human, monkey, bovine, and rat extracellular matrix. *Proc Natl Acad Sci USA*, 80: 6591–6595.

Sano, S., Itami, S., Takeda, K., Tarutani, M., Yamaguchi, Y., *et al.* (1999). Keratinocyte-specific ablation of Stat3 exhibits impaired skin remodeling, but does not affect skin morphogenesis. *Embo J*, 18: 4657–4668.

Shimizu, H., Mitomo, K., Watanabe, T., Okamoto, S. and Yamamoto, K. (1990). Involvement of a NF-kappa B-like transcription factor in the activation of the interleukin-6 gene by inflammatory lymphokines. *Mol Cell Biol*, 10: 561–568.

Soutschek, J., Akinc, A., Bramlage, B., Charisse, K., Constien, R., *et al.* (2004). Therapeutic silencing of an endogenous gene by systemic administration of modified siRNAs. *Nature*, 432: 173–178.

Sutton, A.L. and MacDonald, P.N. (2003). Vitamin D: more than a "bone-a-fide" hormone. *Mol Endocrinol*, 17: 777–791.

Swarthout, J.T., D'Alonzo, R.C., Selvamurugan, N. and Partridge, N.C. (2002). Parathyroid hormone-dependent signaling pathways regulating genes in bone cells. *Gene*, 282: 1–17.

Tarnawski, A. (2000). Molecular mechanisms of ulcer healing. *Drug News Perspect*, 13: 158–168.

Tsumaki, N. and Yoshikawa, H. (2005). The role of bone morphogenetic proteins in endochondral bone formation. *Cytokine Growth Factor Rev*, 16: 279–285.

Urist, M.R. (1965). Bone: formation by autoinduction. *Science*, 150: 893–899.

Urist, M.R., Iwata, H., Ceccotti, P.L., Dorfman, R.L., Boyd, S.D., McDowell, R.M. and Chien, C. (1973). Bone morphogenesis in implants of insoluble bone gelatin. *Proc Natl Acad Sci USA*, 70: 3511–3515.

Urist, M.R. and Strates, B.S. (1971). Bone morphogenetic protein. *J Dent Res*, 50: 1392–1406.

van Bezooijen, R.L., Heldin, C.-H. and ten Dijke, P. (2006). Bone morphogenetic proteins and their receptors. *Encyclopedia of Life Sciences* (published online 27 January 2006).

van der Eerden, B.C., Karperien, M. and Wit, J.M. (2003). Systemic and local regulation of the growth plate. *Endocr Rev*, 24: 782–801.

Wang, E.A., Rosen, V., Cordes, P., Hewick, R.M., Kriz, M.J., *et al.* (1988). Purification and characterization of other distinct bone-inducing factors. *Proc Natl Acad Sci USA*, 85: 9484–9488.

Watanabe, D., Takagi, H., Suzuma, K., Suzuma, I., Oh, H., *et al.* (2004). Transcription factor Ets-1 mediates ischemia- and vascular endothelial growth factor-dependent retinal neovascularization. *Am J Pathol*, 164: 1827–1835.

Witte, M.B. and Barbul, A. (1997). General principles of wound healing. *Surg Clin North Am*, 77: 509–528.

Wozney, J.M., Rosen, V., Celeste, A.J., Mitsock, L.M., Whitters, M.J., Kriz, R.W., *et al.* (1988). Novel regulators of bone formation: molecular clones and activities. *Science*, 242: 1528–1534.

Chapter 5

The extracellular matrix as a biologic scaffold for tissue engineering

Stephen Badylak, Thomas Gilbert and Julie Myers-Irvin

Chapter objectives:

- To understand the composition and organization of extracellular matrix (ECM) in tissues and organs
- To recognize the utility of scaffolds derived from ECM in the field of tissue engineering
- To comprehend the importance of mechanical behavior and mechanical design in the use of ECM scaffolds for tissue repair
- To identify the process by which ECM scaffolds are produced
- To understand some of the mechanisms by which ECM scaffolds promote the formation of functional, site-specific host tissue
- To recognize the rate of degradation of ECM scaffolds *in vivo* and the importance of scaffold degradation on the remodeling process

"An interdisciplinary field that applies the principles of engineering and life sciences toward the development of biological substitutes that restore, maintain, or improve tissue function or a whole organ. Excerpt from a review by Langer and Vancanti (1993) that introduced the concept of tissue engineering to a wide audience"

(Science 260, 920–6; 1993).

5.1 Introduction

Scaffolds to support the constructive remodeling of injured or missing tissues or organs can be composed of synthetic or naturally occurring materials. Such scaffolds can be degradable or non-degradable, and these scaffolds can be engineered to have specific mechanical and material properties that closely approximate those of the tissue to be replaced. Ultimately, the scaffold should facilitate the attachment, migration, proliferation, differentiation and three-dimensional spatial organization of the cell population required for structural and functional replacement of the target organ or tissue. This process has been referred to as functional tissue engineering (Caplan, 2003).

Any scaffold material selected for a tissue engineering application will be subjected to in vivo remodeling. Remodeling in the present context refers to the *in vivo* host response to the scaffold material including the number and type of cells that interact with the scaffold, the degradation and replacement of the scaffold by new host tissue, and the organization and differentiation of the new host tissue in relationship to surrounding structures. The remodeling process is markedly influenced by the microenvironment of each tissue including factors such as blood supply, pH, O_2 and CO_2 concentration, mechanical stressors and the host:scaffold interface. These microenvironmental factors will eventually determine the fate of the scaffold material and the type of remodeling that will occur; i.e., constructive remodeling, destructive remodeling, scarring or something between these extremes. Scarring represents the default mechanism by which adult mammals heal and replace injured or missing tissues. Destructive remodeling refers to chronic inflammation with necrosis and the deposition of dense

fibrous connective tissue. Constructive remodeling refers to the process of scaffold degradation, cellular infiltration, vascularization, and the differentiation and three-dimensional organization of site appropriate tissues.

A naturally occurring scaffold material that has been commonly used in tissue engineering applications is the extracellular matrix (ECM). ECM has been harvested and utilized as a scaffold for numerous tissue engineering applications including repair of musculotendinous structures (Lai *et al.*, 2003; Schlamberg *et al.*, 2004), lower urinary tract reconstruction (le Roux, 2005; Mantovani *et al.*, 2003; Mertsching *et al.*, 2005; van Amerongen *et al.*, 2006), esophageal replacement (Badylak *et al.*, 2000; Badylak *et al.*, 2005; Lopes *et al.*, 2006), skin (chronic wound) healing (Demling *et al.*, 2004), and upper-airway reconstruction (Huber *et al.*, 2003), among others. Decellularized xenogeneic ECM is well tolerated by human hosts because components of the ECM are generally well conserved among species. The decellularization step is an important part of the processing of ECM as xenogeneic and allogeneic cellular antigens are recognized as foreign by the host and can trigger an inflammatory response or tissue rejection (Badylak *et al.*, 2001). The decellularization and preparation process will be discussed later in the chapter.

ECM scaffolds can be seeded with site specific cells, usually autologous cells from the intended recipient before utilization for tissue repair. The requirement for cells to promote a constructive remodeling response scaffold varies depending on the intended application and mode of preparation of the scaffold. If cells are to be seeded onto a matrix, it is necessary to find an autologous cell source and a controlled environment must be available to ensure cell viability when translocating the seeded scaffold to the host recipient. Therefore, by seeding the ECM with cells, the scaffold is no longer available for use as an off-the-shelf product. However, certain applications, such as small diameter tubular structures, may require a cellular component in order to achieve clinical utility (Badylak, 2004).

When implanted into the body, one must consider the *in vivo* physiological stresses and strains that will be placed on the scaffold and affect the remodeling process. It is necessary to manufacture scaffolds that have mechanical properties similar to those of native tissues. Unmodified ECM scaffolds also degrade over time (an important property that will be discussed in the following biologic properties section), changing the strength and mechanical properties of the scaffold. It is therefore important to understand the biomechanical properties of the ECM and some experiments used to do so will be discussed further. Various individual components of the ECM such as collagen and hyaluronic acid have been investigated as scaffold materials with varying degrees of success (Hahn *et al.*, 2006; Lisignoli *et al.*, 2006). The host response to an unmodified ECM is different than that of purified components of the ECM. For example, VEGF will induce angiogenesis and BMP will induce bone formation, but the amount, duration and appropriateness of the response is difficult to control. When ECM is used in its entirety, the response to each component tends to be much more in concert with site appropriate tissue remodeling. The advantages and disadvantages of using individual components will not be further discussed herein. Rather, the present chapter will focus upon the composition, preparation, architecture, mechanical and material properties, and *in vivo* utility of the ECM as a scaffold for tissue engineering applications.

5.2 Extracellular matrix

5.2.1 Definition

The extracellular matrix represents the secreted product of resident cells within each tissue and organ and is composed of a mixture of structural and functional proteins arranged in a unique, tissue specific three-dimensional ultrastructure. These molecules, mainly proteins, provide the mechanical strength required for proper function of each tissue and serve as a conduit for information exchange (i.e. signaling) between adjacent cells and between cells and the ECM itself. The ECM is in a state of dynamic reciprocity (Bissell and Aggeler, 1987) and will change in response to environmental cues such as hypoxia and mechanical loading. In the context of tissue engineering applications therefore, use of the ECM as a scaffold indeed provides structural support, but perhaps more importantly provides a favorable environment for constructive remodeling of tissue and organs.

5.2.2 Composition

The extracellular matrix represents a mixture of structural and functional molecules organized in a three-dimensional architecture that is unique to each tissue. Most of these molecules are well-recognized and they form a complex mixture of proteins, glycosaminoglycans, glycoproteins and small molecules arranged in a unique, tissue specific three-dimensional architecture (Laurie *et al.*, 1989; Baldwin, 1996; Martins-Green and Bissel, 1995). The logical division of the ECM into structural and functional components is not possible because many of these molecules have both structural and functional roles in health and disease. For example, both collagen and fibronectin, molecules that once were considered to exist purely for their 'structural' properties, are now known to have a variety of 'functional' moieties with properties ranging from cell adhesion and motility to promotion of or inhibition of angiogenesis. These 'bimodal' or multifunctional molecules provide a hint of the diverse occult amino acid sequences that exist within certain parent molecules and which, in themselves, harbor biologic activity.

5.2.2.1 *Collagen*
Collagen is the most abundant protein within the mammalian ECM. Greater than 90% of the dry

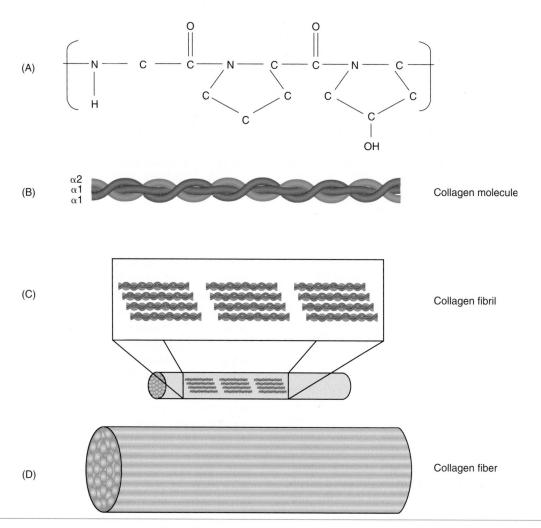

Figure 5.1 Schematic representation of the assembly of collagen I. A, Individual collagen polypeptide chains. They polypeptides consist of the repeating sequence Gly-X-Y, where X is often proline and Y is often hydroxyproline; B, Collagen is made up of three polypeptide strands which are all left-handed helixes and twist together to form a right handed coiled coil. The polypeptides strands are synthesized as precursor chains with propeptides (globular extensions) on the C and N ends. The propeptides are cleaved into short non-helical telopeptides; C, Collagen molecules self assemble into collagen fibrils; D, Collagen fibers are formed by end to end and lateral assembly of collagen fibrils, resulting in a regular banding pattern that is characteristic of collagen.

weight of the ECM from most tissues and organs is represented by collagen (van der Rest and Garrone, 1992). More than twenty distinct types of collagen have been identified, each with a unique biologic function. Type I collagen (Figure 5.1) is the major structural protein present in tissues and is ubiquitous within both the animal and plant kingdoms. Type I collagen is abundant in tendinous and ligamentous structures and provides the necessary strength to accommodate the uniaxial and multiaxial mechanical loading to which these tissues are commonly subjected. These same tissues provide a convenient source of collagen for many medical device applications. Bovine Type I collagen is harvested from Achilles tendon and is perhaps the most commonly used xenogeneic ECM component intended for therapeutic applications.

Other collagen types exist in the ECM of most tissues but typically in much lower quantities. These alternative collagen types provide distinct mechanical and physical properties to the ECM and simultaneously contribute to the population of ligands that interact with the resident cell populations. For example, Type IV collagen is present within the basement membrane of most vascular structures and within tissues that contain an epithelial cell component (Barnard and Gathercole, 1991; Piez, 1984; Yurchenco et al., 1994). The ligand affinity of Type IV collagen for endothelial cells is the reason for its use as a biocompatible coating for medical devices intended to have a blood interface. Type III collagen is found within the submucosal tissue of selected organs such as the urinary bladder; a location in which tissue flexibility and compliance are required for appropriate function as opposed to the more rigid properties required of a tendon or ligament supplied by Type I collagen (Piez, 1984). Type VI collagen is a relatively small molecule that serves as a connecting unit between glycosaminoglycans and larger structural proteins such as Type I collagen, thus providing a gel like consistency to the ECM (Yurchenco et al., 1994). Type VII collagen is found within the basement membrane of the epidermis and functions as an anchoring fibril to protect the overlying keratinocytes

from sheer stresses (Yurchenco et al., 1994). Each type of collagen is of course the result of specific gene expression patterns as cells differentiate and tissues and organs develop and spatially organize (van der Rest and Garrone, 1992; Yurchenco et al., 1994). In nature, collagen is intimately associated with glycosylated proteins, growth factors and other structural proteins such as elastin and laminin to provide unique tissue properties (Yurchenco, 1994). Each of these types of collagen exist within most of the ECM scaffolds that are used for the constructive remodeling of tissues.

5.2.2.2 Fibronectin

Fibronectin is second only to collagen in quantity within the ECM. Fibronectin is a dimeric molecule of 250,000 MW subunits and exists both in soluble and tissue isoforms and possesses ligands for adhesion of many cell types (McPherson and Badylak, 1998; Miyamoto et al., 1998; Schwarzbauer, 1991; 1999) (Figure 5.2). The ECM of submucosal structures, basement membranes, and interstitial tissues all contain abundant fibronectin (McPherson and Badylak, 1998; Schwarzbauer, 1999). The cell friendly characteristics of this protein have made it an attractive substrate for *in vitro* cell culture and for use as a coating for synthetic scaffold materials to promote host biocompatibility. Fibronectin is rich in the Arg-Gly-Asp (RGD) subunit; a tripeptide that is

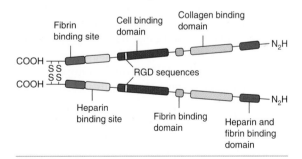

Figure 5.2 Fibronectin is a dimeric molecule joined by two dilsulfide bonds at the carboxy end. Each domain of the fibronectin molecule has binding sites for cell receptors and ECM molecules.

important in cell adhesion via the $\alpha_5\beta_1$ integrin (Yurchenco, 1994). Fibronectin is found at an early stage within the ECM of developing embryos and is critical for normal biologic development, especially the development of vascular structures. Fibronectin was the first 'structural' molecule identified to have a functional motif.

5.2.2.3 Laminin

Laminin is a complex adhesion protein found in the ECM, especially within basement membrane ECMs (Schwarzbauer, 1999). This protein plays an important role in early embryonic development and is perhaps the best studied of the ECM proteins found within embryonic bodies (Li *et al.*, 2002). This trimeric cross-linked polypeptide exists in numerous forms dependent upon the particular mixture of peptide chains (e.g., $\alpha 1$, $\beta 1$, $\gamma 1$) (Timpl, 1996; Timpl and Brown, 1996) (Figure 5.3). The prominent role of laminin in the formation and maintenance of vascular structures is particularly noteworthy when considering the ECM as a scaffold for tissue reconstruction (Ponce *et al.*, 1999; Werb *et al.*, 1999). The crucial role of the β-1 integrin chain in mediating hematopoietic stem cell interactions with fibronectin and laminin has been firmly established (Ponce *et al.*, 1999; Werb *et al.*, 1999). Loss of the β-1 integrin receptors in mice results in intrapartum mortality. This protein appears to be among the first and most critical ECM factors in the process of cell and tissue differentiation. The specific role of laminin in tissue reconstruction when ECM is used as a scaffold for tissue and organ reconstruction in adults is unclear but its importance in developmental biology suggests that this molecule is essential for self assembly of cell populations and for organized tissue development as opposed to scar tissue formation.

5.2.2.4 Glycosaminoglycans

The ECM contains a mixture of glycosaminoglycans (GAGs) depending upon the tissue location of the ECM in the host, the age of the host, and the microenvironment. The GAGs bind growth factors and cytokines, promote water retention, and contribute to the gel properties of the ECM. The heparin binding

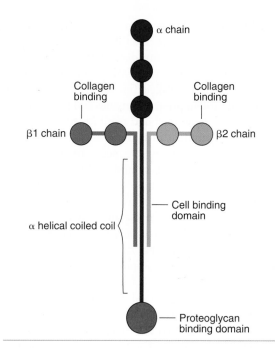

Figure 5.3 Laminin is composed of three polypeptide chains (α, $\beta 1$, and $\beta 2$) organized into the shape of a cross. Laminin has binding domains for heparin, collagen IV, heparin sulfate, and cells.

properties of numerous cell surface receptors and of many growth factors (e.g., fibroblast growth factor family, vascular endothelial cell growth factor) make the heparin-rich GAGs important components of naturally occurring substrates for cell growth. The glycosaminoglycans present in ECM include chondroitin sulfates A and B (Figure 5.4A,B), heparin (Figure 5.5A), heparan sulfate (Figure 5.5B), and hyaluronic acid (Entwistle *et al.*, 1995; Hodde *et al.*, 1996) (Figure 5.6). Hyaluronic acid has been most extensively investigated as a scaffold for tissue reconstruction and as a carrier for selected cell populations in therapeutic tissue engineering applications. The concentration of hyaluronic acid within ECM is highest in fetal and newborn tissues and tends therefore to be associated with desirable healing properties. The specific role, if any, of this GAG upon

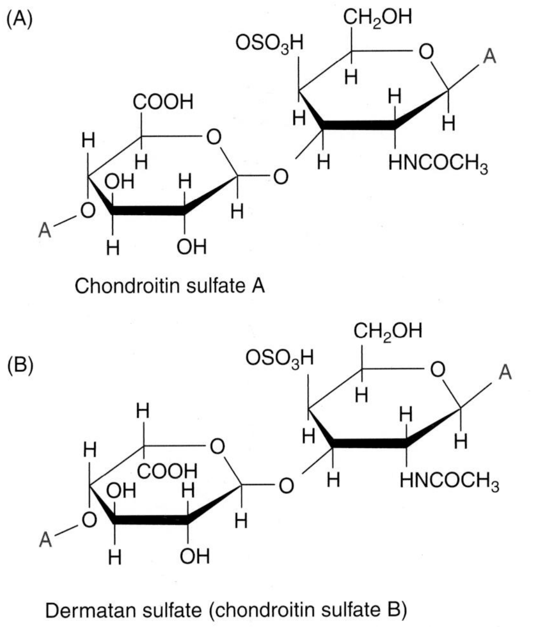

(A)

Chondroitin sulfate A

(B)

Dermatan sulfate (chondroitin sulfate B)

Figure 5.4 Glycosaminoglycans are long unbranched polysaccharides that are composed of a repeating disaccharide unit. The disaccharide unit is composed of one of two modified sugars, either N-acetylgalactosamine or N-acetylglucosamine. A, Chondroitin sulfate A is a major structural component of cartilage whose function depends on the properties of the proteoglycan of which it is part. Dermatan sulfate (Chrondoritin sulfate B) is found mostly in skin and plays a role in wound repair, infection, and coagulation among others.

progenitor cell proliferation and differentiation during adult wound healing is unknown.

5.2.2.5 Growth factors

An important characteristic of the intact ECM that distinguishes it from other scaffolds for tissue reconstruction is its diversity of structural and functional proteins. The bioactive molecules that reside within the ECM and their unique spatial distribution provide a reservoir of biologic signals. Although cytokines and growth factors are present within ECM in very small quantities, they act as potent modulators of cell behavior. The list of growth factors found within ECM is extensive and includes vascular endothelial cell growth factor (VEGF), the fibroblast growth factor (FGF) family, stromal-derived growth factor (SDF-1), epithelial cell growth factor (EGF), transforming growth factor β (TGF-β), keratinocyte growth factor (KGF), hepatocyte growth factor (HGF), platelet derived growth factor (PDGF), and bone morphogenetic protein (BMP) among others (Bonewald, 1999; Kagami et al., 1998; Roberts et al., 1988). These factors tend to exist in multiple isoforms, each with its specific biologic activity. Purified forms of growth factors have been investigated in recent years as therapeutic methods of encouraging blood vessel formation (e.g., VEGF), stimulating deposition of granulation tissue (PDGF), and bone (BMP), and encouraging epithelialization of wounds (KGF). However, this therapeutic approach has struggled because of the difficulty in determining optimal dose and methods of delivery, the ability to sustain and localize the growth factor release at the desired site, and the inability to turn the factor 'on' and 'off' as needed during the course of tissue repair.

An advantage of utilizing the ECM in its native state as a substrate or scaffold for cell growth and differentiation is the presence of all the attendant growth factors (and their inhibitors) in the same relative amounts that exist in nature and perhaps more importantly, in their native three-dimensional ultrastructure. The ECM presents these factors efficiently to resident or migrating cell surface receptors, protects the growth factors from degradation, and modulates their synthesis (Entwistle et al., 1995; Bonewald, 1999; Kagami et al., 1998; Roberts et al., 1988). If one considers the ECM to be a substrate for in vitro and in vivo cell growth, it is reasonable to think of the ECM as a temporary (i.e., degradable) controlled release vehicle for naturally derived growth factors.

5.2.3 Collagen fiber architecture

The biomechanical behavior of an ECM scaffold and the effects of mechanical loading upon the scaffold and

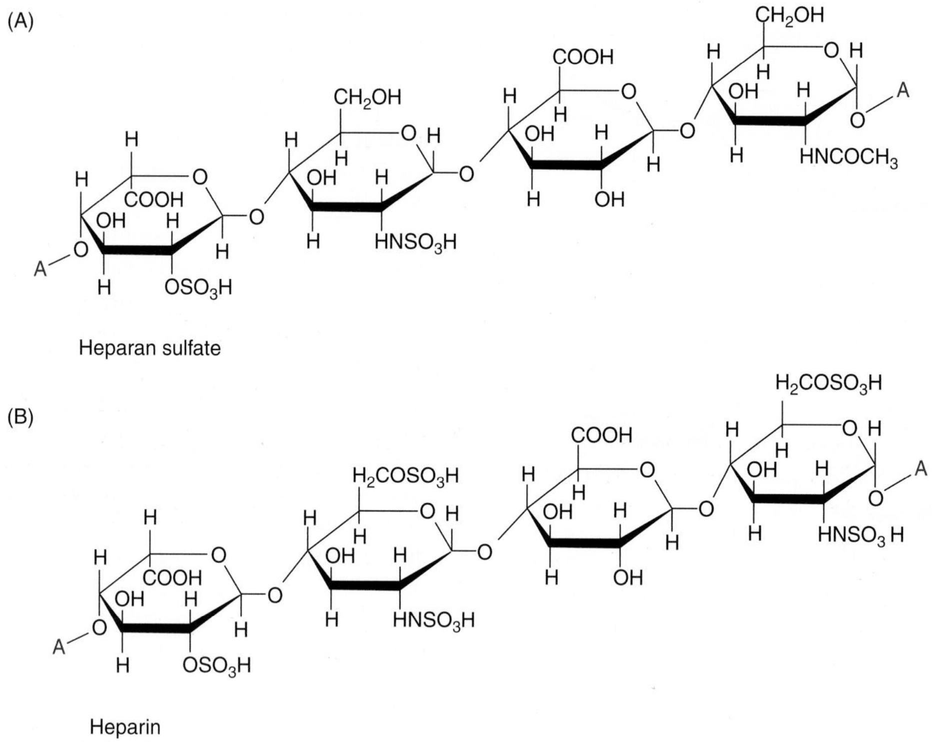

(A)

Heparan sulfate

(B)

Heparin

Figure 5.5 A, Heparan sulfate mediates the interactions between many proteins; B, Heparin, an anticoagulant, has a structure very similar to that of heparan sulfate but is relatively highly sulfated.

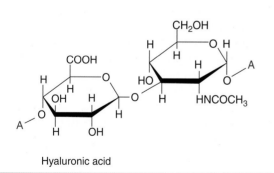

Hyaluronic acid

Figure 5.6 Hyaluronic acid is one of the most extensively studied and the largest glycosaminoglycan, containing between 50 and several thousand disaccharide units.

remodeling process are critically dependent upon the collagen fiber architecture. The alignment and organization of collagen fibers dictate the biomechanical behavior of any tissue, and are highly dependent on the function of the source tissue from which the ECM is derived. For example, it is known that the collagen fibers derived from a ligament or a tendon are highly aligned along the long axis of the tissue to provide the greatest resistance to strain in a load bearing application. Thus, the use of ECM derived from tendons and ligaments is a logical choice for repair of structures such as the anterior cruciate ligament (Cartmell and Dunn, 2000; Harrison and Gratzer, 2005; Woods and Gratzer, 2005). The small intestinal submucosa (SIS), which has been used extensively for tissue engineering applications, also

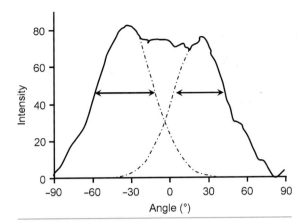

Figure 5.7 The collagen fiber distribution obtained with small angle light scattering (SALS) shows the preferred collagen fiber orientation along the longitudinal axis of the small intestine. The distribution is a composite of two subpopulations of collagen fibers oriented approximately 30° from the preferred fiber direction.

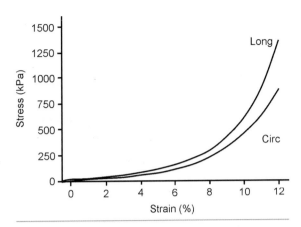

Figure 5.8 Biaxial mechanical behavior of SIS shows the anisotropic behavior with increased modulus and strength in the preferred collagen fiber direction.

has a characteristic collagen fiber organization that is related to its native function. Studies have shown that SIS has a preferred alignment along the longitudinal axis of the small intestine. Using small angle light scattering (SALS) technique (Sacks *et al.*, 1997), it has been shown that this preferred alignment is a composite of two subpopulations of collagen fibers that have alignment approximately 30° from the longitudinal axis of the small intestine (Gloeckner *et al.*, 2000; Sacks and Gloeckner, 1999) (Figure 5.7). This spiral arrangement allows the small intestine to constrict in a manner that promotes the efficient transport of a bolus of biomass. When the SIS-ECM is subjected to biaxial mechanical testing, this preferred fiber orientation leads to an anisotropic biomechanical behavior, with greater strength and tangent modulus along the preferred fiber direction (Sacks and Gloeckner, 1999) (Figure 5.8). Therefore, when SIS is used as an ECM scaffold, certain predictions and expectations can be made regarding its physical and mechanical properties because of the understanding of its collagen fiber architecture. The same principles

can be applied to any scaffold material composed of a tissue derived ECM.

5.2.4 Biomechanical behavior

When ECM scaffolds are used for load bearing applications, such as tendon reconstruction and body wall repair, a single layer sheet of ECM is typically insufficient to accommodate physiologic loads. Multilaminate ECM devices can be designed and manufactured to increase the strength of the scaffolds. Multilayer ECM devices are typically created by mechanically laminating hydrated sheets of ECM under vacuum (Badylak *et al.*, 2005; Freytes *et al.*, 2004). The vacuum-induced compression of the layers of ECM results in the extrusion of water from the tissue and bonds the layers together. The collagen fiber architecture and, ultimately, the biomechanical behavior of each single layer of ECM are design parameters for the arrangement of the multiple layers of ECM in the multilaminate devices for specific load bearing applications.

An example of the need for an ECM scaffold device with significant load bearing capability is repair of the rotator cuff. The rotator cuff has a uniaxial ultimate tensile strength of approximately 1000 N and a complex strain distribution. The Restore™ device, a multilaminate ECM device that is marketed for augmentation of the rotator cuff, consists of 10 layers of SIS-ECM. The 10-layer device is designed and manufactured so that five separate 2-layer bundles of SIS are oriented 72° from the longitudinal axis of the adjacent 2-layer bundles. The result is an isotropic scaffold that has a uniaxial ultimate tensile strength of greater than 1000 N regardless of the direction of loading. Alternatively, when considering repair of the Achilles tendon, an ECM scaffold must be able to withstand some of the greatest forces developed within the body and be subjected to a uniform strain distribution. For this application, a more appropriate design for the SIS-ECM may be to orient all layers along the same axis of the device.

It is important to determine the mechanical behavior of ECM scaffolds under complex loading

regimens that simulate the *in vivo* conditions. For example, body wall and esophageal tissues are subjected to biaxial loading, shear loading, pressure and uneven strain distributions. To address these complex loading environments, several testing methods have been utilized. The ball-burst test subjects the scaffold to biaxial stress in the plane of the scaffold, while applying load in an orthogonal direction to the scaffold. The ball burst testing procedure is an excellent model of body wall loading. A number of studies have utilized this method to examine the mechanical properties of single and multilayered ECM derived scaffolds (Freytes *et al.*, 2004). Pressure volume testing on the other hand is an appropriate test for cylinders or vesicle shaped structures such as the esophagus or urinary bladder (Badylak *et al.*, 2005) (Figures 5.9 and 5.10).

The collagen fiber architecture and kinematics (i.e., the change in fiber architecture when a load is applied) play an important role in defining the mechanical environment to which the cells that eventually populate the ECM scaffold are subjected (Billiar and Sacks, 1997; Gilbert *et al.*, 2006a;

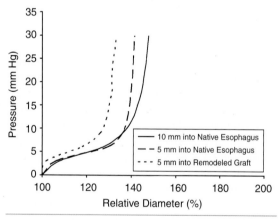

Figure 5.9 The pressure-diameter response curves for remodeled ECM used to repair the canine esophagus and adjacent normal canine esophagus. At 145 days after implantation, the remodeled ECM graft has a similar pressure diameter response as the adjacent normal tissue. The immediately adjacent esophageal tissue appears to have been altered by the presence of the ECM scaffold and the associated remodeling process.

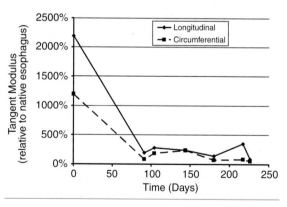

Figure. 5.10 The tangent modulus of the ECM graft during the course of remodeling after placement in full circumferential defect in the esophagus. By 90 days after implantation, the tangent modulus of the graft is very similar to that for native esophageal tissue.

Gloeckner *et al.*, 2000). In turn, the cells play a central role in the remodeling process. Appropriate tension causes fibroblasts to have an elongated, spindle-shaped morphology and the cells are aligned along the direction of load (Bell *et al.*, 1979; Eastwood *et al.*, 1998). As a result, the fibroblasts tend to align the secreted collagen fibers along the lines of stress. Tension also increases collagen synthesis by fibroblasts (Lambert *et al.*, 1992), and that newly secreted collagen tends to be aligned parallel with the orientation of the cell (Wang *et al.*, 2003). These responses of the cells and changes in the new ECM of the remodeled tissue can improve the biomechanical behavior of the scaffold. Conversely, insufficient mechanical loading can increase the synthesis of matrix metalloproteases by cells and result in rapid resorption of a scaffold (Prajapati *et al.*, 2000a, 2002b; Seliktar *et al.*, 2001).

A final consideration for the biomechanical behavior of ECM scaffolds is their degradation rate. It is well established that naturally occurring ECM scaffolds degrade quickly after implantation as long as they are not chemically crosslinked (Badylak *et al.*, 1998; Record *et al.*, 2001). The biologic implications of this rapid degradation will be discussed in more detail later in this chapter. However, the effects of ECM scaffold degradation upon the biomechanical properties of the remodeling scaffold must be understood when considering the design of a scaffold for a given clinical application. For example, an eight layer ECM device was used as a repair scaffold for experimentally created defects in the abdominal wall in a dog model (Badylak *et al.*, 2001). Quantitative analysis showed that the strength of the ECM scaffold decreased by 50% within 10 days of surgical implantation. However, subsequent remodeling resulted in a replacement tissue that eventually possessed twice the strength of the native wall by 24 months. Similarly, when an ECM scaffold was used in a tubular configuration for repair of the esophagus, the material rapidly remodeled such that the pressure diameter response characteristics of the new tissue closely approximated that of the native esophageal tissue within three to four months (Badylak *et al.*,

2005). This data obtained from preclinical studies provides the information needed to engineer a device for clinical application.

In summary, one of the most desirable feature of the use of ECM as a scaffold material for tissue engineering applications is its ability to respond to environmental stressors such as mechanical loading. The remodeling process results in a site specific replacement tissue that has near normal functional and structural characteristics for many clinical applications.

5.3 Preparation of ECM

The preparation of an ECM scaffold derived from native tissues requires a combination of mechanical, physical, and enzymatic processing steps that results in complete decellularization of the source tissue. ECM scaffolds have been prepared from many tissues including small intestine (SIS) (Badylak *et al.*, 1989; 1995; Kropp *et al.*, 1995; Sacks and Gloeckner, 1999), urinary bladder (Chen *et al.*, 1999; Ewalt *et al.*, 1992; Freytes *et al.*, 2004; Gilbert *et al.*, 2005), dermis (Armour *et al.*, 2006; Charge and Rudnicki, 2004), and pericardium (Liang *et al.*, 2004), among others.

Many of the tissues used as a source material for biologic scaffolds are xenogeneic or allogeneic. By definition therefore, the cellular antigens of those intact tissues are recognized as foreign by the host and would induce an adverse inflammatory or an adverse immune mediated response. The proteins of which the ECM is composed, however, are typically conserved across species lines and are well tolerated even by xenogeneic recipients (Bernard *et al.*, 1983a, 1983b; Constantinou and Jimenez, 1991; Exposito *et al.*, 1992). Depending upon the tissue of interest, mechanical methods such as manual or automated delamination of particular layers (e.g. muscle layers) is typically followed by a combination of physical, chemical, and enzymatic methods to achieve complete decellularization (Gilbert *et al.*, 2006b). The ultimate goal of any decellularization process is the removal of all cellular and nuclear material while maintaining the composition, mechanical properties, and biologic activity of the remaining ECM. Physical treatments for decellularization include sonication, massage, or freezing and thawing. Such methods tend to disrupt cell membranes, release the contents of all cells and facilitate the subsequent rinsing and removal of cell contents from the remaining ECM. Enzymatic treatment with agents such as trypsin, and chemical treatment with detergents or ionic solutions further destroy cell membranes and intercellular bonds. Since the ECM of different tissues varies with regard to composition, density and compactness, the combination of methods used to achieve complete decellularization varies among tissues.

Following decellularization, ECM intended for use as a biologic scaffold must be sterilized prior to use as a medical device/implant. Sterilization of biologic materials such as extracellular matrix involves unique considerations such as the shrink temperature of collagen, adverse effects upon any bioactive components, changes in ultrastructure, and changes in surface characteristics. In spite of these considerations, numerous methods for terminally sterilizing biologic scaffolds composed of ECM have been identified. These methods include ionizing radiation such as gamma irradiation and electron beam irradiation, and exposure to ethelyne oxide (ETO) (Freytes and Badylak, 2006) (Figure 5.11).

Some ECM materials are crosslinked by chemicals such as glutaraldehyde, carbodiimide, or photo-oxidizing agents to modify their mechanical, immunogenic, or physical properties. However, such crosslinking

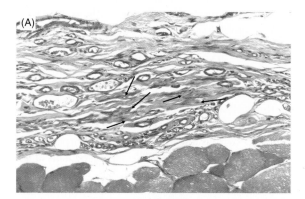

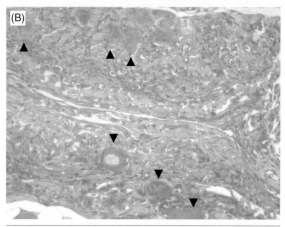

Figure 5.12 Histologic appearance of Restore™ versus CuffPatch™ 16 weeks following implantation as an abdominal wall replacement device. A, The Restore™ device is replaced with vascularized connective tissue and skeletal muscle (arrows); B, The CuffPatch™ device is still present and contains inflammatory cells, including multinucleate giant cells (arrowheads). This figure demonstrates the marked difference in host response to the SIS-ECM (Restore™) when compared to a chemically crosslinked form of SIS-ECM (CuffPatch™).

Figure 5.11 ECM harvested from porcine urinary bladder. This thin (60 μM) sheet of ECM is entirely free of any cellular component, has a multidirectional tensile strength of approximately 40 N, and has not been chemically cross linked.

promotes a foreign body response with fibrous tissue formation, chronic inflammation, and inhibition of both cellular infiltration and scaffold degradation (Badylak, 2002) (Figure 5.12).

Following decellularization, the ECM can be processed into many forms. Depending on the application and intended site of repair, the ECM can be processed into sheets, powder/particulate forms, or shaped into other structures such as tubes.

5.4 Biologic activities of ECM scaffolds

Recent research suggests an important role for biologically active degradation products of the ECM in the healing process. Degradation of the ECM is one of the earliest events following tissue injury and was discussed earlier in relation to its effects upon biomechanical behavior of the scaffolds. However, ECM degradation is a critical and necessary event to

Classical Experiment

Radioactive labeling to track the fate of extracellular matrix scaffolds

Biologic scaffolds composed of ECM are currently being utilized in numerous regenerative medicine applications. Some scaffolds degrade rapidly following implantation and are replaced by host tissue although others are chemically cross-linked to slow degradation. Currently, little is known about the rate of degradation or the fate of the degradation products of the ECM. The present study utilizes ^{14}C labeled ECM to track the degradation products of porcine SIS when used to repair canine bladder wall.

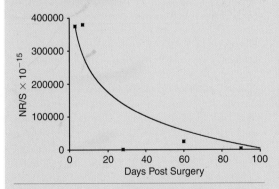

AMS results for remodeled urinary bladder. Normalized ratio of radioactive to stable carbon $(NR/S) \times 10^{-15}$ vs days post surgery. The preoperative devices (n = 3) had an average value of $4,460,000 \pm 275,000$ NR/S $\times 10^{-15}$.

Porcine SIS-ECM was labeled as previously reported (Record et al., 2001) by giving weekly intravenous injection of ^{14}C proline to piglets beginning at 3 weeks of age. Pigs were then sacrificed at 26 weeks of age (i.e., 23 consecutive weeks of injection) and ECM material was prepared from the desired organs. The radioactivity of each prepared ECM material was measured by liquid scintillation counting or accelerator mass spectrometry (AMS).

Following implantation of the labeled scaffold material into the bladder, the scaffold degrades quickly with 40–60% of the scaffold removed from the site of remodeling within 4 weeks. ^{14}C was undetectable by 90 days in the bladder. ^{14}C was however, detected in blood, urine, lung, and liver. Radioactivity was detected in blood and urine almost immediately after repair of the bladder indicating the immediate initiation of the scaffold degradation process. It was only detected in the liver and lung after repair of the bladder and was only detected by AMS (a very sensitive detection method). These results suggest that the primary fate of the ECM is degradation, metabolism, and excretion as urine or exhalation as CO_2.

This study allowed for the determination of the extent of degradation and the fate of degradation products. In the urinary bladder, 40–60% of the ECM is degraded within one month while the remainder of the scaffold is resorbed by 60–90 days and excreted from the body.

realize some important biologic effects. Many growth factors (VEGF, bFGF, etc.) reside as inactive proteins in the intact ECM and can be released as active peptides during degradation. These peptides are released into the wound and can aid in regulating tissue repair. For example, peptides derived from TSP-1 can inhibit angiogenesis and peptides derived from SPARC can stimulate vascular growth (Midwood and Schwarzbauer, 2004). Artificial degradation of the ECM has resulted in the formation of peptides with antimicrobial (Brennan *et al.*, 2006; Sarikaya *et al.*, 2002), chemoattractant (Badylak *et al.*, 2001; Li *et al.*, 2004; Zantop *et al.*, 2006) and angiogenic (Li *et al.*, 2004) properties.

5.4.1 Antimicrobial properties

Naturally occurring antimicrobial peptides are widespread among the animal and plant kingdoms and an online database reports that more than 500 peptides have been identified (Zasloff, 2002). Sarikaya *et al.* have demonstrated that components of ECM scaffolds derived from the small intestinal submucosa (SIS) and urinary bladder submucosa (UBS) from pig contain antibacterial activity against both gram positive and gram negative bacteria (Sarikaya *et al.*, 2002). To examine antimicrobial activity, porcine SIS and UBS was isolated and degraded using high heat and mild acid. The degradation products were mixed with either *E. coli* or *S. aureus* and the growth of bacteria was examined over the course of 24 hours. Both SIS and UBS degradation products demonstrated a strong antibacterial effect against *E. coli* and *S. aureus* for 24 hours (Sarikaya *et al.*, 2002).

Brennen *et al.* (2006) examined the antibacterial activity of ECM scaffolds derived from additional tissues, and further fractionated the ECM degradation products to examine the resultant activity. The ECM materials utilized for these experiments were composed of porcine liver tissue (L-ECM) as well as superficial layers (tunica propria and basement membrane-UBM) of the porcine urinary bladder. Degradation products were produced *in vitro* by the

same methods mentioned above. Products from both UBM and L-ECM demonstrated antibacterial activity against both *S. aureus* and *E. coli*. Fractionation of the degraded ECM resulted in samples with varying antibacterial activity (Figure 5.13).

Interestingly, this antibacterial activity is associated only with degradation products of the ECM (Sarikaya *et al.*, 2002) and not with intact ECM (Holtom *et al.*, 2004). Holtom *et al.* (2004) examined the growth of various bacteria in the presence of a disk of intact SIS. No inhibition of bacterial growth was observed, again supporting the concept that degradation of the ECM is necessary for release of biologically active peptides.

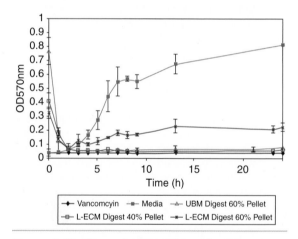

Figure 5.13 Effect of UBM-ECM and L-ECM digest ammonium sulfate fractions on *S. aureus* growth. This graph shows the antibacterial effects against clinical strains of *Staphylococcus aureus* utilizing the degradation products resulting from the acid digestion of ECM scaffolds. The ECM scaffolds were derived from porcine urinary bladder (UBM-ECM) and liver (L-ECM). These biologic scaffolds were digested with acid at high temperatures, fractionated by ammonium sulfate precipitation, and tested for antibacterial activity in a standardized *in vitro* assay. Degradation products from both UBM-ECM and L-ECM demonstrated antibacterial activity against both *S. aureus*. The specific ammonium sulfate fractions that showed antimicrobial activity varied for the two different ECM scaffold types.

State of the Art Experiment

Surgical treatment to repair esophageal disease involves resection or ablation of the tissue and is associated with high morbidity and between 30–40% of procedures present post-surgical complications. Autologous tissue for reconstruction is generally non-existent. To date, tissue engineering applications for esophageal repair have not been extensively studied. Collagen, fibroblasts, and smooth muscle cells have been used *in vitro* to partially reconstruct esophageal tubes. However, successful translation of these constructs to an *in vivo* model have not been reported. Due to previous success utilizing porcine ECM to reconstruct various body tissues, this bioscaffold was used to examine the structural, functional, and biomechanical properties of full circumferential

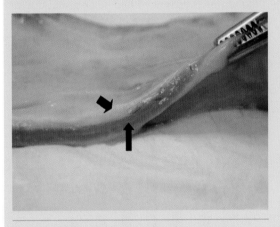

Cross-section of remodeled ECM bioscaffold from group 3 dog (91 day survival). Arrow points out well-organized muscle tissue and arrowhead denotes superficial luminal mucosal lining. Observe the lack of scar tissue.

esophageal tissue following experimental resection and reconstruction with either ECM alone, muscle tissue alone, or ECM plus muscle tissue.

Twenty-two adult dogs were divided into four groups and each dog was subjected to surgical resection of 5 cm of the full circumferential esophageal endomucosa in the mid-cervical region. The groups varied by the amount of skeletal muscle tissue that was subsequently removed from the outer muscular covering of the operated esophageal segment and by the placement or lack of placement of an ECM scaffold as a tubular graft in place of the removed endomucosa. Group 1 was repaired with ECM alone, group 2 with muscle tissue alone, group 3 with ECM plus a partial (30%) covering of muscle tissue, group 4 with ECM plus a complete (100%) covering of muscle tissue. Esophageal function, morphology, and biomechanical properties of the tissue were examined.

Animals in groups 1 and 2 formed intractable esophageal stricture and had to be sacrificed three weeks following surgery. Four of five dogs in group 3 and six of seven dogs in group 4 survived for between 26 and 230 days. These dogs showed constructive remodeling of the esophageal tissue with well organized esophageal layers, little stricture, esophageal motility, and a normal clinical outcome. Mechanical testing of tissue from groups 3 and 4 resulted in tissue with near normal biomechanical properties.

These results suggest that ECM scaffolds composed of porcine urinary bladder plus autologous muscle tissue can facilitate the *in vivo* reconstruction of structurally and functionally viable esophageal tissue, but that ECM scaffolds alone are insufficient for this application.

Structures composed of ECM are resistant to bacterial infections in preclinical and clinical applications. This resistance is seen even in clinical applications at a high risk for bacterial infection (Jernigan *et al.*, 2004; Kim *et al.*, 2000; Mantovani *et al.*, 2003;

Ruiz *et al.*, 2005; Kim *et al.*, 2005) as well as deliberate bacterial infection in preclinical studies (Badylak *et al.*, 1994). In one study, dogs underwent infrarenal aorta replacement with either a synthetic graft or SIS, and were deliberately challenged with *S. aureus* at the time of surgery. Half of the dogs implanted with the synthetic graft had positive bacterial cultures and all dogs showed evidence of persistent infection after 30 days. However, all of the dogs implanted with SIS had negative bacterial cultures and had histologic appearance of graft remodeling and were free of active inflammation after 30 days (Badylak *et al.*, 1994). The comparison of a non-degrading synthetic graft with the SIS-ECM that does undergo degradation again suggests the importance of degradation in the antibacterial properties of ECM.

The short term antibacterial effects of ECM degradation may prevent an implant infection by providing immediate protection while the host inflammatory cell response and humoral immune response become activated. Additionally, the ECM continues to degrade *in vivo* as remodeling occurs and may lead to continued release of antimicrobial peptides providing a sustained antibacterial effect (Brennan *et al.*, 2006).

5.4.2 Chemotactic properties

Degradation of the ECM by physical and chemical methods has produced low molecular weight (LMW) peptides with chemoattractant activity (Li *et al.*, 2004). These peptides, derived from SIS-ECM, range in size from 5–16 kDa and have demonstrated chemotactic activity for primary endothelial cells of mouse heart, liver, and kidney origins. Furthermore, when these ECM degradation samples were utilized in an *in vivo* Matrigel assay, a dose-dependent angiogenic response was observed (Li *et al.*, 2004). An additional *in vivo* experiment revealed that implantation with scaffolds composed of SIS-ECM or UBS-ECM significantly enhances the recruitment of marrow derived cells to the site of tissue repair and these cells participate in the long-term remodeling process (Badylak *et al.*, 2001; Zantop *et al.*, 2006). Recruitment of bone marrow derived cells by ECM may help explain the constructive remodeling that is observed rather than scar tissue formation. These studies demonstrate that ECM is capable of generating LMW peptide degradation products that are able to stimulate

endothelial cell recruitment both *in vitro* and *in vivo* (Figure 5.14).

5.4.3 Angiogenic properties

In addition to antimicrobial and chemotactic properties, degradation products of the ECM also display angiogenic properties (Li *et al.*, 2004). An *in vivo* Matrigel implant assay was performed, utilizing a fraction of degraded porcine derived SIS. Various concentrations of degraded SIS was mixed with Matrigel and heparin and injected subcutaneously into mice. PBS mixed with Matrigel and heparin was used as a negative control. The mice were sacrificed after 7 days and the Matrigel plugs were examined for blood vessels. When degraded SIS was mixed with the Matrigel before implantation, there was a remarkable increase in the number of blood vessels present within the plug, and this increase in blood vessels occurred in a dose dependent manner. There were no blood vessels observed in the PBS control

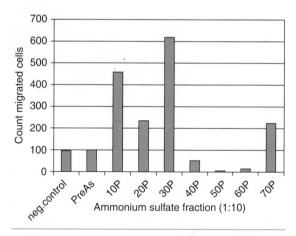

Figure 5.14 Chemotactic response of human umbilical vein cells (HUVECs) to ammonium sulfate fractions of liver ECM (L-ECM) digests. Liver ECM was digested with acid at high temperatures, fractionated by ammonium sulfate precipitation, and tested for chemotactic activity using the Boyden chamber in vitro assay. Specific ammonium sulfate fractions showed chemotactic activity for HUVEC cells.

(Li *et al.*, 2004). Therefore, degradation products of the ECM appear to be able to induce angiogenesis. Taken together, the described experiments demonstrate the importance of ECM degradation products in numerous biologic activities that promote constructive tissue remodeling.

5.5 Commercially available scaffolds composed of extracellular matrix

Scaffold materials composed of ECM are currently being utilized in many tissue engineering applications. Dermatologic applications (e.g. chronic, non-healing ulcers), lower urinary tract reconstruction, gastroesophageal applications, soft tissue repair and musculotendinous tissue reconstruction are among the medical fields currently making widespread use of ECM scaffolds.

An excellent example of the diversity of the use of ECM as a scaffold material involves the commercially available products for use in the field of orthopedic surgery. The Restore™ device is a non-chemically cross-linked material composed of porcine small intestinal submucosa that is marketed for the replacement and augmentation of musculotendinous defects. Cuff Patch™ represents a separate product that is also composed of porcine small intestinal submucosa, but differs from the Restore™ device in that it is chemically cross-linked with carbodiimide. GraftJacket® is a chemically cross-linked form of human dermis ECM marketed for rotator cuff repair. TissueMend® is an ECM derived from fetal bovine skin and Permacol™ is a cross-linked form of porcine dermis; both of which are commercially available for the repair of musculotendinous structures. Each of the above mentioned ECM products is similar in that it is composed of extracellular matrix. However, the host response to each scaffold material differs markedly because of the different methods of processing (Valentin *et al.*, 2006).

Alloderm is a commercially available product in the United States composed of human dermis. Interestingly, this particular product is not regulated as a medical device, but rather as a tissue transplant.

Therefore, terminal sterilization is not required for the material.

A multilaminate form of SIS (Surgisis®) is available for the reconstruction of soft tissues. This biologic scaffold has been successfully used for the repair of acute, traumatic wounds of the abdominal wall, inguinal hernia repair, and thoracic wall repair among other applications.

Table 5.1 summarizes some of the commercially available ECM scaffolds, their source, and tissue engineering application. There are many other scaffold materials available that consist of individual components of the extracellular matrix. However, the focus of this chapter is upon products that use the intact extracellular matrix for tissue repair.

5.6 Future considerations

The use of extracellular matrix scaffold as a biologic scaffold for tissue engineering applications has been remarkably successful when compared with the success of synthetic scaffold materials such as polylactic coglycolic acid (PLGA), purified collagen, or other artificial matrices. However, it is clear that as tissue engineering advances toward 'organ engineering' that the use of generic ECM scaffolds for all applications will likely be insufficient. For example, ECM derived from the small intestinal submucosa (SIS) will likely not be sophisticated enough to support the attachment, proliferation and differentiation of the multiple cell types that compose the liver, kidney, or other complex organs. It is possible, even likely, that extracellular matrix derived from these tissues will be required to support the appropriate site specific differentiation of appropriate cells. Therefore, a second generation of ECM scaffolds should be investigated in anticipation of these needs. Similarly, an understanding of the complex organization of the structural and functional molecules within the ECM would provide an improved roadmap for the *in vitro* production of similar scaffold materials. The ability to manufacture an artificial ECM *in vitro* would minimize variability and allow for 'customization' of ECM scaffolds for

Table 5.1 Commercially available products composed of intact extracellular matrix.

Product	Company	Material	Form		Use
Acellular					
Oasis®	Healthpoint	Porcine small intestinal submucosa (SIS)	Natural	Dry Sheet	Partial and full thickness wounds; superficial and second degree burns
Xelma™	Molnlycke	ECM protein, PGA, water	Gel		venous leg ulcers
AlloDerm	Lifecell	Human skin	Cross-linked	Dry Sheet	Abdominal wall, breast. ENT/head and neck reconstruction, grafting
CuffPatch™	Arthrotek	Porcine small intestinal submucosa (SIS)	Cross-linked	Hydrated Sheet	Reinforcement of soft tissues
TissueMend®	TEI Biosciences	Fetal bovine skin	Natural	Dry Sheet	Surgical repair and reinforcement of soft tissue in rotator cuff
Durepair®	TEI Biosciences	Fetal bovine skin	Natural	Dry Sheet	Repair of cranial or spinal dura
Xenform™	TEI Biosciences	Fetal bovine skin	Natural	Dry Sheet	Repair of colon, rectal, urethral, and vaginal prolapse, pelvic reconstruction, urethral sling
SurgiMend™	TEI Biosciences	Fetal bovine skin	Natural	Dry Sheet	Surgical repair of damaged or ruptured soft tissue membranes
PriMatrix™	TEI Biosciences	Fetal bovine skin	Natural	Dry Sheet	Wound management
Permacol™	Tissue Science Laboratories	Porcine skin	Cross-linked	Hydrated Sheet	Soft connective tissue repair
Graft Jacket®	Wright Medical Tech	Human skin	Cross-linked	Dry Sheet	Foot ulcers
Surgisis®	Cook SIS	Porcine small intestinal submucosa (SIS)	Natural	Dry Sheet	Soft tissue repair and reinforcement
Durasis®	Cook SIS	Porcine small intestinal submucosa (SIS)	Natural	Dry Sheet	Repair dura matter

Stratasis®	Cook SIS	Porcine small intestinal submucosa (SIS)	Natural	Dry Sheet	Treatment or urinary incontinence
OrthADAPT™	Pegasus Biologicals	Horse pericardium	Cross-linked		Reinforcement, repair and reconstruction of soft tissue in orthopedics
DurADAPT™	Pegasus Biologicals	Horse pericardium	Cross-linked		Repair dura matter after craniotomy
Axis™ dermis	Mentor	Human dermis	Natural	Dry Sheet	Pelvic organ prolapse
Suspend™	Mentor	Human fascia lata	Natural	Dry Sheet	Urethral sling
Restore™	DePuy	Porcine small intestinal submucosa (SIS)	Natural	Sheet	Reinforcement of soft tissues
Veritas®	Synovis Surgical	Bovine pericardium		Hydrated Sheet	Soft tissue repair
Dura-Guard®	Synovis Surgical	Bovine pericardium		Hydrated Sheet	Spinal and cranial repair
Vascu-Guard®	Synovis Surgical	Bovine pericardium			Reconstruction of blood vessels in neck, legs, and arms
Peri-Guard®	Synovis Surgical	Bovine pericardium			Pericardial and soft tissue repair
Cellular					
Dermagraft™	Smith & Nephew	Fibroblasts, ECM, bioabsorbable scaffold		Frozen Sheet	Full thickness diabetic foot ulcers
OrCel™	Ortec International	Bovine collagen cultured with human fibroblasts		Hydrated Sheet	Burn wounds
Apligraf®	Organogenesis, Inc.	Human fibroblasts, collagen, secreted ECM		Hydrated Sheet	Venous ulcers, diabetic foot ulcers
Transcyte™	Smith & Nephew	ECM secreted from human fibroblasts		Frozen Sheet	Mid to intermediate partial thickness burns

Jernigan, T.W., Croce, M.A., Cagiannos, C., *et al.* (2004). Small intestinal submucosa for vascular reconstruction in the presence of gastrointestinal contamination. *Ann Surg*, 239(5): 733–738. discussion 738–740.

Kagami, S., Kondo, S., Loster, K., *et al.* (1998). Collagen type I modulates the platelet-derived growth factor (PDGF) regulation of the growth and expression of β1 integrins by rat mesangial cells. *Biochemical & Biophysical Research Communications*, 252(3): 728–732.

Kim, B.S., Baez, C.E. and Atala, A. (2000). Biomaterials for tissue engineering. *World J Urol*, 18(1): 2–9.

Kim, M.S., Hong, K.D., Shin, H.W., *et al.* (2005). Preparation of porcine small intestinal submucosa sponge and their application as a wound dressing in full-thickness skin defect of rat. *International Journal of Biological Macromolecules*, 36(1–2): 54–60.

Kropp, B. P., Eppley, B. L., Prevel, C. D., *et al.* Experimental assessment of small intestinal submucosa as a bladder wall substitute. *Urology*, 46(3), 396–400.

Lai, J.Y., Chang, P.Y. and Lin, J,N. (2003). Body wall repair using small intestinal submucosa seeded with cells. *J Pediatric Surg.*, 38(12): 1752–1755.

Lambert, C.A., Soudant, E.P., Nusgens, B.V., *et al.* (1992). Pretranslational regulation of extracellular matrix macromolecules and collagenase expression in fibroblasts by mechanical forces. *Lab Invest*, 66(4): 444–451.

Langer, and Vancanti, (1993). Tissue engineering. *Science*, 260: 920–926.

Laurie, G.W., Horikoshi, S., Killen, P.D., *et al.* (1989). *In situ* hybridization reveals temporal and spatial changes in cellular expression of mRNA for a laminin receptor, laminin, and basement membrane (type IV) collagen in the developing kidney. *Journal of Cell Biology*, 109(3): 1351–1362.

le Roux, P.J. (2005). Endoscopic urethroplasty with unseeded small intestinal submucosa collagen matrix grafts: a pilot study. *J Urol*, 173(1): 140–143.

Li, F., Li, W., Johnson, S., *et al.* (2004). Low-molecular-weight peptides derived from extracellular matrix as chemoattractants for primary endothelial cells. *Endothelium*, 11(3–4): 199–206.

Li, S., Harrison, D., Carbonetto, S., *et al.* (2002). Matrix assembly, regulation, and survival functions of laminin and its receptors in embryonic stem cell differentiation. *Journal of Cell Biology*, 157(7): 1279–1290.

Liang, H.C., Chang, Y., Hsu, C.K., *et al.* (2004). Effects of crosslinking degree of an acellular biological tissue on its tissue regeneration pattern. *Biomaterials*, 25(17): 3541–3552.

Lisignoli, G.C.S., Piacentini, A., Zini, N., *et al.* (2006). Chondrogenic differentiation of murine and human mesnchymal stromal cells in a hyaluronic acid scaffold: Differences in gene expression and cell morphology. *J Biomed Mater Res*, 77A(3): 497–506.

Lopes, M.F., Cabrita, A., Ilharco, J., *et al.* (2006). Grafts of porcine intestinal submucosa for repair of cervical and abdominal esophageal defects in the rat. *J Invest Surg*, 19(2): 105–111.

Mantovani, F., Trinchieri, A., Castelnuovo, C., *et al.* (2003). Reconstructive urethroplasty using porcine acellular matrix. *Eur Urol*, 44(5): 600–602.

Martins-Green, M. and Bissel, M.F. (1995). Cell-extracellular matrix interactions in development. *Semin Dev Biol*, 6: 149–159.

McPherson, T. and Badylak, S.F. (1998). Characterization of fibronectin derived from porcine small intestinal submucosa. *Tissue Engineering*, 4: 75–83.

Mertsching, H., Walles, T., Hofmann, M., *et al.* (2005). Engineering of a vascularized scaffold for artificial tissue and organ generation. *Biomaterials*, 26(33): 6610–6617.

Midwood, K.S., Williams, L.V. and Schwarzbauer, J.E. (2004). Tissue repair and the dynamics of the extracellular matrix. *The International Journal of Biochemistry and Cell Biology*, 36: 1031–1037.

Miyamoto, S., Katz, B.Z., Lafrenie, R.M., *et al.* (1998). Fibronectin and integrins in cell adhesion, signaling, and morphogenesis. *Ann N Y Acad Sci*, 857: 119–129.

Piez, K.A. (1984). *Molecular and aggregate structures of the collagens*. New York: Elsevier.

Ponce, M.L., Nomizu, M., Delgado, M.C., *et al.* (1999). Identification of endothelial cell binding sites on the laminin gamma 1 chain. *Circ Res*, 84(6): 688–694.

Prajapati, R.T., Chavally-Mis, B., Herbage, D., *et al.* (2000a). Mechanical loading regulates protease production by fibroblasts in three-dimensional collagen substrates. *Wound Repair Regen*, 8(3): 226–237.

Prajapati, R.T., Eastwood, M. and Brown, R.A. (2000b). Duration and orientation of mechanical loads determine fibroblast cyto-mechanical activation: monitored by protease release. *Wound Repair Regen*, 8(3): 238–246.

Record, R.D., Hillegonds, D., Simmons, C., *et al.* (2001). *In vivo* degradation of 14C-labeled small intestinal submucosa (SIS) when used for urinary bladder repair. *Biomaterials*, 22(19): 2653–2659.

Roberts, R., Gallagher, J., Spooncer, E., *et al.* (1988). Heparan sulphate bound growth factors: a mechanism for stromal cell mediated haemopoiesis. *Nature*, 332(6162): 376–378.

Ruiz, C.E., Iemura, M., Medie, S., *et al.* (2005). Transcatheter placement of a low-profile biodegradable pulmonary valve made of small intestinal submucosa: A long-term study in a swine model. *Journal of Thoracic and Cardiovascular Surgery*, 130(2): 477.e471–477.e479.

Sacks, M.S. and Gloeckner, D.C. (1999). Quantification of the fiber architecture and biaxial mechanical behavior of porcine intestinal submucosa. *J Biomed Mater Res*, 46(1): 1–10.

Sacks, M.S., Smith, D.B. and Hiester, E.D. (1997). A small angle light scattering device for planar connective tissue microstructural analysis. *Ann Biomed Eng*, 25(4): 678–689.

Sarikaya, A., Record, R., Wu, C.C., *et al.* (2002). Antimicrobial activity associated with extracellular matrices. *Tissue Engineering*, 8(1): 63–71.

Schwarzbauer, J.E. (1991). Fibronectin: from gene to protein. *Curr Opin Cell Biol*, 3(5): 786–791.

Schwarzbauer, J. (1999). Basement membranes: Putting up the barriers. *Curr Biol*, 9(7): R242–R244.

Sclamberg, S.G., Tibone, J.E., Itamura, J.M., *et al.* (2004). Six-month magnetic resonance imaging follow-up of large and massive rotator cuff repairs reinforced with porcine small intestinal submucosa. *J Shoulder Elbow Surg*, 13(5): 538–541.

Seliktar, D., Nerem, R.M. and Galis, Z.S. (2001). The role of matrix metalloproteinase-2 in the remodeling of cell-seeded vascular constructs subjected to cyclic strain. *Ann Biomed Eng*, 29(11): 923–934.

Timpl, R. (1996). Macromolecular organization of basement membranes. *Current Opinion in Cell Biology*, 8(5): 618–624.

Timpl, R. and Brown, J.C. (1996). Supramolecular assembly of basement membranes. *Bioessays*, 18(2): 123–132.

Valentin, J.E., Badylak, J.S., McCabe, G.P. and Badylak, S.F. (2006). Extracellular Matrix bioscaffolds for Orthopaedic Applications: A Comparative Study. *The Journal of Bone and Joint Surgery*, 88: 2673–2686.

van Amerongen, M.J., Harmsen, M.C., Peterson, *et al.* (2006). The enzymatic degradation of scaffolds and their replacement by vascularized extracellular matrix in the murine myocardium. *Biomaterials*, 27(10): 2247–2257.

van der Rest, M. and Garrone, R. (1992). The collagen family of proteins. *FASEB Journal*, 5: 2814–2823.

Wang, J.H.-C., Jia, F., Gilbert, T.W., *et al.* (2003). Cell orientation determines the alignment of cell-produced collagenous matrix. *J Biomech*, 36(1): 97–102.

Werb, Z., Vu, T.H., Rinkenberger, J.L., *et al.* (1999). Matrix-degrading proteases and angiogenesis during development and tumor formation. *Apmis*, 107(1): 11–18.

Woods, T. and Gratzer, P.F. (2005). Effectiveness of three extraction techniques in the development of a decellularized bone-anterior cruciate ligament-bone graft. *Biomaterials*, 26(35): 7339–7349.

Yurchenco, P., Birk, D.E. and Mecham, R.P. (1994). *Extracellular Matrix Assembly and Structure*. New York: Academic Press.

Zantop, T., Gilbert, T.W. Yoder, M.C. and Badylak, S.F. (2006). Extracellular Matrix Scaffolds Attract Bone Marrow Derived Cells in a Mouse Model of Achilles Tendon Reconstruction. *J Orthop Res*, 24(12): 1299–1309.

Zasloff, M. (2002). Antimicrobial Peptides of Multicellular Organisms. *Nature*, 415: 389–395.

Chapter 6
Natural Polymers in tissue engineering applications

Manuela Gomes, Helena Azevedo, Patrícia Malafaya, Simone Silva, Joaquim Oliveira, Gabriela Silva, Rui Sousa, João Mano and Rui Reis

Chapter objectives:

- To understand the origin, structure and properties of natural polymers used in tissue engineering applications
- To identify the characteristics that make natural polymers interesting for TE applications
- To understand possible factors that may affect cells/tissue response to natural polymers based scaffolds
- To understand the possible specific applications of each natural polymer in the context of tissue engineering

- To understand the processing possibilities of the different natural origin polymers for TE applications
- To recognize the most important achievements in this research field attained by different scientists
- To understand the versatility obtained by combining natural origin polymers with other materials in Tissue Engineering applications

"Perhaps appropriately designed biodegradable templates can be used to regenerate segments of other tissues or organs which have become lost or dysfunctional due to disease or trauma."

Yannas *et al.* (1982)

6.1 Introduction

Life as we know it could not exist without natural polymers. Just think of deoxyribonucleic acid (DNA) and ribonucleic acid (RNA). They are natural polymers essential in so many life processes. In fact, long before there were plastics and synthetic polymers, nature was using natural polymers to make life possible. In the early 1900s, scientists began to understand the chemical makeup of natural polymers and how to make synthetic polymers with properties that complement those of natural materials. Nevertheless, for many purposes, we still don't think of natural polymers in the same way as we think about synthetic polymers. However, that doesn't make natural polymers less important; indeed, it turns out that they are more important in many ways. In fact, after a century of developing synthetic polymers for use as materials, polymer science is turning back toward its roots, as natural polymers show promise in a wide range of biomedical uses, such as scaffolds for growing artificial human tissues, that is, for making life better after injury or disease.

Tissue engineering offers the possibility to help in the regeneration of tissues damaged by disease or trauma and, in some cases, to create new tissues and replace failing or malfunctioning organs. Typically, this is achieved through the use of degradable biomaterials to either induce surrounding tissue and cell ingrowth or to serve as temporary scaffolds for transplanted cells to attach, grow, and maintain differentiated functions. In any case, the role of the biomaterial scaffold is temporary, but still crucial to the success of the strategy. Therefore, the design and production of an appropriate scaffold material is the first, and one of the most important stages, in tissue engineering strategies. In this critical stage, the selection of the most adequate raw material is a primary consideration. Natural polymers were the first to be used as scaffold materials for tissue regeneration. They have frequently been used in tissue engineering applications because they are either components of, or have properties similar to, the natural extracellular matrix.

This chapter provides an overview of the natural origin polymers that are commercially available or currently being studied in different labs for tissue engineering applications, with some emphasis on the most widely studied systems. It describes their chemical structure, main properties and potential applications within the field. Several aspects regarding the development and research status towards their final application are addressed. The main advantages and disadvantages of the use of natural origin polymers as compared to other materials used in tissue engineering scaffolding are also discussed.

6.2 Natural polymers

Natural polymers are derived from renewable resources, namely from plants, animals and microorganisms, and are, therefore, widely distributed in nature. These materials exhibit a large diversity of unique (and in most cases) rather complex structures, and different physiological functions, and may offer a variety of potential applications in the field of tissue engineering due to their various properties, such as pseudoplastic behavior, gelation ability, water binding capacity, biodegradability, among many others. In addition, they possess many functional groups

Invited chapter, **Textbook on tissue engineering**, Clemens van Blitterswijk, Anders Lindahl, Peter Thomsen, David Williams, Jeffrey Hubbell, Rainieri Cancedda (Editors), Elsevier

Classical Experiment

While the term *Tissue Engineering* was still to be 'coined' (this happened only in 1987), researchers were already studying an approach to regenerate skin wounds that resulted in a paper published in *Science* in 1982 (Yannas, Burke *et al.*, 1982). This paper described the prompt and long-term closure of full-thickness skin wounds in guinea pigs and humans, achieved by applying a bilayer polymeric membrane, comprising of a top silicone layer and a bottom layer of a porous cross-linked network of collagen and glycosaminoglycan, seeded with a small number of autologous basal cells before grafting (Yannas, Burke *et al.*, 1982). This study was conducted following three main stages; today recognized by many researchers as the three main phases of a tissue engineering approach: the first corresponds to the development of the material matrix-membranes; following this, in the second stage, the membranes are seeded with cells and finally, in the third stage, the cell seeded membranes are grafted onto the tissue defect. The first stage, i.e. the development of the appropriate membranes based on collagen and glycosaminoglycan, was in fact, extensively explored, as the authors have analyzed a range of different chemical compositions and several methods were compared for preparing membranes with different porosities and pore sizes (Dagalakis, Flink *et al.*, 1980; Yannas and Burke, 1980; Yannas, Burke *et al.*, 1980). Extensive characterization of the materials led to the selection of the most suitable formulation/structure to be used in the further stages of the development of these skin tissue equivalents. Neodermal tissue synthesis occurred in these membranes seeded with autologous cells and there was no evidence of conventional scar formation. New and apparently normal functional skin was generated in less than 4 weeks. It was demonstrated that, although the acellular membranes can also be used to regenerate skin defects, the cell-seeded membranes provide a means for closing the largest full thickness skin wounds in a shorter period of time (Yannas, Burke *et al.*, 1982). This system, mainly based on a 3D polymeric matrix obtained from two natural origin polymers (collagen and glycosaminoglycan), also resulted in the first tissue engineered product to be approved by FDA (in 1996), and clearly opened the way to the concept of tissue engineering, i.e. to the regeneration of tissues using cells seeded onto a 3D matrix, made from natural origin polymers.

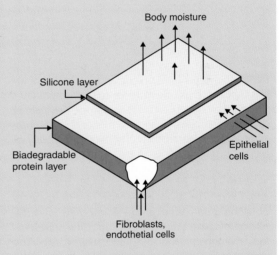

Schematic representation of the polymeric membrane developed by Yannas et al. *(1982) (Wound tissue can utilize a polymeric template to synthesize a functional extension of skin. Science, 215(4529): 174–176) to be used as a template to obtain skin substitutes. The top layer, made of medical grade silicone, is designed to be spontaneously ejected following formation of a confluent neoepidermal layer under it. The bottom layer, a cross-linked network of collagen and chondroitin 6-sulfate, was designed to undergo biodegradation at a controlled rate while is replaced by neodermal tissue.*

State of the Art Experiment

Natural polymers in gene delivery and tissue engineering

Research on natural polymers for the development of matrix-based gene delivery systems has opened the way to new and exciting possibilities to be explored within the field of regenerative medicine (a more detailed review on this subject may be found in Dang and Leong, 2006). In fact, the combination of gene therapy and tissue engineering exploits the potential of genetic cell engineering to provide biochemical signals that direct cell proliferation and differentiation, and simultaneously, the ability of natural polymers to serve as gene carriers and tissue engineering scaffolds. It is true that synthetic polymers and viral carriers have been preferentially used in gene delivery applications, but natural polymers have unique and intrinsic properties that can make them more suitable candidates for this type of application. Such properties include their general biocompatibility, mucoadhesive character and biodegradability. The biocompatibility of natural polymers, for example, allows for cells infiltration into the matrix and transfection can occur as these cells come into contact with the imbedded DNA. The biodegradability of the matrices obtained from natural polymers may also assist the release of gene transfer agents into the surrounding environment and thus affect nearby cells. The current research suggests therefore, that natural polymeric carriers have a different mechanism for intracellular escape and transfection than synthetic polymers. However, a very limited number of studies have focused on the development of matrices based on natural polymers for gene delivery and for cell support. An interesting example is provided by research work from Lim and co-workers (Lim, Liao et al., 2006) in which the authors have investigated a 3D fiber mesh scaffold, based on chitin and alginate, as a way to obtain a better spatial control of plasmid localization, in opposition to other available systems that are based on simple mixture to bond the matrix and the gene delivery elements. In this study, chitin and alginate fibers were formed by polyelectrolyte complexation of the water-soluble polymers, and PEI-DNA nanoparticles containing green fluorescent protein (GFP)-encoding plasmid, were loaded during the fiber drawing process. These fibers were then pro-cessed into a nonwoven fiber mesh scaffold, using a method based on the needle-punching technique. This system was then studied to analyse the tranfectability of human epithelial kidney cells (HEK293) and human dermal fibroblasts (HDF) seeded on the scaffolds. In summary, the results obtained showed that nanoparticles released from the fibers over time retained their bioactivity and successfully transfected cells seeded on the scaffold in a sustained manner. Transgene expression in HEK293 cells and human dermal fibroblasts seeded on the transfecting scaffolds was significant even after 2 weeks of culture compared to 3-day expression in two-dimensional controls. Fibroblasts seeded on scaffolds containing DNA encoding basic fibroblast growth factor (bFGF) demonstrated prolonged secretion of bFGF at levels significantly higher than baseline.

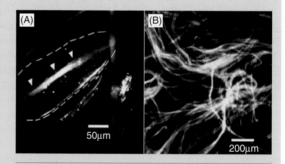

Confocal microscopy images of the fibrous scaffolds and the PEI-DNA nanoparticles (fluorescently stained) encapsulated within the fibers developed by Lim et al. (Nonviral gene delivery from nonwoven fibrous scaffolds fabricated by interfacial complexation of polyelectrolytes. Mol Ther, 13(6): 1163–1172, 2006). (A) Phase microscopy showing the bead region of a single fiber, depicting the nanoparticles dispersed within the bead (dotted lines) and at a higher density in the core fiber segment (arrows); (B) fibers containing nanoparticles.

(amino, carboxylic and hydroxyl groups) available for chemical (hydrolysis, oxidation, reduction, esterification, etherification, cross-linking reactions, etc.) (Kurita, 2001; Xie, Lui *et al.*, 2005) and enzymatic (Broderick, O'Halloran *et al.*, 2005; Chen, Embree *et al.*, 2003) modification and/or conjugation with other molecules, which allows an overwhelming variety of products with tailorable chemistries and properties to be obtained. Protein materials may offer an additional advantage as they are able to interact favorably with cells through specific recognition domains present in their structure. On the other hand, the creation of hybrid materials – by means of combining the advantages of different natural polymers – may constitute a useful approach to mimicking the natural environment of the extracellular matrix and to obtaining scaffolding materials with superior mechanical and biological properties.

An intrinsic characteristic of natural origin polymers is their ability to be degraded by naturally occurring enzymes, which may indicate the greater propensity of these materials to be metabolized by the physiological mechanisms. Another important aspect to consider when using natural polymers, is that they can induce an undesirable immune response due to the presence of impurities and endotoxins (depending on their source), and their properties may differ from batch to batch during large-scale isolation procedures due to the inability to accurately control the processing techniques. Nevertheless, as knowledge about these natural polymers increases, new approaches (including methods for production, purification, controlling material properties and enhancing material biocompatibility) are likely to be developed for designing better scaffolding materials to support the development of more natural and functional tissues.

In summary, both natural and synthetic polymers present important characteristics and, therefore, one must recognize that the best biodegradable polymer for biomedical applications might be found by taking steps towards the development of new biomaterials that combine the most favorable properties of synthetic and natural polymers. Several examples of the combination of natural and synthetic polymers will

be described in further sections. Another approach that will be presented consists of the reinforcement of polymeric matrices with bioactive ceramic materials, such as hydroxyapatite and other calcium phosphates. These fillers have the ability, in most cases, of improving the mechanical properties and the biological behavior simultaneously.

It is well known that living organisms are able to synthesize a vast variety of polymers which can be divided into eight major classes according with their chemical structure: (1) polysaccharides, (2) proteins and other polyamides, (3) polyoxoesters (polyhydroxyalkanoic acids), (4) polythioesters, (5) polyanhydrides (polyphosphate), (6) polyisoprenoids, (7) lignin, and (8) nucleic acids (Steinbüchel and Rhee, 2005). However, only polymers belonging to the three first classes will be described in more detail in this chapter. This is due to their importance as raw-materials in tissue engineering scaffolding. Although most of these natural polymers are obtained from plant (Franz and Blaschek, 1990; Morrison and Karkalas, 1990; Stephen, Churms *et al.*, 1990) and animal (Izydorczyk, Cui *et al.*, 2005; Lezica and Quesada-Allué, 1990) sources or from algae (Percival and McDowell, 1990), there are a large number of microorganisms capable of synthesizing many biopolymers. In fact, with advances in biotechnology, there is an increasing interest in using microorganisms to produce polymers by fermentation (enabling large-scale production, avoiding complex and time-consuming isolation procedures and the risk of animal-derived pathogens) (Naessens, Cerdobbel *et al.*, 2005; Widner, Behr *et al.*, 2005) or *in vitro* enzymatic processes (Kobayashi, Fujikawa *et al.*, 2003). They allow controlling polymer molecular weight, branching patterns and branch chain lengths, crosslinking between chains, altering the fine structure and functional properties of polymers.

6.3 Polysaccharides

Polysaccharides, also known as glycans, consist of monosaccharides (aldoses or ketoses) linked together by O-glycosidic linkages. Each monosaccharide is classified according to the number of carbons in the monosaccharide chain (usually 3 to 9), into trioses (C_3), tetroses (C_4), pentoses (C_5), hexoses (C_6), heptoses (C_7), octoses (C_8), nonoses (C_9). Polysaccharides can be classified as homopolysaccharides or heteropolysaccharides if they consist of one type or more than one type of monosaccharide. Because glycosidic linkages can be made to any of the hydroxyl groups of a monosaccharide, polysaccharides form linear as well as branched polymers (Figure 6.1). Differences in the monosaccharide composition, linkage types and patterns, chain shapes, and molecular weight, dictates their physical properties, including solubility, flow behavior, gelling potential, and/or surface and interfacial properties (Izydorczyk, 2005; Izydorczyk, Cui *et al.*, 2005).

In the living organisms, polysaccharides perform a range of biological functions, such as maintenance and structural integrity (e.g. cellulose, chitin), energy reserve storage (e.g. starch, glycogen) and biological protection and adhesion (e.g. gum exudates, extracellular microbial polysaccharides). These functions can be found in the compilation presented in Table 6.1.

In this chapter, the authors have chosen to focus only on the polymers that have been proposed, by different researchers, for application within the tissue engineering field; namely alginate, dextran, chitosan, cellulose, starch and hyaluronic acid polysaccharides. All of these polymers have been used as scaffold materials and will be described in more detail in the following sections.

6.3.1 Alginate and dextran

Alginate is a biological material derived from sea algae, composed of linear block copolymers of 1-4 linked b-D-mannuronic acid (*M*) and a-L-guluronic acid (*G*) (Figure 6.2). Divalent ions form cross-links in alginate by binding the guluronic residues, inducing a sol–gel transition in the material.

Dextran is a bacterial-derived polysaccharide, consisting essentially of a-1,6 linked D-glucopyranose residues with a few percent of a-1,2-, a-1,3-, or a-1,4-linked side chains (Figure 6.3) synthesized from sucrose by *Leuconostoc mesenteroides streptococcus*.

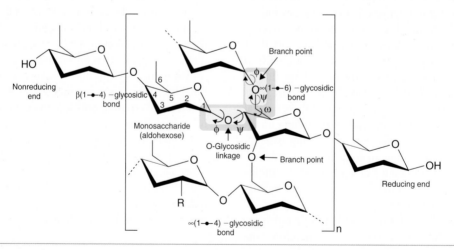

Figure 6.1 General structure of polysaccharides showing their diversity in terms of monosaccharide composition (nature and molar ratios of the monosaccharide building blocks), linkage patterns (linkage positions between the glycosidic linkages and branches), anomeric configuration (α- or β-configuration of the glycosidic linkage), substitutions (position and nature of OH−modifications), degree of freedom in (1→4) and (1→6)-glycosidic bonds. (Izydorczyk, M. (2005). *Understand the Chemistry of Food Carbohydrates. Food Carbohydrates: Chemistry, Physical Properties, and Application* (Cui, S.W. ed.), Boca Raton, CRC Press, Taylor & Francis Group: 1–65.)

Table 6.1 Classification of polysaccharides according with their origin, function, linkage patterns, sequence and composition of sugar units in polysaccharide chains, presence of ionizing groups.

Origin	Polysaccharide	Occurrence/Function	Glycosidic linkage/ Repeating unit	Nature and distribution of the monosaccharide units
Plant	Starch	Starch is synthesized in amyloplasts of green in plants and deposited in the major depots of seeds, tubers and roots in the form of granules. Energy storage material in almost higher plants (corn, rice, potato, wheat, tapioca, etc).	Amylose: α-(1→4)-D-Glc Amylopectin: α-(1→4, 1→6)-D-Glc	Homopolysaccharide, neutral.Amylose: linear Amylopectin: branched
	Cellulose	Structural polysaccharide in the cell walls of higher plants (cotton, wood). Besides the mechanical strength of the plant cell, cellulose is a protective component against external attack by mechanical forces or microorganisms.	β-(1→4)-D-Glc	Homopolysaccharide, neutral, linear.
	Arabinogallactan	Larch arabinogallactan is extracted from the heartwood of the western larch *Larix occidentalis*. It is an exudate gum polysaccharide which is produced on the exterior surfaces of plant usually as a result of trauma or stress (physical injury and/or fungal attack).	Main chain: β-(1→3)-D-Gal Side chains: disaccharides β-D-Gal-(1→6)-β-D-Gal and β-L-Ara-(1→3)-α-L-Ara	Heteropolysaccharide, neutral, branched.
Algal	Alginate	It occurs combined with calcium and other bases in the cell walls and intracellular matrix of brown seaweeds (*Phaeophyceae*), being the main structural component. Contributes to ionic interactions and physical protection.	β-(1→4)-D-ManA-α-(1→4)-L-GulA	Heteropolysaccharide, anionic, linear.
	Agarose	Red algae (*Rhodophycae*). Biological function in algae is antidessication at low tide and to provide mechanical support so that cells do not collapse.	-(1→3)-β-D-Gal-(1→4)-3,6-anhydro-α-L-Gal	Heteropolysaccharide, neutral, linear.
	Carrageenans	Carrageenans are structural polysaccharides of the marine red algae (*Rhodophycae*).	κ-carrageenan: -(1→3)-β-D-Gal-4-sulphate-(1→]4)-3,6-anhydro-α-D-Gal-(1→3)-	Heteropolysaccharide, anionic, linear.

(Continued)

Table 6.2 (*Continued*)

Origin	Polysaccharide	Occurrence/Function	Glycosidic linkage/ Repeating unit	Nature and distribution of the monosaccharide units
			κ-carrageenan: -(1→3)-β-D-Gal-4-sulphate-(1→4)-3,6-anhydro-α-D-Gal-2-sulphate-(1→3)- λ-carrageenan:-(1→3)-β-D-Gal-2-suphate-(1→4)-α-D-Gal-2,6-disulphate-(1→3)-	
Animal	Chitin Chitosan	Chitin is the main component of the exoskeleton of insects and shells of crustaceans (crab, shrimp, lobster etc). Structural/supporting polysaccharide. Chitosan is a chitin derivative obtained by a deacetylation reaction.	Chitin: β-(1→4)-D-GlcNAc Chitosan: β-(1→4)-D-GlcN-β-(1→4)-D-GlcNAc, distributed in a random way depending on the degree of acetylation.	Chitin: homopolysaccharide, neutral, linear.Chitosan: heteropolysaccharide, cationic, linear.
	Hyaluronic acid	Hyaluronan is an important glycosaminoglycan component of connective tissue (cartilage, tendon, skin, and blood vessel walls), synovial fluid (the fluid that lubrificates joints) and the vitreous humor of the eye. It plays a significant role in wound healing.	-β(1→4)-D-GlcUA-β(1→3)-D-GlcNAc-	Heteropolysaccharide, anionic, linear.
Microbial	Dextran	Extracellular polysaccharide produced by the bacterium *Leuconostoc mesenteroides*.	α-(1→2, 1→3, 1→4, 1→6)-Glc	Homopolysaccharide, neutral, branched.
	Gellan gum	Extracellular polysaccharide produced by the bacterium *Sphingomonas elodea*	→3)-β-D-Glc-(1→4)-β-D-GlcUA- (1→4)-β-D-Glc-(1→4)-α-L-Rha-(1→	Heteropolysaccharide, anionic, linear
	Pullulan	Extracellular polysaccharide produced by the fungus *Aureobasidium pullulans*.	α-(1→6)-Maltotriose	Homopolysaccharide, neutral, branched.

Glc – Glucose, Ara – Arabinose, GulA – Guluronic acid, ManA – Mannuronic acid, Gal – Galactose, GlcNAc – N-acetylglucosamine, GlcN – N-glucosamine, GlcUA – Glucuronic acid, Rha – Rhamnose

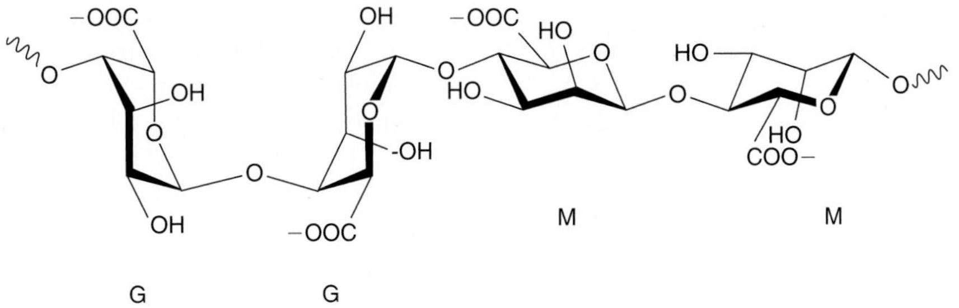

Figure 6.2 Alginate structure. M – mannuronic acid; G – guluronic acid.

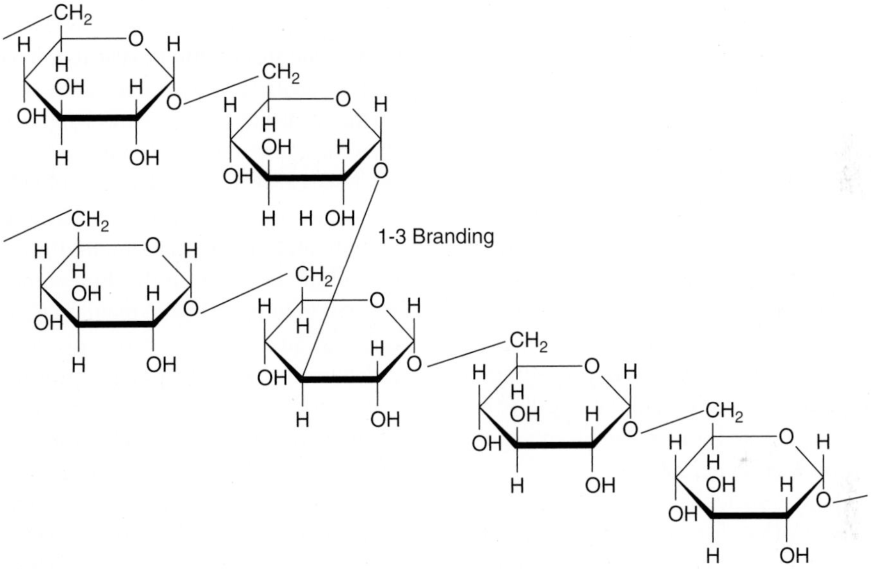

Figure 6.3 Dextran structure evidencing the 1-3 branching.

Because of their biocompatibility, abundance in source, and low prices, they have been widely used in the food industry as thickeners and emulsifying agents. Alginate can be ionically cross-linked by the addition of divalent cations (like Ca^{2+}) in aqueous solution. The gelation and cross-linking of the polymers are mainly achieved by the exchange of sodium ions from the guluronic acids with the divalent cations, and the stacking of these guluronic groups to form the characteristic egg-box structure shown in Figure 6.4.

The cross-links are believed to create a stiff egg-box structure and they impart viscoelastic solid behavior to the material (Grant, Morris et al., 1973; LeRoux, Guilak et al., 1999). The properties of alginate derive from this behavior, and include (Gombotz and Wee, 1998; Lemoine, Wauters et al., 1998; Rowley, Madlambayan et al., 1999): a relatively inert aqueous environment within the matrix; a high gel porosity which allows for high diffusion rates of macromolecules; the ability of control this porosity with simple coating procedures

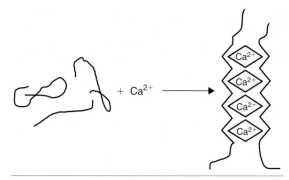

Figure 6.4 Egg box model for alginate gel formation. It is shown the conversion of random coils to buckled ribbonlike structures containing arrays of Ca^{2+} ions ($\diamond$).

and dissolution and biodegradation of the systems under normal physiological conditions and, at room temperature, a mild encapsulation process free of organic solvents. An attractive class of physically cross-linked gels are those where gel formation is not instantaneous, but occurs a certain time after mixing the hydrogel components or after a certain trigger (such as pH or temperature). These systems can be administered by injection as liquid formulation and gellify *in situ* (Kuo and Ma, 2001; Van Tomme, van Steenbergen *et al.*, 2005; Williams, Klein *et al.*, 2005).

It is the latter characteristic that drew the attention of researchers to use alginate for encapsulation of cells as well as bioactive agents. The material to be encapsulated is usually mixed with an alginate solution, and the mixture dripped into a solution containing Ca^{2+} ions, resulting in the instantaneous formation of microparticles that entrap cells or drugs within a three-dimensional lattice (Fundueanu, Nastruzzi *et al.*, 1999). Dextran hydrogels can be created by either physical or chemical cross-linking, taking advantage of the hydroxyl groups present on the a-1,6-linked D-glucose residues. Dextran particles have been widely used as separation matrices, such as Sephadex, as cell microcarriers, such as Cytodex, and as drug delivery vehicles (Levesque, Lim *et al.*, 2005). There has been considerable interest on dextran scaffolds

for tissue engineering applications (Levesque, Lim *et al.*, 2005; Massia and Stark, 2001).

Similarly, alginate cross-linked with Ca^{2+} has been popularized for *in vitro* cell culture (Guo, Jourdian *et al.*, 1989; Smidsrød and Skjak-Bræk, 1990) and tissue engineering applications (Chang, Rowley *et al.*, 2001; Fragonas, Valente *et al.*, 2000; Kataoka, Suzuki *et al.*, 2001; Kuo and Ma, 2001; Miralles, Baudoin *et al.*, 2001; Paige, Cima *et al.*, 1996) primarily because of the ability to immobilize and later recover cells from the culture matrix (LeRoux, Guilak *et al.*, 1999). Alginate has also been used as a bioartificial matrix for cartilage generation and fundamental studies on entrapped chondrocytes (Guo, Jourdian *et al.*, 1989; Paige, Cima *et al.*, 1996). The suspension of cells in a bioartificial matrix, such as alginate, is associated with significant changes in the local physical and mechanical environment of the cells compared to their native extracellular matrix (ECM) (LeRoux, Guilak *et al.*, 1999). ECM plays an important role in tissue engineering because cellular growth and differentiation, in the two-dimensional cell culture as well as in the three-dimensional space of the developing organism, require ECM with which the cells can interact (Cho, Seo *et al.*, 2006). In the artificial culture system, the physical properties of the artificial matrix will govern the deformations and tractions applied to the cells, altering important cell−matrix interactions present in the native system that appear to regulate cell activity in response to mechanical stress (Buschmann, Gluzband *et al.*, 1995; Guo, Jourdian *et al.*, 1989). On the other hand, alginate is well known for forming strong complexes with polycations including, but not limited to, synthetic polymers, proteins and polypeptides. This feature is particularly important when attempting to use alginate as a scaffold for tissue engineering applications such as cartilage and bone, since mechanical constraints limit its applications. Combining alginate with other polymers and ceramic materials has been shown to be able to obviate this feature (Cho, Oh *et al.*, 2005; Ciardelli, Chiono *et al.*, 2005; Li, Ramay *et al.*, 2005; Sivakumar and Panduranga Rao, 2003; Sotome, Uemura *et al.*, 2004).

Alginate has also been widely studied for engineering liver tissue (Dvir-Ginzberg, Gamlieli-Bonshtein et al., 2003; Elkayam, Amitay-Shaprut et al., 2006; Glicklis, Shapiro et al., 2000). The bioartificial liver-assist device or regeneration of the liver-tissue substitutes for liver tissue engineering requires a suitable ECM for hepatocyte culture because hepatocytes are anchorage-dependent cells and are highly sensitive to the ECM milieu for the maintenance of their viability and differentiated functions (Cho, Seo et al., 2006). A potential approach to facilitate the performance of implanted hepatocytes is to enable their aggregation and re-expression of their differentiated function prior to implantation (Glicklis, Shapiro et al., 2000) and alginate has been shown to allow hepatocyte culture and function.

The differentiation and growth of adult stem cells within engineered tissue constructs are believed to be under the influence of cell-biomaterial interactions. Gimble et al. (cited in Awad, Wickham et al., 2004) has shown that alginate-based materials can have an enhancing effect over the differentiation of human adipose-derived adult stem (hADAS) cells, and manipulating the composition of these tissue engineered constructs may have significant effects on their mechanical properties. Additionally, the major role of alginate in tissue engineering has been defined as a vehicle for cell encapsulation and delivery to the site, and attachment of RGD sequences has shown to potentate bone cell attachment and up-regulation of specific bone markers (Alsberg, Anderson et al., 2001).

In summary, the possibility of having an injectable, in situ-gellifying material than can serve as a filler and template for the regeneration/repair of tissues such as cartilage is a very attractive one. Alginate and dextran have shown and continue to show excellent properties for this purpose. Allied to this, the potential of being tailored for several applications — hence a multi-tasking ability for these materials — renders interest from the scientific community.

6.3.2 Chitosan

During the past 30 years, a substantial amount of work has been published on chitosan and its potential use in various pharmaceutical applications (see, for instance, the review by Dodane and Vilivalam, 1998) including tissue engineering. This is due to its similar structure to naturally occurring glycosaminoglycans (GAGs) and its degradability by enzymes in humans (Drury and Mooney, 2003). Figure 6.5 shows a chitosan structure. It is a linear polysaccharide of $(1-4)$-linked D-glucosamine and N-acetyl-D-glucosamine residues derived from chitin, a high molecular weight, the second most abundant natural biopolymer commonly found in arthropod exoskeletons such as shells of marine crustaceans and cell walls of fungi (Di Martino, Sittinger et al., 2005). Chitosan has been proven to be biologically renewable, biodegradable (Chellat, Tabrizian et al., 2000; Chenite, Chaput et al., 2000; Vila, Sanchez et al., 2002), bioadhesive (Bertram and Bodmeier, 2006; Govender, Pillay et al., 2005), and biocompatible (Dodane and Vilivalam, 1998; Chellat, Tabrizian et al., 2000; Park, Lee et al., 2000; Mao, Zhao et al., 2003; Urry, Pattanaik et al., 1998), and used in wound dressing and healing (Adekogbe and Ghanem, 2005; Chellat et al., 2000; Kiyozumi et al., 2006), drug delivery systems (Baran and Reis, 2003; Gupta and Kumar, 2000; Patashnik, Rabinovich et al., 1997; Peniche, Fernandez et al., 2003), and various tissue engineering applications. We focus on these issues in this chapter.

Depending on the source and preparation procedure, chitosan's average molecular weight may range from 50 to 1000 kDa (Francis Suh and Matthew, 2000). The degree of N-deacetylation usually varies from 50% to 90% (Dodane and Vilivalam, 1998). Chitosan is a semi-crystalline polymer and the degree of crystallinity is a function of the degree of deacetylation. Crystallinity is maximum for both chitin (i.e. 0% deacetylated) and fully deacetylated (i.e. 100%) chitosan. Minimum crystallinity is achieved at intermediate degrees of deacetylation. Chitosan is degraded by lysozyme (Haque, Beekley et al., 2001; Vila, Sanchez et al., 2002; Yannas, Burke et al., 1980); the kinetics of degradation is inversely related to the degree of crystallinity. Because of the stable crystalline structure, chitosan is normally insoluble in aqueous solutions above pH 7. However, in dilute acids, the

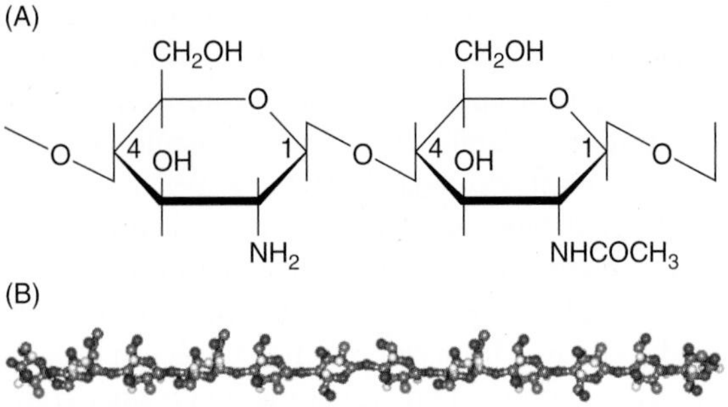

(A)

(B)

Figure 6.5 Structure of chitosan.

free amino groups are protonated and the molecule becomes fully soluble below pH 5. The pH-dependent solubility of chitosan provides a convenient mechanism for processing under mild conditions. Viscous solutions can be extruded and gelled in high pH solutions (Malafaya, Pedro *et al.*, 2005) or baths of non-solvents such as methanol. Such gel forms (particles, fibers or blocks) can be subsequently drawn and dried to form high-strength materials. Much of the potential of chitosan as a biomaterial stems from its cationic nature and high charge density in solution. The charge density allows chitosan to form insoluble ionic complexes or complex coacervates with a wide variety of water-soluble anionic polymers. Chitosan derivatives and blends have also been gelled via glutaraldehyde cross-linking (Simionescu, Lu *et al.*, 2006) and other cross-linking agents such as genipin (Mwale, Iordanova *et al.*, 2005), UV irradiation (Drury and Mooney, 2003), and thermal variations (Di Martino, Sittinger *et al.*, 2005). Besides the referred gelling-based processing method, freeze drying is undoubtedly the most widely processing technology used to process chitosan shapes. Furthermore, the cationic nature of chitosan is primarily responsible for electrostatic interactions with anionic glycosaminoglycans (GAG), proteoglycans and other negatively charged molecules. This property is of great interest because a large number of cytokines/growth factors are linked to GAG, and a scaffold incorporating a chitosan–GAG complex may retain and concentrate growth factors secreted by colonizing

cells (Di Martino, Sittinger *et al.*, 2005). Several researchers have examined the host tissue response to chitosan-based materials. In general, these materials evoke a minimal foreign body reaction, with little or no fibrous encapsulation (VandeVord, Matthew *et al.*, 2002). Formation of normal granulation tissue associated with accelerated angiogenesis, appears to be the typical course of the healing response. This immunomodulatory effect has been suggested to stimulate the integration of the implanted material by the host (Suh and Matthew, 2000).

Due to its promising properties, chitosan as been applied in tissue engineering applications targeting several tissues as it was possible to review in Table 6.2. The most commonly aimed tissues are bone, cartilage and skin, but others such as liver or trachea have applied chitosan as scaffolds to support the temporary cell functions. Due to its easy processability, chitosan as been moulded in a range of shapes including porous scaffolds, injectable gels, membranes, tubular systems, particles as described in Table 6.2. Chitosan scaffolds for bone tissue engineering have been widely investigated and shown to enhance bone formation both *in vitro* and *in vivo*, mainly in the presence of other polymers such as gelatin (Simionescu, Lu *et al.*, 2006), alginates (Li, Ramay *et al.*, 2005). When one considers cartilage tissue engineering applications, in particular, chitosan seems to be a good candidate given the importance of GAGs in stimulating the chondrogenesis (Ong, Trabbic-Carlson *et al.*, 2006), the use of GAGs or GAG analogs such as chitosan as components of a cartilage tissue scaffold appears to be a logical approach for enhancing chondrogenesis as shown by several papers (Ong, Trabbic-Carlson *et al.*, 2006; Suh and Matthew, 2000; Xia, Liu *et al.*, 2004; Yamane, Iwasaki *et al.*, 2005). It thus shares some characteristics with various GAGs and hyaluronic acid present in articular cartilage (Francis Suh and Matthew, 2000).

At present, chitosan is one of the most promising natural-origin polymers for tissue engineering. In particular, its chemical versatility and the possibility to generate structures with predictable pore sizes and degradation rates make chitosan a promising

Table 6.2 Chitosan-based scaffolds for different tissue engineering applications

Material	Scaffold Structure	Processing methodology	Cell type (source)	TE application	References
Chitosan	3D fibber meshes	Wet spinning	Osteoblast-like SAOS-2 (human osteosarcoma cell line)	Bone	Tuzlakoglu, Bolgen et al., 2005
Chitosan	3D porous blocks	Freeze-drying	Osteoblast-like ROS (rat osteosarcoma cell line)	Bone	Ho, Kuo et al., 2004; Ho, Wang et al., 2005
Chitosan/polyester	3D fibber meshes	Fibber extrusion	MSC's (human bone marrow, primary culture)	Bone	Correlo, Boesel et al., 2005a; 2005b; Pinto, Correlo et al., 2005
Chitosan/alginate	3D porous cylinders	Freeze-drying	Osteoblast-like MG63 (human osteosarcoma cell line)	Bone	Li, Ramay et al., 2005
Chitosan/alginate	Injectable gel	Gelation by sonication	MSC's (rat bone marrow, primary culture)	Bone	Park, Choi et al., 2005
Chitosan/(HA)	3D porous cylinders	Particle aggregation	ADAS cells (human adipose tissue, primary culture)	Bone	Malafaya, Pedro et al., 2005
Chitosan/(HA)	3D porous cuboids	3D-Printing	Osteoblasts (human calvaria, primary culture)	Bone	Takata, Wang et al., 2001
Chitosan/nano-HA	3D porous blocks	Freeze-drying	Osteoblast-like MC3T3-E1 (newborn mouse calvaria cell line)	Bone	Kong, Gao et al., 2005
Chitosan/β-TCP	3D porous blocks	Freeze-drying	Osteoblast-like MG63 (human osteosarcoma cell line)	Bone	Zhang and Zhang, 2001; Zhang, Ni et al., 2003
Chitosan/coralline	3D porous cylinders	Freeze-drying	MSC's CRL-12424 (mouse bone marrow cell line)	Bone	Gravel, Gross et al., 2006
Chitosan/gelatin/ HA	3D porous disks	Freeze-drying	Osteoblasts (neonatal rats calvaria, primary culture)	Bone	Zhao, Yin et al., 2002
Chitosan/gelatin/ (HA)	3D porous disks	Freeze-drying	MSC's (human bone marrow, primary culture)	Bone	Zhao, Grayson et al., 2006

(Continued)

Table 6.2 (*Continued*)

Material	Scaffold Structure	Processing methodology	Cell type (source)	TE application	References
Chitosan	3D porous disks	Freeze-drying	Chondrocytes (pig knee and dogs shoulder joint, primary culture)	Cartilage	Kim, Park et al., 2003; Subramanian and Lin 2005; Ong, Trabbic-Carlson et al., 2006
Chitosan	3D porous cylinders	Particle aggregation	ADAS cells (human adipose tissue, primary culture)	Cartilage	Malafaya, Pedro et al., 2006
Chitosan/polyester	3D fibber meshes	Fibber extrusion	Chondrocytes (bovine knee, primary culture)	Cartilage	Correlo, Boesel et al., 2005a; 20005b; Oliveira, Correlo et al., 2005
Chitosan/gelatin	3D porous cylinders	Freeze-drying	Chondrocytes (rabbit knee and pig auricular cartilage, primary culture)	Cartilage	Xia, Liu et al., 2004; Guo, Zhao et al., 2006
Chitosan/GP	Injectable gel	Gelation by polyol salts	Chondrocytes (calf knee, primary culture)	Cartilage	Chenite, Chaput et al., 2000; Hoemann, Sun et al., 2005
Chitosan/hyaluronan	3D fibers sheets	Wet spinning	Chondrocytes (rabbits knee, hip, and shoulder joints, primary culture)	Cartilage	Yamane, Iwasaki et al., 2005
Chitosan/alginate	3D porous cylinders	Freeze-drying	Chondrocyte-like HTB-94 (human bone chondrosarcoma cell line)	Cartilage	Ciardelli, Chiono et al., 2005
Chitosan/(HA)	Bilayered	Particle aggregation	ADAS cells (human adipose tissue, primary culture)	Osteochondral	Malafaya, Pedro et al., 2005

Chitosan/hyaluronan	Bilayered with PLA	Freeze-drying	In vivo (rabbit femoral chondyle)	Osteochondral	Frenkel, Bradica et al., 2005
Chitosan	Porous membranes	Freeze-drying	-	Skin	Adekogbe and Ghanem, 2005
Chitosan/gelatin	Bilayered porous membranes	Freeze-drying	Co-culture of fibroblasts and keratinocytes (human skin, primary culture)	Skin	Mao, Zhao et al., 2003; Mao, Zhao et al., 2003
Chitosan/collagen	Porous membranes	Freeze-drying	Co-culture of fibroblasts and keratinocytes (human foreskin, primary culture)	Skin	Black, Bouez et al., 2005
Chitosan	Tubular system	Wire-heating and freeze-drying	-	Neural	Huang, Huang et al., 2005
Chitosan	Porous hollow conduits	Thermal-induced phase separation	Neuro-2a cells (mouse neuroblastoma cell line)	Neural	Chandy, Rao et al., 2003; Wang, Wu et al., 2005; Ao, Wang et al., 2006
Chitosan derivative	Tubular system	Crab tendon treatment	In vivo (rat sciatic nerve)	Neural	Itoh, Matsuda et al., 2005
Chitosan/hyaluronan	3D fibers sheets	Wet spinning	Fibroblasts (rabbit patellar tendon, primary culture)	Ligament	Funakoshi, Majima et al., 2005
Chitosan/collagen	3D porous blocks	Freeze-drying	Hepatocytes (rat liver, primary culture)	Liver	Wang, Bogracheva et al., 1998
Chitosan/gelatin	3D hydrogel cylinders	Solvent casting	Respiratory epithelial cells (human tissue, primary culture)	Tracheal	Risbud, Endres et al., 2001

(): with or without; HA: hydroxylapatite; β-TCP: β-tricalcium phosphate; GP: glycerophosphate disodium salt; PLA: Polylactic acid; MSC's: mesenchymal stem cells; ROS: rat osteosarcoma; ADAS cells: adipose derived adult stem cells

candidate scaffold for these applications. In fact, the combination of good biocompatibility, intrinsic antibacterial activity, and ability to bind to growth factors renders this material as a good potential for several tissue engineering applications.

6.3.3 Cellulose

Cellulose is the main component of plant cell walls. It also constitutes the most abundant, renewable polymer resource available today, existing mainly in lignocellulosic material in forests, with wood being the most important source. The primary structure of this linear polymer consists of up to 15,000 D-glucose residues linked by $\beta(1\rightarrow4)$-glycosidic bond (Voet, 1995) (Figure 6.6). The fully equatorial conformation of β-linked glucopyranose residues stabilizes the chain structure, minimizing its flexibility. It is the ability of these chains to hydrogen-bond together into fibres (microfibrils) that gives cellulose its unique properties of mechanical strength and chemical stability, leading also to insoluble materials with small degradability *in vivo* (Entcheva, Bien *et al.*, 2004). The biodegradability of cellulose is considered to be limited, if it occurs at all, because of the

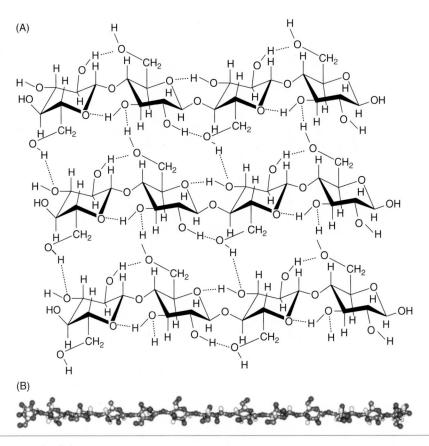

Figure 6.6 Structure of cellulose.

absence of hydrolases that attack the β(1→4) linkage (Beguin and Aubert, 1994). This fact, together with the difficult processing, is the most limited factor for the use of cellulose in tissue engineering applications. However, some partial degradation in processed cellulose sponges *in vivo* was reported (Martson, Viljanto *et al.*, 1999). The modification of the highly regular structural order of cellulose may also improve and tailor its degradation, as well as its tissue response (Miyamoto, Takahashi *et al.*, 1989). For example, *in vitro* and *in vivo* studies on acetyl-cellulose and ethyl-cellulose sponges allowed to conclude that, for the first case, a gradual degradation overtime could be detected, consistent with the observation on implanted sponges in Wistar rats (Elcin, 2006).

Cellulose and its derivatives have been employed with success as biomaterials, and there are some indications that they could be an adequate source for tissue engineering applications. In orthopaedic applications, it has been shown that cellulose sponges could support bone tissue ingrowths, suggesting that it could be used in bone tissue engineering (Martson, Viljanto *et al.*, 1998). Takata and co-workers (Takata, Wang *et al.*, 2001) compared different membranes for guided bone regeneration and, among other materials, cellulose which also exhibited the ability to induce cell migration. Cellulose acetate and cellulose scaffolds also showed to be interesting for cardiac tissue regeneration, as they could promote cardiac cell growth, enhancing cell connectivity and electrical functionality (Entcheva, Bien *et al.*, 2004).

Bacterial cellulose (BC) is a biotechnological method for producing pure nanofibrilar cellulose structures that have high mechanical strength, high water content and high crystallinity (see for instances refs in Backdahl, Helenius *et al.*, 2006). BC is excreted extracellularly by the *Acetobacter xylinum* bacteria, and the pellicle formed has been proposed to be used in tissue engineering related applications. The biocompatibility of BC was confirmed *in vivo* where subcutaneous implantations in rats did not show substantial inflammatory response (Helenius, Backdahl *et al.*, 2006). BC was shown to be able to support the proliferation of bovine-derived chondrocytes, and thus suggested the potential for use in tissue engineering of cartilage (Svensson, Nicklasson *et al.*, 2005). Moreover, the adequate mechanical properties of BC pellicles and the fact that smooth muscle cells adhere to and proliferate onto it suggested that BC could also be attractive for tissue engineering of blood vessels (Backdahl, Helenius *et al.*, 2006).

The properties of cellulose can be highly altered with chemical modification (e.g. through the substitution of the hydroxyl groups) allowing expansion and tailoring of the physical features and the response to tissues of this material. A few cellulose derivatives have been specifically proposed for tissue engineering purposes. For example, 2,3-dialdehydecellulose porous membranes were prepared from methylcellulose combining water-induced phase separation and salt leaching techniques; this material is biodegradable and has been used as a drug carrier. Human neonatal skin fibroblast cells attached and spread on these membranes (RoyChowdhury and Kumar, 2006). Hydroxypropyl methylcellulose grafted with silanol groups was developed as a injectable and self-setting hydrogel, that could be used to deliver and fix cells into a site through a non-evasive procedure (Vinatier, Magne *et al.*, 2005). Chondrocytes from two different origins were found to maintain their viability and to proliferate when cultured into the hydrogel (Vinatier, Magne *et al.*, 2005), indicating that cellulose derivatives could be also used as a carrier of chondrocytes in cartilage tissue engineering. Cellulose sulphate was found to be biocompatible and non-immunogenic and also showed to be adequate to encapsulate cells, to be used, for example, in protecting pancreatic xenogeneic cells from the immune system as a potentially curative treatment option for diabetes (Schaffellner, Stadlbauer *et al.*, 2005).

Cellulose and its derivatives may be seen as a potential source of natural based materials in tissue engineering applications. There is, however, more work to be done in order to enhance the degradation rate and to found more processing routes to produce scaffolds with controlled architectures.

6.3.4 Starch

Starch is the dominant carbohydrate reserve material of higher plants, being found in the leaf chloroplasts and in the amyoplasts of storage organs such as seeds and tubers (Wang, Bogracheva *et al.*, 1998). Although there is a broad range of possible origins of native starch, most of the starch utilized world-wide comes from a relatively small number of crops, the most important being corn, potato, wheat, and tapioca with smaller amounts from rice, sorghum, sweet potato, arrowroot, sago, and mung beans (Wang, Bogracheva *et al.*, 1998).

As a natural polymer, it has received great attention as a possible alternative to synthetic polymers in several applications, mainly for being one of the cheapest biopolymers, being totally biodegradable into carbon dioxide and water (Galliard, 1987), and abundantly available (Galliard, 1987; Reis and Cunha, 2001; Xie, Yu *et al.*, 2006).

Native starch is composed of granules of variable sizes and shapes depending on the source of the starch. Chemically starch is a polysaccharide consisting only of homoglucan units (Galliard, 1987; Reis and Cunha, 2001). Starch is constituted by α-D-glucose units, which can be organized to form two distinct molecules, amylose and amylopectin (Avérous, 2004; Beery and Ladisch, 2001; Galliard, 1987; Eagles, Lesnoy *et al.*, 1996; Wang, Bogracheva *et al.*, 1998). The typical structure of amylose consists of a linear, very sparsely branched, polymer linked basically by 1→4 bonds (Avérous, 2004; Eagles, Lesnoy *et al.*, 1996; Galliard, 1987; Reis and Cunha, 2001; Willett, Jasberg *et al.*, 1995). On the contrary, amylopectin is highly branched on multiple points of the backbone, and contains not only 1→4 bonds, but also 1→6 branching points, that tend to appear each 25–30 glucose units (Galliard, 1987; Willett, Jasberg *et al.*, 1995; Eagles, Lesnoy *et al.*, 1996; Avérous, 2004). The correspondent molecular weights are around 10^5 to 10^6 for amylose and 10^7 to 10^9 g/mol for amylopectin (Beery and Ladisch, 2001; Galliard, 1987). The distinct molecular weights and degrees of branching of both molecules are responsible for the quite different properties of starch isolated from sources with diverse amylose/amylopectin relative ratios (Avérous, 2004; Galliard, 1987; Reis and Cunha, 2001) (Figure 6.7). Besides its two basic macromolecular constituents, traces of lipids, proteins and minerals (mainly phosphates) may also be found on native starch (Avérous, 2004).

One of the most important properties of native starch is its semi-crystallinity. Depending on the source and the moisture content, the degree of crystallinity in native starch ranges between 15% to 50% (Yilmaz, Jongboom *et al.*, 2006). The crystallinity of starch is due to amylose and amylopectin, but mostly depends on amylopectin. Even though amylopectin has a branched structure, the branches form double helices between branches. The starch granule has been found to have alternating crystalline and amorphous concentric layers. The amorphous may be due to areas where the α-1,6 branch points form the chains, while the crystalline regions arise when the joined α-1,4 joined branches intertwine with each other and form double helices (Beery and Ladisch, 2001), resulting in the formation of parallel crytalline lamellae (Yilmaz, Jongboom *et al.*, 2006).

Besides its use as a filler material, native starch must be modified by destructuring of its granular structure to find other applications. The destructuring agent is usually water. The disruption of the granule organization obtained with the combination of water and heat is termed 'gelatinization' and is characterized by the swelling of starch, forming a viscous past with destruction of most inter-molecular hydrogen links (Avérous, 2004; Reis and Cunha, 2001).

To be able to make a thermoplastic starch (TPS) that can be processed by conventional processing techniques such as extrusion or injection moulding, it is necessary to disrupt the granule and melt the partially crystalline nature of starch in the granule (Avérous, 2004; Da Róz, Carvalho *et al.*, 2006; Willett, Jasberg *et al.*, 1995). For granular starch the glass transition temperature (T_g) is above the T_d of the polymer chains due to the strong interactions by hydrogen bonding of the chains. Several authors have estimated the T_g of dry starch to be in the 230−250°C (Avérous, 2004). Therefore, plasticizers have to be added to lower the T_g beneath the T_d. Very important factors that will determine the final properties of TPS

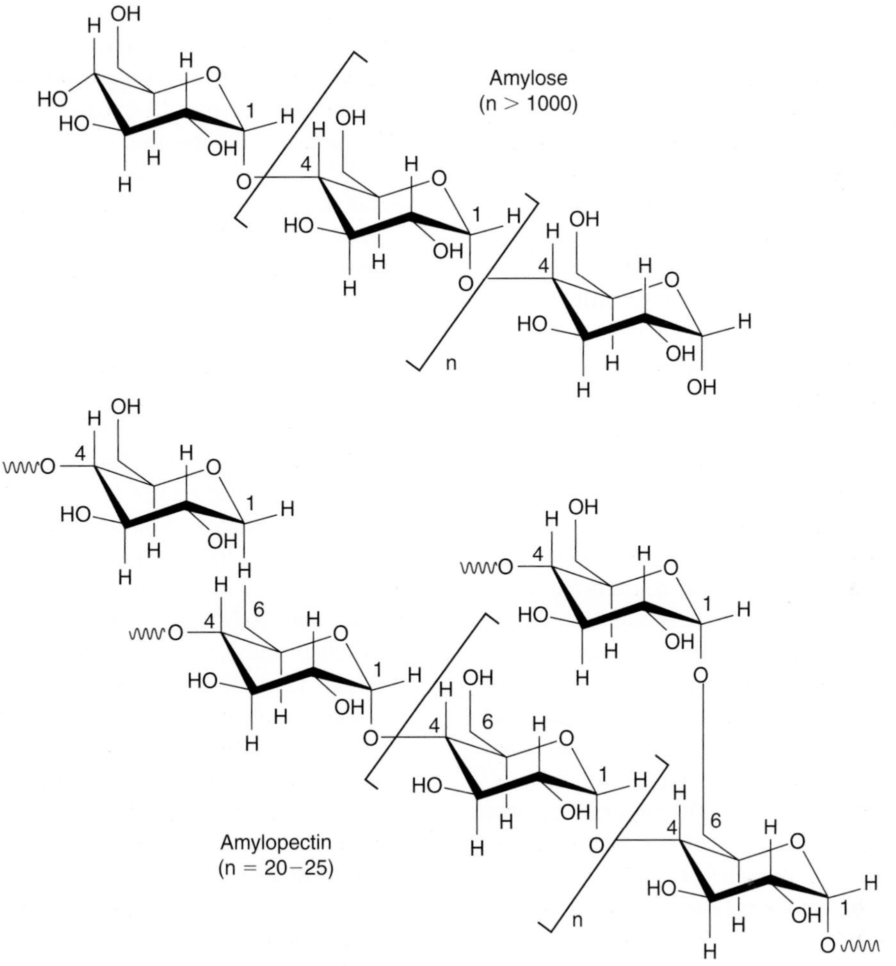

Starch
10–20% amylose (1–4-α-D-glucopyranoside) polymer
and
80–90% amylopectin (1–4, 1–6'-α-D-glucopyranoside)
>100 branched, interconnecting chains of 20–25 glucose units

Figure 6.7 Structure of the two molecules that constitute starch, amylase and amylopectin. The amylose content of tarch can vary between 10% to 20% and the amylopectin content from 80% to 90%. The different ratios of amylase/amylopectin found in starch isolated from different sources, determine its properties.

products are, among others, the type and amount of used plasticizers, the amylose/amylopectin ratio, the molecular weight of the starch (both mainly depend on the plant of origin), and the final crystallinity of the products (Avérous, 2004; Reis and Cunha, 2001; Willett, Jasberg *et al.*, 1995). Important plasticizers are water, and several polyols such as glycerol and glycol (Avérous, 2004; Reis and Cunha, 2001).

Nevertheless, the application of unblended TPS is limited because of the thermal sensitivity and degradation of starch due to water loss at elevated temperatures (Galliard, 1987; Reis and Cunha, 2001). Generally for temperatures exceeding 180−190°C, rapid degradation occurs during processing of TPS. The behavior of TPS is glassy and materials can only be processed by the addition of water, other plasticizers or melt flow accelerators.

To overcome difficulties associated with the limited applicability of unblended TPS, while the starch is being destructurized in the extruder, it is possible to add, together with the plasticizers and other additives, other polymers in order to create biodegradable blends that will confer a more thermoplastic nature to the TPS. Other aimed properties are a better resistance to thermo-mechanical degradation, meaning that the blends are more readably processable, have a less brittle nature, and enhanced resistance to water and ageing, as compared to fully starch thermoplastics. This has led to the development of a large range of starch based thermoplastic blends for several different applications, including in the biomedical field.

Reis and others (Gomes, Bossano *et al.*, 2006; Gomes, Godinho *et al.*, 2002; Gomes, Reis *et al.*, 2001; Gomes, Ribeiro *et al.*, 2001; Gomes, Sikavitsas *et al.*, 2003; Mano, Vaz *et al.*, 1999; Marques, Reis *et al.*, 2002; Mendes, Bezemer *et al.*, 2003; Oliveira, Malafaya *et al.*, 2003; Oliveira and Reis, 2004; Salgado, Coutinho *et al.*, 2004; Salgado, Figueiredo *et al.*, 2005; Salgado, Gomes *et al.*, 2002; Tuzlakoglu, Bolgen *et al.*, 2005) have developed an extensive work using blends of corn starch (in amounts varying from 30 up to 50%wt) with several different synthetic polymers, namely, poly(ethylene vinyl alcohol) (SEVA-C), cellulose acetate (SCA), poly(ε-caprolactone) (SPCL) and poly-lactic acid (SPLA) (Bastiolli, Belloti *et al.*, 1993; Bastiolli, 1995; Reis, Mendes *et al.*, 1997). These polymers can be designed into distinct structural forms and/or properties by tailoring the synthetic component of the starch-based blend, their processing methods, and the incorporation of additives and reinforcement materials. These polymeric blends are degraded by hydrolytic processes and several enzymes

(Araújo, Vaz *et al.*, 2001; Azevedo, Gama *et al.*, 2003) can also be involved in the process, mainly alpha-amylase, beta-amylase, alpha-glucosidase and other debranching enzymes (Azevedo, Gama *et al.*, 2003). The biocompatibility and non-immunogenecity of starch-based polymers has been well demonstrated by several *in vitro* (Gomes, Reis *et al.*, 2001; Marques, Reis *et al.*, 2002; Salgado, Coutinho *et al.*, 2004) and in vivo studies (Marques, Reis *et al.*, 2002; Mendes, Reis *et al.*, 2001). For all these reasons, starch-based polymers have been suggested for a wide range of biomedical applications, such as partially degradable bone cements (Boesel, Fernandes *et al.*, 2004; Boesel, Mano *et al.*, 2004; Boesel and Reis, 2004; Espigares, Elvira *et al.*, 2002; Pereira, Cunha *et al.*, 1998) as systems for controlled release of drugs (Gomes, Ribeiro *et al.*, 2001; Elvira, Mano *et al.*, 2002; Oliveira, Malafaya *et al.*, 2003; Pereira, Cunha *et al.*, 1998) as bone substitutes in the orthopaedic field (Reis, Cunha *et al.*, 1996; Reis, Cunha *et al.*, 1997; Sousa, Kalay *et al.*, 2000; Sousa, Mano *et al.*, 2002; Sousa, Reis *et al.*, 2003) or as scaffolds for tissue engineering (Gomes, Godinho *et al.*, 2002a, 2002b; Gomes, Holtorf *et al.*, 2006; Gomes, Ribeiro *et al.*, 2001; Gomes, Salgado *et al.*, 2002; Gomes, Malafaya *et al.*, 2004; Pereira, Gomes *et al.*, 1998; Salgado, Coutinho *et al.*, 2004; Salgado, Gomes *et al.*, 2002). A wide range of starch based scaffolds have been developed exhibiting different properties and porous architectures, using several different processing methodologies, from conventional melt based technologies, such as extrusion and injection moulding using blowing agents (Gomes, Godinho *et al.*, 2002; Gomes, Ribeiro *et al.*, 2001) to innovative techniques, such as microwave baking (Gomes, Ribeiro *et al.*, 2001). Some of these scaffolds have been successfully used in bone tissue engineering studies using human osteoblasts (Salgado, Coutinho *et al.*, 2004; Salgado, Gomes *et al.*, 2002) and rat bone marrow stromal cells (Gomes, Holtorf *et al.*, 2006; Gomes, Sikavitsas *et al.*, 2003; Mendes, Bezemer *et al.*, 2003). These and some other examples of tissue engineering studies performed using of starch based polymers as scaffold materials are summarized in Table 6.3.

Table 6.3 Examples of the use of starch based polymers in tissue engineering research.

Polymer	Scaffold	Processing methodology	Cells/animal model	TE Application	Reference
Starch/ polycaprolactone	2D porous scaffold	Selective laser sintering	NIH-3T3 mouse fibroblasts	not specified	(Ciardelli, Chiono et al., 2005)
Starch/dextran/ gelatin	3D porous cylindrical	Rapid prototyping technologies (3D printing)	not shown	not specified	(Lam, Mo et al., 2002)
Starch/ polycaprolactone	Fiber mesh	Fiber bonding	rat marrow stromal cells	bone, cartilage	(Gomes, Sikavitsas et al., 2003; Tuzlakoglu, Bolgen et al., 2005; Gomes, Bossano et al., 2006; Gomes, Holtorf et al., 2006)
			micro- and macrovascular endothelial cells	bone	(Santos, Fuchs et al., 2007)
	Nano-micro fiber mesh	Fiber bonding + electrospinning	rat marrow stromal cells	bone	(Tuzlakoglu, Bolgen et al., 2005)
Starch/ethylene-vinyl alcohol	3D porous	Extrusion/injection moulding with blowing agents	Human osteoblast-like cells (SaOS-2)	bone	(Gomes, Reis et al., 2001; Gomes, Ribeiro et al., 2001; Gomes, Godinho et al., 2002; Salgado, Gomes et al., 2002; Salgado, Coutinho et al., 2004; Salgado, Figueiredo et al., 2005)

6.3.5 Hyaluronan

Hyaluronan, formely known as hyaluronic acid (HA), is a natural and highly hydrophilic polysaccharide, which has been found to be a key constituent of native extracellular matrix (ECM) and tissues. HA belongs to the family of glycosaminoglycans and is synthesized as a large, negatively charged and linear polysaccharide of varying chain length (2−25 μm) composed of repeating disaccharide units (Figure 6.8) (Toole, 2004).

It is believed that interactions between HA with other ECM macromolecules and chondrocytes on the one hand, and its hydrodynamic characteristics, especially its viscosity and ability to retain water on the other, are critical for the maintenance of both cartilage homeostasis and biomechanical integrity. The characteristics of HA are, to a great extent, responsible for the regulation of the porosity and malleability of these matrices. Moreover, in the human body, HA is cleaved by enzymes called hyaluronidases (Menzel and Farr, 1998), showing that the cells of the host or the ones present in the engineered tissue, may regulate

the local clearance of the material, while the new tissue is being formed. Moreover, it has been demonstrated that HA is biocompatible (Ehlers, Behrens *et al.*, 2001) and has a greater bacteriostatic effect when compared with other matrices namely, collagen type I, poly(lactide-co-glycolide) (PLGA) and hydroxyapatite (Carlson, Dragoo *et al.*, 2004).

Ehlers and others (Hegewald, Ringe *et al.*, 2004; Zou, Li *et al.*, 2004) reported a double action of HA on chondrocytes, when added to the culture media. Results showed that chondrocytes presented a great tendency to differentiate and a higher rate of proliferation. These findings are corroborated by the works of other authors, which demonstrated that HA also stimulates bone marrow stromal cells proliferation and differentiation. These results are quite interesting since it is known that usually a differentiation-inducing stimuli leads to a lower cell proliferation. Therefore, HA possess some of the required features that are in the basis of choosing a material suitable for tissue engineering scaffolding; although, little is known about the mechanical properties of the HA molecules. The pioneering work of Fujii and co-workers (Fujii, Sun *et al.*, 2002) has demonstrated that the persistence length of single hyaluronan molecules is about 4.5 nm. This is important data for designing tissue engineering scaffolds, where the mechanical component is essential.

It was also found that HA interacts with cell surfaces in two ways, by binding to specific cell-surface receptors such as the hyaluronan receptor CD44 (van der Kraan, Buma *et al.*, 2002) and RHAMM (receptor for hyaluronan-mediated motility), and sustained transmembrane interactions with its synthetases (Toole, 2004). The binding of chondrocytes to HA through the CD44 receptor greatly affects the functioning of these cells, thus cartilage homeostasis. In fact, the blocking of the CD44 receptors of chondrocytes results in the degradation of cartilage matrix. Although, the physiological role of HA is not only restricted to its participation in the synovial fluid of joints, umbilical cord and vitreous body of the eye (Menzel and Farr, 1998) but also has been described to be involved on the co-regulation of cell behavior

(A)

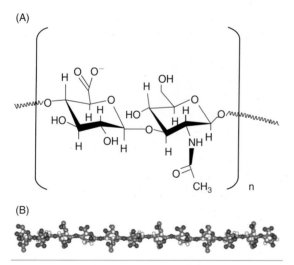

(B)

Figure 6.8 Structure of hyaluronan, which is composed of a repeating disaccharide of (1→3) and (1→4)-linked β-D-glucuronic acid and N-acetyl-β-D-glucosamine units.

during embryonic lung development, angiogenesis, wound healing processes, and inflammation (Toole, 2004). In the last few years, HA and its derivatives have been showing interesting results when used in medicine, namely in the treatment of several soft and hard tissue defects such as skin (Horch, Kopp *et al.*, 2005; Pianigiani, Andreassi *et al.*, 1999), blood vessels (Amarnath, Srinivas *et al.*, 2006; Peattie, Nayate *et al.*, 2004), eye (Cho, Chung *et al.*, 2003), ear (Campoccia, Doherty *et al.*, 1998), and bone (Gao, Dennis *et al.*, 2002) tissue. HA could certainly find other applications but the water solubility and rapid resorption preclude many clinical applications. To circumvent some of HA limitations, several authors have been proposing the modification of the HA molecular structure, inan attempt to obtain more stable HA-based materials. Covalent cross-linking (Liu, Shu *et al.*, 2005; Prestwich, Marecak *et al.*, 1998; Shu, Liu *et al.*, 2003), partial or total esterification of its free carboxylic groups (Campoccia, Doherty *et al.*, 1998), and annealing (Frenkel, Bradica *et al.*, 2005) are all ways to obtain a modified and stable form of HA. Most of hyaluronan-based polymers that can be obtained by cross-linking are water-insoluble gels or hydrogels (hylans), and much of which has still to be explored. In fact, HA and its derivatives offers a wide range of features that allows its prevalent use in tissue engineering as scaffolds since it can be used in the form of gels (Park, Park *et al.*, 2005; Shu, Liu *et al.*, 2004), sponges (Halbleib, Skurk *et al.*, 2003; Solchaga, Temenoff *et al.*, 2005), films (Liu, Shu *et al.*, 2005; Shu, Liu *et al.*, 2003), fibers (Funakoshi, Majima *et al.*, 2005; Milella, Brescia *et al.*, 2002) and microparticles (Lee, Park *et al.*, 2001). These materials met some of the criteria for its successful application not only in tissue engineering scaffolding (Halbleib, Skurk *et al.*, 2003) but also in drug delivery applications (Cho, Chung *et al.*, 2003; Lee, Park *et al.*, 2001). Despite this, a great deal of attention has been given to the development of alternative high-quality scaffolds. In this context, several authors have been proposing the blending with other polymers to utilize the benefits of each biomaterial. Hyaluronan has been combined with fibrin glue (useful cell delivery matrix) (Park,

Park *et al.*, 2005), alginate (Oerther, Payan *et al.*, 1999), collagen (Segura, Anderson *et al.*, 2005), gelatin (Shu, Liu *et al.*, 2003), laminin (Hou, Xu *et al.*, 2005), chitosan (Denuziere, Ferrier *et al.*, 1998; Yamane, Iwasaki *et al.*, 2005), polyesters (Yoo, Lee *et al.*, 2005) and calcium phosphates (Gao, Dennis *et al.*, 2002) to develop composite scaffolds for regeneration of several damaged tissues (Table 6.4).

It has been known for some time that HA also plays a key role the interactions with tumour cells (Knudson, Biswas *et al.*, 1989). In fact, there is an association of high levels of HA with malignancy of tumours (Toole, Wight *et al.*, 2002). These observations highlight the biological role of HA, demonstrating that this molecule can be a viable therapeutic target, which might be useful for developing more effective therapeutic strategies in the coming years.

6.4 Proteins

Proteins are the most abundant organic molecules within the cell extracellular and intracellular medium, where they ensure multiple biological functions such as transport, regulation of pathways, protection against foreign molecules, structural properties, protein storage, as well as being the catalyst for a great diversity of reactions, acting as biocatalysts (enzymes).

In a molecular perspective, proteins may be considered as polymer structures composed by 20 distinct amino acids linked by amide (or peptide) bonds. Amino acids are therefore the building blocks of polypeptides and proteins, which consist of a central carbon linked to an amine group, a carboxyl group, a hydrogen atom, and a side chain (R groups). R groups can be classified as nonpolar groups, uncharged polar groups or charged polar groups, which their distribution along the protein backbone renders proteins with distinct characteristics.

The structure of a protein is not, however, as simple as a polysaccharide, or other polymer. Generally, the protein structure is described on four levels. The *primary structure* of a protein is its amino acid

Table 6.4 Applications of hyaluronan and its derivatives in tissue engineering

Polymer	TE Application	References
Hyaluronan/fibrin glue	Articular cartilage	Park, Park et al., 2005
Hyaluronan/alginate	Articular cartilage	Oerther, Payan et al., 1999; Dausse, Grossin et al., 2003
Hyaluronan/chitosan	Articular cartilage and Skin	Denuziere, Ferrier et al., 1998; Funakoshi, Majima et al., 2005; Yamane, Iwasaki et al., 2005
Hyaluronan/collagen crosslinked via a poly(ethyleneglycol) diepoxide	Not specified	Segura, Anderson et al., 2005
Hyaluronan/collagen crosslinked via pyridinoline	Articular cartilage	Abe, Takahashi et al., 2005
Hyaluronan modified with methacylic anhydride	Articular cartilage	Burdick, Chung et al., 2005
Hyaluronan/elastin	Articular cartilage	Shu, Liu et al., 2003; Goodstone, Cartwright et al., 2004
Fibronectin-coated ACP™ (hyaluronan-based sponge)	Osteochondral	Solchaga, Temenoff et al., 2005
Hyaluronan/calcium phosphates	Bone and Osteochondral	Gao, Dennis et al., 2002
Hyaluronan/PLGA	Osteochondral and Articular cartilage	Solchaga, Temenoff et al., 2005; Yoo, Lee et al., 2005
Hyaluronan/laminin	Brain	Hou, Xu et al., 2005
Hyaff® (hyaluronan derivative obtained by esterifying the free carboxylic group)	Skin, cartilage, trachea and other soft tissue	Campoccia, Doherty et al., 1998; Milella, Brescia et al., 2002; Ziegelaar, Aigner et al., 2002; Halbleib, Skurk et al., 2003; Tonello, Zavan et al., 2003; Hemmrich, von Heimburg et al., 2005; Zavan, Brun et al., 2005; Arrigoni, Camozzi et al., 2006
Laserskin® (hyaluronan 100% esterified with benzyl alcohol)	Skin	Pianigiani, Andreassi et al., 1999; Horch, Kopp et al., 2005
Dissulfide-cross linked hyaluronan	Soft tissues	Liu, Shu et al., 2005
Hyaluronan-graft-poloxamer	Eye	Cho, Chung et al., 2003
Hylans (hydrogels based on cross-linked hyaluronan)	Vascular and Aortic heart valves	Peattie, Nayate et al., 2004; Ramamurthi and Vesely 2005; Amarnath, Srinivas et al., 2006
Non-woven Hyaff (esterified Hyaluronan)	Vascular	Turner, Kielty et al., 2004; Arrigoni, Camozzi et al., 2006

sequence, whereas the *secondary structure* refers to the local spatial arrangement of the polypeptide's backbone atoms without regard to the conformation of its side chains. The folding of the polypeptide chain is responsible for putting in close contact different parts of the chain to create binding sites to the substrate, etc. The *tertiary structure* is related with the three-dimensional structure of the entire polypeptide. When proteins are composed of more than one polypeptide chain (referred as subunits), the resultant spatial arrangement of its subunits is known as the protein's *quaternary structure.*

The configuration assumed by a protein, and thus the one which determines its properties, is the one which minimizes the molecule's free energy. Protein conformation is determinant for protein bioactivity, being known that a certain three-dimensional structure is essential for protein functionality. Most of the forces that stabilize the protein structure are weak (hydrogen bonding, ionic and hydrophobic interactions, van der Waals forces) giving some flexibility to the macromolecule. In general, non-polar amino acid side chains (e.g. phenylalanine, leucine, tryptophan, valine, etc) are located in the interior of the protein away from the aqueous solvent. The hydrophobic effects that promote this distribution are largely responsible for the three-dimensional structure of native proteins. On the contrary, ionized side chains tend to be on the surface of the molecule to interact with the aqueous solvent. In addition, the polypeptide chains of larger proteins tend to exist in structural domains independently folded and connected by segments of peptide chains.

Taking into account the low stabilities of protein conformations, these molecules are easily susceptible to denaturation by changing the balance of the weak interactions that maintain the native conformation. Proteins can be denatured by a variety of conditions and substances as heating, extreme pH, chaotropic agents, detergents, adsorption to certain surfaces, etc. (Bailey and Ollis, 1986).

We now focus on the most important proteins that have been studied for tissue engineering applications, such as collagen, elastin, soybean and silk fibroin.

6.4.1 Collagen

Collagen is the most abundant protein in mammalian tissues (cornea, blood vessels, skin, cartilage, bone, tendon and ligament) and is the main component of the extracellular matrix (ECM) (Gelse, Poschl *et al.*, 2003; Yang, Hillas *et al.*, 2004). Its main function is to maintain the structural integrity of vertebrates and many other organisms. However, collagens also exert important functions in the cell microenvironment and are involved in the storage and release of cell mediators, like growth factors (Gelse, Poschl *et al.*, 2003). More than 20 genetically distinct collagens have been identified (Friess, 1998; Gelse, Poschl *et al.*, 2003; Lee, Singla *et al.*, 2001; Yang, Hillas *et al.*, 2004), but the basic structure of all collagen is composed of three polypeptide chains, which wrap around one another to form three-stranded rope structure (triple helix, Figure 6.9B). Close-packing of the chains near the central axis imposes the requirement that glycine (Gly) occupies every third position, generating a $(X-Y-Gly)_n$ repeating sequence. Proline (Prol) and 4-hydroxyproline (Hyp), which in collagens constitute about 20% of all residues, are found almost exclusively in the X and Y positions, respectively. Therefore, the most common triplet in collagen is Prol-Hyp-Gly which accounts for about 10% of the total sequence (Bella, Eaton *et al.*, 1994) (Figure 6.9A). Peptides that contain Gly as every third residue and have large amounts of Prol and Hyp behave as triple helices in solution.

The individual triple helices are arranged to form fibrils which are of high tensile strength and can be further assembled and cross-linked (collagen fibrils are stabilized in the ECM by the enzyme lysyl oxidase).

In tissues that have to resist shear, tensile, or pressure forces, such as tendons, bone, cartilage, and skin, collagen is arranged in fibrils, with a characteristic 67 nm axial periodicity, which provides the tensile strength (Rosso, Marino *et al.*, 2005). Only collagen types I, II, III, V, and XI self-assemble into fibrils (Gelse, Poschl *et al.*, 2003). The fibrils are composed of collagen molecules, which consist of a triple helix of approximately 300 nm in length and 1.5 nm in

(A) (B)

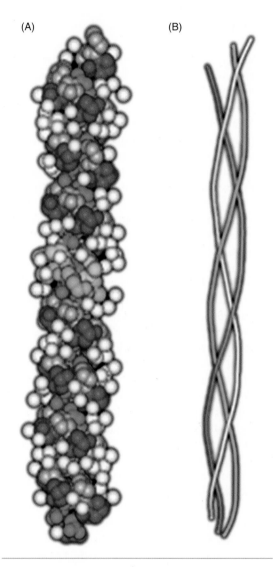

Figure 6.9 (A) CPK model of the structure of the triple-helical collagen-like peptide (1QSU, retrieved from Protein Data Bank at http://www.rcsb.org) showing Gly residues in green, Prol resides in grey and Hyp (magenta). All residues are exposed to the solvent (water molecules are displayed in white); (B) Schematic representation of the collagen-like peptide showing the triple helix. The structures were generated using the WebLab ViewerLite 3.7 program (Molecular Simulations Inc, USA).

diameter (Rosso, Marino *et al.*, 2005). Collagen fibril formation is an extracellular process, which occurs through the cleavage of terminal procollagen peptides by specific procollagen metalloproteinases. An article by Stevens and George (2005) provides an interesting schematic diagram showing the natural assembly of collagen fibres. Further reading about collagen fibril formation and molecular packing in collagen fibrils may be found in previous publications (Kadler, Holmes *et al.*, 1996; Pachence, 1996; Fratzl, 2003).

During the 1970s and 1980s, academics and commercial researchers began to use collagen as a biomaterial in a variety of connective tissue applications because of its excellent biocompatibility (Friess, 1998; Lee, Singla *et al.*, 2001), low antigenicity (Friess, 1998; Lee, Singla *et al.*, 2001), high biodegradability (Friess, 1998; O'Cearbhaill, Barron *et al.*, 2006; van Amerongen, Harmsen *et al.*, 2006), and good haemostatic and cell-binding properties (Boccafoschi, Habermehl *et al.*, 2005; Friess, 1998).

The primary sources of industrial collagens are from animal tissues (porcine and calf skin, bovine tendon, rat tail, etc.). It may readily be purified from animal tissues with enzyme treatment and salt/acid extraction. However, the use of animal-derived collagen raises concerns over the possible transmission of infectious agents such as viruses and prions (Yang, Hillas *et al.*, 2004). Transmissible bovine spongiform encephalopathy (BSE) is one of the most difficult contaminating agents to detect and remove from animal tissues. Therefore, different attempts have been made to find new and safer sources of collagen, namely from marine sources (e.g. jellyfish collagen) (Song, Yeon Kim *et al.*, 2006) or by producing human recombinant collagen (rhC) for clinical use (Yang, Hillas *et al.*, 2004) using different expression systems. The use of recombinant sources of human collagen provides a reliable, predictable and chemically defined source of purified human collagens that is free of animal components (please see for instances the review of Yang, Hillas *et al.*, 2004, for more details about the application of recombinant human collagen in tissue engineering). The triple-helical collagens made by recombinant technology have the same amino acid sequence as human tissue-derived collagen. In addition, collagen products can be purified from fibres, from molecules reconstituted as fibres, or from specific recombinant polypeptides with specific composition and conformation.

A feature common to all of these collagen materials is the need for stable chemical cross-linking to control the mechanical properties and the residence time in the body, and to some extent their potential immunogenicity. This can be achieved via chemical (glutaraldehyde, formaldehyde, carbodiimides, diphenylphosphoryl azide), physical (UV radiation, freeze drying, heating, thermaldehydration) and enzymatic cross-linking. These cross-linking agents react with specific amino acid residues on the collagen molecule imparting individual biochemical, thermal, and mechanical stability to the biomaterial. Collagen may be an ideal scaffold material, as it is the major component of the extracellular matrix, and because it can be processed into a wide variety of structures and shapes (sponges, fibres, films, 3D gels, fleeces — Table 6.5). Furthermore, collagen substrates can modify the morphology, migration and in certain cases the differentiation of cells due to the presence of cell adhesion sequences present on its structure (e.g. RGD). Collagen is naturally degraded by matrix metalloproteinases, specifically collagenase and serine proteases (Pachence, 1996). These enzymes are secreted by neutrophils during the foreign body reaction, allowing the collagen degradation to be controlled by the cells present at the implantation site (van Amerongen, Harmsen et al., 2006). However, its low thermal stability, due to its protein nature, does not allow collagen to be processed by melt-based techniques, limiting its processing to solvent-based methods and consequently the final properties of the scaffolds, normally characterized by poor mechanical strength. Another drawback related with the use of collagen materials is the requirement of additional chemical or physical cross-linking to confer mechanical strength and enzymatic resistance. Intermolecular cross-linking reduces the degradation rate by making collagen less susceptible to enzymatic attack. Collagen

has long been known to elicit minimal inflammatory and antigenic responses and has been approved by the Unites States Food and Drug Administration (FDA) for many types of medical applications, including wound dressings and artificial skin (Kim, Baez et al., 2000; Pachence, 1996). These properties of collagen emphasize its significance in tissue regeneration and its value as a scaffold material, being currently used in a great number of tissue engineering applications. Table 6.5 presents some examples of TE applications using collagen scaffolds. In addition, composites of collagens with glycosaminoglycans, as well as with synthetic biodegradable polymers and ceramics, have also been extensively studied for their potential application as scaffolds for tissue engineering. A vast number of publications can be found in the literature, covering a diversity of clinical applications, such as general surgery, orthopaedics, cardiovascular, dermatology, otorhinolaryngology, urology, dentistry, ophthalmology, plastic and reconstructive surgery (Pachence, 1996).

Although these examples offer encourage applications of collagen in tissue engineering, its low mechanical properties, the risk of viral infection, its antigenicity potential and fast biodegradation when implanted in the human body are, to some extent, limiting the clinical applications of this natural biomaterial.

6.4.2 Elastin

Elastin is another of the key structural proteins found in the extracellular matrices (ECM) of connective tissues (e.g. blood vessels, oesophagus, skin) that need to stretch and retract following mechanical loading and release (Joddar and Ramamurthi, 2006; Li, Mondrinos et al., 2005). It is found predominantly in the walls of arteries, lungs, intestines and skin, as well as other elastic tissues. However, unlike type I collagen, elastin has found little use as a biomaterial, due to two main reasons (Daamen, Nillesen et al., 2005; Daamen, van Moerkerk et al., 2003): (i) elastin preparations have a strong tendency to calcify upon

Table 6.5 Examples of application of collagen scaffolds in tissue engineering research.

Type of Scaffold	Processing methodology	TE application	Ref.
Collagen sponge with 11 mm in diameter and 2 mm in thickness and pore volume fraction of 97.5%.	Freeze-drying. Crosslinking by thermaldehydration.	Tooth tissue engineering Guided Tissue Regeneration (GTR) in Dentistry	Sumita, Honda *et al.*, 2006
Collagen membranes.	Conversion of rhCl monomers into oligomers and reconstitution into collagen fibrils. The resulting fibrillar networks were subsequently crosslinked with ethyl-3-(3-dimethyl aminopropyl)carbodiimide (EDC).		Yang, Hillas *et al.*, 2004
Scaffolds with predefined and reproducible internal channels with widths of 135 μm.	Rapid prototyping (solid freeform fabrication technology and critical point drying technique).	Cardiovascular (aortic valve, blood vessel) tissue engineering	Sachlos, Reis *et al.*, 2003; Taylor, Sachlos *et al.*, 2006
Scaffolds of 6 mm in diameter and 0.75 mm in thickness.	Freeze-drying, crosslinked with hexamethylenediisocyanate,		van Amerongen, Harmsen *et al.*, 2006
Flat sheets of collagen type I.	Solvent evaporation.		Boccafoschi, Habermehl *et al.*, 2005
Porous tubular scaffolds with an inner diameter of 3 mm, an outer diameter of 6 mm and a length of 4 cm.	Freeze-drying of a suspension of type I insoluble collagen and insoluble elastin. Crosslinked with a carbodiimide.		Engbers-Buijtenhuijs, Buttafoco *et al.*, 2006
Collagen-gel tubular constructs.	Polymerization into glass test tubes.		Seliktar, Black *et al.*, 2000

Type	Processing	Application	Reference
Type I collagen sponge with interconnected pores.	Discs cored from sheets of Ultrafoam® collagen hemostat (Davol Inc., Cranston RI).	Bone tissue engineering Bone graft substitutes	Meinel, Karageorgiou et al., 2004
Recombinant collagen sponges (porous micromatrice structures interconnected by homogenous thin sheets of recombinant human collagen I fibrils).	In-mold fibrillogenesis/crosslinking process followed by lyophilization. Crosslinking with EDC.		Yang, Hillas et al., 2004
Collagen gel (Atelocollagen gel from Koken Co., Tokyo, Japan). Dome shape of 0.8 cm diameter and 0.2 cm top height.	Gelation at 37 °C for 60 min.	Cartilage tissue engineering	Kino-Oka, Maeda et al., 2005
Matriderm®. 3-D structure made of purified collagen I of bovine epidermis and small amounts of elastine.	Freeze drying. Crosslinking with a carbodiimide.		Stark, Suck et al., 2006
Collagen-based wound dressings (membranes, fibres, sponges).	Several.	Dermal tissue engineering (artificial skin, skin substitutes).	Pachence 1996
Type I collagen contracted gels (discs).	Gelation at 37°C for 60 min.		Feng, Yamato et al., 2003
Fibrous scaffolds	Electrospinning. Crosslinking with 1,6-diisocyanatohexane.	TE applications in general	Li, Mondrinos et al., 2005

implantation, probably because of the microfibrillar components (mainly fibrillin) within the elastic fibre that are difficult to remove; (ii) the purification of elastin is complex (Daamen, Nillesen *et al.*, 2005). The insoluble nature of elastin as also limited in its use in traditional reconstituted matrix fabrication techniques (Berglund, Nerem *et al.*, 2004) and when applied only poorly defined elastin preparations have been used (Daamen, van Moerkerk *et al.*, 2003).

Elastin consists of several repetitive amino acid sequences, including VPGVG, APGVGV, VPGFGVGAG AND VPGG (Haider, Megeed *et al.*, 2004). Highly insoluble and extensively cross-linked, mature elastin is formed from tropoelastin, its soluble precursor (Debelle, Alix *et al.*, 1998). Tropoelastin is secreted from elastogenic cells as a 60-kDA monomer that is subjected to oxidation by lysyl oxidase. Subsequent protein-protein associations give rise to massive macroarrays of elastin (Li, Mondrinos *et al.*, 2005). The structure of tropoelastin consists of an alteration of hydrophobic regions, responsible for elasticity, and cross-linking domains. Additionally, it ends with a hydrophilic carboxy-terminal sequence containing its only two cysteine residues (Debelle, Alix *et al.*, 1998). As a consequence, elastin is a substantially insoluble protein network (Debelle, Alix *et al.*, 1998; Li, Mondrinos *et al.*, 2005). Soluble material is typically derived either as a fragmented elastin in the form of alpha- and kappa-elastin or preferable through expression of the natural monomer tropoelastin (Li, Mondrinos *et al.*, 2005). In the production of alpha elastin, bovine ligament elastin is treated (Leach, Wolinsky *et al.*, 2005) with a mild acid hydrolysis to yield a high molecular weight digest that retains the amino acid composition of native elastin. Despite structural heterogeneities resulting from the hydrolysis, α-elastin retains several key physicochemical properties of the nascent elastin. Nevertheless, the development of α-elastin based biomaterials is still a quite unexplored area (Figure 6.10).

Recombinant protein technologies have allowed the synthesis of well-defined elastin-derived polypeptides, which have driven insightful structure-function studies of tropoelastin, as well as several discrete elastin

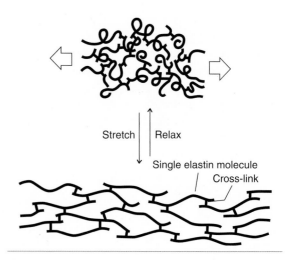

Figure 6.10 Structure of Elastin.

domains. Elastin-like polypeptides (ELPs) are artificial polypeptides with unique properties that make them attractive as biomaterial for tissue-engineering, as it has been demonstrated by the work of Urry and co-workers (Urry, Hugel *et al.*, 2002; Urry and Parker, 2002; Urry, Pattanaik *et al.*, 1998). ELPs consist of oligomeric repeats of the pentapeptide sequence Val-Pro-Gly-Xaa-Gly (Xaa is any amino acid except proline), a naturally occurring sequence in the protein elastin. ELPs are soluble in aqueous solution bellow their transition temperature (Tt) but when the solution temperature is raised above their Tt , the polymers start a complex self assembly process that leads to an aggregation of the polymer chains, initially forming nano- and micro-particles, which segregates from the solution (Betre, Setton *et al.*, 2002; Ong, Trabbic-Carlson *et al.*, 2006). This 'smart' nature may not be of particular interest for the final application of ELPs as ECM, but it is extremely important to simplify several steps in the production of ELPS and preparation of the ECM (Girotti, Reguera *et al.*, 2004). ELPs have also demonstrated an outstanding biocompatible behavior. Apparently, the immune response system of the human body does not differentiate the ELPs from endogenous elastin. Moreover, because of their protein nature, their bioabsorption is carried out by conventional metabolic routes, yielding just natural amino acids (Girotti, Reguera *et al.*, 2004). In addition, the matrices resulting from cross-linking of ELPs show a mechanical response quite similar to the natural elastin (Lee, Macosko *et al.*, 2001). This characteristic is very important for their application in tissue engineering, as the scaffold (artificial ECM) has to properly transmit the forces from the surrounding environment to the attached cells so that they can build new tissue that can eventually replace the artificial ECM (Girotti, Reguera *et al.*, 2004).

However, the broad application of these materials is limited by the inherent challenges of synthesizing recombinant proteins (e.g. residual endotoxin, capital cost and expertise, scale-up).

Table 6.6 gives several examples of tissue engineering studies in which elastin-based scaffolds were used.

6.4.3 Soybean

Soybeans belong to the legume family and can be processed into three kinds of protein-rich products: soy flour (Chabba, Mattews *et al.*, 2005; Swain, Biswal *et al.*, 2004), soy concentrate (Alibhai, Mondor *et al.*, 2006; Swain, Biswal *et al.*, 2004), and soy isolate (Schmidt, Giacomelli *et al.*, 2005; Swain, Biswal *et al.*, 2004). Soy protein, the major component of the soybean (30−45%) is readily available from renewable resources, is economically competitive, and presents good water resistance as well as storage stability (Hinds, Rowe *et al.*, 2006). About 90−95% of the soy is storage protein, with two subunits, namely 35% conglycinin (7S) and 52% glycinin (11S) (Kim, Nikolovski *et al.*, 1999). Due to its low cost and surface active properties, soy protein is of great importance to the food industry; specially as it provides stability against phase separation in food systems (Elizalde, Bartholomai *et al.*, 1996). Nevertheless, the combination of its properties with a similarity to tissue constituents and a reduced susceptibility to thermal degradation, makes soy an ideal template for use in biodegradable polymer for biomedical applications (Mano, Vaz *et al.*, 1999). Membranes, microparticles

and thermoplastics-based soy materials have been developed for tissue regeneration (Vaz, Graaf *et al.*, 2002). Biodegradable soy plastics have been developed by melt-based methods such as extrusion and injection moulding (Mano, Vaz *et al.*, 1999). Mano and co-workers (Mano, Vaz *et al.*, 1999) reported that soy protein-based thermoplastics presented a suitable range of mechanical and dynamical properties that might allow their use as biomaterials, namely in controlled release applications.

Soy protein has many reactive groups, such as $-NH_2$, $-OH$ and $-SH$, that are susceptible to chemical and physical modifications (Swain, Rao *et al.*, 2004). Some studies reported that the combination of soy protein with other proteins (e.g. wheat gluten (Were, Hettiarachchy *et al.*, 1999), casein (Mano, Vaz *et al.*, 1999) and polysaccharides such as cellulose (Xie, Yu *et al.*, 2006), dialdehyde starch (Rhim, Gennadios *et al.*, 1998) and chitosan (Silva, Elvira *et al.*, 2004; Silva, Santos *et al.*, 2005) in film form may promote physical and chemical interactions which improve some properties. Silva and co-workers (Silva, Oliveira *et al.*, 2006) have reported that, by means of combining a sol-gel process with the freeze-drying technique, it was possible to develop cross-linked porous structures based on chitosan and soy protein. It was demonstrated that the developed porous structures possess a suitable porosity and adequate interconnectivity. Furthermore, tetraethylorthosilicate (TEOS) can be used to introduce specific interactions in the interfaces between chitosan and soy protein, and improve its mechanical stability and degradability. Therefore, this work has shown that these structures have great potential for tissue engineering of cartilage.

6.4.4 Silk fibroin

Silk fibroin is a highly insoluble fibrous protein produced by domestic silk worms *(Bombyx mori)* containing up to 90% of the amino acids glycine, alanine, and serine leading to antiparallel β-pleated sheet formation in the fibers (Jin, Fridrikh *et al.*, 2005). Fibroin is a structural protein of silk

Table 6.6 Examples of application of elastin based scaffolds in tissue engineering research

Polymer	Scaffold	Processing methodology	Cells/animal model	TE Application	Reference
α-elastin	films	Crosslinking	Bovine aortic smooth muscle cells	Vascular tissue	Leach, Wolinsky et al., 2005
Elastin and tropoelastin	fibers	Electrospinning followed by crosslinking	HEPM	Not specified	Li, Mondrinos et al., 2005
Aortic Elastin	3D porous structure	Cyanogenbromide treatment for decellularization and removal of collagen and other ECM components.	3T3 mouse fibroblast cell line (ATCC)	Not specified	Lu, Ganesan et al., 2004
Collagen/elastin/ PLGA	Electrospun fiber meshes	Electrospinning	Bovine endothelial and smooth muscle cells	Vascular tissue	Stitzel, Liu et al., 2006
Elastin/collagen	3D structure composed of thin sheets and fibrils (collagen) and thick fibres (elastin)	Liophilization followed by crosslinking	Sprague-Dawley rats (subcutaneous pockets	Not specified	Daamen, Nillesen et al., 2005
Aortic Elastin	3D porous structure	Cyanogenbromide treatment (CNBr)	Sprague-Dawley rats (Subdermal implantation)	Vascular tissue	Simionescu, Lu et al., 2006
Collagen and elastin (1:1)	Tubular porous structures	Freeze-drying followed by crosslinking	Human smooth muscle cells	Vascular tissue	Buttafoco, Engbers-Buijtenhuijs et al., 2006; Engbers-Buijtenhuijs, Buttafoco et al., 2006
Elastin-like polypeptides (ELPs)	Injectable scaffolds	Gene design and synthesis	Pig chondrocytes	cartilage	Betre, Setton et al., 2002; Ong, Trabbic-Carlson et al., 2006

PLGA: poly(D,L-lactide-co-glycolide) HEPM: Human embryonic palatal mesenchyme

fibers and sericins are the water-soluble glue-like proteins that bind the fibroin fibers together (Altman, Diaz *et al.*, 2003) (see Figure 6.11). High purity silk fibroin fiber can be obtained easily from degummed silk (boiling-off), which refers to partial or complete removal of the sericin. Removal of the sericin coating before use removes the thrombogenic and inflammatory response of silk fibroin (Santin, Motta *et al.*, 1999). *Bombyx mori* silk fibroin can be dissolved with neutral salt solutions such as lithium bromide (LiBr), lithium thiocyanate (LiSCN), hexafluroisopropryl alcohol (HFIP) and calcium nitrate-methanol [Ca(NO$_3$)$_2$-MeOH] (Ha, Park *et al.*, 2003). Their mixtures are dialyzed to get pure fibroin solution, which can be used to prepare silk fibroin membranes, fiber, hydrogel, scaffolds and others types of materials (Altman, Diaz *et al.*, 2003). Traditionally, silk fibroin has been used for decades as suture material (Altman,

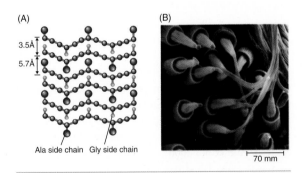

(A)

3.5Å

5.7Å

Ala side chain   Gly side chain

(B)

70 mm

Figure 6.11 Structure of silk. The fibers used to make silk cloth or a spider web are made up of the protein fibroin. (A) Fibroin consists of layers of antiparallel β sheets rich in Ala (purple) and Gly (yellow) residues. The small side chains interdigitate and allow close packing of each layered sheet, as shown in this side view; (B) Strands of fibroin (blue) emerge from the spinnerets of a spider in this colorized electron micrograph. Source: Nelson, D.L. and Cox, M.M. (2003). Lehninger Principles of Biochemistry, 3rd edn, Worth Publishers, New York, NY, p. 174. (From Nelson, D.L. and Cox, M.M. (2003). Lehninger Principles of Biochemistry, 3rd edn. Worth Publishers, New York, NY.)

Diaz *et al.*, 2003). Nowadays, several studies demonstrate the utility of silk matrices in films (Altman, Diaz *et al.*, 2003; Santin, Motta *et al.*, 1999; Servoli, Maniglio *et al.*, 2005), nanofibers (Ayutsede, Gandhi *et al.*, 2005; Jin, Chen *et al.*, 2004; Jin, Fridrikh *et al.*, 2005), hydrogels (Motta, Migliaresi *et al.*, 2004) and porous matrices (Kim, Park *et al.*, 2005; Nazarov, Jin *et al.*, 2004) for biomaterials and tissue engineering with stem cells for cartilage and bone applications. These applications of silk fibroin are related to its permeability to oxygen and water, cell adhesion and growth characteristics, slow degradability, low inflammatory response, and high tensile strength with flexibility (Altman, Diaz *et al.*, 2003). Porous three-dimensional scaffolds with silk fibroin have been obtained using various processing techniques (Table 6.7); these include salt leaching (Kim, Kim *et al.*, 2005; Kim, Park *et al.*, 2005; Nazarov, Jin *et al.*, 2004), electrospinning (Ayutsede, Gandhi *et al.*, 2005; Jin, Chen *et al.*, 2004; Li, Vepari *et al.*, 2006), freeze-drying (Gobin, Froude *et al.*, 2005; Li, Lu *et al.*, 2001; Li, Wu *et al.*, 2001; Li, Zhang *et al.*, 2002; Nazarov, Jin *et al.*, 2004), and gas-foaming processing (Nazarov, Jin *et al.*, 2004). Li and co-workers (Li, Lu *et al.*, 2001; Li, Wu *et al.*, 2001; Li, Zhang *et al.*, 2002) reported a series of studies on preparation conditions of porous silk fibroin materials and its relationship between the structure and properties. These materials were prepared by means of freeze-drying. A new process to form a silk fibroin spongy porous 3-D structure with both good porous structures and mechanical properties has also been reported (Tamada, 2005). This process involves freezing and a thawing fibroin aqueous solution in the presence of a small amount of water-miscible organic solvent. It requires no freeze-drying, no cross-linking chemicals, or the aid of other materials. In general, the silk scaffolds produced by different methods described here presented good porosity and mechanical properties which can be controlled by silk fibroin concentration, freezing temperature and particle size of salt used in the process. Other approaches to form silk scaffolds involved the blending of polymers, such as poly(ethylene oxide) (Jin, Fridrikh *et al.*, 2005), chitosan (Gobin,

Table 6.7 Examples of application of soy and silk based materials in tissue engineering studies

Polymer	Scaffold	Processing methodology	TE Application	Reference
Soy protein/				
Chitosan blend	not specified	Freeze-drying and sol-gel process	cartilage	Silva, Oliveira *et al.*, 2006
Silk fibroin	not specified	Freeze-drying	not specified	Li, Lu *et al.*, 2001; Li, Wu *et al.*, 2001; Li, Zhang *et al.*, 2002; Gobin, Froude *et al.*, 2005
Silk fibroin	not specified	Salt leaching	Cartilage, Bone	Meinel, Hofmann *et al.*, 2004; Kim, Kim *et al.*, 2005; Meinel, Fajardo *et al.*, 2005
Silk fibroin	not specified	Gas-foaming	not specified	Nazarov, Jin *et al.*, 2004
Silk fibroin	nanofiber	Electrospinning	Bone	Li, Vepari *et al.*, 2006

Froude *et al.*, 2005), or the surface modification of synthetic polymers such as poly(ε-caprolactone) (Chen, Zhou *et al.*, 2004) and polyurethane (Petrini, Parolari *et al.*, 2001) with silk fibroin coating in order to improve their collective properties, especially processability, mechanical properties and the biocompatibility, respectively.

With respect to using silk fibroin for cell culture, many researchers have investigated the effects of the silk matrices in nanofiber and porous matrix obtained by methodologies described previously on the culture of osteoblasts-like cells (Unger, Wolf *et al.*, 2004), human mesenchymal stem cells (Jin, Chen *et al.*, 2004; Kim, Kim *et al.*, 2005; Li, Vepari *et al.*, 2006; Meinel, Fajardo *et al.*, 2005; Meinel, Hofmann *et al.*, 2004) have shown very promising results regarding their application in cartilage and bone tissue engineering. Altman and co-workers (Jin, Chen *et al.*, 2004) concluded that electrospun silk matrices support bone marrow mesenchymal stem cells attachment, spreading and growth *in vitro*. Meinel and co-workers (Meinel, Fajardo *et al.*, 2005) reported the feasibility of silk-based implants with engineered bone for the (re-) generation of bone tissues. Recently, the potential of electrospun silk fibrous scaffold for bone formation from human bone marrow-derived mesenchymal stem cells (hMSCs) was explored by combining the unique structural features

generated by eletrospinning with functional factors, such as bone morphogenic protein-2 (BMP-2) and nanohydroxyapatite particles (Li, Vepari *et al.*, 2006).

6.5 Polyhydroxyalkanoates

Polyhydroxyalkanoates (PHAs) are naturally occurring biodegradable polymers. PHAs are synthesized and stored as water-insoluble inclusions in the cytoplasm of several bacteria and used as carbon and energy reserve materials (Anderson and Dawes, 1990; Sudesh, Abe *et al.*, 2000). The first PHA to be identified was poly(3-hydroxybutyric acid) (P[HB]). This homopolymer is the most abundant bacteria synthesized polyester and its 3-hydroxybutyrate (HB) monomer was thought to be the unique PHA constituent in bacteria (Dawes and Senior, 1973; Sudesh, Abe *et al.*, 2000). Further research (Wallen and Rohwedder, 1974) reported heteropolymers in chloroform extracts of activated sewage sludge, like 3-hydroxyvalerate (HV) among others. The introduction of other units in the PHA chain (besides 3HB) has a significant effect on mechanical behavior of the polyester (Barham, 1990; Barham, Barker *et al.*, 1992). The homopolymer of PHB is a brittle material, while the increase in HV content turns the HB-co-HV copolymer more ductile (Bauer and Owen, 1988; Mitomo, Barham *et al.*, 1988;

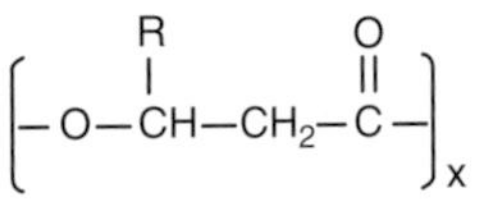

$$\left[-O-CH-CH_2-C-\right]_x$$

with R on the CH carbon and O double-bonded to C.

Figure 6.12 Structure of polyhydroxyalkanoates.

Owen, 1985). The mechanical behavior of PHAs depends on both the length of the pendant groups and the distance between ester linkages. PHAs with short pendant groups are prone to crystallization but exhibit stiff and brittle behavior, while PHAs with longer pendant groups are ductile (Williams, Martin *et al.*, 1999). The wide performance range of PHA copolymers justified additional scientific and industrial interest, which led to the discovery of further bacterial PHAs. PHAs can be synthesized in molecular weights which depend on the growth conditions and on the microorganism species – between, 200,000 to 3,000,000 Da (Sudesh, Abe *et al.*, 2000) (Figure 6.12).

The wide range of mechanical properties (Barham and Keller, 1986; Bauer and Owen, 1988; Gassner and Owen, 1996; Hobbs, 1998; Hobbs and Barham, 1998; Hobbs and Barham, 1999; Ishikawa, Kawaguchi *et al.*, 1991; Knowles, 1993; Owen, 1985; Scandola, Focarete *et al.*, 1997) coupled with the biodegradable (Gassner and Owen, 1996; Knowles, 1993; Chaput, Yahia *et al.*, 1995; Freier, Kunze *et al.*, 2002; Pouton and Akhtar, 1996; Scandola, Focarete *et al.*, 1997) and the biocompatible behaviors (Freier, Kunze *et al.*, 2002; Gogolewski, Jovanovic *et al.*, 1993; Shangguan, Wang *et al.*, 2006; Volova, Shishatskaya *et al.*, 2003; Wang, Yang *et al.*, 2005) of PHAs makes them potential biomedical candidates including drug delivery and tissue engineering applications (Pouton and Akhtar, 1996; Williams, Martin *et al.*, 1999). The biocompatibility assessment of PHAs has indicated that cell response also depends on the type of polyester. In a research study, the viable cell number of mouse fibroblasts (cell line L929) on polyhydroxybutyrate (PHB) films have increased more than two orders of magnitude upon blending with poly(hydroxybutyrate-co-hydroxyhexanoate) (PHBHH) (Yang, Zhao *et al.*, 2002). The influence of PHB content on mechanical behavior is also evident from the strong ductility increase which occurs with the introduction of PHBHH in PHBHH/PHB blends (Zhao, Deng *et al.*, 2003). Several studies (Deng, Zhao *et al.*, 2002; Rivard, Chaput *et al.*, 1996; Sodian, Hoerstrup *et al.*, 2000a, 2000b; Sodian, Sperling *et al.*, 2000; Wang, Wu *et al.*, 2004; Zheng, Bei *et al.*, 2005) reported the investigation of PHAs as potential scaffold materials in diverse range of tissue engineering applications. Sodian and co-workers developed a trileaflet heart valve from a porous PHA scaffold produced by salt leaching. Constructs were produced using vascular cells harvested from an ovine carotid artery and placed into a pulsatile flow bioreactor (Sodian, Hoerstrup *et al.*, 2000; Sodian, Sperling *et al.*, 2000). Results indicated that cells were mostly viable and grew into scaffolds pores. The formation of connective tissue between the inside and the outside of the porous heart valve scaffold was also observed. Other studies have focussed the assessment of PHA scaffolds for bone and cartilage tissue engineering. A study by Rivard and co-workers (Rivard, Chaput *et al.*, 1996) investigated the proliferation of ovine chondrocytes and osteoblasts in poly(β-hydroxybutyrate-β-hydroxyvalerate) scaffolds. Another study assessed the performance of porous PHBHH/PHB scaffolds, produced by salt leaching method, as matrices for three-dimensional growth of chondrocytes. Cell densities were higher for PHBHH/PHB scaffolds as compared to PHB scaffolds alone. The authors explained this discrepancy based on eventual differences in crystalline and amorphous arrangements between PHB and PHBHH/PHB scaffolds, as the presence of PHB crystalline domains may reduce oxygen permeability (Deng, Zhao *et al.*, 2002). Another study (Zheng, Bei *et al.*, 2005) has shown that PHB/PHBHH blends with 1:1 ratio have higher surface free energy as compared to PHB alone, which maximizes chondrocytes adhesion. Furthermore, polarity of the PHA on the scaffold seems to play an important role on what concerns cell morphology. In the case of PHBHH/PHB substrates, PHB content affects blend polarity which has an important effect on cell shape. Polarity increases with decreasing blend crystallinity, which affected chondrocyte shape, by altering it from spherical to flat.

6.6 Future developments

Several scaffolds based on natural origin polymers have been widely studied for tissue engineering application. Many of them exhibit unique advantageous features concerning intrinsic cellular interaction and degradability. However, these materials do also exhibit some disadvantages that limit their widespread use. Therefore, it is necessary to increase the knowledge about these natural polymers in order to enable the development of new approaches, including methods for production, purification, controlling material properties (molecular weight, mechanical, degradation rate) and for enhancing material biocompatibility (for instances by using non-animal derived production), in order to design better and more versatile scaffolding materials.

Tissue engineering scaffolding will also benefit from advances in recombinant protein technologies, which have proven to be a very powerful tool for the design and production of complex protein polymers with well-defined molecular weights, monomer compositions, sequences and stereochemistry. Very little has been explored within this new class of polymers and therefore much remains to be investigated about their versatility and possibilities of obtaining tailored properties for target applications. Accordingly, special interest has emerged for the use of these protein-based polymers for tissue engineering and other biomedical applications.

Further studies are expected to widen the range of natural origin materials (and combination of these with synthetic polymers) and the tailoring of their properties in order to make them even further suitable for applications within tissue engineering.

6.7 Summary

1. A wide range of natural origin polymers have frequently been used, and might in future be potentially useful in tissue engineering.
2. Tissue engineering scaffolds comprised of naturally derived macromolecules have potential advantages of biocompatibility, cell-controlled degradability, and intrinsic cellular interaction.
3. However, they may exhibit batch variations and, in many cases, exhibit a narrow and limited range of mechanical properties. In many cases, they can also be difficult to process by conventional methods.
4. In contrast, synthetic polymers can be prepared with precisely controlled structures and functions. However, many synthetic polymers do not degrade as desired in physiological conditions, and the use of toxic chemicals in their synthesis or processing may require extensive purification steps. Many of them are also not suitable for cell adhesion and proliferation.
5. The combination of natural origin polymer with synthetic polymers and the further development in emerging methodologies such as recombinant protein technologies is expected to lead to outstanding developments towards the development of improved materials to be used in tissue engineering application.
6. No material alone will satisfy all design parameters in all applications within the tissue engineering field, but a wide range of materials can be tailored for discrete applications.

References

Abe, M., Takahashi, M., *et al.* (2005). The effect of hyaluronic acid with different molecular weights on collage cross-link synthesis in cultured chondrocytes embedded in collagen gels. *Journal of Biomedical Materials Research*, 75A: 494–499.

Adekogbe, I. and Ghanem, A. (2005). Fabrication and characterization of DTBP-cross-linked chitosan scaffolds for skin tissue engineering. *Biomaterials*, 26(35): 7241.

Alibhai, Z., Mondor, M., *et al.* (2006). Production of soy protein concentrates/isolates: traditional and membrane technologies. *Desalination*, 191: 351–358.

Alsberg, E., Anderson, K.W., *et al.* (2001). Cell-interactive alginate hydrogels for bone tissue engineering. *Journal of Dental Research*, 80(11): 2025–2029.

Altman, G.H., Diaz, F., *et al.* (2003). Silk-based biomaterials. *Biomaterials*, 24(3): 401–416.

Amarnath, L.P., Srinivas, A., *et al.* (2006). *In vitro* hemo-compatibility testing of UV-modified hyaluronan hydrogels. *Biomaterials*, 27(8): 1416.

Anderson, A.J. and Dawes, E.A. (1990). Occurrence, metabolism, metabolic role, and industrial uses of bacterial polyhydroxyalkanoates. *Microbiological Reviews*, 54(4): 450–472+iii.

Ao, Q., Wang, A., *et al.* (2006). Manufacture of multimicrotubule chitosan nerve conduits with novel molds and characterization *in vitro*. *Journal of Biomedical Materials Research Part A*, 77A(1): 11–18.

Araújo, M., Vaz, C., *et al.* (2001). *In-vitro* degradation behaviour of starch/EVOH biomaterials. *Polymer Degradation and Stability*, 73: 237–244.

Arrigoni, C., Camozzi, D., *et al.* (2006). The effect of sodium ascorbate on the mechanical properties of hyaluronan-based vascular constructs. *Biomaterials*, 27(4): 623.

Avérous, L. (2004). Biodegradable multiphase systems based on lasticized starch: a review. *Journal of Macromolecular Science*, 44(3): 231–274.

Awad, H.A., Wickham, M.Q., *et al.* (2004). Chondrogenic differentiation of adipose-derived adult stem cells in agarose, alginate, and gelatin scaffolds. *Biomaterials*, 25(16): 3211–3222.

Ayutsede, J., Gandhi, M., *et al.* (2005). Regeneration of Bombyx mori silk by electrospinning. *Part 3: characterization of electrospun nonwoven mat. Polymer*, 46(5): 1625–1634.

Azevedo, H.S., Gama, F.M., *et al.* (2003). *In vitro* assessment of the enzymatic degradation of several starch based biomaterials. *Biomacromolecules*, 4(6): 1703–1712.

Backdahl, H., Helenius, G., *et al.* (2006). Mechanical properties of bacterial cellulose and interactions with smooth muscle cells. *Biomaterials*, 27(9): 2141–2149.

Bailey, J.E. and Ollis, D.F. (1986). *Biochemical Engineering Fundamentals*. Singapore: McGraw-Hill.

Baran, E. T., & Reis, R. L. (2003). *Development and in vitro evaluation of chitosan and soluble starch-chitosan nano-microparticles to be used as drug delivery vectors*. Society for Biomaterials 29th Annual Meeting Transactions, Reno, USA.

Barham, P.J. (1990). Physical properties of poly(hydroxybutyrate) and poly(hydroxybutyrate-co-hydroxyvalerate). *Novel Biodegradable Microbial Polymers*, : 81–96.

Barham, P.J., Barker, P., *et al.* (1992). Physical properties of poly(hydroxybutyrate) and copolymers of hydroxybutyrate and hydroxyvalerate. *FEMS Microbiology Reviews*, 103 (2–4): 289–298.

Barham, P.J. and Keller, A. (1986). Relationship between microstructure and mode of fracture in polyhydroxybutyrate. *Journal of Polymer Science. Part A-2, Polymer Physics*, 24(1): 69–77.

Bastiolli, C. (1995). Starch-polymer composites. In *Degradable Polymers* (Scott, G. and Gilead, D., eds). London: Chapman anmd Hall, pp. 112–137.

Bastiolli, C., Belloti, C., *et al.* (1993). Mater-Bi: properties and biodegradability. *J Env Polym Deg*, 1: 181–191.

Bauer, H. and Owen, A.J. (1988). Some structural and mechanical properties of bacterially produced poly-beta-hydroxybutyrate-co-beta-hydroxyvalerate. *Colloid and Polymer Science*, 266(3): 241–247.

Beery, K.E. and Ladisch, M.R. (2001). Chemistry and properties of starch based desiccants. *Enzyme Microb Technol*, 28(7-8): 573–581.

Beguin, P. and Aubert, J.P. (1994). The Biological Degradation of Cellulose. *Fems Microbiology Reviews*, 13(1): 25–58.

Bella, J., Eaton, M., *et al.* (1994). Crystal-Structure and Molecular-Structure of a Collagen-Like Peptide at 1.9-Angstrom Resolution. *Science*, 266(5182): 75–81.

Berglund, J.D., Nerem, R.M., *et al.* (2004). Incorporation of intact elastin scaffolds in tissue-engineered collagen-based vascular grafts. *Tissue Eng*, 10(9-10): 1526–1535.

Bertram, U. and Bodmeier, R. (2006). *In situ* gelling, bioadhesive nasal inserts for extended drug delivery: *In vitro* characterization of a new nasal dosage form. *European Journal of Pharmaceutical Sciences*, 27(1): 62.

Betre, H., Setton, L.A., *et al.* (2002). Characterization of a genetically engineered elastin-like polypeptide for cartilaginous tissue repair. *Biomacromolecules*, 3(5): 910–916.

Black, A.F., Bouez, C., *et al.* (2005). Optimization and characterization of an engineered human skin equivalent. *Tissue Engineering*, 11(5–6): 723–733.

Boccafoschi, F., Habermehl, J., *et al.* (2005). Biological performances of collagen-based scaffolds tor vascular tissue engineering. *Biomaterials*, 26(35): 7410–7417.

Boesel, L.F., Fernandes, M.H., *et al.* (2004). The behavior of novel hydrophilic composite bone cements in simulated body fluids. *J Biomed Mater Res*, 70B(2): 368–377.

Boesel, L.F., Mano, J.F., *et al.* (2004). Optimization of the formulation and mechanical properties of starch based partially degradable bone cements. *J Mater Sci Mater Med*, 15(1): 73–83.

Boesel, L.F. and Reis, R.L. (2004). Hydrophilic matrices to be used as bioactive and degradable bone cements. *J Mater Sci Mater Med*, 15(4): 503–506.

Broderick, E.P., O'Halloran, D.M., *et al.* (2005). Enzymatic stabilization of gelatin-based scaffolds. *Journal of Biomedical Materials Research Part B-Applied Biomaterials*, 72B(1): 37–42.

Burdick, J.A., Chung, C., *et al.* (2005). Controlled degradation and mechanical behavior of photopolymerized hyaluronic acid networks. *Biomacromolecules*, 6: 386–391.

Buschmann, M.D., Gluzband, Y.A., *et al.* (1995). Mechanical compression modulates matrix biosynthesis in chondrocyte/agarose culture. *J Cell Sci*, 108: 1497–1508.

Buttafoco, L., Engbers-Buijtenhuijs, P., *et al.* (2006). Physical characterization of vascular grafts cultured in a bioreactor. *Biomaterials*, 27(11): 2380–2389.

Campoccia, D., Doherty, P., *et al.* (1998). Semisynthetic resorbable materials from hyaluronan esterification. *Biomaterials*, 19(23): 2101.

Carlson, G.A., Dragoo, J.L., *et al.* (2004). Bacteriostatic properties of biomatrices against common orthopaedic pathogens. *Biochemical and Biophysical Research Communications*, 321(2): 472.

Chabba, S., Mattews, G.F., *et al.* (2005). Green composites using cross-linked soy flour and flax yarns. *Green Chemistry*, 7: 576–581.

Chandy, T., Rao, G.H., *et al.* (2003). The development of porous alginate/elastin/PEG composite matrix for cardiovascular engineering. *J Biomater Appl*, 17(4): 287–301.

Chang, S.C., Rowley, J.A., *et al.* (2001). Injection molding of chondrocyte/alginate constructs in the shape of facial implants. *Journal of Biomedical Materials Research*, 55(4): 503–511.

Chaput, C., Yahia, L.H. *et al.* (1995). *Natural poly(hydroxybutyrate-hydroxyvalerate) Polymers as degradable Biomaterials.* Materials Research Society Symposium - Proceedings, San Francisco, CA, USA, Materials Research Society.

Chellat, F., Tabrizian, M., *et al.* (2000). *In vitro* and *in vivo* biocompatibility of chitosan-xanthan polyionic complex. *Journal of Biomedical Materials Research*, 51(1): 107–116.

Chen, G., Zhou, P., *et al.* (2004). Silk fibroin modified porous poly(E-caprolactone) scaffold for human fibroblast culture *in vitro. Journal of Materials Science-Materials in Medicine*, 15(6): 671–677.

Chen, T.H., Embree, H.D., *et al.* (2003). Enzyme-catalyzed gel formation of gelatin and chitosan: potential for *in situ* applications. *Biomaterials*, 24(17): 2831–2841.

Chenite, A., Chaput, C., *et al.* (2000). Novel injectable neutral solutions of chitosan form biodegradable gels *in situ. Biomaterials*, 21(21): 2155.

Chenite, A., Chaput, C., *et al.* (2000). Novel injectable neutral solutions of chitosan form biodegradable gels *in situ. Biomaterials*, 21(21): 2155–2161.

Cho, C.S., Seo, S.J., *et al.* (2006). Galactose-carrying polymers as extracellular matrices for liver tissue engineering. *Biomaterials*, 27(4): 576.

Cho, K.Y., Chung, T.W., *et al.* (2003). Release of ciprofloxacin from poloxamer-graft-hyaluronic acid hydrogels *in vitro. International Journal of Pharmaceutics*, 260(1): 83.

Cho, S.H., Oh, S.H., *et al.* (2005). Fabrication and characterization of porous alginate/polyvinyl alcohol hybrid scaffolds for 3D cell culture. *Journal of Biomaterials Science-Polymer Edition*, 16(8): 933–947.

Ciardelli, G., Chiono, V., *et al.* (2005). Blends of poly-(epsilon-caprolactone) and polysaccharides in tissue engineering applications. *Biomacromolecules*, 6(4): 1961–1976.

Correlo, V.M., Boesel, L., *et al.* (2005). Hydroxyapatite Reinforced Chitosan and Polyester Blends for Biomedical Applications. *Macromolecular Materials Engineering*, 290(12): 1157–1165.

Correlo, V.M., Boesel, L., *et al.* (2005). Properties of Melt Processed Chitosan and Aliphatic Polyester Blends. *Materials Science and Engineering A*, 403(1–2): 57–68.

Da Róz, A.L., Carvalho, A.J.F., *et al.* (2006). The effect of plasticizers on thermoplastic starch compositions obtained by melt spinning. *Carbohydrate Polymers*, 63: 417–424.

Daamen, W.F., Nillesen, S.T., *et al.* (2005). Tissue response of defined collagen-elastin scaffolds in young and adult rats with special attention to calcification. *Biomaterials*, 26(1): 81–92.

Daamen, W.F., van Moerkerk, H.T., *et al.* (2003). Preparation and evaluation of molecularly-defined collagen-elastin-glycosaminoglycan scaffolds for tissue engineering. *Biomaterials*, 24(22): 4001–4009.

Dagalakis, N., Flink, J., *et al.* (1980). Design of an artificial skin. Part III. Control of pore structure. *J Biomed Mater Res*, 14(4): 511–528.

Dang, J.M. and Leong, K.W. (2006). Natural polymers for gene delivery and tissue engineering. *Adv Drug Deliv Rev*, 58(4): 487–499.

Dausse, Y., Grossin, L., *et al.* (2003). Cartilage repair using new polysaccharidic biomaterials: macroscopic, histological and biochemical approaches in a rat model of cartilage defect. *Osteoarthritis and Cartilage*, 11(1): 16.

Dawes, E.A. and Senior, P.J. (1973). The role and regulation of energy reserve polymers in micro-organisms. *Advances in Microbial Physiology*, 10: 135–266.

Debelle, L., Alix, A.J., et al. (1998). The secondary structure and architecture of human elastin. *Eur J Biochem*, 258(2): 533–539.

Deng, Y., Zhao, K., et al. (2002). Study on the three-dimensional proliferation of rabbit articular cartilage-derived chondrocytes on polyhydroxyalkanoate scaffolds. *Biomaterials*, 23(20): 4049–4056.

Denuziere, A., Ferrier, D., et al. (1998). Chitosan-chondroitin sulfate and chitosan-hyaluronate polyelectrolyte complexes: biological properties. *Biomaterials*, 19(14): 1275.

Di Martino, A., Sittinger, M., et al. (2005). Chitosan: A versatile biopolymer for orthopaedic tissue-engineering. *Biomaterials*, 26(30): 5983.

Dodane, V. and Vilivalam, V.D. (1998). Pharmaceutical applications of chitosan. *Pharmaceutical Science & Technology Today*, 1(6): 246–253.

Drury, J.L. and Mooney, D.J. (2003). Hydrogels for tissue engineering: scaffold design variables and applications. *Biomaterials*, 24(24): 4337.

Dvir-Ginzberg, M., Gamlieli-Bonshtein, I., et al. (2003). Liver tissue engineering within alginate scaffolds: Effects of cell-seeding density on hepatocyte viability, morphology, and function. *Tissue Engineering*, 9(4): 757–766.

Eagles, D., Lesnoy, D., et al. (1996). Starch fibers: processing and characteristics. *Textile Research Journal*, 66(4): 277–282.

Ehlers, E.M., Behrens, P., et al. (2001). Effects of hyaluronic acid on the morphology and proliferation of human chondrocytes in primary cell culture. *Annals of Anatomy – Anatomischer Anzeiger*, 183(1): 13.

Elcin, A.E. (2006). *In vitro* and *in vivo* degradation of oxidized acetyl- and ethyl-cellulose sponges. *Artificial Cells Blood Substitutes and Biotechnology*, 34(4): 407–418.

Elizalde, B.E., Bartholomai, G.B., et al. (1996). The effect of pH on the relationship between hydrophilic/lipophilic characteristics and emulsification properties of soy proteins. *Food Science and Technology-Lebensmittel-Wissenschaft & Technologie*, 29(4): 334–339.

Elkayam, T., Amitay-Shaprut, S., et al. (2006). Enhancing the drug metabolism activities of C3A – A human hepatocyte cell line – By tissue engineering within alginate scaffolds. *Tissue Engineering*, 12(5): 1357–1368.

Elvira, C., Mano, J.F., et al. (2002). Starch-based biodegradable hydrogels with potential biomedical applications as drug delivery systems. *Biomaterials*, 23(9): 1955–1966.

Engbers-Buijtenhuijs, P., Buttafoco, L., et al. (2006). Biological characterization of vascular grafts cultured in a bioreactor. *Biomaterials*, 27(11): 2390–2397.

Engbers-Buijtenhuijs, P., Buttafoco, L., et al. (2006). Biological characterization of vascular grafts cultured in a bioreactor. *Biomaterials*, 27(11): 2390–2397.

Entcheva, E., Bien, H., et al. (2004). Functional cardiac cell constructs on cellulose-based scaffolding. *Biomaterials*, 25(26): 5753–5762.

Espigares, I., Elvira, C., et al. (2002). New partially degradable and bioactive acrylic bone cements based on starch blends and ceramic fillers. *Biomaterials*, 23(8): 1883–1895.

Feng, Z., Yamato, M., et al. (2003). Investigation on the mechanical properties of contracted collagen gels as a scaffold for tissue engineering. *Artificial Organs*, 27(1): 84–91.

Fragonas, E., Valente, M., et al. (2000). Articular cartilage repair in rabbits by using suspensions of allogenic chondrocytes in alginate. *Biomaterials*, 21(8): 795.

Francis Suh, J.K. and Matthew, H.W.T. (2000). Application of chitosan-based polysaccharide biomaterials in cartilage tissue engineering: a review. *Biomaterials*, 21(24): 2589.

Franz, G. and Blaschek, W. (1990). Cellulose. *Methods in plant biochemistry. Carbohydrates* (Dey, P.M., ed.). London: Academic Press Limited, Volume 2: 291–322.

Fratzl, P. (2003). Cellulose and collagen: from fibres to tissues. *Current Opinion in Colloid & Interface Science*, 8(1): 32–39.

Freier, T., Kunze, C., et al. (2002). *In vitro* and *in vivo* degradation studies for development of a biodegradable patch based on poly(3-hydroxybutyrate). *Biomaterials*, 23(13): 2649–2657.

Frenkel, S.R., Bradica, G., et al. (2005). Regeneration of articular cartilage – Evaluation of osteochondral defect repair in the rabbit using multiphasic implants. *Osteoarthritis and Cartilage*, 13(9): 798.

Friess, W. (1998). Collagen – biomaterial for drug delivery. *European Journal of Pharmaceutics and Biopharmaceutics*, 45(2): 113–136.

Fujii, T., Sun, Y.-L., et al. (2002). Mechanical properties of single hyaluronan molecules. *Journal of Biomechanics*, 35(4): 527.

Funakoshi, T., Majima, T., et al. (2005). Novel chitosan-based hyaluronan hybrid polymer fibers as a scaffold in ligament tissue engineering. *Journal of Biomedical Materials Research Part A*, 74A(3): 338–346.

Funakoshi, T., Majima, T., *et al.* (2005). Novel chitosan-based hyaluronan hybrid polymer fibers as a scaffold in ligament tissue engineering. *Journal of Biomedical Materials Research*, 74A: 338–346.

Fundueanu, G., Nastruzzi, C., *et al.* (1999). Physico-chemical characterization of Ca-alginate microparticles produced with different methods. *Biomaterials*, 20(15): 1427–1435.

Galliard, T. (1987). *Starch: Properties and potential*. New York: John Wiley.

Gao, J., Dennis, J.E., *et al.* (2002). Repair of osteochondral defect with tissue-engined two-phase composite material of injectable calcium phosphate and hyaluronan sponge. *Tissue Engineering*, 8: 827–837.

Gassner, F. and Owen, A.J. (1996). Some properties of poly(3-hydroxybutyrate) – Poly(3-hydroxyvalerate) blends. *Polymer International*, 39(3): 215–219.

Gelse, K., Poschl, E., *et al.* (2003). Collagens – structure, function, and biosynthesis. *Advanced Drug Delivery Reviews*, 55(12): 1531–1546.

Girotti, A., Reguera, J., *et al.* (2004). Design and bioproduction of a recombinant multi(bio)functional elastin-like protein polymer containing cell adhesion sequences for tissue engineering purposes. *J Mater Sci Mater Med*, 15(4): 479–484.

Glicklis, R., Shapiro, L., *et al.* (2000). Hepatocyte behavior within three-dimensional porous alginate scaffolds. *Biotechnology and Bioengineering*, 67(3): 344–353.

Gobin, A.S., Froude, V.E., *et al.* (2005). Structural and mechanical characteristics of silk fibroin and chitosan blend scaffolds for tissue regeneration. *Journal of Biomedical Materials Research Part A*, 74A(3): 465–473.

Gogolewski, S., Jovanovic, M., *et al.* (1993). Tissue response and *in vivo* degradation of selected polyhydroxyacids: Polylactides (PLA), poly(3-hydroxybutyrate) (PHB), and poly(3- hydroxybutyrate-co-3-hydroxyvalerate) (PHB/VA). *Journal of Biomedical Materials Research*, 27(9): 1135–1148.

Gombotz, W.R. and Wee, S.F. (1998). Protein release from alginate matrices. *Advanced Drug Delivery Reviews*, 31(3): 267–285.

Gomes, M., Godinho, J., *et al.* (2002). Design and processing of starch based scaffolds for hard tissue engineering. *J Appl Med Polym*, 6(2): 75–80.

Gomes, M., Salgado, A., *et al.* (2002). Bone tissue engineering using starch based scaffolds obtained by different methods. In *Polymer Based Systems on Tissue Engineering, Replacement and Regeneration* (Reis, R. and Cohn, D., eds). Amsterdam, Kluwer Academic Publishers: pp. 221–249.

Gomes, M.E., Bossano, C.M., *et al.* (2006). *In vitro* localization of bone growth factors in constructs of biodegradable scaffolds seeded with marrow stromal cells and cultured in a flow perfusion bioreactor. *Tissue Engineering*, 12(1): 177–188.

Gomes, M.E., Godinho, J.S., *et al.* (2002). Alternative tissue engineering scaffolds based on starch: processing methodologies, morphology, degradation and mechanical properties. *Materials Science & Engineering C-Biomimetic and Supramolecular Systems*, 20(1-2): 19–26.

Gomes, M.E., Godinho, J.S., *et al.* (2002). Alternative tissue engineering scaffolds based on starch: processing methodologies, morphology, degradation and mechanical properties. *Materials Science and Engineering: C*, 20(1-2): 19–26.

Gomes, M.E., Holtorf, H.L., *et al.* (2006). Influence of the porosity of starch-based fiber mesh scaffolds on the proliferation and osteogenic differentiation of bone marrow stromal cells cultured in a flow perfusion bioreactor. *Tissue Engineering*, 12(4): 801–809.

Gomes, M.E., Malafaya, P.B., *et al.* (2004). Methodologies for processing biodegradable and natural origin scaffolds for bone and cartilage tissue-engineering applications. *Methods in Molecular Biology Series* (Hollander, A. and Hatton, P., eds). Totowa, USA, The Humana Press Inc. 238: 65–76.

Gomes, M.E., Reis, R.L., *et al.* (2001). Cytocompatibility and response of osteoblastic-like cells to starch-based polymers: effect of several additives and processing conditions. *Biomaterials*, 22(13): 1911–1917.

Gomes, M.E., Ribeiro, A.S., *et al.* (2001). A new approach based on injection moulding to produce biodegradable starch-based polymeric scaffolds: morphology, mechanical and degradation behaviour. *Biomaterials*, 22(9): 883–889.

Gomes, M.E., Sikavitsas, V.I., *et al.* (2003). Effect of flow perfusion on the osteogenic differentiation of bone marrow stromal cells cultured on starch-based three-dimensional scaffolds. *J Biomed Mater Res A*, 67: 87–95.

Goodstone, N.J., Cartwright, A., *et al.* (2004). Effects of high molecular weight hyaluronan on chondrocytes cltured within a resorbable gelatin sponge. *Tissue Engineering*, 10: 621–631.

Govender, S., Pillay, V., *et al.* (2005). Optimization and characterization of bioadhesive controlled release tetracycline microspheres. *International Journal of Pharmaceutics*, 306(1–2): 24.

Grant, G.T., Morris, E.R., *et al.* (1973). Biological interactions between polysaccharides and divalent cations: the egg-box model. *FEBS Letters*, 32: 195–198.

Gravel, M., Gross, T., *et al.* (2006). Responses of mesenchymal stem cell to chitosan-coralline composites microstructured using coralline as gas forming agent. *Biomaterials*, 27(9): 1899.

Guo, J., Jourdian, G.W., *et al.* (1989). Culture and growth characteristics of chondrocytes encapsulated in alginate beads. *Connective Tissue Research*, 19: 277–297.

Guo, T., Zhao, J., *et al.* (2006). Porous chitosan-gelatin scaffold containing plasmid DNA encoding transforming growth factor-[beta]1 for chondrocytes proliferation. *Biomaterials*, 27(7): 1095.

Gupta, K.C. and Kumar, M.N.V.R. (2000). Trends in controlled drug release formulations using chitin and chitosan. *Journal of Scientific & Industrial Research*, 59(3): 201–213.

Ha, S.W., Park, Y.H., *et al.* (2003). Dissolution of Bombyx mori silk fibroin in the calcium nitrate tetrahydrate-methanol system and aspects of wet spinning of fibroin solution. *Biomacromolecules*, 4(3): 488–496.

Haider, M., Megeed, Z., *et al.* (2004). Genetically engineered polymers: status and prospects for controlled release. *J Control Release*, 95(1): 1–26.

Halbleib, M., Skurk, T., *et al.* (2003). Tissue engineering of white adipose tissue using hyaluronic acid-based scaffolds. I: *In vitro differentiation of human adipocyte precursor cells on scaffolds*. *Biomaterials*, 24(18): 3125.

Haque, M.I., Beekley, A.C., *et al.* (2001). Bioabsorption qualities of chitosan-absorbable vascular templates(1). *Curr Surg*, 58(1): 77–80.

Hegewald, A.A., Ringe, J., *et al.* (2004). Hyaluronic acid and autologous synovial fluid induce chondrogenic differentiation of equine mesenchymal stem cells: a preliminary study. *Tissue and Cell*, 431–438.

Helenius, G., Backdahl, H., *et al.* (2006). *In vivo* biocompatibility of bacterial cellulose. *Journal of Biomedical Materials Research Part A*, 76A(2): 431–438.

Hemmrich, K., von Heimburg, D., *et al.* (2005). Implantation of preadipocyte-loaded hyaluronic acid-based scaffolds into nude mice to evaluate potential for soft tissue engineering. *Biomaterials*, 26(34): 7025.

Hinds, M.T., Rowe, R.C., *et al.* (2006). Development of a reinforced porcine elastin composite vascular scaffold. *J Biomed Mater Res A*, 77: 458–469.

Ho, M.-H., Kuo, P.-Y., *et al.* (2004). Preparation of porous scaffolds by using freeze-extraction and freeze-gelation methods. *Biomaterials*, 25(1): 129.

Ho, M.-H., Wang, D.-M., *et al.* (2005). Preparation and characterization of RGD-immobilized chitosan scaffolds. *Biomaterials*, 26(16): 3197.

Hobbs, J.K. (1998). The fracture of poly(hydroxybutyrate): Part I Fracture mechanics study during ageing. *Journal of Materials Science*, 33(10): 2509–2514.

Hobbs, J.K. and Barham, P.J. (1998). The fracture of poly(hydroxybutyrate): Part II Fracture mechanics study after annealing. *Journal of Materials Science*, 33(10): 2515–2518.

Hobbs, J.K. and Barham, P.J. (1999). The fracture of poly(hydroxybutyrate): Part III Fracture morphology in thin films and bulk systems. *Journal of Materials Science*, 34(19): 4831–4844.

Hoemann, C.D., Sun, J., *et al.* (2005). Tissue engineering of cartilage using an injectable and adhesive chitosan-based cell-delivery vehicle. *Osteoarthritis and Cartilage*, 13(4): 318.

Horch, R.E., Kopp, J., *et al.* (2005). Tissue engineering of cultured skin substitutes. *J. Cell. Mol. Med*, 9(3): 592–608.

Hou, S., Xu, Q., *et al.* (2005). The repair of brain lesion by implantation of hyaluronic acid hydrogels modified with laminin. *Journal of Neuroscience Methods*, 148(1): 60.

Huang, Y.-C., Huang, Y.-Y., *et al.* (2005). Manufacture of porous polymer nerve conduits through a lyophilizing and wire-heating process. *Journal of Biomedical Materials Research Part B: Applied Biomaterials*, 74B(1): 659–664.

Ishikawa, K., Kawaguchi, Y., *et al.* (1991). Plasticization of bacterial polyester by the addition of acylglycerols and its enzymatic degradability. *Kobunshi Ronbunshu*, 48(4): 221–226.

Itoh, S., Matsuda, A., *et al.* (2005). Effects of a laminin peptide (YIGSR) immobilized on crab-tendon chitosan tubes on nerve regeneration. *Journal of Biomedical Materials Research Part B: Applied Biomaterials*, 73B(2): 375–382.

Izydorczyk, M. (2005). Understand the Chemistry of Food Carbohydrates. *Food Carbohydrates: Chemistry, Physical Properties, and Applications* (Cui, S.W., ed.). Boca Raton: CRC Press Taylor & Francis Group, pp. 1–65.

Izydorczyk, M., Cui, S.W., *et al.* (2005). Polysaccharide Gums: Structures, Functional Properties, and Applications. *Food Carbohydrates: Chemistry, Physical Properties, and Applications* (Cui, S.W., ed.). Boca Raton: CRC Press Taylor & Francis Group, pp. 263–307.

Jin, H.J., Chen, J.S., *et al.* (2004). Human bone marrow stromal cell responses on electrospun silk fibroin mats. *Biomaterials*, 25(6): 1039–1047.

Jin, H.J., Fridrikh, S.V., et al. (2005). Electrospinning Bombyx mori Silk with Poly(ethylene oxide). *Biomacromolecules*, 3: 1233–1239.

Joddar, B. and Ramamurthi, A. (2006). Fragment size- and dose-specific effects of hyaluronan on matrix synthesis by vascular smooth muscle cells. *Biomaterials*, 27(15): 2994–3004.

Kadler, K.E., Holmes, D.F., et al. (1996). Collagen fibril formation. *Biochemical Journal*, 316: 1–11.

Kataoka, K., Suzuki, Y., et al. (2001). Alginate, a bioresorbable material derived from brown seaweed, enhances elongation of amputated axons of spinal cord in infant rats. *Journal of Biomedical Materials Research*, 54(3): 373–384.

Kim, B.-S., Nikolovski, J., et al. (1999). Engineered Smooth Muscle Tissues: Regulating Cell Phenotype with the Scaffold. *Experimental Cell Research*, 251(2): 318–328.

Kim, B.S., Baez, C.E., et al. (2000). Biomaterials for tissue engineering. *World Journal of Urology*, 18(1): 2–9.

Kim, H.J., Kim, U.J., et al. (2005). Influence of macroporous protein scaffolds on bone tissue engineering from bone marrow stem cells. *Biomaterials*, 26(21): 4442–4452.

Kim, S.E., Park, J.H., et al. (2003). Porous chitosan scaffold containing microspheres loaded with transforming growth factor-[beta]1: Implications for cartilage tissue engineering. *Journal of Controlled Release*, 91(3): 365.

Kim, U.J., Park, J., et al. (2005). Three-dimensional aqueous-derived biomaterial scaffolds from silk fibroin. *Biomaterials*, 26(15): 2775–2785.

Kino-Oka, M., Maeda, Y., et al. (2005). A kinetic modeling of chondrocyte culture for manufacture of tissue-engineered cartilage. *Journal of Bioscience and Bioengineering*, 99(3): 197–207.

Kiyozumi, T., Kanatani, Y., et al. (2006). Medium (DMEM/F12)-containing chitosan hydrogel as adhesive and dressing in autologous skin grafts and accelerator in the healing process. *Journal of Biomedical Materials Research Part B: Applied Biomaterials*, 79B(1): 129–136.

Knowles, J.C. (1993). Development of a natural degradable polymer for orthopaedic use. *Journal of Medical Engineering and Technology*, 17(4): 129–137.

Knudson, K., Biswas, C., et al. (1989). The role and regulation of tumour-associated hyaluronan. *The Biology of Hyaluronan*, Ciba Foundation Symposium, 143: 150–159.

Kobayashi, S., Fujikawa, S., et al. (2003). Enzymatic synthesis of chondroitin and its derivatives catalyzed by hyaluronidase. *Journal of the American Chemical Society*, 125(47): 14357–14369.

Kong, L., Gao, Y., et al. (2005). Preparation and characterization of nano-hydroxyapatite/chitosan composite scaffolds. *Journal of Biomedical Materials Research Part A*, 75A(2): 275–282.

Kuo, C.K. and Ma, P.X. (2001). Ionically cross-linked alginate hydrogels as scaffolds for tissue engineering: Part 1. Structure, gelation rate and mechanical properties. *Biomaterials*, 22(6): 511–521.

Kurita, K. (2001). Controlled functionalization of the polysaccharide chitin. *Progress in Polymer Science*, 26(9): 1921–1971.

Lam, C.X.F., Mo, X.M., et al. (2002). Scaffold development using 3D printing with a starch-based polymer. *Materials Science and Engineering: C*, 20(1-2): 49–56.

Leach, J.B., Wolinsky, J.B., et al. (2005). Cross-linked [alpha]-elastin biomaterials: towards a processable elastin mimetic scaffold. *Acta Biomaterialia*, 1(2): 155–164.

Lee, C.H., Singla, A., et al. (2001). Biomedical applications of collagen. *International Journal of Pharmaceutics*, 221(1–2): 1–22.

Lee, J.-E., Park, J.-G., et al. (2001). Preparation of collagen modified hyaluronan microparticles as antibiotic carrier. *Yonsei Medical Journal*, 42(3): 291–298.

Lee, J., Macosko, C.W., et al. (2001). Mechanical properties of cross-linked synthetic elastomeric polypentapeptides. *Macromolecules*, 34(17): 5968–5974.

Lemoine, D., Wauters, F., et al. (1998). Preparation and characterization of alginate microspheres containing a model antigen. *International Journal of Pharmaceutics*, 176(1): 9–19.

LeRoux, M.A., Guilak, F., et al. (1999). Compressive and shear properties of alginate gel: Effects of sodium ions and alginate concentration. *Journal of Biomedical Materials Research*, 47(1): 46–53.

Levesque, S.G., Lim, R.M., et al. (2005). Macroporous interconnected dextran scaffolds of controlled porosity for tissue-engineering applications. *Biomaterials*, 26(35): 7436–7446.

Lezica, R.P. and Quesada-Allué, L. (1990). Chitin. *Methods in plant biochemistry. Carbohydrates* (Dey, P.M., ed.). London: Academic Press Limited, Volume 2: pp. 443–481.

Li, C., Vepari, C., et al. (2006). Electrospun silk-BMP-2 scaffolds for bone tissue engineering. *Biomaterials*, 27: 3115–3124.

Li, M., Mondrinos, M.J., et al. (2005). Electrospun protein fibers as matrices for tissue engineering. *Biomaterials*, 26(30): 5999–6008.

Li, M.Z., Lu, S.Z., *et al.* (2001). Study on porous silk fibroin materials. I. Fine structure of freeze dried silk fibroin. *Journal of Applied Polymer Science*, 79(12): 2185–2191.

Li, M.Z., Wu, Z.Y., *et al.* (2001). Study on porous silk fibroin materials. II. Preparation and characteristics of spongy silk fibroin materials. *Journal of Applied Polymer Science*, 79(12): 2192–2199.

Li, M.Z., Zhang, C.S., *et al.* (2002). Study on porous silk fibroin materials: 3. Influence of repeated freeze-thawing on the structure and properties of porous silk fibroin materials. *Polymers for Advanced Technologies*, 13(8): 605–610.

Li, Z., Ramay, H.R., *et al.* (2005). Chitosan-alginate hybrid scaffolds for bone tissue engineering. *Biomaterials*, 26(18): 3919.

Lim, S.H., Liao, I.C., *et al.* (2006). Nonviral gene delivery from nonwoven fibrous scaffolds fabricated by interfacial complexation of polyelectrolytes. *Mol Ther*, 13(6): 1163–1172.

Liu, Y., Shu, X.Z., *et al.* (2005). Biocompatibility and stability of dissulfide-cross-linked hyaluronan films. *Biomaterials*, 26: 4737–4746.

Lu, Q., Ganesan, K., *et al.* (2004). Novel porous aortic elastin and collagen scaffolds for tissue engineering. *Biomaterials*, 25(22): 5227–5237.

Malafaya, P.B., Pedro, A., *et al.* (2005). Chitosan particles agglomerated scaffolds for cartilage and osteochondral tissue engineering approaches with adipose tissue derived stem cells. *Journal of Materials Science: Materials in Medicine*, 16(12): 1077.

Mano, J.F., Vaz, C.M., *et al.* (1999). Dynamic mechanical properties of hydroxyapatite-reinforced and porous starch-based degradable biomaterials. *J Mater Sci Mater Med*, 10(12): 857–862.

Mao, J., Zhao, L., *et al.* (2003a). Study of novel chitosan-gelatin artificial skin *in vitro*. *Journal of Biomedical Materials Research*, 64A(2): 301–308.

Mao, J.S., Zhao, L.G., *et al.* (2003b). Structure and properties of bilayer chitosan-gelatin scaffolds. *Biomaterials*, 24(6): 1067–1074.

Marques, A.P., Reis, R.L., *et al.* (2002). The biocompatibility of novel starch-based polymers and composites: *in vitro* studies. *Biomaterials*, 23(6): 1471–1478.

Martson, M., Viljanto, J., *et al.* (1998). Biocompatibility of cellulose sponge with bone. *European Surgical Research*, 30(6): 426–432.

Martson, M., Viljanto, J., *et al.* (1999). Is cellulose sponge degradable or stable as implantation materialβ An *in*

vivo subcutaneous study in the rat. *Biomaterials*, 20(21): 1989–1995.

Massia, S. and Stark, J. (2001). Immobilized RGD peptides on surfaces-grafted dextran promote biospecific cell attachment. *Journal of Biomedical Materials Research*, 56: 390–399.

Meinel, L., Fajardo, R., *et al.* (2005). Silk implants for the healing of critical size bone defects. *Bone*, 37(5): 688–698.

Meinel, L., Hofmann, S., *et al.* (2004). Engineering cartilage-like tissue using human mesenchymal stem cells and silk protein scaffolds. *Biotechnology and Bioengineering*, 88(3): 379–391.

Meinel, L., Karageorgiou, V., *et al.* (2004). Bone tissue engineering using human mesenchymal stem cells: Effects of scaffold material and medium flow. *Annals of Biomedical Engineering*, 32(1): 112–122.

Mendes, S.C., Bezemer, J., *et al.* (2003). Evaluation of two biodegradable polymeric systems as substrates for bone tissue engineering. *Tissue Engineering*, 9 Suppl 1: S91–S101.

Mendes, S.C., Reis, R.L., *et al.* (2001). Biocompatibility testing of novel starch-based materials with potential application in orthopaedic surgery: a preliminary study. *Biomaterials*, 22(14): 2057–2064.

Menzel, E.J. and Farr, C. (1998). Hyaluronidase and its substrate hyaluronan: biochemistry, biological activities and therapeutic uses. *Cancer Letters*, 131: 3–11.

Milella, E., Brescia, E., *et al.* (2002). Physico-chemical properties and degradability of non-woven hyaluronan benzylic esters as tissue engineering scaffolds. *Biomaterials*, 23(4): 1053.

Miralles, G., Baudoin, R., *et al.* (2001). Sodium alginate sponges with or without sodium hyaluronate: *In vitro* engineering of cartilage. *Journal of Biomedical Materials Research*, 57(2): 268–278.

Mitomo, H., Barham, P.J., *et al.* (1988). Temperature dependence of mechanical properties of poly(β-hydroxybutyrate-β-hydroxyvalerate). *Polymer communications Guildford*, 29(4): 112–115.

Miyamoto, T., Takahashi, S., *et al.* (1989). Tissue Biocompatibility of Cellulose and Its Derivatives. *Journal of Biomedical Materials Research*, 23(1): 125–133.

Morrison, W.R. and Karkalas, J. (1990). Starch. In *Methods in plant biochemistry. Carbohydrates* (Dey, P.M., ed.). London: Academic Press Limited. Volume 2: 323–352.

Motta, A., Migliaresi, C., *et al.* (2004). Fibroin hydrogels for biomedical applications: preparation, characterization and *in vitro* cell culture studies. *Journal of Biomaterials Science-Polymer Edition*, 15(7): 851–864.

Mwale, F., Iordanova, M., *et al.* (2005). Biological evaluation of chitosan salts cross-linked to genipin as a cell scaffold for disk tissue engineering. *Tissue Engineering*, 11(1-2): 130–140.

Naessens, M., Cerdobbel, A., *et al.* (2005). Leuconostoc dextransucrase and dextran: production, properties and applications. *Journal of Chemical Technology and Biotechnology*, 80(8): 845–860.

Nazarov, R., Jin, H.J., *et al.* (2004). Porous 3-D scaffolds from regenerated silk fibroin. *Biomacromolecules*, 5(3): 718–726.

O'Cearbhaill, E.D., Barron, V., *et al.* (2006). Characterization of a collagen membrane for its potential use in cardiovascular tissue engineering applications. *Journal of Materials Science-Materials in Medicine*, 17(3): 195–201.

Oerther, S., Payan, E., *et al.* (1999). Hyaluronate-alginate combination for the preparation of new biomaterials: investigation of the behaviour in aqueous solutions. *Biochimica et Biophysica Acta*, 1426: 185–194.

Oliveira, A.L., Malafaya, P.B., *et al.* (2003). Sodium silicate gel as a precursor for the *in vitro* nucleation and growth of a bone-like apatite coating in compact and porous polymeric structures. *Biomaterials*, 24(15): 2575–2584.

Oliveira, A.L. and Reis, R.L. (2004). Pre-mineralization of starch/polycrapolactone bone tissue engineering scaffolds by a calcium-silicate-based process. *J Mater Sci Mater Med*, 15(4): 533–540.

Oliveira, J.T., Correlo, V.M., *et al.* (2005). *Chitosan-Polyester Scaffolds Seeded with Bovine Articular Chondrocytes for Cartilage Tissue Enginnering Applications*. Bologna: Artificial Organs.

Ong, S.R., Trabbic-Carlson, K.A., *et al.* (2006). Epitope tagging for tracking elastin-like polypeptides. *Biomaterials*, 27(9): 1930–1935.

Owen, A.J. (1985). Some dynamic mechanical properties of microbially produced poly-beta-hydroxybutyrate/beta-hydroxyvalerate copolymers. *Colloid and Polymer Science*, 263(10): 799–803.

Pachence, J.M. (1996). Collagen-based devices for soft tissue repair. *Journal of Biomedical Materials Research*, 33(1): 35–40.

Paige, K.T., Cima, L.G., *et al.* (1996). De novo cartilage generation using calcium alginate−chondrocyte constructs. *Plastic and Reconstructive Surgery*, 97: 168–180.

Park, D.-J., Choi, B.-H., *et al.* (2005). Injectable bone using chitosan-alginate gel/mesenchymal stem cells/BMP-2 composites. *Journal of Cranio-Maxillofacial Surgery*, 33(1): 50.

Park, S.-H., Park, S., *et al.* (2005). Tissue-engineered cartilage using fibrin/hyaluronan composite gel and its *in vivo* implantation. *Artificial Organs*, 29(10): 838–845.

Park, Y.J., Lee, Y.M., *et al.* (2000). Controlled release of platelet-derived growth factor-BB from chondroitin sulfate-chitosan sponge for guided bone regeneration. *J Control Release*, 67(2-3): 385–394.

Patashnik, S., Rabinovich, L., *et al.* (1997). Preparation and evaluation of chitosan microspheres containing bisphosphonates. *Journal of Drug Targeting*, 4(6): 371–380.

Peattie, R.A., Nayate, A.P., *et al.* (2004). Stimulation of *in vivo* angiogenesis by cytokine-loaded hyaluronic acid hydrogel implants. *Biomaterials*, 25: 2789–2798.

Peniche, C., Fernandez, M., *et al.* (2003). Drug delivery systems based on porous chitosan/polyacrylic acid microspheres. *Macromolecular Bioscience*, 3(10): 540–545.

Percival, E. and McDowell, R.H. (1990). Algal polysaccharides. In *Methods in plant biochemistry. Carbohydrates* (Dey, P.M., ed.). London: Academic Press Limited, Volume 2: 523–547.

Pereira, C., Cunha, A., *et al.* (1998). New starch-based thermoplastic hydrogels for use as bone cements or drug-delivery carriers. *J Mater Sci: Mater Med*, 9: 825–833.

Pereira, C., Gomes, M.E., *et al.* (1998). Hard Cellular Materials in the Human Body: Properties and Production of Foamed Polymers for Bone Replacement. *NATO/ASI Series, Kluwer Press, Drodercht*, (1998). (Rivier, N. and Sadoc, J., eds). Drodercht: Kluwer Press, pp. 193–204.

Petrini, P., Parolari, C., *et al.* (2001). Silk fibroin-polyurethane scaffolds for tissue engineering. *Journal of Materials Science-Materials in Medicine*, 12(10-12): 849–853.

Pianigiani, E., Andreassi, A., *et al.* (1999). A new model for studying differentiation and growth of epidermal cultures on hyaluronan-based carrier. *Biomaterials*, 20(18): 1689.

Pinto, A.R., Correlo, V.M. *et al.* (2005). *Behaviour of Human Bone Marrow Mesenchymal Stem Cells Seeded on Fiber Bonding Chitosan Polyester based for Bone Tissue Engineering Scaffolds*. 8th TESI Annual Meeting, Shanghai.

Pouton, C.W. and Akhtar, S. (1996). Biosynthetic polyhydroxyalkanoates and their potential in drug delivery. *Advanced Drug Delivery Reviews*, 18(2): 133–162.

Prestwich, G.D., Marecak, D.M., *et al.* (1998). Controlled chemical modification of hyaluronic acid: synthesis, applications, and biodegradation of hydrazide derivatives. *Journal of Controlled Release*, 53(1-3): 93.

Ramamurthi, A. and Vesely, I. (2005). Evaluation of the matrix-synthesis potential of cross-linked hyaluronan gels for tissue engineering of aortic heart valves. *Biomaterials*, 26(9): 999.

Reis, R. and Cunha, A., *et al.* (2001). Starch and starch based thermoplastics. In *Encyclopedia of Materials Science and Technology* (Jurgen, K.H., *et al.* eds.). Amsterdam: Elsevier Science, p. 8810.

Reis, R.L., Cunha, A.M., *et al.* (1996). Mechanical behavior of injection-molded starch-based polymers. *Polymers for Advanced Technologies*, 7(10): 784–790.

Reis, R.L., Cunha, A.M., *et al.* (1997). Structure development and control of injection-molded hydroxyl-apatite-reinforced starch/EVOH composites. *Advances in Polymer Technology*, 16(4): 263–277.

Reis, R.L., Mendes, S.C., *et al.* (1997). Processing and *in vitro* degradation of starch/EVOH thermoplastic blends. *Polymer International*, 43(4): 347–352.

Rhim, J.W., Gennadios, A., *et al.* (1998). Soy protein isolate dialdehyde starch films. *Industrial Crops and Products*, 8(3): 195–203.

Risbud, M., Endres, M., *et al.* (2001). Biocompatible hydrogel supports the growth of respiratory epithelial cells: Possibilities in tracheal tissue engineering. *Journal of Biomedical Materials Research*, 56(1): 120–127.

Rivard, C.H., Chaput, C., *et al.* (1996). Bioabsorbable synthetic polyesters and tissue regeneration: A study on the three-dimensional proliferation of ovine chondrocytes and osteoblasts [Polyesters biosynthetiques absorbables et regeneration tissulaire. Etude de la proliferation tridimensionnelle de chondrocytes et osteoblastes ovins]. *Annales de Chirurgie*, 50(8): 651–658.

Rosso, F., Marino, G., *et al.* (2005). Smart materials as scaffolds for tissue engineering. *Journal of Cellular Physiology*, 203(3): 465–470.

Rowley, J.A., Madlambayan, G., *et al.* (1999). Alginate hydrogels as synthetic extracellular matrix materials. *Biomaterials*, 20(1): 45–53.

RoyChowdhury, P. and Kumar, V. (2006). Fabrication and evaluation of porous 2,3-dialdehydecellulose membrane as a potential biodegradable tissue-engineering scaffold. *Journal of Biomedical Materials Research Part A*, 76A(2): 300–309.

Sachlos, E., Reis, N., *et al.* (2003). Novel collagen scaffolds with predefined internal morphology made by solid freeform fabrication. *Biomaterials*, 24(8): 1487–1497.

Salgado, A.J., Coutinho, O.P., *et al.* (2004). Novel starch-based scaffolds for bone tissue engineering: cytotoxicity, cell culture, and protein expression. *Tissue Engineering*, 10(3-4): 465–474.

Salgado, A.J., Figueiredo, J.E., *et al.* (2005). Biological response to pre-mineralized starch based scaffolds for bone tissue engineering. *J Mater Sci Mater Med*, 16(3): 267–275.

Salgado, A.J., Gomes, M.E., *et al.* (2002). Preliminary study on the adhesion and proliferation of human osteoblasts on starch-based scaffolds. *Materials Science and Engineering: C*, 20(1-2): 27–33.

Santin, M., Motta, A., *et al.* (1999). *In vitro* evaluation of the inflammatory potential of the silk fibroin. *Journal of Biomedical Materials Research*, 46(3): 382–389.

Santos, M.I., Fuchs, S., *et al.* (2007). Response of micro- and macrovascular endothelial cells to starch-based fiber meshes for bone tissue engineering. *Biomaterials*, 28(2): 240–248.

Scandola, M., Focarete, M.L., *et al.* (1997). Polymer blends of natural poly(3-hydroxybutyrate-co-3-hydroxyvalerate) and a synthetic atactic poly(3-hydroxybutyrate). Characterization and biodegradation studies. *Macromolecules*, 30(9): 2568–2574.

Schaffellner, S., Stadlbauer, V., *et al.* (2005). Porcine islet cells microencapsulated in sodium cellulose sulfate. *Transplantation Proceedings*, 37(1): 248–252.

Schmidt, V., Giacomelli, C., *et al.* (2005). Soy protein isolate based films: influence of sodium dodecyl sulfate and polycaprolactone-triol on their properties. *Macromolecular Symposia*, 229: 127–137.

Segura, T., Anderson, B.C., *et al.* (2005). Cross-linked hyaluronic acid hydrogels: a strategy to functionalize and pattern. *Biomaterials*, 26(4): 359.

Seliktar, D., Black, R.A., *et al.* (2000). Dynamic mechanical conditioning of collagen-gel blood vessel constructs induces remodeling *in vitro*. *Annals of Biomedical Engineering*, 28(4): 351–362.

Servoli, E., Maniglio, D., *et al.* (2005). Surface properties of silk fibroin films and their interaction with fibroblasts. *Macromolecular Bioscience*, 5(12): 1175–1183.

Shangguan, Y.Y., Wang, Y.W., *et al.* (2006). The mechanical properties and *in vitro* biodegradation and biocompatibility of UV-treated poly(3-hydroxybutyrate-co-3-hydroxyhexanoate). *Biomaterials*, 27(11): 2349–2357.

Shu, X.Z., Liu, Y., *et al.* (2003). Disulfide-cross-linked hyaluronan-gelatin hydrogel films: a covalent mimic of the extracellular matrix for *in vitro* cell growth. *Biomaterials*, 24: 3825–3834.

Shu, X.Z., Liu, Y., *et al.* (2004). *In situ* cross-linkable hyaluronan hydrogels for tissue engineering. *Biomaterials*, 25(7–8): 1339.

Silva, R.M., Elvira, C., *et al.* (2004). Influence of Beta-Radiation Sterilization in Properties of New Chitosan/Soybean Protein Isolate Membranes for Guided Bone Regeneration. *Journal of Materials Science-Materials in Medicine*, 15: 523–528.

Silva, S.S., Oliveira, J.M., *et al.* (2006). *Physicochemical Characterization of Novel Chitosan-Soy Protein/TEOS Porous Hybrids for Tissue Engineering Applications.*

Silva, S.S., Santos, M.I., *et al.* (2005). Physical properties and biocompatibility of chitosan/soy blended membranes. *Journal of Materials Science-Materials in Medicine*, 16(6): 575–579.

Simionescu, D.T., Lu, Q., *et al.* (2006). Biocompatibility and remodeling potential of pure arterial elastin and collagen scaffolds. *Biomaterials*, 27(5): 702–713.

Sivakumar, M. and Panduranga Rao, K. (2003). Preparation, characterization, and *in vitro* release of gentamicin from coralline hydroxyapatite-alginate composite microspheres. *Journal of Biomedical Materials Research*, 65: 222–228.

Smidsrød, O. and Skjak-Bræk, G. (1990). Alginate as immobilization matrix for cells. *Trends in Biotechnology*, 8: 71–78.

Sodian, R., Hoerstrup, S.P., *et al.* (2000). Early *in vivo* experience with tissue-engineered trileaflet heart valves. *Circulation*, 102(19).

Sodian, R., Hoerstrup, S.P., *et al.* (2000). Tissue engineering of heart valves: *In vitro* experiences. *Annals of Thoracic Surgery*, 70(1): 140–144.

Sodian, R., Sperling, J.S., *et al.* (2000). Fabrication of a trileaflet heart valve scaffold from a polyhydroxyalkanoate biopolyester for use in tissue engineering. *Tissue Engineering*, 6(2): 183–188.

Solchaga, L.A., Temenoff, J.S., *et al.* (2005). Repair of osteochondral defects with hyaluronan- and polyester-based scaffolds. *Osteoarthritis and Cartilage*, 13(4): 297.

Song, E., Yeon Kim, S., *et al.* (2006). Collagen scaffolds derived from a marine source and their biocompatibility. *Biomaterials*, 27(15): 2951–2961.

Sotome, S., Uemura, T., *et al.* (2004). Synthesis and *in vivo* evaluation of a novel hydroxyapatite/collagen-alginate as a bone filler and a drug delivery carrier of bone morphogenetic protein. *Materials Science & Engineering C-Biomimetic And Supramolecular Systems*, 24(3): 341–347.

Sousa, R., Reis, R., *et al.* (2003). Processing and properties of bone-analogue biodegradable and bioinert polymeric composites. *Composite Sci Tech*, 63: 389–402.

Sousa, R.A., Kalay, G., *et al.* (2000). Injection molding of a starch/EVOH blend aimed as an alternative biomaterial for temporary applications. *Journal of Applied Polymer Science*, 77(6): 1303–1315.

Sousa, R.A., Mano, J.F., *et al.* (2002). Mechanical performance of starch based bioactive composite biomaterials molded with preferred orientation. *Polymer Engineering and Science*, 42(5): 1032–1045.

Stark, Y., Suck, K., *et al.* (2006). Application of collagen matrices for cartilage tissue engineering. *Experimental and Toxicologic Pathology*, 57(4): 305–311.

Steinbüchel, A. and Rhee, S.K. (2005). *Polysaccharides and Polyamides in the Food Industry. Properties, Production and Patents*. Weinheim: WILEY-VCH Verlag GmbH & Co. KGaA.

Stephen, A.M., Churms, S.C., *et al.* (1990). Exudate Gums. In *Methods in plant biochemistry. Carbohydrates* (Dey, P. M., ed.). London: Academic Press Limited, pp. 483–522. Volume 2

Stevens, M.M. and George, J.H. (2005). Exploring and engineering the cell surface interface. *Science*, 310(5751): 1135–1138.

Stitzel, J., Liu, J., *et al.* (2006). Controlled fabrication of a biological vascular substitute. *Biomaterials*, 27(7): 1088–1094.

Subramanian, A. and Lin, H.-Y. (2005). Cross-linked chitosan: Its physical properties and the effects of matrix stiffness on chondrocyte cell morphology and proliferation. *Journal of Biomedical Materials Research Part A*, 75A(3): 742–753.

Sudesh, K., Abe, H., *et al.* (2000). Synthesis, structure and properties of polyhydroxyalkanoates: Biological polyesters. *Progress in Polymer Science (Oxford)*, 25(10): 1503–1555.

Suh, J.K.F. and Matthew, H.W.T. (2000). Application of chitosan-based polysaccharide biomaterials in cartilage tissue engineering: a review. *Biomaterials*, 21(24): 2589.

Sumita, Y., Honda, M.J., *et al.* (2006). Performance of collagen sponge as a 3-D scaffold for tooth-tissue engineering. *Biomaterials*, 27(17): 3238–3248.

Svensson, A., Nicklasson, E., *et al.* (2005). Bacterial cellulose as a potential scaffold for tissue engineering of cartilage. *Biomaterials*, 26(4): 419–431.

Swain, S.N., Biswal, S.M., *et al.* (2004). Biodegradable soy-based plastics: Opportunities and challenges. *Journal of Polymers and the Environment*, 12(1): 35–42.

Swain, S.N., Rao, K.K., *et al.* (2004). Biodegradable Polymers. III. Spectral, Thermal, Mechanical, and Morphological Properties of Cross-linked Furfural−Soy Protein Concentrate. *Journal of Applied Polymer Science*, 93: 2590–2596.

Takata, T., Wang, H.L., *et al.* (2001). Migration of osteo-blastic cells on various guided bone regeneration membranes. *Clinical Oral Implants Research*, 12(4): 332–338.

Tamada, Y. (2005). New process to form a silk fibroin porous 3-D structure. *Biomacromolecules*, 6(6): 3100–3106.

Taylor, P.M., Sachlos, E., *et al.* (2006). Interaction of human valve interstitial cells with collagen matrices manufactured using rapid prototyping. *Biomaterials*, 27(13): 2733–2737.

Tonello, C., Zavan, B., *et al.* (2003). *In vitro* reconstruction of human dermal equivalent enriched with endothelial cells. *Biomaterials*, 24(7): 1205.

Toole, B., Wight, T., *et al.* (2002). Hyaluronan-cell interactions in cancer and vascular disease. *Journal of Biological Chemistry*, 277: 4593–4596.

Toole, B.P. (2004). Hyaluronan: From extracellular glue to pericellular cue. *Nature review: Cancer*, 4: 528–539.

Turner, N.J., Kielty, C.M., *et al.* (2004). A novel hyaluronan-based biomaterial (Hyaff-11(R)) as a scaffold for endothelial cells in tissue engineered vascular grafts. *Biomaterials*, 25(28): 5955.

Tuzlakoglu, K., Bolgen, N., *et al.* (2005). Nano- and microfiber combined scaffolds: a new architecture for bone tissue engineering. *J Mater Sci Mater Med*, 16(12): 1099–1104.

Unger, R.E., Wolf, M., *et al.* (2004). Growth of human cells on a non-woven silk fibroin net: a potential for use in tissue engineering. *Biomaterials*, 25(6): 1069–1075.

Urry, D.W., Hugel, T., *et al.* (2002). Elastin: a representative ideal protein elastomer. *Philos Trans R Soc Lond B Biol Sci*, 357(1418): 169–184.

Urry, D.W. and Parker, T.M. (2002). Mechanics of elastin: molecular mechanism of biological elasticity and its relationship to contraction. *J Muscle Res Cell Motil*, 23(5-6): 543–559.

Urry, D.W., Pattanaik, A., *et al.* (1998). Elastic protein-based polymers in soft tissue augmentation and generation. *J Biomater Sci Polym Ed*, 9(10): 1015–1048.

van Amerongen, M.J., Harmsen, M.C., *et al.* (2006). The enzymatic degradation of scaffolds and their replacement by vascularized extracellular matrix in the murine myocardium. *Biomaterials*, 27(10): 2247–2257.

van der Kraan, P.M., Buma, P., *et al.* (2002). Interaction of chondrocytes, extracellular matrix and growth factors: relevance for articular cartilage tissue engineering. *Osteoarthritis and Cartilage*, 10(8): 631.

Van Tomme, S.R., van Steenbergen, M.J., *et al.* (2005). Self-gelling hydrogels based on oppositely charged dextran microspheres. *Biomaterials*, 26(14): 2129–2135.

VandeVord, P.J., Matthew, H.W.T., *et al.* (2002). Evaluation of the biocompatibility of a chitosan scaffold in mice. *Journal of Biomedical Materials Research*, 59(3): 585–590.

Vaz, C.M., Graaf, L.A., *et al.* (2002). Soy protein-based systems for different tissue regeneration applications. In *Polymer Based Systems on Tissue Engineering, Replacement and Regeneration* (Reis, R. and Cohn, D., eds), Amsterdam, Kluwer Academic Publishers: pp. 93–110.

Vila, A., Sanchez, A., *et al.* (2002). Design of biodegradable particles for protein delivery. *J Control Release*, 78(1-3): 15–24.

Vinatier, C., Magne, D., *et al.* (2005). A silanized hydroxy-propyl methylcellulose hydrogel for the three-dimensional culture of chondrocytes. *Biomaterials*, 26(33): 6643–6651.

Voet, D. (1995). *Biochemistry*. John Wiley & Sons, Inc.

Volova, T., Shishatskaya, E., *et al.* (2003). Results of biomedical investigations of PHB and PHB/PHV fibers. *Biochemical Engineering Journal*, 16(2): 125–133.

Wallen, L.L. and Rohwedder, W.K. (1974). Poly B hydroxyalkanoate from activated sludge. *Environmental Science and Technology*, 8(6): 576–579.

Wang, T.L., Bogracheva, T.Y., *et al.* (1998). Starch: as simple as A, B, C? *Journal of Experimental Botany*, 49(320): 481–502.

Wang, Y.W., Wu, Q., *et al.* (2004). Attachment, proliferation and differentiation of osteoblasts on random biopolyester poly(3-hydroxybutyrate-co-3-hydroxyhexanoate) scaffolds. *Biomaterials*, 25(4): 669–675.

Wang, Y.W., Wu, Q., *et al.* (2005). Evaluation of three-dimensional scaffolds made of blends of hydroxyapatite and poly(3-hydroxybutyrate-co-3-hydroxyhexanoate) for bone reconstruction. *Biomaterials*, 26(8): 899–904.

Wang, Y.W., Yang, F., *et al.* (2005). Effect of composition of poly(3-hydroxybutyrate-co-3-hydroxyhexanoate) on growth of fibroblast and osteoblast. *Biomaterials*, 26(7): 755–761.

Were, L., Hettiarachchy, N.S., *et al.* (1999). Properties of cysteine-added soy protein-wheat gluten films. *Journal of Food Science*, 64(3): 514–518.

Widner, B., Behr, R., *et al.* (2005). Hyaluronic acid production in Bacillus subtilis. *Applied and Environmental Microbiology*, 71(7): 3747–3752.

Willett, J.L., Jasberg, B.K., *et al.* (1995). Rheology of thermoplastic starch: effects of temperature, moisture content and additives on melt viscosity. *Polymer Engineering and Science*, 35(2): 202–210.

Williams, G.M., Klein, T.J., *et al.* (2005). Cell density alters matrix accumulation in two distinct fractions and the mechanical integrity of alginate-chondrocyte constructs. *Acta Biomaterialia*, 1(6): 625–633.

Williams, S.F., Martin, D.P., *et al.* (1999). PHA applications: Addressing the price performance issue I. *Tissue engineering. International Journal of Biological Macromolecules*, 25(1–3): 111–121.

Xia, W., Liu, W., *et al.* (2004). Tissue engineering of cartilage with the use of chitosan-gelatin complex scaffolds. *Journal of Biomedical Materials Research Part B: Applied Biomaterials*, 71B(2): 373–380.

Xie, F., Yu, L., *et al.* (2006). Starch modification using reactive extrusion. *Starch/Starke*, 58: 131–139.

Xie, S.X., Liu, Q., *et al.* (2005). Starch Modification and Applications. In *Food Carbohydrates: Chemistry, Physical Properties, and Applications* (Cui, S.W., ed.). Boca Raton: CRC Press Taylor & Francis Group, pp. 357–405.

Yamane, S., Iwasaki, N., *et al.* (2005). Feasibility of chitosan-based hyaluronic acid hybrid biomaterial for a novel scaffold in cartilage tissue engineering. *Biomaterials*, 26(6): 611.

Yang, C.L., Hillas, P.J., *et al.* (2004). The application of recombinant human collagen in tissue engineering. *Biodrugs*, 18(2): 103–119.

Yang, X., Zhao, K., *et al.* (2002). Effect of surface treatment on the biocompatibility of microbial polyhydroxyalkanoates. *Biomaterials*, 23(5): 1391–1397.

Yannas, I.V. and Burke, J.F. (1980). Design of an artificial skin. I. Basic design principles. *J Biomed Mater Res*, 14(1): 65–81.

Yannas, I.V., Burke, J.F., *et al.* (1980). Design of an artificial skin. II. Control of chemical composition. *J Biomed Mater Res*, 14(2): 107–132.

Yannas, I.V., Burke, J.F., *et al.* (1982). Wound tissue can utilize a polymeric template to synthesize a functional extension of skin. *Science*, 215(4529): 174–176.

Yilmaz, G., Jongboom, R.O.J., *et al.* (2006). Thermoplastic starch as a biodegradable matrix for encapsulation and controlled release. In *Handbook of biodegradable Polymeric materials and their applications* (Mallapragada, S. and Narasimhan, B., eds). California: American Scientific Publishers, p. 2.

Yoo, H.S., Lee, E.A., *et al.* (2005). Hyaluronic acid modified biodegradable scaffolds for cartilage tissue engineering. *Biomaterials*, 26(14): 1925.

Zavan, B., Brun, P., *et al.* (2005). Extracellular matrix-enriched polymeric scaffolds as a substrate for hepatocyte cultures: *in vitro* and *in vivo* studies. *Biomaterials*, 26(34): 7038.

Zhang, Y., Ni, M., *et al.* (2003). Calcium phosphate-chitosan composite scaffolds for bone tissue engineering. *Tissue Engineering*, 9(2): 337–345.

Zhang, Y. and Zhang, M. (2001). Synthesis and characterization of macroporous chitosan/calcium phosphate composite scaffolds for tissue engineering. *Journal of Biomedical Materials Research*, 55(3): 304–312.

Zhao, F., Grayson, W.L., *et al.* (2006). Effects of hydroxyapatite in 3-D chitosan-gelatin polymer network on human mesenchymal stem cell construct development. *Biomaterials*, 27(9): 1859.

Zhao, F., Yin, Y., *et al.* (2002). Preparation and histological evaluation of biomimetic three-dimensional hydroxyapatite/chitosan-gelatin network composite scaffolds. *Biomaterials*, 23(15): 3227.

Zhao, K., Deng, Y., *et al.* (2003). Polyhydroxyalkanoate (PHA) scaffolds with good mechanical properties and biocompatibility. *Biomaterials*, 24(6): 1041–1045.

Zheng, Z., Bei, F.F., *et al.* (2005). Effects of crystallization of polyhydroxyalkanoate blend on surface physicochemical properties and interactions with rabbit articular cartilage chondrocytes. *Biomaterials*, 26(17): 3537–3548.

Ziegelaar, B.W., Aigner, J., *et al.* (2002). The characterization of human respiratory epithelial cells cultured on resorbable scaffolds: first steps towards a tissue engineered tracheal replacement. *Biomaterials*, 23(6): 1425.

Zou, X., Li, H., *et al.* (2004). Stimulation of porcine bone marrow stromal cells by hyaluronan, dexamethasone and rhBMP-2. *Biomaterials*, 25(23): 5375.

Chapter 7
Degradable polymers for tissue engineering

Riemke van Dijkhuizen-Radersma, Lorenzo Moroni, Aart van Apeldoorn, Zheng Zhang and Dirk Grijpma

Chapter objectives:

- To understand the requirements for degradable polymers to be suitable as scaffold material in tissue engineering
- To recognize how polymers can be synthesized and the advantages of copolymerization
- To understand the importance and to know examples of labile bonds in degradable polymers
- To identify the most important degradation mechanisms

- To be aware of the effect of chain scission and oligomer dissolution on molecular weight and mass loss
- To understand the effect of tissue response on the (*in vivo*) degradation rate
- To recognize examples of degradable polymers broadly used in tissue engineering

"The whole is more than the sum of its parts"

Aristotle (384 BC–322 BC), *Metaphysica*

"It is a mistake to think you can solve any major problems just with potatoes"

Douglas Adams (1952–2001), *Life, the Universe and Everything* (2005) Published by Del Ray.

7.1 Introduction and background

A general definition of tissue engineering can be found in *The Williams Dictionary for Biomaterials* (Williams, 1999): 'Tissue engineering is the persuasion of the body to heal itself through the delivery to appropriate sites of molecular signals, cells and supporting structures.' Although very short, this definition addresses the three main components of tissue engineering:

1. Molecular signals, such as growth factors, that stimulate the proliferation and differentiation of cells *in vitro* and/or *in vivo*, the infiltration by surrounding tissue,
2. Cells, to regenerate the lost or damaged tissue, and
3. Scaffolds and matrices, i.e. porous supporting structures and/or gels that allow cell attachment and tissue ingrowth.

The technology of tissue engineering is being applied to (re)generate many human tissues and organs. Examples are skin, cartilage, tendon, liver, esophagus, trachea, urothelial tissue, cardiovascular structures, intestine and bone. As an indication of the magnitude of the problem that tissue engineering addresses, worldwide organ replacement therapies utilizing standard metallic or polymeric devices consume 8% of medical spending, or approximately €300 billion per year (McIntire, 2002). The potential market world-wide for tissue-engineered products is estimated at nearly €100 billion per year.

In this chapter, we will not discuss the specific roles of cells and molecular signals in tissue engineering, as they are treated in other chapters of this book. Instead, we will focus here on basic aspects of polymers used in the preparation of polymeric scaffolds and matrices.

An important requirement for a tissue engineering scaffold (see Chapter 14), is that it degrades and resorbs at a rate that matches the formation of new tissue. The advantage is two-fold: the supportive structure does not hinder the development and growth of the tissue and any detrimental long-term tissue reactions of the body are prevented.

In addition to this, other important aspects regarding the polymeric material that is to be used are:

- Biocompatibility is of utmost importance (see Chapter 9) to prevent an adverse tissue reaction of the immune system.
- It should be possible to tune the rate of degradation of the polymer. For instance, the material should not degrade too quickly in load bearing areas, as it may cause a significant decrease of the mechanical properties and its consequent premature failure. On the other hand, degradation should not be too slow as it may prevent tissue regeneration.
- The resulting degradation products should be non-toxic and, therefore, should not induce an inflammatory reaction. They should dissolve in body fluids and, after transportation via the lymphatic system, the kidneys should be able to excrete them from the body.

To reach these goals, scaffolds for tissue engineering have been prepared from biologically derived polymers, such as collagen, hyaluronic acid and chitosan. However, the number of biologically-derived polymers that can be used, as well as the possible modifications to improve their mechanical properties and degradability, are limited. Natural polymers are often difficult to process and pathogenic risks may be associated with materials of animal or human origin.

Alternatively, the use of synthetic polymers of different chemistries enables the design of scaffolds with specific mechanical and biological properties, and degradation rates (Vacanti and Vacanti, 2000). Synthetic polymers can be produced cheaply and reproducibly and can be easily processed into devices of virtually any shape and form.

Much of the research on degradable polymers in tissue engineering has been focused on hybrid cell/scaffold constructs using degradable polymers such as poly(lactic acid) (PLA), poly(glycolic acid) (PGA) and their copolymers (PLGA), which are well-known in the medical field. These polymers degrade via hydrolysis of the main chain ester bonds (Reed and Gilding, 1981) and are resorbable *in vivo* as their degradation products (lactic- and glycolic acid) are part of the Krebs metabolic cycle. Besides PLA, PGA, and their copolymers, which are considered the golden standard in tissue engineering and regenerative medicine applications, novel polymers are continuously being developed, characterized and used to fabricate scaffolds and medical devices. Some of these polymers will be outlined in the following paragraphs of this chapter. From a degradation mechanism point of view, it is also important to differentiate the resorption behavior of polymers from that of ceramics: in the former case soluble compounds are generated upon degradation and scission of the polymer chain, while in the latter case the scaffolds resorb by slow dissolution of the constituent ions into intracellular or extracellular fluids (see Chapter 14). In this chapter we will discuss how polymers are synthesized and how their properties can be characterized and controlled. Particular emphasis will be put on degradation mechanisms of these biomaterials.

7.2 Synthesis and properties of polymers

7.2.1 Polymer synthesis

Polymers are long-chain, high molecular weight macromolecules formed by reaction of monomers.

A polymer chain consists of many covalently bound repeating units, formed upon sequential reactions of the monomers (Allcock *et al.*, 2003; Cowie, 1997; Odian, 2004). These monomers are multi-functional molecules that can react with each other to form homopolymers or with monomers of a different type to form copolymers. Important reaction mechanisms through which polymers are prepared are *addition polymerizations, step-growth polymerizations* and *ring-opening polymerizations*.

In *addition polymerizations*, double bond containing monomers such as ethylene, vinyl chloride, styrene and methyl methacrylate are polymerized by action of radicals or ionic species. In these polymerizations distinct steps can be discerned: initiation and the formation of reactive radicals or ionic compounds, chain propagation and chain termination. Once a reactive species is formed, successive additions of large numbers of monomers will lead to the formation of a polymer chain. Each monomer addition regenerates a reactive center. This chain reaction will continue until termination occurs. Overall, typical addition polymerizations can be presented as:

$$n\ H_2C=CH_2 \longrightarrow -[H_2C-CH_2]_n-$$
$$\text{ethylene} \qquad\qquad \text{polyethylene (or PE)}$$

and

$$n\ H_2C=C(CH_3)COOCH_3 \longrightarrow$$
$$\text{methyl methacrylate}$$

$$-[H_2CC(CH_3)COOCH_3]_n-$$
$$\text{poly(methyl methacrylate) (or PMMA)}$$

As a rule, addition polymers are not considered degradable polymers, as no labile bonds are present in the main chain. However, prolonged implantation of PE and PMMA and other such polymers will result in some chain scission due to cellular action in which radical species are formed.

Step growth polymerizations are characterized by the stepwise reaction of the functional groups of the reactants. These reactions are analogous to the simple reactions of functional groups. The size of the polymer molecule increases relatively slowly as dimers, trimers, tetramers, pentamers, etc. are formed until eventually large-sized molecules are formed. These reactions occur between any of the different-sized species present in the reaction system.

In *step growth* polymerization reactions, often a condensation reaction takes place in which a small molecule is liberated. For example, lactic acid is a molecule that contains two functional groups; a hydroxyl group (-OH) and a carboxylic group (-COOH). Upon removal of water by heating and application of vacuum, lactic acid can be polymerized to form a polyester:

$$nHOCH(CH_3)COOH \longrightarrow$$
lactic acid

$$-[OCH(CH_3)CO]_n- + (n-1)H_2O$$
poly(lactic acid) (or PLA)

But in the *step growth* preparation of polyurethanes, in which a diisocyanate reacts with a (polymeric) diol, a small molecule is not liberated and condensation does not take place:

$$OCN-(CH_2)_6-NCO + HO-[CH_2CH_2O]_m-H \longrightarrow$$
Hexamethylene   polyethylene glycol
diisocyanate

$$-[O(CH_2CH_2O)_mCONH(CH_2)_6NHCO]-$$
a polyurethane

A third mechanism by which polymers can be prepared is *ring opening polymerization* (ROP) of cyclic monomers. As an example, the ring opening of ethylene oxide is shown:

ethylene oxide $\longrightarrow$ $-[CH_2CH_2O]_n-$ Poly (ethylene oxide) (or PEO)

Besides the monomer used and the way in which it has been polymerized, the molecular weight and the molecular weight distribution of the polymer obtained are important factors that determine its properties.

7.2.2 Polymer and copolymer properties

The nature of the monomer(s) used to prepare the polymer will determine the repeating units within the polymer chain, and, therefore, will determine to a large extent the physical, chemical, and biological properties of the polymeric material as an implant or tissue engineering scaffold.

Depending on the intended application, polymeric materials with widely differing physical properties (e.g. thermal properties, mechanical properties, and hydrophilicity) and degradation characteristics are required (Ratner *et al.*, 2004). Parameters that often need to be optimized are the melting and glass transition temperatures, the tensile strength, the elastic modulus or stiffness, and the (surface) hydrophilicity. Therefore, copolymerization, i.e. the preparation of polymers from two or more types of monomers, is often employed to tune the material properties (Odian, 2004).

Copolymers can be classified as *random-, alternating-, block-* or *graft copolymers. Random copolymers* have a statistical arrangement of different monomer units in their backbone. In *alternating copolymers*, the different monomer units are arranged in such a way that they alternate along the polymer chain. The physical and degradation properties depend strongly on the monomers nature and composition. Examples of how these properties can be tuned are the statistical copolymerizations of L- and D-lactide to influence the crystallinity of the polymer (Grijpma and Pennings, 1994), lactide and ε-caprolactone copolymerizations to affect the stiffness of the copolymers (Grijpma *et al.*, 1991), and lactide and glycolide copolymerizations to adjust hydrolysis rates of the copolymers (Reed and Gilding, 1981). In general, random and alternating copolymer structures have characteristics which are intermediate between the properties of the different parent homopolymers, while block- and graft copolymers possess properties of both homopolymers (Allcock *et al.*, 2003; Cowie, 1997; Odian, 2004). The preparation of biodegradable polyesters by ring opening polymerization and the effect of copolymerization on their properties is described in Box 1.

Box 1 Biodegradable polyesters synthesized by ring opening polymerization

In the past years, biodegradable aliphatic polyesters, such as poly(lactide) [poly(lactic acid), PLA], poly(glycolide) [poly(glycolic acid), PGA], poly(ε-caprolactone) (PCL), and their copolymers, have attracted much attention (for reviews see Albertsson and Varma, 2002; Sodergard and Stolt, 2002; Ulrich *et al.*, 1999).

   Although polyesters can be synthesized by polycondensation of hydroxyl acids such as lactic acid, it is difficult to achieve high molecular weights and control the molecular weight, molecular weight distribution, and architecture of the polymer. In most cases, biodegradable polyesters are synthesized by ring-opening polymerization (ROP) of cyclic esters (or lactones). Structures of some of these lactones often used are:

Lactide   Glycolide ε-caprolactone

Structures of cyclic esters used in ring-opening polymerizations. Lactide has two optically active carbon atoms in the ring.

   ROP of cyclic esters proceeds through cationic polymerization, anionic polymerization or coordination-insertion reactions. The coordination-insertion reactions (Kowalski *et al.*, (2000); Kricheldorf *et al.*, 2000; Nijenhuis *et al.*, 1992) are often initiated by metal alkoxides, which contain a covalent metal-oxygen bond and have a weak Lewis acid character. The propagation proceeds by coordination of the monomer to the metal alkoxide, followed by the insertion of the monomer into the metal-oxygen bond. A scheme of the propagation step of this mechanism is shown above. It should be noted that the growing chains contain the metal-oxygen bond during the propagation, i.e., they remain active.

Scheme illustrating the propagation step of lactone ring opening polymerization via a coordination-insertion mechanism.

   Tin (II) 2-ethylhexanoate (stannous octoate or SnOct$_2$) is the most often used catalyst for the ROP of cyclic esters. SnOct$_2$ is a highly active catalyst that is soluble in the monomer melts, allowing the ROP to be performed in bulk. Although there is some concern about the cytotoxicity of SnOct$_2$, the American Food and Drug Administration (FDA) has approved many degradable implants in which the polymer has been prepared using small amounts of this catalyst.

   The mechanism of the ROP process using SnOct$_2$ as a catalyst has been studied in depth by Kowalski *et al.* (2000) and Kricheldorf *et al.* (2000). For the ROP of L-lactide and ε-CL, Kowalski and co-workers proposed the following mechanism: first, a tin-alkoxide bond is formed via a reversible reaction between SnOct$_2$ and a compound containing a hydroxyl group (a). The monomer is then coordinated to and inserted into the formed tin-alkoxide bond. Subsequent coordination and insertion of a next monomer then propagates the polymerization (b). Chain transfer reactions between the growing polymer chains (in which the tin-alkoxide bonds are still present) and compounds containing hydroxyl groups can also occur, leading to an increase in the molecular weight distribution (c).

(a) Sn(Oct)$_2$ + ROH $\rightleftharpoons$ Oct−Sn−OR + OctH

(b) −Sn−OR + n Cyclic Ester $\longrightarrow$ −Sn−(O~$\overset{\overset{\displaystyle O}{\|}}{C}$)$_n$ −OR

(c) −Sn−(O~$\overset{\overset{\displaystyle O}{\|}}{C}$)$_n$ −OR + ROH $\rightleftharpoons$ −Sn−OR + H−(O~$\overset{\overset{\displaystyle O}{\|}}{C}$)$_n$ −OR

Ring-opening polymerization of cyclic esters using SnOct$_2$ as a catalyst: (a) formation of a tin-alkoxide bond by reaction of an alcohol with SnOct$_2$ (b) propagation by monomer coordination and insertion; (c) chain transfer reaction.

The compounds containing hydroxyl groups (above) can be very small amounts of impurities present in the monomer, catalyst or reaction environment. If stringent purification and reaction conditions are applied, very high molecular weight polyesters (such as PLA with M_n of 310×10^3 g/mol) can be synthesized. These compounds can also be added on purpose to control molecular weight, composition and architecture of the polymer that is to be synthesized. When *e.g.* monomethoxy-PEG (mPEG) or a PEG-diol is used together with SnOct$_2$, the corresponding diblock or triblock copolymers of PEG and polyesters can be synthesized (Jeong *et al.*, 1997).

Lactides and polylactides
Of the degradable polyesters used in medical applications, PLA has received special interest due to the presence of asymmetrical carbon atoms in the lactide monomers. Lactide compounds exist in three different stereoisomeric forms; D-(R-) lactide (DLA), L-(S-) lactide (LLA), and meso-lactide. It should be noted that with D,L-lactide (DLLA), an equimolar racemic mixture of LLA and DLA isomers is meant. The structures of these different compounds are shown below.

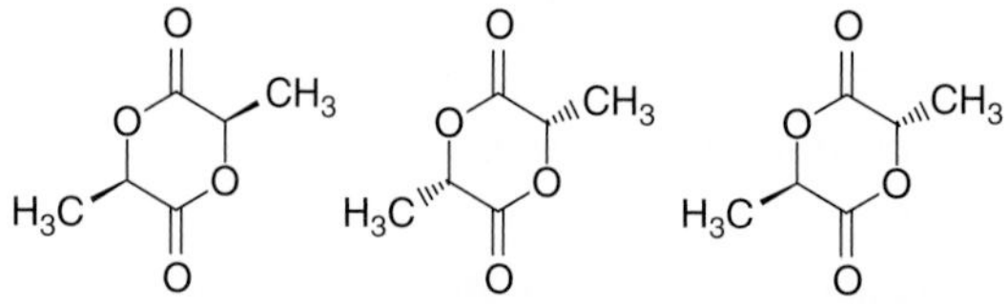

D-lactide   L-lactide   meso-lactide

Structures of lactide stereoisomers. D,L-lactide is an equimolar racemic mixture of LLA and DLA.

In the ROP with SnOct$_2$, the asymmetric carbon atoms in the lactide monomers can retain their conformation. Therefore, PLA polymers with different stereo-regularities (isotactic poly(LLA) (PLLA), poly(DLA) (PDLA), and atactic poly(DLLA) (PDLLA)) can be prepared.

PLLA and PDLA are semi-crystalline polymers with a glass transition temperature (T_g) of approximately 60°C, and a peak melting temperature (T_m) of approximately 180°C. PDLLA is amorphous with a T_g of approximately 55°C. Both semi-crystalline PLLA and amorphous PDLLA polymers are rigid materials. Their modulus of elasticity and stress at break (σ_{break}) values are close to, respectively, 3.5 GPa and 65 MPa. However, these polymers are relatively brittle with an elongation at break (ε_{break}) less than 6%.

Interestingly, mixtures of PLLA and PDLA can form so called stereo-complexes with even higher melting temperatures of up to 230°C.

Both polymers degrade by hydrolysis in which naturally occurring lactic acid is formed. Degradation of the polymers starts with water uptake, followed by random cleavage of the ester bonds in the polymer chain. The degradation is throughout the bulk of the material. Upon degradation, the number of carboxylic end group increases, which leads to a decrease in pH and autocatalytic hydrolytic degradation (Li et al., 1990c). During the degradation of semi-crystalline PLLA, crystallinity of the residual material increases as hydrolysis preferentially takes place in the amorphous domains. In general, the rate of degradation and erosion of amorphous PDLLA is faster than that of PLLA.

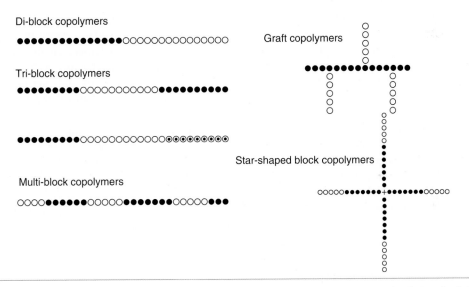

Figure 7.1 Schematic representation of block copolymer structures: ● A-monomer unit, ○-B unit, ◉ C-unit.

Block copolymers are composed of connected polymeric or oligomeric segments built up of different monomer units. Examples are diblock copolymers, triblock copolymers and multiblock copolymers. In *graft copolymers*, a comb-like structure is obtained by attaching (co)polymer segments to a polymer backbone at various places. Several types of block copolymer architectures can be synthesized for use as biomaterials. They are schematically represented in Figure 7.1 and can be categorized into linear and non-linear copolymers.

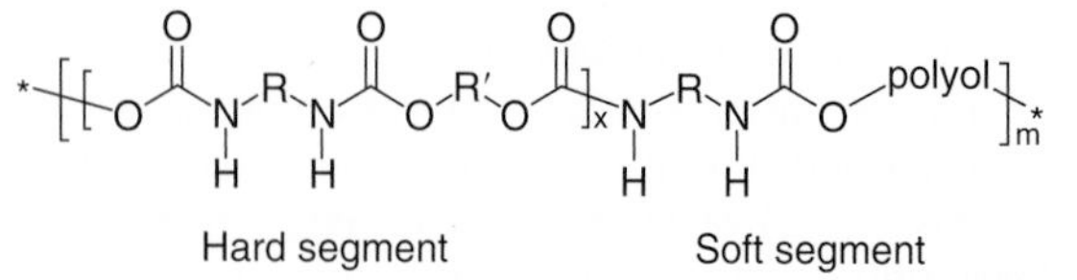

Hard segment   Soft segment

Figure 7.2 General structure of a segmented poly(urethane) prepared from a diisocyanate, OCN-R-NCO; chain extender, HO-R'-OH; and polyol building blocks.

In blends (or mixtures) of high molecular weight polymers the miscibility of the constituents is usually quite poor and macroscopic phase separation occurs, which leads to poor mechanical properties. This is due to the absence of (specific) interactions and large differences in their solubility parameters. In contrast, the constituent segments in block copolymers are covalently bound, so that microphase-separated structures are often only formed; block copolymers possess attractive properties that combine the intrinsic properties of the different building blocks.

An important class of phase-separated block copolymers is that of multi-block copolymers containing alternating soft- and hard segments (Bates and Fredrickson, 1996). Frequently, a hard phase (with high melting temperature or glass transition temperature) dispersed in a continuous soft phase (with a glass transition temperature below room temperature) can be distinguished. These hard segments form physical cross-linkages between the polymeric chains adding strength and toughness to a flexible and elastic thermoplastic elastomeric material.

Well-known examples of multi-block copolymers are segmented poly(urethane)s. Building blocks of segmented poly(urethane)s include diisocyanates, polyols, and chain extenders. Usually the hard segment originates from the diisocyanate and chain extender molecules, and the soft segment contains the polyol (a hydroxyl group terminated polymer or oligomer) moiety. A general structure of such a poly(urethane) is shown in Figure 7.2.

7.2.3 Chemically cross-linked polymer networks

Besides the *physical* cross-linkages between the polymeric chains as described above, polymer networks can be obtained by *chemical* cross-linking. Bi-functional monomers such as methyl methacrylate and lactic acid yield linear polymers (which are often soluble in solvents and can be processed at elevated temperatures above their glass transition temperature or melting temperature). Monomers with a functionality larger than two, e.g. glycerol which contains three reactive hydroxyl groups, can be used to prepare branched polymers and cross-linked network structures when polymerized with a diacid monomer.

In chemically cross-linked polymers a network structure is created by the formation of covalent linkages between the polymer chains (Allcock *et al.*, 2003; Cowie, 1997; Odian, 2004). In contrast to linear and branched polymers, chemically cross-linked polymers do not dissolve in solvents, but only swell in them. The extent of the polymer network swelling by a solvent is dependent on the density of cross-linkages: the higher the cross-linking density, the lower the degree of swelling. As a chemically cross-linked polymer is in fact a single giant molecule, it cannot be processed at temperatures above its melting temperature or glass transition temperature either.

Important examples of cross-linked polymers are hydrogels. These polymers are based on hydrophilic, water soluble oligomeric or polymeric chains that have been reacted with each other to form a covalent network. In one preparation route, polyethylene glycol (PEG) is functionalized at both termini with acrylic groups that in a later stage are linked by radical photopolymerization reactions (Sawhney *et al.*, 1993). To prepare a degradable and resorbable polymeric scaffold for tissue engineering, it will be necessary to have cleavable bonds in the polymer main chain or in the cross-linkages between the polymer chains.

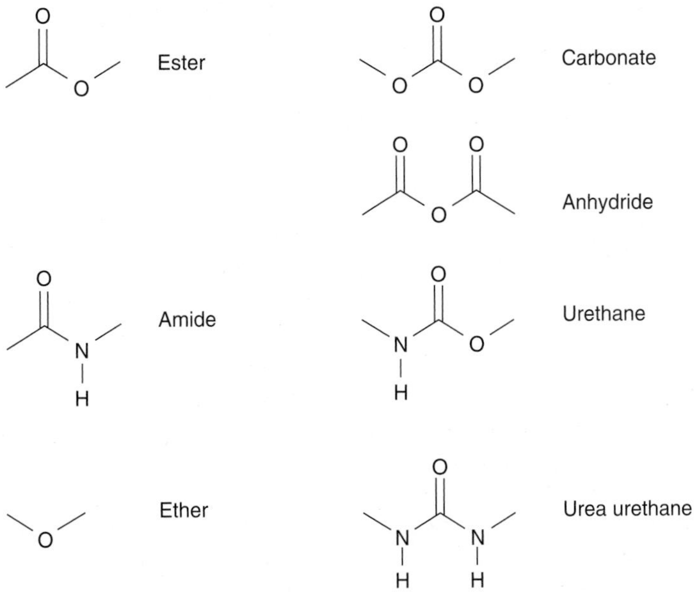

Figure 7.3 Overview of important liable bonds in the main chain of biodegradable polymers.

7.3 (Bio)degradable polymers

7.3.1 Labile bonds

Significant efforts have been devoted to prepare bio-degradable polymers and use them in medical applications as implants, drug delivery devices and tissue engineering scaffolds (Uhrich *et al.*, 1999). Polymer degradation and decrease of its molecular weight occurs through scission of the main chain. Although degradation reactions can occur during processing at elevated temperatures or upon sterilization with high energy radiation, for a polymeric material to resorb in the body it is necessary that the main chain of the macromolecule contains labile bonds. Such bonds can then be scissioned by hydrolysis or oxidation reactions (see section 7.2, also through enzymes and cellular activity) to yield (soluble) compounds of lower molecular weight (Göpferich, 1996; Tamada and Langer, 1993). Figure 7.3 gives an overview of important cleavable bonds present in biodegradable polymers.

7.3.2 Biodegradable polymers in tissue engineering

Of the currently employed biomaterials, synthetic polymers possess most characteristics required to fabricate scaffolds for tissue engineering applications. The most often used synthetic polymers, their properties and applications area are summarized in Table 7.1.

Of the synthetic polymers, linear aliphatic polyesters like poly(lactic acid) (PLA), poly(glycolic acid) (PGA) and copolymers (PLGA) are very well known in the medical field and have therefore been broadly used in tissue engineering. In general they elicit a minimal or mild foreign body reaction, and as such are considered to be biocompatible. By varying their molecular weight and in case of PLGA also

Table 7.1 Principal properties of synthetic polymeric biomaterials used in tissue engineering

Degradable polymer		Biocompatibility	Biodegradation	Bulk Mechanical Stiffness: E-Modulus (GPa)	Tissue Engineering applications
poly(lactic acid)	PLA	(Enzymatic) hydrolysis	Bulk–5 months to ~5 years	2–3	Skin, Cartilage, Bone, Ligament, Tendon, Vessels,
poly(glycolic acid)	PGA	Degradation products in metabolic pathway Localized Inflammation	Bulk–1 to ~12 months	5–7	Nerve, Bladder, Liver
poly(lactic-co-glycolic acid)	PLGA		Bulk–1 to ~12 months	2–7	
poly(ε-caprolactone)	PCL	Hydrolysis Minimal inflammation	Bulk–more than 3 years	0.4	Skin, Cartilage, Bone, Ligament, Tendon, Vessels, Nerve
multi block copolymers comprising poly(ethylene oxide) and butylene terephtalate units	PEOT/PBT	Hydrolysis Mild foreign body reaction No inflammation	Bulk–1 month to ~5 years	0.01–0.1	Skin, Cartilage, Bone, Muscle
polyphospho-esters	PPEs	Hydrolysis Minimal foreign body reaction Minimal inflammation	Erosion–1 to more than 3 years	0.4–0.7	Cartilage, Bone, Nerve, Liver
polyphosphazenes	PPAs	Hydrolysis Minimal foreign body reaction Minimal inflammation	Erosion–1 week to more than 3 years	$0.02*10^{-3}$– $0.2*10^{-3}$	Bone, Nerve
polyanhydrides	PAs	Hydrolysis Minimal foreign body reaction Minimal inflammation	Erosion–within 1 month	$0.2* 10^{-3}$– $6*10^{-3}$	Bone
polyorthoesters	POEs	Degradation products in metabolic pathway Minimal inflammation	Erosion–1week to ~16 months	$0.012*10^{-3}$–4	Bone
poly(propylene fumarate)-diacrylate networks	PPF-DA	Hydrolysis Minimal foreign body reaction Mild inflammation	Bulk–6 months to more than 3 years	0.002–0.12	Bone, Vessels
poly(ethylene glycol)-diacrylate networks	PEG-DA	Hydrolysis Minimal foreign body reaction	Varies with composition	$0.032*10^{-3}$– $0.5*10^{-3}$	Skin, Vessels, Nerve

their copolymer ratio, the biodegradation rate and the mechanical properties can be tailored. They are suited for tissue engineering applications (Anderson and Langone, 1999; Babensee *et al.*, 2000; Chu *et al.*, 1997; Freed *et al.*, 1993; Honda *et al.*, 2000; Sarazin *et al.*, 2004), as their degradation products (lactic and glycolic acids) obtained resulting from hydrolysis are normally present in the metabolic pathways of the human body. However, their bulk degradation behavior (see section 7.4) may lead to the formation and accumulation of relatively large amounts of acidic degradation products that cannot be easily disposed of. This can result in local inflammation (Bostman *et al.*, 1989; Fu *et al.*, 2000). Furthermore, in semi-crystalline polymers, the crystalline residues may take a long time to resorb completely. Another linear aliphatic polyester commonly used in tissue engineering is poly(ε-caprolactone) (PCL). This polymer has found many applications as a result of its good biocompatibility and favorable mechanical properties. It degrades at a much lower rate than PLA, PGA, and PLGA, making it attractive when long-term implants are desired (Choi and Park, 2002; Honda *et al.*, 2000; Hutmacher *et al.*, 2001; Wang, 1989).

Another family of thermoplastic polymers that has been developed recently for tissue engineering and drug delivery is multi blockcopolymers comprising poly(ethylene oxide) (PEO) and poly(butylene terephthalate) (PBT) (PEOT/PBT)) They are prepared by polycondensation of (dimethyl) terephtalate (T), butanediol (B) and polyethylene glycol (PEG). These poly(ether ester) copolymers (Figure 7.4) exhibit good physical properties like elasticity, toughness and strength in combination with easy processability (Bezemer *et al.*, 1999). Their properties result mainly from a phase separated morphology in which soft, hydrophilic PEO segments are physically cross-linked by hard, semi crystalline PBT segments. In contrast to chemically cross-linked materials, these cross-links are reversible and will be disrupted at temperatures above their glass transition- or melting point, giving the polymer material its good processability.

The interest arisen in tissue engineering applications is due to the fact that by varying the molecular weight of the starting PEG segments and the weight ratio of PEOT and PBT blocks it is possible to tailor properties, such as wettability (Olde Riekerink *et al.*, 2003), swelling (Bezemer *et al.*, 1999; Deschamps *et al.*, 2002; van Dijkhuizen-Radersma *et al.*, 2002), biodegradation rate (Deschamps *et al.*, 2002), protein adsorption Mahmood *et al.*, 2004), and mechanical properties (Leong *et al.*, 1985; Moroni *et al.*, 2005). PEOT/PBT copolymers have shown to be biocompatible both *in vitro* and *in vivo* in skin, cartilage, and bone regeneration experiments (Bakker *et al.*, 1998; Beumer *et al.*, 1994a, 1994b; van Blitterswijk *et al.*, 1993) and they have reached clinical application (PolyActive™, IsoTis Orthopaedics SA) as cement stoppers and bone fillers in orthopedic surgery (Bulstra *et al.*, 1996; Mensik *et al.*, 2002). Being poly(ether ester)s, degradation occurs in aqueous media by hydrolysis and by oxidation. Degradation rate is very slow for multi block-copolymers with high PBT contents and more rapid contents of PEOT and longer PEO segments (Bezemer *et al.*, 1999; Deschamps *et al.*, 2002). A further modulation in degradation rate can be achieved by substituting

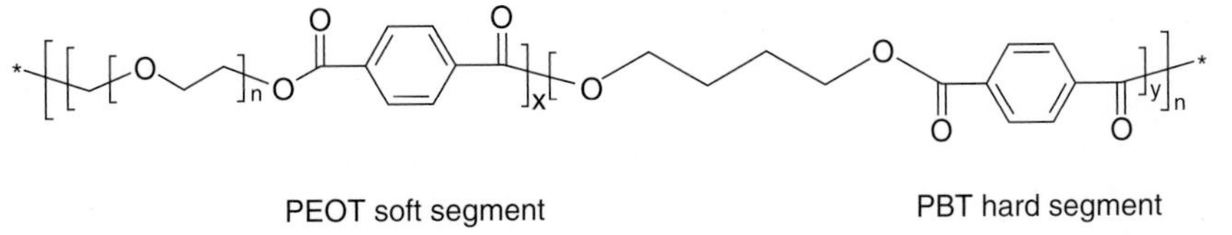

PEOT soft segment   PBT hard segment

Figure 7.4 Chemical structure of PEOT/PBT multi-block copolymers.

the terephtalate moieties domains with succinic acid derivatives during the copolymerization reaction (van Dijkhuizen-Radersma *et al.*, 2003, 2005).

It can be stated that PLA, PGA, PLGA copolymers, PCL, and PEOT/PBT copolymers are equally well suited for many tissue engineering applications. They are attractive biomaterials to fabricate scaffolds and have been widely used with promising results in regenerative medicine.

Among the multitude of other synthetic polymers that have been investigated for controlled release and tissue engineering applications, interesting classes are polyphospho-esters (PPEs) (Wang *et al.*, 2001a, 2001b), polyphosphazenes (PPAs) (Aldini *et al.*, 2001; Ambrosio *et al.*, 2002; Caliceti *et al.*, 2000; Cohen *et al.*, 1993), polyanhydrides (PAs) (Leong *et al.*, 1986) and polyortho-esters (POEs) (Choi and Heller, 1978). These polymers have shown to resorb via a surface erosion mechanism (see section 7.2) (Andriano *et al.*, 1999; Burkoth *et al.*, 2000), which is also known to affect the stability of the scaffolds to a lesser extent and to elicit a milder *in vivo* inflammatory response, as compared to polyesters and poly(ether ester)s previously considered. PPEs have adequate mechanical properties that make them suitable for hard tissue engineering. Although PPAs, PAs, and POEs have also been used for hard tissue repair, they might be more suitable for soft tissue engineering due to their lower rigidities.

An alternative to the preparation of tissue engineering scaffolds by processing polymers from melts or solutions is making use of injectable polymers. This class of polymeric materials is very attractive as they can be used in minimally invasive surgery such as arthroscopy, which is highly beneficial to the patient. Furthermore, they can fill irregularly shaped defects (Burdick *et al.*, 2001; Fisher *et al.*, 2001; Kui and Ma, 2001), and cells and bioactive agents can easily be incorporated (Elbert *et al.*, 2001; Mann *et al.*, 2001a, 2001b). In particular, photopolymerizable systems based on poly(propylene fumarate)-diacrylates (PPF-DA) and poly(ethylene glycol)-diacrylates (PEG-DA), have been much investigated. They even can be hardened transdermally by applying light (Behravesh *et al.*, 1999; Elbert and Hubbell, 2001; He *et al.*, 2000a,

2000b). Alternatively, chemically curable polymers also based on PPF have also been studied (Behravesh *et al.*, 1999). Despite the advantages in using these polymers, some issues may still rise from their low mechanical properties and possible cytotoxicity remains due to the acrylic groups.

7.4 Mechanisms of polymer degradation and erosion

7.4.1 Definitions

In polymer degradation, chain scission occurs and oligomers, monomers and other low molecular weight species are formed (Tamada and Langer, 1993). The degradation of polymers can be induced by thermal activation, oxidation, photolysis, radiolysis, or hydrolysis processes (Gopferich, 1996). When degradation is affected by the biological environment, the term biodegradation can be used. As a consequence of degradation, *erosion* of the material can occur. Erosion specifically refers to the loss of material by monomers and oligomers leaving the polymer mass. Obviously, if polymer erosion and resorption is to occur, first soluble components need to be formed.

7.4.2 Chain scission and polymer degradation

For polymeric biomaterials, the most important degradation reaction is hydrolysis (Figure 7.6). There are several factors that influence the velocity rate of this reaction: *the nature of the chemical bond, the pH, the copolymer composition and the extend of water uptake* are the most relevant. Of these factors, it is mainly the type of bond in the polymer backbone that determines the rate of hydrolysis. Anhydrides and ortho-ester bonds are the most reactive ones. In water, anhydrides are more reactive than esters and amides. Such ranking must be viewed with circumspection. However, reactivity of the bonds is much dependent on catalysis and steric or electronic effects.

In organic chemistry, ester hydrolysis of organic compounds occurs by reaction of water with the ester bond. However, in the presence of acidic or basic environments, the hydrolysis reactions are catalyzed and the rates are much higher (Smith, 2001). The hydrolytic degradation of poly(CL), poly(DLLA) and related aliphatic polyesters involves the generation of carboxylic end groups that are also able to catalyze the hydrolysis reaction (Pitt *et al.*, 1981a, 1981b). Based on a first-order kinetic model, Pitt and co-workers (Pitt *et al.*, 1981a) related the rate of chain scission of an aliphatic polyester autocatalyzed by the generated carboxylic acid end groups to the decrease of the number average molecular weight in time:

$$ln(\bar{M}n) = ln(\bar{M}n^0) - kt$$

where is the average number molecular weight at the start of the hydrolysis, k is the rate constant and t is the degradation time.

If the hydrolysis is not autocatalytic, chain cleavage will follow the rate law (Pitt and Gu, 1987):

$$\frac{1}{\bar{M}n} = \frac{1}{\bar{M}n^0} + k't$$

As discussed before (section 7.2.2), the composition can have a large influence on the chemical and physical properties of a (co)polymer. The rate of degradation of a polymer is dependent on the ease of hydrolysability as well as on the accessibility of the cleavable main chain bonds to enzymes and water. Besides the nature of these labile bonds, the hydrophilicity of the material, the morphology and crystallinity of the polymer and its molecular weight are important parameters in determining the degradability of a polymer.

In the hydrolysis of a polyester, which initially is hydrophobic and not soluble in water, the average molecular weight of the polymer will decrease in time. It will require a certain degree of main chain hydrolysis before appreciable amounts of water soluble oligomers and low molecular compounds are formed.

7.4.3 Bulk- and surface-erosion

Polymer degradation processes can result in erosion, i.e. loss of mass of the materials, when degradation products diffuse and dissolve into the degradation environment. Polymer erosion can be a more complex process as it depends on many other processes besides degradation, such as morphological changes and characteristics of the oligomers formed (Göpferich, 1996).

In polyesters there are four main factors which determine the erosion diffusion and dissolution phenomena (Vert, 2005):

1. The hydrolysis rate constant of the ester bond.
2. The diffusion coefficient of water within the polymer matrix.
3. The diffusion coefficient of chain fragments within the polymer matrix.
4. The solubility of the degradation products (oligomers) in the surrounding medium.

Erosion of polymers can be classified as a bulk erosion process or surface erosion process (Göpferich, 1993, 1996; von Burkersroda *et al.*, 2002). In bulk erosion, the polymer chain scission occurs throughout the specimen. The molecular weight and the mechanical strength of the specimens decrease in time. The decrease in molecular weight occurs essentially from the beginning of the degradation process, whereas loss of mass is much delayed. The external dimensions of the polymer specimens remain essentially unchanged, until the specimens disintegrate at a critical time point. It should be noted that the loss in mechanical properties of the material precedes the loss in mass. The process of bulk erosion is schematically described in Figure 7.5a.

In surface erosion, loss of material is confined to the surface of the polymer device only. Size and

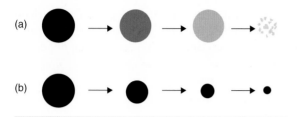

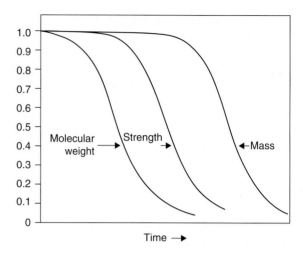

Figure 7.5 Scheme illustrating degradation of materials by bulk erosion (a) and surface erosion (b) processes.

mass of the device decrease in time, whereas molecular weight and mechanical properties of the polymer device remain unchanged. In surface erosion, the rate of mass loss is proportional to the surface area. Surface erosion is schematically depicted in Figure 7.5b.

Most biodegradable polyesters that are currently available degrade by a bulk erosion process that predominantly involves simple hydrolysis of main chain ester bonds (Middleton and Tipton, 2000). First, small amounts of water diffuse into the bulk of the material preferentially attacking the ester bonds present in the amorphous domains of the polymer. Initially, this decrease in molecular weight will not affect the mechanical properties of the device as physical cross-linkages due to entanglements and regions of crystallinity maintain the structure of the material. At a certain point, however, the reduction in molecular weight will lead to a decrease in the mechanical properties of the material. As more carboxylic acid end groups are formed, the rate of the hydrolysis reaction will increase and loss of mass will occur when the degradation products become soluble in water. Figure 7.6a shows a graphic representation of these effects.

In a surface erosion process (Figure 7.6b), the rate of hydrolysis of labile bonds is relatively fast in comparison to the diffusion rate of water into the bulk and the degradation reactions are limited to the surface of the polymer material. The molecular weight of the polymer, and therefore its mechanical properties,

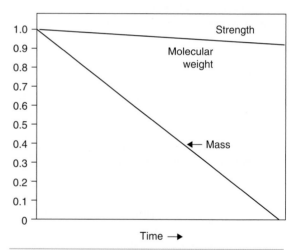

Figure 7.6 (a) Representation of the bulk erosion process of degradable polymers; (b) Representation of the surface erosion process of degradable polymers.

remain more or less unchanged with time, while the mass of the device decreases at an appreciable rate. A surface-eroding polymer could provide constant and well controlled release rates in drug delivery applications, but only few polymers show surface erosion characteristics: examples are polyanhydrides (copolymers of sebacic acid (SA), 1,3-bis(p-carboxyphenoxy)propane (CPP), and 1,3-bis(p-carboxyphenoxy)hexane (CPH)) (Leong *et al.*, 1985), poly(adipicanhydride) (Albertsson and Eklund, 1995), and poly(orthoesters) (Heller, 1990). These polymers have hydrolytically very labile bonds in their main chains, which react with water rapidly. Therefore, the degradation process occurs at the polymer surface resulting in a surface eroding material.

7.4.4 *In vivo* degradation

In the development of a novel degradable polymer, most degradation experiments are performed *in vitro* by incubating the polymeric specimen in phosphate buffered saline (PBS) at body temperature (37°C) to study the degradation mechanism. However, *in vitro* degradation behaviour can considerably differ from *in vivo* degradation. In most cases *in vivo* degradation is faster than *in vitro* degradation (Fulzele *et al.*, 2003; Lu *et al.*, 2000; Mainil-Varlet *et al.*, 1997; Menei *et al.*, 1993; Spenlehauer *et al.*, 1989; Stoll *et al.*, 2001; Tracy *et al.*, 1999). The faster *in vivo* degradation can be attributed to the tissue response (Ali *et al.*, 1994; Stoll *et al.*, 2001). Acute inflammation, due to the injury of implantation, and foreign body reaction, due to the presence of the implant, induce the migration of polymorphonuclear leucocytes and macrophages to the implant site (Anderson, 1994; Anderson and Shive, 1997). Reactive species, like super oxide and hydrogen peroxide species, produced by these inflammatory cells can oxidize the polymer chains (Deschamps *et al.*, 2002; Goedemoed *et al.*, 1998). Therefore, the *in vivo* decrease in molecular weight values is not only due to hydrolytic degradation caused by the extracellular fluid, but probably also by the influence of the oxygen free radicals and of the other species generated by the inflammatory cells (Ali *et al.*, 1994). In some cases however, such as in hydrogel-like polymer systems, a slower *in vivo* degradation has been reported compared to the *in vitro* degradation in phosphate buffer (Changez *et al.*,

2003). The presence of less fluid near the implant site retards swelling and subsequently inhibits degradation. For copolymers only degrading by hydrolysis in the bulk, the *in vitro* degradation will mimic the *in vivo* degradation (Deschamps *et al.*, 2003; Suggs *et al.*, 1998).

Besides the increased chain scission rate, the erosion rate *in vivo* differs from the *in vitro* situation. Mass loss *in vivo* can be based on either *passive transport* of monomers and oligomers by dissolution and diffusion (similar to *in vitro*) or by *active transport* by phagocytes or a combination of these processes. The solubility of the oligomers may be increased *in vivo* due to the presence of lipids, resulting in a faster mass loss. The dissolved degradation products will be removed from the implantation site via the lymphatic system and subsequently secreted from the body by the kidneys. The *in vivo* mass loss can be further increased by mechanical stresses and cellular activity. Various studies demonstrated the fragmentation of the biomaterial in time. Implants based on PEOT/PBT copolymers have been evaluated extensively in various implantation sites and extensive fragmentation of the implants was visible within 4 weeks, thereby significantly enlarging the implant surface (Lu *et al.*, 2000; Menei *et al.*, 1993). Higher fragmentation rates and smaller average particle sizes were observed at intramuscular implantation sites as compared to subcutaneous sites (Menei *et al.*, 1993). This could be attributed to the higher stresses and higher extents of tissue vascularization in muscular tissue. Ultimately, the implant breaks down into particles smaller than 10 μm, which undergo macrophage phagocytosis (Anderson and Shive, 1997). Transmission Electron Microscopy (TEM) and Raman spectroscopy and imaging showed phagocytes containing intracellular polymeric particles (Beumer *et al.*, 1994; van Blitterswijk *et al.*, 1993; van Loon, 1995). (A clear example of polymer phagocytosis can be found in Box 2 'Examples of *in vivo* polymer degradation'.)

Besides the site of implantation, other parameters that affect the *in vivo* degradation are the *size* and the *shape* of the implant. As described in Chapter 6, the tissue response towards a biomaterial depends to a large extent on the shape of the implant (Anderson, 1994; Matlaga *et al.*, 1976). Being a surface related process, a higher surface area (due to fragmentation or porosity) can affect the tissue response. In addition, sharp angular shapes can induce a higher response than more rounded shapes (Matlaga *et al.*, 1976). The formation of a fibrous capsule as part of the foreign body response may also affect the rate of degradation. Furthermore, one would expect that larger implants require longer degradation times. However, this may depend upon auto or normal catalysis of the degradation process. In the case of poly(lactide-co-glycolide) (PLGA); for example, the *in vivo* degradation is autocatalyzed due to formation of acidic degradation products (see

Classical Experiment). Various studies showed that the autocatalytic effect can be affected by the size of the implant (Grizzi *et al.*, 1995; Göpferich and Tessmar, 1999; Lu *et al.*, 1999; Therin *et al.*, 1992a). Comparison of PLGA implants of various sizes, indicated that a critical thickness can be determined above which hydrolytic degradation is accelerated due to accumulation of acidic degradation products in the implant (Therin *et al.*, 1992a).

In conclusion, the *in vivo* degradation of a biodegradable polymer is affected by many factors, which are influenced by the implant characteristics. This complex situation stresses the importance of performing relevant *in vivo* experiments when selecting a biodegradable polymer for a specific tissue engineering application.

Box 2 Examples of *in vivo* polymer degradation

One of the key issues in successful application of biomaterials for tissue engineering and drug delivery is a well-characterized *in vivo* (bio)degradation behavior. Scientists have access to several techniques by which this can be studied. These vary from nuclear magnetic resonance (NMR), a destructive analysis technique, to simple light microscopy, which is a non-destructive technique. Once a degradable (bio)material is implanted, a number of processes are initiated that lead to the breakdown of the implant. By using combinations of analysis techniques, information on chemical and morphological changes caused by degradation can be gathered. A simple approach to study the effects of degradation is to design an *in vitro* experiment first. This model investigates the basic changes in the material at hand. The information obtained form this experiment can than later be used to interpret the data obtained during an *in vivo* experiment, which is the ultimate performance test for a biomaterial. In this part, two examples of studies on the degradation of biomaterials are described. The first describes the degradation of several PEOT/PBT multiblock copolymers. The second describes the intra cellular degradation of poly(lactic-co-glycolic acid) microspheres.

Example 1
As a model for degradation, 300PEOT55PBT45, 300PEOT70PBT30 and 1000PEOT70PBT30 copolymers (indicated as **a**PEOT**b**PBT**c** in which **a** is the PEO molecular weight, **b** the weight percentage of PEOT, and **c** (=100 − **b**) the weight percentage of PBT) were analyzed after subcutaneous implantation in rats. At 12 weeks, only 1000PEOT70PBT30 polymer discs showed clear signs of degradation, as a loss in PEOT content was observed. For 300PEOT70PBT30 and 300PEOT55PBT45, no changes in composition were observed. Moreover, Histological examination by light microscopy revealed that macrophage-like cells clearly play an active role in the removal of these materials through phagocytosis and possible further degradation of the polymer.

Light micrograph of 300PEOT55PBT45 polymer at 1 week (left) and 12 weeks of implantation (right). Besides minor changes in the surface, no effects on polymer morphology can be observed after 12 weeks of implantation. Scalebars are 100 μm. Reproduced with permission from JACS: van Apeldoorn, A.A., van Manen, H.J., et al. (2004). Raman imaging of PLGA microsphere degradation inside macrophages. J Am Chem Soc, 126(41): 13226–13227.

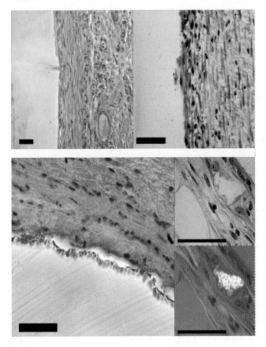

Light micrographs of 300PEOT70PBT30 polymer at 1 (top left) 4 (top right) and 12 (bottom left) weeks of implantation. Bottom right: small polymer fragments can be observed by regular light microscopy (bottom right top) and by polarized light microscopy (bottom right bottom) at 12 weeks. The dark blue granulation at the polymer surface at 12 weeks, indicates calcification of the polymer. Scalebars are 100 μm. Reproduced with permission from JACS: van Apeldoorn, A.A., van Manen, H.J., et al. (2004). Raman imaging of PLGA microsphere degradation inside macrophages. J Am Chem Soc, 126(41): 13226–13227.

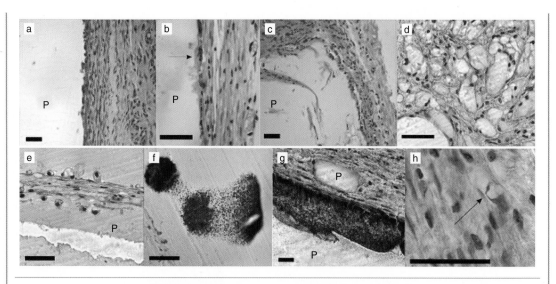

Light micrographs of 1000PEOT70PBT30 polymer at 1 (a), 2 (b), 4 (c) and 12 weeks (e–h) of implantation. In (a) the onset of beginning lacunae formation on the polymer (p) surface is shown, which progressed after 2 weeks (b), macrophage-like cells (arrow) occupying the lacunae can be observed. After 4 weeks of implantation (c) large cracks have formed throughout the polymer and are filled with connective tissue. After 12 weeks of implantation severe fragmentation of the polymer discs was observed (d and e), and also signs of extensive calcification (dark blue and purple granulation) – this occurs not only at the surface (g) but also within the polymer (f); (g) and (h) not only show that a large number of small polymer fragments were present in the connective tissue surrounding the implant, but in the cytoplasm of macrophage-like cells as well (arrow). Scalebars are 100 μm. Reproduced with permission from JACS: van Apeldoorn, A.A., van Manen, H.J., et al. (2004). Raman imaging of PLGA microsphere degradation inside macrophages. J Am Chem Soc, 126(41): 13226–13227.

Based on these histological results several conclusions can be drawn. The 1000PEOT70PBT30 polymer shows the most pronounced signs of degradation, while the other two compositions show only minor surface changes. In the case of 300PEOT70PBT30 some small fragments detaching from the surface and indications of calcification could be also observed.

When following the degradation of the 1000PEOT70PBT30 polymers in time, several processes can be described. Degradation of the sample starts with a mild erosion at the surface, where macrophage-like cells are involved in lacunae formation (4B). In time this surface effect progresses towards fragmentation of the polymer sample (4C) and cracks start to appear throughout the sample. These fissures become infiltrated by connective tissue. Then the sample falls apart into small fragments surrounded by macrophage like cells and calcification can be observed. Upon closer examination, the macrophage-like cells seem to contain small polymer fragments within their cytoplasm; this is indicative of the active role these cells play in the *in vivo* degradation of these materials.

The above example shows that, based on histological observations, one can already obtain a wealth of information on the degradation process of biomaterials. An key observation is that cells apparently play a great part in the degradation process, which stresses the importance of performing *in vivo* studies.

Example 2

Poly(lactic-*co*-glycolic acid) (PLGA) has been used for a wide variety of medical applications, ranging from resorbable sutures to bone screws and microspheres for drug delivery applications. From literature it is known that the degradation of PLGA *in vivo* and *in vitro* mainly takes place through either hydrolysis of the ester linkages (Edlund and Albertsson, 2002; Schliecker *et al.*, 2003).

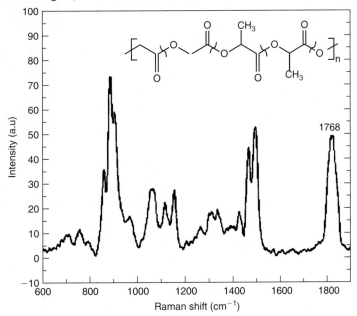

Raman spectrum and chemical structure of PLGA. In the structure, two glycolic acid units are shown on the left, whereas lactic acid units are displayed on the right. The ester bonds (marked '~'), which are hydrolyzed upon degradation, are characterized by a carbonyl stretching vibration at 1768 cm⁻¹. Reproduced with permission from JACS: van Apeldoorn, A.A., van Manen, H.J., et al. (2004). Raman imaging of PLGA microsphere degradation inside macrophages. J Am Chem Soc, *126(41): 13226–13227.*

We also know from literature that devices composed of PLGA degrade in a heterogeneous manner called autocatalytic degradation (Therin *et al.*, 1992b).

In 1995, Park proposed a possible model describing this process which was based on previous work done by Vert and co-workers on these materials. They described a model in which the degradation proceeds more rapidly in the center than at the exterior of the specimen (Li *et al.*, 1990a, 1990c; Park, 1995; Therin *et al.*, 1992) (see Classical Experiment). This model involves the autocatalytic action of the carboxylic acid end groups of the degrading material trapped within the internal environment of the specimen. It has been questioned whether this model is also valid when dealing with small implants or microspheres, since these are so small in size that entrapment of these acid end group cannot take place (Li, 1999). It has been postulated that for small implants and thin films, the degradation process is more homogeneous (Grizzi *et al.*, 1995).

In this case chemical information might provide insight on the value of the degradation models described in the literature. After phagocytosis, degradation may continue within an organelle called the 'phagosome'. While the *in vivo* response to an implanted biomaterial can to a great extent be studied by histology, no information in relation to the chemical composition of the degrading materials can be obtained.

Raman spectroscopy can be used to study the chemical bonds involved in degradation of polymers by detecting intensity and wavelength changes in the vibrational bands of these bonds. By implementing this technology into a confocal (very high resolution) microscope, it is possible to simultaneously study the chemistry and the morphology of samples of interest.

By using an *in vitro* cell culture model and confocal Raman microscopy we now can study the degradation of PLGA microspheres, after macrophage phagocytosis *in vitro*, through non-resonant confocal Raman spectroscopy and imaging. The cells were cultured for 1 and 2 weeks at which point they were fixed. The samples were washed and placed in PBS for Raman measurements, which were performed using a home-built confocal Raman microscope (Uzunbajakava *et al.*, 2003).

Light microscopic observation indicates that the internalized microspheres show signs of intracellular degradation already after 1 week of culture. We can see clear cavity formation in the center of the microspheres at 1 and 2 weeks of cell culture (right). In contrast, PLGA microspheres incubated under identical conditions, but in the absence of macrophages (insets in c and d), did not show any signs of degradation. Confocal Raman spectroscopy and imaging of degraded PLGA spheres complete these observations. Raman images, created in the 1768 cm^{-1} band specific for the ester bonds in PLGA as well show a low intensity of this band in the internal area of several particles already after 1 week (B) of cell culture. This indicates the loss of PLGA ester bonds from the center of these microspheres.

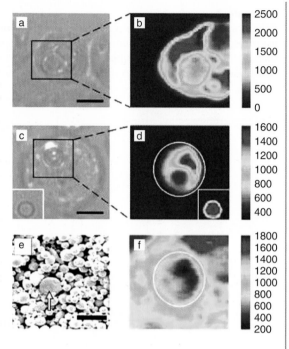

In (a) and (c), light micrographs of macrophages containing phagocytosed PLGA microspheres cultured for 1 and 2 weeks are shown (scale bar 5 μm). The Raman images (7.5×7.5 μm²) of these beads are shown in (b) and (d). The scale in (b) and (d) shows the relative intensity of the 1768 cm^{-1} band used for imaging (specific for PLGA ester bonds). Insets in (c) and (d) show controls (image size 7.5×7.5 μm²). The Raman image in (f) depicts the intensity of the 1527–1714 cm^{-1} region, which contains bands from cytoplasmic species. (e) Electron micrograph of PLGA microspheres after freezing and crushing in liquid nitrogen. A cross-section of a broken sphere, showing the solid center, can be seen (arrow) (scale bar 10 μm). Reproduced with permission from JACS: van Apeldoorn, A.A., van Manen, H.J., et al. (2004). Raman imaging of PLGA microsphere degradation inside macrophages. J Am Chem Soc, 126(41): 13226–13227.

By using the spectroscopic information we now can obtain information on the chemical composition of several locations within these samples. In order to quantify the data, spectra of PLGA and averaged spectra taken from the high- and low-intensity regions of the PLGA microsphere (see spectra below) to the 875 cm^{-1} band were scaled. This lactic acid band is not affected by hydrolysis (Geze *et al.*,

1999). By subtracting the reference data from the experimental data, we can find that 2 weeks after phagocytosis a $\pm 30\%$ reduction in the ester bond intensity in both the low-and high-intensity area of the internalized microsphere has occurred. This is demonstrated by the negative band in A and B, at $1768 \, cm^{-1}$ after subtraction of the scaled spectra of pure PLGA. Noteworthy is the fact that the difference of spectrum in (A) shows Raman positive bands which are specific for the cell cytoplasm.

Based on the presence of Raman bands at $1004 \, cm^{-1}$ (phenylalanine) and $1662 \, cm^{-1}$ (amide I) and the $1440 \, cm^{-1}$ band assigned to CH_2 groups predominantly found in lipids, we can conclude that both proteins and lipids are present in the degradation-induced hole in the microspheres. Proteins and lipids have probably diffused through one or more pores formed by degradation connecting the cavity to the phagosomal environment. The collapsed microsphere in the top right corner (B) of the Raman image, shows that this might be a possible explanation.

The degradation of PLGA causing concentric cavities in the center of the microspheres after uptake by macrophages, as described here, indicates that also in microspheres an autocatalytic hydrolysis effect occurs. At the same time, the analysis of the spectroscopic data suggests that degradation takes place throughout the microsphere. Furthermore, degradation takes place by hydrolysis of the ester bonds preferentially related to the glycolic acid block component in the polymer, as indicated by the negative bands in A and B. Interestingly, this finding is more related to the suggestion done by Li on a homogeneous degradation model when dealing with small samples such as microspheres (Li, 1999).

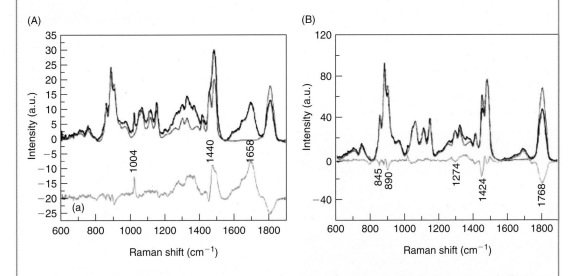

Raman spectra (in black) of the low- (A) and high-intensity (B) regions of the PLGA microsphere shown above. Pure PLGA is shown in red, and difference spectra (after scaling to the $875 \, cm^{-1}$ band) in green. Note the negative band at $1768 \, cm^{-1}$ (indicating a decrease in the number of ester bonds) and the negative bands specific for glycolic acid at 845, 890, 1274, and $1424 \, cm^{-1}$. Reproduced with permission from JACS: van Apeldoorn, A.A., van Manen, H.J., et al. (2004). Raman imaging of PLGA microsphere degradation inside macrophages. J Am Chem Soc, 126(41): 13226–13227.

Existing models do not take into account the influence of the cell on the microsphere degradation process, therefore they must differ from phagosomal degradation. We have seen that PLGA degradation shown here is a cell-mediated process, caused by either the low pH (≤5.5) and/or the presence of hydrolytic enzymes in the phagosome. This follows from the results of the control samples; in this case macrophages were not present and no degradation was observed after 1 and 2 weeks (refer back to insets in previous figure).

Classical Experiment

Amorphous, non-crystallizable copolymers based on L-lactide, D-lactide and glycolide show a surface to center differentiation upon hydrolysis, resulting in a hollowing out of the specimens during degradation (Grizzi *et al.*, 1995; Li *et al.*, 1990a, 1990b). This phenomenon is related to the autocatalytic hydrolysis of the ester bonds in the main chain.

Until the publications of Vert and coworkers, the extent of degradation of these polymers both *in vitro* and *in vivo* was investigated by visual observation, viscometry and changes in specimen mass. Polymer degradation was considered to be a homogeneous process where water absorption throughout the polymer is followed by hydrolysis reactions. However, careful analysis of the molecular weight of the degrading polymer in 2 mm thick specimens showed a bimodal molecular

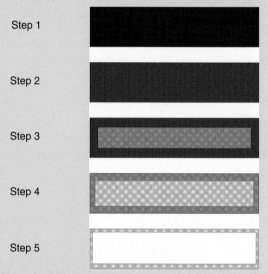

Step 1

Step 2

Step 3

Step 4

Step 5

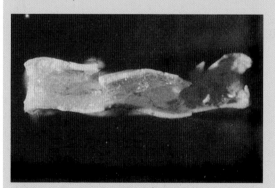

Cross-section of a PDLLA specimen degraded for 5 weeks in saline buffer. Reproduced with permission from Li, S.M., Garreau, H. and Vert, M. (1990). J Mater Sci Mater Med, 1: 131.

Schematic representation of the different steps of the degradation of PDLLA specimens in aqueous medium (Reproduced with permission from Li, S.M., Garreau, H. and Vert, M. (1990). J Mater Sci Mater Med 1: 131). Step 1: initial specimen; Step 2: water absorption, start of ester bond cleavage, and decrease in molecular weight; Step 3: differentiation between surface and center, with dramatic decrease in molecular weight in inner part of the specimen; Step 4: diffusion of oligomers through thinning surface when molecular weight is low enough to allow solubilization in the medium; Step 5: hollow shell remaining after release of oligomers and slow degradation of the shell.

weight distribution. It was found that in such poly(lactide) specimens the inner part of the specimen degrades at a higher rate than the outside (Grizzi et al., 1995; Vert 2005).

When the specimen is placed in the aqueous medium, water penetrates into the material. Hydrolytic cleavage of the polymer chains occurs and as carboxylic acid groups are generated autocatalysis occurs as well. Within the specimen, degradation of the still insoluble polymer chains proceeds homogenously via an autocatalytic mechanism. As soon as the molecular weight of the oligomers formed becomes low enough to allow solubility in the surrounding aqueous medium, these oligomers diffuse to the surface of the specimen and into the surrounding medium as they continue to degrade. This process, which combines hydrolytic degradation, diffusion and solubilization, results in a differentiation between the rates of degradation at the surface and at the interior of the polymer specimen. As a result, hollow specimens can be formed during degradation. The scheme shown here illustrates this process.

State of the Art Experiment

Poly(trimethylene carbonate)
Poly(trimethylene carbonate) (PTMC) is flexible, biodegradable and biocompatible, showing attractive surface erosion behavior *in vivo* (Zhang et al., 2006). It is synthesized by ring-opening polymerization of TMC monomer. The ring-opening polymerization of TMC is illustrated here.

Trimethylene carbonate

Poly(Trimethylene carbonate) (or PTMC)

Ring opening polymerization of trimethylene carbonate (TMC).

High molecular weight PTMC polymer is an amorphous material with a glass transition temperature of approximately $-20°C$; the polymer is in the rubbery state at room temperature and under physiological conditions. It is a hydrophobic polymer with a static water contact angle of approximately 74° and the equilibrium water uptake in phosphate buffered saline (pH = 7.4) is approximately 1.3 wt%. These values are close to those of PCL and PDLLA.

PTMC is a very good substrate for the culturing of cells (Pêgo et al., 2003). Excellent adhesion and proliferation behavior of human Schwann cells, human umbilical vein endothelial cells, and rat cardiomyocytes on the surface of high molecular weight PTMC polymers was observed.

Surface Erosion Behavior of poly(trimethylene carbonate)
Poly(trimethylene carbonate) (PTMC) degrades very slowly *in vitro* by hydrolysis of the carbonate linkages. Contrary to the slow degradation *in vitro*, the degradation of high molecular weight PTMC ($M_n = 320 \times 10^3$ g/mol) *in vivo* in the back of rats was strikingly fast. After implantation, the mass of 600 µm thick PTMC samples decreased linearly in time and degradation was nearly complete in 3 weeks. The specimen dimensions changed accordingly, while the molecular weight remained constant; which suggests that degradation *in vivo* had occurred by an enzymatic surface erosion process (Zhang et al., 2006).

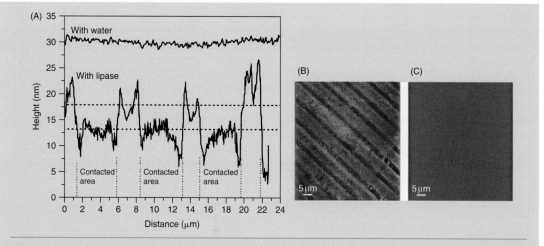

Contact mode AFM height profiles of the surfaces of PTMC films after µCP using lipase solution and using water as control (A), and friction images of PTMC films after µCP using lipase solution (B) and water (C).

When PTMC was incubated in lipase solutions (from *Thermomyces Lanuginosus*) similar behavior could also be observed (Zhang *et al.*, 2005).

Micro-patterned structures

The possibility of utilizing the rapid enzymatic surface erosion to form a micro-patterned structure on PTMC films by micro contact printing (µCP) using lipase solutions was investigated (Zhang *et al.*, 2005). Patterned poly(dimethyl siloxane) (PDMS) stamps, consisting of 2 µm wide recessed lines spaced by 5 µm wide protruding lines were impregnated with the lipase solution. After removing the excess liquid, the patterned structure was carefully pressed onto the PTMC film for 30 s, subsequently the PTMC film was extensively rinsed with water and blown-dry.

The resulting pattern on the surface of the PTMC films was probed using contact mode AFM to obtain height and friction images. From height images, alternating and parallel 2- and 5-µm wide lines could be distinguished. The 5-µm wide lines, which had been contacted with the lipase solution on the PDMS stamp, were lower in height. A typical height profile across

the patterned surface of the film is shown (a). The figure shows that the 5-µm wide lines are approximately 5 nm lower than the 2-µm wide lines. Simultaneously, the AFM friction image after µCP was obtained. (b) and (c) shows that the pattern of the friction image corresponds very well with that of the height profile. Alternating lines with relatively high friction are approximately 5-µm wide and can readily be distinguished from relatively low friction lines 2 µm in width. The 5-µm wide lines with higher friction correspond to the area that has been contacted with lipase solution, their higher friction is possibly due to a higher hydrophilicity in these lines.

Control experiments showed that after µCP of the PTMC films with only water, a patterned surface was not obtained; the height profile (a) and the friction image (c) did not show any features.

These results demonstrate that by µCP using lipase solutions, a micro-patterned PTMC surface with differences in height and in friction properties can be readily formed. These micro-patterned PTMC surfaces could have great potential for applications where cell patterning is required.

7.5 Future perspectives

Current research on degradable polymers in tissue engineering focuses on balancing the required mechanical (elasticity, strength, etc.) properties with the degradation rate. A polymeric structure should disappear without leaving inert (crystalline) residues behind. The optimal mechanical behavior characteristics and resorption rates are specific for each specific application. With reference to the definition given in *The Williams Dictionary for Biomaterials* (Williams, 1999), future research on degradable polymers will not only focus on the supporting task of these materials. Introduction of biologically active functional groups in the polymeric chain will result in materials that attract (stem) cells and/or growth factors in the body that are required for tissue regeneration. In addition, specific molecular signals may be incorporated into the degradable polymer, to be released during their application.

7.6 Summary

1. Degradable polymers are suitable as scaffold material for tissue engineering as interference with the development and growth of new tissue and unwanted long-term reactions are prevented.
2. Degradable polymers for use in tissue engineering should be biocompatible, and the resulting degradation products should be non-toxic. The time required to complete degradation resorption depends on the intended application and preferably matches the formation of functional tissue.
3. Polymers are long-chain macromolecules formed by covalent coupling of monomers. Copolymerization, i.e. the preparation of polymers from two or more types of monomers, is often employed to tune their properties.
4. For a polymeric material to be degradable in the body it is necessary that the main chain of the macromolecule contains labile degradable bonds, such as ester, carbonate, or anhydride bonds.
5. The most important degradable polymers are polyesters. In the presence of water, chain scission occurs by hydrolysis of ester bonds, which lowers the molecular weight. The resulting oligomers dissolve in the surrounding environment or undergo further hydrolysis.
6. The soluble degradation products are transported from the implantation site via the lymphatic system to the kidneys which excrete them from the body.
7. Mass loss of the polymeric device, due to removal of the oligomers is referred to as erosion. Distinction is made between *surface* and *bulk erosion* processes.
8. Often, *in vivo* degradation is faster than *in vitro* degradation. This is due to the tissue response (inflammation, foreign body response) and the presence of enzymes.
9. Linear aliphatic polyesters such as poly(lactic acid) (PLA), poly(glycolic acid) (PGA) and their copolymers (PLGA) have been broadly used in tissue engineering. These polyesters degrade via hydrolysis of the main chain ester bonds. Their degradation products (lactic- and glycolic acid) are part of the Krebs metabolic cycle.
10. More advanced degradable polymers will have an optimal balance between mechanical properties and degradation rate. In addition, functional groups attract cells and/or growth factors required for tissue formation.

References

Albertsson, A.-C. and Eklund, M. (1995). Influence of molecular structure on the degradation mechanism of degradable polymers: In vitro degradation of poly(trimethylenecarbonate), poly(trimethylenecarbonate-co-caprolactone), and poly(adipicanhydride). *J Appl Polym Sci*, 57: 87–108.

Albertsson, A.-C. and Varma, I.K. (2002). polyesters: Synthesis, Properties and Applications. *Adv Polym Sci*, 157: 1–40.

Aldini, N.N., Caliceti, P., Lora, S., *et al.* (2001). Calcitonin release system in the treatment of experimental osteoporosis. Histomorphometric evaluation. *J Orthop Res*, 19(5): 955–961.

Ali, S.A.M., Doherty, P.J. and Williams, D.F. (1994). Molecular biointeractions of biomedical polymers with extracellular exudates ad inflammatory cells and their effects on the biocompatibility, *in vivo. Biomaterials*, 15(10): 779–785.

Allcock, H.R., Lampe, F.W. and Mark, J.E. (2003). *Contemporary topics in polymer chemistry* (3rd edn.). Upper Saddle River: Pearson Education.

Ambrosio, A.M., Allcock, H.R., Katti, D.S. and Laurencin, C.T. Degradable polyphosphazene/poly(alpha-hydroxyester) blends: degradation studies. *Biomaterials*, 23(7): 1667-1672.

Anderson, J.M. (1994). *In vivo* biocompatibility of implantable delivery systems and biomaterials. *Eur J Pharm Biopharm*, 40: 1–8.

Anderson, J.A. and Shive, M.S. (1997). Biodegradation and biocompatibility of PLA and PLGA microspheres. *Adv Drug Del Rev*, 28: 5–24.

Anderson, J.M. and Langone, J.J. (1999). Issues and perspectives on the biocompatibility and immunotoxicity evaluation of implanted controlled release systems. *J Control Release*, 57(2): 107–113.

Andriano, K.P., Tabata, Y., Ikada, Y. and Heller, J. (1999). *In vitro* and *in vivo* comparison of bulk and surface hydrolysis in absorbable polymer scaffolds for tissue engineering. *J Biomed Mater Res*, 48(5): 602–612.

Babensee, J.E., McIntire, L.V. and Mikos, A.G. (2000). Growth factor delivery for tissue engineering. *Pharm Res*, 17(5): 497–504.

Bakker, D., van Blitterswijk, C.A., Hesseling, S.C. and Grote, J.J. (1988). Effect of implantation site on phagocyte/polymer interaction and fibrous capsule formation. *Biomaterials*, 9(1): 14–23.

Bates, F.S. and Fredrickson, G.H. (1996). Block Copolymer Thermo-dynamics: Theory and Experiment. In *Thermoplastic Elastomers* (Holden., G., Legge, N.R., Quirk., R.P. and Schroeder, H.E., eds). Cincinnati: Hanser and Gardner.

Behravesh, E., Yasko, A.W., Engel, P.S. and Mikos, A.G. (1999). Synthetic biodegradable polymers for orthopaedic applications. *Clin Orthop Relat Res* (367 Suppl): S118–S129.

Beumer, G.J., van Blitterswijk, C.A. and Ponec, M. (1994a). Degradative behaviour of polymeric matrices in (sub)dermal and muscle tissue of the rat: a quantitative study. *Biomaterials*, 15(7): 551–559.

Beumer, G.J., van Blitterswijk, C.A. and Ponec, M. (1994b). Biocompatibility of a biodegradable matrix used as a skin substitute: an *in vivo* evaluation. *J Biomed Mater Res*, 28(5): 545–552.

Bezemer, J.M., Grijpma, D.W., Dijkstra, P.J., *et al.* (1999). A controlled release system for proteins based on poly(ether ester) block-copolymers: polymer network characterization. *J Control Release*, 62(3): 393–405.

Bostman, O., Hirvensalo, E., Vainionpaa, S., *et al.* (1989). Ankle fractures treated using biodegradable internal fixation. *Clin Orthop Relat Res* (238): 195–203.

Bulstra, S.K., Geesink, R.G., Bakker, D., *et al.* (1996). Femoral canal occlusion in total hip replacement using a resorbable and flexible cement restrictor. *J Bone Joint Surg Br*, 78(6): 892–898.

Burdick, J.A., Peterson, A.J. and Anseth, K.S. (2001). Conversion and temperature profiles during the photoinitiated polymerization of thick orthopaedic biomaterials. *Biomaterials*, 22(13): 1779–1786.

Burkersroda, , von, F., Sched, L. and Göpferich, A. (2002). Why degradable polymers undergo surface erosion or bulk erosion. *Biomaterials*, 23: 4221.

Burkoth, A.K., Burdick, J. and Anseth, K.S. (2000). Surface and bulk modifications to photocross-linked polyanhydrides to control degradation behavior. *J Biomed Mater Res*, 51(3): 352–359.

Caliceti, P., Veronese, F.M. and Lora, S. (2000). Polyphosphazene microspheres for insulin delivery. *Int J Pharm*, 211(1–2): 57–65.

Changez, M., Koul, V., Krishna, B., *et al.* (2005). Studies on biodegradation and release of gentamcin sulphate from interpenetrating network hydrogels based on poly(acrylic acid) and gelatin: *in vitro* and *in vivo. Biomaterials*, 25(1): 139–146.

Choi, N.S. and Heller, J. (1978) Drug Delivery devices manufactured from poly(orthoesters) and poly (orthocarbonates). US Patent 4.093.709.

Choi, S.H. and Park, T.G. (2002). Synthesis and characterization of elastic PLGA/PCL/PLGA tri-block copolymers. *J Biomater Sci Polym Ed*, 13(10): 1163–1173.

Chu, C.R., Dounchis, J.S., Yoshioka, M., *et al.* (1997). Osteochondral repair using perichondral cells. A 1-year study in rabbits. *Clin Orthop Relat Res* (340): 220–229.

Cohen, S., Bano, M.C., Cima, L.G., *et al.* (1993). Design of synthetic polymeric structures for cell transplantation and tissue engineering. *Clin Mater*, 13(1–4): 3–10.

Cowie, J.M.G. (1997). *Polymers: Chemistry and Physics of Modern Materials* (2nd ed.). London: Blackie Academic and Professional.

Deschamps, A.A., Claase, M.B. and Sleijster, W.J. *et al.* (2002). Design of segmented poly(ether ester) materials and structures for the tissue engineering of bone. *J Control Release*, 78(1-3): 175–186.

Deschamps, A.A., Van Apeldoorn, A.A., De Bruijn, J.D., *et al.* (2003). Poly(ether ester amide)s for tissue engineering. *Biomaterials*, 24: 2643–2652.

Edlund, U. and Albertsson, A.C. (2002). Degradable polymer microspheres for controlled drug delivery. *Adv Polym Sci*, 157: 67–112.

Elbert, D.L. and Hubbell, J.A. (2001). Conjugate addition reactions combined with free-radical cross-linking for the design of materials for tissue engineering. *Biomacromolecules*, 2(2): 430–441.

Elbert, D.L., Pratt, A.B., Lutolf, M.P., *et al.* (2001). Protein delivery from materials formed by self-selective conjugate addition reactions. *J Control Release*, 76(1–2): 11–25.

Fisher, J.P., Holland, T.A., Dean, D., *et al.* (2001). Synthesis and properties of photocross-linked poly(propylene fumarate) scaffolds. *J Biomater Sci Polym Ed*, 12(6): 673–687.

Freed, L.E., Marquis, J.C., Nohria, A., *et al.* (1993). Neocartilage formation *in vitro* and *in vivo* using cells cultured on synthetic biodegradable polymers. *J Biomed Mater Res*, 27(1): 11–23.

Fu, K., Pack, D.W., Klibanov, A.M. and Langer, R. (2000). Visual evidence of acidic environment within degrading poly(lactic-co-glycolic acid) (PLGA) microspheres. *Pharm Res*, 17(1): 100–106.

Fulzele, S.V., Satturwar, P.M. and Dorle, A.K. (2003). Study of the biodegradation and *in vivo* biocompatibility of novel biomaterials. *Eur J Pharm Sci*, 29(10): 53–61.

Geze, A., Chourpa, I., Boury, F., *et al.* (1999). *Analyst*, 124: 37–42.

Goedemoed, J.H., Hennink, W.H., Bezemer, J.M. *et al.* (1998) Polyetherester copolymers as drug delivery matrices. European Patent Application 97202533.2.

Göpferich, A. (1996). Mechanisms of polymer degradation and erosion. *Biomaterials*, 17: 103.

Göpferich, A. and Langer, R. (1993). Modeling of Polymer Erosion. *Macromolecules*, 26: 4105.

Göpferich, A. and Tessmar, J. (2002). Polyanhydride degradation and erosion. *Adv Drug Del Rev*, 54: 911–931.

Grijpma, D.W. and Pennings, A.J. (1994). Copolymers of L-lactide 1. Synthesis, thermal properties and hydrolytic degradation. *Macromol Chem Phys*, 195: 1633–1647.

Grijpma, D.W., Zondervan, G.J. and Pennings, A.J. (1991). High molecular weight copolymers of L-lactide and ε-caprolactone as biodegradable elastomeric implant materials. *Polym Bull*, 25: 327–333.

Grizzi, I., Garreau, H., Li, S. and Vert, M. (1995). Hydrolytic degradation of devices based on poly(DL-lactic acid) size dependence. *Biomaterials*, 16(4): 305–311.

He, S., Yaszemski, M.J. and Yasko, A.W. (2000a). Synthesis of biodegradable poly(propylene fumarate) networks with poly(propylene fumarate)-diacrylate macromers as cross-linking agents and characterization of their degradation products. *Polymer*, 42: 1251–1260.

He, S., Yaszemski, M.J., Yasko, A.W., *et al.* (2000b). Injectable biodegradable polymer composites based on poly(propylene fumarate) cross-linked with poly (ethylene glycol)-dimethacrylate. *Biomaterials*, 21(23): 2389–2394.

Heller, J. (1990). Development of poly(ortho esters): a historical overview. *Biomaterials*, 11: 659–665.

Honda, M., Yada, T., Ueda, M. and Kimata, K. (2000). Cartilage formation by cultured chondrocytes in a new scaffold made of poly(L-lactide-epsilon-caprolactone) sponge. *J Oral Maxillofac Surg*, 58(7): 767–775.

Hutmacher, D.W., Schantz, T., Zein, I., *et al.* (2001). Mechanical properties and cell cultural response of polycaprolactone scaffolds designed and fabricated via fused deposition modeling. *J Biomed Mater Res*, 55(2): 203–216.

Jeong, B., Bae, Y.H., Lee, D.S. and Kim, S.W. Biodegradable block copolymers as injectable drug-delivery systems, *Nature*, 388: 860.

Kowalski, A., Duda, A. and Penczek, S. (2000). Kinetics and mechanism of cyclic esterspolymerization initiated with tin(II)octoate. 3. Polymerizationofl,l-dilactide. *Macromolecules*, 33: 7359.

Kricheldorf, H.R., Kreiser-Saunders, I. and Stricker, A. (2000). Polylactones 48. SnOct2-initiated polymerizations of lactide: A mechanistic study. *Macromolecules*, 33: 702.

Kuo, C.K. and Ma, P.X. (2001). Ionically cross-linked alginate hydrogels as scaffolds for tissue engineering: part 1. Structure, gelation rate and mechanical properties. *Biomaterials*, 22(6): 511–521.

Leong, K.W., Brott, B.C. and Langer, R. (1985). Bioerodible polyanhydrides as drug-carrier matrices. *J Biomed Mater Res*, 19: 941–955.

Leong, K.W., Kost, J., Mathiowitz, E. and Langer, R. (1986). Polyanhydrides for controlled release of bioactive agents. *Biomaterials*, 7(5): 364–371.

Li, S.M. (1999). Hydrolytic degradation characteristics of aliphatic polyesters derived from lactic and glycolic acids. *J Biomed Mater Res*, 48: 342–353.

Li, S.M., Garreau, H. and Vert, M. (1990a). Structure-property relationships in the case of the degradation of massive poly(α-hydroxy acids) in aqueous media. *J Mater Sci Mater Med*, 1: 131.

Li, S.M., Garreau, H. and Vert, M. (1990b). Structure-property relationships in the case of the degradation of massive poly(α-hydroxy acids) in aqueous media. *J Mater Sci Mater Med*, 1: 198.

Li, S.M., Garreau, H. and Vert, M. (1990c). Structure-property relationships in the case of the degradation of massive poly(α-hydroxy acids) in aqueous media. *Mater Sci Mater Med*, 1: 123.

Lu, L., Garcia, C.A. and Mikos, A.G. (1999). *In vitro* degradation of thin poly(DL-lactic-co-glycolic acid) films. *J Biomed Mat Res*, 46: 236–244.

Lu, L., Peter, S.J., Lyman, M.D., *et al.* (2000). *In vitro* and *in vivo* degradation of porous poly(DL-lactic-co-glycolic acid) foams. *Biomaterials*, 21: 1845–1873.

Mahmood, T.A., de Jong, R., Riesle, J., *et al.* (2004). Adhesion-mediated signal transduction in human articular chondrocytes: the influence of biomaterial chemistry and tenascin-C. *Exp Cell Res*, 301(2): 179–188.

Mainil-Varlet, P., Curtis, R. and Gogolewski, S. (1997). Effect of *in vivo* and *in vitro* degradation on molecular and mechanical properties of various low-molecular-weight polylactides. *J Biomed Mater Res*, 36: 360–380.

Mann, B.K., Gobin, A.S., Tsai, A.T., *et al.* (2001a). Smooth muscle cell growth in photopolymerized hydrogels with cell adhesive and proteolytically degradable domains: synthetic ECM analogs for tissue engineering. *Biomaterials*, 22(22): 3045–3051.

Mann, B.K., Schmedlen, R.H. and West, J.L. (2001b). Tethered-TGF-beta increases extracellular matrix production of vascular smooth muscle cells. *Biomaterials*, 22(5): 439–444.

Matlaga, B.F., Yasenchak, L.P. and Salthouse, T.N. (1976). Issue response to implanted polymers: the significance of sample shape. *J Biomed Mater Res*, 10: 391–397.

McIntire, L.V. (2002). *WTEC Panel Report on Tissue Engineering Research*. International Technology Research Institute, World Technology (WTEC) Division.

Menei, P., Daniel, V., Montero-Menei, C., *et al.* (1993). Biodegradation and brain tissue reaction to poly(D,L-lactide-co-glycolide) microspheres. *Biomaterials*, 14(6): 470–478.

Mensik, I., Lamme, E.N., Riesle, J. and Brychta, P. (2002). Effectiveness and Safety of the PEGT/PBT Copolymer Scaffold as Dermal Substitute in Scar Reconstruction Wounds (Feasibility Trial). *Cell Tissue Bank*, 3(4): 245–253.

Middleton, J.C. and Tipton, A.J. (2000). Synthetic biodegradable polymers as orthopedic devices. *Biomaterials*, 21: 2335–2346.

Moroni, L., de Wijn, J.R. and van Blitterswijk, C.A. (2005). Three-dimensional fiber-deposited PEOT/PBT copolymer scaffolds for tissue engineering: Influence of porosity, molecular network mesh size, and swelling in aqueous media on dynamic mechanical properties. *J Biomed Mater Res A*, 75: 957–965.

Nijenhuis, A.J., Grijpma, D.W. and Pennings, A.J. (1992). Lewis acid catalyzed polymerisation of L-lactide. Kinetics and mechanism of esters of bulk polymerization. *Macromolecules*, 25: 6419–6424.

Odian, G. (2004). *Principles of Polymerization.* (4th edn). Hoboken: Wiley Interscience.

Olde Riekerink, M.B., Claase, M.B., Engbers, G.H., *et al.* (2003). Gas plasma etching of PEO/PBT segmented block copolymer films. *J Biomed Mater Res A*, 65(4): 417–428.

Park, T.G. (1995). Degradation of Poly(lactic-co-glycolic acid) Microspheres: Effect of Copolymer Composition. *Biomaterials*, 16: 1123–1130.

Pêgo, A.P., Vleggeert-Lankamp, C.L.A.M., Deenen, M., *et al.* (2003). *J Biomed Mater Res*, 67A: 876.

Pitt, C.G. and Gu, Z. (1987). Modification of the rates of chain cleavage of poly(ε-caprolactone) and related polyesters in the solid state. *J Control Rel*, 4: 283–292.

Pitt, C.G., Chasalow, F.I., Hibionada, D.M., Klimas, D.M. and Schindler, A. (1981a). Aliphatic polyesters. I. The degradation of poly(ε-caprolactone) *in vivo*. *J Appl Polym Sci*, 26: 3779–3787.

Pitt, C.G., Gratzl, M.M., Kimmel, G.L., Surles, J. and Schindler, A. (1981b). Aliphatic polyesters II. The degradation of poly(DL-lactide), poly (ε-caprolactone), and their copolymers *in vivo*. *Biomaterials*, 2: 215–220.

Ratner, B.D., Hoffman, A.S., Schoen, F.J. and Lemons, J.E., (eds) (2004). *Biomaterials Science, An introduction to materials in medicine* (2nd edn). Amsterdam: Academic Press, Elsevier.

Reed, A.M. and Gilding, D.K. (1981). Biodegradable polymers for use in surgery-poly(glycolic)/poly(lactic acid) homo and copolymers, 2. *In vitro* degradation. *Polymer*, 22: 494–498.

Sarazin, P., Roy, X. and Favis, B.D. (2004). Controlled preparation and properties of porous poly(L-lactide) obtained from a co-continuous blend of two biodegradable polymers. *Biomaterials*, 25(28): 5965–5978.

Sawhney, A.S., Pathak, C.P. and Hubbell, J.A. (1993). Bioerodible hydrogels based on photopolymerized poly(ethylene glycol)-co-(alpha-hydroxy acid) diacrylate macromers. *Macromolecules*, 26: 581–587.

Schliecker, G., Schmidt, C., Fuchs, S., *et al.* (2003). *Int J Pharm*, 266: 39–49.

Smith, M.B. and March, J. (2001). In *March's Advanced Organic Chemistry* (March, J., ed.). (5th edn). New York: Wiley Interscience.

Sodergard, A. and Stolt, M. (2002). Properties of lactic acid based polymers and their correlation with composition. *Prog. Polym. Sci*, 27: 1123–1163.

Spenlehauer, G., Vert, M., Benoit, J.P. and Boddaert, A. (1989). *In vitro* and *in vivo* degradation of poly(D,L lactide/glycolide) type microspheres made by solvent evaporation method. *Biomaterials*, 10: 557–563.

Stoll, G.H., Nimmerfall, F., Acemoglu, M., *et al.* (2001). Poly(ethylene carbonate)s, part II: degradation mechanisms and parenteral delivery of bioactive agents. *J Control Rel*, 76: 209–225.

Suggs, L.J., Krishnan, R.S., Garcia, C.A., *et al.* (1998). *In vitro* and *in vivo* degradation of poly(propylene fumarate-co-ethylene glycol) hydrogels. *J Biomed Mat Res*, 42: 312–320.

Tamada, A. and Langer, R. (1993). Erosion kinetics of hydrolytically degradable polymers. *Proc Natl Acad Sci USA*, 90: 552–556.

Therin, M., Christel, P., Li, S., *et al.* (1992a). *Biomaterials*, 13: 594–600.

Therin, M., Christel, P., Li, S., *et al.* (1992b). *In vivo* degradation of massive poly(α-hydoxy acids): validation of *in vitro* findings. *Biomaterials*, 13(9): 594–600.

Tracy, M., Ward, K.L., Firouzabadian, L., *et al.* (1999). Factors affecting the degradation rate of poly(lactide-co-glycolide) microspheres *in vivo* and *in vitro*. *Biomaterials*, 20: 1057–1062.

Uhrich, K.E., Cannizzaro, S.M., Langer, R.S. and Shakesheff, K.M. (1999). Polymeric systems for controlled drug release. *Chem Rev*, 99: 3181–3198.

Uzunbajakava, N., Lenferink, A., Kraan, Y., *et al.* (2003). *Biophys J*, 84: 3968–3981.

Vacanti, C.A. and Vacanti, J.P. (2000). The Science of Tissue Engineering. *Orthop Clin, North America*, 31: 351–355.

van Blitterswijk, C.A., van den Brink, J., Leenders, H. and Bakker, D. (1993). The effect of PEO ratio on degradation, calcification and bone bonding of PEO/PBT copolymer (PolyActive). *Cell and Materials*, 3: 23–26.

van Dijkhuizen-Radersma, R., Peters, F.L., Stienstra, N.A., *et al.* (2002). Control of vitamin B12 release from poly(ethylene glycol)/poly(butylene terephthalate) multiblock copolymers. *Biomaterials*, 23(6): 1527–1536.

van Dijkhuizen-Radersma, R., Roosma, J.R., Kaim, P., *et al.* (2003). Biodegradable poly(ether-ester) multiblock copolymers for controlled release applications. *J Biomed Mater Res A*, 67(4): 1294–1304.

van Dijkhuizen-Radersma, R., Metairie, S., Roosma, J.R., *et al.* (2005). Controlled release of proteins from degradable poly(ether-ester) multiblock copolymers. *J Control Release*, 101(1-3): 175–186.

van Loon (1995). Thesis, The Netherlands: University of Leiden.

Vert, M. (2005). Aliphatic polyesters: great degradable polymers that cannot do everything. *Biomacromolecules*, 6: 538–546.

Wang, P.Y. (1989). Compressed poly(vinyl alcohol)-polycaprolactone admixture as a model to evaluate erodible implants for sustained drug delivery. *J Biomed Mater Res*, 23(1): 91–104.

Wang, J., Mao, H.Q. and Leong, K.W. (2001a). A novel biodegradable gene carrier based on polyphosphoester. *J Am Chem Soc*, 123(38): 9480–9481.

Wang, S., Wan, A.C., Xu, X., *et al.* (2001b). A new nerve guide conduit material composed of a biodegradable poly(phosphoester). *Biomaterials*, 22(10): 1157–1169.

Williams, D.F. (1999). *The Williams Dictionary of Biomaterials*. Liverpool: Liverpool University Press.

Zhang, Z., Zou, S., Vancso, G.J., *et al.* (2005). *Biomacromolecules*, 6: 3404.

Zhang, Z., Kuijer, R., Bulstra, S.K., *et al.* (2006). *Biomaterials*, 27: 1741–1748.

Chapter 8
Degradation of bioceramics

Florence Barrère, Ming Ni, Pamela Habibovic, Paul Ducheyne and Klaas de Groot

Chapter objectives:

- To consider what bioceramics, glasses, and ceramics are
- To understand how the body reacts towards biomaterials
- To consider how bioceramics degrade
- To understand the application of degradable ceramics in bone regeneration
- To recognize how to tailor the degradation to stimulate tissue regeneration

Table 8.1 Calcium phosphates of biological interest

Name	Formula	Abbreviation	Ca/P
Dicalcium phosphate anhydrate	$CaHPO_4$	DCPA	1.00
Dicalcium phosphate dehydrate	$CaHPO_4.2H_2O$	DCPD	1.00
Octacalcium phosphate	$Ca_8(PO_4)_4(HPO_4)_2.5H_2O$	OCP	1.33
Tricalcium phosphate	$Ca_3(PO_4)_2$	TCP	1.50
Hydroxyapatite	$Ca_{10}(PO_4)_6(OH)_2$	OHAp	1.67

solubility in physiological solutions, and hence, its bioactivity and resorbability. For example, BG45S5, the first glass composition tested contains (in weight percentage) 45% of SiO_2, 24.5% CaO, 24.5% Na_2O and 6% P_2O_5. This glass was first studied in the late 1960s and early 1970s (Beckham *et al.*, 1971; Greenlee *et al.*, 1972; Hench *et al.*, 1971). Although mechanisms were not understood, it was suggested that this glass type formed a chemical bond with bone. Later on, it has been established that glasses do not bond to bone if the Ca/P ratio is substantially lower than 5:1, as seen in Figure 8.1. Glass-ceramics were developed in order to increase the mechanical properties of these glasses. To this end, bioactive glasses are subjected to thermal treatments that alter the material's microstructure. Some investigators showed that when bioactive glass in the system MgO—CaO—SiO_2—P_2O_5 was thermally treated, the glass transformed into apatite/wollastonite (A-W) glass-ceramic (38% apatite, 34% wollastonite, and 28 wt% residue glassy phase) (El-Ghannam and Ducheyne, 2005). The first A-W glass-ceramic were developed by Kokubo in the 1980s and contains oxyfluorapatite $Ca_{10}(PO_4)_6(OH,F_2)$ and wollastonite (CaO · SiO_2) (Kokubo, 1993). These changes in

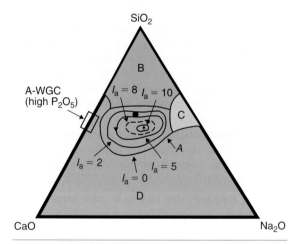

Figure 8.1 Compositional dependence (in wt%) of bone bonding and soft tissue bonding of bioactive glasses and glass-ceramics. All compositions in region A are bioactive and bond to bone. Compositions in region B are bioinertand lead to formation of a nonadherantfibrous capsule. Compositions in region C are resorbable. Region D is restricted by technical factors. Region E (soft tissue bonding) is inside the dashed line (Hench, L.L. (1991). *J. Amer Ceram Soc*, 74(7): 1487–1510.)

bioactive glass microstructure influence its bioactivity. A special form of BG 45S5 is BioGran®, a bioactive glass granules with a narrow size range (300–355 μm) (Schepers *et al.*, 1991, 1993; Schepers and Ducheyne, 1997). These materials have been used clinically as bone-regenerative material in dental and orthopedic applications. Clinical applications are mainly focused on non-load bearing sites. These are sites that enable mechanically stable insertion such as for middle ear, ossicle and maxillofacial reconstruction, periodontal repair and spinal repair (Wang, 2004).

8.2 Degradation mechanisms of calcium phosphate ceramics

In artificial or natural aqueous environments bioceramics can degrade via (i) solution-mediated mechanisms leading to physicochemical dissolution of the ceramic with possibly phase transformation, (ii) cell-mediated mechanisms via macrophages and osteoclasts, and (iii) loss of mechanical integrity as a result of the aforementioned mechanisms. In biological systems, degradation of bioceramics is a combination of non-equilibrium processes that occur simultaneously or in competition with each other.

8.2.1 Physicochemical degradation

The physicochemical degradation, or dissolution, of calcium phosphate ceramics can be described as a dissolution-reprecipitation cascade which is the result of exchanges at a solid-liquid interface. In artificial or natural aqueous environments these bioceramics dissolve, this physicochemical process is typical of inorganic substrates, i.e. having dominant ionic features. It is the result of a multi-component dynamic process that cannot be mimicked *in vitro*. However, *in vitro* experiments simplifying the biological environment have led to conclusions that, in general, fit with *in vivo* observations.

From a thermodynamic point of view, most calcium phosphates are sparingly soluble in water, and some are very insoluble, but all dissolve in acids.

Their solubility, defined as the amount of dissolved solute contained in a saturated solution when particles of solute are continually passing into solution (dissolving) while other particles are returning to the solid solute phase (growth) at exactly the same rate (Wu and Nancollas, 1998), decreases with the increase of pH (de Groot, 1983). Figure 8.2a shows the solubility of various calcium phosphate phases as a function of the supersaturation and as a function of pH under equilibrium conditions at 37°C (Tang *et al.*, 2001). Supersaturation can be defined as either a metastable or unstable state of a solution. Supersaturated solutions contain more of the dissolved material than could be dissolved by the water under equilibrium conditions. Crystal growth has not yet occurred, but that can be initiated by introducing seeds (Figure 8.2b).

From a surface reactivity point of view, physicochemical dissolution can be seen as ionic transfer from he solid phase to the aqueous liquid via surface hydration of calcium (Ca^{2+}), phosphate species (PO_4^{3-}, HPO_4^{2-}, $H_2PO_4^-$) and possible impurities, e.g. carbonate (CO_3^{2-}), fluoride (F^-), chloride (Cl^-) present in the biomaterial. Under physiological conditions (pH = 7.4 and 37°C), this dissolution process is highly dependent on the nature of the calcium phosphate substrate (Barrere *et al.*, 2003; Dhert *et al.*, 1993; Doi *et al.*, 1999; Ducheyne *et al.*, 1993; Okazaki *et al.*, 1982; Radin and Ducheyne, 1993) on one hand, and the environment, i.e. composition and supersaturation of the liquid *in vitro* (Hyakuna *et al.*, 1990; Raynaud *et al.*, 1998; Tang *et al.*, 2001) or implantation site *in vivo* (Barralet *et al.*, 2000; Daculsi *et al.*, 1990), on the other hand. As a result of these ionic transfers, phase transformations occur at the ceramic surface. A phase transformation occurs for calcium phosphate phases, which are unstable under physiological conditions, e.g. octacalcium phosphate, dicalcium phosphate dehydrate, tricalcium phosphate or amorphous calcium phosphate. Similar phase transformations also occur for apatite-based ceramics, since apatitic structures have a strong ability to adapt to their environment by hosting foreign ions and subsequently to sustain atomic rearrangements (Cazalbou *et al.*, 2004). However, crystalline

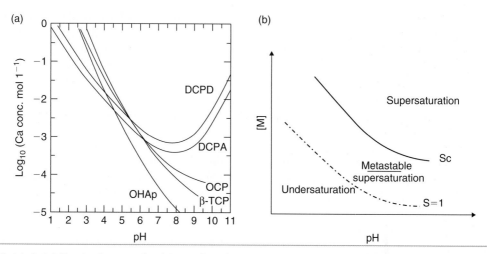

Figure 8.2 (a) Solubility isotherms of calcium phosphates at 37°C (J.C. Elliot, 1994, *Structure and Chemistry of the Apatites and other Calcium Orthophosphates*, Amsterdam: Elsevier) Structure and chemistry of the apatites and other calcium orthophosphates), (b) Solubility isotherm of salts in solution as a function of the pH. This diagram represents the three different saturation states of an ionic solution susceptible of precipitation.

hydroxyapatitic substrates are often too stable to transform.

In vitro and under physiological conditions, the newly formed crystals are comparable to bone mineral with respect to structure and inorganic composition, i.e. nanocrystalline carbonated apatite. They are both issued from the nucleation and growth process from supersaturated extracellular fluids (Barrere *et al.*, 2003; Daculsi *et al.*, 1989, 1990; Heughebaert *et al.*, 1988; Johnsson *et al.*, 1991; Radin and Ducheyne, 1993) and, in particular, in the presence of magnesium and carbonate (Elliot, 1994; Furedi-Milhofer *et al.*, 1976; LeGeros, 1991). *In vivo*, this newly formed mineral phase is also associated with organic matrix (Figure 8.3) (Barrere *et al.*, 2003; de Bruijn *et al.*, 1995; Heughebaert *et al.*, 1988).

The proteins from the surrounding fluids are also involved in the ionic exchanges mechanisms as observed *in vitro*. Their interaction with calcium phosphate substrates depend on the bioceramics characteristics (phase, crystallinity, composition, texture etc.) (El-Ghannam *et al.*, 1999; Rouahi *et al.*, 2006; Sharpe *et al.*, 1997), the proteins' feature (conformation, isoelectric point, composition, etc.), their

concentration, and whether they interact with the salts in the surrounding fluids or on substrates (Combes and Rey 2002; Hunter *et al.*, 1996; Johnsson *et al.*, 1991; Koutsopoulos and Dalas, 2000; Ofir *et al.*, 2004). Firstly, in suspension, proteins can inhibit or support calcium phosphate nucleation and growth (Boskey and Paschalis, 2001; Hunter *et al.*, 1996; Johnsson *et al.*, 1991). Regarding phosphorylated proteins, such as collagen, osteopontin, osteonectin, bone sialoprotein or osteocalcin, phosphorylated entities have demonstrated their ability to nucleate and grow calcium phosphate crystals (Boskey and Paschalis, 2001). However, not all of the phosphorylated proteins induce calcium phosphate formation; osteopontin, particularly, is a strong crystallization inhibitor (Hunter *et al.*, 1996). Secondly, when proteins adsorb onto calcium phosphate substrates, their charge, their concentration and the presence of calcium in the surrounding fluids, influence the kinetics and pattern of the surface coverage with time (Kawasaki *et al.*, 2003). These adsorbed proteins can also influence the new formation of calcium phosphate crystals by blocking or not the substrate's nucleation sites (Combes and Rey, 2002; Koutsopoulos and Dalas, 2000; Ofir *et al.*,

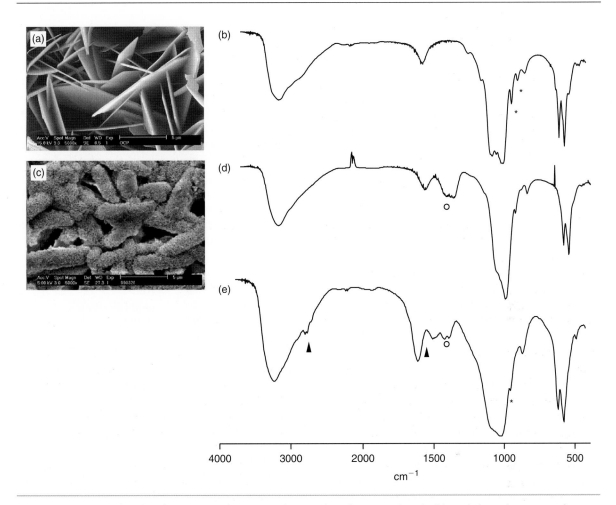

Figure 8.3 Morphology and structure of an octacalcium phosphate coating (a, b) and the subsequent changes due to the dissolution/reprecipitation process *in vitro* after immersion in a-MEM for 2 weeks (c, d), and *in vivo* after 4 weeks of subcutaneous implantation in rat (e). The morphology of the substrate is represented by Scanning Electronic Microscopy (SEM) pictures at magnification 5000 (a, c). The structure is characterized by Fourier Transform Infra-Red (FTIR) spectroscopy (b, d, e). The initial OCP substrate is composed of sharp elongated crystals (a) and its crystallized structure is characterized by sharp phosphate (P—O) bands between 1100 and 1000 cm^{-1} and at around 600 and 560 cm^{-1} (b). Two OCP typical bands can be noted at 906 and 852 cm^{-1} (*). *In vitro*, the substrate exhibits smaller crystals that seem to have grown along the initial OCP crystals (c), the FTIR structure of these crystals have lost their OCP features and exhibit broader and less defined bands (d). In addition, new bands typical of carbonate groups at 1460 and 1416 cm^{-1} (O) have appeared. This FTIR spectrum indicates a carbonate apatitic structure (d). *In vivo* (subcutaneous implantation in rat for 4 weeks), the FTIR structure of the coating has evolved in a similar way as *in vivo*: the coating transformed into a carbonated apatite (e). In addition, organic compounds (▲) were found in the transformed substrate.

2004), irrespective to the protein's isoelectric point (Ofir *et al.*, 2004). These results obtained from *in vitro* experiments are highly dependent on the chosen model. In contrast, *in vivo*, hundreds of proteins are present in biological fluids and their global effect on calcium phosphate reactivity is not sufficiently understood. But when present in the surrounding fluids, proteins are detected in close association with the nanocrystalline carbonated apatite formed on the surface of calcium phosphate bioceramics (Barrere *et al.*, 2003; de Bruijn *et al.*, 1992, 1994; Heughebaert *et al.*, 1988; Johnsson *et al.*, 1991; Radin *et al.*, 1998). Consequently, the nature, the quantity and the conformation of these proteins at the biomaterial surface will determine the cellular activity (El-Ghannam *et al.*, 1999; Rouahi *et al.*, 2006).

The physicochemical dissolution behavior *in vitro* and *in vivo* (under physiological conditions) of calcium phosphate ceramics can be affected by:

1. **The crystalline features**: for a given calcium phosphate crystalline phase the presence of defaults in the crystal lattice (due to the processing technique, for example) accelerates the dissolution. Eventually amorphous biomaterials dissolve faster than crystalline ones (Barrere *et al.*, 2000; de Bruijn *et al.*, 1992; Elliot, 1994; LeGeros, 1991; Porter *et al.*, 2001).
2. **The thermodynamical solubility**: the physicochemical dissolution observed *in vivo* and *in vitro* follows, in general, the thermodynamical solubility of calcium phosphate (Barrere *et al.*, 2000; Ducheyne *et al.*, 1993; Elliot, 1994; LeGeros, 1991; Radin and Ducheyne, 1993; Tang *et al.*, 2001), i.e. from the least soluble to the most soluble:

$$OHAp > TCP > OCP > DCPD$$

3. **The presence of additives**: the presence of some additives of mineral origins within the calcium phosphate structure can affect the crystal lattice, and therefore can accelerate the dissolution, e.g. carbonate, silicate or strontium in hydroxylapatite. On the other hand, some additives reduce the dissolution process, e.g fluoride in hydroxylapatite,

magnesium or zinc in beta-tricalcium phosphate *in vitro* and *in vivo* (Christoffersen *et al.*, 1997; Dhert *et al.*, 1993; Elliot, 1994; Ito *et al.*, 2002; LeGeros, 1991; Okazaki *et al.*, 1982; Porter *et al.*, 2004). The incorporation of organic molecules (proteins, amino acids, sugars) reduces also the physicochemical dissolution mechanism (Liu *et al.*, 2003)

4. **The structure**: last but not least, the structure plays an important role in the physicochemical dissolution process of calcium phosphates. The larger the exposed surface to its environment, the faster the biomaterial dissolves, simply because larger quantities of exchanges can take place (Radin and Ducheyne, 1994).

8.2.2 Cellular degradation

The typical cellular degradation of bioceramics is mediated by osteoclasts. Osteoclasts are derived from hematopoietic stem cells that differentiate along the monocyte/macrophage lineage. They are responsible for bone resorption by acidification of bone mineral leading to its dissolution and by enzymatic degradation of demineralized extracellular bone matrix. The mature osteoclast is a functionally polarized cell that attaches via its apical pole to the mineralized bone matrix by forming a tight ring-like zone of adhesion, the sealing zone. This attachment involves the specific interaction between the cell membrane and some bone matrix proteins via integrins (transmembrane adhesion proteins mediating cell-substratum and cell-cell interactions). In the resorbing compartment, situated under the cell and delimited by the sealing zone, osteoclasts generate a milieu acidification resulting in the dissolution of bone mineral. This osteoclastic acidification is mediated by the action of carbonic anhydrase that produces carbon dioxide, water and protons that are extruded across the cell membrane into the resorbing compartment (Kartsogiannis and Ng, 2004). In Figure 8.4, osteoclasts detected at bioceramics surfaces are represented.

Heymann *et al.* (1999) have extensively addressed the cellular degradation of calcium phosphate ceramics *in vitro* and *in vivo*. After colonization of the substrate by monocytes/macrophages that are recruited during the inflammatory reaction following surgical act (Basle *et al.*, 1993; Heymann *et al.*, 1999), osteoclasts are responsible for bone mineral degradation, i.e. bone resorption. *In vitro* and *in vivo*, it has been observed that osteoclasts degrade calcium phosphate ceramics in a similar way as bone mineral: osteoclasts attach firmly to the substrate sealing zone. In the center of this sealing zone, they secrete H^+ leading to a local pH = 4–5. *In vivo*, osteoclasts partially participate to the degradation of calcium phosphate ceramics (Gauthier *et al.*, 1999; Lu *et al.*, 2002; Ooms *et al.*, 2002; Wenisch *et al.*, 2003; Zerbo *et al.*, 2005). Resorbing osteoclasts have been found at the surface of the bioceramics in association with newly formed bone (Gauthier *et al.*, 1999; Ooms *et al.*, 2002; Wenisch *et al.*, 2003). However, the *in vivo* degradation of calcium phosphate materials is also associated with a dissolution phenomenon (Gauthier *et al.*, 1999; Lu *et al.*, 2002). The degree of osteoclastic activity and dissolution process of a calcium phosphate ceramic depends on the nature of the calcium phosphate (cement, bulk ceramic, particles, highly soluble tricalcium phosphate, sparingly soluble hydroxyapatite, etc.) (Gauthier *et al.*, 1999; Lu *et al.*, 2002). In the case of the degradation of highly soluble tricalcium phosphate ceramics *in vivo*, Zerbo *et al.* (2005) have shown that the degree of physicochemical dissolution was higher than osteoclastic resorption. Calcium phosphate particles can also be phagocyotosed by osteoclasts, i.e. they are incorporated in the cytoplasm, and thereafter dissolved by acid attack or enzymatic process. These phagocytosed particles are generated as the result of either local dissolution at grain boundaries or by mechanical stress generating debris (Heymann *et al.*, 1999). The osteoclastic activity depends on intrinsic properties of the calcium phosphate ceramics. However, some general trends can be outlined:

1. **The physicochemical dissolution kinetics of the biomaterial**: calcium phosphate ceramics do not all interact in the same way with osteoclasts. The release of calcium ions from the biomaterial seems to play a critical role in the osteoclastic activity; above a critical range calcium ions level, osteoclastic resorption is inhibited (de Bruijn *et al.*, 1994; Doi *et al.*, 1999; Yamada *et al.*, 1997). Together with the dissolution behavior, the structure of the calcium phosphate ceramic the crystallinity influence and the osteoclastic activity (de Bruijn *et al.*, 1994; Leeuwenburgh *et al.*, 2001; Yamada *et al.*, 1997).

2. **Carbonates and other mineral ions**: carbonated apatitic salts – e.g. dentine, bone mineral, synthetic carbonated apatite, and calcium carbonate structures, i.e. aragonite, calcite – are resorbed by osteoclasts. It has been proposed that the carbonate content may stimulate the carbonic anhydrase activity known to promote the osteoclastic acidic secretion *in vitro* (Doi *et al.*, 1999; Leeuwenburgh *et al.*, 2001). On the other hand zinc and fluoride included in calcium phosphate biomaterials have shown also an inhibiting effect on osteoclastic resorption *in vitro* (Ito *et al.*, 2002; Sakae *et al.*, 2000) and *in vivo* (Kawamura *et al.*, 2003; Sakae *et al.*, 2003) for specific concentration with regard to zinc-containing calcium phosphates.

3. **The surface energy of the calcium phosphate biomaterial**: the polar component of the surface energy was found to modulate the osteoclastic adhesion *in vitro*. However, the further spreading and resorption was not influenced anymore by surface energy differences between substrates (Redey *et al.*, 1999).

4. **The surface roughness**: known in general to influence the cell attachment *in vitro*, including osteoclasts attachment. Rough apatitic surfaces appear to enhance osteoclastic attachment compared with smooth ones (Gomi *et al.*, 1993).

8.2.3 Mechanical degradation

From a mechanical point of view, the calcium phosphate ceramics are brittle polycrystalline materials for which mechanical properties are governed by

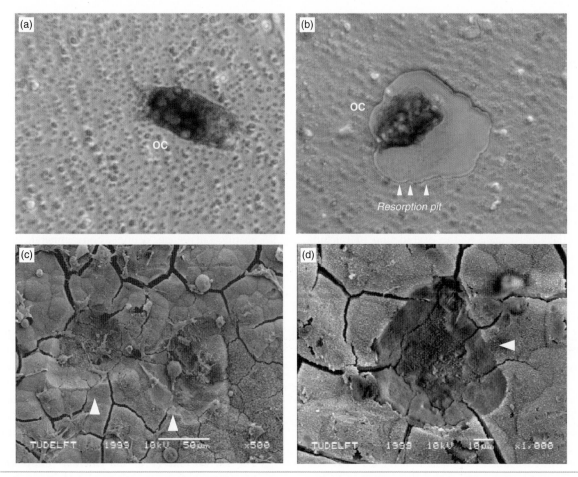

Figure 8.4 Osteoclasts on calcium phosphate substrates in vitro (a–e) and in vivo (f–i). (a, b) The light-microscopy pictures represent tartrate-resistant acid phosphatase (TRAP)-stained rabbit osteoclasts (OC). They respectively represent (a) a non-resorbing and (b) a resorbing-state. The nuclei staining by 4'-6'-diamino-2-phenylindiole (DAPI) shows that osteoclasts (OC) are multinucleated (Courtesy Dr S.G. Perez); (c, d) The Scanning Electronic Microscopy pictures represent resorption pits on carbonate apatite substrate by rat osteoclasts after their removal; (e) Osteoclast on an octacalcium phosphate substrate. Actin ring staining by F-phalloidin, magnification 200; (f, g) Light microscopy pictures of an explanted calcium phosphate: (f) staining for TRAP identified osteoclasts (→) localized at the calcium phosphate surface (*) and the adjacent bone (b) (bm = bone marrow; scale bar 50 μm), (g) higher magnification of an osteoclast associated with the implant (*) (scale bar 10 μm) (from Wenisch *et al.* (2003), *J Biomed Mater Res*, 67A: 713–718); (h) Transmission electronic overview of an osteoclast that is closely associated with the ceramic surface. The osteoclast is cuboidalin shape and displays intended nuclei (N), which are located toward the base of the cell. Clearly defined plasma membrane domains, such as the ruffled border (r), the sealing zone (s), and the dorsal microvilli (mv), can be identified. Large vacuoles filled with longer slender crystals are preferentially localized close to the ruffled border (→) whereas smaller vacuoles enclosing sphericparticles of high electron density are located below (▼). Golgi complexes (G) are restricted to the areas immediately around the nuclei. Note the alterations of electron density of calcium phosphate crystals that are located close to the ruffled border, which is indicative for the intense resorption activity of the osteoclast

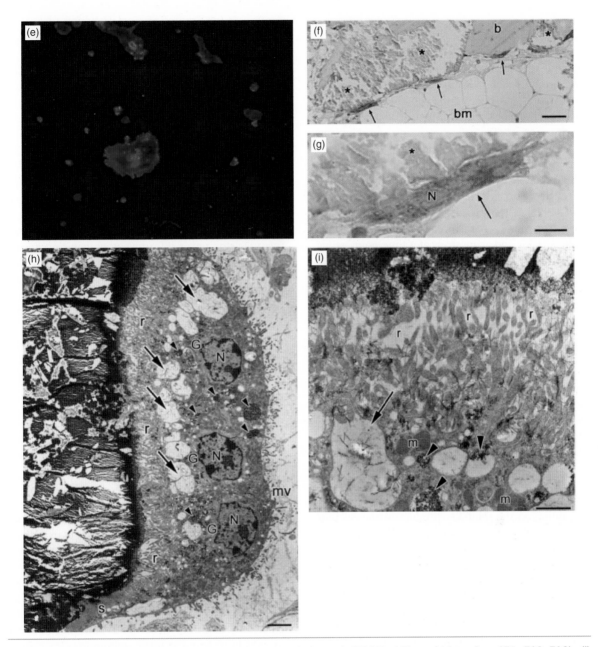

Figure 8.4 (Continued) (scale bar = 2 μm) (from Wenisch *et al.* (2003). *J Biomed Mater Res*, 67A: 713–718); (i) Ultratructural details of the osteoclastic ruffled border facing the implant surface. Mitochondria (m) and numerous vacuoles lie adjacent to the ruffled border (r) showing calcium phosphate phagocytosis. Large vacuoles containing long slender crystals (→) can be differentiated from smaller vacuoles filled with electron dense material (▼) (scale bar 1 μm). (a, b, e, Courtesy Dr. S.G. Perez; c, d are from de Leeuwenburg *et al.*, 2001, *J Biomed Mat Res*, p. 213; f, g, h, *in vivo* images are from de Wenisch *et al.* (2003). *J Biomed Mat Res*, 67a: 713–718).

the grain size, grain boundaries and porosity. They have a compression strength between 79 and 106 GPa (LeGeros, 2002) Under humid conditions, e.g. in liquids or physiological fluids, and as a consequence of the physicochemical dissolution mechanisms, calcium phosphate ceramics undergo a decrease of mechanical strength (de Groot, 1983; Mirtchi *et al.*, 1989; Pilliar *et al.*, 2001; Raynaud *et al.*, 1998) and of resistance to fatigue (de Groot, 1983; Raynaud *et al.*, 1998). The mechanical strength of a material can be seen as its resistance to fracture formation under specific and acute stress at a time point, while failure by fatigue includes an additional parameter which is the long-duration strength. For biomedical applications, long-duration stress is important. For example, if a ceramic rod breaks at a stress σ in a short-time, an identical rod, when stressed to about $0.75\,\sigma$, would fracture after an about 100 times longer period. Generally, decrease of (tensile) strength of brittle ceramic materials is caused by slow or subcritical crack growth, occurring under stress, sometimes assisted by environment factors (de Groot, 1983). With regard to monophasic dense ceramics, the mechanical strength may be affected either (i) by grain decohesion governed by a higher solubility of nanophases present at grain boundaries due to the processing technique (Raynaud *et al.*, 1998), or (ii) by a uniform physicochemical dissolution of grains, depending on the solubility product of the calcium phosphate phase (Pilliar *et al.*, 2001). With regard to pluriphasic dense calcium phosphate ceramics, the various sensitivities towards physicochemical dissolution affect the mechanical properties measured after aging under humid conditions. On one hand, the most soluble phase destabilizes the bulk ceramic inducing a decrease in mechanical strength (Mirtchi *et al.*, 1989; Raynaud *et al.*, 2002). On the other hand, the presence of another phase (for example, tricalcium phosphate combined with hydroxyapatite, i.e. BCP) or a compound (fluoride added in the hydroxyapatite lattice) may provoke grains densification, which mechanically stabilizes the ceramic (Barinov *et al.*, 2003; Raynaud *et al.*, 2002). With regard to total porosity, mechanical strength failure depends

directly upon the macroporosity and microporosity (de Groot, 1983; Nilsson *et al.*, 2003; Pilliar *et al.*, 2001). The higher the micro- and macroporosity of a ceramic, the more surface area is exposed to the fluids and therefore the more intensively the cracks and fractures are propagating though the ceramic.

Since the beginning of the earlier mechanical investigations of calcium phosphate ceramics in the 1920s, and despite some improvement of their mechanical characteristics, it has been shown that calcium phosphate ceramics as such are not appropriate for permanent skeleton repair, if loading other than compressive occurs (de Groot, 1983). Composite materials, e.g. calcium phosphate coated metals or calcium phosphate-polymeric blends have been developed to combine their biological properties with the mechanical properties of other biomaterials.

Parameters influencing the mechanical strength degradation *in vitro* and *in vivo* are directly related to the parameters influencing the physicochemical dissolution.

8.3 Degradation mechanisms of bioactive glasses

8.3.1 Physicochemical degradation

When a bioactive glass is immersed in an aqueous solution, a series of reactions takes place at its surface. The composition of the medium plays an important role, as the surface reaction layer that forms is dependent on the medium. The surface layer in its turn determines the protective effect against further corrosion. This can be generalized as three processes including leaching, dissolution and precipitation. Briefly, in its simplest form, sodium ions leach from the surface and are replaced by hydrogen ions through an ion-exchange reaction. This depletion of sodium leads to the formation of a silica-rich layer. An amorphous calcium phosphate layer forms on the silica-rich layer. Calcium is present in the solid state, but is equally drawn from solution. In simple electrolyte solutions (which are not a good simulation of *in vivo* behavior), the amorphous layer crystallizes to form carbonated hydroxyapatite (c-Ap) with properties akin to the mineral phase of bone. This simple reaction sequence that brings about the deposition of c-Ap was first proposed by in 1971 (Clark *et al.*, 1976; Hench, 1991; Hench *et al.*, 1971). Hench described five reaction stages:

Stage 1: Leaching and formation of silanols

Stage 2: Loss of soluble silica and formation of silanols

Stage 3: Polycondensation of silanols to form a hydrated silica gel

Stage 4: Formation of an amorphous calcium phosphate layer

Stage 5: Crystallization of c-Ap layer (Figure 8.5a).

The dissolution of bioactive glasses has been studied for over 40 years. Many different factors influence the mechanism and rate of the dissolution of bioactive glasses, such as glass composition, glass particle size and glass powder type. The results of *in vitro* experiments are also dependent on pH and ionic strength of the solution, type of solution and BG surface area to solution volume ratio (Bunker, 1994; Bunker *et al.*, 1983).

Radin *et al.* (1997) studied surface reactions on bioactive glass (BG) 45S5 under various modeling conditions. They used one-parametric variations of solution composition. The solutions used were 0.05 M tris hydroxymethyl aminomethane/HCl (tris buffer), tris buffer complemented with plasma electrolyte and/or serum, and serum. After a short period of immersion (3 h), the reacted surfaces of BG were analyzed using Fourier transform infrared (FTIR). All the FTIR spectra of BG after immersion for 3 h in the various solutions (Figure 8.5b) showed changes in their appearance in comparison to the spectrum

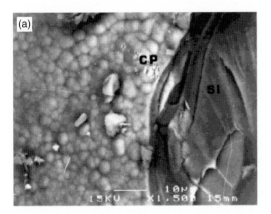

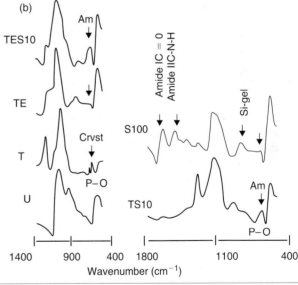

Figure 8.5 Phase transformations of bioglasses *in vitro*. (a) SEM view of a fractured granule showing the Ca-P (CP) surface layer and underlying silica (Si) (Radin *et al.* (2000). *J Biomed Mater Res*, 49: 264–272; (b) FTIR spectra of unreacted BG 45S55 prior to (U) and after immersion for 3 h into 0.05 M trisbuffer (T), T + plasma electrolyte (TE), TE + 10% serum (TES-10), T + 10% serum (TS-10), and 100% serum (S-100). Notice the appearance of split P—O bend (indicating formation of a crystalline (Cryst) phase) after immersion in T and undivided P—O bend (indicating amorphous (Am) phosphate phase) after immersion in other solutions (Radin, *et al.* (1997). *J Biomed Mat Res*, 37: 363–375).

of BG 45S5 before immersion. At this time period, a crystalline calcium phosphate ceramic (CPC) phase was only formed on BG in tris buffer as indicated by the P—O bend peak splitting. In contrast, an amorphous phosphate phase was formed in modeling solutions containing either plasma electrolyte (TE) or serum (TS-10), or both (TES-10, S-100), as the P—O bend peak appeared undivided. Thus, the formation of a crystalline HA surface was delayed in the presence of plasma electrolyte and serum.

Bioactive glass granules can be internally hollowed out leading to shells of calcium-phosphate in which osteogenesis can be observed (Figure 8.6) (Schepers *et al.*, 1991). BioGran® is a bone graft material made of bioactive glass granules (300–355 μm). Animal studies (Schepers *et al.*, 1991; Schepers and Ducheyne, 1997) showed this material to be replaced by new bone. When BioGran® is implanted in bone tissue, it reacts to form internal silica-gel cores with a calcium phosphate-rich surface. After a period of time, the internal silica-gel core degrades and leaves an external calcium phosphate shell. Schepers *et al.* (1993) demonstrated cellular basis for this phenomenon. This internal excavation of BG granules was simulated by *in vitro* immersion experiments (Radin *et al.*, 2000). Bioactive glass granules were immersed under integral (no solution exchange during the experiment, thereby simulating stagnant fluid conditions) or differential conditions (conditions simulating continuous fluid flow *in vivo*) in tris buffered solution complemented with either plasma electrolyte (TE) or with electrolyte and 10% serum (TES-10). Only when the solution was continuously replenished, thereby avoiding Si saturation in solution, and only when the solution contained serum proteins, was full Si dissolution from the core of the granules observed. This study supports the hypothesis that there is a physico–chemical mechanism of Si transport through the Ca-P-rich layer followed by Si dissolution. In addition to cellular resorption, this mechanism may be operative *in vivo* and thereby may contribute to the observed *in vivo* excavation.

Cerruti *et al.* (2005) studied the effects of pH and ion strength of the dissolution solution on the

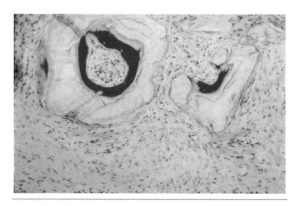

Figure 8.6 Bone tissue formation in excavated bioactive glass particles of narrow size range. Bone tissue is stained red. Also note the channels connecting the interior of the particle with the surrounding milieu (Ducheyne and Qui (1999). 20: 2287–2303).

reactivity of BG 45S5. BG 45S5 particles with size of 2 μm were immersed in water and different TRIS buffers. The pH values of TRIS buffers were varied from 6.9 to 8.8. After a period of immersion (from 0.5 to 2880 min), it showed that only at pH 7.4 could c-Ap be formed, as described by Hench. At higher pH, calcium phosphate precipitation occurred immediately after the immersion and prevented any further dissolution of BG 45S5. Calcium carbonate was formed more than c-Ap in these conditions. At lower pH, no c-Ap was observed within the first two days of reaction, and a total breakdown of glassy network occurred. Therefore, pH is an important factor that influences bioactive glass degradation.

Sepulveda *et al.* (2002) showed that the type of glass powders influences the dissolution rates. Two types of powders were used in their study, the melt-derived 45S5 glass powders and 58S gel-glass powders. The 58S gel-glass exhibits a highly mesoporous texture (pores in the range of 2–50 nm) and high surface area (126–164 m²/g), whereas melt-derived 45S5 glass exhibits low porosity (pores in the range of 1.63–2.13 nm) and low surface area (0.15–2.7 m²/g). It was found that 58S glass had a much higher degradation rate compared to that of melt-derived 45S5.

State of the Art Experiment
Guidelines to study the degradation of bioceramics

Importance of well characterizing the initial materials: Besides the classical discrepancies based on different experimental settings to assess the degradation mechanisms, important deviations in the nature of the starting materials have to be addressed with regard to: (i) the preparation of the bioceramic which induce inherent properties and affect their dissolution behavior, (ii) the difficulty of a precise chemical characterization of the end product (for example Raynaud *et al.* (2002) have shown that a 1% deviation in the calcium-to-phosphorus ratio can affect the physicochemical and biological profile of calcium phosphate ceramics), and (iii) the choice of the implantation site (ectopic versus orthotopic, the fluid circulation flow, the degree of motion of the implant at the site, the presence or absence of osteoclasts, inflammation response of the host tissue that may lead to a decrease of the local pH).

Models to evaluate physicochemical dissolution of calcium phosphate ceramics: *In vitro* dissolution can be assessed by the measurement of calcium and phosphate ions concentration by immersion in aqueous solutions at 37°C at several time points. Various composition and condition alternatives can be chosen: (i) buffering conditions: at pH = 7.4 (physiological conditions) or at pH=5.0 (inflammation pH) (Barrere *et al.*, 2000), (ii) ionic strength and mineral composition comparable to blood plasma (Barrere *et al.*, 2000; Radin *et al.*, 2000), (ii) presence of proteins in the medium (Radin *et al.*, 2000). Phase transformation can be assessed by typical structural analyses (infra red spectroscopy, X-ray diffraction).

In vivo, the weight of the implants can be measured prior ectopic implantation and after explantation by scale or by thermogravimetric analyses (Barralet *et al.*, 2000; Heughebaert *et al.*, 1988). The introduction of the 45Ca radiotracer in the calcium phosphate material (den Hollander *et al.*, 1991; Le Huec *et al.*, 1998) or the titration of silicon (Si) by atomic adsorption spectrophotometry (Lai *et al.*, 2005) can be used to trace the elements after implantation in the surrounding and distant tissues. Generally, back-scattering electronic microscopy or light microscopy of the explants is commonly used to evaluate the volume loss (Barrere *et al.*, 2003). Phase transformation may be also assed by structural analyses (thermogravimetric analysis, infra red spectroscopy, X-ray diffraction (Habibovic *et al.*, 2006; Heughebaert *et al.*, 1988)) or by phase contrast (back-scattering electronic microscopy (de Bruijn *et al.*, 1994), transmission electron microscopy (Daculsi *et al.*, 1990; Neo *et al.*, 1993)).

Models to evaluate osteoclast degradation

For *in vitro* experiments, osteoclasts are generally obtained from rodent bone marrow biopsies (Doi *et al.*, 1999; Leeuwenburgh *et al.*, 2001; Monchau *et al.*, 2002; Sakae *et al.*, 2000; Yamada *et al.*, 1997) but also from humans (Monchau *et al.*, 2002). The osteoclastic activity is assessed with light microscopy after the staining of osteoclast-specific gene (Tartrate Resistant Acid Phosphatase, TRAP) (Wenisch *et al.*, 2003; Zerbo *et al.*, 2005) or actin-ring (Monchau *et al.*, 2002; Redey *et al.*, 1999) as shown in Figure 8.6. Scanning electronic microscopy after a cell-fixation step (de Bruijn *et al.*, 1994; Doi *et al.*, 1999; Monchau *et al.*, 2002) or cell removal (de Bruijn *et al.*, 1994; Leeuwenburgh *et al.*, 2001) are also classical methods to perform qualitative observation of the resorption pits (size, density, depth, surface area) and of the cells (size, morphology, presence of filopodia) (de Bruijn *et al.*, 1994; Monchau *et al.*, 2002; Redey *et al.*, 1999; Yamada *et al.*, 1994, 1997). *In vivo*, the degradation of bioceramics can be evaluated on sections by histomorphometry (Dhert *et al.*, 1998; Lu *et al.*, 2002; Ooms *et al.*, 2002), counting resorption lacunae (Dhert *et al.*, 1998), location of osteoclasts in contact with the implant by transmission electronic microscopy (Wenisch *et al.*, 2003) or after TRAP staining of cells by light microscopy (Wenisch *et al.*, 2003; Zerbo *et al.*, 2005).

Models to evaluate mechanical degradation

Ceramics and glass ceramics should not be used as such in load-bearing applications. Fatigue test in humid environment may be the most representative *in vitro* model (Pilliar *et al.*, 2001; Raynaud *et al.*, 1998; Shinn-Jyh Dinga 2005). *In vivo,* interface strength between an implant and the surrounding bone can be evaluated by using the push (Spivak *et al.*, 1994) and pull-out test (Dhert *et al.*, 1991).

Therefore, modifying the micro-structure of bioactive glass can alter the degradation rate.

Most immersion studies, as mentioned previously, were performed in acellular, serum free solution. There are two kinds of solution commonly used: TRIS buffer and simulated body fluid (SBF). TRIS buffer provides a pH similar to the physiological condition. SBF is an ionic solution similar to that found in human plasma. When in the presence of proteins and serum, the degradation of bioactive glass becomes complex.

As Radin *et al.* (1997) has demonstrated, and Mahmood and Davies (2000) has documented, the amino acids contained in the culture medium were incorporated within the growing calcium phosphate rich surface reaction layer of bioactive glass. BG 45S5 discs were immersed in the standard SBF-K9 solution or tissue culture medium α-MEM for up to 3 days. The discs were examined by scanning electron microscopy (SEM) and X-ray photoelectron spectroscopy (XPS). XPS is one of the most widely used surface analytical techniques which provides valuable information about the chemical surface composition of approximately the top 80 Å of a sample surface (Ratner and Castner, 1997). XPS depth profiling showed the presence of nitrogen throughout the depth of the surface layer of the sample incubated in α-MEM for 72 h, whereas an insignificant amount of nitrogen was found at the surface of any sample immersed in SBF. The nitrogen signal most likely came from cysteine, an amino acid contained in α-MEM. Their work further confirmed that SBF can only be used as a first solution to study the biological behavior of bioactive glass. In fact, SBF does not contain any organic constituents of body fluid and these organic components, such as amino acids, can alter the composition, structure and stability of the surface layers on bioactive glasses.

In another study, Kaufmann *et al.* (2000) demonstrated that the presence of serum significantly delayed the formation of c-Ap at the surface of bioactive glass. BG 45S5 discs were immersed in two solutions: SBF or Dulbecco's Modified Eagle's minimum essential medium supplemented with 10% Nu-Serum TM, 2 mM of L-glutamine, and 50 U/ml of penicillin/streptomycin. After immersion, the disks were rinsed and air-dried and examined by atomic force microscopy (AFM) and Rutherford backscattering spectrometry (RBS). Both AFM and RBS are powerful surface analysis tools. The results demonstrated that the reaction layer composition, thickness, morphology, and kinetics of formation were different when serum was present. This is because the uniform and rapid adsorption of serum proteins on the surface may serve to protect the surface from further direct interaction with aqueous media, slowing down the transformation reactions.

In conclusion, immersion in TRIS buffer or SBF is only a first step, and by itself is not an appropriate model of *in vivo* growth of c-Ap on bioactive glass. In fact, bioactive glass is exposed to a complex environment of electrolytes, proteins and cells, when implanted *in vivo*. The present data suggest that surface dissolution and precipitation reactions occur in tandem with protein adsorption. Therefore, we propose that the following events take place in an integrated and overlapping manner at the interface between bioactive glass implant and tissue: serum protein adsorption prior to, and in parallel with, solution mediated reactions leading to the formation of an amorphous silicon-containing surface with accumulated calcium phosphate phases; selective adsorption of attachment molecules such as fibronectin; humoral and cell-mediated c-Ap formation; cell proliferation, differentiation, and extracellular, matrix formation (Ducheyne and Qiu, 1999).

8.3.2 Cellular degradation

The effects of bioactive glass on osteoclast were not well studied. Osteoclast is a multinuclear cell, which is responsible for bone resorption. When osteoclasts resorb bone or biomaterials, the local pH will be lowered. As mentioned previously, bioactive glass is able to be dissolved completely at lower pH.

Yamada et al. (1994) studied osteoclastic resorption on wollastonite-containing glass ceramic (A-W GC). Two types of substrates were used: plain A-W GC and pre-conditioned A-W GC. The pre-conditioned A-W GC was immersed in SBF for 5 days to form a bone-like apatite layer on A-W GC surface. Their results showed that for 2-day culture, osteoclasts with a non-motile appearance formed no lacunae on the plain A-W GC, whereas actively moving osteoclasts made many track-like resorption lacunae on the pre-conditioned A-W GC. Their results demonstrate that the osteoclastic activity is much higher on apatite surface than on plain glass-ceramic surface.

Vaahtio et al. (2005) investigated the osteoclast response to different silica and carbonate containing calcium phosphate (CaP) layers on bioactive glass S53P4. Bioactive glass S53P4 was immersed in conventional C-SBF and revised R-SBF. A revised SBF contains ion concentrations, including those of Cl^- and HCO_3^-, equal to those of the human blood plasma. It was found that in R-SBF, the CaP layer formed faster compared to C-SBF, and the CaP layer formed in R-SBF was amorphous compared to the poorly crystalline bonelike c-Ap formed in C-SBF. The optimal surface for osteoclast activity was an amorphous calcium phosphate having mesoporous nanotopography and proper dissolution rate of calcium and silica. Once again, active osteoclasts were

found on the calcium phosphate layer instead of plain bioactive glass.

Vogel et al. (2004) investigated the presence of multinuclear cells (MNC) during the degradation of BG 45S5 particles using a rabbit animal model. It was found that BG 45S5 particles degraded either in silicon-rich remnants or in calcium phosphate (CaP)-shells. Although they observed both foreign body giant cell type and osteoclast-like type cells, exclusively osteoclast-like cells developed on resorbable substrates. They also found that the absolute number of MNCs depended on the time after implantation and the solubility of the implant. Bone bonding, however, only occurred on Ca- and P-rich surfaces. Osteoclast-like cells were detected on the particles after transformation in CaP-shells. Cellular degradation of CaP-shells is similar to that of calcium phosphate ceramics and has been described in details in previous section.

Both in vivo and in vitro studies (Vaahtio et al., 2005; Vogel et al., 2004; Yamada et al., 1994) demonstrate that active osteoclasts are only detected on the calcium phosphate layer. As mentioned previously, the calcium phosphate layer can be formed on bioactive glass surface. Osteoclasts may be responsible to resorb this calcium phosphate layer afterwards. However, there is no direct evidence of cellular mediated degradation of silicon. The silicon content of bioactive glass can be safely excreted from urine when implanted in vivo (Lai et al., 2005).

8.3.3 Mechanical degradation

Although bioactive glasses can form a strong interfacial bond with bone, the mechanical properties of these glasses are limited. The failure load of BG 45S5 after 8 weeks implantation was only $2.75 \pm 1.80\,kg$ (Thompson and Hench, 1998). Previous studies show that BG 45S5 does not resist cyclic loading or cracking (Thompson and Hench, 1998). Therefore, bioglass itself can only be used for non-load bearing sites. The use of bioactive glass has been restricted mainly to powder, granules, or small monoliths.

Box 1 Theoretical background on ceramics, glasses and glass ceramics

Inorganic non-metallic solids can be classified as ceramic, glass or glass-ceramics.

The word *ceramic* is derived from the Greek word *keramikos*, 'having to do with pottery'. The American Society for Testing and Materials (ASTM) defines a ceramic article as '*an article having a … body of crystalline or partly crystalline structure, or of glass, which body is produced from essentially inorganic, nonmetallic substances and either is formed from a molten mass which solidifies on cooling, or is formed and simultaneously or subsequently matured by the action of the heat*' (ASTM 1988). Ceramic materials are ionic or covalently-bonded materials. They can be crystalline or amorphous. The crystalline and amorphous states are typical solid states which represent the degree of order between ions, atoms or molecules. The *crystalline state* is characterized by a definite internal molecular structure (O'Bannon, 1984). This structure is periodically repeated in a specific tri-dimensional pattern (a). The cohesion in crystals is insured by the binding energy between the atoms in the case of covalent solids, or by electrostatic forces existing between anions and cations. On the contrary, *amorphous state* is characterized by no order between ions, atoms or molecules (b), therefore the values of binding energy between entities greatly vary within the solid. The dissolution of ceramics is strongly dependent on these binding energies.

Crystalline solids are different from the theoretical crystals. First, these crystals present generally imperfections: (i) additional entities are inserted in the crystal lattice (c), or (ii) some entities can be

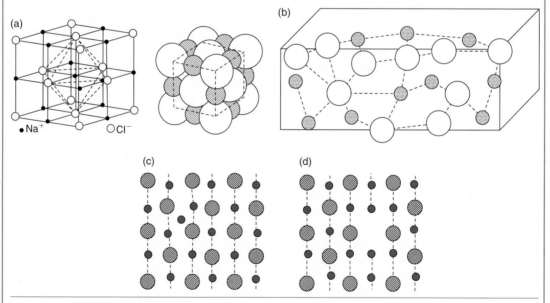

(a) Crystalline state (sodium chloride, NaCl). Three-dimensional sketch of a sodium chloride (NaCl) crystal lattice (Paul Arnaud; Bordas-Dunod)

(b) Bi-dimensional representation of an amorphous state showing the unorganized structure between the entities; 3-dimensional effect of unorganized structure

(c) and (d) Bi-dimensional representation of common crystal defaults. (c) Insertion, and (d) vacancy (Paul Arnaud; Bordas-Dunod).

absent in the crystal, i.e. vacancies (d). Second, crystalline solids are composed of several crystals packed together in a more or less organized structure. These imperfections result from the materials' processing (presence of impurities, kinetic of crystallization, etc …). The crystallinity of a solid is determined by X-ray diffraction, and in particular the width spectrum peaks. In short, the broader the peaks are, the less crystalline the solids are. However, it is important to realize that a low crystallinity can result either from a solid totally composed of crystals of very small dimensions, or from a mixture of crystals with an amorphous phase (LeGeros, 1991). Although their apparent crystallinity may appear similar, their dissolution can greatly differ (Barrere *et al.*, 2000). Therefore, the concept of crystallinity may be used with precautions.

One of the specific forms of amorphous inorganic solid is ***glass***. Glass formation is the result of the very rapid cooling of a viscous molten material to a solid state without crystallizing (ASTM 1988). Silica (SiO_2) is the common base for glasses. A slower and controlled cooling down of a glass can induce the formation of glass-ceramics, which are partly crystalline and partly glassy (ASTM, 1988). An important change in the glass microstructure, which usually precedes crystallization, is the glass-in-glass phase separation. After phase separation, the glass no longer has a homogenous composition, but rather, consists of two or more immiscible glassy phases of different chemical constituents. Unlike crystallization, phase separation in glass might not be visible by optical microscopy and in most cases can be detected only by electron microscopy. Previous work revealed that thermal treatment of bioactive glass at 550°C resulted in glass-in-glass phase separation and minor crystallization (El-Ghannam and Ducheyne, 2005).

The mechanical properties of bioactive glass can be improved by simply inducing crystalline phase (Thompson and Hench, 1998). The crystalline phase is produced in the glassy BG by heat treatment. The increase of density and the appearance of microcrystalline enhance the mechanical properties. After heat treatment, the ceramic-glass still retains its bioactivity, which means it still has the ability to form a calcium phosphate layer when immersed in SBF.

The mechanical properties of bioactive glass can also be improved by reinforcing it with a second phase. It has been shown that BG with 60% volume fraction of metal fibers increased its bending strength from 42 MPa to 340 MPa (Ducheyne and Hench, 1982).

A polymeric bioceramic composite was developed to provide long-term mechanical support for fracture healing. This concept was first introduced by Bonfield *et al.* (1981), who developed a substance made from two materials that matched the natural component of bone: hydroxyapatite (HA) and collagen. The collagen can be replaced by high-density polyethylene (PE) or other polymers and the HA can be replaced by synthetically produced HA or bioactive glass. Based on this concept, bioactive glass/PE composite was prepared and its mechanical properties were improved (Wang *et al.*, 1998).

In summary, the limited mechanical properties of bioactive glasses limit their applications to non-load bearing sites. The mechanical properties of bioactive glass can be improved by increasing crystalline phase, reinforcing with fibers, or preparing polymeric bioceramic composites.

8.4 Translation to bone tissue engineering systems

The principles of the degradation mechanisms specific to bioceramics are directly linked to important parameters of biomaterials. As bone tissue engineering aims to regenerate tissue, the scaffold should eventually degrade entirely to be replaced by the regenerated bone. Therefore, tuning the scaffold degradability is highly desired for the development of bioceramic scaffolds. In addition, the scaffold degradation is currently used as a tool to stimulate specific cellular activity and/or bone formation.

8.4.1 The bone-bonding ability

Dissolution/precipitation phenomena are associated with reactivity towards bone bonding, i.e. the formation

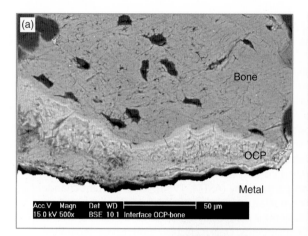

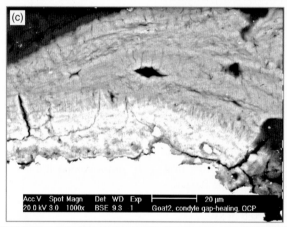

Figure 8.7 Back-scattering Electronic Microscopy picture of an octacalcium phosphate (OCP) coating on a metallic porous scaffold implanted for 12 weeks in the femoral condyle (goat) at different magnification. (a, b) Between the octacalcium phosphate coating and the newly formed bone, an interfacial phase (arrow) that can be attributed to superficial phase transformation is clearly visible. However, on the same implant, this interfacial layer is not always visible (c).

of an interfacial mineralized layer between bioceramics and bone tissue that insures their cohesion (Figure 8.7). Structurally, this layer is comparable to the films grown *in vitro* by dissolution-precipitation mechanisms, i.e. nanocrystals of carbonated apatite, in simulated body fluids that mimic the mineral composition of blood plasma. When formed in the presence of osteogenic cells experiments, this mineralized layer is comparable to the cement lines present in bone

(Davies, 1996; de Bruijn *et al.*, 1995). *In vivo* (osseous and non-osseous environment), physicochemical and crystallographic continuity are observed between the calcium phosphate implant and the newly formed mineralized layer (Daculsi *et al.*, 1989; de Bruijn *et al.*, 1992; Dhert *et al.*, 1993; Neo *et al.*, 1993). Its occurrence and thickness are related to the reactivity (dissolution/precipitation) of the calcium phosphate substrate (Neo *et al.*, 1993), so-called bioactivity

(Hench and Wilson, 1984). This mineralized interface insures a physicochemical and mechanical cohesion between the implant and the host bone. It is particularly relevant for load-bearing applications, i.e. hip metallic prostheses coated with calcium phosphate which layer improve undoubtedly the mechanical stability of the implant by augmenting and accelerating the bone apposition (Dhert *et al.*, 1993; Geesink *et al.*, 1987; Rahbek *et al.*, 2004).

8.4.2 Osteoinduction by calcium phosphates

Osteoinductive biomaterials are biomaterials which have intrinsic ability to induce bone formation in non-osseous environment (Figure 8.8), i.e. in an environment where bone cells are initially absent, such as muscle or subcutis. Although the mechanism of osteoinduction by biomaterials is not completely unraveled at the time of writing, the dissolution/precipitation behavior of calcium phosphate scaffolds has been pointed out as a relevant parameter (Habibovic *et al.*, 2005). By intramuscular implantation in goats of two macroporous calcium phosphate ceramics identical in composition, crystallinity and porosity but with different microporosities, Habibovic *et al.* (2005, 2006) have demonstrated that an elevated microporosity was responsible for ectopic bone formation. This positive effect of increased microporosity on ectopic bone formation could be direct or indirect. Increased microporosity is directly related to the change of surface topography, i.e. increased surface roughness, which might affect cellular differentiation. In addition, an increased microporosity means a larger surface that is exposed to body fluids, leading to a more elevated dissolution/reprecipitation phenomenon as compared to non-microporous surfaces. This phenomenon is most pronounced in the areas with the least body fluid replenishment, i.e. in the macropores, where stable critical free calcium- and orthophosphate-ion levels and/or consequent formation of a bone like mineral layer might trigger cell differentiation into the osteogenic

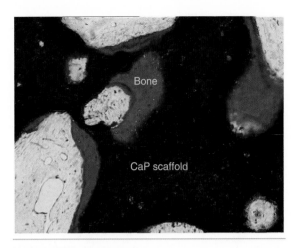

Figure 8.8 Light-microscopy pictures a macroporous calcium phosphate (CaP) scaffold implanted in goat muscle for 12 weeks representing the ectopic formation of bone (dark purple) (magnification × 20).

lineage. Another possibility is that the formation of this bone-like mineral layer is accompanied by the coprecipitation of endogenous osteogenic factors such as Bone Morphogenetic Proteins, which thus indirectly triggers the differentiation of cells into the osteogenic lineage followed by ectopic bone formation (Ripamonti, 1996).

Another property which has been shown to be relevant for osteoinduction by biomaterials is their mechanical stability as is illustrated in Figure 8.9. This figure represents a macroporous osteoinductive BCP ceramic after 12 weeks of implantation in goat muscle. Sporadic mechanical degradation occurred through the implants and interestingly, ectopic bone formation was solely observed in the non-degraded zones of the implant.

8.4.3 Osteogenic stimulation via dissolution/reprecipitation mechanisms

8.4.3.1 Calcium-phosphate
The radiolabeling of calcium (45Ca) in beta tricalcium phosphate ceramics has shown that dissolution

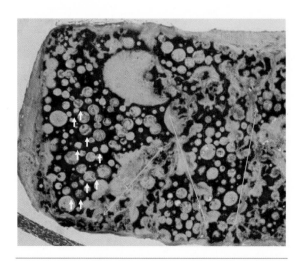

Figure 8.9 Digital picture a macroporous CaP scaffold (cross-section) after implantation in goat muscle for 12 weeks. (a) The white arrow represents the ectopic formation of bone (dark purple) and the yellow arrows represent the cracks within the ceramic illustrating the mechanical degradation of the ceramic and the infiltration of fibrous tissues.

products originating from the implants were partially detected in the surrounding bone tissue (Le Huec *et al.*, 1998). *In vitro*, the medium concentration in calcium and phosphate may increase or decrease when calcium phosphate ceramics are immersed, depending on the reactivity of the biomaterial (Doi *et al.*, 1999; Hyakuna *et al.*, 1989). These calcium and phosphate level modifications in the cell culture medium induced by the presence of a calcium phosphate substrate are reported to affect cell proliferation (Anselme *et al.*, 1997; Arinzeh *et al.*, 2005; Knabe *et al.*, 2004; Midy *et al.*, 2001), differentiation (Arinzeh *et al.*, 2005; Midy *et al.*, 2001) and may even induce cell death (Hyakuna *et al.*, 1989). In the absence of calcium phosphate substrates, increasing or decreasing calcium or phosphate contents in culture medium directly affect the osteoblastic activities (Dvorak *et al.*, 2004; Meleti *et al.*, 2000). On one hand, alkaline phosphatase activity was maximized for calcium concentration between 1.2 and 1.8 mM in the culture medium; above and below this window, the alkaline phosphatase activity was inhibited (Dvorak *et al.*, 2004). On the other hand, a content of inorganic phosphate higher than 5 mM induced osteoblasts apoptosis (Meleti *et al.*, 2000).

De Bruijn *et al.* (1992, 1994) demonstrated *in vitro* and *in vivo* (in small animal models) a clear link between dissolution rate of calcium phosphates and early bone formation, and therefore the influence of dissolution products on bone formation. In larger animal models (goats), no significant differences with respect to bone contact could be established between the calcium phosphate substrates having different dissolution rates in the course of the first month of implantation. However, a significantly higher osteoid production (matrix produced by the osteoblasts) was detected on the more soluble calcium phosphate substrate (Dhert *et al.*, 1998). Recently, in a bone tissue engineering approach, human mesenchymal stem cells were cultured on various biphasic calcium phosphate ceramics of comparable macroporosity ranging from 100% hydroxyapatite to 100% tricalcium phosphate. In a mouse ectopic model, by altering the composition of HA/TCP to 20% HA/80% TCP, hMSC bone induction occurred at the fastest rate *in vivo* over the other formulations of the more stable 100% HA, HA/TCP (76/24, 63/37, 56/44), and the fully degradable, 100% TCP (Arinzeh *et al.*, 2005). However, no further surface characterization, e.g. microporosity and specific surface area measurements, were performed on the scaffolds, leaving doubts as to whether degradation rate of the various BCPs ceramics play a role or not.

In view of the dissolution properties of calcium phosphates, several groups have used calcium phosphate ceramics as delivery systems for gene (Shen *et al.*, 2004) or drugs (Kroese-Deutman *et al.*, 2005; Lebugle *et al.*, 2002; Liu *et al.*, 2005; Stigter *et al.*, 2004). Their association with calcium phosphates can be performed by (i) adsorption on powder followed by compaction, (ii) co-precipitation, or (iii) addition in the cement paste applied in bone regeneration field and related fields of. With regard to the stimulation of bone regeneration, specific proteins have been administered via calcium phosphate carriers. Bone morphogenic proteins (BMP, especially BMP-2) adsorbed onto ceramics (Kroese-Deutman *et al.*, 2005; Urist *et al.*, 1987) or co-precipitated with carbonated apatite coatings (Liu *et al.*, 2005) induce more bone formation than ceramics alone *in vivo*. Recently, incorporation of silicate and zinc ions, respectively in tricalcium phosphate and hydroxyapatite ceramics, were reported to have a significant influence on osteogenesis *in vitro* and *in vivo* (Ikeuchi *et al.*, 2003; Kawamura *et al.*, 2000; Porter *et al.*, 2004).

8.4.3.2 Bioactive glass
From *in vitro* cell culture studies, bioactive glasses have shown stimulatory effects on osteoblast (bone forming cells) differentiation. Ducheyne *et al.* (1994) documented the characteristics of the BG substrate that markedly affect the synthesis of extracellular matrix by cells expressing the osteoblastic phenotype. After conditioning, the BG disks were placed in 60 mm diameter Petri dishes, wetted with tissue culture medium and inoculated with about 1×10^6 neonatal rat calvaria osteoblasts. After 1 week in culture, SEM micrograph (Figure 8.10) shows that the BG surface was covered by collagen fibrils and bone-like tissue. El-Ghannam *et al.* (1997) showed that when bioactive glass surfaces are modified by a first treatment with tris buffered solution complemented with plasma electrolyte (TE) to form a layer of carbonated-hydroxyapatite and then adsorption of serum proteins, a high number of MC3T3 cells colonize the bioactive glass surfaces and express high alkaline phosphatase activity. Their results suggest that the enhancement of osteoblastic phenotype expression may be due to the adsorption of a high quantity of fibronectin from serum onto the reacted bioactive glass surface. This finding is consistent with other data of conditioned BG surfaces on the effect of osteoblast adhesion, proliferation and differentiation (El-Ghannam *et al.*, 1997, 1999).

The effects of fibronectin (Fn) adsorption and glass surface reaction stage on the attachment of osteoblast-like cells (ROS 17/2.8) to bioactive glass were studied by García *et al.* (Garcia *et al.*, 1998). Bioactive glass disks were pretreated in a simulated

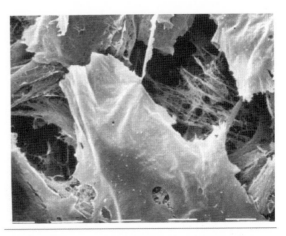

Figure 8.10 Scanning electron micrograph of the bioactive glass surface inoculated with neonatal rat calvaria osteoblasts for 1 week. The glass surface was covered by collagen fibrils and bone-like tissue. × 1,000 (Ducheyne *et al.* (1994). *Journal of Cellular Biochemistry*, 56: 162–167).

physiologic solution to produce three reaction layers: unreacted glass (BG0), amorphous calcium phosphate (BG1d), and carbonated hydroxyapatite (BG7d). Synthetic hydroxyapatite (sHA) and nonreactive borosilicate glass (CG) were used as controls. A spinning disk device which applied a linear range of forces to attached cells while maintaining uniform chemical conditions at the interface was used to quantify cell adhesion. The number of adherent cells decreased in a sigmoidal fashion with applied force, and the resulting detachment profile provided measurements of adhesion strength. For the same amount of adsorbed Fn, cell adhesion was higher on surface-reacted bioactive glasses (BG1d and BG7d) than on BG0, CG, and sHA (Figure 8.11). These data clearly demonstrate that the surface treatment of bioactive glass results in enhanced fibronectin-mediated osteoblast-like cell attachment. Since the surface treatment did not affect the amount of adsorbed fibronectin, these data suggested that the observed increase in attachment strength resulted from differences in fibronectin conformation.

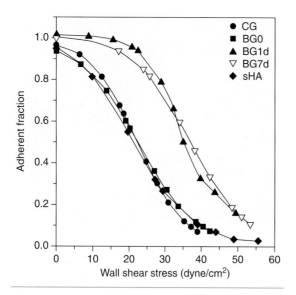

Figure 8.11 Cell detachment profiles for substrates coated with 0.1 mg/mLFn (10 dyne/cm^2 = 1N/m^2). Cell adhesion was significantly higher on reacted bioactive (BG1d, BG7d) than on unreacted bioactive (BG0) and control (CG) glasses and stoichiometric hydroxyapatite (sHA) (Garcia *et al.* (1998). *J Biomed Mat Res*, 40: 48–56).

Dissolution products of bioactive glass play an important role in stimulation of osteogenesis. Xynos *et al.* (2000a, 2000b) investigated the relationship between stimulation of osteogenesis and ionic products dissolved from bioactive glass 45S5. They observed that these ionic products increased osteoblast proliferation and up-regulated gene expression of various factors regulating osteoblast growth and differentiation. Specifically, insulin-like growth factor II (IGF-II) was increased to 300% by exposure of the osteoblasts to the bioactive glass stimuli. IGF-II is the most abundant growth factor in bone and is also a known inducer of osteoblast proliferation *in vitro*. In another study, Yao *et al.* (2005) also observed solution-mediated enhancement of osteoblast differentiation. This effect was even more pronounced when cells were present on the bioceramic surfaces from which the dissolution products

originated. In summary, bioactive glass has a stimulatory effect on osteoprogenitor cells.

8.5 Future developments: tailoring the resorption kinetic of bioceramics for optimal bone regeneration

Degradation mechanisms of bioceramics play a pivotal role in the bone-bonding ability and in the excellent biocompatibility of these implants as they are resorbed by osteoclasts like bone mineral. In critical-sized defects, i.e. defects that are too large in size to be healed by natural bone healing process, adjusting the degradation kinetics with bone formation rate remains a main objective in bone tissue regeneration.

8.5.1 Pre-clinical and clinical evidences reports

Coral – a natural porous mineral composed of calcium carbonate – loaded with mesenchymal stem cells have shown a well orchestrated degradation-bone formation kinetics, while unloaded coral degraded entirely without bone formation (Petite *et al.*, 2000; Zhu *et al.*, 2006). However, regarding synthetic biomaterials, a synchronized and efficient degradation/bone formation rate is not yet achieved. After 6 years implantation of a tissue engineered constructs based on human mesenchymal stem cells and hydroxyapatite scaffolds in three patients, Matsrogiacomo *et al.* (2005) reported that the scaffold is still visible by X-ray (see Figure 8.12). Although the penetration and the presence of bone through the scaffold can toughen the ceramic (Chistolini *et al.*, 1999), a brittle body in bone may cause instability. Besides the low solubility of hydroxyapatite which could explain its presence after many years of implantation, it has often been observed that, as soon as bone has formed on the surface of the calcium phosphate ceramics, the degradation process is decreased (de Bruijn *et al.*, 1994; Dhert *et al.*, 1993, 1998; Rahbek *et al.*, 2004).

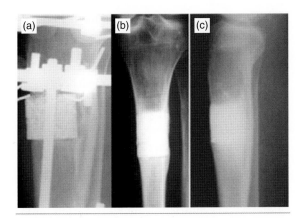

Figure 8.12 Tibia repair by tissue engineering approach in a human subject. Radiographs obtained immediately post (a), 18 months (b), and 5.5 years after surgery (Mastrogiacomo *et al.* (2005). *Orthod canio Facial Res*, 8: 277–284).

8.5.2 Tailoring the dissolution properties of calcium phosphate

In principle: (i) mixing at various ratios a low solubility (hydroxyapatite) phase with a highly soluble phase (amorphous, tricalcium phosphate) resulting in biphasic calcium phosphates ceramics (BCP) (Legeros *et al.*, 2003), (ii) including additives (Mg, F, CO$_3$) in a given crystalline phase (Dhert *et al.*, 1993), (iii) selecting different calcium phosphate phases (amorphous, DCPD, OCP, HA, TCP) (Barrere *et al.*, 2000), or (iv) changing the crystallinity and pore size of the ceramic are the options for tailoring the degradation kinetics of calcium ceramics from a few weeks to a few years. In theory and in practice, one can change the resorption kinetics of calcium phosphate ceramics. However, no universal degradable scaffolds have yet been developed, and their degradation properties should be designed depending on their application. For example, in non-loading situations, calcium phosphates are used as granules or cement. Their composition can virtually indefinitely vary with regard to their structure, the incorporation of additives and the mixture of phases. The relevance of this tuning is illustrated in

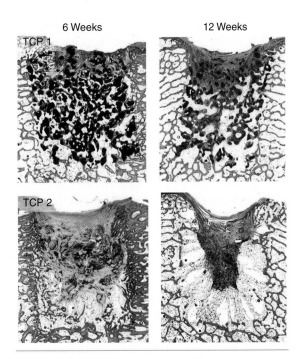

Figure 8.13 Light microscopy pictures of macroporous calcium phosphate ceramics implanted in goat's femoral condyle (magnification ×2). This figure represents the tunability of degradation/bone formation of two ceramics of similar composition (TCP), but different synthesis method. TCP 1 degrades slower than TCP2, allowing a continuous bone growth from the host bone towards the center of the implant, suggesting a full bone regeneration in the above defect. While the too fast resorbing TCP2 does not support bone ingrowth, instead the defect is filled with fibrous tissues (Courtesy of Dr. Yuan Huipin).

Figure 8.13. Two macroporous tricalcium phosphates, TCP-1 and TCP-2 obtained by different fabrication procedures were implanted in the femoral condyle of goats. After 6 weeks of implantation, the resorption of TCP-1 and TCP-2 were significantly different. While TCP-1 was still visible in the bone defect and bone was found through the macroporous structure, TCP-2 was fully resorbed and only fibrous tissue observed in the bone defect. After 12 weeks, TCP-1 had resorbed compared to the first implantation time in favor of bone growth, while only fibrous tissue was observed in the bone defect where TCP-2 was implanted.

In the case of hip implants, in load bearing applications, metals are the sole current option to sustain mechanical loading. Calcium phosphates applied as coatings on metallic prostheses have been recorded as one of the most highly successful clinical methods with regard to hip arthroplasty (Epinette and Manley, 2004). These coatings significantly accelerate bone growth on to the metallic implant, improve their fixation, and extend the prostheses longevity. Extensive animal studies have been conducted taking into account bone formation rate versus resorption rate and versus mechanical stability. By changing the coating's parameters (technique, temperature, composition) one can change the coating's degradation characteristics. With regard to hip prostheses, the calcium phosphate coating's resorption kinetic holds a certain paradigm. Soluble coatings enhance bone formation at the early implantation stage, inducing a fast early fixation; but soluble coatings may in a second stage lead to mechanical instability between the metallic implant and the surrounding bone. Accordingly, insoluble coatings delay bone fixation, i.e. the first stage mechanical stability, but insoluble coatings stabilize the long-term fixation of the implant with the surrounding tissues (de Bruijn *et al.*, 1994; Dhert *et al.*, 1993, 1998; Rahbek *et al.*, 2004). So far, there is therefore no rationale on: (i) what an optimal dissolution rate should be, and (ii) how we should design such scaffold.

8.5.3 Cellular contribution for tailoring the degradation properties

By changing the composition of the ceramics, it is possible to estimate their dissolution rate. *In vitro*, it has been observed that an elevated release of calcium ions, i.e. an accelerated dissolution rate, may inhibit osteoclastic activity (Doi *et al.*, 1999). Regarding the *in vivo* degradation of the highly soluble tricalcium phosphate ceramic, the observations of Zerbo *et al.* (2005) confirm the dominant dissolution mechanism

versus the osteoclastic resorption mechanism. By using appropriate techniques of osteogenic immunolocalization with TRAP activity staining of osteoclasts, we could get an insight into this 'competition' between cell-mediated and fluid-mediated degradation and bone formation (Zerbo *et al.*, 2005). However, to date, there is a lack of knowledge on the cellular contribution of ceramic degradation *in vivo* of well-characterized materials with well-defined dissolution properties. Coincidently, Habibovic *et al.* (2005) observed multinucleated cells in contact with an osteoinductive material after 6 weeks of implantation in goat's muscle (Figure 8.14). Further investigations on the role of bone resorbing cells on the mechanisms of bone formation (ectopically and orthotopically) should be emphasized: (i) on one hand, this could give more insight on the osteoinduction mechanism of bioceramics; (ii) on the other

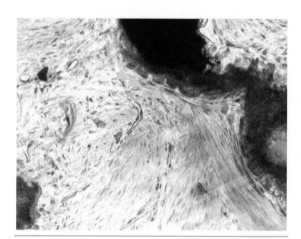

Figure 8.14 Light microscopy picture of an osteoinductive scaffold (octacalcium phosphate coated porous titanium scaffold) after 6 weeks of implantation in goat's muscle (magnification ×20). In this implant, bone was found in contact with the scaffold. Resorbing multinucleatedcells (▲) were found at the surface of the octacalcium phosphate coating. No specific staining has been performed, enabling more complete identification of these cells. Would resorption of an osteoinductive material be a key factor in the osteoinduction mechanism?

hand, the coordination of bone resorbing and bone forming cells could play a pivotal role in a well-orchestrated degradation scheme combining bone formation and scaffold resorption.

8.6 Summary

1. From a material science point of view, bone is a composite material composed of collagen fibers along which are organized calcium phosphates nanocrystals. Like bone mineral, calcium phosphate ceramics and bioglasses contain calcium and phosphate ions.
2. The typical degradation mechanisms of calcium phosphate ceramics and bioglasses occur via physicochemical dissolution and osteoclastic resorption. The mechanical degradation of these bioceramics is the result of both previous degradation mechanisms.
3. Associated with the surface dissolution, the bioceramic's surface undergoes a precipitation mechanism characterized by the formation of a carbonate apatite layer. By classical material characterization techniques (infra-red or X-ray diffraction), the structure of this layer is comparable to bone mineral. Consequently, physicochemical degradation does not mean necessarily loss of mass.
4. Calcium phosphates and bioglasses are too brittle and subjected to fatigue to be used as such in load bearing applications like hip prostheses or dental implants.
5. Calcium Phosphates and bioglasses have unique bone-bonding properties related to their superficial phase transformation mechanisms involving dissolution and precipitation.
6. What happens *in vitro* does not necessarily represents what happen *in vivo*, and vice versa.
7. Osteoclasts initiate bone remodeling process, i.e removal of old bone and formation of new bone.
8. 1% deviation in the biomaterial's characteristics, e.g. silica content regarding bioglasses or calcium-to-phosphorus ratio, can affect the physicochemical and biological profile of the implant.
9. For bone tissue engineering, we aim to design degradable biomaterials that would disappear simultaneously with bone formation, without creating a mechanical instability in the defect until full bone regeneration.
10. From an engineering point of view, we can vary the bioceramics composition indefinitely and in a controlled-fashion to tailor their dissolution rate. However, living bodies and their parts do not work in such a controlled-fashion. This makes it difficult to establish rules on the degradation kinetics of a degradable biomaterial.

References

Anselme, K., Sharrock, P., *et al.* (1997). *In vitro* growth of human adult bone-derived cells on hydroxyapatite plasma-sprayed coatings. *J Biomed Mater Res*, 34(2): 247–259.

Arinzeh, T.L., Tran, T., *et al.* (2005). A comparative study of biphasic calcium phosphate ceramics for human mesenchymal stem-cell-induced bone formation. *Biomaterials*, 26(17): 3631–3638.

ASTM, Ed. (1988). General products, *chemical specialties and en used products*: Glass; Ceramic Whitewares. Philadelphia: Annual book of ASTM standards.

Barinov, S.M., Tumanov, S.V., *et al.* (2003). Environment effect on the strength of hydroxy- and fluorohydroxyapatite ceramics. *Inorganic Materials*, 39(8): 877–880.

Barralet, J., Akao, M., *et al.* (2000). Dissolution of dense carbonate apatite subcutaneously implanted in Wistar rats. *J Biomed Mat Res*, 49(2): 176–182.

Barrere, F., Stigter, M., *et al.* (2000). *In vitro* dissolution of various calcium-phosphate coatings on Ti6Al4V. Bioceramics Key Engineering Materials.

Barrere, F., van der Valk, C.M., *et al.* (2003). *In vitro* and *in vivo* degradation of biomimetic octacalcium phosphate and carbonate apatite coatings on titanium implants. *J Biomed Mater Res A*, 64A(2): 378–387.

Basle, M.F., Chappard, D., *et al.* (1993). Osteoclastic resorption of Ca-P biomaterials implanted in rabbit bone. *Calcif Tissue Int*, 53(5): 348–356.

Beckham, C.A., Greenlee, T.K., Jr. *et al.* (1971). Bone formation at a ceramic implant interface. *Calcified Tissue Research*, 8: 165–171.

Bonfield, W., Grynpas, M.D., *et al.* (1981). Hydroxyapatite Reinforced Polyethylene – A Mechanically Compatible

Implant Material for Bone-Replacement. *Biomaterials*, 2(3): 185–186.

Boskey, A. and Paschalis, E. (2001). Matrix proteins and biomineralization. Bone Engineering. J. Davies, Toronto, em square: 44–61.

Bunker, B.C. (1994). Molecular Mechanisms for Corrosion of Silica and Silicate-Glasses. *Journal of Non-Crystalline Solids*, 179: 300–308.

Bunker, B.C., Arnold, G.W., *et al.* (1983). Mechanisms for Alkali Leaching in Mixed-Na-K Silicate-Glasses. *Journal of Non-Crystalline Solids*, 58(2–3): 295–322.

Cazalbou, S., Combes, C., *et al.* (2004). Adaptative physicochemistry of bio-related calcium phosphates. *J Mater Chem*, 14(14): 2148–2153.

Cerruti, M., Greenspan, D., *et al.* (2005). Effect of pH and ionic strength on the reactivity of Bioglass ® 45S5. *Biomaterials*, 26(14): 1665–1674.

Chistolini, P., Ruspantini, I., *et al.* (1999). Biomechanical evaluation of cell-loaded and cell-free hydroxyapatite implants for the reconstruction of segmental bone defects. *J Mater Sci Mater Med*, 10(12): 739–742.

Christoffersen, J., Christoffersen, M.R., *et al.* (1997). Effects of strontium ions on growth and dissolution of hydroxyapatite and on bone mineral detection. *Bone*, 20(1): 47–54.

Clark, A.E., Hench, L.L., *et al.* (1976). The influence of surface chemistry on implant interface histology: a theoretical basis for implant materials selection. *J Biomed Mater Res*, 10: 161–174.

Combes, C. and Rey, C. (2002). Adsorption of proteins and calcium phosphate materials bioactivity. *Biomaterials*, 23(13): 2817–2823.

Daculsi, G., Legeros, R.Z., *et al.* (1990). Formation of Carbonate-Apatite Crystals after Implantation of Calcium-Phosphate Ceramics. *Calcif Tissue Int*, 46(1): 20–27.

Daculsi, G., Legeros, R.Z., *et al.* (1989). Transformation of Biphasic Calcium-Phosphate Ceramics Invivo – Ultrastructural and Physicochemical Characterization. *J Biomed Mater Res*, 23(8): 883–894.

Damien, C.J. and Parsons, J.R. (1991). Bone graft and bone graft substitutes: a review of current technology and applications. *J Appl Biomater*, 2(3): 187–208.

Davies, J. (1996). *In vitro* modeling of the bone/implant interface. *Anatomical Record*, 245(2): 426–445.

de Bruijn, J.D., Bovell, Y.P., *et al.* (1994). Osteoclastic resorption of calcium phosphates is potentiated in post-osteogenic culture conditions. *J Biomed Mater Res*, 28(1): 105–112.

de Bruijn, J.D., Bovell, Y.P., *et al.* (1994). Structural arrangements at the interface between plasma sprayed calcium phosphates and bone. *Biomaterials*, 15(7): 543–550.

de Bruijn, J.D., Davies, J.E., *et al.* (1992). *Biological responses to calcium phosphate ceramics. Bone bonding biomaterials* (Ducheyne, P., Kokubo, T. and van Blitterswijk, C.A., eds). Leiderdorp: Reed Healthcare communications, pp. 57–72.

de Bruijn, J.D., Klein, C.P., *et al.* (1992). The ultrastructure of the bone-hydroxyapatite interface *in vitro. J Biomed Mater Res*, 26(10): 1365–1382.

de Bruijn, J.D., van Blitterswijk, C.A., *et al.* (1995). Initial bone matrix formation at the hydroxyapatite interface *in vivo. J Biomed Mater Res*, 29(1): 89–99.

de Groot, K. (1983). Ceramics of calcium phosphates: preparation and properties. In *Bioceramics of calcium phosphate* (de Groot, K., ed.). CRC Press Inc, pp. 100–111.

den Hollander, W., Patka, P., *et al.* (1991). Macroporous calcium phosphate ceramics for bone substitution: a tracer study on biodegradation with 45Ca tracer. *Biomaterials*, 12(6): 569–573.

Dhert, W.J., Klein, C.P., *et al.* (1993). A histological and histomorphometrical investigation of fluorapatite, magnesiumwhitlockite, and hydroxylapatite plasma-sprayed coatings in goats. *J Biomed Mater Res*, 27(1): 127–138.

Dhert, W.J., Klein, C.P., *et al.* (1991). A mechanical investigation of fluorapatite, magnesiumwhitlockite, and hydroxylapatite plasma-sprayed coatings in goats. *J Biomed Mater Res*, 25(10): 1183–1200.

Dhert, W.J., Thomsen, P., *et al.* (1998). Integration of press-fit implants in cortical bone: a study on interface kinetics. *J Biomed Mater Res*, 41(4): 574–583.

Doi, Y., Iwanaga, H., *et al.* (1999). Osteoclastic responses to various calcium phosphates in cell cultures. *J Biomed Mater Res*, 47(3): 424–433.

Ducheyne, P., El-Ghannam, A., *et al.* (1994). Effect of bioactive glass templates on osteoblast proliferation and *in vitro* synthesis of bone-like tissue. [Review]. *Journal of Cellular Biochemistry*, 56: 162–167.

Ducheyne, P. and Hench, L. (1982). The processing and static mechanical-properties of meta fiber reinforced bioglass. *Journal of Materials Science*, (2): 595–606.

Ducheyne, P. and Qiu, Q. (1999). Bioactive ceramics: the effect of surface reactivity on bone formation and bone cell function. *Biomaterials*, 20(23/24): 2287–2303.

Ducheyne, P., Radin, S., *et al.* (1993). The effect of calcium-phosphate ceramic composition and structure on *in vitro* behavior.1. Dissolution. *J Biomed Mater Res*, 27(1): 25–34.

Dvorak, M.M., Siddiqua, A., *et al.* (2004). Physiological changes in extracellular calcium concentration directly control osteoblast function in the absence of calciotropic hormones. *Proc Natl Acad Sci USA*, 101(14): 5140–5145.

El-Ghannam, A. and Ducheyne, P. (2005). Biomaterials. In *Basic orthopaedic biomechanics and mechanobiology* (Mow, V.C. and Huiskes, R., eds). Philadelphia: Lippincott Williams and Wilkins, pp. 511–519.

El-Ghannam, A., Ducheyne, P., *et al.* (1997). Formation of surface reaction products on bioactive glass and their effects on the expression of the osteoblastic phenotype and the deposition of mineralized extracellular matrix. *Biomaterials*, 18: 295–303.

El-Ghannam, A., Ducheyne, P., *et al.* (1997). Porous bioactive glass and hydroxyapatite ceramic affect bone cell function *in vitro* along different time lines. *J Biomed Mater Res*, 36(2): 167–180.

El-Ghannam, A., Ducheyne, P., *et al.* (1999). Effect of serum proteins on osteoblast adhesion to surface-modified bioactive glass and hydroxyapatite. *J Orthop Res*, 17(3): 340–345.

Elliot, J.C. (1994). *Structure and Chemistry of the Apatites and other Calcium Orthophosphates*. Amsterdam: Elsevier.

Epinette, J.A. and Manley, M.T. Eds. (2004). Fifteen Years of Clinical Experience with Hydroxyapatite Coatings in Joint Arthroplasty. Paris: Springer.

Furedi-Milhofer, H., Brecevic, L., *et al.* (1976). Crystal growth and phase transformation in the precipitation of calcium phosphates. *Faraday Discuss Chem Soc*, (61): 184–190.

Garcia, A.J., Ducheyne, P., *et al.* (1998). Effect of surface reaction stage on fibronection-mediated adhesion of osteoblast-like cells to bioactive glass. *J Biomed Mater Res*, 40: 48–56.

Gauthier, O., Bouler, J.M., *et al.* (1999). Kinetic study of bone ingrowth and ceramic resorption associated with the implantation of different injectable calcium-phosphate bone substitutes. *J Biomed Mater Res*, 47(1): 28–35.

Geesink, R.G., de Groot, K., *et al.* (1987). Chemical implant fixation using hydroxyl-apatite coatings. The development of a human total hip prosthesis for chemical fixation to bone using hydroxyl-apatite coatings on titanium substrates. *Clin Orthop Relat Res*, (225): 147–170.

Gomi, K., Lowenberg, B., *et al.* (1993). Resorption of sintered synthetic hydroxyapatite by osteoclasts *in vitro*. *Biomaterials*, 14(2): 91–96.

Greenlee, T.K. Jr., Beckham, C.A., *et al.* (1972). Glass ceramic bone implants. *J Biomed Mater Res*, 6: 235–244.

Habibovic, P., Li, J., *et al.* (2005). Biological performance of uncoated and octacalcium phosphate-coated Ti6Al4V. *Biomaterials*, 26(1): 23–36.

Habibovic, P., Sees, T.M., *et al.* (2006). Osteoinduction by biomaterials-Physicochemical and structural influences. *J Biomed Mater Res A*, 77(4): 747–762.

Habibovic, P., Yuan, H., *et al.* (2005). 3D microenvironment as essential element for osteoinduction by biomaterials. *Biomaterials*, 26(17): 3565–3575.

Hench, L. (2001). Ethical issues of implant. In *Science, Faith and Ethics* (Hench, L., ed.). London: Imperial College Press, pp. 84–118.

Hench, L.L. (1991). Bioceramics: from concept to clinic. J Am Cer Soc, 74(7): 1487–1510.

Hench, L.L., Splinter, R. J., *et al.* (1971). Mechanisms of interfacial bonding between ceramics and bone. *J Biomed Mater Res Symp*, 2: 117–141.

Hench, L.L. and Wilson, J. (1984). Surface-active biomaterials. *Science*, 226(4675): 630–636.

Heughebaert, M., LeGeros, R.Z., *et al.* (1988). Physicochemical characterization of deposits associated with HA ceramics implanted in nonosseous sites. *J Biomed Mater Res*, 22(3 Suppl): 257–268.

Heymann, D., Pradal, G., *et al.* (1999). Cellular mechanisms of calcium phosphate ceramic degradation. *Histol Histopathol*, 14(3): 871–877.

Hunter, G.K., Hauschka, P.V., *et al.* (1996). Nucleation and inhibition of hydroxyapatite formation by mineralized tissue proteins. *Biochem J*, 317(Pt 1): 59–64.

Hyakuna, K., Yamamuro, T., *et al.* (1989). The influence of calcium phosphate ceramics and glass-ceramics on cultured cells and their surrounding media. *J Biomed Mater Res*, 23(9): 1049–1066.

Hyakuna, K., Yamamuro, T., *et al.* (1990). Surface reactions of calcium phosphate ceramics to various solutions. *J Biomed Mater Res*, 24(4): 471–488.

Ikeuchi, M., Ito, A., *et al.* (2003). Osteogenic differentiation of cultured rat and human bone marrow cells on the surface of zinc-releasing calcium phosphate ceramics. *J Biomed Mater Res A*, 67(4): 1115–1122.

Ito, A., Kawamura, H., *et al.* (2002). Resorbability and solubility of zinc-containing tricalcium phosphate. *J Biomed Mater Res*, 60(2): 224–231.

Johnsson, M.S.A., Paschalis, E., *et al.* (1991). Kinetics of mineralization, demineralization and transformation of calcium phosphates at mineral and protein surfaces. In *The Bone-Biomaterial Interface* (Davies, J. E., ed.). Toronto, Canada: University of Toronto Press, pp. 68–75.

Kartsogiannis, V. and Ng, K.W. (2004). Cell lines and primary cell cultures in the study of bone cell biology. *Mol Cell Endocrinol*, 228(1-2): 79–102.

Kaufmann, E.A., Ducheyne, P., *et al.* (2000). Initial events at the bioactive glass surface in contact with protein containing solutions. *J Biomed Mater Res*, 52(825): 830.

Kawamura, H., Ito, A., *et al.* (2000). Stimulatory effect of zinc-releasing calcium phosphate implant on bone formation in rabbit femora. *J Biomed Mater Res*, 50(2): 184–190.

Kawamura, H., Ito, A., *et al.* (2003). Long-term implantation of zinc-releasing calcium phosphate ceramics in rabbit femora. *J Biomed Mater Res A*, 65(4): 468–474.

Kawasaki, K., Kambara, M., *et al.* (2003). A comparison of the adsorption of saliva proteins and some typical proteins onto the surface of hydroxyapatite. *Colloids Surf B-Biointerfaces*, 32(4): 321–334.

Knabe, C., Berger, G., *et al.* (2004). Effect of rapidly resorbable calcium phosphates and a calcium phosphate bone cement on the expression of bone-related genes and proteins *in vitro*. *J Biomed Mater Res A*, 69(1): 145–154.

Kokubo, T. (1993). Bioactivity of glasses and glass ceramics. In *Bone-bonding Biomaterials* (Ducheyne, P., Kokubo, T. and van Blitterswijk, C.A., eds). Leiderdorp, The Netherlands: Reed Healthcare Communications, pp. 31–46.

Koutsopoulos, S. and Dalas, E. (2000). The effect of acidic amino acids on hydroxyapatite crystallization. *J Crystal Growth*, 217(4): 410–415.

Kroese-Deutman, H.C., Ruhe, P.Q., *et al.* (2005). Bone inductive properties of rhBMP-2 loaded porous calcium phosphate cement implants inserted at an ectopic site in rabbits. *Biomaterials*, 26(10): 1131–1138.

Lai, W., Garino, J., *et al.* (2005). Excretion of resorption products from bioactive glass implanted in rabbit muscle. *J Biomed Mater Res Part A*, 75A(2): 398–407.

Le Huec, J.C., Clement, D., *et al.* (1998). Evolution of the local calcium content around irradiated beta-tricalcium phosphate ceramic implants: *in vivo* study in the rabbit. *Biomaterials*, 19(7–9): 733–738.

Lebugle, A., Rodrigues, A., *et al.* (2002). Study of implantable calcium phosphate systems for the slow release of methotrexate. *Biomaterials*, 23(16): 3517–3522.

Leeuwenburgh, S., Layrolle, P., *et al.* (2001). Osteoclastic resorption of biomimetic calcium phosphate coatings *in vitro*. *J Biomed Mater Res*, 56(2): 208–215.

LeGeros, R. (1991). *Calcium Phopshates in oral biology and medicine*. San Fransisco: Calif, Karger.

LeGeros, R.Z. (2002). Properties of osteoconductive biomaterials: calcium phosphates. *Clin Orthop Relat Res*, (395): 81–98.

Legeros, R.Z., Lin, S., *et al.* (2003). Biphasic calcium phosphate bioceramics: preparation, properties and applications. *J Mater Sci Mater Med*, 14(3): 201–209.

Liu, Y., de Groot, K., *et al.* (2005). BMP-2 liberated from biomimetic implant coatings induces and sustains direct ossification in an ectopic rat model. *Bone*, 36(5): 745–757.

Liu, Y., Hunziker, E.B., *et al.* (2003). Proteins incorporated into biomimetically prepared calcium phosphate coatings modulate their mechanical strength and dissolution rate. *Biomaterials*, 24(1): 65–70.

Lu, J., Descamps, M., *et al.* (2002). The biodegradation mechanism of calcium phosphate biomaterials in bone. *J Biomed Mater Res*, 63(4): 408–412.

Mahmood, T.A. and Davies, J.E. (2000). Incorporation of amino acids within the surface reactive layers of bioactive glass *in vitro*: an XPS study. *Journal of Materials Science-Materials in Medicine*, 11(1): 19–23.

Mastrogiacomo, M., Muraglia, A., *et al.* (2005). Tissue engineering of bone: search for a better scaffold. *Orthod Craniofac Res*, 8(4): 277–284.

Meleti, Z., Shapiro, I.M., *et al.* (2000). Inorganic phosphate induces apoptosis of osteoblast-like cells in culture. *Bone*, 27(3): 359–366.

Midy, V., Dard, M., *et al.* (2001). Evaluation of the effect of three calcium phosphate powders on osteoblast cells. *J Mater Sci Mater Med*, 12(3): 259–265.

Mirtchi, A.A., Lemaitre, J., *et al.* (1989). Calcium phosphate cements: study of the beta-tricalcium phosphate-monocalcium phosphate system. *Biomaterials*, 10(7): 475–480.

Monchau, F., Lefevre, A., *et al.* (2002). *In vitro* studies of human and rat osteoclast activity on hydroxyapatite, beta-tricalcium phosphate, calcium carbonate. *Biomol Eng*, 19(2–6): 143–152.

Neo, M., Nakamura, T., *et al.* (1993). Transmission electron microscopic study of apatite formation on bioactive ceramics *in vivo*. In *Bone-bonding Materials* (Ducheyne, P., Kokubo, T. and van Blitterswijk, C.A., eds). Leiderdorp: Reed Healthcare Communications, pp. 111–120.

Nilsson, M., Fernandez, E., *et al.* (2003). The effect of aging an injectable bone graft substitute in simulated body fluid. *Bioceramics 15 Key Engineering Materials* (B. Ben Nissan, D. Sher, and W. Walsh, eds), pp. 403–406.

OBannon, L. S. (1984). *Dictionary of Ceramic Science and Engineering*. New-York: Plenum Press.

Ofir, P.B.Y., Govrin-Lippman, R., *et al.* (2004). The influence of polyelectrolytes on the formation and phase transformation of amorphous calcium phosphate. *Crystal Growth & Design*, 4(1): 177–183.

Okazaki, M., Takahashi, J., *et al.* (1982). Crystallinity, solubility, and dissolution rate behavior of fluoridated CO3 apatites. *J Biomed Mater Res*, 16(6): 851–860.

Ooms, E.M., Wolke, J.G., *et al.* (2002). Trabecular bone response to injectable calcium phosphate (Ca-P) cement. *J Biomed Mater Res*, 61(1): 9–18.

Petite, H., Viateau, V., *et al.* (2000). Tissue-engineered bone regeneration. *Nat Biotechnol*, 18(9): 959–963.

Pilliar, R.M., Filiaggi, M.J., *et al.* (2001). Porous calcium polyphosphate scaffolds for bone substitute appli-cations – *in vitro* characterization. *Biomaterials*, 22(9): 963–972.

Porter, A.E., Botelho, C.M., *et al.* (2004). Ultrastructural comparison of dissolution and apatite precipitation on hydroxyapatite and silicon-substituted hydroxyapatite *in vitro* and *in vivo*. *J Biomed Mater Res A*, 69(4): 670–679.

Porter, A.E., Patel, N., *et al.* (2004). Effect of sintered silicate-substituted hydroxyapatite on remodelling processes at the bone-implant interface. *Biomaterials*, 25(16): 3303–3314.

Porter, N.L., Pilliar, R.M., *et al.* (2001). Fabrication of porous calcium polyphosphate implants by solid freeform fabrication: a study of processing parameters and *in vitro* degradation characteristics. *J Biomed Mater Res*, 56(4): 504–515.

Radin, S., Ducheyne, P., *et al.* (1998). Effect of serum proteins and osteoblasts on the surface transformation of a calcium phosphate coating: a physicochemical and ultrastructural study. *J Biomed Mater Res*, 39(2): 234–243.

Radin, S., Ducheyne, P., *et al.* (2000). *In vitro* transformation of bioactive glass granules into Ca-P shells. *J Biomed Mater Res*, 49(2): 264–272.

Radin, S., Ducheyne, P., *et al.* (1997). The effect of *in vitro* modeling conditions on the surface reactions on bioactive glass. *J Biomed Mater Res*, 37: 363–375.

Radin, S.R. and Ducheyne, P. (1993). The effect of calcium phosphate ceramic composition and structure on *in vitro* behavior. II. Precipitation. *J Biomed Mater Res*, 27(1): 35–45.

Radin, S.R. and Ducheyne, P. (1994). Effect of bioactive ceramic composition and structure on *in vitro* behavior. III. Porous versus dense ceramics. *J Biomed Mater Res*, 28(11): 1303–1309.

Rahbek, O., Overgraad, S., *et al.* (2004). Calcium phosphate coatings for implant fixation. In *Fifteen Years of Clinical Experience with Hydroxyapatite Coatings in Joint Arthroplasty* (Epinette, J.A. and Manley, M.T., eds). Paris: Springe, pp. 35–51.

Ratner, B.D. and Castner, D.G. (1997). *Surface Analysis – The Principal Techniques V*. J.C. New York: John Wiley & Sons Ltd. pp. 43–98.

Raynaud, S., Champion, E., *et al.* (1998). Dynamic fatigue and degradation in solution of hydroxyapatite ceramics. *J Mater Sci Mater Med*, 9(4): 221–227.

Raynaud, S., Champion, E., *et al.* (2002). Calcium phosphate apatites with variable Ca/P atomic ratio III. Mechanical properties and degradation in solution of hot pressed ceramics. *Biomaterials*, 23(4): 1081–1109.

Redey, S.A., Razzouk, S., *et al.* (1999). Osteoclast adhesion and activity on synthetic hydroxyapatite, carbonated hydroxyapatite, and natural calcium carbonate: relationship to surface energies. *J Biomed Mater Res*, 45(2): 140–147.

Ripamonti, U. (1996). Osteoinduction in porous hydroxyapatite implanted in heterotopic sites of different animal models. *Biomaterials*, 17(1): 31–35.

Rouahi, M., Champion, E., *et al.* (2006). Physicochemical characteristics and protein adsorption potential of hydroxyapatite particles: influence on *in vitro* biocompatibility of ceramics after sintering. *Colloids Surf B Biointerfaces*, 47(1): 10–19.

Sakae, T., Hoshino, K., *et al.* (2000). *In vitro* interactions of bone marrow cells with carbonate and fluoride containing apatites. *Bioceramics Key Engineering Materials* 192–191, Bologna.

Sakae, T., Ookubo, A., *et al.* (2003). Bone formation induced by several carbonate- and fluoride-containing apatite implanted in dog mandible. *Bioceramics Key Engineering Materials* 240–242, Sidney.

Schepers, E., Declercq, M., *et al.* (1991). Bioactive glass particulate material as a filler for bone lesions. *J Oral Rehab*, 18: 439–452.

Schepers, E. and Ducheyne, P. (1997). Bioactive glass granules of a narrow size range for the treatment of oral and bony defects: A twenty-four month animal experiment. *Journal of Oral Rehabilitation*, 24(3): 171–181.

Schepers, E.J.G., Ducheyne, P., *et al.* (1993). Bioactive glass particles of narrow size range: A new material for the repair of bone defects. *Implant Dentistry*, 2: 151–156.

Sepulveda, P., Jones, J.R., *et al.* (2002). *In vitro* dissolution of melt-derived 45S5 and sol-gel derived 58S bioactive glasses. *J Biomed Mater Res*, 61(2): 301–311.

Sharpe, J.R., Sammons, R.L., *et al.* (1997). Effect of pH on protein adsorption to hydroxyapatite and tricalcium phosphate ceramics. *Biomaterials*, 18(6): 471–476.

Shen, H., Tan, J., *et al.* (2004). Surface-mediated gene transfer from nanocomposites of controlled texture. *Nature Materials*, 3(8): 569–574.

Shinn-Jyh Dinga, C.-W.W., David Chan-Hen Chenb and Hsien-Chang Chang, (2005). *In vitro* degradation behavior of porous calcium phosphates under diametral compression loading. *Ceramics International*, 31(5): 691–696.

Spivak, J.M., Neuwirth, M.G., *et al.* (1994). Hydroxyapatite enhancement of posterior spinal instrumentation fixation. *Spine*, 19(8): 955–964.

Stigter, M., Bezemer, J., *et al.* (2004). Incorporation of different antibiotics into carbonated hydroxyapatite coatings on titanium implants, release and antibiotic efficacy. *J Control Release*, 99(1): 127–137.

Tang, R.K., Nancollas, G.H., *et al.* (2001). Mechanism of dissolution of sparingly soluble electrolytes. *J Am Chem Soc*, 123(23): 5437–5443.

Thompson, I.D. and Hench, L.L. (1998). Mechanical properties of bioactive glasses, glass-ceramics and composites. Proceedings of the Institution of Mechanical Engineers Part H. *Journal of Engineering in Medicine*, 212(H2): 127–136.

Urist, M.R., Nilsson, O., *et al.* (1987). Bone regeneration under the influence of a bone morphogenetic protein (BMP) beta tricalcium phosphate (TCP) composite in skull trephine defects in dogs. *Clin Orthop Relat Res*, (214): 295–304.

Vaahtio, M., Peltola, T., *et al.* (2005). Osteoclast response of biomimetically processed silica and carbonate containing calcium phosphate layers on bioactive glass S53P4. *Key Engineering Materials* 284-286: 549–552.

Vogel, M., Voigt, C., *et al.* (2004). Development of multinuclear giant cells during the degradation of Bioizlass (R) particles in rabbits. *J Biomed Mater Res Part A*, 70A(3): 370–379.

Wang, M. (2004). Bioactive Materials and Processing. In *Biomaterials and Tissue Engineering* (Shi, D., ed.). Berlin/New York: Springer, pp. 1–87.

Wang, M., Hench, L.L., *et al.* (1998). Bioglass (R) high density polyethylene composite for soft tissue applications: Preparation and evaluation. *J Biomed Mater Res*, 42(4): 577–586.

Wenisch, S., Stahl, J.P., *et al.* (2003). *In vivo* mechanisms of hydroxyapatite ceramic degradation by osteoclasts: fine structural microscopy. *J Biomed Mater Res A*, 67(3): 713–718.

Williams, D.F. (1999). *The Williams Dictionary of Biomaterials*. Liverpool University Press.

Wu, W.J. and Nancollas, G.H. (1998). The dissolution and growth of sparingly soluble inorganic salts: A kinetics and surface energy approach. *Pure and Applied Chemistry*, 70(10): 1867–1872.

Xynos, I.D., Edgar, A.J., *et al.* (2000a). Ionic products of bioactive glass dissolution increase proliferation of human osteoblasts and induce insulin-like growth factor II mRNA expression and protein synthesis. *Biochem Biophys Res Commun*, 276: 461–465.

Xynos, I.D., Hukkanen, M.V.J., *et al.* (2000a). Bioglass ® 45S5 stimulates osteoblast turnover and enhances bone formation *in vitro*: Implications and applications for bone tissue engineering. *Calcif Tis Int*, 67(4): 321–329.

Yamada, S., Heymann, D., *et al.* (1997). Osteoclastic resorption of calcium phosphate ceramics with different hydroxyapatite/beta-tricalcium phosphate ratios. *Biomaterials*, 18(15): 1037–1041.

Yamada, S., Nakamura, T., *et al.* (1994). Osteoclastic Resorption of Apatite Formed on Apatite-Containing and Wollastonite-Containing Glass-Ceramic by A Simulated Body-Fluid. *J Biomed Mater Res*, 28(11): 1357–1363.

Yao, J., Radin, S., *et al.* (2005). The effect of bioactive glass content on synthesis and bioactivity of composite poly (lactic-co-glycolic acid)/bioactive glass substrate for tissue engineering. *Biomaterials*, 26: 1935–1943.

Zerbo, I.R., Bronckers, A.L.J.J., *et al.* (2005). Localisation of osteogenic and osteoclastic cells in porous [β]-tricalcium phosphate particles used for human maxillary sinus floor elevation. *Biomaterials*, 26(12): 1445–1451.

Zhu, L., Liu, W., *et al.* (2006). Tissue-Engineered Bone Repair of Goat Femur Defects with Osteogenically Induced Bone Marrow Stromal Cells. *Tissue Eng.*, 12(3): 423–433.

Chapter 9
Biocompatibility

David Williams

Chapter objectives:

- To understand the basic concepts of biocompatibility in general
- To understand the evolution of conventional biomaterials on the basis of their biocompatibility
- To understand why the mechanisms of biocompatibility of tissue engineering scaffolds differ from those of implantable biomaterials

- To appreciate the differences between scaffolds and matrices on the basis of their biocompatibility characteristics
- To understand the evolution and potential uses of current scaffold and matrices

Edited by Van Blitterswijk, Lindahl, Thomsen, Williams, Hubbell and Cancedda.

9.1 Introduction

One of the most widely discussed factors that controls the overall performance of a tissue engineering product is the biocompatibility of the scaffold or matrix that is used. The main problem with the understanding of this particular subject has been the fact that biocompatibility has, until now, been largely concerned with implantable devices (and the performance of the biomaterials used in their construction) within the human body, where the intention has usually been related to the long term replacement of tissues. Within the context of tissue engineering, however, the specifications of the biomaterials used as scaffolds and matrices are very different to those for long term implantable devices. The precise requirements and characteristics of their biocompatibility will also be substantially different, although this has rarely been discussed. Indeed, it is a matter of concern that the materials used for the majority of tissue engineering scaffolds and matrices in recent years have been similar to those used for some implantable devices and drug delivery systems, largely on the basis that these materials have had prior regulatory approval with respect to such systems and have, putatively, been demonstrated to display biological safety in these other situations.

The most important characteristic that distinguishes a biomaterial from any other material is its ability to exist in contact with the relevant tissues, of the human body, or components of tissues, without causing an unacceptable degree of harm to those tissues or components. It has become clear that there are many ways in which materials and the components of tissues can interact such that this co-existence may be compromised, and the search for biomaterials that are able to provide for the best performance in devices has been based upon the acquisition of knowledge and understanding about these interactions.

These interactions are usually discussed in the broad context of the subject of biocompatibility. Biocompatibility is a word that is used extensively within biomaterials science, but there still exists a great deal of uncertainty about what it actually means and about the mechanisms that are subsumed within the phenomena that collectively constitute biocompatibility. As biomaterials are being used in increasingly diverse and complex situations, with applications now involving tissue engineering, as well as the longer established implantable medical devices, invasive sensors and drug delivery systems, this uncertainty over the mechanisms of biocompatibility is becoming a serious impediment to the development of these new techniques.

9.2 The evolution of current concepts of biocompatibility

Biocompatibility has traditionally been concerned with implantable devices that have been intended to remain within an individual for a long time. To those who were developing and using the first generation of implantable devices, during the years between 1940 and 1980, it was becoming increasingly obvious that the best performance biologically would be achieved with materials that were the least reactive chemically. Thus, within metallic systems the plain carbon steels, which demonstrated overt corrosion, were replaced by increasingly superior stainless steels, then by the strongly passivated cobalt-chromium alloys, titanium alloys and the platinum group metals. With polymers, the readily available and versatile nylons and polyesters were replaced by the more degradation resistant polyethylene, fluorocarbon polymers, acrylics and silicones. Consistent with this approach, the selection criteria for implantable biomaterials evolved as a list of events that had to be avoided, most of these originating from those events associated with the release of some products of corrosion or degradation and their biological activity, either locally or systemically. Materials were therefore selected, or occasionally developed, on the basis that they would be non-toxic, non-immunogenic, non-thrombogenic, non-carcinogenic, non-irritant and so on, such a list of negatives becoming, by default, the definition of biocompatibility.

Three factors initiated a re-evaluation of this position. The first was that it became obvious that the response to specific individual materials could vary from one situation to another. Thus biocompatibility could not solely be dependent on the material characteristics but also had to be defined by the situation in which the material is used (Williams, 2003). Secondly, an increasing number of applications required that the material should specifically react with the tissues rather than be ignored by them, as in the case of an inert material (Williams, 1999a). Thirdly, and in a similar context, some applications required that the material should degrade over time in the body rather than remain indefinitely (Williams, 1982), a point which has gained increasing relevance in drug delivery systems and tissue engineering scaffolds.

It was therefore considered that the requirement that biocompatibility, which was considered to be equated with biological safety, meant that the material should do the patient no harm, was no longer a sufficient pre-requisite, even for long term implantable devices. Accordingly, biocompatibility was redefined in 1987 as follows:

> Biocompatibility refers to the ability of a material to perform with an appropriate host response in a specific situation (Williams, 1999b)

This definition, which clearly places the word in the category of a concept rather than a practical descriptor of a process, is based on the three assumptions that a material has to perform and not simply exist in the tissues, that the response which it evokes has to be appropriate for the application, and that the nature of the response to a specific material and its appropriateness may vary from one situation to another.

This definition is of course very general and it may not be of any real help in advancing knowledge of biocompatibility. Indeed, it is true that it has not led to a greater understanding of specific mechanisms and individual processes. Moreover, it is likely that one concept cannot apply to all material-tissue interactions that pertain to widely varying applications, ranging from a drug eluting stent to a tissue engineering cartilage construct or an invasive biosensor. It is with this diversity in mind, and the wide ranging

potential mechanisms of interactions based both in materials science and biology, that a different paradigm of biocompatibility can be devised.

9.3 The agents of biocompatibility

The paradigm of biocompatibility outlined in this chapter involves the separate, but potentially interrelated, responses of the two phases of the biomaterial–tissue construct and the interfacial phenomena that come into play when they meet. Probably the most important underlying principle is that the mechanisms by which materials and human tissues respond to each other are not unique to this particular use but are merely variations of natural processes that occur within materials and biological sciences. Thus, in general, the response of a material to implantation in the human body will not involve totally new mechanisms not found in other environments, and the cellular and humoral responses of the body do not involve the cells and extracellular constituents performing in ways which are entirely non-physiological. The key to understanding biocompatibility is the determination of which chemical, biochemical, physiological, physical or other mechanisms become operative (and why), under the highly specific conditions associated with contact between biomaterials and the tissues or cells of the body, and what are the consequences of these interactions.

It is worth noting that there are several mediators of the biocompatibility of a material other than the characteristics of the material itself. Of great significance is the nature and quality of the clinical intervention that places the material into contact with the tissues. For implantable medical devices, the characteristics of the individual in or on whom the device is placed are also of considerable importance and it is to be anticipated that wide patient-to-patient variability will be seen. Age, sex, general health, physical mobility, lifestyle features and pharmacological status all contribute to this variation. The design of the device and the physical relationship between the device and the body play a significant role, as do the presence or absence of micro-organisms and endotoxins. Within the context of a tissue engineering scaffold, the source of the cells and the physical and mechanical conditions within a bioreactor, are expected to have significant effects.

In Box 1, the major material characteristics that may influence the host response are listed. These can be divided into characteristics of the bulk material and those of the surface. The majority of these characteristics are self evident although obviously some subsume a number of features. The elastic constants, for example, include Young's Modulus, shear and bulk moduli and Poisson's ratio. Crystallinity in polymers includes the degree of crystallinity and the nature of the molecular symmetry, whilst in metals it includes crystal structure and grain size.

Box 1 Material variables in biocompatibility

Bulk material composition, microstructure, morphology
Crystallinity and crystallography
Elastic constants, compliance
Surface chemical composition, chemical gradient, molecular mobility
Surface topography and porosity
Water content, hydrophobic–hydrophilic balance, surface energy
Corrosion parameters, ion release profile, metal ion toxicity
Polymer degradation profile, degradation product toxicity
Leachables, catalysts, additives, contaminants
Ceramic dissolution profile
Wear debris release profile, particle size
Sterility and endotoxins

When placed in or on the tissues of the body, a number of reactions to a material may be seen over time, and these are listed in Box 2. Some of these may constitute important determinants of the host response, whilst others are of greater importance in the functioning of the device. Within the host, in the majority of circumstances, we may envisage a sequence of events, potentially involving the interaction between proteins and other physiological macromolecules with the material surface, the initiation of inflammatory and/or immune responses, and then the repair and/or regeneration processes that may lead to stable equilibrium between material and host. This is the classical biocompatibility paradigm that has been discussed in one form or another over the last couple of decades (Williams, 2008).

We may try to rationalize the actual observations of biocompatibility phenomena on both clinical and experimental levels. A perception has developed recently that the evolution of biomaterials for these devices has been following a pattern in which so-called bio-inertness has been displaced by new concepts of bioactivity. Indeed, much has been written about the development of second and third generations of biomaterials on the basis of the desirability of intentional interactivity of the material with the host, either to assist in the incorporation of the device into the host or to achieve some specific functional activity. The introduction of bioactivity into the specification of biomaterials has to be predicated on mechanisms whereby specific biomaterials characteristics (such as identified in Box 1) control specific host responses (identified in Box 2) and that refinement of the former should lead to improvement of the latter and the production of better biocompatibility-based performance.

The actual evidence, however, would suggest otherwise. An analysis of the performance of clinical devices over several decades points unequivocally to

Box 2 Host response characteristics in biocompatibility

Protein adsorption and desorption characteristics
Complement activation
Platelet adhesion, activation and aggregation
Activation of intrinsic clotting cascade
Neutrophil activation
Fibroblast behavior and fibrosis
Microvascular changes
Macrophage activation, foreign body giant cell production
Osteoblast/osteoclast responses
Endothelial proliferation
Antibody production, lymphocyte behavior
Acute hypersensitivity/anaphylaxis
Delayed hypersensitivity
Genotoxicity, reproductive toxicity
Tumour formation

the conclusion that the best performances are seen with the use of materials that are as inert as possible and that most attempts to induce bioactivity or to intentionally or unintentionally deviate from inertness have led to poorer clinical performance. Over the years, most significant developments in biomaterials specifications have been concerned with the improvement to inertness, or the optimization of functional properties (e.g. mechanical or physical properties) without decreasing inertness.

Thus we can see that the successful long term implantable devices today use a smaller group of acceptable biomaterials than twenty years ago, and this group of materials has emerged as the preferred options for many, varied, applications. The majority of total joint replacement prostheses utilize cobalt–chromium alloys, titanium alloys, ultra high molecular weight polyethylene and alumina. A minority will have some component with a surface layer of hydroxyapatite and some prostheses are attached to bone by the use of a polymethylmethacrylate cement. The majority of mechanical heart valves involve the same two types of alloy with either a polyester or polytetrafluoroethylene-based sewing ring and a carbon or carbon coated leaflet or disc. Synthetic vascular grafts use the same polyester and polytetrafluoroethylene. Implanted

microelectronic devices use titanium for the can and either cobalt–chromium alloys or platinum group alloys for the leads and electrodes, with either silicone elastomer or polyurethane insulation. Intra-ocular lenses and other ophthalmological devices use either acrylics or silicones. Breast implants still just use silicone polymers.

In some situations, the search for optimal inertness has had to be tempered with the need for highly specific functional properties, such as the need to control the expansion and mechanical properties of intravascular stents, leading to the use of nickel–titanium shape memory alloys in some devices, and the need to optimize electrical insulation in certain pacemaker and implantable defibrillator components, where questions have been raised about the environmental stability of two of the better functional materials, polyurethanes and polyimides, under some circumstances.

With this history in mind, we can see that the vast majority of implantable devices utilize cobalt–chromium, titanium or platinum group metal-based alloys, polyethylene, acrylic or fluoropolymer-based thermoplastics, polyurethane or silicone-based thermoplastic elastomers, aromatic polyester or fluoropolymer textiles, aluminium oxide ceramic and carbon. Their use has been almost entirely predicated on the well proven

In the Discussion paragraph of this chapter an overview of factors of attention with regard to biocompatibility testing in tissue engineering is given. Tissue-biomaterial interactions, cell-biomaterial surface interactions, mechanical and molecular signaling, and incorporation of the material into the host are largely dependent on physico-chemical and structural properties of the material. Section 9.4 gives an overview of the scaffolds that can be used in tissue engineering. They all largely

Monomer composition	RA Day 6	Day 6 24 h RA pulse	Day 6;No RA	RA Day 1	Day 1 No RA:
100% 1					
70% 1, 30%*					
100% 3					
70% 3, 30% 1					
70% 3, 30% 18					
70% 3, 30% 21					
100% 6					
100% 13					
100% 7					
70% 7, 30% 4					
70% 7, 30%*					
100% 11					
70% 11, 30% 1					
70% 11, 30% 21					
100% 12					
70% 12, 30% 3					
70% 12, 30% 21					
100% 18					
70% 18, 30%*					
70% 18, 30% 13					
100% 21					
100% 23					
70% 23, 30% 1					
70% 23, 30% 21					

Polymer effects on human embryonic stem (hES) cell attachment, growth and proliferation. (a) Four million hES cell embryoid body day-6 cells were grown on the 'hit' polymer arrays in the presence of retinoic acid, the absence of retinoic acid and with a 24-h pulse of retinoic acid for 1 or 6 d. Cells were then stained for cytokeratin 7 (green), vimentin (red) and DNA (blue). Representative images at each time point for each polymer are shown for all conditions. RA, retinoic acid. (From Anderson D.G., Levenberg, S. and Langer, R. (2004). Nanoliter-scale synthesis of arrayed biomaterials and application to human embryonic stem cells, Nature Biotech, *22(7): 863–866.)*

differ in their chemical composition and degradation behavior leaving us with a complex array of parameters which influence biocompatibility of the tissue-engineered construct and determine its clinical success. This array is further complicated by the macrostructural (overall geometry, porosity, etc.) and microstructural (surface topography, microporosity, roughness, etc.) modification of the scaffold material. Making a choice of the appropriate tissue engineering scaffold for each application is therefore difficult, if not impossible by using the existing assays. That is why reliable high-throughput assays are needed for fast screening of large number of candidate materials. An example of the system that could be useful for this purpose is described in the paper by Anderson et al (2004). They have described an approach for rapid, nanoliter-scale synthesis of polymeric biomaterials and characterization of their interaction with cells. Over 1700 material-cell interactions were simultaneously screened which allowed for identification of a host of unexpected material effects and control over cell behavior. The described study was performed on polymeric materials in contact with human embryonic stem cells, but similar systems could be applied for other material types and relevant cells. Although high-throughput *in vitro* assays as the one described here can give relevant information on cell-material interactions, a future challenge lies in designing smart *in vivo* high-throughput systems as the difficulty to precisely mimic complex *in vivo* environments *in vitro* still limits the predictive value of the *in vitro* assay systems.

performances that balance appropriate mechanical or physical functionality with chemical and biological inertness. Almost every time a material with less than optimal inertness, either chemical or biological, has been introduced, it has produced poorer biocompatibility and, usually inferior clinical performance.

It is therefore possible to restate the paradigm of materials selection here as 'when selecting materials for long term implantable devices, choose the material that optimizes the functional properties of the device, consistent with maximum chemical and biological inertness'. Moreover, we can redefine biocompatibility specifically for the situation of long term implantable devices as:

The biocompatibility of a long term implantable medical device refers to the ability of the device to perform its intended function, with the desired degree of incorporation in the host, without eliciting any undesirable local or systemic effects in that host.

It is also noted, and particularly relevant, that a number of applications of biomaterials have required specifically and intentionally degradable materials for the completion of full function. These include those devices where it is required that dissolution or degradation occurs after they have successfully completed a temporary function, such as assisting in wound healing, and those where the degradation process is actually involved in the performance of the principal function, especially in the control of drug release. The specific requirements of degradation here are quite similar to those for long term implantable devices, however, since the principal objective is to degrade in a controlled manner without inflicting any detrimental effect on the surrounding tissues. The materials used for resorbable surgical sutures or degradable bone fracture plates are not intended to actively influence the wound healing process itself. There are, of course, some situations where it is intended that a degrading material should have some specific beneficial effect on the surrounding or newly developing tissue, especially with respect to the formation of new bone on the surface of an implantable device, which shall be discussed later in the context of tissue engineering constructs. In general, therefore, the desirable biocompatibility characteristics of intentionally biodegradable materials for implantation are not too different from those of long term stable materials, and may be considered to be as follows:

The biocompatibility of a biodegradable material for implantation in a host refers to the ability of the

Classical Experiment
In vivo biocompatibility studies

In the early days of implantable materials, determining the biocompatibility was generally based on histological and morphological observations and was directed towards the understanding of the cellular responses adjacent to the implant. The biocompatibility of a material within tissue was in general described in terms of acute and chronic inflammatory responses and fibrous capsule formation. Histological observation of tissue adjacent to the implanted material as a function of implantation time, such as in studies described by Gourley *et al.* (1978) was the commonly used method for investigating biocompatibility. A second approach was enzyme histochemical investigation to determine cell function at implantation site, as was introduced by Salthouse and Matlaga (1981). These approaches were in particular applicable for testing implantable devices that were intended to remain within an individual for a long time. While bioinert, stable materials were gradually replaced by the materials with the ability to exert a certain function upon implantation and degrade in time, different ways of testing biocompatibility were required. Not only the tissue and blood response to the biomaterial needed to be tested, but also the material response to cells, proteins and other factors which may be present at the material surface.

In 1983, Marchant and coworkers (Marchant *et al.*, 1983) designed a cage implantation system which could be used to appreciate the dynamic nature of cell response at implantation site. The design was such that resorbable polymer of interest, that is inserted into a porous stainless steel cage could be subcutaneously implanted. Such a system provided a means of taking samples of exudate surrounding the material within the cage by a syringe, without need of sacrificing the animal. This exudate could then be tested for quantitative and differential white cell counts (polymorphonuclear leukocytes, monocytes, macrophages), extracellular enzyme activity (alkaline phosphatase, acid phosphatase, prostatic acid phosphatase, etc.) and protein analyses. In addition to the exudate analysis, upon explantation, both changes on the implanted material within the cage and the fibrous tissue formed around the implant could be investigated. This system therefore served as a short-term method to evaluate selected aspects of the biocompatibility of the material, i.e. the influence of cells and enzymes on the material and the influence of the material on cells and enzymes in the inflammatory response.

device to perform its intended function, including an appropriate degradation profile, without eliciting any undesirable local or systemic effects in that host.

Much of biomaterials science research over the last couple of decades has been concerned with materials surfaces and their modification for improved biocompatibility. This attention to the surface is not illogical since it will be the surface of the device that contacts the host and should, *a priori*, have the most influence. However, it has to be said again that the evidence suggests that surface modification does not usually result in better performance of any implantable device, whatever its intended function and duration, unless it is directly devoted to improving chemical or biological

inertness. Surface treatments to metallic materials that enhance surface oxide stability (for example, through anodizing) may reduce metal ion release under steady state conditions, which can only be a good outcome.

The question has to be asked, however, whether the addition of any different functional groups to the surface layer of a material can have any influence on the interactions that will be taking place continuously over many years in the interfacial region. This is an immensely important question at the heart of the discussion of biocompatibility mechanisms. If we take into account the obvious fact the host compartment in the region of interest is an enormously complex, dynamic, living milieu of structural and metabolically active macromolecules that provide the support for many different types of cell, each with different functions and life cycles, all contained within an aqueous fluid medium of precise ionic content and concentration, we have to ask how a 'functionalized' material surface, which can only physically interact with a minute proportion of these host compartment entities, and which will either be covered with and passivated by, or actively digested or degraded by, certain of these macromolecules almost immediately, can possibly influence the response of the host in the long term. The answer has to be that in the vast majority of circumstances, material surface modifications have no influence at all on long term biocompatibility.

There are some exceptions to this based on the mechanistic possibility that the initial interaction with the surface will control subsequent events. Before discussing these, however, let us consider how material characteristics could possibly influence the features of the host response. It is particularly important to reflect on how any component of the tissue could recognize and be influenced by the characteristics of the material. Let us start with the chemical nature of the biomaterial. It would be entirely illogical to assume that any of the host components could distinguish whether they were in contact with a sample of polyethylene, polypropylene, polyetheretherketone or any other synthetic thermoplastic polymer after implantation. Nor could they distinguish any one oxide ceramic from another. The same arguments would apply to the possibility that host components could discriminate between samples of titanium, tantalum, zirconium, platinum and so on. Within all of these classes of materials, there appears to be no mechanistic possibility that any structural or morphological feature of the bulk material, such as the degree of crystallinity of a polymer or the crystal structure or grain size of metals or ceramics, could seriously affect the outcome of the host response evolution. There is one characteristic of the bulk material that can have an influence, and indeed is one that is rarely discussed except in the context of stress shielding in bone, and that is the elastic moduli. In the dynamic environment of the human body, tissue and organs are subjected to constant movement and the relative movement between tissues

and synthetic materials caused by differential stiffness can influence some parts of the response.

With all of these bulk characteristics per se seeming to be so irrelevant, we are left with the surface as the mediator of biocompatibility. It is very important to distinguish two separate aspects here, one being the physico-chemical characteristics of the surface layer itself and the other being the fact that the material surface provides the portal through which material and host components can be exchanged.

The former characteristics (many of which are of course inter-related) include surface charge, surface energy, the hydrophobic/hydrophilic balance, molecular mobility, and surface topography. Importantly the topography includes macro-, micro-, meso-, and nano-topography and embraces surface porosity. The host components that could conceivably be influenced by variations in these surface properties include proteins, which can theoretically undergo adsorption, desorption or conformational change processes on foreign surfaces, and cells which, depending on their nature and function, may adhere to surfaces (which may depend on the adsorbed protein layers), following which their functions may be modulated, or indeed following which they may change their phenotype. There is indeed a plethora of possible biological reactions that may take place at biomaterials surfaces. Whether or not these take place at the surfaces of implanted biomaterials is a different matter. The biomaterials literature contains many reports of attempts to quantitatively correlate parameters of material surfaces with the behavior of proteins and cells but while there are individual success stories in which some relationship between a parameter and a response can be identified under a narrow range of experimental conditions, no generic relationships of any relevance to long term biomaterials performance have emerged. There is no such thing as an optimal surface charge, energy, roughness, porosity or any other parameter which can be reliably used to modulate or even predict any macromolecular or cellular response at an implant surface. It is an immensely important question as to whether this will be any different for a tissue engineering scaffold material.

The second surface feature that controls biocompatibility is the role of that surface as a portal, or interface, between the bulk of the material and the host. For there to be any chemical or biological interaction between material and host there has to be an exchange of components, usually although not always involving some atoms, ions, molecules or larger fragments being released from the material, via the surface, into the host. The ability of the surface, or surface layer, to permit the passage of such species may be the most important factor in biocompatibility, and may be partially controlled by the deliberate modification of surfaces to make them more resistant to such passage, for example through deliberate oxidation, or more amenable to such passage, for example by increasing hydrophilicity.

9.4 Tissue engineering scaffolds and matrices

We now come to the considerations of biocompatibility of materials specifically used in tissue engineering scaffolds and matrices. Let us start by differentiating between these two types of structure, in the context of the requirements for a material within the tissue engineering process. The now classical tissue engineering paradigm involves deriving cells from an appropriate source, be that autologous or allogeneic, and be they fully differentiated or stem cells, and manipulating them *ex vivo* to produce the optimal cell characteristics, and then seeding these cells into a material substrate where they are stimulated to produce the desired new tissue. It should be borne in mind here that the material has a number of functions, which can be divided into three main types:

- To provide the overall shape to the construct in order to facilitate the generation of a useful size and volume of tissue.
- To facilitate the delivery of signals, both molecular and mechanical to the cells.
- To support the cells and optimize their function within the scaffold.

In many circumstances, although not necessarily all of them, the material should degrade at some time, with appropriate kinetics and metabolic consequences.

Of course not all tissue engineering processes follow this simple paradigm and there is a substantial move towards *in vivo* rather than *in vitro/ex vivo* regeneration. In other words, far more of the tissue regeneration will occur within the host rather than in an *in vitro* bioreactor. The principles of biocompatibility will, however, be much the same.

During the last couple of decades, two types of supporting structure have emerged as possible forms for tissue engineering. First, there is the scaffold, this being a microporous, usually three dimensional, structure, within which a cell suspension can be placed and where the cells themselves have the opportunity to attach to the free surfaces of the material, and the media, with nutrients and other agents, can circulate. Secondly there are the matrices, which are gels, and more specifically hydrogels, involving a network of structural, usually cross-linked, molecules, within a water-based viscous matrix. The principal advantages with the porous scaffold approach are the potential control over the morphology and mechanical characteristics, while the matrix has the very considerable advantage of resembling the extracellular matrix within which most cells normally reside. It should be borne in mind that a porous scaffold does not resemble the normal extracellular matrix and that those cells which are capable to expressing new tissue do not normally do so under these quite artificial conditions. It should also be noted that a three dimensional scaffold may well absorb fluid from the cell suspension medium and slowly form a gel, so that the distinction is not always clear-cut.

In general, it is possible to define a scaffold as a three dimensional microporous structure within which cells are cultivated for the purpose of generating new tissue, while a matrix is a three dimensional gel within which cells are cultivated for the purpose of generating new tissue. There are some circumstances in tissue engineering where a membrane rather than a three dimensional supporting structure is used, these being discussed a little later since they tend to have different functions and specifications.

When considering the specific functions of three dimensional supporting structures mentioned above, the provision of shape is not really a biocompatibility issue, but the other three functions can easily be seen as significantly related to biocompatibility.

First and foremost is the need for the material to directly influence the cells, persuading them to function in an optimal manner with respect to their expression of new extracellular matrix. Depending on cell type, this may be best achieved by some form of contact guidance, where the cell interaction with the material surface, including processes of adhesion, spreading and proliferation, may be precursors and determinants of processes of matrix production. With respect to stem cell populations, an extremely important characteristic would be the ability to influence their differentiation and the subsequent maintenance of the desired phenotype.

It would seem intuitively obvious that optimal guidance of cell behavior in this respect will not normally be achieved through chemical and biological inertness and immediately we see a conflict between the biocompatibility characteristics of materials for tissue engineering scaffolds and for implantable devices. An immensely important conundrum also becomes apparent here. If the internal material surfaces of a scaffold play a significant positive role in the control of cell behavior in a tissue engineering process, then it follows that the contact area between cells and surfaces should be maximized. If a material scaffold has a porosity in excess of 80%, as some do, then it is unlikely that the cells in suspension within the scaffold will come into direct contact with these surfaces to any significant extent. On the other hand, if the porosity is reduced in order to increase internal surface area, this may place unacceptable interference with fluid (including nutrients and biomolecules) flow, and with the generation of new tissue.

From a scaffold functionality perspective, therefore, it would seem that optimal performance will be achieved with the appropriate balance between the specific biological activity of the scaffold material and

the morphology and architecture of the porous structure. With respect to a hydrogel, it should be possible for each cell to be surrounded by the type of environment that it is naturally used to, such that there can be signalling to the cell via the normal molecular or morphological processes. This suggests that the gel should have a physical and molecular structure that mostly resembles its normal environment, which de facto becomes a biocompatibility specification.

The second biocompatibility-related factor is the ability to support molecular and mechanical signalling. With the matrix type of support, this is very similar to the qualities mentioned in the previous paragraph, with the additional caveats that the gel should be able to control the delivery to the cells of any biomolecules required for tissue expression, including growth factors, and should have the appropriate mechanical properties to allow relevant strains to be transmitted to cell membranes. With a scaffold, it is far more difficult for the intrinsic material properties to directly influence mechanical signalling. Indeed it is more relevant here to consider that these biocompatibility factors are more controlled by fluid flow within the porous structure than any other characteristic.

Thirdly, we have to consider the degradation profile of those scaffolds and matrices that are intended to biodegrade during or after the tissue generation process. This particularly concerns the avoidance of any adverse effects associated with the degradation processes, which will be largely controlled by inflammation and immune responses. In principle, the mechanisms here are quite straightforward. As a material degrades within a physiological environment, which could be either *in vivo* or *in vitro*, several different types or formats of degradation product could be generated. Assuming for a moment that the biomaterial is a synthetic polymer, depending on its precise chemical and morphological nature and on the precise conditions of the environment, the degradation products could include monomers, oligomers or fragments of the polymer, or all three, as well as any residual processing or other additives. The degradation profile could be such that depolymerization was the dominant event, in which case monomeric products are most likely, which may be metabolized. If the process were random chain scission, oligomers and small molecular fragments may be first produced, which could themselves be degraded further and metabolized.

Many degradation processes result in the fragmentation of the polymer at some stage, yielding voluminous microscopic particles. The result of the degradation process is, therefore, that there will be a temporal sequence in the release of molecular or particulate components into the cellular or tissue environment of the engineered construct. Many of these components can have an influence on cell behavior. In the extreme they may be cytotoxic, with the capacity to kill cells, or alter cell functions such as respiration, metabolism and proliferation, or have some other effect on cell structure and properties. The very sensitive processes in tissue engineering associated with the optimization of the behavior of those cells intended to express new matrix may be seriously impaired by interference from these degradation products. Equally, if an *in vitro* tissue engineering process results in the formation of a construct that consists of part scaffold and part new tissue which is then placed in patient, or if the whole of the tissue engineering process occurs *in vivo*, then the characteristics of the regenerated tissue may be radically altered by the response to the degradation products. In particular, if the degradation products are pro-inflammatory, either by virtue of the chemical nature of the molecules, or their pH, or the physical form of any particulates, then the host response may be overwhelmed by macrophages and foreign body giant cells, which will, by definition, be vastly different to the required functional tissue of, say, skin, cartilage or bone.

This inflammation scenario is a serious concern for many synthetic biodegradable polymers. As noted below, the main alternative scaffold materials are natural biopolymers or tissue derived matrices, where it is often assumed that the biocompatibility will be superior, either because they introduce some desirable bioactive effect or because they tend to be less pro-inflammatory. This may well be true, but the use of natural products and their derivatives is not without its own biocompatibility issues, especially in view of the

fact that such substances or their eventual metabolic products, may well be immunogenic. This is especially possible for some protein-based scaffold materials.

In view of the above analysis of the potential interactions between tissue engineering scaffolds and matrices, we may reconsider the requirements for biocompatibility in these applications. It is clear that when designing materials for tissue engineering scaffolds and matrices, all the lessons of the mechanisms of biocompatibility of materials for long term implantable devices should be remembered, but that totally new paradigms for the design of materials that can simultaneously or sequentially deliver molecular signals and mechanical signals to the relevant cells during the dynamic interchange between material degradation and tissue development are required. Indeed, biocompatibility may be redefined in the context of tissue engineering as:

> The biocompatibility of a scaffold or matrix for a tissue engineering product refers to the ability to perform as a substrate that will support the appropriate cellular activity, including the facilitation of molecular and mechanical signalling systems, in order to optimize tissue regeneration, without eliciting any undesirable local or systemic responses in the eventual host.

Clearly the use of tissue engineering scaffolds and matrices is relatively new and there is little clear evidence about precise mechanisms and performance with respect to biocompatibility. A surprisingly wide range of materials has already been proposed for this use, each employing a range of fabrication technologies and morphologies and architectures. These materials and their forms are summarized in Table 9.1. The following comments may be made about the current status of knowledge about the biocompatibility of these materials.

9.4.1 Synthetic degradable polymers

It is not surprising that most of the so-called first generation of tissue engineering substrate materials were

Table 9.1 Classification of materials for tissue engineering scaffolds and matrices

Class	Examples
Synthetic Degradable Polymers	Polylactides/glycolides
	Polycaprolactone
	Polyhydroxyalkanoates
	Poly(propylene fumarates)
	Polyurethanes
Natural Biopolymers	Proteins
	Collagen
	Elastin
	Fibrin/fibrinogen
	Silk
	Polysaccharides
	Alginates
	Chitosan
	Hyaluronic acid
Bioactive ceramics	Calcium phosphates
	Bioactive glasses
Composites	Synthetic polymers/ bioactive ceramics
	Biopolymers/bioactive ceramics
Tissue derived ECM	Small intestine submucosa
	Skin ECM

synthetic polymers since several of these already had had clinical experience within the fields of implantable devices and drug delivery systems where flexibility and degradability were important properties. Some of the more significant of these are discussed briefly below in relation to their biocompatibility.

9.4.1.1 *Polylactic acid, polyglycolic acid and their copolymers*

Perhaps the most widely used type of synthetic polymer here is the polyester family based on polylactic acid and polyglycolic acid. The different isomers of these polymers (especially poly(L-lactide) and poly(DL-lactide)), and the different copolymers that can be formed (poly(lactide-co-glycolide), PLGA), provides a very wide spectrum of degradation profile

and physical and chemical properties. The time to complete degradation can vary from a few weeks up to several years, most attention in tissue engineering being paid to PLGA of around 75/25 molar ratio, DL-lactide to glycolide. Although in theory these polymers look quite suitable, they do have two drawbacks related to their biocompatibility. The first is that they are relatively hydrophobic and non-wettable, such that cell interactions with the surface are very limited. Secondly, the degradation products are quite pro-inflammatory. It is clear that macrophages and foreign body giant cells are associated with the degradation and the response to the degradation (Xia *et al.*, 2006) and this results in a significant disadvantage with respect to the optimization of the eventual host response. Thus, although they are still used in tissue engineering, both experimentally and clinically, they are far from ideal.

In view of the hydrophobicity problem, there have been many attempts to alter the surface characteristics of these polymers by a number of physical or chemical techniques (Khang *et al.*, 2002). Some of these, through the reduction in contact angle, will result in more significant cellular interactions, for example adhesion, but it is still as yet uncertain how these treatments can be made effective in a clinical context.

9.4.1.2 *Polyhydroxyalkanoate-based materials*

Polyhydroxyalkanoates are polyesters that are naturally produced by some microorganisms and utilized by them for energy storage. It has been known for some time that it is possible to culture certain bacteria under industrial fermentation conditions and harvest and purify the relevant molecules. It was assumed that since these bacteria synthesize the polymer intracellularly and then degrade it enzymatically when energy is required, this degradation process would be replicated in other biological systems. One of these polyesters with the most potential was considered to be polyhydroxybutyric acid, PHB, and it was investigated for use in medical applications as a naturally occurring biodegradable polymer. It quickly became known, however, that this degradation mechanism was not reproduced within mammalian systems and by mammalian enzymes and PHB did not degrade very rapidly in the human body (Miller and Williams, 1987). It was appreciated then that the degradation profile could be altered through co-polymerization and control of polymer morphology. Since then (Chen and Wu, 2005), a variety of these polymers and their derivatives have been used clinically in a wide variety of applications, and their use considered for tissue engineering, including tissue engineered heart valves (Sodian *et al.*, 2000). Copolymers of hydroxybutyrate with hydroxyhexanoate (PHBHHx) look particularly attractive, with apparently good cell adhesion and functionality, and a controllable degradation profile with little evidence of any cytotoxicity and subsequent inflammation.

9.4.1.3 Polycaprolactone

Polycaprolactone is another polyester with a substantial history of exploration as a degradable scaffold material, being easily fabricated by a variety of routes and capable of surface modification if necessary and with a moderate degradation rate. Porous polycaprolactone scaffolds, with pore sizes ranging from 100 to 500 μm appear to be able to support good cell growth for a variety of cells, although with varying optimal pores sizes depending on cell phenotype (Oh *et al.*, 2007). As noted in general terms earlier, the cells need the pores to be large enough to allow cell migration, nutrient supply and waste removal, whilst small enough to provide a sufficiently high surface area to facilitate cell–polymer interactions. Although some limited inflammation may be seen during the degradation process *in vivo*, this does not appear to be detrimental to tissue development, although it is not clear whether this has been followed to the point of complete degradation.

Polyfumarate-based materials

Although most materials investigated within tissue engineering have been considered rather generically, some have been developed with specific end-points in mind. One of these is poly(propylene fumarate) (PPF), which has been explored for use within bone tissue engineering, and especially for *in vivo* bone tissue engineering, where the biodegradable scaffolds

are intended to be implanted into defect sites for the support of the migration, proliferation, differentiation and new bone matrix production by osteoblasts (Fisher *et al.*, 2002). It is notable that *in vivo* animal experiences show that although the porous polymer 'supports' bone ingrowth in the sense of guiding the bone formation, the material alone does not induce bone formation. It degrades slowly, with some erosion and fragmentation being seen over the course of many months (Hedberg *et al.*, 2005) with increasing evidence of inflammation associated with the release of insoluble particulates and soluble acidic products.

9.4.1.4 Polyethylene glycol terephthalate/ polybutylene terephthalate

In a similar manner to the design of PPF for bone tissue engineering, a family of segmented block co-polymers of polyethylene glycol terephthalate and polybutylene terephthalate has been designed for cartilage tissue engineering (Malda *et al.*, 2004). Here, there has been considerable variability in the ability of the polymers to support cell growth *in vitro* (Classe *et al.*, 2003) although in animal models neovascularization has been shown to occur when used as a dermal substitute (Ring *et al.*, 2006). The degradation profile is one of a very slow process, with fragments of polymer still being seen a couple of years after construct implantation (El Ghalbzouri *et al.*, 2004). From the biocompatibility perspective, it is difficult to see that this material will provide an optimal solution.

Polyurethanes

Polyurethanes comprise a remarkably varied group of polymers that have had many different uses in medicine. The advantages conferred by polyurethanes include very versatile mechanical properties and excellent biocompatibility under many circumstances, including very good blood compatibility. However, there have always been questions over the resistance to degradation of this class of polymer because of the intrinsic instability of the urethane bond and over the toxicity of breakdown products following the degradation of some of the materials. The search for truly biostable polyurethane biomaterials with good overall biocompatibility has been of considerable importance in medical device technology. On the other hand, the potential degradability of polyurethanes has tempted many groups to seek to develop an intentionally bio-degradable material with good biocompatibility from this family. Several possibilities have emerged, including a porous polyesterurethane (Saad *et al.*, 2000) and lysine di-isocyanate polyurethanes (Zhang *et al.*, 2002) both of which have been investigated with respect to tissue engineering scaffolds. The evidence would once again suggest that these types of polymer are able to support cellular activity under some conditions but without actively promoting any particular behavior and that the degradation profile can be adjusted to give a clinically acceptable degradation rate but without, as yet, any guarantee of good, inflammation-free, long term biocompatibility.

Interestingly, the general versatility of the chemical formulation of polyurethanes has allowed for room temperature synthesis under some situations, leading to the concept of an injectable, *in situ* curing scaffold (Bonzani *et al.*, 2007). This would facilitate the process of *in vivo* tissue engineering, but the success would critically depend on the lack of toxicity associated with the curing process as well as the chronic resistance to inflammation.

9.4.2 Synthetic non-degradable polymers

For completeness, it should be mentioned that a tissue engineering process does not necessarily have to incorporate a degradable material. If it is acceptable that the regenerated tissue contains the residue of a non degradable material, it might still be possible to obtain the relevant functionality, in which case the biocompatibility problems associated with the degradation process and the release of pro-inflammatory products would be obviated.

9.4.3 Synthetic hydrogels

We now moving from porous polymeric scaffolds to hydrogels. A wide range of properties is possible, bearing in mind that hydrogels can contain anything from, typically 30% to 90% of water, and with a huge variety of forms of polymeric network holding the water (Drury and Mooney, 2003). Generally it is sensible to consider these as either synthetic polymers or biopolymers, although probably more attention has been paid to blends and mixtures of two or more types than individual polymers. With respect to the synthetic hydrogels, two stand out as being studied more than any other. These are the closely related poly(ethylene oxide) and poly(ethylene glycol); it is worth noting that many blends of synthetic and natural hydrogels used in medicine involve one or other of these as a major component.

Generally these chemically simple hydrogels are non-cytotoxic, without any stimulation of the inflammatory or immune processes, but equally they do not have any intrinsic biological functionality and do not, by themselves promote any particular cellular function. Synthetic hydrogels tend to be degraded through hydrolysis of ester linkages but it is possible to synthetically introduce other linkages to facilitate enzymatic cleavage of the molecules. By definition, hydrogels do not yield solid fragments as they degrade, nor should they necessarily yield acidic by-products and so significant inflammation is not expected.

9.4.4 Protein-based biopolymers

Since the structure and function of most tissues and organs is critically dependent on the presence of certain proteins, it is logical to consider that proteins themselves, or their analogues, could provide for effective scaffolds or matrices for tissue engineering. Several of these have now been developed to a considerable extent and point the way towards optimization of biocompatibility.

9.4.4.1 *Collagen and elastin*
There are many different types of naturally occurring collagens, and equally many different methods to prepare biomaterials-based collagen products. It is therefore difficult to be precise over the general biocompatibility characteristics of these materials and their usefulness as tissue engineering scaffolds

and matrices. Nevertheless there is some consistency in the overall picture that collagen sponges and gels can give good performance, and probably better than that seen with synthetic polymers. One study compared collagen extracted from porcine skin that was cross-linked by a physical process to a polyester in the context of tooth tissue engineering, showing higher initial cell attachment, a greater formation of calcified tissue and more relevant tooth morphology with the collagen (Sumita *et al.*, 2006). The suitability of collagen type I for a tissue engineered heart valve from the biocompatibility perspective has also been demonstrated (Taylor *et al.*, 2006), while equine derived collagen fleece has acceptable compatibility with bladder smooth muscle cells (Danielsson *et al.*, 2006).

Collagen may be made available in many different forms, including sponges, granules and gels. Their variable physical form can be controlled by the arrangement of the collagen strands and the degree and nature of the cross-linking induced between them. Collagen is slowly degraded in the body, largely through the activity of the metalloprotease collagenase and some serine proteases, such degradation therefore being variable depending on the precise nature of the immediate cellular environment. Collagen has one major advantage over all synthetic and most natural scaffolds in that it is a natural extracellular matrix protein to which the receptors on many cell membranes can be attracted, creating the possibility of directly influencing cell adhesion and function. How much this actually occurs in practice is not entirely clear but there is a growing body of evidence to support this possibility. Collagen (and indeed elastin) scaffolds prepared from the porcine aorta have been shown to support *in vivo* cellular repopulation and de novo extracellular matrix synthesis (Simionescu *et al.*, 2006). The role of collagen in skin tissue engineering has also been well demonstrated in a small human clinical trial of a product based in human dermal fibroblasts are cultured for up to seven weeks in a fibrinogen rich medium, wherein they autosynthesize collagen, the resulting 'living skin equivalent' being placed on damaged skin (Boyd *et al.*, 2007). This collagen not only supports the original seeded fibroblasts but also appears to support the proliferation of infiltrating keratinocytes.

Most of the collagens used in tissue engineering applications are derived from human or animal sources. Immunological problems and the possibility of disease transmission have always to be considered. In order to minimize risk, and indeed to optimize performance, it has often been suggested that recombinant collagen would be more appropriate (Yang *et al.*, 2004). It is possible to provide chemically defined purified human collagens that are free of animal components and furthermore the recombinant technologies allow for modification of collagen structure in order to optimize properties. It is not known whether recombinant collagen offers any superiority compared to natural animal derived materials, but this remains an attractive possibility.

Elastin is another structural protein that could offer potential as a tissue engineering scaffold, although by definition the rather special mechanical properties associated with the elasticity of this protein limit the types of tissue that could be involved. Vascular tissue is the more obvious target, including scaffolds for engineered arteries (Leach *et al.*, 2005). The essential characteristics of the biocompatibility of elastin scaffolds should not be too different from those of collagen, although it would be expected that these two proteins would preferentially support the activity of different cells, and the highly insoluble nature of elastin should influence degradability.

9.4.4.2 *Fibrin*

Fibrin glues have been used in a number of clinical applications over many years and have often been proposed for tissue engineering. Under normal circumstances, fibrin preparations are soluble and unstable due to degradation by fibrinolysis, and whilst this may be acceptable in some short term situations such as adhesion and sealing, this is not ideal for a tissue engineering application when the construct is required to retain shape over many weeks. This has resulted in the development of more stable fibrin gels (Eyrich *et al.*, 2007) using fibrinogen, thrombin and calcium ions to alter the structure, some of these formulations being stable, and retaining strength for several weeks. With a specific ambition of using these 'long term stable' gels in cartilage tissue engineering, it has been seen that chondrocytes are able to proliferate very well within some of these gels, with the expression of extracellular matrix components such as collagen II and glycosaminoglycans.

Silk

Silk has been extensively investigated in recent years, this material being derived from several different natural sources such as the spider and silkworm and prepared by several different routes. Silk was for very many years used as a surgical suture material, but this use was largely discontinued because of variable degradation rates and hypersensitivity reactions. Their use in tissue engineering scaffolds has been predicated on superior purification methods. The *in vitro* and *in vivo* biocompatibility of silk scaffolds has been investigated and summarized by Kaplan's group (Meinel *et al.*, 2005). Little in the way of inflammatory responses were seen on implantation of tissue generated by bone marrow-derived mesenchymal stem cells within silk scaffolds and improved purification techniques that have resulted in the removal of the immunologically troublesome sericins has resulted in minimal antigenicity. Silk appears to have little negative effect on cell viability and function, although whether there is any active support for proliferation or tissue generation from either silk surfaces or degradation products is unclear.

9.4.5 Polysaccharide-based biopolymers

Comprising the other main class of biopolymer are the polysaccharides, several of which have been extensively used in medical applications including tissue engineering, primarily because of a beneficial combination of biological properties, including biodegradation.

9.4.5.1 *Hyaluronic acid*

Hyaluronic acid is a relatively simple glycosaminoglycans, a linear polysaccharide of repeating β-D-glucuronic acid and N-acetyl-β-D glucosamine units.

It is found in virtually all tissues of adult mammals and is particularly involved in wound healing and in lubrication in synovial joints. There are many ways in which the hyaluronic acid can be modified to produce relevant and practical structures, including cross-linking and esterification. These products range from water soluble molecules, to hydrogels and to films, foams, sponges and textiles. The hyaluronic acid itself can be degraded by hyaluronidase, but the differing physical forms result in quite widely varying degradation profiles, with complete degradation taking place in times ranging from a few days to several months (Segura *et al.*, 2005). There are many indications that the hyaluronic acid does play a role in the positive stimulation of cell function in tissue engineering processes. The degradation process is variable, and there are examples of very unremarkable responses to the degrading material but also some evidence of inflammation in other situations.

9.4.5.2 *Chitosan*

Chitosan is a deacetylated derivative of chitin, a common natural biopolymer found in the shells of crustaceans and the cell wall of fungi. It is also a linear polysaccharide, comprised of glucosamine and N-acetyl glucosamine segments, the ratio of which determines the properties, especially the solubility/degradability. It can be formed as porous structures by many different processes and is used in several different types of medical application, ranging from drug and DNA delivery systems to wound dressings and anti-microbial systems. It has also been considered for the material of tissue engineering scaffolds, including cartilage and the intervertebral disc (de Martino *et al.*, 2005). In many of the applications for which chitosan has been proposed, the conflicting biological characteristics that it displays under different conditions, and the quite varying degradation profile, have led to a poor translation of the potential to actual clinical use. With respect to tissue engineering, it is still not clear whether any specific biological stimulation is provided by chitosan surfaces or gels on cells and it is equally unclear exactly how the degradation process is related to the ultimate host response.

9.4.5.3 *Alginate*

Somewhat similar to chitosans, the alginates are gels that have several potential medical applications but with a variability in biocompatibility characteristics and biological performance. Part of the problem in the past has related to the presence of contaminants, including endotoxins, which has adversely impacted on biocompatibility, and the situation is a little clearer with the introduction of highly purified materials. Again with a polysaccharide we see how the biological performance is dependant on the precise composition, and in this case with the mannuronic acid to guluronic acid ratio (Klock *et al.*, 1997). Many alginates are highly pro-inflammatory, and potent stimulators of cells such as macrophages and lymphocytes, but this can be significantly controlled by optimization of the composition. The same applies to the degradation profile, where both the mannuronic acid to guluronic acid ratio and the nature and degree of cross-linking will control solubility. Ionically cross-linked alginates slowly dissolve while some other derivatives will hydrolyse. The influence on the overall biocompatibility of alginate tissue engineering gels is similar to that seen with chitosans.

9.4.6 Extracellular matrix derived materials

The topic of the extracellular matrix as a biological scaffold material has been extensively discussed by Badylak (2007). As noted earlier, the functional and structural molecules of the natural extracellular matrix should allow cells to communicate with each other and with their environment in a near-normal manner. There are many commercially available products obtained from both human and animal origins, including porcine small intestine mucosa, fetal bovine skin, bovine pericardium, human dermis and human fascia lata. Obviously the biocompatibility characteristics will vary with source and with preparation method. The main biocompatibility issues appear to have been based on the ability (or not) of these products to facilitate the constructive remodelling of tissues and on the recipient immune response.

In contrast to the situation with most other scaffolds, either synthetic polymers or biopolymers, some of the extracellular matrix materials do appear to influence the tissue remodelling after implantation, where angiogenesis, infiltration and mitogenesis of host cells and the deposition and reorganization of new extracellular matrix may all be seen. Exactly why this occurs is not clear but it is likely due to the release of growth factors during the matrix degradation, which is itself mediated by enzymatic processes. It is likely that unmodified as opposed to cross-linked tissue will be superior since cross-linking may resist or retard these degradation processes.

With respect to the recipient immune response to these extracellular matrix scaffolds, which are obviously either xenogeneic or allogeneic, the situation is likely to vary depending on source and processing route, but it would appear that recipients of some of these scaffolds may well recognize the material as 'non-self' and produce antibodies but that these fail to initiate a classical immune response, a situation backed up by considerable clinical experience with some of these products.

9.4.7 Bioactive glasses

It has to be recognized that, because of the quite different mechanical properties required from the forming tissue, the scaffold for successful bone tissue engineering may have to be quite different to that used in soft tissue applications. It is generally considered that one of the so-called bioactive ceramics or glass-ceramics would be most appropriate, especially since they can be made with the type of porosity that is normally considered a prerequisite for bone ingrowth and maintenance, that is a pore size in excess of 500 μm and an interconnection size in the region of 200 μm, because they can allow apatite formation on their surface and because they can be degraded.

Several bioactive glasses come into this category, of which there several varieties with different formulations and ionic concentrations. There is no doubt that under *in vitro* conditions, some of these glasses do facilitate apatite deposition on their surfaces, and that they can facilitate new bone formation at their surfaces *in vivo*. The biocompatibility of these glasses is straightforward but it is not yet certain just how the surface activities will translate into effective performance in the regeneration of bone in clinically relevant situations. The evidence would suggest that extracellular matrix formation and mineralization can be induced within the porosity of bioactive glasses under some circumstances (Jones *et al.*, 2007).

9.4.8 Calcium phosphate ceramics

The situation with calcium phosphate ceramics is somewhat similar. They contain the same ionic species as found in bone mineral, and experience over several decades with the various forms of the bioceramic family, including hydroxyapatite, tricalcium phosphate and several biphasic materials, show that they can facilitate bone regeneration *in vivo*. The overall performance of calcium phosphates depends on the exact formulation, including the presence of different atomic species and crystallographic form, and the consequential effects on both mineralization and the rate of degradation or dissolution.

Porous calcium phosphate ceramics have been shown to be capable of supporting cell function *in vitro* and also of supporting bone formation *in vivo*. For example, adipose-derived stromal cells can differentiate into osteoblasts within such scaffolds and such constructs, after suitable culture, can stimulate the formation of bone *in vivo*, both ectopically and within bony defects Lin *et al.*, 2007). However, it is rarely obvious just how much of this stimulation and inducement is directly due to the material per se. Generally it can be assumed that calcium phosphates have good overall biological safety and facilitate bone formation under some circumstances, but just how much real biological activity there is remains to be seen. One of the more interesting developments has been that of the injectable, porous calcium phosphate, which may be described as either a cement or a scaffold (Hockin *et al.*, 2006).

9.4.9 Polymer–polymer composites/blends and ceramic–polymer composites

It is a commentary on the performance of each of the individual materials mentioned in the above sections that far more attention is currently being paid to composites and blends of these materials than the 'pure' materials themselves. Within the synthetic polymers, for example, PLGA is often used as a copolymer in order to modify the degradation rate of other polymers, including polycaprolactone and PHB. With proteins, collagen, elastin and fibrin have been mixed together on many times and there have been large numbers of hyaluronic acid, chitosan and alginate blends. Similarly, partly in order to alter mechanical properties but mainly in the hope of modifying bioactivity, many synthetic and natural polymers have been formed as composites with the calcium phosphates, bioactive glasses and indeed other bioceramics. Within this group are very many composites that involve nanoscale dispersed particles, that is particles less than 100 nm in size, especially nanoparticles of hydroxyapatite and carbon nanotubes. The classical rationale for the design of composites is that it should be possible to obtain combinations of properties that could not be achieved by either phase alone and it is therefore possible that composite material scaffolds could be effective from the biocompatibility perspective, although there is little evidence that this is being achieved in practice. It should be noted that there is an inherent biocompatibility concern over the use of the so-called nanocomposites in that the fate of released nanoparticles, and especially their translocation, or systemic distribution, is not understood.

9.4.10 Modified surfaces

It will be obvious that none of the first generation of synthetic materials discussed in the above sections have all of the requisite properties to be considered as ideal with respect to biocompatibility, especially in the context of facilitation of relevant cellular behavior at or near their surfaces. In view of this, there have been many attempts to modify these surfaces in order to improve cell function. In some situations, this has involved physico-chemical processes that alter surface energy and wettability, for example through plasma treatments. More importantly, it is recognized that the adhesion of anchorage-dependent cells to surfaces is mediated by integrins and that the promotion of cell-surface interactions can be achieved by the attachment of peptide sequences onto surfaces in order to facilitate these integrin-controlled events. The RGD sequence has been most commonly used for this purpose, but many others are possible (Santiago *et al.*, 2006) and there is no doubt that cellular behavior can be altered by these processes.

9.4.11 Membranes

The above discussions have been based on the tissue engineering model of cells expressing extracellular matrix within a biomaterial scaffold or matrix, either *in vitro* or *in vivo*, where the biocompatibility is determined by the interactions between those materials and cells and subsequent host tissues. There are other tissue engineering models which have to be mentioned, several of which involve membranes of biomaterials rather than porous scaffolds or three dimensional gels. The most important of these is associated with so-called cell-sheet engineering in which cells are cultured on thermally sensitive polymers and then released from the material surface by a change in temperature, forming an independent an integral sheet of cells which is used by itself, without attached material, in the host. Here the biocompatibility is concerned solely with the lack of cytotoxicity of the material in culture and the ability of cells to adhere to the polymer. Membranes are also used for cell encapsulation techniques, which are not considered here.

9.5 General discussion of biocompatibility in tissue engineering

Within the context of the general tissue engineering paradigm (Figure 9.1), we can see that biocompatibility,

as defined in section 9.3, may be influential to the overall process in a number of ways. However, the precise manner in which the biocompatibility of the material used in the scaffold or matrix is not yet clear.

The following would appear to be the important factors:

1. As with all biomaterial – tissue interactions, the first events that take place are those associated with the initial adsorption of proteins onto the material surface. In theory this should be influential in determining subsequent cellular behavior at that interface but it is far from clear how much the physico-chemical characteristics of the surface that should play some role in the protein adsorption process or how much the nature of the adsorbed layer that should play a role in subsequent cell adhesion actually influence the cellular events in tissue generation.

2. Whether mediated by an adsorbed protein layer or not, the surface of the scaffold will interact with the cells to some extent. It is important to know whether the surfaces themselves have any positive role in:

   - Facilitating the desired differentiation of stem or progenitor cells.
   - Maintaining the phenotype of the target cells.
   - Facilitating the adhesion of target cells to the surface.
   - Facilitation of the proliferation of target cells.
   - Facilitating the expression of extracellular matrix.
   - Facilitating the organization of the new extracellular matrix.

It is again far from clear what characteristics of the surfaces dictate these possibilities and indeed it is highly likely that any given material surface physico-chemical characteristic will influence these cellular events in different ways. There is no guarantee, for example, that the features of a surface that facilitate cell adhesion and spreading will also favor cell proliferation, although this is a common assumption in the literature. One study has shown

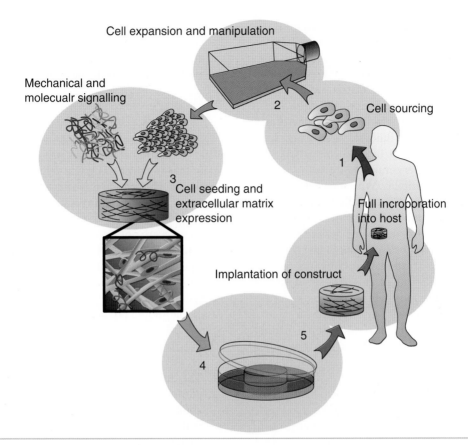

Figure 9.1 The central tissue engineering paradigm.

that there is considerable inconsistency in bone cell attachment and proliferation in contact with chitosan samples of different surface characteristics (Hamilton *et al.*, 2007). It may well be that the cellular behavior, including the control of stem cell differentiation, is more influenced by the surface nanotopography that the surface chemistry (Dalby *et al.*, 2007). There is a suggestion from the available evidence that natural biopolymers should be more effective at controlling cell behavior than synthetic polymers.

3. Very significantly, it is highly unlikely that any single material, synthetic or natural biopolymer, will, unaided, be able to fully and effectively modulate the behavior of the cells used in tissue engineering such that they meet all expectations of production of extracellular matrix. Even the mixing and blending of several components into a composite does not resolve this basic problem, although improvements to the baseline can be made by some physico-chemical surface treatments, where changes to wettability can affect the phenotype of cells at the surface (Lim *et al.*, 2005). The behavior and performance of cells within biomaterial constructs can, however, be radically altered by the surface attachment of adhesion peptides, or their incorporation into hydrogels, with particular emphasis on the RGD group, which can, for example, influence osteogenesis with bone marrow stromal cells in a dose dependent manner (Yang *et al.*, 2005).

4. The central, and indeed pivotal point, in most tissue engineering processes involves the delivery of the appropriate mechanical and molecular signals to the target cells so that they are persuaded to express the appropriate matrix, under conditions where they would not normally do so. As far as mechanical signals are concerned, these may involve structural or fluid stresses, which have little to do with the intrinsic biocompatibility of the material, but very much to do with the biomechanical compatibility. With molecular signals, this is where we see a potentially huge difference between scaffold materials. The molecules themselves may be indigenous, being derived from the relevant tissue itself, or exogenous, being delivered to the site from an 'external' location. It is unlikely that a synthetic scaffold material could play a role in the indigenous delivery of a biomolecule, but this is exactly what an extracellular matrix material will do, and also what some natural biopolymers may do. The delivery of exogenous molecular signals is concerned with the scaffold or matrix acting as a controlled drug delivery system.

5. If a favorable interaction between cells and biomaterial can be engineered, and an acceptable shape, size and functionality of regenerated tissue produced, the clinical benefit will only be derived if that construct can be fully incorporated into the host. This usually involves the complete and safe degradation and removal of the biomaterial, without inflammation and stimulation of the immune system, and the development of the appropriate tissue characteristics such as vascularity and innervation. Virtually all synthetic degradable polymers have the propensity to stimulate inflammation, which, depending on the degradation profile, is detrimental to processes such as cell growth and angiogenesis (Sung *et al.*, 2004). This can be modulated but not eliminated. Clearly, cells are influenced by the nature of the biomaterial environment, and global gene expression associated with cells within these environments will depend on the precise nature of that environment (Klapperich and Bertozzi, 2004) and so the possibility does exist to control this behavior. There

is evidence that natural biopolymers may have better all-round biocompatibility properties, but also that none of these have yet matched the biological performance of the tissue derived matrices.

9.6 Future perspectives

Tissue engineering brings a new challenge within the field of biocompatibility: the use of non-inert materials that interact and guide cells toward a desired response. This results in an increased complexity as material biocompatibility can influence the tissue engineering process in many ways. Unfortunately, these ways are not yet clear and it will be an important task to precisely identify them. In this respect, the possibility to evaluate the biocompatibility of many different biomaterials within a short time using high-throughput *in vitro* assays represents a valuable asset that should not be neglected.

9.7 Summary

1. The most important characteristic that distinguishes a biomaterial from any other material is its ability to exist in contact with the relevant tissues without causing an unacceptable degree of harm to those tissues.
2. The overall performance of a tissue engineering product is the biocompatibility of the scaffold or matrix that is used.
3. Implanted materials interact in many ways with tissue components. The acquisition of knowledge and understanding about these interactions is the basis for finding biomaterials with the best performances and is usually discussed in the broad context of biocompatibility.
4. The key to understanding biocompatibility is the determination of which chemical, biochemical, physiological, physical or other mechanisms become operative, (and why), under the highly specific conditions associated with contact between biomaterials and the tissues or cells of the body, and what are the consequences of these interactions.
5. There are several mediators of the biocompatibility of a material other than the characteristics

of the material itself. The nature and quality of the clinical intervention, the characteristics of the individual (age, sex, general health, physical mobility, lifestyle features and pharmacological status), the design of the device, the presence or absence of micro-organisms and endotoxins, all contribute to this variation.

6. For long-term implantable devices, the best performances are seen with the use of materials that are as inert as possible.

7. On the contrary, in tissue engineering the material is designed to support and influence cells in an optimal manner, therefore conflicting with biological inertness. In addition, materials used in tissue engineering should degrade at some time, resulting in degradation products that may cause an inflammation scenario and impaired biocompatibility.

8. Biocompatibility may be influential to the tissue engineering process in a number of ways that are unfortunately not yet clear.

References

Anderson, D.G., Levenberg, S. and Langer, R. (2004). Nanoliter-scale synthesis of arrayed biomaterials and application to human embryonic stem cells. *Nature Biotech*, 22(7): 863–866.

Badylak, S.F. (2007). The extracellular matrix as a biologic scaffold material. *Biomaterials*, 28: 3587–3593.

Bonzani, I.C., Adhikari, R., Houshyar, S., *et al.* (2007). Synthesis of two-component injectable polyurethanes for bone tissue engineering. *Biomaterials*, 28: 423–433.

Boyd, M., Flasza, M., Johnson, P.A., *et al.* (2007). Integration and persistence of an investigational human living skin equivalent in human surgical wounds. *Regen Med*, 2: 369–376.

Chen, G.-Q. and Wu, Q. (2005). The application of polyhydroxyalkanoates as tissue engineering materials. *Biomaterials*, 26: 6565–6578.

Classe, M.B., Grijpma, D.W., Mendes, S.C., *et al.* (2003). Porous PEOT/PBT scaffolds for bone tissue engineering; preparation, characterization, and *in vitro* bone marrow cell culturing. *J Biomed Mater Res*, 64A: 291–300.

Dalby, M.J., Gadegaard, N., Curtis, A.S.G. and Oreffo, R. O.C. (2007). Nanotopographical control of human osteoprogenitor differentiation. *Current Stem Cell Research and Therapy*, 2: 129–138.

Danielsson, C., Rualt, S., Basset-Dardare, A. and Frey, P. (2006). Modified collagen fleece, a scaffold for transplantation of human bladder smooth muscle cells. *Biomaterials*, 27: 1054–1060.

de Martino, A., Sittinger, M., Risbud, M.V. (2005). Chitosan: a versatile biopolymer for orthopaedic tissue engineering. *Biomaterials*, 26: 5983–5990.

Drury, J.L. and Mooney, D.J. (2003). Hydrogels for tissue engineering; scaffold design variables and applications. *Biomaterials*, 24: 4337–4351.

El Ghalbzouri, A., Lamme, E.N., van Blitterswijk, C., *et al.* (2004). The use of PEGT/PBT as a dermal scaffold for skin tissue engineering. *Biomaterials*, 25: 2987–2996.

Eyrich, D., Brabdl, F., Appel, B., *et al.* (2007). Long term stable fibrin gels for cartilage engineering. *Biomaterials*, 28: 55–65.

Fisher, J.P., Vehof, J.W.M., Dean, D., *et al.* (2002). Soft and hard tissue response to photocrosslinked poly(propylene fumarate) scaffolds in a rabbit model. *J Biomed Mater Res*, 59: 547–556.

Gourlay, S.J., Rice, R.M., Hegyeli, A.F., *et al.* (1978). Biocompatibility testing of polymers: *In vivo* implantation studies. *J Biomed Mater Res*, 12: 219–232.

Hamilton, V., Yuan, Y., Rigney, D.A., *et al.* (2007). Bone cell attachment and growth on well-characterized chitosan films. *Polymer International*, 56: 641–647.

Hedberg, E.L., Krosse-Deutman, H.C., Shih, C.K., *et al.* (2005). *In vivo* degradation of porous poly(propylene fumarate)-poly(DL-lactic–co-glycolic acid) composite scaffolds. *Biomaterials*, 26: 4616–4623.

Hockin, H.K.X., Weir, M.D., Burguera, E.F. and Fraser, A.M. (2006). Injectable and macroporous calcium phosphate cement scaffold. *Biomaterials*, 27: 4279–4287.

Jones, J.R., Tsigkou, O., Coates, E.E., *et al.* (2007). Extracellular matrix formation and mineralization on a phosphate free porous bioactive glass scaffold using primary human osteoblast cells. *Biomaterials*, 28: 1653–1663.

Khang, G., Choee, J.-H., Rhee, J.M. and Lee, H.B. (2002). Interaction of different types of cells on physicochemically treated poly(L-lactide-co-glycolide) surfaces. *J Appl Poly Sci*, 85: 1253–1262.

Klapperich, C.M. and Bertozzi, C.R. (2004). Global gene expression of cells attached to a tissue engineering scaffold. *Biomaterials*, 25: 5631–5641.

Klock, G., Pfeffermann, A., Ryser, C., *et al.* (1997). Biocompatibility of mannuronic acid-rich alginates. *Biomaterials*, 18: 707–713.

Leach, J.B., Wolinsky, J.B., Stone, P.J. and Wong, J.Y. (2005). Cross-linked elastin biomaterials: towards a prcessable elastin mimetic scaffold. *Acta Biomaterialia*, 1: 155–164.

Lim, J.Y., Taylor, A.F., Li, Z., *et al.* (2005). Integrin expression and osteopontin regulation in human fetal osteoblastic cells mediated by substratum surface characteristics. *Tissue Engineering*, 11: 19–29.

Lin, Y., Wang, T., Wu, L., *et al.* (2007). Ectopic and *in situ* bone formation of adipose tissue derived stromal cells in biphasic calcium phosphate nanocomposite. *J Biomed Mater Res*, 81: 900–910.

Malda, J., Woodfield, T.B.F., van der Vloodt, F., *et al.* (2004). The effect of PEGT/PBT scaffold architecture on oxygen gradients in tissue engineered cartilaginous constructs. *Biomaterials*, 25: 5773–5780.

Marchant, R., Hiltner, A., Hamlin, C., *et al.* (1983). *In vivo* biocompatibility studies. I. The cage implant system and a biodegradable hydrogel. *J Biomed Mater Res*, 17: 301–325.

Meinel, L., Hofmann, S., Karageorgiou, V., *et al.* (2005). The inflammatory responses to silk films *in vitro* and *in vivo*. *Biomaterials*, 26: 147–155.

Miller, N.D. and Williams, D.F. (1987). On the biodegradation of poly beta-hydroxybutyrate homopolymer and poly beta-hydroxybutyrate-hydroxyvalerate copolymers. *Biomaterials*, 8: 129–137.

Oh, S.H., Park, I.K., Kim, J.M. and Lee, J.H. (2007). *In vitro* and *in vivo* characteristics of PCL scaffolds with pore size gradient fabricated by a centrifugation method. *Biomaterials*, 28: 1664–1671.

Ring, A., Langer, S., Homann, H.H., *et al.* (2006). Analysis of neovascularisation of PEGT/PBT copolymer dermis substitutes in balb/c mice. *Burns*, 32: 35–41.

Saad, B., Kuboki, Y., Welti, M., *et al.* (2000). DegraPol-foam: a degradable and highly porous polyester urethane foam as a new substitute for bone formation. *Artif Organs*, 24: 939–945.

Salthouse, T.N. and Matlaga, B.F. (1981). Enzyme histochemistry of the cellular response in implants. In: *Fundamental Aspects of Biocompatibility*, (Williams, D.F., ed.). Boca Raton, FL: CRC Press, pp. 233–257. Vol. II.

Santiago, L.Y., Nowak, R.W., Rubin, J.P. and Marra, K.G. (2006). Peptide-surface modification of polycaprolactone with laminin-derived sequences for adipose-derived stem cell applications. *Biomaterials*, 27: 2962–2969.

Segura, T., Anderson, B.C., Chung, P.H., *et al.* (2005). Cross-linked hyaluronic acid hydrogels; a strategy to functionlize and pattern. *Biomaterials*, 26: 359–371.

Simionescu, D.T., Lu, Q., Song, Y., *et al.* (2006). Biocompatibility and remodelling potential of pure arterial elastin and collagen scaffolds. *Biomaterials*, 27: 702–713.

Sodian, R., Hoerstrup, S.P., Sperling, J.S., *et al.* (2000). Early *in vivo* experience with tissue engineered trileaflet heart valves. *Circulation*, 102(Suppl): 22–29.

Sumita, Y., Honda, M.J., Ohara, T., *et al.* (2006). Performance of collagen sponge as a 3D scaffold for tooth tissue engineering. *Biomaterials*, 27: 3238–3248.

Sung, H.-J., Meredith, C., Johnson, C. and Galis, Z.S. (2004). The effect of scaffold degradation rate on three-dimensional cell growth and angiogenesis. *Biomaterials*, 25: 5735–5742.

Taylor, P.M., Sachlos, E., Dreger, S.A., *et al.* (2006). Interaction of human valve interstitial cells with collagen matrices manufactured using prototyping. *Biomaterials*, 27: 2733–2737.

Williams, D.F. (1982). *Degradation of surgical polymers*. *Journal of Materials Science*, 17: 1233–1246.

Williams, D.F. (1999a). Bioinertness: an outdated concept. In *Tissue Engineering of Vascular Prosthetic Grafts* (Zilla, P. and Greissler, H., eds). Georgetown, Texas: R. G. LandesGeorgetown, Texas, pp. 459–462.

Williams, D.F. (1999b). *The Williams Dictionary of Biomaterials*. Liverpool University Press.

Williams, D.F. (2003). The inert-bioactivity conundrum. In *The Implant Tissue Interface* (Ellingsen, J.-E., ed.). Boca Raton: CRC Press, pp. 407–430.

Williams, D. F. (2007). On the mechanisms of biocompatibility. *Biomaterials*, To be published.

Xia, Z., Huang, Y., Adamopoulos, I.E., *et al.* (2006). Macrophage-mediated biodegradation of poly(DL-lactide-co-glycolide) *in vitro*. *J Biomed Mater Res*, 79: 582–590.

Yang, C., Hillas, P.J., Baez, J.A., *et al.* (2004). The application of recombinant human collagen in tissue engineering. *BioDrugs*, 18: 103–119.

Yang, F., Williams, C.G., Wang, D.A., *et al.* (2005). The effect of incorporating RGD adhesive peptide in polyethylene glycol diacrylate hydrogel on osteogenesis of bone marrow stromal cells. *Biomaterials*, 26: 5991–5998.

Zhang, J.Y., Beckman, E.J., Hu, J., *et al.* (2002). Synthesis, biodegradability and biocompatibility of lysine diisocyanate glucose polymers. *Tissue Engineering*, 8: 771–785.

Chapter 10
Cell source

Paolo Bianco, Pamela Gehron Robey, Giuseppina Pennesi and Ranieri Cancedda

Chapter objectives:

- To know which types of post-natal tissues contain stem cells that could be useful in tissue engineering and regenerative medicine
- To understand the concept of a stem cell niche within a tissue and how the niche controls stem cell activity
- To be familiar with basic methodologies of isolation and characterization of post-natal stem cells, and in some cases how they are manipulated *in vitro*
- To understand the notion of post-natal stem cell 'plasticity' and the issues surrounding it
- To be aware of the issues related to the use of autologous, allogenic and xenogenic cells

"Omnis cellula e cellula."

(Every cell stems from another cell)

Rudolf Virchow (1821–1902),

"Die Cellularpathologie in ihrer Begründung auf physiologische und pathologische Gewebelehre"

August Hirschwald (1858)

10.1 Evidence for the presence of stem cells in adult tissues

10.1.1 A discovery and its link to history

The ability of blood cells to self-renew throughout the organism's lifespan was one of the historical observations leading, at the turn of the XIX century, to the formulation of the hypothesis that stem cells could exist (reviewed in Baserga and Zavagli, 1981). Interestingly, 60 years and two World Wars separate the formulation of the hypothesis from the first experimental hint as to the actual existence of hematopoietic stem cells (HSCs). It was perhaps the tragic evidence of the effects of nuclear weapons that brought to the attention the need for developing radioprotection measures of strategic importance. This fueled research on hematopoietic stem cells, leading to the classic experiments of Till and McCullough (1961) and to the subsequent development of bone marrow transplantation, the first and most successful application of stem cell technology to clinical medicine worldwide. In a way, the discovery of HSCs in the western world (and of skeletal/mesenchymal stem cells in the Soviet Union) is a product of hot (nuclear) war, harvested during the subsequent cold war.

10.1.2 Adult stem cells (SCs) are undifferentiated cells that are found in differentiated adult tissues

Adult SCs have been derived from many animal and human tissues such as: bone marrow, peripheral blood, brain, spinal cord, dental pulp, blood vessels, skeletal muscle, heart, epidermis, mucosa of the digestive system, cornea, liver, and pancreas (Table 10.1). This list includes tissues that, due to their limited turnover, were not previously thought to harbour them, such as the CNS.

Adult SCs are undifferentiated cells with a long-term self-renewal capacity that can differentiate to mature cell types with specialized functions (Figure 10.1). The main role of adult stem cells is to maintain tissue homeostasis, and to replace cells lost due to normal tissue turnover, injury or disease. Typically, SCs generate intermediate cell types (progenitors and precursors) before they reach the fully differentiated state of mature cells. Progenitors and precursor cells are considered as cells committed toward differentiation along a specific cellular pathway.

10.1.3 Deterministic and stochastic models of self-renewal

Two alternative concepts can explain self-renewal and asymmetric cell division (Figure 10.2). In one, a stem cell is determined to divide asymmetrically by specific spatial and functional constraints imposed by a 'niche'. In this view, it is a spatial environmental cue that generates asymmetry of division and self-renewal of a stem cell, as long as the cell is subjected to the cue, and therefore physically retained in the space where the cue is active. This is believed to be the case with respect to the HSC (Morrison and Weissman, 1994). In the

Table 10.1 Adult SC types and their developmental plasticity (modified from Lemoli *et al.* (2005). *Hematologica*, 90: 360–381)

Tissue of origin	Cell type	Species	Tissue damage	Tissue formed	
				In vitro	*In vivo*
Muscle	Satellite cells	-	-	-	-
Skin	Epithelial SC	-	-	-	-
CNS	Neural SC	Mice	TBI	-	BM, Muscle
Liver	Hepatocytes, oval cells***	Rat	STZ*	Pancreas	Pancreas
Kidney	Renal SC	-	-	-	-
	MSC	Human	-	BM, Liver	-
Pancreas	Pancreatic SC	Rat	-	Liver	-
Heart	Cardiac SC	Mice	TBI**	-	Endothelium
	MAPC	Mice, Rat, Human	None or TBI	Multiple tissues of the 3 germinal layers	Multiple tissues of the 3 germinal layers
	MSC	Mice, Rat, Human	None or TBI	Cardiomyocytes Skeletal muscle Osteoblasts Chondroblasts Adipocytes Neural Cells	Cardiomyocytes Skeletal muscle Osteoblasts Chondroblasts Adipocytes Neural Cells
Bone marrow	Unfractionated	Mice Rat Human Dog	TBI CCl$_4$		Pancreas Liver Skin Intestine Epithelium CNS Skeletal muscle Cardiac muscle

* STZ, Streptozotocin; ** TBI, Total body irradiation; *** Only oval cells showed the capacity to generate pancreatic endocrine hormone – producing cells, *in vitro* and *in vivo*.

alternative concept, a stem cell can stochastically fluctuate between an asymmetric or symmetric division mode. In the latter case, both daughter cells can enter clonal expansion and differentiation, or, alternately, both cells can be restrained from expansion. This implies that stem cells can stochastically self-renew, or consume, or expand (Watt and Hogan, 2000).

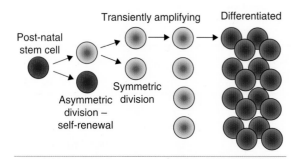

Figure 10.1 Although direct evidence is lacking in mammalian cells, it is presumed that post-natal stems are deeply quiescent, but when they divide, they do so asymmetrically, with one daughter remaining a stem cell, and another daughter that is more restricted. This daughter transiently divides via symmetric division and these amplified cells then differentiate.

10.2 Hemopoietic stem cell niche(s)

10.2.1 Stemness – functional properties and phenotype of HSCS

A single hematopoietic stem cell can reconstitute the hemato-lymphoid system of a lethally irradiated animal; that is, it can regenerate all hemato-lymphoid cell types for the lifespan of an animal in which lethal irradiation caused irreversible marrow aplasia (Spangrude, Heimfeld *et al.*, 1988) (Figure 10.3). From the bone marrow of reconstituted animals, stem cells can again be harvested and used for serial transplantation of additional irradiated animals (4–5 in a series, in the mouse) (Harrison and Astle, 1982). These experimental data prove the long-term self-renewal, and multipotency, of HSCs; that is their 'stemness.' In the mouse, 'hematopoietic stemness' is ascribed to cells with the surface phenotype $CD34^+Sca-1^+Thy-1^{low}$ Lin^-. That is, the cells highly express the cell surface proteins CD34 and Sca-1 and low levels of the cell surface protein Thy-1, and lack markers of more differentiated hematopoietic cells (lineage negative) which defines HSCs (Osawa, Hanada *et al.*, 1996; Spangrude,

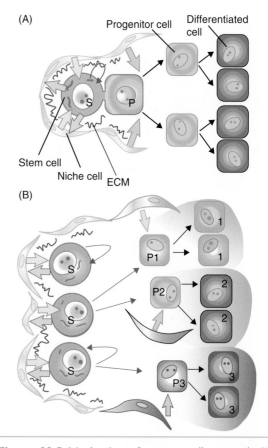

Figure 10.2 Mechanisms for stem cell renewal. (A) Deterministic – A stem cell (S) gives rise to a progenitor (P), which has a more restricted proliferation capacity, and differentiates in response to extrinsic cues. The stem cell is held in a quiescent state by immediate local signals (blue arrows). When these constraints are altered (green arrow), the cell rarely proliferates to reform itself (a stem cell) and a daughter cell that will transiently amplify, and differentiate; (B) Stochastic – There is no corral in which a stem cell is hurdled that dictates its activity. A stem cell could divide to give rise to daughter cells that are stem cells, or else more committed progenitors that differentiate along different pathways (P1, P2, P3), depending on the combination of extrinsic factors to which they are exposed, and both modalities at the same time. (From Watt, F.M. and Hogan, B.L. (2000). Out of Eden: stem cells and their niches. *Science*, 287(5457): 1427–1430.)

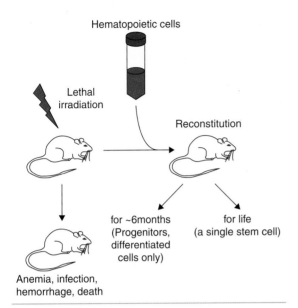

Hematopoietic cells

Lethal
irradiation

Reconstitution

for ~6months
(Progenitors,
differentiated
cells only)

for life
(a single stem cell)

Anemia, infection,
hemorrhage, death

Figure 10.3 Upon lethal irradiation, mice will succumb due to anemia, infection and hemorrhage. If committed hematopoietic progenitors are infused, mice will survive for approximately six months, but if infused with progenitors containing at least one stem cell, the hematopoiesis is reconstituted for life.

Heimfeld et al., 1988). Definition of the phenotype of human HSCs is still incomplete.

10.2.2 Development and migration of HSCs

In mammals, definitive hematopoietic stem cells are generated during embryonic life (~10.5 pc in the mouse) in the ventral floor of the dorsal aorta, in a specific region called the aorta-gonad-mesonephron (AGM) (Sanchez, Holmes et al., 1996). In this region, where HSCs are generated, hematopoiesis does not develop, but instead HSCs migrate via the bloodstream to the liver and spleen, and, as the bone marrow cavity is excavated in developing bone, to the bone marrow. In contrast to the AGM, active hematopoiesis is established at all of these sites. The bone marrow will become the predominant (the only,

in humans) hematopoietic organ in the post-natal organism (reviewed in Tavian and Peault, 2005).

In the bone marrow, HSCs can self-renew, but also proliferate, commit to specific hematopoietic lineages, and differentiate. HSCs circulate in the peripheral blood (Udomsakdi, Lansdorp et al., 1992), and therefore can colonize other tissues, which however do not provide a suitable environment for self-renewal or differentiation. In the bone marrow, this environment is provided by stromal cells, which make the bone marrow stroma unique in its ability to support HSC self-renewal and differentiation (Bianco and Riminucci, 1998). HSCs are enriched in fetal blood; hence the ease with which cell populations enriched in HSCs can be obtained from the cord blood (reviewed in Broxmeyer, 2005).

10.2.3 The HSC 'niche'

Recent studies have supported the view that a specific HSC niche exists in the bone marrow (in which HSCs are retained), may self-renew, and are protected from entering a differentiative or apoptotic fate. Different experimental approaches have been employed to anatomically dissect the 'niche', including the study of genetically engineered mice. There is no conclusive evidence as to the precise anatomical identity of the niche. In one view, bone surfaces and the cells associated with them (osteoblasts) represent the niche (reviewed in Zhu and Emerson, 2004) (Figure 10.4). In another, the niche is localized to the walls of bone marrow blood vessels (sinusoids) (reviewed in Kopp, Avecilla et al., 2005). In reality, the structure of the bone marrow is so complex that vascular walls exist in very close proximity to bone surfaces, and may in turn be lined, at least on one aspect, by bone cells. In addition, cells of osteogenic lineage exist, in the bone marrow stroma, also away from bone surfaces. Hence, a too strict assignment of the niche to one specific histological structure is probably not warranted. The possibility remains though, that retention within the niche may determine the quiescence or self-renewal of HSCS, while egress from the niche may determine the exposure of HSCs

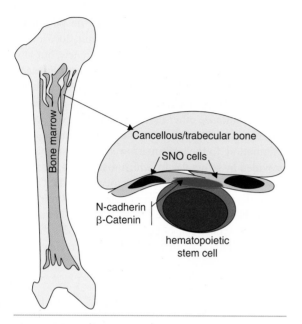

Figure 10.4 It has recently been suggested that the hematopoietic niche is formed by association of a hematopoietic stem cell with a particular type of cell found on the surfaces of trabecular bone (Spindle-shaped, N-caderin+ Osteoblastic cell, SNO). The association may be due to the co-expression of the adherens junction proteins, N-cadherin and β-catenin by both cell types. (From Zhang, J., Niu, C. *et al.* (2003). Identification of the haematopoietic stem cell niche and control of the niche size. *Nature*, 425(6960): 836–841.)

to changing environmental clues, dictating their proliferation and differentiation into a hematopoietic cell progeny rather than quiescence or self-renewal.

10.3 Epithelial stem cell and their niches

10.3.1 Self renewal of epithelia depends upon the presence of epithelial stem cells

The human epidermis is a stratified epithelium of cells named keratinocytes. The epidermis is organized in different cell layers: basal, spinosum, granulosum and corneum. Only keratinocytes in the basal layer can divide. Basal keratinocytes that become committed to terminal differentiation stop dividing,

and are displaced upward to the stratum corneum, this layer being composed of anucleate corneocytes continuously shedding into the environment. While evolving through the different stages of maturation and completing the process of keratinization, the keratinocytes sequentially acquire the phenotypes characteristic of the different cell layers (Figure 10.5).

For each corneocyte shed from the epidermal surface, a basal keratinocyte undergoes a round of cell division to replace the lost cell. As a result of this process, the epidermis is renewed approximately every month. The self-renewal process is possible only because epithelial stem cells exist in the epidermis. Skin keratinocyte stem cells are located in the bulge-containing region of the hair follicle (Figure 10.6). Separate populations of epithelial stem cells have been shown to be present also in interfollicular epidermis, hair follicles and sebaceous glands (Alonso and Fuchs, 2003; Gambardella and Barrandon, 2003). However, the bulge-containing region of the hair follicle is considered the main reservoir of skin keratinocyte stem cells (Oshima, Rochat *et al.*, 2001). It is well-known that in patients with partial-thickness burns (i.e. total epidermal destruction, but maintenance of skin appendages), healing can be obtained starting from keratinocytes migrating from hair follicles. It is likely that epithelial stem cells migrating from the hair follicle bulge restore the stem cells normally present in the basal layer of the epidermis.

A comparable condition also exists in other human stratified epithelia such as certain mucosae (oral, vaginal, urethral, etc.) and the ocular limbal-corneal epithelium. In particular, the corneal epithelium is a transparent and flat stratified squamous epithelium composed of cuboidal cells lying on the avascular corneal stroma. The stem cells of the corneal epithelium are located exclusively in the basal layer of the limbal epithelium, which is located in the narrow transitional zone between the bulbar conjunctiva and the cornea (Pellegrini, Golisano *et al.*, 1999) (Figure 10.7). When limbal stem cells become committed to terminal differentiation, they migrate from the periphery towards the center of the corneal epithelium. Cell shedding occurs from the central region of the epithelium. The corneal epithelium is fully replaced approximately every year.

10.3.2 Transient amplifying cells: an intermediate step between stem and fully differentiated epithelial cells

Keratinocyte stem cells generate transient amplifying (TA) cells; i.e., non-stem daughter cells that rapidly divide, but perform only limited cell duplications before their terminal differentiation. Three types of keratinocytes with different proliferation potential are present in primary cultures of human keratinocytes from both epidermis and limbal epithelium: holoclones, meroclones and paraclones (Barrandon and Green, 1987) (Figure 10.8). Cells giving rise to holoclones are considered to be the keratinocyte stem cells. These cells can undergo more than 180 cell doublings, have telomerase activity, long telomeres and generate a mature epithelium when transplanted *in vivo*. The meroclone is most likely a reservoir of TA cells. The paraclones are formed by TA cells, which can perform a maximum of 15 cell divisions and generate aborted colonies containing only terminally differentiated cells. The number of cells that can be obtained from a single holoclone is incredibly

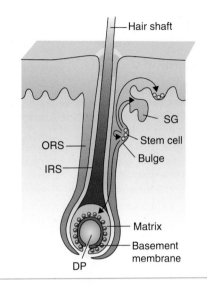

Figure 10.6 Location of epithelial stem cells in the bulge-containing region of the hair follicle. This region is the main reservoir of sweat glands, hair follicles, and epidermal stem cells. ORS, outer root sheath; IRS, internal root sheath; DP, follicular papilla; SG, sebaceous gland.

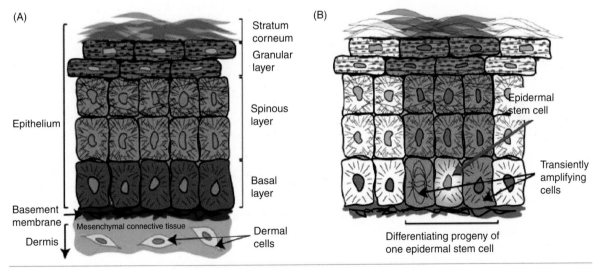

Figure 10.5 Stratified structure of human epidermis. (A) In this tissue epidermal stem cells are located in the basal layer of the epidermis; (B) Each stem cell gives rise to a progeny that pass through different stages of maturation. While progressing toward the terminally differentiated corneocytes, cells are displaced from the basal layer to the stratum corneum layer.

State of the Art Experiment

Demonstration of stem cells in hair follicles

Rochat, Kobayashi and Barrandon (1994) examined the growth capacity of keratinocytes isolated from human scalp hair follicles. Some of them had extensive growth potential, as they were able to undergo at least 130 doublings. Therefore, the hair follicle, like the epidermis, contains keratinocytes with the expected property of stem cells (holoclones). The same authors also examined the distribution of clonogenic keratinocytes within the hair follicle. Several hundred colony-forming cells were concentrated at a region below the midpoint of the follicle and outside the hair bulb. This region lies deeper than the site of insertion of the erector pili muscle, which corresponds with the position of the bulge when the latter can be identified. In contrast, few colony-forming cells were present in the hair bulb, where most of the mitotic activity is observed during the active growth phase of the follicle.

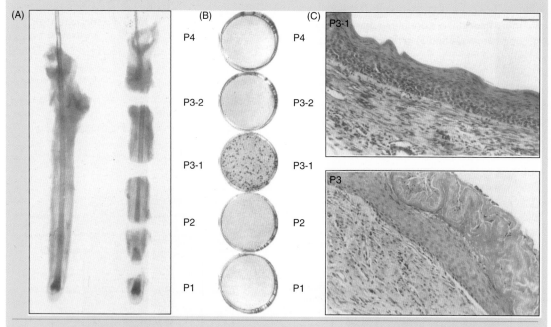

(A) A single human hair follicle was removed and sectioned into different regions and used to establish clonogenic cultures of epidermal cells. (B) Colony-forming units were only found in cultures established from the P3-1 region. (C) When grafted onto the skin of immunocompromised mice, cells from the P3 region were able to regenerate epidermis, although the transplants liked a well defined stratum granulosum (modified from Rochat, A., Kobayashi, K. and Barrandon, Y. (1994). Location of stem cells of human hair follicles by clonal analysis. Cell, 76: 1063–1073).

high. In fact, a single holoclone can produce as many as 10^{45} keratinocytes before senescence. A human individual has approximately 8×10^{10} epidermal keratinocytes scattered over a body surface of approximately $2\,m^2$. Thus, in principle one could obtain enough keratinocytes from a single clone to cover 1.25×10^{34} human individuals (coresponding to a body surface area of approximately $2.5 \times 10^{28}\ Km^2$).

10.3.3. Sheets of stratified epithelium can be obtained *in vitro*

Human epidermal keratinocytes propagated in culture form sheets of stratified epithelium maintaining the

characteristics of authentic epidermis (Figure 10.9). Epidermis produced *in vitro* is currently used to permanently restore severely damaged epidermis surfaces, as in the case of burn patients (Gallico, O'Connor *et al.*, 1984). Similarly, corneal epithelial grafts, aimed at replacing the damaged corneal epithelium, have been successfully obtained starting from small limbal biopsies taken from the uninjured eye (Pellegrini, Traverso *et al.*, 1997).

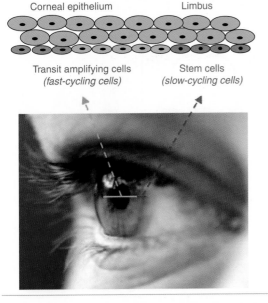

Figure 10.7 The stem cells of the corneal epithelium are located exclusively in the basal layer of limbal epithelium, the transitional zone between the bulbar conjunctiva and the cornea.

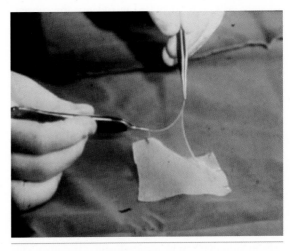

Figure 10.9 Sheet of cultured epidermis produced *in vitro* starting from a small skin biopsy.

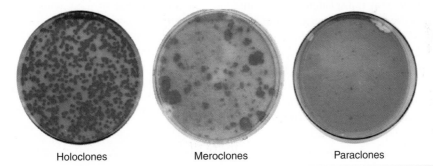

Figure 10.8 Colonies produced by different clonal type of epidermal cells. The cells were allowed to grow for 12 days before their fixation and staining with rhodamine. The holoclone has the greatest reproductive capacity. The meroclones contain a mixed population of cells with different growth potential. The paraclones contain cells with a reproductive capacity of not more than 15 cell generations.

10.4 Neuronal stem cell and their niches

10.4.1 Localization of neurogenesis and neural stem cells in the post-natal organism

Although post-natal neural tissue had long been thought to be unable to regenerate, the work of Altman in the 1960s revealed that in rodents, there is a constant birth of new neurons throughout life (Altman and Das, 1965). Subsequently, two primary locations of post-natal neurogenesis, fueled by post-natal neural stem cells, were identified in rodents: the subventricular zone of the lateral ventricle wall (SVZ) and the subgranular zone of the hippocampal dentate gyrus (SGZ) (Figure 10.10A), and the olfactory bulb and hippocampus in humans (reviewed in Doetsch, 2003; Ma, Ming *et al.*, 2005). Both sites of active neurogenesis are in close proximity to blood vessels (Palmer, Willhoite, *et al.* 2000), and within these regions, the extracellular matrix components, fibronectin and laminins 1 and 2, along with their cognate β1-containing integrin receptors have been implicated in controlling neurogenesis (Campos, 2005). Unlike in the hematopoietic stem cell field, there are no definitive markers for isolation of post-natal neuronal stem cells. Proof of their existence has relied heavily on the generation of proliferative cells *in vitro* that can be induced with a variety of culture conditions to form the three types of cells found in neural tissue: neurons, astrocytes and oligodendrocytes. Importantly, *in vivo* transplantation of undifferentiated proliferative neural cells isolated from both the SVZ and the SGZ have the ability to form neurons and supportive glial cells, typical of their site of origin, but also will form appropriate cell types when transplanted into another region of the brain. These studies demonstrate the multipotency of neural stem cells, and the ability of the local environment to dictate their pattern of differentiation (reviewed in Gage, 2000).

10.4.2 The neural stem cell and its niche

The character of the neural stem cell and its niche has been hotly pursued, and although there is

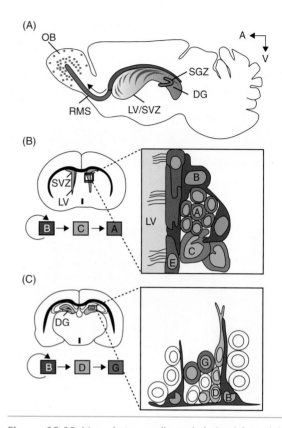

Figure 10.10 Neural stem cells and their niches. (A) In the adult mouse brain, there are two sites of active neurogenesis, the subventricular zone (SVZ) located in the lateral wall of the ventricle (LV), giving birth to neuroblasts that migrate along the rostral migratory stream (RMS), and the subgranular zone (SGZ) of the dentate gyrus (DG); (B) In the SVZ, astrocytes (B) are in close contact with ependymal cells (E), and extend processes into the lateral ventricle (LV). They form transiently amplifying C-type cells (C), which generate A-type neuroblasts; (C) In the SGZ, astrocytes (B) also form precursor cells (D) which give rise to new subgranular neurons (G). (Figure modified from Doetsch, F. (2003). The glial identity of neural stem cells. *Nat Neurosci*, 6(11): 1127–3411.)

controversy, it appears that a certain type of glial cell serves not only as the niche, but also as the stem cell (Doetsch, 2003). While often thought of as 'housekeeping cells' by removing metabolic waste and providing nutritional support, astroglial cells also function by providing signals that regulate axon guidance and synapse formation, and in the SVZ, by forming a tunnel that guides the migration of neuroblasts along the rostral migratory stream (RMS) to the olfactory bulb. In the SVZ, radial glial-like astrocytes (type B, positive for glial fibrillary acidic protein (GFAP), but negative for the usual astrocyte marker, S100β), give rise to type C cells (transiently amplifying cells) that subsequently form type A neuroblasts (Figure 10.10B). In the SGZ, radial astrocytes that are adjacent to blood vessels send projections that intimately associate with the granular cell layer. These cells are GFAP positive, but a subpopulation also express the immature neural precursor marker, nestin, and surround clusters of neuroblasts (Figure 10.10C), and may in fact give rise to them (reviewed in Doetsch, 2003; Ma, Ming et al., 2005).

10.4.3 Repair by post-natal neural stem cells

Based on current understandings of post-natal neural stem cells, two avenues of regeneration can be envisioned, either by encouraging endogenous stem cells to bring about repair via the administration of appropriate growth factors and regulators (Lie, Song et al., 2004), or by transplantation of a neural stem cell population (Bjorklund, 2000). It is evident that following a variety of insults, endogenous neurogenesis in the SVZ and the SGZ is enhanced, as well as in other parts of the nervous system characterized by more dormant neural stem cells. However, there is limited evidence that the generation of new neurons results in functional repair above a certain (very low) threshold, perhaps due to inappropriate migration, differentiation into the necessary cell types and formation of circuits. In some cases, local neurogenesis may actually contribute to pathological processes. This highlights even further the obstacles to be overcome in the use of exogenous neural stem cells in neural repair.

10.5 Mesenchymal stem cells and their niches

10.5.1 Skeletal (mesenchymal) stem cells can be derived from bone marrow

In addition to HSCs, a second population of stem cells exists in the bone marrow. Originally identified as bone marrow stromal stem cells (Owen and Friedenstein, 1988), these cells were later renamed 'mesenchymal stem cells.' Stromal (mesenchymal) stem cells stand to the stromal system as HSCs stand to the hematopoietic system. A single stromal stem cell can give rise to all differentiated cell types in the stromal system, which in essence coincides with the range of different connective tissues found in the skeleton (bone, cartilage, fibrous tissue, adipocytes, myelosupportive stroma; i.e., skeletal stem cells, SSCs) (Figure 10.11). As their original name indicates, the stromal stem cells are non-hematopoietic in nature, origin and differentiation potential; they are physically comprised within the stromal framework upon which HSCs and their differentiating progeny proliferate and mature (reviewed in Bianco and Robey, 2004).

SSCs are best identified as part of a subset of clonogenic stromal cells noted for the ability to establish discrete colonies (each derived from a single colony-forming cell, called the colony forming unit-fibroblastic, CFU-F) when plated in culture at low density (Figure 10.11). There are ~1 CFU-F/10,000 nucleated bone marrow cells. Just as the ability to reconstitute the entire hemato-lymphoid system of a lethally irradiated mouse defines the HSC, it is the ability to generate the complete range of skeletal tissues upon transplantation *in vivo* that defines the SSC (reviewed in Bianco and Robey, 2004). However, whereas HSCs cannot be efficiently grown *in vitro*, and are usually transplanted directly following isolation *ex vivo*, SSCs cannot yet be prospectively purified. As a result, they have never been used in any *in vivo* transplantation experiment as purified, uncultured cells. They must be established in culture in order to obtain a cell number high enough for *in vivo* transplantation. SSCs isolated from the bone marrow can generate bone,

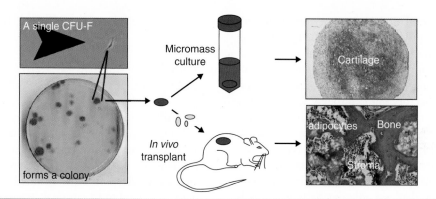

Figure 10.11 When single cell suspensions of bone marrow are plated at low density, a colony is formed by a single stromal cell, the Colony Forming Unit-Fibroblast (CFU-F). Approximately 10% of the individual colonies are multipotent skeletal stem cells as demonstrated by their formation of cartilage when placed in micromass cultures, and by the formation of a complete heterotopic ossicle composed of bone, hematopoiesis-supportive stroma and marrow adipocytes when transplanted in conjunction with hydroxyapatite/tricalcium phosphate ceramic particles into immunocompromised mice.

cartilage, fat, fibrous tissue and myelosupportive stroma *in vivo*. CFU-Fs vary in their growth ability and *in vivo* differentiation potential. It is thought that CFU-Fs whose progeny establish a complete heterotopic 'ossicle' (bone and hematopoiesis-supporting stroma) represent multipotent SSCs, whereas CFU-Fs that are only able to establish heterotopic bone but not bone marrow are seen as committed osteogenic progenitors (Kuznetsov, Krebsbach *et al.*, 1997).

Although the precise identity of SSCs in situ has not been determined, it is thought that they may be associated with the bone marrow vasculature. Definition of the precise in situ identity of SSCs will allow determination of whether a true 'niche' exists in the bone marrow for SSCs. The existence of a functional 'niche' for SSCs is suggested by the fact that they are mostly quiescent (G_0) cells *in vivo*, and only resume growth *in vitro* following prolonged exposure to high concentrations of serum or cocktails of growth factors.

10.5.2 *In vitro* and *in vivo* testing of skeletal (mesenchymal) stem cells multipotency

In vitro differentiation assays are commonly used as surrogates for the stringent *in vivo* transplantation assays. These assays probe the ability of cultured stromal cells to undergo adipogenic, osteogenic, or chondrogenic differentiation when exposed to specific inductive stimuli. The *in vitro* assays, however, must be conducted on clonal cell strains if meant to prove multipotency, are somewhat artificial, and are not as significant as the *in vivo* assays. When transplanted *in vivo* under specific conditions (subcutaneous transplantation with an osteoconductive carrier such as hydroxyapatite, or intraperitoneal transplantation within cell-impermeant diffusion chambers), SSCs demonstrate native (i.e., induction-independent) osteogenic, adipogenic and chondrogenic potential (Bianco, Kuznetsov *et al.*, 2006).

10.5.3 Multipotent adult progenitor cells (MAPCs)

The isolation of a population of cells from marrow, able to form virtually every type of cell in the body, has been reported (Jiang, Jahagirdar *et al.*, 2002). These cells, termed multipotent adult progenitor cells or MAPCs, differentiate, at the single cell level, not only into mesenchymal cells, but also into cells with visceral mesoderm, neuroectoderm and endoderm characteristics *in vitro*. When injected into an early blastocyst, single MAPCs contribute to all somatic cell types (Figure 10.12). It has been suggested that these cells with such remarkable capabilities may in fact be residual embryonic stem cells (Prindull, 2005).

In principle, as MAPCs proliferate extensively *in vitro* without apparent senescence or loss of differentiation potential, they may be an ideal cell source for therapy of inherited and/or degenerative diseases. Nevertheless, given the difficulties encountered by several laboratories in reproducing these results, some limitations may exist to their wide use in cell therapy.

10.5.4 Cells with properties similar to those exhibited by bone marrow-derived skeletal (mesenchymal) stem cells may exist in other connective tissues

Conceivably, cells with properties similar to those exhibited by bone marrow-derived SSCs may exist in other connective tissues, and a line of thought tends to encompass a variety of cell strains isolated from a variety of tissue sources under the umbrella label of 'mesenchymal stem cells.' Such putative cells have been described in adipose tissue, cord blood, peripheral blood of post-natal animals, synovial tissue, periosteum, and other sources. However, as far as we are aware, a direct comparison of the biological properties of MSCs from such different sources has not been conducted. Similarities have been established based on marker expression in culture, but this is not stringent. Assays of clonogenicity are also biased by the different nature of the starting material, which includes a vast majority of non-adherent, non-stromal cells in the bone marrow, and conversely a vast majority of adherent, stromal cells (in the absence of a hematopoietic component) in other connective tissues. As a result, whether in fact 'mesenchymal' stem cells obtained from the bone marrow and from different sources represent similar or different populations remain to be determined.

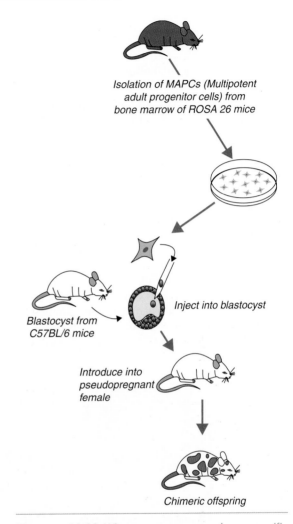

Isolation of MAPCs (Multipotent adult progenitor cells) from bone marrow of ROSA 26 mice

Inject into blastocyst

Blastocyst from C57BL/6 mice

Introduce into pseudopregnant female

Chimeric offspring

Figure 10.12 When grown under specific culture, adult bone marrow from mice has been demonstrated to give rise to multipotent adult progenitor cells (MAPCs). MAPCs were isolated from the Rosa 26 mouse that constitutively express LacZ as a marker, and were injected into blastocysts and placed into the uterus of pseudo-pregant females. The resulting chimeric offspring demonstrated LacZ positive cells throughout the entire mouse, indicating that MAPCs contribute to most, if not all, somatic cell types. As MAPCs proliferate extensively without obvious senescence or loss of differentiation potential, they may be an ideal cell source for therapy of inherited or degenerative diseases.

Classical Experiment

The discovery of multipotent stromal cells in the bone marrow

Starting in the late 1960s, Alexander J. Friedenstein and coworkers (Friedenstein, Piatetzky-Shapiro *et al.*, 1966) began to unravel the mysterious relationships between bone and its marrow, composed of hematopoietic cells and the stroma that supports them. In his first series of experiments, Friedenstein determined that bone marrow fragments and single cell suspensions derived from them form bone and cartilage and less frequently, fat and connective tissue, but did not support hematopoiesis when transplanted in a closed system (diffusion chamber) where in growth of recipient tissue is prevented (A, E). He subsequently determined that the bone forming cells were derived from the rapidly adherent, non-hematopoietic stromal cells, which he originally termed 'mechanocytes' due to their fibroblast-like character (B). When single cell

suspensions of bone marrow were plated at low density, discrete colonies were formed by these cells, which he later termed the Colony Forming Unit-Fibroblast (C, E). Next, he took individual colonies adsorbed into collagen sponges and transplanted them in an open system under the renal capsule, where recipient blood vessels could invade the donor tissue (E). In this case, a complete bone/marrow organ was formed by some (but not all) of the individual colonies (D, E). Using genetic markers, it was determined that the bone, stroma and marrow adipocytes were of donor origin, whereas the hematopoietic tissue was of recipient origin. This is the reverse of what is seen in bone marrow transplantation, where stroma is of recipient origin and hematopoietic tissue is of donor origin (reviewed in Friedenstein, 1990). These experiments established for the first time that bone marrow stromal cells transfer the

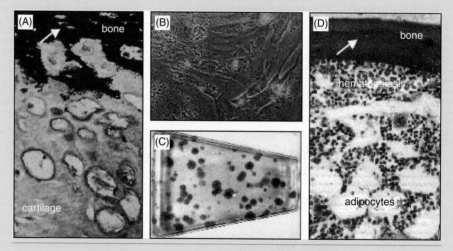

(A) Bone and cartilage formed by a single cell suspension of bone marrow in a diffusion chamber (closed system). (B) Rapidly adherent, non-hematopoietic 'mechanocytes' (bone marrow stromal cells) obtained in vitro from single cell suspension of bone marrow. (C) Establishment of colonies by the Colony Forming Units-Fibroblast. (D) Formation of a complete bone/marrow organ following transplantion of bone marrow stromal cells derived from a single under the kidney capsule.

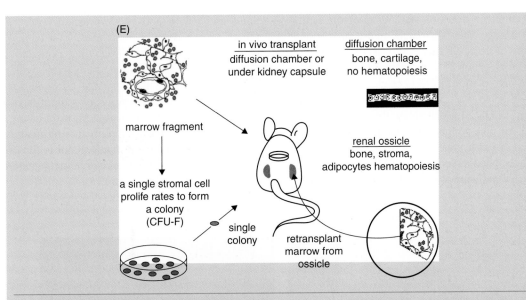

(E) Post-natal bone marrow stroma contains a self-renewing, multipotent stem cell – proof of principle. When either marrow fragments are transplanted in diffusion chambers, bone and cartilage are formed in the absence of hematopoiesis. When single cell suspensions of bone marrow are plated at low density, the Colony Forming Unit-Fibroblast (CFU-F) rapidly adheres and generates a colony. Upon in vivo transplantation underneath the renal capsule, 10–20% of these colonies form a bone/marrow organ. By re-transplanting the marrow from the ectopic ossicle, another complete bone marrow organ is formed, demonstrating self-renewal of the bone marrow stromal stem cell (skeletal stem cell).

hematopoietic microenvironment (Friedenstein, Chailakhyan et al., 1974). By taking the marrow from the ectopic ossicle and transplanting it again under the kidney capsule, another ectopic ossicle was formed (E). Based on these experiments, along with others performed by Dr. Friedenstein with his collaborators, Dr. Maureen Owen and coworkers (reviewed in Friedenstein, 1990), they hypothesized that post-natal bone marrow is home not only for the hematopoietic stem cell, but also what they termed a 'bone marrow stromal stem cell', capable of self-renewal, and with the ability to form all skeletal connective tissues (bone, cartilage, hematopoiesis-supportive stroma and marrow adipocytes). Furthermore, based on the fact that osteogenesis can be induced outside of the skeletal system (e.g., in skeletal muscle) by transitional urinary epithelium and demineralized bone matrix (both of which contain bone morphogenetic proteins), Friedenstein proposed that inducible osteogenic precursor cells (IOPCs) must exist in many tissues, that may be similar to those in marrow, the determined osteogenic precursor cells (DOPCs) (reviewed in Friedenstein, 1990). In fact, recent studies suggest that these types of 'mesenchymal' stem cells exist in many connective tissues.

10.5.5 Muscle derived stem cells (MDSCs)

Cells with stem cell-like properties derived from various tissues now include cells from the skeletal muscle compartment. Researchers have identified two types of stem cells in skeletal muscle: satellite cells and multipotent mesenchymal stem cells. Although multipotent mesenchymal stem cells and satellite cells could represent different stages of maturation of the same stem cell lineage, they also could represent distinct populations of stem cells that exist in skeletal muscle.

The research group of Johnny Huard in Pittsburgh developed an isolation protocol that selects for cells with a low adherence capacity. The muscle-derived stem cells (MDSCs) isolated by this method and expanded for 300 population doublings showed no indication of replicative senescence. MDSCs preserved their phenotype (Sca1+/CD34+/desmin^{low}; that is, high expression levels of the surface molecules Sca-1 and CD34, and low expression levels of the muscle protein desmin) for at least 200 cell doublings. After expansion to this level, they exhibited high skeletal muscle regeneration comparable with that exhibited by minimally expanded cells when transplanted into the skeletal muscle of mutant mice with a single base substitution within an exon of the dystrophin gene, which causes premature termination of the polypeptide chain (mdx mice) and is considered a model of Duchenne muscular dystrophy (Deasy, Gharaibeh *et al.*, 2005) (Figure 10.13).

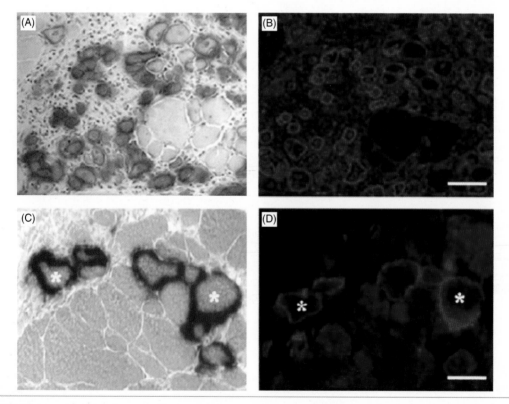

Figure 10.13 Muscle-derived stem cells from mdx mice were stably transfected with a plasmid encoding both Lac Z and dystrophin genes and injected intramuscularly into hind limb (panels A-B) or i.v. (panels C-D) of mdx mice. After 7 days the animals were sacrified. Several newly-formed Lac Z positive myofibers (panels A, C asterisks) co-expressing dystrophin (panels B, D asterisks) were observed in the limb muscle of the mice (from Lee, J.Y., Qu-Petersen, Z. *et al.* (2000). Clonal isolation of muscle-derived cells capable of enhancing muscle regeneration and bone healing. *J Cell Biol*, 150: 1085).

Although researchers have initially focused on the utilization of these cells with regard to their myogenic capacity, it is becoming increasingly clear that such cells may be used in the treatment of non-myogenic disorders such as cartilage, bone, tendon, hematopoietic disorders.

10.6 Adult stem cells can cross lineage-specific boundaries

10.6.1 'Transgermal' plasticity of bone marrow-derived skeletal (mesenchymal) stem cells

As the list of tissues that contain post-natal stem cell has expanded, so has the number of reports suggesting the highly controversial notion that tissue specific stem cells may have the ability to cross lineage specific boundaries; that is, that they are 'plastic' (reviewed in Lakshmipathy and Verfaillie, 2005; Quesenberry, Dooner et al., 2005). In particular, stem cells present in bone marrow have emerged as exceptionally interesting in this regard. Using whole marrow or purified (or enriched) hematopoietic stem cells (HSCs) or bone marrow stromal cells (BMSCs) (a subset of which are skeletal stem cells, 'mesenchymal' stem cells), cells of donor origin have been identified in a large number of recipient tissues, including those of endodermal and ectodermal origin, following their infusion either locally or systemically (reviewed in Grove, Bruscia et al., 2004). However, the level of 'conversion' appears to vary. The variability noted from tissue to tissue and from one study to another may reflect the nature of the model system employed; e.g., in an intact organism, versus ones in which disease or injury are present (Quesenberry, Dooner et al., 2005). In intact animals, the level of 'converted' bone marrow cells of donor origin is vanishing small. Marrow cells have been found to engraft at low levels at the sites of injury in skeletal muscle and to produce dystrophin (Ferrari, Cusella-De Angelis et al., 1998). In a study utilizing mice with tyrosinemia type 1, which ultimately leads to liver failure, highly enriched HSCs engrafted in the degenerating liver (by up to 50%) and restored liver function (Lagasse, Connors et al., 2000). Following lethal irradiation, the progeny of a single HSC not only repopulated marrow, but also were found in multiple epithelial tissues including the lungs, the gastrointestinal tract, the liver and skin (Krause, Theise et al., 2001). Bone marrow cells (unfractionated, HSCs and BMSCs) have been suggested in a number of studies to have the ability to form cardiomyocytes (reviewed in Menasche, 2005), and neuronal tissue (reviewed in Ortiz-Gonzalez, Keene et al., 2004).

10.6.2 Stem cell plasticity: fact, fusion or fiction?

While these studies are highly intriguing, proof of true plasticity by rigorous criteria has been lacking in most (Lakshmipathy and Verfaillie, 2005), and some observations have not been reproduced. True plasticity must be defined by the ability of the progeny of a single cell to not only express a certain number of markers characteristic of another cell type, but to actually function as that cell type. It can be seen that plasticity could occur in three different ways.

In the case of transdifferentiation, a cell with a particular phenotype converts to another phenotype without cell division, as exemplified by the conversion of bone marrow stromal cells to marrow adipocytes during aging. Transdetermination is characterized by the conversion of one type of stem cell into another, which may be the mode by which a purified HSC is made able to reform all types of nerve cells when placed in the brain. Plasticity may also be evoked by dedifferentiation of cell to a more primitive cell type that then differentiates into another phenotype, as has been demonstrated in the regeneration of a newt limb (reviewed in Fang, Alison et al., 2004). Irrespective of the way that cells become plastic, it is still an open question as to whether such truly plastic stem cells, capable of transgermal differentiation exist in the adult mammalian organism, or whether they are created by *ex vivo* manipulation.

It has often been stated that for proof of principle, plasticity of a given cell must be 'robust and persistent' (Lakshmipathy and Verfaillie, 2005). While this is most certainly necessary from a tissue regeneration point of view, conversion as a biological phenomenon is of scientific interest, irrespective of its frequency. Determination of the factors that regulate plasticity will undoubtedly lead to better ways to manipulate various populations of cells. It should be noted that some studies have challenged the concept of stem cell trans-differentiation potential by showing that cell fusion is a main mechanism by which HSC acquire the function of mature hepatocytes (Wang, Willenbring *et al.*, 2003; Vassilopoulos, Wang *et al.*, 2003). True plasticity would not be a function of donor cell fusion with a recipient cell type of another phenotype. Therefore, investigators should test, when possible, whether cell fusion is responsible. However, within the context of tissue regeneration, that is not to say that fusion is not of interest, as fusion is a normal event in many tissues, including liver, skeletal muscle and Purkinje cells. Furthermore, fusion of a donor stem cell may rejuvenate a recipient cell via nuclear reprogramming and encourage regeneration. Whether current observations are due to true plasticity or cell fusion, further work is needed on the technological aspects of the phenomenon for effective translation into clinical strategies for tissue regeneration.

10.7 Expansion of the stem cell compartment through cell culture

10.7.1 Stem cells versus progenitor cells

The ability to purify post-natal stem cells for direct application to ensure long-term tissue regeneration remains a major goal, with the exception of HSCs. However, reconstitution of two- and three-dimensional tissues will undoubtedly require larger numbers of cells than can be easily purified. Post-natal stem cells cannot be amplified in number, as can embryonic stem cells *in vitro*. The decrease in frequency of stem cells during culture must be taken into consideration. Due to the postulated asymmetric division of post-natal stem cells, attempts to expand them *ex vivo* dilute the population with more committed transiently amplifying and differentiated cells (Bianco, Kuznetsov *et al.*, 2006) (see Figure 10.1). Post-natal stem cells are not immortal, consequently, in tissues with a high rate of turnover, over dilution can affect the long-term efficacy. Two- and three-dimensional tissues present another problem. Small neighborhoods may be effectively regenerated, by a single stem cell, but recovery of vast territories may not be possible due to constraints on migration of cells through solid structures. Nonetheless, *ex vivo* expanded populations, such as they are, may be very effective in restoring tissues with low rates of turnover (Robey and Bianco, 2004). Furthermore, the paracrine effects of factors produced by transiently amplifying and differentiating cells on their parent stem cells cannot be overlooked, and are likely involved, at least in part, in stem cell maintenance.

10.7.2 Stem cell isolation and purification: methods and caveats

In all systems other than the hematopoietic system, isolation of stem cells means isolation in culture. In the hematopoietic system, stem cells cannot be grown in culture (although it is claimed that they can be partially activated to a limited mitotic activity by the use of defined cytokine cocktails). Therefore, in the hematopoietic system isolation of stem cells and purification of stem cells coincide with one another. One can only isolate a purified population, or nothing at all. The ability to purify a stem cell population coincides with the ability to prospectively isolate stem cells from the parent tissue. Prospective isolation of stem cells in turn requires the definition of a surface phenotype suited for use in immunoselection procedures. The HSC is the only stem cell type that has been effectively purified, and the mouse is the only species in which this has been actually accomplished (Figure 10.14). Purification of murine

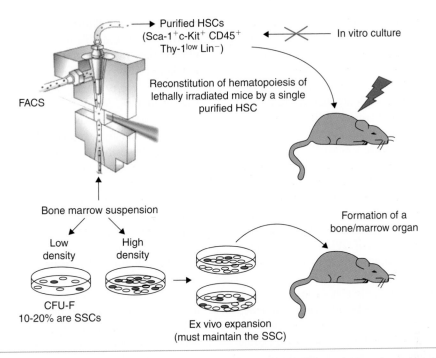

Figure 10.14 Using single cell suspensions of bone marrow, it is possible to isolate hematopoietic stem cells (HSCs) via FACS by virtue of their cell surface markers (Sca-1+ cKit+ CD45+ Thy-1^{low} Lin$^-$). This is the only system in which stem cells can be 'purified' (from mice), as demonstrated by the FACS isolated cells' ability to reconstitute hematopoiesis in lethally irradiated mice. HSCs cannot be grown or maintained in culture. Conversely, skeletal stem cells within the bone marrow stromal population can not be purified by cell surface markers to date. Plating bone marrow at low density results in purification of the Colony Forming Unit-Fibroblasts (CFU-Fs), of which 10–20% are multi-potent skeletal stem cells (SSCs). In order to obtain enough cells for tissue regeneration, these cells must be expanded in culture, but keeping in mind that due to asymmetric division and the presence of transiently amplifying cells, SSCs will become diluted with increasing passage number.

HSCs has been made possible by the recognition of a defined surface phenotype, which has been missing so far both for human HSCs, and for all other kinds of putative stem cells in other systems.

The rationale for pursuing prospective isolation of purified stem cell fractions is twofold. On the one hand, it is of theoretical importance to define the characteristics of the true stem cell subset as opposed to the varied range of committed progenitors that are found in any lineage. On the other hand, it is assumed that the availability of a purified population of stem cells would potentiate all applicative uses – present and established, as well as future and

to be developed – of stem cells in regenerative medicine. For example, a much lower number of cells is required for bone marrow reconstitution, and reliable and predictable long-term reconstitution can be predicted if enriched (in humans) or purified (in the mouse) hematopoietic stem cells are at hand.

10.7.3 *In vitro* culture: a critical step for future stem/progenitor cell applications

Hematopoietic stem cells are isolated through immunoselection only; neural stem cells and epithelial stem cells are isolated through culture only; mesenchymal stem cells are isolated through culture (Figure 10.14), or through a combination of immunoselection and culture. Even when immunoselection is used (based, for example, on the use of antibodies against STRO-1, a mouse monoclonal antibody that has been used to identify all CFU-F, but whose epitope is currently unknown), mesenchymal stem cells are not necessarily purified. In the best-case scenario, it is the whole of the clonogenic population (which is highly heterogeneous in itself) that can be purified. More importantly, selection by adherence to plastic in primary cultures established at clonal density by itself results in the isolation of CFU-Fs, and there is no evidence that higher numbers of CFU-Fs can be established in culture by immunoselection with STRO-1 compared to selection by adherence.

For all kinds of stem cells that cannot be transplanted by infusion into the bloodstream (like HSCs can be), cell number is critical for all prospective applicative uses. Hence, all kinds of stem cells other than the HSCs need to be 'expanded' in culture, since the absolute number of cells obtained even from a significant tissue volume is regularly insufficient for applicative use (Figure 10.14). This circumstance makes the purification of stem cells less important in solid phase tissues than in blood – since cells will have to be cultured anyway, isolation through culture will be good enough in most cases. A second important caveat moderating the excessive obstination in pursuing prospective purification of non-hematopoietic stem cells emanates from the fact that hierarchically 'lower' progenitors do contribute to tissue repair in all prospective applications to cell therapy. Hence, the actual benefit of handling a 'pure' population of stem cells remains to be determined.

10.7.4 Primary cultures versus cell lines

Continuous cell lines have often been used as model systems of post-natal stem cell activity (e.g., C3H10T1/2 cells as a model for 'mesenchymal' stem cells). Studies have focused on defining the factors that direct these cell lines, which are considered to be uncommitted, and able to differentiate into multiple phenotypes. However, in many cases, these studies are flawed based on the lack of verification of differentiation capacity by the gold standard assay, *in vivo* transplantation. Many lines are derived from non-human species and have been of particular interest in cardiomyocyte regeneration, but would obviate the need for immune suppression (Xiao, Min *et al.*, 2004). While it has been a long-standing dream of tissue engineers to generate cell lines that can be used 'off the shelf' (Conrad and Huss, 2005), it is not at all clear that non-autologous stem cells, and more importantly, their differentiated progeny, would escape from immune surveillance of the recipient without induction of tolerance or immunosuppression. Furthermore, by definition, a cell line is continuous; that is, immortalized either by spontaneous mutation or by design. Immortalization is not always reflective of tumorigenicity, however, it has recently been reported that bone marrow stromal cells transfected with the catalytic subunit of telomerase, hTERT, do form tumors (Serakinci, Guldberg *et al.*, 2004). Current techniques for immortalization cannot ensure that the behavior of even a patient's own cells could be appropriately controlled.

10.8 Can we use allogeneic or xenogeneic stem cells?

10.8.1 Autografts versus allografts

In terms of genetic disparity between donor and recipient, grafts can be classified as autografts, isografts, allografts or xenografts. Autografts, from one part of the body to another of the same individual, do not elicit rejection. Isografts between genetically identical individuals, such as monozygotic twins, do not express antigen foreign to the recipient and do not activate a rejection response. Allograft is when one individual donates an organ or tissue to a genetically different individual belonging to the same species, having allelic variants of genes of the major histocompatibility complex (MHC). In this case, the cells of the allograft will express alloantigens recognized as foreign by the recipient. Xenograft is when donor and recipient belong to different species. In this case, the maximal genetic disparity occurs, and a xenograft is generally rapidly rejected.

Graft rejection can be the final outcome of the activation of various mechanisms of humoral and cellular immunity, both specific and non-specific (Figure 10.15). These responses can be reduced by selection of donor and recipient combinations matched for their MHC molecules, and immunosuppressive agents can be used to block transplant rejection. Allogeneic stem cell transplantation, in particular hematopoietic stem cells, has a significant potential for the treatment of malignancy, autoimmunity (Ikehara, 1998), genetic disorder (Staba, Escolar et al., 2004), and can also be used to facilitate gene therapy and solid organ transplantation (Seung, Mordes et al., 2003).

In allogeneic bone marrow grafts, a special situation may occur in which immunocompetent T lymphocytes transplanted with the bone marrow can attack the recipient unable to reject them for immaturity or immunosuppression: the graft-versus-host disease (GVHD). A beneficial graft-versus-leukemia or a graft-versus-tumor effect, analogous to GVHD effect, could be generated in the treatment of hematological and non-hematological malignancies using allogeneic hematopoietic stem cell transplant, limiting radiation and chemotherapy to doses sufficient for donor stem cell engraftment and reduction of toxicity (Kolb, Schmid et al., 2004).

In the field of organ transplantation, the therapeutic potential of allogeneic hematopoietic transplant lives in its capacity to induce chimerism (the coexistence of cells from genetically different individuals) in the host, leading to the induction of central immunotolerance, which is the most robust state of donor-specific transplantation tolerance known (Seung, Mordes et al., 2003).

MHC-mismatched cord blood should be considered an acceptable source of hematopoietic stem cell grafts for adults in the absence of an MHC-matched adult donor. Cord-blood grafts from unrelated donors

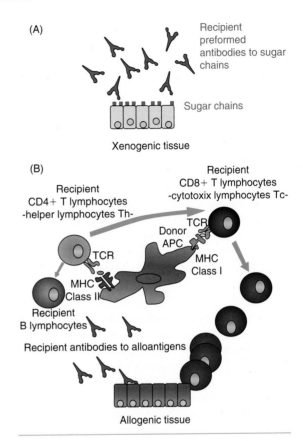

Figure 10.15 Mechanisms of xenograft and allograft rejection. Rejection of a xenograft is caused by recipient's circulating natural antibodies against sugar on the cell membrane of the xenogenic tissue (A). Allograft rejection involves the direct action of CD8+ cytotoxic lymphocytes recognizing alloantigen presented by MHC class I molecules (direct recognition) and the indirect action of CD4+ T lymphocytes recognizing alloantigens presented by MHC class II molecules and activating CD8+ clones and antibody-producing B lymphocytes (B). APC = Antigen Presenting Cell; TCR = T Cell Receptor.

have been used successfully in children, where these grafts reconstitute hematopoiesis more slowly than do bone marrow grafts, but the incidence and severity of GVHD are not excessive, and graft-versus leukemia effects are well maintained. A major limitation to the use of cord-blood grafts in adults is the concern that these grafts have an insufficient number of precursor cells as compared with bone-marrow grafts (Laughlin, Eapen *et al.*, 2004).

Despite the problems related to allograft reaction, issues related to cell harvesting and culturing, quality control and safety makes the use of allogenic stem cells a feasible approach for therapy of disparate pathologies. In humans, allogeneic transplantation of mesenchymal precursors (BMSC) in patients affected by osteogenesis imperfecta resulted in some engraftment of donor osteoblasts (Horwitz, Prockop *et al.*, 1999).

10.8.2 Stem cells can modulate the immune response

The immune response is subject to a variety of control mechanisms. These controls restore the immune system to a resting state when responsiveness to a given antigen is no longer required. Cell-mediated immune responses are controlled by molecular interactions leading to inhibition of activation signals. There are also different down-regulatory pathways involving cytokines resulting in suppression of the immune response, Transforming growth factor β (TGFβ) and Interleukin 10 (IL-10) among the others (Khaled and Durum, 2002). Next to elements of the immune system (Shevach, 2002), other cell types (Caspi, Roberge *et al.*, 1987), including self-renewing progenitor cells isolated from the mammalian central nervous system or bone marrow (Hori, Ng *et al.*, 2003), can block lymphocyte activation and proliferation.

The reason why cells such as BMSCs have a direct role in regulating the immune response is still unclear; it might be the effect of their place in the bone marrow. In adults, the cellular elements of the blood, including lymphocytes and the other cells of the immune system, are produced in the bone marrow. The development of the bone marrow cavity is a coordinated process in which blood precursors migrate and colonize spaces carved out of embryonic bone and cartilage. Thus, an intimate physical association between bone cells

and blood cells is established early in life. Several animal models strongly implicate mesenchymal precursor cells in hematopoiesis by virtue of creating a niche (Taichman, 2005). In the bone marrow, hematopoietic precursors reside close to endosteal surfaces, and it has been shown that osteoblasts produce many factors essential for the survival, renewal, and maturation of hematopoietic stem cells. Co-transplant of BMSC may improve the outcome of allogeneic hematopoietic stem cell transplantation by promoting hematopoietic engraftment and limiting GVHD in patients (Le Blanc, Rasmusson *et al.*, 2004). Moreover, the disruption or perturbation of normal osteoblastic function has a profound and central role in defining the operational structure of the hematopoietic niche (Taichman, 2005). In mouse models, bone marrow-derived multipotent stem cells have been used as immosuppressive agents also for the treatment of autoimmune disease, particularly experimental autoimmune encephalitis (Zappia, Casazza *et al.*, 2005).

10.8.3 Mechanisms by which BMSCs can modulate the immune response are only partially known

The mechanisms by which stem cells exert their role in the immune system are pleiotropic and redundant, involving pathways such as the activation of the Program death 1 (PD1) pathway in a cell-to cell contact fashion, and the release of TGFβ1 (Figure 10.16). Disparate types of cells are the potential targets, among them T and B-lymphocytes, but other elements cannot be excluded (Di Nicola, Carlo-Stella *et al.*, 2002; Krampera, Glennie *et al.*, 2003; Potian, Aviv *et al.*, 2003; Augello, Tasso *et al.*, 2005; Maccario, Podesta *et al.*, 2005). It has been also suggested that stem cells might be able to educate T regulatory lymphocytes that are non-immunologically restricted, that are included in the whole population of spleen cells, and that are able to continue the inhibitory activity initiated by stem cells. Targeted lymphocytes were rendered unresponsive to proliferation stimuli by down regulation of the expression of Stat5b, a transducer of the activation

cells in children with osteogenesis imperfecta. *Nat Med*, 5(3): 309–313.

Ikehara, S. (1998). Autoimmune diseases as stem cell disorders: normal stem cell transplant for their treatment. *Int J Mol Med*, 1(1): 5–16.

Jiang, Y., Jahagirdar, B.N., *et al.* (2002). Pluripotency of mesenchymal stem cells derived from adult marrow. *Nature*, 418(6893): 41–49.

Khaled, A. and Durum, S. (2002). Lymphocide: cytokines and the control of lymphoid homeostasis. *Nat Rev Immunol*, 2(11): 817–830.

Kolb, H., Schmid, C., *et al.* (2004). Graft-versus-leukemia reactions in allogeneic chimeras. *Blood*, 103(3): 767–776.

Kopp, H.-G., Avecilla, S.T., *et al.* (2005). The bone marrow vascular niche: home of HSC differentiation and mobilization. *Physiology*, 20(5): 349–356.

Krampera, M., Glennie, S., *et al.* (2003). Bone marrow mesenchymal stem cells inhibit the response of naive and memory antigen-specific T cells to their cognate peptide. *Blood*, 101(9): 3722–3729.

Krause, D.S., Theise, N.D., *et al.* (2001). Multi-organ, multi-lineage engraftment by a single bone marrow-derived stem cell. *Cell*, 105(3): 369–377.

Kuznetsov, S.A., Krebsbach, P.H., *et al.* (1997). Single-colony derived strains of human marrow stromal fibroblasts form bone after transplantation *in vivo. J Bone Miner Res*, 12(9): 1335–1347.

Lagasse, E., Connors, H., *et al.* (2000). Purified hematopoietic stem cells can differentiate into hepatocytes *in vivo. Nat Med*, 6(11): 1229–1234.

Lakshmipathy, U. and Verfaillie, C. (2005). Stem cell plasticity. *Blood Rev*, 19(1): 29–38.

Laughlin, M., Eapen, M., *et al.* (2004). Outcomes after transplantation of cord blood or bone marrow from unrelated donors in adults with leukemia. *N Engl J Med*, 351(22): 2265–2275.

Le Blanc, K., Rasmusson, I., *et al.* (2004). Treatment of severe acute graft-versus-host disease with third party haploidentical mesenchymal stem cells. *Lancet*, 363(9419): 1439–1441.

Lee, J.Y., Qu-Petersen, Z., *et al.* (2000). Clonal Isolation of Muscle-derived Cells Capable of Enhancing Muscle Regeneration and Bone Healing. *J Cell Biol*, 150(5): 1085–1100.

Lie, D.C., Song, H., *et al.* (2004). Neurogenesis in the adult brain: new strategies for central nervous system diseases. *Annu Rev Pharmacol Toxicol*, 44: 399–421.

Ma, D.K., Ming, G.L., *et al.* (2005). Glial influences on neural stem cell development: cellular niches for adult neurogenesis. *Curr Opin Neurobiol*, 15(5): 514–520.

Maccario, R., Podesta, M., *et al.* (2005). Interaction of human mesenchymal stem cells with cells involved in alloantigen-specific immune response favors the differentiation of CD4+ T-cell subsets expressing a regulatory/suppressive phenotype. *Haematologica*, 90(4): 516–525.

MacKenzie, D., Hullett, D., *et al.* (2003). Xenogeneic transplantation of porcine islets: an overview. *Transplantation*, 76(6): 887–891.

Menasche, P. (2005). Stem cells for clinical use in cardiovascular medicine: current limitations and future perspectives. *Thromb Haemost*, 94(4): 697–701.

Morrison, S.J. and Weissman, I.L. (1994). The long-term repopulating subset of hematopoietic stem cells is deterministic and isolatable by phenotype. *Immunity*, 1(8): 661–673.

Ortiz-Gonzalez, X.R., Keene, C.D., *et al.* (2004). Neural induction of adult bone marrow and umbilical cord stem cells. *Curr Neurovasc Res*, 1(3): 207–213.

Osawa, M., Hanada, K., *et al.* (1996). Long-term lymphohematopoietic reconstitution by a single CD34-low/negative hematopoietic stem cell. *Science*, 273(5272): 242–245.

Oshima, H., Rochat, A., *et al.* (2001). Morphogenesis and renewal of hair follicles from adult multipotent stem cells. *Cell*, 104(2): 233–245.

Owen, M. and Friedenstein, A.J. (1988). Stromal stem cells: marrow-derived osteogenic precursors. *Ciba Found Symp*, 136: 42–60.

Palmer, T.D., Willhoite, A.R., *et al.* (2000). Vascular niche for adult hippocampal neurogenesis. *J Comp Neurol*, 425(4): 479–494.

Pellegrini, G., Golisano, O., *et al.* (1999). Location and clonal analysis of stem cells and their differentiated progeny in the human ocular surface. *J Cell Biol*, 145(4): 769–782.

Pellegrini, G., Traverso, C.E., *et al.* (1997). Long-term restoration of damaged corneal surfaces with autologous cultivated corneal epithelium. *Lancet*, 349(9057): 990–993.

Potian, J., Aviv, H., *et al.* (2003). Veto-like activity of mesenchymal stem cells: functional discrimination between cellular responses to alloantigens and recall antigens. *J Immunol*, 171(7): 3426–3434.

Prindull, G. (2005). Hypothesis: cell plasticity, linking embryonal stem cells to adult stem cell reservoirs and metastatic cancer cells? *Exp Hematol*, 33(7): 738–746.

Quesenberry, P.J., Dooner, G., *et al.* (2005). Stem cell biology and the plasticity polemic. *Exp Hematol*, 33(4): 389–394.

Robey, P.G. and Bianco, P. (2004). Stem cells in tissue engineering. In *Handbook of Adult and Fetal Stem Cells* (Lanza, R.P., ed.). San Diego: Academic Press, pp. 785–792.

Rochat, A., Kobayashi, K. and Barrandon, Y. (1994). Location of stem cells of human hair follicles by clonal analysis. *Cell*, 76(6): 1063–1073.

Sanchez, M.J., Holmes, A., *et al.* (1996). Characterization of the first definitive hematopoietic stem cells in the AGM and liver of the mouse embryo. *Immunity*, 5(6): 513–525.

Serakinci, N., Guldberg, P., *et al.* (2004). Adult human mesenchymal stem cell as a target for neoplastic transformation. *Oncogene*, 23(29): 5095–5098.

Seung, E., Mordes, J., *et al.* (2003). Hematopoietic chimerism and central tolerance created by peripheral-tolerance induction without myeloablative conditioning. *J Clin Invest*, 112(5): 795–808.

Shevach, E. (2002). CD4+ CD25+ suppressor T cells: more questions than answers. *Nat Rev Immunol*, 2(6): 389–400.

Spangrude, G.J., Heimfeld, S., *et al.* (1988). Purification and characterization of mouse hematopoietic stem cells. *Science*, 241(4861): 58–62.

Staba, S., Escolar, M., *et al.* (2004). Cord-blood transplants from unrelated donors in patients with Hurler's Syndrome. *N Engl J Med*, 350(19): 1960–1969.

Taichman, R. (2005). Blood and bone: two tissues whose fates are intertwined to create the hematopoietic stem-cell niche. *Blood*, 105(7): 2631–2639.

Tavian, M. and Peault, B. (2005). Embryonic development of the human hematopoietic system. *Int J Dev Biol*, 49(2–3): 243–250.

Till, J.E. and McCulloch, E.A. (1961). A direct measurement of the radiation sensitivity of normal mouse bone marrow cells. *Radiat Res*, 14: 213–222.

Udomsakdi, C., Lansdorp, P.M., *et al.* (1992). Characterization of primitive hematopoietic cells in normal human peripheral blood. *Blood*, 80(10): 2513–2521.

Vassilopoulos, G., Wang, P.R., *et al.* (2003). Transplanted bone marrow regenerates liver by cell fusion. *Nature*, 422(6934): 901–904.

Wang, X., Willenbring, H., *et al.* (2003). Cell fusion is the principal source of bone-marrow-derived hepatocytes. *Nature*, 422(6934): 897–901.

Watt, F.M. and Hogan, B.L. (2000). Out of Eden: stem cells and their niches. *Science*, 287(5457): 1427–1430.

Xiao, Y.F., Min, J.Y., *et al.* (2004). Immunosuppression and xenotransplantation of cells for cardiac repair. *Ann Thorac Surg*, 77(2): 737–744.

Zappia, E., Casazza, S., *et al.* (2005). Mesenchymal stem cells ameliorate experimental autoimmune encephalomyelitis inducing T cell anergy. *Blood*, 04: 1496.

Zhang, J., Niu, C., *et al.* (2003). Identification of the haematopoietic stem cell niche and control of the niche size. *Nature*, 425(6960): 836–841.

Zhu, J. and Emerson, S.G. (2004). A new bone to pick: osteoblasts and the haematopoietic stem-cell niche. *Bioessays*, 26(6): 595–599..

Chapter 11

Cell culture: harvest, selection, expansion, and differentiation

Tommi Tallheden

Chapter objectives:

- To obtain knowledge in principles and methods for clinical cell-based tissue engineering
- To recognize different methods for isolation of primary cells
- To understand principles for selection and purification of primary cells
- To be aware of how the choice of culture conditions influences proliferation and differentiation of cells

- To understand how expansion *in vitro* changes the differentiation state of the cells
- To identify methods and conditions for differentiation of cells *in vitro*

"The human body is a machine…"

> This statement was made by the French philosopher, mathematician, and scientist, René
> Descartes, March 31, 1596–February 11, 1650

11.1 Introduction

Descartes' view of the human body was at that time was based on primary, although not scientifically proven, observations and was probably considered as total nonsense by many. However, the Descartes view of the body as a machine, with parts that could be replaced or removed temporarily from the body

for refurbishing, is today becoming a reality through techniques applied in clinical tissue engineering (TE).

Tissue engineering utilizes a challenging and promising method for treatment of human organ malfunctions and disorders. Several TE strategies include a combination of human cells with biomaterials (scaffolds) before the construct are implanted into the host (Figure 11.1). The logic of using cells instead of only

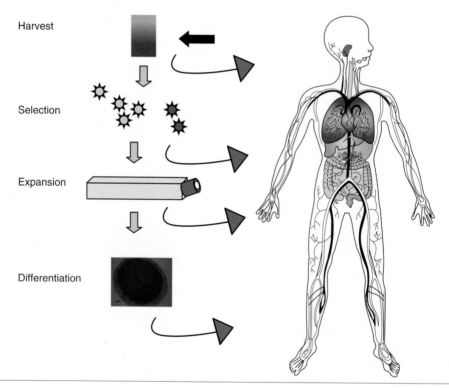

Figure 11.1 In most tissue engineering applications aimed at restoring lost function of organs or tissue, cells are harvested, processed and expanded in culture before being transplanted back into same patient (autologous) or into another patient (allogenic). Even though this scheme demonstrates the most common principle, some techniques do not require expansion of the cell number and consists, therefore, only of the purification of isolated cells from aspirates or biopsies, before re-implantation takes place.

scaffolds is that the cells will function as continuous producers of the missing components and thereby restoring the malfunction of the tissue. The implanted cells could also take an active part in restoring the function of the inherent cells of the tissue or organ.

The cells are isolated with a variety of methods from either the patient (autologous) or from donated organs or tissues (allogenic). In some strategies, the cells are implanted back into the patient after being isolated; more commonly, the cells are transported to a specialized laboratory for further processing. At the laboratory, the cells can be purified and/or expanded before being sent back for implantation. In some cases, pieces of tissue (biopsies) are harvested and from these, the cells of interest are isolated. After the isolation, the cell number can be further multiplied *in vitro*, although, it is of greatest importance that the cells are treated in the proper way in order to keep their therapeutic potential. The general aim of this chapter is to give an overview of different methods for cell harvest, *in vitro* culture expansion and differentiation of cells for clinical TE. Particular focus will be drawn on methods describing the use of primary human cells.

11.2 Harvest

The initial harvest of cells or pieces of tissues can be made using a range of different methods and procedures. Optimal harvest methods differ depending on the cell source and cell type used in a given clinical application. For some applications, it is best to make an aspiration under local anesthesia, whereas other applications require biopsies to be harvested with a fiberoptic endoscope. Most cell harvests originate from some kind of diagnostic procedure. The methods, therefore, are most often developed because of the high demands of patient safety (especially sterility). However, in many cases, the material harvested for diagnostic purposes will not be implanted back into the patient but will instead be processed for histological examination. In contrast, for materials harvested for implantation, special care has to be taken during handling, storing and transportation of the cells to the laboratory for further processing.

11.2.1 Bone marrow aspirates

A common source of cells for tissue engineering applications is the bone marrow. This is because marrow is easily accessible, can be harvested autologous and contain several different cell types of interest for TE, including mesenchymal stem cells. These cells have been hypothesized to have the ability to differentiate into most of the mesenchymal tissues (Caplan, 1991; Lewthwaite, Bastow et al., 2006) and are therefore of great interest for clinical TE. The bone marrow is normally harvested under local anesthesia from the upper part of the hip (the iliac crest), but it can also be isolated from the sternum. To access the bone marrow, a strong needle (16 gauges) is led through the skin and the outer part of the bone until it reaches the softer, central part of the bone (bone marrow) (Figure 11.2A). From this softer part, the bone marrow is sucked out with a heparinized syringe. The heparin is necessary for preventing coagulation, which would make it impossible to harvest single cells. During the aspiration, the first milliliters of the marrow will contain a mixture of hematopoetic cells, adipocytes, endothelial progenitor cells and ostoeprogenitor cells. However, if more than 2 ml of marrow is harvested from a given location, peripheral blood is drawn into the syringe, diluting the number of mesenchymal stem cells harvested (Muschler, Boehm et al., 1997).

11.2.2 Uroepithelial cell flush-out

For harvesting autologous uroepithelial cells for tissue engineering of the bladder, it is possible to use a flush-out method (the cells are released and collected from the bladder wall by repeated filling and emptying with an isotonic salt solution) (Figure 11.2B). This method is especially useful for harvesting cells from children as it does not require any anesthesia (Fossum, Gustafson et al., 2003).

(A)

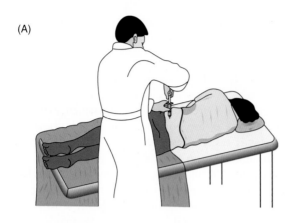

(B)   Bladder and sphincter muscles

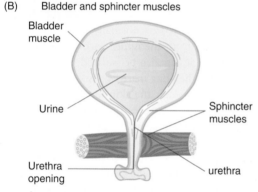

(C)

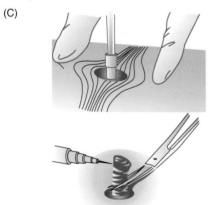

Figure 11.2 Cells can be harvested by using different methods. (A) Aspiration of bone marrow from iliac crest under local anaesthesia; (B) Harvest of epithelial cells from the bladder by flushing with saline; (C) Dermal core skin biopsy.

11.2.3 Tissue biopsies (skin, cartilage, lungs, liver, kidney and heart)

For obtaining starting materials for skin TE (keratinocytes, melanocytes and dermal fibroblasts) a punch biopsy can be used. The biopsy is taken under local anesthetic with a small, sharp, normally round tool (punch). The punch is placed over the actual area to be harvested, pushed down, and slowly rotated to remove a circular piece of skin (Figure 11.2C). The skin sample is then lifted with forceps or a needle, and then cut from the tissue below it. Sometimes, when larger skin samples are harvested, sutures may be needed to close any remaining hole after the biopsy.

The most common way of harvesting articular cartilage for cartilage repair procedures is isolating biopsies during diagnostic arthroscopic examinations. During this procedure, the surgeon is able to define the size and severity of the defect and by the use of a suction curette or a raspatorium isolate biopsies. The choice of instrument, as well as the experience of the surgeon, influences the size of the biopsies. In a recent comparative study between five different experienced surgeons at three different centers, the average size of the harvested articular cartilage biopsies differed between 176–239 mg corresponding to 1800–3300 cells/mg (Sjögren Jansson, Tallheden *et al.*, 2004).

In the harvest of nasal and auricular cartilage, it is necessary to remove the skin before accessing the cartilage. Normally this is made through opening the skin and then accessing the cartilage; although successful harvesting of auricular cartilage by using a 12-gauge core biopsy needle through the skin has also been described (Megerian, Weitzner *et al.*, 2000). For several internal organs such as the lungs, liver, kidney and heart, biopsies can be harvested under fluoroscopic control (a combined fiberoptic endoscope and an X-ray machine). Viable cells can be isolated from the biopsies after mechanical and/or enzymatic disruption (see below) (Seferovic, Ristic *et al.*, 2003).

11.2.3.1 Cell retrieval from harvested tissues
From marrow aspirates and bladder flush outs, cells are retrieved in suspension, which facilitates further

State of the Art Experiment

Intraoperative cell based tissue engineering

Cell-based TE applications using autologous cells based on *in vitro* cell expansion and/or differentiation have been used in the clinic for over 25 years (O'Connor, Mulliken *et al.*, 1981; Brittberg, Lindahl *et al.*, 1994). During these years, the basic knowledge of the potential of using autologous cells has increased and led to improvements and modifications of the original protocols. Another reason for these modifications is that the original protocols are logistically complicated, relatively expensive and are therefore limited to a small patient population.

One recent method for articular cartilage resurfacing that has kept the original principle of using autologous cells, and is utilizing the human body itself as a bioreactor, has been described by Lu *et al.* (2006). The method is principally based on small biopsies that have been harvested from a non-weight-bearing area of the articular cartilage, processed within the operating room, and transplanted back into the same patient under the same operational séance. Before re-transplanting the cartilage, it is washed with PBS, minced into small fragments (~1 mm^3) and imbedded into a polyglycolide/polylactide (PGA/PLA) or polyglycolide/polycaprolactone (PGA/PCL) foam reinforced with polydiaxanone (PDS) mesh.

This method is currently under international multicenter clinical trials and these studies will demonstrate the usefulness of the method in human.

This approach is promising because it uses current potential knowledge to make use of autologous cells in a state-of-the-art scaffold. It also simplifies the current cell-based procedures for cartilage repair, and eliminates the logistical problems as well as dramatically decreases the costs. This, in turn, should lead to the method becoming available for larger patient populations.

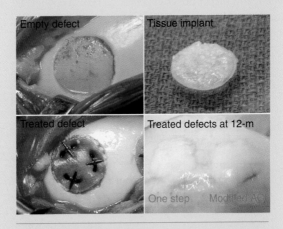

Empty full thickness defects in the trochlear groove (upper left); tissue implant containing cells and covered with fibrin glue (upper right); defect treated with implant and stapled with resorbable PGA/PDS fixatives (lower left); and treated defect after 12 months (lower right) (courtesy of Dr Lu and Dr Francois, DePuy Biologics).

processing of cells. In contrast, for tissue biopsies cells have to be dissociated from each other and from the surrounding extracellular matrix (ECM). Cell release from the ECM is normally started by mechanical disruption. The instrumentation and method depends on the amount of tissue and the type of ECM. For example, for adipose tissues, vortexing in digestion buffer is sufficient, while for other tissues such as cartilage, dicing with a scalpel is needed. The disrupted biopsies can then be directly implanted (Lu, Dhanaraj *et al.*, 2006), further processed with enzymes or placed in culture (explant culture). If the disrupted biopsies are placed in culture, cells will start migrating from the explants and these cells can be further expanded in culture.

Some cell–cell adhesions are mediated by cell adhesion molecules that are calcium dependent (cadherins) and are, therefore, sensitive for chelating agents such as ethylenediaminetetraacetic acid (EDTA). In this case, cells can be released by placing the biopsy in EDTA containing buffer. For most biopsies, enzymes such as protease, collagenase, dispase and trypsin have to be used for disrupting the biopsy and releasing the cell from their extracellular matrix (ECM). These enzymes are specific to certain proteins and are normally used as single separate steps or in combination after each other.

However, because there is no consensus in nomenclature between the companies selling the enzymes, they have their own labeling systems for the preparations that are partially purified. For example, SIGMA indicates their crudest collagenase preparation as type I, and their pure collagenase as type XI, whereas Boeringer Mannhaim refers to the crudest as type A and the purest type D. It is important, therefore, to read the information provided by the companies in order to match the intended use. It has been shown that the most effective islet recovery from the human pancreas can be obtained with purified combinations of collagenase and elastase in specific ratios (Gill, Chambers *et al.*, 1995; Olack, Swanson *et al.*, 1999).

11.2.4 Portions of organs (liver, kidney, and pancreas)

The methods described above are used mainly for isolating small numbers (10^3–10^6) of autologous

---

Classical Experiment

Treatment of burn wounds with autologous cultured cells

In the middle of the 1970s, Rheinwald and Green (1975) published the clonal expansion of human epidermal cells *in vitro*. Five years later, the basic protocol of this method was used for treatment of severe burn wounds with autologous culture expanded epithelial cells (O'Connor, Mulliken *et al.*, 1981; Gallico, O'Connor *et al.*, 1984). The method is based on the isolation of a small biopsy form a patient and after treatment with enzymes the epidermal layer was separated from the dermal layer. The epidermal cells were then expanded in the presence of lethally irradiated 3T3 cells until the cells reached confluence, normally after 2 to 3 weeks. The confluent cells were transferred on to a Vaseline-impregnated gauze (2 × 2 cm) which was then transplanted back to wounded area of the patient. Since its introduction, this method has saved the lives of thousands of people and is still today a backbone for severe burn wound treatment.

One of the major drawbacks with the original method is that is takes several weeks for the cells to grow and during that time the patients will be suffering from their burn wounds. Therefore, methods for applying subconfluent keratinocytes, either in suspension together with carriers such as type I collagen, fibrin glue or hyaluronic acid, or mixed in with synthetic (polyurthetan, Teflon or Poly(hydroxyethyl Methacrylate)) or natural (acellular porcine dermis) biomaterials have been explored. The procedure has also been explored commercially by several companies under various brand names.

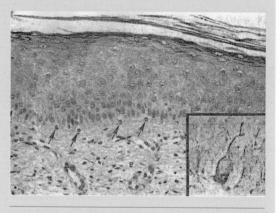

Microscopical appearance of epidermis 14 weeks after transplantation of culture expanded keratinocyte. The section is showing all layers but rete ridges are lacking (adopted from Gallico, G.G. et al. (1984). J New Eng J Med, 311(7): 448–451.

cells. In other cases, larger numbers of allogeneic cells can be isolated from a whole or part of a donated large organ such as a kidney, pancreas or the liver. As the highest priority of organs that become available from brain-dead, beating-heart donors is to use them for organ transplantation, only organs rejected by these programs then become available for investigators in academic or industrial settings. The reason for rejection could be because of infection, a high percentage of fat, or prolonged ischemia. All of these organs are highly vascularized and their metabolism is highly dependent on oxygen. A prolonged time of ischemia could therefore be detrimental for the viability and functionality of the cells.

Digestion of smaller pieces or whole organs can be made by perfusing the organ with buffers and enzymes (Reese and Byard, 1981; Strom, Jirtle et al., 1982). This is normally done by aseptically connecting silicon tubing to one of the larger arteries or veins (liver) into which the disrupting solutions can be delivered with a peristaltic pump. Most protocols start with perfusion of the organ with calcium- and magnesium-free buffers and EDTA. This step is then followed by enzymatic digestion after a cell suspension containing a mixture of different cell types (hematopoetic, epithelial, endothelial and stromal) is obtained. Most often, the cell suspension obtained after single rounds of enzymatic digestion results in a non-homogenous mixture of small cell aggregates and single cells. For some downstream processes, single cells (i.e. cell sorting) are requested and the cell suspension could therefore be further mechanical and/or enzymatic dissociated.

11.3 Selection

The harvested cells from most tissues contain a mixture of cell types and usually the downstream processing starts and ends with mixed cell populations, achieving only enrichment and not 'purification'. However, enrichment for specific cell types can be achieved in a number of ways.

11.3.1 Selective adhesion during *in vitro* culture

When isolated cells are placed in culture, some cells will adhere to the surface while others will stay in suspension. The adherence process is highly dependent on the culture conditions, expression of cell surface receptors and on the type of surface (see below). By modification of the surface or special coatings, specific cells can be isolated. One protein used for coating plastic surfaces for selective adherence is fibronectin. This protein is highly abundant in fetal and embryonic tissues but is also found in the extra cellular matrix of adult skin and cartilage. The receptor for fibronectin is called beta 1 integrin and is highly expressed on progenitor populations of keratinocytes (Jones and Watt, 1993) and chondrocytes (Dowthwaite, Bishop et al., 2004). The progenitor population can therefore be isolated from the original mixture of cells by brief plating on fibronectin coated dishes.

Another way to select a specific cell type is to adopt the time for cell adherence. By seeding a mixture of primary isolated cells onto plastic dishes and at given time intervals remove the non-adhering cells, a purified stem cell population can be isolated from a skeletal muscle (Jankowski, Haluszczak et al., 2001). The procedure is called the pre-plating method and is relatively time-consuming, but does not require any specific adherence proteins and can therefore be performed in a total autologous procedure.

The difference in adherence properties can also be used for purification of culture expanded cells before implantation. This is especially efficient for the removal of fibroblasts from adherent cultures of skeletal myocytes and is based on the fibroblasts' faster adherence to plastic (Chachques, Herreros et al., 2004). A clear advantage of this method is that it does not require any staining or special equipment; however, it is limited to highly active cells.

11.3.2 Gradient centrifugation

Another way to separate specific cell types, based on the size and density of the cells, is by density gradient centrifugation. There are two forms of density gradient centrifugation: rate zonal and isopycnic. The rate zonal gradient is generally created by layering solutions (i.e. sucrose) of varying densities (concentrations) in centrifugation tubes with the solution with the highest density at the bottom. The cells are then loaded on top of the gradient and the size difference between the cells affects their separation along with the density of the cells. Small cells move faster through the gradient than large cells, if the density of the cells is greater than the density of the medium in the tube at all points during the separation. The run is then terminated before the separated cells reach the bottom of the tube.

In the isopycnic separation of cells, the density range of the gradient medium in the tube encompasses all densities of the different cell types and is often made with polyvinylpyrrolidone coated silica beads (Percoll). The gradient is created by high-speed centrifugation ($10,000 - 25,000\times$ g depending on osmolarity adjustment medium) for 15 minutes. During that time the gradient material, with a pre-chosen range of density, is distributed accordingly in the tube. The cells are then loaded onto the preformed gradient and centrifuged at a lower speed ($400\times$ g) for 20–30 minutes

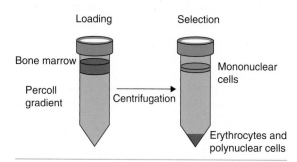

Figure 11.3 Isolation of bone marrow derived mesenchymal stem cells by density gradient centrifugation. The bone marrow is carefully loaded onto the premade Percoll gradient and centrifuged for 30 min at $400\times$ g. This procedure separates the mononuclear cells in which monocytes and the mesenchymal stem cells from the erythrocytes and polynuclear blood cells.

(Figure 11.3). During the centrifugation, the cells will sediment to an equilibrium position where the gradient density is equal to the density of the cell (isopycnic position). Thus, in this type of separation, the particles are separated solely on the basis of differences in density, irrespective of size.

It is, however, important to remember that the osmolality of the medium in the tube can significantly alter the size and apparent buoyant density of cells. A high external osmolality will cause cells to shrink, while a low osmolality in the medium will cause the particles to swell. Other important factors are pH of the medium, rotor type and the angle of the tube during centrifugation.

11.3.3 Antibody driven separation

A more specific cell selection can be obtained by using the principle of an antibody detecting a specific target antigen, preferably on the cell surface. To manage this, a fluorescent dye is coupled to an antibody of interest and then, after labeling of the cells with the antibody, the cells can be separated in an electronic fluorescence activated cell sorter (FACS). In this method, single cells are passed through a laser beam and the fluorescence of the cells is momentarily captured. After the exposure the single cells are trapped within small droplets, created by a small vibrating nozzle and the droplets are given a positive or negative electrical charge depending on whether the droplet contains a fluorescent cell or not. The electrical charge is used together with a strong electrical field to sort the positive and negative droplets into separate containers. If several cells are trapped within the same droplet, the droplet is left uncharged and sorted into the waste bucket. By using this method, a few specific cells can be separated from several thousands in a few seconds.

However, since it is hard to sterilize all the tubing within the machine, the FACS technique has clear limitations for aseptic production of cells for clinical implantation. Another separation technique that also utilizes specific antibodies, and is potentially more suitable for clinical use, is magnetic activated cell sorting (MACS). With this technique, cells become magnetic after being labeled with antibodies to which small magnetic beads are coupled. The magnetic cells, positive to the target surface molecule, are then separated from the negative cells by exposure to a strong magnetic field. After the positively sorted cells have passed through the magnet, they can be run again to increase the purity of the selection. To avoid implanting the magnetic beads they can be detached with anti-fab antiserum, or simply by using negative selection, e.g. elimination of the contaminating cells (Rasmussen, Smeland *et al.*, 1992).

Another method for isolation of specific cell types with the use of antibodies is called panning (Wysocki and Sato, 1978). In this method, the culture dishes are coated with the appropriate antibody for a specific marker. The cell mixture is then added to the dish and cells expressing the antigen matching the antibody will attach to the dish while other cells are easily washed away. The method works if the surface to which the unique antibody is coated has low binding capacity of the unwanted cells; otherwise the discrimination of the different cell types will be reduced. This method has, for example, been used to successfully isolate epithelial (Castleman, Northrop *et al.*, 1991) and hematopoetic stem cells (Cardoso, Watt *et al.*, 1995) from mixtures of cells.

11.4 Expansion

Because a high number of cells are needed for most TE, it is common for the cells to be expanded *in vitro* before clinical use. The expansion is normally made in tissue culture vessels in the presence of a culture medium supplemented with growth factors, serum and other specific additives (see further in Chapter 12, Cell Nutrition).

11.4.1 Anchorage dependent cells

The most common way of expanding cells is to culture the cells as a single cell layer (monolayer) on the bottom of a culture dish (Figure 11.4A). Historically, the culture dish was made of glass and known as 'in vitro culture' (*in vitro* = in glass). Today, the glass has been replaced by disposable polystyrene plastic bottles and dishes. The culture in monolayer is based on the fact that the cells adhere to the bottom of the culture dish and get their nutrition and oxygen from the surrounding culture medium. The typical cell types which are expanded in monolayer are fibroblasts, skeletal myocytes, chondrocytes, etc. (i.e. anchor dependent cells). These cells must adhere to a surface before they start to proliferate.

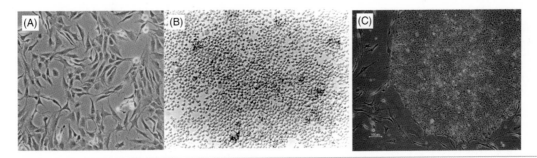

Figure 11.4 (A) Monolayer culture of human articular chondrocytes; (B) Suspension culture of hematopoetic progenitor cells stimulated with trombopoetin (adopted from Birkmann, J. *et al.* (1997). *Stem Cells*, 15: 18–32); and (C) Co-culture of human embryonic stem cell SA502s and mouse embryonic fibroblasts (on the right) (courtesy of Cellartis II AB, Sweden).

11.4.1.1 Tissue culture plastic

During the manufacturing of the plastic culture dishes and bottles, the polystyrene is molded into the desired shape and then treated with energy and gas to create the surface roughness and hydrophilicity that is necessary for protein adsorption and cell attachment (Figure 11.5). This special treatment of the surface reduces the aromatic groups and increases the oxygen containing functional groups of the polystyrene backbone. By this treatment, the water-contact angle decreases which gives to higher wettability of the surface. Higher wettability leads to increased adherence and spreading which results in increased cell proliferation (van Kooten, Spijker *et al.*, 2004).

11.4.1.2 Surface coatings

For some cell types, it is necessary to increase the ability of cell binding to the surface by applying coatings (for example, gelatin) or more specific ECM proteins such as collagen, laminin or fibronectin. By applying a thin layer with ECM proteins before seeding the desired cell type, the cells are able to adhere, stretch out and proliferate faster compared to uncoated surfaces. It has further been shown that,

by regulating the concentration of ECM proteins in the coatings, it is possible to direct the cellular fate by turning them into a proliferative phase (Mooney, Hansen *et al.*, 1992). Traditionally, most of the protein coatings have an animal (bovine or porcine) source origin, and are therefore potential carriers of animal-derived diseases, which make them less interesting for clinical TE. Although there are several recombinant proteins on the market, these tend to be very expensive.

11.4.1.3 Microcarriers

One way to increase the surface for culture of anchor dependent cells is to use microcarriers (MCs). The microcarrier technology has emerged from the need for mass production of protein synthesizing animal cells in large reactors (up to several hundred liters in volume). The technology is based on inoculation of 5–10 cells per bead (~0.2 mm) and culture of the cells in magnetically stirred (spinner) flasks (Figure 11.6). The mixed bioreactor culture systems are necessary to provide sufficient mass transfer for high density

Spinner bottle

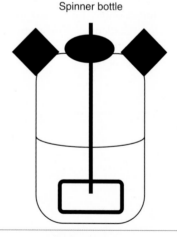

Figure 11.6 The spinner bottle is used for placing the bottle on a magnetic stirrer inside or outside a tissue culture incubator. The continuous stirring prevents cells from adhering to the inner surface of the bottle and improves cell nutrition of larger cell aggregates or tissue constructs.

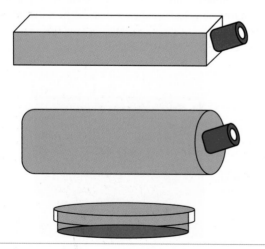

Figure 11.5 Tissue culture plastic can be molded into a variety of different forms (square and rounded culture flasks, Petri dishes) to meet a wide range of cull culture applications.

microcarrier cultures (see further Chapter 12, Cell Nutrition and Chapter 16, Bioreactors). After a few days, when the cells reach confluence, normally 200–300 cells can be found on each bead. Traditionally, microcarriers have been manufactured from different synthetic materials including dextran, polyacrylamide and polystyrene, although, new biodegradable microcarriers suitable for human cell therapy, have been developed (Xu and Reid, 2001). Additionally, by increasing the size of the microcarriers, having a porous structure and thereby utilizing the interior, the surface can be further increased. Examples of these macroporous microcarriers are the CultiPher-S beads and the Cytopore 1 and 2.

11.4.2 Suspension cultures

Certain cell types, especially blood cells, do not need to adhere to a surface to proliferate and differentiate. Instead, these cells grow in 3D environments in semisolid suspension (i.e. suspension cultures) (Mizrahi and Moore, 1971; Kubota and Preisler, 1982). In these cultures, the cells are seeded within the semisolid culture media, such as agar, methyl cellulose or collagen and the culture media provides the cells with nutrition and growth factors (refer back to Figure 11.4B). Depending on the solidness of the semisolid media medium, changes and additional culture medium are made either by diluting the medium or by replacing the medium on top of the substrate. The suspension cultures are also a useful system for analyzing 3D growth and sprouting.

11.4.3 Culture on feeder cells

In many organs, different cell types function side-by-side as a team by helping and supporting each other. In one example, co-cultures are made by seeding a rapidly growing cell line (so-called 'feeder cells') on Petri dishes and then arresting their growth when the dishes reach the desired confluence. Feeder cell growth is arrested either by irradiation or by treating the cells with Mitomycin (a mitosis arresting drug). The feeder cells can be any rapidly growing anchorage dependent cell line and one of the most common feeder cell lines is the mouse NIH 3T3 (refer back to Figure 11.4C).

The desired slowly growing cells are then seeded on top of the growth-arrested feeder cell monolayer. After a few days, or even weeks, the cells of interest will establish their typical clonal growth appearance. Cell types which are commonly expanded in feeder cell co-cultures are keratinocytes, heamatopoetic stem cells, and embryonic stem cells.

The functions of the feeder cells are to detoxify, process nutrients, activate, deactivate hormones, growth factors and provide a supporting cell matrix (Figure 11.7). During the cultures there is intensive cross-talk and cell-migration between the different cell types and many cell types could not expand without the use of feeder cells. However, by careful studies of the function of feeder cells, new culture medium formulations have been made. These new mediums have, to some extent, replaced the feeder cells (Petzer, Hogge *et al.*, 1996).

From a clinical point of view, the feeder cell co-culture method has clear disadvantages. The extra work with feeder cells is time-consuming and makes the production of large amounts of cells more difficult compared to feeder-free cultures. Also, cells used for co-culture, or animal origin derived serum supplements, contain high amounts of the sialic acid Neu5Gc, which is incorporated in high amounts to exposed human cells during culture (Martin, Muotri *et al.*, 2005). Due to this, many humans have antibodies towards Neu5Gc the exposed cells will be immunogenic when transplanted. This information confirms the standpoint that all human cell lines expanded on animal feeder cells should be considered as xenogenic and work should be made towards purified human clinical cell culture systems (Skottman, Dilber *et al.*, 2006).

11.4.4 Change of phenotype during expansion

During expansion *in vitro*, several cell types change their phenotype towards fibroblast like cells and start to express primitive embryonic markers (Schnabel,

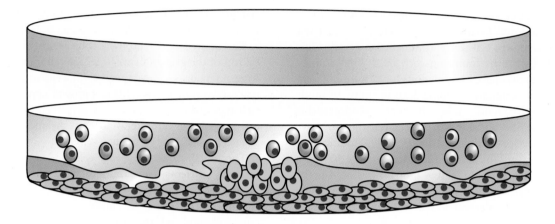

Figure 11.7 Coculture can either be made as a direct cell-to-cell contact by seeding the cell type onto a premade feeder cell layer or indirectly by separating the different cell types by placing one cell type within an insert. Here we have feeder cells on the bottom of the plate, yellow cells in direct contact with the feeders and other cells in suspension culture.

Marlovits *et al.*, 2002; Diaz-Romero, Gaillard *et al.*, 2005). This transformation is called 'dedifferentiation' and is a common phenomena observed in many types of cells and tissues during expansion (Reischl, Prelle *et al.*, 1999; Baker, Carfagna *et al.*, 2001). The degree of dedifferentiation has been correlated to the number of cell divisions or passages (Hardingham, Tew *et al.*, 2002; Dell'Accio, De Bari *et al.*, 2003; Schulze-Tanzil, Mobasheri *et al.*, 2004) and to the method for expansion (Frondoza, Sohrabi *et al.*, 1996). This is a typical character as demonstrated in studies of the myogenic differentiation of human articular chondrocytes (Dell'Accio, De Bari *et al.*, 2003). In that study, no differentiation into myocytes was obtained when cells from a low passage were used, while with later passages cartilage derived skeletal muscle fibers were obtained.

Complementary to the dedifferentiation theory, the change in phenotype during *in vitro* culture is that the culture conditions favor cell subpopulations, such as transit amplifying cells, because these cells are able to respond to the culture conditions more rapidly. This could be because that particular subpopulation expresses receptors for serum containing growth factors and, when exposed to these *in vitro*, respond more rapidly (Shimazu, Nah *et al.*, 1996). However,

even if it is not clear whether the end result from *in vitro* expansion is true dedifferentiation, the change in functionality is presumed to cause problems for different TE applications. This is probably due to the fact that the objective of cell expansion is to create a more functional cell count to restore the lost function of the target tissue or organ. In this perspective, the functionality of the cells could be defined as the ability to quickly, and on a long-term basis, incorporate with host cells and tissue. The quick incorporation is needed to start the production of the missing components, and the long-term basis is to keep this function going. However, it is most likely that these functions do not come from a single cell but rather from a differentiated cell and a stem cell, respectively. The ultimate cell population for transplantation is a mixture of cells with different functionalities, adapted to the needs and characteristics of the target tissue or organ.

11.5 Differentiation

If the expansion phase is considered to result in a change in phenotype towards a proliferative nonprotein secreting cell, the differentiation phase is the opposite. During differentiation, the cells slow down or stop their cell division and mature into a functional protein synthesizing phenotype. Generally, *in vitro* the differentiation is accomplished by eliminating or lowering the serum concentration and culturing the cells in a 3D environment. The reduction in serum reduces the concentrations of proliferation driving growth factors (Kruse and Tseng, 1993; Tallheden, Van der Lee *et al.*, 2005; Zhang, Li *et al.*, 2005), and 3D culture environment increases cell-to-cell contact, which can lead to increased differentiation (Benya and Shaffer, 1982; Colburn *et al.*, 1992; Hall and Miyake, 1992; Li, Tacchetti, Tavella *et al.*, 1992).

11.5.1 Monolayer

When cells are cultured in monolayers and reach growth inhibition due to space limitations, most cell types stop proliferating and start to differentiate. One simple way of inducing cell differentiation is, therefore, to seed the cells at very high cell densities and allow them to grow in multiple layers, i.e. over-confluence. Differentiation can also be increased by the removal of serum and/or the addition of differentiation factors such as dexamethasone, retinoic acid, and ascorbic acid (see Box 1).

Box 11.1 Differentiation of fat cells (adipocytes)

Differentiation of preadipoctes or stem cells into lipid filled adipocytes can be induced by supplementing the culture medium in high density cultures with dexamethasone, 1-methyl-3-isobutylxanthine and insulin. This cocktail of synthetic glucocorticoid, an inhibitor of nucleotide phosphodiesterase (increases the amounts of cAMP and cGMP) and the pancreatic hormone (activates PI3-kinase), induces the expression of the transcription factors PPAR gamma (peroxisome proliferator activated receptor gamma), which is essential for adipocyte differentiation. The grade of differentiation into adipocytes has further been shown to be increased if the cells are cultured on fibronectin coated dishes (Hemmrich, von Heimburg *et al.*, 2005). This is of special importance when autologous preadipocytes are for use in clinical implantation. For clinical use, it is of great practical importance that the cells can be cultured in 3D environments and that it is possible to induce differentiation in the presence of autologous human serum instead of fetal calf serum. In this way, the cells and serum can be harvested from the same patient and, after expansion and subsequent differentiation *in vitro* under stimulation of the serum, implanted back to the patient as a functional patch of tissue.

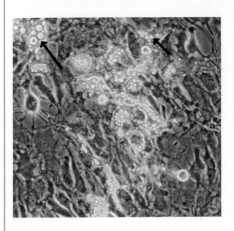

Differentiation of mesenchymal stem cells into adipogenic lineage in monolayer cultures. The differentiation was made after addition of dexamethasone, 1-methyl-3-isobutylxanthine and insulin to the cultures for 21 days. Arrows point out the formation of large lipid vesicles.

By coating the bottom of the culture vessel with ECM proteins, the monolayer cultured cells can be given specific signals that initiate differentiation (Reid, 1990). For example, by adding an additional layer of collagen coating on top of high-density seeded hepatocytes in monolayer, a collagen gel sandwich culture is formed. This type of culture changes the monolayer culture into a 3D culture, allowing the cells to have apical orientation and thereby a differentiated state of the cells can be induced and preserved for longer culture periods.

11.5.2.3 Culture

One simple way of inducing differentiation in mono-layer expanded cells is to allow the spontaneous formation of spherical cell aggregates in 96-well plates or on non-adherent bacterial culture dishes (Li, Colburn *et al.*, 1992; Korff and Augustin, 1998; Conley, Young *et al.*, 2004). By regulating the number of seeded cells, the size of the spheres can be controlled. Another way to induce differentiation is to force cells to aggregate by centrifugation (100–500 g). These cell pellets will initially be discoid, but will eventually become spherical due to cell migration and/or ECM synthesis (Figure 11.8B).

Another way to mimic a 3D culture environment is to use different forms of water-rich natural or synthetic gels or scaffolds. One kind of biomaterial that consists of about 60 per cent of water is the hydrogel. Of the naturally occurring hydrogels, agarose, alginate, hyaluronic acid, and methyl cellulose have been shown to be able to provide cells with a 3D environment in which the cells have mechanical support and relatively good diffusion of gases and nutrients. Synthetic hydrogels have also been shown to be able to support cell differentiation and eliminate the variance between the biological hydrogels (different scaffolds for tissue engineering are described more thoroughly in other chapters).

11.5.3 Co-culture

As previously discussed in this chapter, certain human cell types such as heamatopoetic, embryonic, epidermal and neuronal stem cells, currently need support from other cells in order to proliferate. The function of feeder cells for stem cell differentiation has also been explored with neuronal stem cells (Lou, Gu *et al.*, 2003). These cells are able to differentiate into neurons, astrocytes and oligodendrocytes, three different cell types of the brain. When the stem cells were cultured on a feeder layer made from bone

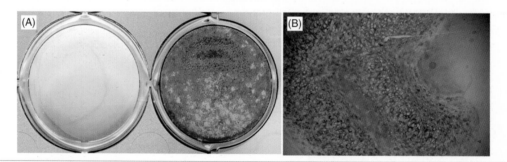

Figure 11.8 Differentiation of a cell can either be made in 2D or 3D. (A) Negative and positive von Kossa staining of monolayer cultures of human mesenchymal stem cells differentiated towards osteogenic lineage. The brown stain indicates mineralization induced by the osteogenic culture medium supplements; (B) The red type II Collagen staining of a histological section of culture expanded human articular chondrocytes indicate that the cells are redifferentiating in 3D high-density pellet model under serum-free conditions.

marrow stromal cells (BMSC), exclusive differentiation into neurons was observed. To clarify whether the cell to cell contact, certain membrane proteins or factors secreted in the culture medium, were inducing the differentiation, new stem cell cultures were made with BMSC conditioned medium or membrane fragments of live or paraformaldehyde fixed BMSCs. The only condition that resulted in the formation of neurons was the conditioned medium, indicating that the differentiating effects of the feeder cell co-culture were due either to the secretion or processing of factors in the culture medium.

11.5.4 Biomechanical signaling

Mechanical stress is an important modulator of cell physiology and tissue development, and specific mechanical signals may promote cell differentiation into a particular phenotype (see Box 2). For example, in studies in which skeletal myofibers were cultured

---

Box 11.2 Biomechanical signaling during embryonic development

One example of the importance of biomechanical load for proliferation and differentiation is the embryonic development of the synovial joints. During this process the hyaluronan secretion from the interzonal cells plays an important role in the cavitation of the joint. The secretion is especially seen from the cells in a presumptive new surface of the joint and has been shown to be regulated by the movement of the joint (Dowthwaite, Flannery et al., 2003). The cells in the surface also express elevated levels of CD44 (the major hyaluronan-binding protein) and it has been shown that oligosaccharide-induced blockade of hyaluronan and CD44 interaction blocks joint formation (Dowthwaite, Edwards et al., 1998), which additionally highlights the central role for hyaluronan in formation of the joint cavity.

In addition, the enzymatic synthesis of hyaluronan is specifically regulated through the mitogen-activated, protein kinase cascade (Raf/MEK/ERK) (Dowthwaite, Ward et al., 1999) (Lewthwaite, Bastow et al., 2006) (Bastow, Lamb et al., 2005). This data has been generated by using experimental model of in ovo immobilization of chick embryos and by using a selective inhibitor (PD98059) of ERK activation in vitro (Lewthwaite, Bastow et al., 2006).

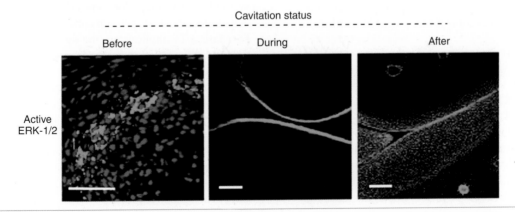

Strong expression of phosphorylated ERK-12 (green) before and during cavitation of synovial joints. The expression is retained at congruent joint surfaces after cavitation (adopted from Barstow, E.R. et al. (2005). J Biol Chem, 280: 11749–11758).

in the presence of cyclic mechanical stretch, increased differentiation has been shown (Vandenburgh and Kaufman, 1979). Signaling from the changes in the physical environment is thought to enhance differentiation through special mechanoreceptors. These receptors sense the environment and regulate the intracellular responses (Muller and Littlewood-Evans, 2001).

In TE, several models and bioreactors have been developed for applying biomechanical signaling to cells and tissue constructs. Generally, most of these models have shown that, when a biomechanical *in vivo* situation is mimicked, cells respond by driving themselves towards differentiation and protein synthesis. One cell type which has been shown to respond to intermittent cyclic uniaxial unconfined compression with increased synthesis of proteoglycans is articular chondrocytes (Kisiday, Jin *et al.*, 2004). A similar response to loading has also been reported for pulsatile shear stress of endothelial cells (Elhadj, Mousa *et al.*, 2002).

11.6 Future developments

Most of the current clinical cell based TE applications include processing of cells with GMP facilities, which are very expensive. The future challenge is to minimize the costs and find new innovating ways to use the body as a reservoir for spare parts. The following list includes some future potential solutions.

1. Intra-operative treatment. Methods for extracting and purifying cells and proteins for implantation back to the patient during the same surgical séance will be developed.
2. Cellular plasticity. Cells will be isolated from one tissue and used for implantation into another, i.e. hepatocytes for treatment of insulin deficiency.
3. Universal stem cells. HLA matched allogenic stem cells either from embryos or adults will be used as an alternative source to the autologous cells and tissues.
4. The body is the best bioreactor. Methods for using the body for proliferation and/or differentiation of cells and tissue constructs will be developed.

The TE field is still young and unexplored and nobody yet knows if cell-based therapies will be an alternative for the future. Nonetheless, ongoing work with human cells will increase our knowledge of which factors dictate the proliferation and differentiation of human cells which will, in turn, pave the way for the development of new drug targets and therapeutic principles.

11.7 Summary

1. Autologous or allogenic cells can be utilized alone or in combination with biomaterials to restore lost function of tissue and organs.
2. Cells can be harvested from different locations of the human body and, depending on the tissue to be restored, processed for selection, expansion of cell number or re-implanted directly into the patient. In all cases, care has to be taken to handle the harvested cells in a proper way from the beginning.
3. When a special cell type is needed for implantation, cells can be isolated from the initial harvest by targeting special characteristics of the cells such as size, surface markers, growth behavior and/or functional properties.
4. The most common way of expanding the initial cell number is by placing the cells in monolayer cultures; although for some cell types such as hematopoetic cells suspension cultures are needed.
5. For clinical TE applications, the selection, expansion and differentiation are made in specially designated laboratories.
6. A potential problem with serial expansion in monolayer is that the cells gradually change their phenotype and that contaminating fast growing cells could become too dominant.
7. When the required cell number is achieved, the cells can be differentiated by exposing them to specific feeder cell lines, growth factors and/or to extra cellular matrix proteins.
8. The cells can be combined with synthetic scaffolds, biomaterials, forced into high density cultures or exposed to physical forces to additionally stimulate differentiation.

9. Most of the current clinical cell-based TE therapies include expansion of the cell number in *in vitro* cultures in special GMP laboratories. Future therapies will be aimed at shortcutting this logistic and relatively expensive procedures.

References

Baker, T.K., Carfagna, M.A., *et al.* (2001). Temporal gene expression analysis of monolayer cultured rat hepatocytes. *Chem Res Toxicol*, 14(9): 1218–1231.

Bastow, E.R., Lamb, K.J., *et al.* (2005). Selective Activation of the MEK-ERK Pathway Is Regulated by Mechanical Stimuli in Forming Joints and Promotes Pericellular Matrix Formation. *J Biol Chem*, 280(12): 11749–11758.

Benya, P.D. and Shaffer, J.D. (1982). Dedifferentiated chondrocytes reexpress the differentiated collagen phenotype when cultured in agarose gels. *Cell*, 30(1): 215–224.

Brittberg, M., Lindahl, A., *et al.* (1994). Treatment of deep cartilage defects in the knee with autologous chondrocyte transplantation. *N Engl J Med*, 331(14): 889–895.

Caplan, A.I. (1991). Mesenchymal stem cells. *J Orthop Res*, 9(5): 641–650.

Cardoso, A.A., Watt, S.M., *et al.* (1995). An improved panning technique for the selection of CD34+ human bone marrow hematopoietic cells with high recovery of early progenitors. *Exp Hematol*, 23(5): 407–412.

Castleman, W.L., Northrop, P.J., *et al.* (1991). Replication of parainfluenza type-3 virus and bovine respiratory syncytial virus in isolated bovine type-II alveolar epithelial cells. *Am J Vet Res*, 52(6): 880–885.

Chachques, J.C., Herreros, J., *et al.* (2004). Autologous human serum for cell culture avoids the implantation of cardioverter-defibrillators in cellular cardiomyoplasty. *Int J Cardiol*, 95(Suppl 1): S29–S33.

Conley, B.J., Young, J.C., *et al.* (2004). Derivation, propagation and differentiation of human embryonic stem cells. *Int J Biochem Cell Biol*, 36(4): 555–567.

DellAccio, F., De Bari, C., *et al.* (2003). Microenvironment and phenotypic stability specify tissue formation by human articular cartilage-derived cells *in vivo*. *Exp Cell Res*, 287(1): 16–27.

Diaz-Romero, J., Gaillard, J.P., *et al.* (2005). Immunophenotypic analysis of human articular chondrocytes: Changes in surface markers associated with cell expansion in monolayer culture. *J Cell Physiol*, 202(3): 731–742.

Dowthwaite, G.P., Bishop, J.C., *et al.* (2004). The surface of articular cartilage contains a progenitor cell population. *J Cell Sci*, 117(Pt 6): 889–897.

Dowthwaite, G.P., Edwards, J.C., *et al.* (1998). An essential role for the interaction between hyaluronan and hyaluronan binding proteins during joint development. *J Histochem Cytochem*, 46(5): 641–651.

Dowthwaite, G.P., Flannery, C.R., *et al.* (2003). A mechanism underlying the movement requirement for synovial joint cavitation. *Matrix Biol*, 22(4): 311–322.

Dowthwaite, G.P., Ward, A.C., *et al.* (1999). The effect of mechanical strain on hyaluronan metabolism in embryonic fibrocartilage cells. *Matrix Biol*, 18(6): 523–532.

Elhadj, S., Mousa, S.A., *et al.* (2002). Chronic pulsatile shear stress impacts synthesis of proteoglycans by endothelial cells: effect on platelet aggregation and coagulation. *J Cell Biochem*, 86(2): 239–250.

Fossum, M., Gustafson, C.J., *et al.* (2003). Isolation and *in vitro* cultivation of human urothelial cells from bladder washings of adult patients and children. *Scand J Plast Reconstr Surg Hand Surg*, 37(1): 41–45.

Frondoza, C., Sohrabi, A., *et al.* (1996). Human chondrocytes proliferate and produce matrix components in microcarrier suspension culture. *Biomaterials*, 17(9): 879–888.

Gallico, G.G. III., OConnor, N.E., *et al.* (1984). Permanent coverage of large burn wounds with autologous cultured human epithelium. *N Engl J Med*, 311(7): 448–451.

Gill, J.F., Chambers, L.L., *et al.* (1995). Safety testing of Liberase, a purified enzyme blend for human islet isolation. *Transplant Proc*, 27(6): 3276–3277.

Hall, B.K. and Miyake, T. (1992). The membranous skeleton: the role of cell condensations in vertebrate skeletogenesis. *Anat Embryol (Berl)*, 186(2): 107–124.

Hardingham, T., Tew, S., *et al.* (2002). Tissue engineering: chondrocytes and cartilage. *Arthritis Res*, 4(Suppl 3): S63–S68.

Hemmrich, K., von Heimburg, D., *et al.* (2005). Optimization of the differentiation of human preadipocytes *in vitro*. *Differentiation*, 73(1): 28–35.

Jankowski, R.J., Haluszczak, C., *et al.* (2001). Flow cytometric characterization of myogenic cell populations obtained via the preplate technique: potential for rapid isolation of muscle-derived stem cells. *Hum Gene Ther*, 12(6): 619–628.

Jones, P.H. and Watt, F.M. (1993). Separation of human epidermal stem cells from transit amplifying cells on the basis of differences in integrin function and expression. *Cell*, 73(4): 713–724.

Kisiday, J.D., Jin, M., *et al.* (2004). Effects of dynamic compressive loading on chondrocyte biosynthesis in self-assembling peptide scaffolds. *J Biomech*, 37(5): 595–604.

Korff, T. and Augustin, H.G. (1998). Integration of endothelial cells in multicellular spheroids prevents apoptosis and induces differentiation. *J Cell Biol*, 143(5): 1341–1352.

Kruse, F.E. and Tseng, S.C. (1993). Serum differentially modulates the clonal growth and differentiation of cultured limbal and corneal epithelium. *Invest Ophthalmol Vis Sci*, 34(10): 2976–2989.

Kubota, K. and Preisler, H.D. (1982). Comparison of agar and methylcellulose culture methods for human erythroid colony formation. *Exp Hematol*, 10(3): 292–299.

Lewthwaite, J.C., Bastow, E.R., *et al.* (2006). A specific mechanomodulatory role for p38 MAPK in embryonic joint articular surface cell MEK-ERK pathway regulation. *J Biol Chem*, 281(16): 11011–11018. Epub 2006 Feb 7.

Li, A.P., Colburn, S.M., *et al.* (1992). A simplified method for the culturing of primary adult rat and human hepatocytes as multicellular spheroids. *In vitro Cell Dev Biol*, 28A(9–10): 673–677.

Lou, S., Gu, P., *et al.* (2003). The effect of bone marrow stromal cells on neuronal differentiation of mesencephalic neural stem cells in Sprague-Dawley rats. *Brain Res*, 968(1): 114–121.

Lu, Y., Dhanaraj, S., *et al.* (2006). Minced cartilage without cell culture serves as an effective intraoperative cell source for cartilage repair. *J Orthop Res*, 24(6): 1261–1270.

Martin, M.J., Muotri, A., *et al.* (2005). Human embryonic stem cells express an immunogenic nonhuman sialic acid. *Nat Med*, 11(2): 228–232. Epub 2005 Jan 30.

Megerian, C., Weitzner, B., *et al.* (2000). Minimally invasive technique of auricular cartilage harvest for tissue engineering. *Tissue Engineering*, 6(Feb): 69–74.

Mizrahi, A. and Moore, G.E. (1971). Role of sodium carboxymethyl cellulose and hydroxyethyl starch in hematopoietic cell line cultures. *Appl Microbiol*, 21(4): 754–757.

Mooney, D., Hansen, L., *et al.* (1992). Switching from differentiation to growth in hepatocytes: control by extracellular matrix. *J Cell Physiol*, 151(3): 497–505.

Muller, U. and Littlewood-Evans, A. (2001). Mechanisms that regulate mechanosensory hair cell differentiation. *Trends Cell Biol*, 11(8): 334–342.

Muschler, G.F., Boehm, C., *et al.* (1997). Aspiration to obtain osteoblast progenitor cells from human bone marrow: the influence of aspiration volume. *J Bone Joint Surg Am*, 79(11): 1699–1709.

OConnor, N.E., Mulliken, J.B., *et al.* (1981). Grafting of burns with cultured epithelium prepared from autologous epidermal cells. *Lancet*, 1(8211): 75–78.

Olack, B.J., Swanson, C.J., *et al.* (1999). Improved method for the isolation and purification of human islets of langerhans using Liberase enzyme blend. *Hum Immunol*, 60(12): 1303–1309.

Petzer, A.L., Hogge, D.E., *et al.* (1996). Self-renewal of primitive human hematopoietic cells (long-term-culture-initiating cells) *in vitro* and their expansion in defined medium. *Proc Natl Acad Sci USA*, 93(4): 1470–1474.

Rasmussen, A.M., Smeland, E.B., *et al.* (1992). A new method for detachment of Dynabeads from positively selected B lymphocytes. *J Immunol Methods*, 146(2): 195–202.

Reese, J.A. and Byard, J.L. (1981). Isolation and culture of adult hepatocytes from liver biopsies. *In vitro*, 17(11): 935–940.

Reid, L.M. (1990). Stem cell biology, hormone/matrix synergies and liver differentiation. *Curr Opin Cell Biol*, 2(1): 121–130.

Reischl, J., Prelle, K., *et al.* (1999). Factors affecting proliferation and dedifferentiation of primary bovine oviduct epithelial cells *in vitro*. *Cell Tissue Res*, 296(2): 371–383.

Rheinwald, J.G. and Green, H. (1975). Serial cultivation of strains of human epidermal keratinocytes: the formation of keratinizing colonies from single cells. *Cell*, 6(3): 331–343.

Schnabel, M., Marlovits, S., *et al.* (2002). Dedifferentiation-associated changes in morphology and gene expression in primary human articular chondrocytes in cell culture. *Osteoarthritis Cartilage*, 10(1): 62–70.

Schulze-Tanzil, G., Mobasheri, A., *et al.* (2004). Loss of chondrogenic potential in dedifferentiated chondrocytes correlates with deficient Shc-Erk interaction and apoptosis. *Osteoarthritis Cartilage*, 12(6): 448–458.

Seferovic, P.M., Ristic, A.D., *et al.* (2003). Diagnostic value of pericardial biopsy: improvement with extensive sampling enabled by pericardioscopy. *Circulation*, 107(7): 978–983.

Shimazu, A., Nah, H.D., *et al.* (1996). Syndecan-3 and the control of chondrocyte proliferation during endochondral ossification. *Exp Cell Res*, 229(1): 126–136.

Sjögren Jansson, E., Tallheden, T., *et al.* (2004). *The Eurocell Experience of Autologous Chondrocyte Implantation*, Oswestry: The Institute of Orthopaedics (Oswestry) Publishing Group.

Skottman, H., Dilber, M.S., *et al.* (2006). The derivation of clinical-grade human embryonic stem cell lines. *FEBS Lett*, 580(12): 2875–2878. Epub 2006 Apr 7.

Strom, S.C., Jirtle, R.L., *et al.* (1982). Isolation, culture, and transplantation of human hepatocytes. *J Natl Cancer Inst*, 68(5): 771–778.

Tacchetti, C., Tavella, S., *et al.* (1992). Cell condensation in chondrogenic differentiation. *Exp Cell Res*, 200(1): 26–33.

Tallheden, T., Van der Lee, J., *et al.* (2005). Human Serum for Culture of Articular Chondrocytes. *Cell Transplantation*, 14: 1–11.

van Kooten, T.G., Spijker, H.T., *et al.* (2004). Plasma-treated polystyrene surfaces: model surfaces for studying cell-biomaterial interactions. *Biomaterials*, 25(10): 1735–1747.

Vandenburgh, H. and Kaufman, S. (1979). *In vitro* model for stretch-induced hypertrophy of skeletal muscle. *Science*, 203(4377): 265–268.

Wysocki, L.J. and Sato, V.L. (1978). Panning for lymphocytes: a method for cell selection. *Proc Natl Acad Sci USA*, 75(6): 2844–2848.

Xu, A.S. and Reid, L.M. (2001). Soft, porous poly(D,L-lactide-co-glycotide) microcarriers designed for *ex vivo* studies and for transplantation of adherent cell types including progenitors. *Ann NY Acad Sci*, 944: 144–159.

Zhang, E., Li, X., *et al.* (2005). Cell cycle synchronization of embryonic stem cells: effect of serum deprivation on the differentiation of embryonic bodies *in vitro*. *Biochem Biophys Res Commun*, 333(4): 1171–1177.

Chapter 12
Cell nutrition

Jos Malda, Milica Radisic, Shulamit Levenberg, Tim Woodfield, Cees Oomens, Frank Baaijens, Peter Svalander and Gordana Vunjak-Novakovic

Chapter objectives:

- To learn about nutrient requirements of cells *in vitro*
- To understand the role of mass transport in providing nutrients to the cells
- To learn about the bioreactor design features that can enhance local control of oxygen and nutrients in the cell microenvironment
- To understand how mathematical modeling can enhance the understanding of nutrient limitation in tissue engineering

"The great book of nature can be read only by those who know the language in which it was written. And this language is mathematics."

Galileo Galilei (1623), quoted in Bialek and Botstein, 2004, Introductory science and mathematics education for 21st-century biologists, *Science* 303(5659): 788–790.

12.1 Introduction

Blood is the main carrier of nutrients and waste products in the human body. The blood actively transports nutrients to all sites in the body, where they will diffuse between a capillary lumen and the cell membrane. Each cell within our body, with the exception of cells within avascular tissues, such as articular cartilage, is close to a blood vessel (Figure 12.1). Should there be a lack of nutrients, most

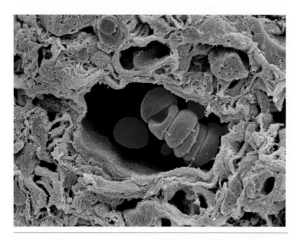

Figure 12.1 Colored scanning electron micrograph (SEM) of a cross-section through a small, thin-walled artery known as an arteriole. Red blood cells can be seen in the central space (lumen) and connective tissue surrounds the arteriole. The inner wall (tunica intima) of the lumen is composed of an endothelial lining and a thin elastic layer that stretches. Magnification: ×4000.

cells located further away than 200 µm would die (Muschler *et al.*, 2004). Within the field of tissue engineering, adequate supplies of nutrients and the removal of waste products poses a significant challenge. *In vitro*-created tissue-engineered constructs lack vascularity and the cells receive nutrients by diffusion over distances that are often in the order of millimetres. Because diffusion can supply oxygen to the cells only up to a distance of approximately 200 µm, the interior of these constructs will be void of viable cells. Thus, the lack of a direct blood supply *in vitro* limits the size and characteristics of engineered tissues. Keeping the graft viable upon implantation *in vivo* is an additional challenge. The graft must be vascularized immediately to allow for cell survival and the integration with the host vasculature. However, the host's vascularization is not sufficient to keep the construct viable in the first period following implantations (Nomi *et al.*, 2002; Ostrander *et al.*, 2001).

This chapter will focus on nutrient supplies to engineered tissues in some detail. Obviously, before addressing possible improvements, a thorough understanding of the problem is a necessity. Therefore, we will firstly explain the nutrient requirements of the cells and discuss the components of the cell culture medium. Specific attention will be paid to how the medium composition can regulate cellular proliferation and differentiation. Secondly, the underlying theory describing the development of nutrient limitation and the basics of mass transfer will be described. Using some practical examples, the development of nutrient gradients within tissue-engineered constructs will be explained.

Finally, possible approaches to decrease nutrient limitation will be described. These include increasing the external mass transport by convective mixing of culture medium around the constructs, increasing the internal mass transfer by medium perfusion through cultured tissues *in vitro*, and vascularization of the cultured tissues *in vivo*. In addition, modeling will be discussed, as a tool to further understand the nutrient limitation problems in engineered tissues.

12.2 Cell culture media

12.2.1 Types of cell culture media

Culture media are generally designed to function as an *in vitro* substitute for a particular biological fluid, present in the normal physiological environment of the corresponding cells *in vivo*. Development of techniques for *in vitro* culture of cells is a good example of 'empirical science' in which current knowledge is based on over a century of painstaking work by many cell biologists in various application fields. Culture media formulation can be categorized as *simple, complex* and *chemically defined*.

Simple media, also known as 'balanced' or 'physiological' salt solutions, are ionic buffer solutions usually supplemented with energy substrates and protein (serum or albumin). Simple media do not support prolonged survival of cells and are therefore mainly used for cell isolation and handling.

Complex media are based on simple media supplemented with amino acids, vitamins, trace metals, serum or albumin and a variety of nutrient substances. Complex media contains all essential components for prolonged cell growth *in vitro*.

Chemically defined media are sometimes called 'serum-free' or 'protein-free'. The first successful cell culture techniques used biological fluids of unknown chemical composition (Arnold, 1887). In his pioneering work on tissues removed from the body, Ringer (1883) was the first to acknowledge the need for chemically defined media. The reasons for using chemically defined media are that they enable reproducibility and avoid interference by unknown substances. To this day, there are very few truly defined culture media available, and the common practice is still to supplement media with sera or purified albumin that contain unknown components.

When certain cell culture applications are being addressed, it is often better to talk about the *cell culture system* rather than *cell culture medium* (which is only one part of it), since many physical and chemical factors (e.g. temperature, chemical environment) and contact materials (such as culture plastics and/or bioreactor materials, pipettes, micro filters, and tubing) can mediate the overall effectiveness of culture medium. A multitude of factors can influence the culture system and a good culture medium may not work in the hands of an inexperienced cell biologist working in a poor laboratory. Therefore, it is essential to consider the entire cell culture system wherein the medium is a pivotal factor and indeed, often the limiting performance factor. The scope of this section is to discuss, in general terms, the various components of cell culture media and their function.

12.2.2 Medium components

12.2.2.1 Water

The major component of all culture media and a major potential source of contamination that can block or alter cell growth is water. A common practice for the production of water for cell culture media is to start with a good source of ground water and to use reverse osmosis and ultra filtration as purification methods. Surface water is avoided since it frequently contains fertilizers and chemicals, which put high demands on the water purification process. Distillation may be needed as a step in the purification of water for media preparation. The Pharmacopoeia specification for purified water used for production of human infusion liquids, so called WFI (water for injection), demands distillation in Europe (PhEur), but this is not a universal requirement in the USA (USP). Cell biologists frequently use purified, low conductivity, so called 18.2 MegaOhm water, for media production. The low conductivity is reached by 'polishing' ultra purified water by ion exchange so that all charged molecules are removed. Since many cytotoxic components, like bacterial endotoxin and leachates from plastic cartridges, tubing and filters are non-charged organic molecules; their presence in the water will remain undetected by the measurement of resistance. TOC (total organic carbon) measurement is commonly used to detect uncharged organic molecules.

12.2.2.2 Balanced salts

The role of balanced salts diluted in purified water is to provide a physiologic ionic environment that promotes cell metabolism and maintains intra- and extra-cellular osmotic balance. They typically consist of various components and concentrations of the following: $CaCl_2$, KCl, KH_2PO_4, $MgCl_2$, $MGSO_4$, NaCl, $NaHCO_3$, Na_2HPO_4, NaH_2PO_4, Citrate and Glucose (Dulbecco and Freeman, 1959; Eagle 1959; Earle, 1943; Hanks and Wallace, 1949). Certain ions, such as Na^+ and K^+, are crucial for cell membrane pump function since they serve in ionic counter pump processes across the cell membrane to regulate the concentration of other ions such as for example NH_4+. Osmolality is an important physiological variable (normally around 285 mOsm/kg H_2O) and it is normally adjusted by varying the amount of NaCl when making up the medium.

12.2.2.3 Energy substrates

Glucose, pyruvate and lactate are the most commonly used energy substrates in growth media, because they are used in the chain of metabolism (glycolysis, citric acid cycle and oxidative phosphorylation). The relative concentrations of energy substrates in media are important. For example, high levels of glucose inhibit oxidative phosphorylation (and oxygen consumption) known as the Crabtree effect (Crabtree, 1929). High levels of lactic acid and pyruvate suppress oxidation of glucose. Thus, adequate levels of energy substrate and adequate oxygen supply are critical for driving the metabolism in the right physiologic direction. Otherwise, compensatory metabolic mechanisms come into play and the culture system can no longer be called physiologic. The goal is to avoid metabolic derangements in where, for example, detrimental free oxygen radicals are produced. The term carbohydrate is not ideal for describing energy substrates since modern media may contain carbohydrates in the form of macromolecules serving other purposes than energy substrate.

12.2.2.4 Buffers

Optimal growth of cells is dependent on maintaining a physiologic pH. The most commonly used culture system employs incubation in a high carbon dioxide (CO_2) environment typically 5%. Based on the well-known Henderson-Hasselbach equation (Equation 12.1), bicarbonate has to be present in the medium at a concentration of about 25 mM to reach a pH of about 7.4.

$$pH = pK_a + \log\left(\frac{[base]}{[acid]}\right) \qquad (12.1)$$

In complex media many other components (such as amino acids) also act as buffers. Therefore, in order to reach the desired pH, bicarbonate must be adjusted and the pH titrated. It should be noted that the CO_2 level is not only essential for maintaining correct pH. Earlier studies have shown that CO_2 can have substantial stimulatory effect on metabolism (Fanestil *et al.*, 1963; Hastings and Longmore, 1965; McLimans, 1972). The disadvantage with bicarbonate buffered media is that they require incubation in a high CO_2 environment before use in order to attain physiologic pH. When handling cells outside the incubator, bicarbonate is not an adequate buffer since the pH of the media will rise to non-physiological levels. Therefore, when handling of cells outside the incubator in ambient air, other buffers with appropriate pKa levels are used, for example HEPES (N-[2-Hydroxyethyl] piperazine-N′-[2-ethanesulfonic acid]) or MOPS (3-Morpholinopropanesulfonic acid).

There are some disadvantages of HEPES buffered media. It has been reported that toxic metabolites are formed when HEPES is exposed to intense light, for example under the microscope, which may be detrimental to certain cells (Zigler *et al.*, 1985). Furthermore, should a HEPES-buffered medium be used in a CO_2 incubator, pH may drop to acidic levels. It is advised that the cell culture is validated in respect to required pH since the optimal pH varies with the cell type. For example, human embryos are extremely sensitive to pH changes, whereas fibroblasts are much more resilient. This is because these very different cell types exhibit different sets of membrane ion pumps that maintain intra-cellular pH and homeostasis under stress *in vitro*.

Phenol red is frequently used in media to function as a visual aid in monitoring pH. However, it should be noted, that: (1) phenol red may contain impurities that may affect cellular behavior, for example via estrogenic effects, and (2) pH and temperature both change quickly when a culture dish is taken outside the incubator. Therefore, validation of pH and temperature changes is strongly recommended when optimizing a new cell culture protocol.

12.2.2.5 Anti-fungal/bacterial substances

Antibiotics (e.g. penicillin G, streptomycin, amphotericin B, gentamicin) are frequently used in culture media to avoid contamination. Ideally, it is preferable to use strict aseptic working techniques, air handling systems and incubator cleaning routines, which indeed may exclude the need for antibiotics in cell culture media. When in doubt, it is wise to run bio burden tests, using contact plates, sedimentation plates and air sampling of the laboratory environment and staff, for microbiological screening detecting both bacteria and moulds. In preparation and isolation of cells from biological specimens, it is good practice to transiently use antibiotics in cell preparation and handling media in order to avoid unintentional introduction of microbes into the cell culture system.

Needless to say, the sensitivity to antibiotics varies with each cell type and should be investigated as part of the process of setting up a new culture system.

12.2.2.6 Protein and amino acid supplementation

Proteins in media serve typically as membrane stabilizing agents and carriers (as well as a source) of essential molecules; for example, fatty acids. Proteins also serve as 'anti-sticking' agents for the handling and culture of non-attaching cell types. Furthermore, it has been hypothesized that proteins could help detoxify media by binding contaminants from plastic culture dishes. The most common sources of protein are serum and purified albumin. Amino acids are an essential part of complex media formulations. They are divided into essential and non-essential amino acids. Amino acids are building blocks in *de novo* protein synthesis, but they may also serve as energy substrates, pH regulators and osmolytes (regulators of water balance). Amino acids, in particular L-glutamine, are, together with vitamins, the most labile components of media as a function of time and temperature. When incubated at 37°C, amino acids spontaneously break down and release ammonia, which may inhibit cell growth should the concentration become too high. It is therefore important to limit the exposure of complex media to 37°C as well as light. Regular changing of media is needed to avoid build up of breakdown products.

12.2.2.7 Vitamins

Vitamins are essential components of various complex media formulations; for example, D-MEM, F-12, RPMI 1640, NCTC). The most abundant vitamins in complex media belong to the water-soluble group, in particular B-vitamins (thiamin, riboflavin, niacin, pyridoxin, folic acid, cyanocobalamin, pantothenic acid, biotin). It can be anticipated that fat-soluble vitamins are provided through the protein supplements used; for example, serum. Vitamins are components that contribute strongly to the light sensitivity of culture media.

12.2.2.8 *Cytokines, growth factors and hormones*

Molecules in this component group exert their action in minute quantities by binding to specific receptors inducing signaling pathways that change cell behavior, e.g. differentiation, migration, secretion, uptake or growth. These molecules are highly cell specific and their use must be investigated for each cell type. Certain factors that are required for optimal growth of a particular cell type may be toxic to another cell type. Their practical use, biological activity and accessibility to the cells may be dependent on carrying molecules such as albumin. The use of factors in media frequently create challenges since many of them degrade rapidly at 37°C, attach to plastic ware, or may be lost in micro filters used for sterile-filtration of media. It is advised that the biological activity of factors is assayed in each cell culture system before experimentation commences. The effect of these components are to a large extent cell specific factors although general growth promoting factors, such as insulin, are widely used in complex media.

12.2.2.9 *Other components*

Attachment factors (e.g. collagen, fibronectin, laminin) are often used for attachment-dependent cell types, such as fibroblasts and epithelial cells. They are normally required for attachment, spreading and polarity of certain cell types to attain their natural phenotype *in vitro*. For example, since epithelial cells are polarized *in vivo* (i.e. exhibit an apical – secretory – side and a basal – anchoring – side that takes up nutrients from the blood stream), these cells need to be cultured using dual chamber membrane culture systems, along with proper media. Otherwise, epithelial cells will dedifferentiate *in vitro* and lose their epithelial characteristics.

Lipids are also essential for cell growth, such as linoleic acid, which is necessary for the synthesis of cell membrane.

Conditioned media, obtained from cultures of specific cells, are sometimes used to enhance the maintenance of a specific cell phenotype or to induce cell differentiation in a desired direction. The use of conditioned media is highly empiric and it is beneficial only when there is no knowledge about the specific factors that need to be supplemented to the basic medium.

Serum is frequently used to provide for a multitude of undefined components, and is still needed to grow most cell types. The serum source can be either autologous (from the patient to be treated) or allogeneic (from another donor, human or animal). FBS (fetal bovine serum) is the most commonly used medium supplement. Unfortunately, the use of FBS does not help clarify the nutrient needs of cells and carries an inherent safety risk. Recent outbreaks of bovine spongiform encephalopathy (BSE) show that future access to pathogen-free herds cannot be guaranteed indefinitely (Belay and Schonberger, 2005).

12.2.3 Medium preparation

Cell biologists know that most media work best when they are freshly prepared, or refrigerated and kept in the dark for only short periods of time. Although most basic components in media are stable, there are several components that are labile (for example, bicarbonate, amino acids, growth factors and vitamins). This presents challenges when it comes to the logistics of media preparation and storage. Media are typically sterile-filtered through a 0.22 μm pore-size filter (sterile filter). However, should a previous growth of micro organisms have occurred, the media may have been depleted of critical components, or toxic components may have been added (for example, endotoxins). It is good practice to minimize the formulation time and sterile-filter the components or the medium immediately after preparation.

It is reasonable to assume that all currently used media formulations contain only a subset of the necessary components found in the cell environment *in vivo*. When selecting a suitable medium for a particular application, it is best to start asking questions about the normal milieu in which the cells are normally growing which implies, among other things, nutrition, osmolality, oxygenation, macromolecular environment and cell regulating factors.

The common view of cell culture media implies a watery red (if containing Phenol red) fluid, which is indeed very non-physiological since there is no such fluid in the body. On the contrary, the physiologic environment *in vivo* is highly viscous and contains known and unknown macromolecules contributing to the osmotic pressure affecting water transport across the cell membrane. Unfortunately, osmotic pressure is frequently not even measured in culture media.

The creation of *in vivo*-like conditions *in vitro* requires an in-depth analysis on the physiologic environment. Appropriate substitution of a standard medium such as, for example, Eagle's minimal essential medium (MEM) with additives specific for the cell type to be cultured is a good start. Furthermore, it is essential that physical-chemical factors are considered and biological assays used before even starting to culture cells. When the culture system and its possible limitations are known to some extent, cell culture work can be done with fewer artefacts. Table 12.1 shows an overview of the composition of some classic culture media compositions.

12.3 Directing cellular behavior by culture medium composition

12.3.1 Stages of cell culture

Once the appropriate cells are isolated, cell culture encompasses three different stages, as shown in Figure 12.2 (Strehl *et al.*, 2002): (1) expansion of the cells; (2) initiation of differentiation; (3) maintenance of differentiation. The first stage drives the appropriate cells through the growth phase as quickly as possible to obtain sufficient numbers of cells within a short period of time. The second stage aims at reducing proliferation and inducing tissue-specific (re)differentiation. During the third stage, the tissue differentiation is stabilized in order to establish and maintain tissue-specific cell characteristics. These same stages are inherent for tissue engineering by cultivation of cells on a biomaterial scaffold.

During each of the three stages, the cells will have different nutrient requirements. In tissue engineering approaches, the formation of a functional, differentiated tissue on a biomaterial scaffold is often preceded by the multiplication of the cells (stage 1). The expansion phase is typically performed in a two-dimensional (2D) fashion, e.g. in a T-flask or a Petri dish. Cell culture media that have historically been optimized for the maximum rate of cell proliferation are also used for the expansion of cells for tissue engineering. Often these media contain fetal bovine serum or adult serum.

During the differentiation phase, cells require another spectrum of nutrients and/or stimulatory agents and are often cultured in media containing low levels of serum or no serum. Further development for chemically defined media that are adapted to the requirements of cells differentiating into specific lineages is necessary.

12.3.2 Growth factors

Exogenous factors modulate cell behavior, *in vitro* and *in vivo*. For example, bone formation by human bone marrow stromal cells can be enhanced by dexamethasone (Mendes, 2002). Similarly, it has been demonstrated that cartilage formation by expanded human chondrocytes could be enhanced by specific combinations of growth factors and hormones, including transforming growth factor β (TGF-β) and insulin-like growth factor 1 (IGF-1), during three-dimensional culture (Yaeger *et al.*, 1997).

Growth factors and hormones can sequentially be applied to promote cells to first de-differentiate into a proliferative state and then re-differentiate and regenerate full tissues (Jakob *et al.*, 2001; Martin *et al.*, 2001). For the regeneration of cartilage tissue, exposure of isolated chondrocytes to fibroblast growth factor-2 (FGF-2) *in vitro* causes the cells to proliferate and de-differentiate, but the ability to undergo chondrogenic differentiation and form cartilage tissue is maintained (Martin *et al.*, 1999). Moreover, the cell's potential to respond to regulatory

Table 12.1 Examples of classic media compositions

Component	EBSS[1] g/L	MEM[2] g/L	DMEM[3] g/L	RPMI 1640[4] g/L	Ham's F10[5] g/L	Medium 199[6] g/L
Inorganic Salts						
NaCl	6.8	6.8	6.4	6.0	7.4	6.8
$NaHCO_3$	2.2	2.2	3.7	2.0	1.2	2.2
KCl	0.4	0.4	0.4	0.4	0.285	0.4
$CaCl_2 \times 2H_2O$	0.265	0.265	0.265	–	0.0441	0.265
$Ca(NO_3) \times 4H_2O$	–	–	–	0.1	–	–
NaH_2PO_4	0.122	0.122	0.109	0.8	0.1537	0.122
KH_2PO_4	–	–	–	–	0.083	–
$MgSO_4$	0.09767	0.09767	0.09767	0.04884	0.07464	0.09767
$CuSO_4$	–	–	–	–	0.0000025	–
$FeSO_4$	–	–	–	–	0.000834	–
$Fe(NO_3)_3 \times 9H_2O$	–	–	–	–	–	0.00072
$ZnSO_4$	–	–	–	–	0.00000288	–
Na-Acetate	–	–	–	–	–	0.05
Amino acids						
L-Alanine	–	0.0089	–	–	0.009	–
DL-Alanine	–	–	–	–	–	0.05
L-Arginine	–	0.126	0.084	0.2	0.211	0.07
L-Asparagine	–	0.015	–	0.05	0.01501	–
L-Aspartic Acid	–	0.0133	–	0.02	0.0133	–
DL-Aspartic Acid	–	–	–	–	–	0.06
L-Cysteine	–	–	–	–	0.035	0.00011
L-Cystine	–	0.0313	0.0626	0.0652	–	0.026
L-Glutamic Acid	–	0.0147	–	0.02	0.0147	–
DL-Glutamic Acid	–	–	–	–	–	0.1336
L-Glutamine	–	0.292	0.584	0.3	0.146	0.1
Glycine	–	0.0075	0.030	0.01	0.00751	0.05
L-Histidine	–	0.042	0.042	0.015	0.021	0.02188
L-Isoleucine	–	0.052	0.105	0.05	0.0026	–

DL-Isoleucine	–	–	–	–	–	0.04
L-Leucine	–	0.052	0.105	0.05	0.0131	–
DL-Leucine	–	–	–	–	–	0.12
L-Lysine	–	0.0725	0.146	0.04	0.0293	0.07
L-Methionine	–	0.015	0.030	0.015	0.00448	–
DL-Methionine	–	–	–	–	–	0.03
L-Phenylalanine	–	0.032	0.066	0.015	0.00496	–
DL-Phenylalanine	–	–	–	–	–	0.05
L-Proline	–	0.0115	–	0.02	0.0115	0.04
Hydroxy-L-Proline	–	–	–	0.02	–	0.01
L-Serine	–	0.0105	0.042	0.03	0.0105	–
DL-Serine	–	–	–	–	–	0.05
L-Threonine	–	0.048	0.095	0.02	0.00357	–
DL-Threonine	–	–	–	–	–	0.06
L-Tryptophan	–	0.01	0.016	0.005	0.0006	–
DL-Tryptophan	–	–	–	–	–	0.02
L-Tyrosine	–	0.0519	0.10379	0.02883	0.00261	0.05766
L-Valine	–	0.046	0.094	0.02	0.0035	–
DL-Valine	–	–	–	–	–	0.05
Vitamins						
L-Ascorbic acid	–	–	–	–	–	0.0000566
D-Biotin	–	0.001	–	0.0002	0.000024	0.00001
Calciferol	–	–	–	–	–	0.0001
Choline chloride	–	0.001	0.004	0.003	0.000698	0.0005
Folic acid	–	0.001	0.004	0.001	0.00132	0.00001
Menadione (sodium bisulfite)	–	–	–	–	–	0.000016
myo-Inositol	–	0.002	0.0072	0.035	0.000541	0.00005
Niacinamide	–	0.001	0.004	0.001	0.000615	0.000025
Nicotinic Acid	–	–	–	–	–	0.000025
p-Amino Benzoic Acid	–	–	–	0.001	–	0.00005
D-Pantothenic acid	–	0.001	0.004	0.00025	0.000715	0.00001

(Continued)

Table 12.1 (Continued)

Component	EBSS[1] g/L	MEM[2] g/L	DMEM[3] g/L	RPMI 1640[4] g/L	Ham's F10[5] g/L	Medium 199[6] g/L
Pyridoxal	–	0.001	–	–	–	0.000025
Pyridoxine	–	–	0.004	0.001	0.000206	0.000025
Retinol Acetate	–	–	–	–	–	0.00014
Riboflavin	–	0.0001	0.0004	0.0002	0.000376	0.00001
DL–alpha–Tocopherol Phosphate	–	–	–	–	–	0.00001
Thiamin	–	0.001	0.004	0.001	0.001	0.00001
Thioctic acid	–	–	–	–	0.00021	–
Vitamine B–12	–	–	–	0.000005	0.00136	–
Other						
D–Glucose	1.0	1.0	1.0	2.0	1.1	1.0
Sodium Pyruvate	–	–	0.11	–	0.11	–
Tween 80	–	–	–	–	–	0.02
Glutathione (reduced)	–	–	–	0.001	–	0.00005
Thymidine	–	–	–	–	0.00073	–
Hypoxanthine	–	–	–	–	0.00408	0.0003
Adenine Sulphate	–	–	–	–	–	0.01
Adenosine Triphosphate	–	–	–	–	–	0.001
Adenosine Monophosphate	–	–	–	–	–	0.0002385
Cholesterol	–	–	–	–	–	0.0002
Deoxyribose	–	–	–	–	–	0.0005
Guanine	–	–	–	–	–	0.0003
Ribose	–	–	–	–	–	0.0005
Thymine	–	–	–	–	–	0.0003
Uracil	–	–	–	–	–	0.0003
Xanthine	–	–	–	–	–	0.000344

[1]EBSS = Earle's Balanced Salt Solution (Earle, 1943). [2]MEM = Minimum Essential Medium Eagle (Eagle, 1956). [3]DMEM = Dulbecco's Modified Eagle's Medium (Dulbecco and Freeman, 1959). [4]RPMI-1640 = Roswell Park Memorial Hospital Medium (Moore, 1967). [5]Ham's F10 (Ham, 1963). [6]Medium 199 (Morgan, 1950).

	Step 1	Step 2	Step 3
Goal	Expansion of cells	Initiation of differentiation	Maintenance of differentiation
Epithelia		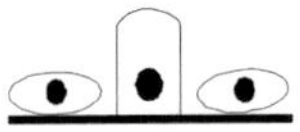	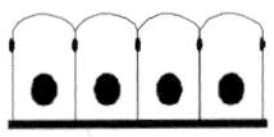
Connective tissue	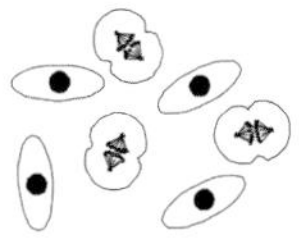	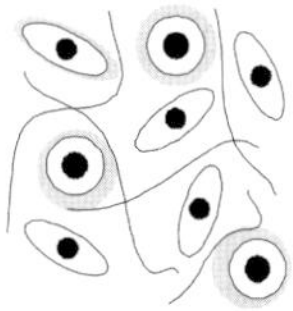	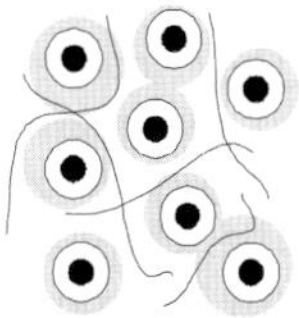
Culture technique	Static culture	Perfusion culture	Perfusion culture
Stimulus	Growth factors FCS in medium	Morphogens Serum-free media	Electrolyte-adapted, Serum-free media
Tissue reaction	Rapid cell division cycle	Reduced cell division cycle	Postmitotic phase Arrest in interphase
Mitotic stress	High	Low	Low
Differentiation	Low	Upregulated	High

Figure 12.2 Cell culture generally involves a three-step protocol. In step 1, the cell number is expanded in conventional culture media often containing growth factors, fetal bovine serum or adult human serum. In step 2, proliferation is reduced and tissue-specific differentiation is induced. In general, the developing tissue is maintained in a bioreactor and continuously supplied with fresh and preferably serum and growth-factor-free culture medium. During step 3, tissue differentiation is stabilized in order to maintain tissue-specific characteristics. Reprinted from Strehl, R., Schumacher, K. *et al.* (2002). Proliferating cells versus differentiated cells in tissue engineering. *Tissue Eng*, 8(1): 37–42, with permission from the author and M.A. Liebert.

molecules upon transfer into a three-dimensional (3D) environment was also enhanced. In the presence of bone morphogenetic protein-2 (BMP-2) the formation of engineered cartilage on polyglycolic acid (PGA) scaffolds was further improved (Martin *et al.*, 2001).

Supplementation of culture media with exogenous growth factors in specific combinations and temporal sequences can be used to promote cells to first proliferate and subsequently re-differentiate and form the specific tissue of interest.

12.4 Mass transport

To engineer a tissue *in vitro*, it is essential to provide access to substrate molecules, such as oxygen, glucose, and amino acids. Also, clearance of products of metabolism, such as CO_2, lactate, and urea, are critical to cell survival. The movement of these molecules in and out of the construct is collectively referred to as *mass transport*. Mass transport occurs by *convection* and by *diffusion*. *Convection* is the mass transfer as a result of bulk fluid motion, whereas *diffusion* is

Box 1 Oxygen gradients in tissue engineering

Within tissue-engineered grafts, oxygen concentrations have been demonstrated to rapidly decrease by moving from the exterior to the interior and intensify during the early stages of tissue development (Kellner *et al.*, 2002; Malda *et al.*, 2004; Radisic, *et al.*, 2006). For example, using optical sensitive foils containing luminescent oxygen-sensitive indicator dyes, 2D oxygen distributions have been measured in engineered cartilage constructs grown for 3 weeks using fibrous polyglycolic acid scaffolds and bovine chondrocytes, and anoxic conditions were observed in the center of these constructs (Kellner *et al.*, 2002). In addition, oxygen gradients have been measured using microelectrodes in cylindrical cartilage constructs created from porous poly(ethylene glycol)-terephthalate/poly(butylene terephthalate) (PEGT/PBT) block co-polymer scaffolds (Malda *et al.*, 2004a, 2004b). It was shown that from 3 to 14 days onward the oxygen gradient became steeper. This effect was associated with

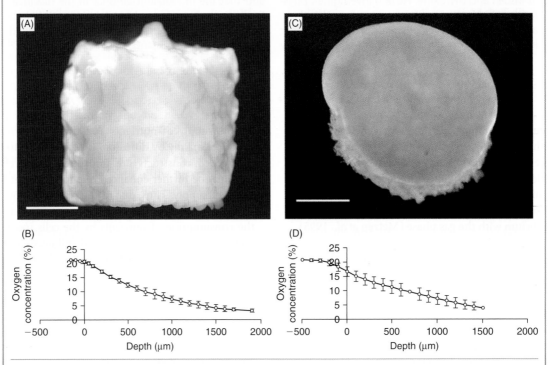

Light micrograph of a cylindrical (A) polymer-cartilaginous and (C) collagen-cardiac construct. Cartilage constructs were created from porous poly(ethylene glycol)-terephthalate/poly(butylene terephthalate) (PEGT/PBT) block co-polymer scaffolds (as previously described Malda, J., Rouwkema, J. et al. (2004). Oxygen gradients in tissue-engineered Pegt/Pbt cartilaginous constructs: Measurement and modeling. Biotechnology and Bioengineering, 86(1): 9–18) seeded with 3.0 × 10⁶ immature bovine chondrocytes and cultured for 14 days in magnetically stirred flasks. Cardiac constructs were created by seeding 3.0 × 10⁶ neonatal rat cardiomyocytes on collagen sponges, and cultured statically for 16 days. Oxygen concentrations rapidly decreased from the construct exterior to the interior (Malda et al., 2004) as measured with a oxygen microelectrode. Scale bar is 1 mm.

vascularization will be addressed. Upon implantation, the tissue-engineered graft must be rapidly vascularized to allow for the survival of the implanted cells. Induction of the blood vessel formation *in vitro* prior to implantation will likely improve the integration of the vascular network of the host. Thirdly, the incorporation of capillary networks into the scaffolds will be discussed. Current fabrication techniques allow the formation of organized networks of channels that can be perfused with culture medium, mimicking vascularized tissue. Finally, this section will describe the use of artificial oxygen carriers to increase the oxygen capacity of the culture medium by mimicking the role of hemoglobin.

12.6.1 Bioreactor culture

In general, a bioreactor is a vessel in which biochemical reactions take place. Based on this definition, even a tissue culture flask or a dish can be regarded as a bioreactor. Within the field of tissue engineering, bioreactors are generally defined as devices in which biological and/or biochemical processes develop under closely monitored and tightly controlled environmental and operating conditions (e.g. pH, temperature, pressure, nutrient supply and waste removal, and in many cases hydrodynamic shear, mechanical and electrical signals) (Martin *et al.*, 2004). Ideally, a bioreactor should provide an *in vitro* environment for rapid and orderly tissue development starting from isolated cells and three-dimensional scaffolds. Bioreactors are designed to perform one or more of the following functions:

- Establish spatially uniform concentrations of cells seeded onto clinically sized biomaterial scaffolds.
- Control conditions in culture medium (e.g. temperature, pH, osmolality, levels of oxygen, nutrients, metabolites, regulatory molecules).
- Provide physiologically relevant physical signals (e.g. interstitial fluid flow, shear, pressure, compression, stretch).
- Facilitate mass transfer between the cells and the culture environment.

In this section, we focus on the use of bioreactors to improve cell nutrition. Bioreactor operation is optimized to provide sufficient amounts of a limiting nutrient. In most cases, due to its high consumption rate and low solubility and diffusivity, that nutrient is oxygen. For example, let's consider non-contracting cardiomyocytes. Under common culture conditions (21% atmospheric oxygen at 37°C), the maximum concentration of oxygen in the culture medium will be 220 μM, while the oxygen diffusion coefficient and the consumption rate will be 2.0×10^{-5} cm^2 s^{-1} and 3.7×10^{-17} mol s^{-1} cell^{-1}, respectively (Casey and Arthur, 2000). On the other hand, the concentration of glucose in the culture medium can be as high as 25000 μM (in 'high glucose' DMEM) while the diffusivity and the consumption rate remain comparable to that of oxygen (1.0×10^{-5} cm^2 s^{-1} and 3.5×10^{-17} mol s^{-1} cell^{-1}, respectively (Casey and Arthur, 2000)). Clearly, for most mammalian cells (cardiomyocytes are a good example) oxygen is the critical nutrient and its mass transport will essentially determine cell survival and wellbeing.

Tissue engineering bioreactors that are used most extensively include static and mixed flasks, rotating vessels and perfusion bioreactors. These bioreactors offer three distinct flow conditions (static, turbulent, and laminar) and, hence, differ significantly in the rate of nutrient supply to the surface of the tissue construct. The medium composition is maintained either by periodic medium exchange or by constant renewal (e.g. by perfusion). Thereby, nutrients such as glucose are replenished, and waste products such as lactate are removed from the bioreactor system. In dishes, flasks and rotating bioreactors, mass transport within the constructs is by passive diffusion only. Convective mixing (e.g. in orbitally mixed dishes, stirred flasks, rotating vessels) improves the rate of external mass transport, while the internal mass transport remains diffusion controlled. Interstitial flow *through the tissue construct* is needed to enhance mass transport to the cells.

Due to their simplicity, *static dishes* remain a widely used set-up for the development of tissue-engineered

State of the Art Experiment

Numerous attempts at improving nutrient supply within engineered tissue constructs have been described. A state of the art example is the work on cardiac tissue engineering by Radisic *et al.* (2005). To mimic the capillary network, neonatal rat heart cells were cultured on poly(glycerol-sebacate) (PGS) scaffolds with a parallel array of channels made using a laser cutting/engraving system and perfused with culture medium. To mimic oxygen supply by hemoglobin, culture medium was supplemented by an oxygen carrier, a perfluorocarbon (PFC) emulsion (Oxygent™); constructs perfused with un-supplemented culture medium served as controls. Constructs were subjected

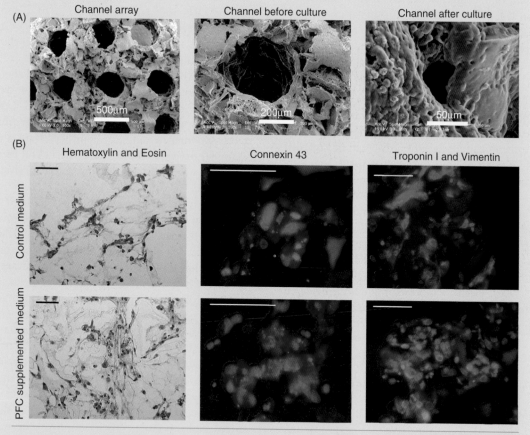

Scanning electron micrograph of the laser-bored channels in poly(glycerol sebacate) scaffolds (a, b) before and (c) after cell culture in the presence of PFC. Upon cultivation the channels remain open and the cells are present in the scaffold pores (c). Histological and Immunofluorescent assessment of tissue constructs cultured (d, e, f) without PFC and (g, h, i) with PFC. Stains are (d, g) hematoxilin and eosin, (e, f, h, i) Nuclei-blue (DAPI), Connexin-43 green (punctuated), Troponin-I green (larger continuous areas), Vimentin-red.

to unidirectional medium flow at a flow rate of 0.1 mL/min (linear flow velocity of 500 μm/s). As the medium flowed through the channel array, oxygen was depleted from the aqueous phase of the culture medium by diffusion into the construct space where it was used for cell respiration. Depletion of oxygen in the aqueous phase acted as a driving force for the diffusion of dissolved oxygen from the PFC particles, thereby contributing to the maintenance of higher oxygen concentrations in the medium. Due to the small size of PFC particles, diffusion of dissolved oxygen from the PFC phase into the aqueous phase was very fast, and estimated not to be a rate-limiting step in this system. For comparison, in un-supplemented culture medium, oxygen was depleted faster since there is no oxygen carrier phase that acts as a reservoir.

PFC-supplemented medium, the decrease in the partial pressure of oxygen in the aqueous phase was only 50% of that in control medium. Consistently, constructs cultivated in the presence of PFC had higher amounts of DNA, troponin I and Cx-43, and significantly better contractile properties as compared to control constructs. In both groups, cells were present at the channel surfaces as well as within constructs. Improved constructs properties were correlated with the enhanced supply of oxygen to the cells within constructs. Mathematical modeling indicated that the addition of 3–6 vol% of PFC emulsion increased the effective diffusivity of the culture medium by 9–18%. In addition, convective term was increased by 60–120% without increasing the average fluid velocity in the channel array. Therefore, both the transport in axial and radial direction is improved compared to that in regular culture medium. The presence of PFC emulsion increased volume averaged and minimum oxygen concentration in the tissue space as well as the bulk oxygen concentration at the outlet from channel lumen. Taken together, these studies showed that mimicking the convective-diffusive transport of oxygen in native tissue, by utilization of channeled scaffolds and oxygen carriers, markedly improved the structure and contractile properties of engineered cardiac constructs.

grafts. In static dishes, both external and internal mass transfer are governed by diffusion (Carrier et al., 2002; Martin et al., 2004). Delivery of oxygen and nutrients to construct surfaces can be enhanced by orbital mixing, but diffusion remains the main mechanism of mass transport within the tissue. In both cases the interior of the scaffold often remains acellular (Radisic et al., 2006). An additional limitation of the culture in dishes is that the bottom surface of the construct is in contact with the bottom of the dish and thereby lacks proper oxygenation and nutrients, and yielding asymmetric cell distribution with compact tissue mostly on the top surface.

To improve cell survival and assembly on all surfaces of the engineered tissue, constructs can be cultivated suspended in culture medium in *spinner flasks*. Within spinner flasks, scaffolds are attached to needles hanging from the lid of the flask. Convective flow, generated by a magnetic stirrer bar, allows continuous mixing of the media surrounding the constructs. Because of the relatively simple setup, spinner flasks are also widely used for seeding cells on scaffolds (Vunjak-Novakovic et al., 1998), the culture of cell/polymer constructs for cartilage (Malda et al., 2005; Vunjak-Novakovic et al., 1996), skin (Wang et al., 2004), cardiac muscle (Carrier et al., 1999) and bone tissue regeneration (Goldstein et al., 2001; Sikavitsas et al., 2002). However, the flow conditions in the spinner flasks are turbulent. As a result, fibrous capsule forms on the construct surfaces, a common cell response to non-physiologic flow.

In order to enhance external mass transfer under laminar flow conditions, the tissue engineered constructs can be cultivated in *rotating vessels* (Carrier et al., 1999; Freed and Vunjak-Novakovic 2002; Papadaki *et al.*, 2001). In the rotating vessels up to 12 tissue constructs can be suspended without fixation in the annular space between two cylinders, gas is exchanged via a silicone membrane of the inner cylinder, and mixing is provided by construct settling in the rotating flow. Depending on the size of the tissue construct, the rotational speed has to be adjusted in order to keep the construct settling in the fluid flow. Dynamic laminar flow of rotating bioreactors generally improves properties of the peripheral tissue layer, fibrous capsule do not form, but the limitations of the diffusional transport of oxygen to the construct interior still remains.

In an attempt to enhance mass transport within cultured constructs, *perfusion bioreactors* (Bancroft *et al.*, 2003; Carrier *et al.*, 2002; Radisic *et al.*, 2003; Sikavitsas *et al.*, 2003) have been developed. They provide interstitial medium flow through the cultured construct at velocities similar to those in native tissues (~400–500 μm/s) (Figure 12.6). In such systems, oxygen and nutrients are supplied to the construct interior by both diffusion and convection. The flow rate can be optimized in respect of the limiting nutrient, in most cases oxygen, due to its low solubility in culture medium.

The minimum flow rate in perfused cartridges can be determined from a simple mass balance of oxygen supply by culture medium and consumption by the cells.

$$F(C_{in} - C_{out}) = \bar{r}_o N \tag{12.9}$$

where F [mol s^{-1}] is the flow rate of culture medium, C_{in} is the inlet oxygen concentration and in most cases corresponds to 21% oxygen (220 μmol L^{-1}) and C_{out} is the outlet oxygen concentration, r_o is the average oxygen consumption [mol cell^{-1} s^{-1}] rate and N the total number of cells. When the flow rate is calculated for the maximum allowed drop in the oxygen partial pressure it is necessary to check the shear stress that will be imposed onto the cells as medium flows through the interstitial space. The shear stress will be a function of construct diameter D and should be below the level that induces damage for a given cell type. In most cases the desired shear stress is below 1 dyn/cm².

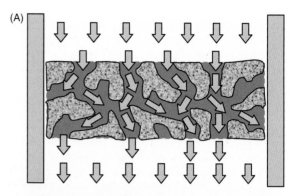

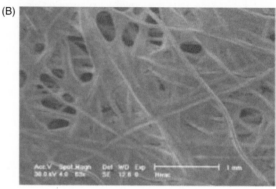

Figure 12.6 In perfusion culture the culture medium is forced through the internal porous network of the scaffold (A). This can mitigate internal diffusion limitations present in three-dimensional scaffolds to enhance nutrient delivery and waste removal from the cultured cells. Consequently, a porous structure is formed (B) when, for example osteoblast are cultured on titanium mesh scaffolds in a perfusion bioreactor for 16 days. Reprinted (A) from Bancroft G.N., Sikavitsas, V.I. *et al.* (2003). Design of a flow perfusion bioreactor system for bone tissue-engineering applications. *Tissue Eng*, 9(3): 549–554, with permission from the authors and M.A. Liebert, and (B) from Sikavitsas, V.I., Bancroft, G.N. *et al.* (2003). Mineralized matrix deposition by marrow stromal osteoblasts in 3D perfusion culture increases with increasing fluid shear forces. *Proc Natl Acad Sci USA*, 100(25): 14683–14688, with permission from the authors and The National Academy of Sciences of the United States of America.

> **Box 2   Cardiac tissue engineering in spinner flasks, rotating vessels and perfusion bioreactors**
>
> The culture of cardiac tissue constructs based on neonatal rat cardiomyocytes and PGA scaffolds has been performed in different culture systems. Beneficial effects on cell number and metabolic activity were reported for 2-week constructs cultured in *spinner flasks* in comparison to *static flasks* (Carrier *et al.*, 1999). Mixing maintained medium gas and pH levels within the physiological range, yielded a more aerobic glucose metabolism in mixed flasks as compared to static flasks (Carrier *et al.*, 1999). Constructs cultured in spinner flasks contained a peripheral tissue-like region (50–70 μm thick) in which cells stained positive for tropomyosin (Bursac *et al.*, 1999). Electrophysiological studies were conducted using a linear array of extracellular electrodes and showed that the peripheral layer of these constructs sustained macroscopically continuous impulse propagation on a centimeter-size scale (Bursac *et al.*, 1999). However, construct interiors remained empty due to the diffusional limitations of the oxygen transport within the bulk tissue, and the density of viable myocytes was orders of magnitude lower than that in the neonatal rat ventricles (Bursac *et al.*, 1999).
>
> Culture in *rotating vessels* enabled better maintenance of pH and oxygen concentration and resulted in mostly aerobic cell metabolism (Carrier *et al.*, 1999). The outer layer of viable tissue was up to 160 μm thick (Papadaki *et al.*, 2001), electrophysiological properties were improved (Bursac *et al.*, 2003) and the metabolic activity of cells increased into the range of values measured for neonatal rat ventricles, significantly higher than in mixed flasks. However, the construct cellularity still remained 2–6 times lower than in native heart ventricles. Also, cells expressed cardiac-specific markers (e.g. tropomyosin, gap junction protein connexin-43, creatin kinase-MM, sarcomeric myosin heavy chain) at levels that were lower than in neonatal rat ventricles but higher than in constructs cultured in spinner flasks (Papadaki *et al.*, 2001).
>
> In order to improve the supply of oxygen to the cardiomyocytes, which are extremely sensitive to hypoxia, cells were cultured on scaffolds in a *perfusion bioreactor*. In this particular study it was demonstrated that constructs seeded in dishes had most cells located in the ~100 μm thick layer at the top surface; whereas constructs seeded in a perfusion bioreactor had physiologically high and spatially uniform cell density throughout the perfused volume of the construct. Throughout the cultivation, the number of live cells in perfused constructs was significantly higher than in dish-grown constructs. Importantly, the final cell viability in perfused constructs was not significantly different than the viability of the freshly isolated cells and was almost twice as high as in dish-grown constructs (Radisic *et al.*, 2004). Consistently, the molar ratio of lactate produced to glucose consumed was ~1 for perfused constructs, indicating aerobic cell metabolism. In dishes, the ratio increased progressively from ~1 to ~2, indicating a transition to anaerobic cell metabolism. Clearly, medium perfusion during seeding and cultivation was key for engineering thick constructs with high densities of viable cells, presumably due to enhanced transport of oxygen within the construct.

12.6.2 Formation of vasculature *in vitro*

In vivo, the construct must be rapidly vascularized to allow for the survival of the construct and, later on, its integration with surrounding host tissues. However, the host's vascularization is often not sufficient to feed the implant, especially not for keeping the construct viable in the first period following implantation and ensuing inflammatory response and fibrin clot formation (Nomi *et al.*, 2002; Ostrander *et al.*, 2001). Oxygen supply is limiting as cells more than approximately 200 μm from blood vessel suffer from hypoxia and die (Carmeliet and Jain, 2000; Colton, 1995). The vasculature of the

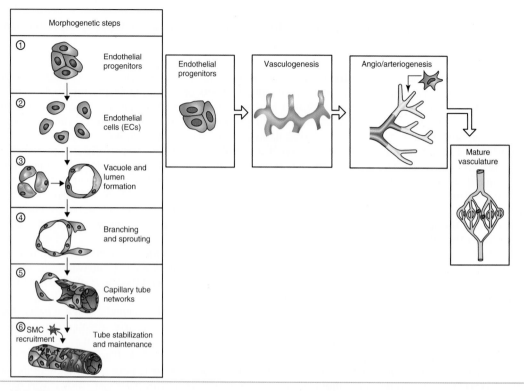

Figure 12.7 The steps and accompanying cellular processes and ECM regulators involved in forming a mature vascular network. Adapted from Lutolf, M.P. and Hubbell, J.A. (2003). Synthetic biomaterials as instructive extracellular microenvironments for morphogenesis in tissue engineering. *Nat Biotechnol*, 23(1): 47–55, and Carmeliet P. (2003). Angiogenesis in health and disease. *Nat Med*, 9(6): 653–660.

host will ultimately penetrate macroporous matrices via angiogenesis, whereby new blood vessels sprout from those that already exist and populate the nearby non-vascularized regions of the construct. This spontaneous angiogenic response, however, is often insufficient to support the large cell masses required to provide therapeutic support.

In the body, the vascular network branches in a hierarchical fashion from large diameter arteries/veins to small diameter capillary beds, and is organized spatially to provide adequate nutrients to the cells of all organs and supporting structures by diffusion and convection. The walls of the vessels are composed of endothelial and mural cells, which are embedded in an extracellular matrix.

There are three distinct mechanisms of vessel growth: *vasculogenesis* resulting from *de novo* blood vessel formation by endothelial progenitors; *angiogenesis* resulting from the stabilization of pre-existing blood vessels; and *arteriogenesis* resulting from the sprouting of blood vessels by mural cells. During vasculogenesis or angiogenesis, endothelial progenitor cells differentiate to arterial and venous endothelial cells (ECs) resulting in primitive vacuole and lumen formation. Subsequent vessel branching and sprouting results in capillary network formation, with vessels eventually becoming stabilized by the recruitment of smooth muscle cells (SMCs) differentiating from their progenitors. The vascular network must mature, both at the level of the vessel wall and at the network level to obtain its complex hierarchical organization of branching arteries, arterioles and capillaries (Figure 12.7).

Various methods to administer vasculogenic and angiogenic growth factors (such as VEGF, FGF and PDGF) and facilitate the angiogenic response are currently being developed. At the time of writing, this was done via slow release from the scaffolds or by genetically modifying the cells to express higher levels of the factors (von Degenfeld *et al.*, 2003; Zisch *et al.*, 2003). Another exciting approach is to vascularize the tissue in vitro, in an attempt to restore cell viability, induce structural organization, and promote integration upon implantation.

Endothelial cells have been incorporated into cardiac valve leaflets, liver tissue, and engineered skin, and we list here some representative examples of these studies. In cardiac valves, engineered leaflets were created using PGA scaffolds seeded with endothelial cells and fibroblasts from bovine arteries (Shinoka *et al.*, 1996). After surgical implantation into the right posterior leaflet of the pulmonary valve for eight weeks, the transplanted autologous cells generated a proper matrix on the polymer. One study showed that when human aortic endothelial cells and hepatocytes were co-cultured in double layered sheets using thermoresponsive culture dishes, the hepatocyte function (i.e. albumin expression) was better maintained than in hepatocytes grown in culture alone (Harimoto *et al.*, 2002). Another type of vascularized tissue construct was fabricated by co-culturing human keratinocytes, dermal fibroblasts, and umbilical vein endothelial cells together in a collagen matrix (Black *et al.*, 1998). This vascularized skin equivalent demonstrated capillary formation and extracellular matrix production.

Endothelial cells co-cultured with skeletal myoblasts on 3D polymer scaffolds could and organize into vessel-like structures throughout the differentiating myotubes. Prevascularization of the engineered muscle promoted vacularization of the implant resulting in increased survival of the implanted muscle construct (Levenberg *et al.*, 2005). These studies indicate that engineered tissues can be fabricated with endothelial cells to closely mimic native tissue function and to improve tissue survival (Figure 12.8).

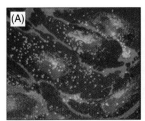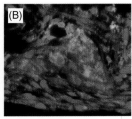

Figure 12.8 (A) Endothelial cells stained with endothelial markers: anti human CD31 antibodies on the membrane (red) and anti-von willibrand factor in the cytoplasm (green), and with DAPI for nuclear staining (blue); (B) Vascularization of engineered muscle tissue. Co-culture of endothelial cells with muscle cells on 3D scaffolds. Endothelial cells (red) form tube structures (red) in between the muscle cells (green) and with DAPI for nuclear staining (blue) (Levenberg, S. Rouwkema, J. *et al.* (2005). Engineering vascularized skeletal muscle tissue. *Nat Biotechnol*, 23(7): 879–884.)

Endothelial cells and nascent vessels (even prior to blood vessel function) were shown to provide inductive signals that are critical for liver and pancreatic development (Lammert *et al.*, 2001; Matsumoto *et al.*, 2001). Therefore, having endothelial cells in the scaffolds and the formation of a blood vessel network in the scaffolds could help support the cell organization into desired tissues. To engineer complex tissue structure with endothelial network formation, human embryonic stem cells were seeded on polymer scaffolds and supplemented with growth factors to direct their differentiation (Levenberg *et al.*, 2003). It was found that during differentiation of the cells into specific epithelial or mesenchymal cells and formation of specific tissue structures (affected by addition of specific growth factors), some of the cells also differentiated into endothelial cells. Moreover, these endothelial cells organized into vessel-like structures throughout the tissue, indicating that three-dimensional culture of the embryonic cells can promote formation of massive three-dimensional vascular networks that closely interacted with the surrounding tissue (Figure 12.9A). Upon implantation into SCID mice, the donor endothelial cells within the implants appeared

limitations due to matrix accumulation, significantly affect metabolite distributions? The figure here shows a typical result for two of the simulated configurations. System A is a model of a Petri dish, where no mechanical stimulation is applied to the construct. In system B, the bioreactor is designed to apply a compressive loading to the construct. C and D shows the calculated lactate accumulation in the medium and construct after 48 h of cultivation. The same simulations supply data on glucose and oxygen distributions and enable the tissue engineer to optimize the design of bioreactor systems.

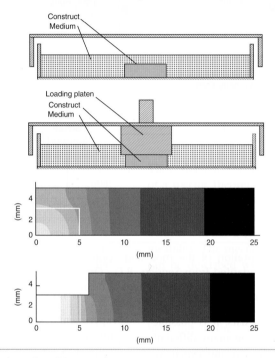

Culture configurations and predicted lactate distributions. (A) Petri Dish; (B) Compression set-up. Lactate distributions are given for the static case after 48 h. Because of symmetry only the right half of the domain is depicted, for; (C) the Petri dish; (D) the ucompression set-up (Adapted from Sengers, B.G., van Donkelaar, C.C. et al. (2005). Computational study of culture conditions and nutrient supply in cartilage tissue engineering. Biotechnol Prog, *21(4): 1252–1261).*

To fully understand the processes in a bioreactor, an integrated approach is required, where the mechanical as well as the biochemical environment are included. In the characterization of cell behavior, computational models can contribute to the interpretation of experimental results, by providing a quantitative metabolic structure, accounting for time dependency and spatial distribution. Short-term studies can be used to identify metabolic pathways. Longer-term studies are required to determine to what extent cellular behavior is representative for tissue engineering applications. An adequate nutrient supply is a prerequisite for tissue engineering, which has to be ensured first so that further biomechanical stimulation can be most effective. This includes not only maintaining viability, but also conditions that favour matrix synthesis. Clearly, computational models can be employed to provide holistic understanding of the culture process and aid the bioreactor design and the optimization of culture protocols.

Acknowledgments

The authors gratefully acknowledge the support for the studies described in this chapter, provided in part by IsoTis Orthobiologics, NIH (P41 EB002520-01, R01 HL076485-01 and RO1 DE016525-01 to GV-N).

References

Arnold, J. (1887). Über Theilungsvorgänge an der Wanderzellen: ihre progressiven und regressiven Metamorphosen. *Arch Mikrosk Anat*, 30: 205–310.

Bancroft, G.N., Sikavitsas, V.I., *et al.* (2003). Design of a flow perfusion bioreactor system for bone tissue-engineering applications. *Tissue Engineering*, 9(3): 549–554.

Belay, E.D. and Schonberger, L.B. (2005). The public health impact of prion diseases. *Annu Rev Public Health*, 26: 191–212.

Bialek, W. and Botstein, D. (2004). Introductory science and mathematics education for 21st-Century biologists. *Science*, 303(5659): 788–790.

Bibby, S.R., Jones, D.A., *et al.* (2005). Metabolism of the intervertebral disc: effects of low levels of oxygen, glucose, and pH on rates of energy metabolism of bovine nucleus pulposus cells. *Spine*, 30(5): 487–496.

Black, A.F., Berthod, F., *et al.* (1998). *In vitro* reconstruction of a human capillary-like network in a tissue-engineered skin equivalent. *Faseb J*, 12(13): 1331–1340.

Bursac, N., Papadaki, M., *et al.* (1999). Cardiac muscle tissue engineering: toward an in vitro model for electrophysiological studies. *Am J Physiol*, 277(2 Pt 2): H433–H444.

Bursac, N., Papadaki, M., *et al.* (2003). Cultivation in rotating bioreactors promotes maintenance of cardiac myocyte electrophysiology and molecular properties. *Tiss Eng*, 9(6): 1243–1253.

Carmeliet, P. (2003). Angiogenesis in health and disease. *Nat Med*, 9(6): 653–660.

Carmeliet, P. and Jain, R.K. (2000). Angiogenesis in cancer and other diseases. *Nature*, 407(6801): 249–257.

Carrel, A. and Lindbergh, C. (1935). The culture of whole organs. Science, 81(2112): 621–623.

Carrier, R., Papadaki, M., *et al.* (1999). Cardiac tissue engineering: cell seeding, cultivation parameters, and tissue construct characterization. *Biotechnol Bioeng*, 64(5): 580–589.

Carrier, R., Rupnick, M., *et al.* (2002). Perfusion improves tissue architecture of engineered cardiac muscle. *Tiss Eng*, 8(2): 175–188.

Casey, T.M. and Arthur, P.G. (2000). Hibernation in non-contracting mammalian cardiomyocytes. *Circulation*, 102(25): 3124–3129.

Colton, C. (1995). Implantable biohybrid artificial organs. *Cell Transplant*, 4(4): 415–436.

Crabtree, H. (1929). Observations on the carbohydrate metabolism of tumours. *Biochem J*, 23: 536–545.

Doran, P. (2000). *Bioprocess Engineering Principles*. London: Academic Press.

Dulbecco, R. and Freeman, G. (1959). Plaque production by the polyoma virus. *Virology*, 8: 396–397.

Eagle, H. (1956). myo-Inositol as an Essential Growth Factor for Normal and Malignant Human Cells in Tissue Culture. *J Biol Chem*, 214: 845–847.

Eagle, H. (1959). Amino acid metabolism in mammalian cell cultures. *Science*, 130: 432–437.

Earle, W. (1943). Production of Malignancy In Vitro. IV. The mouse fibroblast cultures and changes seen in the living cells. *JNCI*, 4: 165–169.

Ehrbar, M., Djonov, V.G., *et al.* (2004). Cell-demanded liberation of VEGF121 from fibrin implants induces local and controlled blood vessel growth. *Circ Res*, 94(8): 1124–1132.

Fanestil, D., Hastings, A., *et al.* (1963). Environmental CO_2 stimulation of mitochondrial adenosine triphosphatase activity. *J Biol Chem*, 238: 836–842.

Fidkowski, C., Kaazempur-Mofrad, M.R., *et al.* (2005). Endothelialized microvasculature based on a biodegradable elastomer. *Tiss Eng*, 11(1-2): 302–309.

Freed, L. E., Marquis, J.C., *et al.* (1994). Composition of Cell-Polymer Cartilage Implants. *Biotech Bioeng*, 43(7): 605–614.

Freed, L. E. and Vunjak-Novakovic, G. (2002). Spaceflight bioreactor studies of cells and tissues. *Adv Space Biol Med*, 8: 177–195.

Goldstein, A., Juarez, T., *et al.* (2001). Effect of convection on osteoblastic cell growth and function in biodegradable polymer foam scaffolds. *Biomaterials*, 22(11): 1279–1288.

Ham, R. (1963). An improved Nutrient Solution for Diploid Chinese Hamster and Human Cell Lines. *Exp Cell Res*, 29: 515–526.

Ham, R. (1981). Cell growth requirements - the challenge we face. In *The growth requirements of vertebrate cells in vitro* (Waymouth, C., Ham, R. and Chapple, P., eds). Cambridge: Cambridge University Press, pp. 1–15.

von Degenfeld, G., Banfi, A., *et al.* (2003). Myoblast-mediated gene transfer for therapeutic angiogenesis and arteriogenesis. *Br J Pharmacol*, 140(4): 620–626.

Vunjak-Novakovic, G., Freed, L., *et al.* (1996). Effects of mixing on tissue engineered cartilage. *AIChE J*, 42: 850–860.

Vunjak-Novakovic, G., Obradovic, B., *et al.* (1998). Dynamic cell seeding of polymer scaffolds for cartilage tissue engineering. *Biotechnol Prog*, 14(2): 193–202.

Wang, H.J., Pieper, J., *et al.* (2004). Stimulation of skin repair is dependent on fibroblast source and presence of extracellular matrix. *Tiss Eng*, 10(7-8): 1054–1064.

White, P. (1946). Cultivation of animal tissues in vitro in nutrients of known concentrations. *Growth*, 10: 231–289.

White, P. (1949). *J Cell Comp Physiol*, 34: 221–241.

Woodfield, T.B.F., Bezemer, J.M., *et al.* (2002). Scaffolds for tissue engineering of cartilage. *Crit Rev Eukaryot Gene Expr*, 12(3): 209–236.

Yaeger, P., Masi, T., *et al.* (1997). Synergistic action of transforming growth factor-beta and insulin-like growth factor-I induces expression of type II collagen and aggrecan genes in adult human articular chondrocytes. *Exp Cell Res*, 237(2): 318–325.

Zigler, J.S., Lepe-Zuniga, J.L., Jr *et al.* (1985). Analysis of the cytotoxic effects of light-exposed HEPES-containing culture medium. *In Vitro Cell Dev Biol*, 21(5): 282–287.

Zisch, A.H., Lutolf, M.P., *et al.* (2003). Cell-demanded release of VEGF from synthetic, biointeractive cell ingrowth matrices for vascularized tissue growth. *Faseb J*, 17(15): 2260–2262.

Zisch, A.H., Lutolf, M.P., *et al.* (2003). Biopolymeric delivery matrices for angiogenic growth factors. *Cardiovasc Pathol*, 12(6): 295–310.

Chapter 13
Cryobiology

Lilia Kuleshova and Dietmar Hutmacher

Chapter objectives:

- To understand the difference between vitrification and rapid cooling based on the freezing concept as well as in the application to cryopreservation of TECs and stem cells, and their advantages
- To obtain knowledge on the variety of cryoprotectants and the diversity of their properties
- To obtain a background in technologies based on slow cooling such as controlled and uncontrolled cooling
- To understand how to determine vitrification properties and toxicity of cryoprotectant solutions
- To become familiar with sample holders used in vitrification protocols
- To obtain general knowledge about methods and protocols for 'safe' cryopreservation
- To understand requirements that should be taken into consideration during the design of vitrification protocols for TECs
- To obtain knowledge about current status of cryopreservation; in particular, vitrification of cells, tissues, organs and TECs

"Cryopreservation plays an important role in tissue banking and will assume even greater importance when tissue engineering becomes an everyday reality"

David Pegg

"Cryobiology usually thought of as the study of the effects of subfreezing temperatures on biological systems stands at the interface between physics and biology"

Gregory Fahy

Abstract

Cryobiology is the science of living organisms, organs, biological tissue or biological cells at low temperatures. From an historical perspective, the word cryobiology literally means the science of life in icy temperatures. In practice, this field comprises the study of any biological material or system (e.g. proteins, cells, tissues, or organs) subjected to any temperature below the physiological range (ranging from cryogenic temperatures to moderately hypothermic conditions).

Cryopreservation plays an important role in cell and tissue banking (i.e. long-term storage at cryogenic temperatures). It will attain even greater importance in the future within the field of tissue engineering (TE) as off-the-shelf products are a prerequisite for routine clinical applications. The promises of TE, however, depend on the ability to physically distribute the products of regenerative medicine to patients in need and to produce these products in a way that allows for acceptable cost, reproducibility, inventory control and quality assurance. For this reason, the ability to cryogenically preserve cells, tissues, and even whole laboratory-produced organs may be indispensable.

Low temperature preservation is a science/technology whereby cells, tissues and organs are preserved at cryogenic temperatures (typically of liquid nitrogen or liquid nitrogen vapors) and restored to original living states with a sufficient survival rate, viability and functionality. In general, the major directions of research and technology currently under investigations for cell and tissue cryopreservation can be classified as freezing and vitrification.

Given the technology available at present, special supporting chemicals (namely cryoprotectants) are essential for achieving cryopreservation of cells/tissue. Cryopreserved suspension of living cells are frequently used in biomedical sciences. In regenerative medicine, the vitrification of tissue engineered constructs (TEC) is of major interest. This chapter will provide an overview of the different techniques currently available for cryopreservation, with special emphasis on vitrification. The prospects in this direction are promising since successful vitrification of stem cells, cell-matrix and cell-scaffold constructs have already been achieved, as will be illustrated in this chapter.

13.1 Introduction to fundamentals of cryobiology

In general, cryobiology covers three broad areas: (i) the study of cold-adaptation of plants and animals; (ii) cryosurgery (a minimally invasive approach for destruction of unhealthy tissue); and (iii) cryopreservation. Cryopreservation, which deals with the storage of biological materials at low temperature, is of particular interest for tissue engineering (TE).

The most common concept underlying TE is to combine a scaffold (cellular solids) or matrix (hydrogels) with living cells to form a 'tissue engineered construct' (TEC) to promote the repair and regeneration of tissues. The scaffold and matrix is expected to support cell colonization, migration, growth and differentiation, and to guide the development of the required tissue. The promises of TE, however, depend on the ability to physically distribute the products of regenerative medicine to patients in need and to produce these products in a way that allows for adequate cost, reproducibility, inventory control and quality assurance. For this reason, the ability to cryogenically preserve cells, tissues, and even whole laboratory-produced organs may be indispensable.

In this chapter the authors try to define the needs versus the wants of vitrifying TEC, with particular emphasis on the cryoprotectant properties in general, and attempt to define some broad constraints for the properties of suitable materials and morphology. Some examples of such difficulties are described. It is concluded that the formation of ice, through both direct and indirect effects, is probably fundamental to these difficulties, and this is why vitrification seems to be the most likely way forward. However, two major problems still to be overcome are cryoprotectant toxicity and recrystallization during rewarming. Less obvious, and certainly less well understood is chilling injury – damage caused by reduction in temperature per se; this may yet turn out to be of fundamental importance. We subsequently review the current state of the art in vitrification techniques applied to TEC in the context of these constraints on cell and tissue survival.

13.1.1 General principles of cryopreservation procedures

Low temperature preservation plays an important role in TE. The goal of any improved cryopreservation protocol is to minimize sudden intracellular or extracellular formation of ice crystals that could result in ultrastructural damage, and thus maintain cell viability and metabolic activity on warming. Generally, cryopreservation of cell suspensions is achievable without much difficulty. However, it is a great challenge to preserve multicellular living objects, and 3D structures such as TEC or biological tissue at low temperature, because cryopreservation of tissue requires not only the preservation of cells but also the morphological structure and tissue architecture; both are vital for mechanical support as well as to aid in

critical biological functions. As the usefulness of stem cells in regeneration medicine has been explored and realized, the development of advanced protocols for effective cryopreservation of stem cells for TE purposes, as well as for a broad range of TEC, has become one of the main focuses in the research of TE.

In general, the major directions of research and technology currently under investigations for cell and tissue preservation can be classified as follows:

1. Cryopreservation
   - freezing
   - vitrification.
2. Freeze-drying (lyophilization).
3. Hypothermia (+4/28–33°C).

Freezing and vitrification allow long-term preservation of cells, tissues and organs with certain efficacy, and will be described in detail in this chapter. Freeze-drying has been employed successfully in prolonging shelf life of certain biologics such as pharmaceutical and food products. The success in preserving primary cells and biological tissue using this method is rather limited, however; hence it will not be covered in this chapter. Hypothermia, on the other hand, allows short-term preservation of cells and tissues (up to 72 h); it is mainly applied when the method of cryopreservation is not feasible or for transportation purposes. Therefore, it is also not a subject of our consideration in this chapter.

One of the primary issues in TE is to minimise the duration of exposure of cells/tissue to potentially damaging conditions *in vitro*. Development of protocols for effective cryopreservation of cells and TEC and even prospective engineered organs is an important area of research in TE. In relation to such considerations, one of the central issues is whether freezing or vitrification approaches, which have been developed in cryobiology, satisfy the fundamental principle of minimizing damage of cells and TEC during cryostorage.

To prevent cells/tissue or an organ from damage during the exposure to low temperature, cryoprotective agents are employed. During cryopreservation procedure, cells should be equilibrated with cryoprotectants

solutions and living cells that completely avoids ice crystal formation during cooling and warming. This is a key advantage for cells/tissue cryopreservation. The avoidance of ice formation during warming is of particular importance for the recovery of cells for biological applications. Retrospectively, the phenomenon of vitrification was first investigated and described in the nineteenth century (Tammann, 1898). Strictly speaking, solidification as a glass had been possible some forty years earlier. The founder of cryobiology, Basil Luyet, recognized the potential of achieving an ice-free, structurally arrested state for cryopreservation over sixty years ago and described it in his classical studies (Luyet, 1937; Luyet and Gehenio, 1940). In Luyet's concept, living systems could be cooled so quickly that ice would not have time to form. Subsequently, it became apparent that solutions with sufficient concentration play a special role in successful vitrification of living cells and tissues. The occurrence and usefulness of such amorphous solutions were investigated and described in several fundamental studies (Fahy, 1981, 1989; MacFarlane and Forsyth, 1990). Their proposal was to use a combination of high concentrations of cryoprotectants and rapid cooling rates to allow vitrification throughout the sample. If used correctly this would avoid both intra- and extra-cellular ice

formation. It was assumed that these solutions would be more efficient for the preservation of cells than solutions that crystallize. An early attempt of cryopreservation using the vitrification approach involved an electron microscopic observation of rapidly cooled red blood cells in the presence of high concentration of glycerol (Rapatz and Luyet, 1968a). Cryopreservation by direct immersion into liquid nitrogen in a vitrifying solution was first achieved in 1985 (Rall and Fahy, 1985). Figure 13.2 schematically illustrates the route of vitrification versus slow freezing.

Another advantage of vitrification is that it is less time-consuming since it usually involves rapid cooling by direct immersion into liquid nitrogen. It is important to note that 'vitrification' is not synonymous to 'rapid /ultrarapid freezing'. Fahy has clearly underlined that 'vitrification is the solidification of a liquid into an amorphous state opposed to solidification into crystalline or partially crystalline state, the latter process being known as freezing' (Fahy, 1989). The state of vitrification in cells/tissue will not be achieved by direct immersion into liquid nitrogen using any cryoprotectant solution; certain criteria need to be fulfilled in order for a solution to vitrify. These criteria/characteristics of vitrification solutions will be described in detail in Section 3.

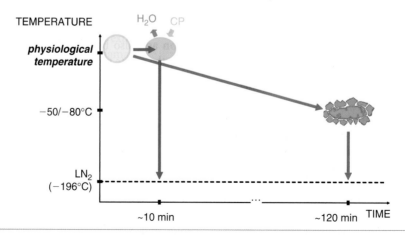

Figure 13.2 Schematic diagram illustrating vitrification and slow cooling. The red arrows represent conventional slow cooling protocols (freezing approach). Green arrows represent vitrification.

13.1.5 Cryoprotectants

The role of cryoprotectants is to minimize/prevent cells/tissue or an organ from being damage during exposure to low temperatures. Although the cryoprotective properties of first penetrating agent (glycerol) were discovered in 1949 (Polge *et al.*, 1949), the role of component selection became crucial only with the development of vitrification technology since the employment of high concentration of chemicals is a prerequisite to using this method. Various agents were tested and new formulations developed for the application of vitrification.

Cryoprotectants can de divided into two main groups based on their role in cryopreservation:

1. penetrating agents
2. non-penetrating agents

These cryoprotectant agents can be further classified by their molecular weight and other properties:

1. Low molecular weight agents
2. Sugars
3. Polymers – high molecular weight agents

The classification is described in Table 13.1.

Generally, the nature of the cryoprotectant action is multifold. A number of factors form their protective properties. Adding a cryoprotectant leads to a lowering in the freezing point of the solution, which decreases with increasing concentration of cryoprotectant. Although at low solute concentrations (10–15%) the effect is insignificant since the freezing point of such solutions drops only up to 3°C, cryoprotectants at high concentration help to minimize damage from ice crystals by encouraging glass formation rather than ice crystallization. The basis of their protective properties lies in their ability to form hydrogen bonds with water molecules thereby minimizing/preventing water-to-water hydrogen bonding that is the basis of ice formation. Almost all penetrating cryoprotectants have a high degree of solubility in water. An exothermic reaction during mixing of cryoprotectants with water or culture medium shows that they change water structure by breaking hydrogen bonds. It also appears that cryoprotectants reduce negative impact caused by high levels of salt. This important property is useful for cell freezing while the cells are dehydrated and surrounded by concentrated salt, particularly during slow cooling.

13.2 Technology based on the freezing concept

13.2.1 Slow cooling

13.2.1.1 Uncontrolled cooling

The concept of slow cooling has developed gradually over the past 50 years. As this method has been used

Table 13.1 Classification of cryoprotectants

Penetrating cryoprotectants	Non-penetrating cryoprotectants	
Low-molecular weight agents MW < 100 Da	Sugars 180 < MW < 594 Da	High-molecular weight agents MW > 1000 Da
Ethylene glycol*	Glucose	Ficoll
Dimethyl sulfoxide (DMSO)*	Fructose	Polyvinyl pyrrolidone (PVP)
Propylene glycol*	Lactose	Dextran
Glycerol*	Sucrose	Polyethylene glycol (PEG)
1,2-/2,3-butandiol	Trehalose	Polymer polyvinyl alcohol (PVA)
Formamid, acetamid**	Raffinose	Hydroxyethyl starch (HES)

*Preferable **Proven to be almost unsuitable due to high toxicity.

most successfully in cryopreserving cells, the procedure described will be based on experiments carried out on cell suspensions. Basically, two different methods, both involving uncontrolled cooling, are applied. In the first method, cells are introduced in a container (e.g. a cryovial) filled with cryopreservation solutions for a short equilibration time. After that the container is transferred to a −20°C freezer for 2 hours followed by placing in a −80/86°C freezer overnight before transferring to liquid nitrogen tank for storage. This protocol is not suitable for more sensitive cells, such as the majority of mammalian cells used in TE.

The second method, which involves cooling in liquid nitrogen vapor, is convenient and easy to apply. This procedure requires a foam box with thick walls and a lid which is able to maintain the temperature of liquid nitrogen for several hours. By adjusting the distance and position of a container from the level of liquid nitrogen it is possible to achieve a cooling rate of 10°C/min or higher. The containers are usually kept on a cooling platform for 5–20 minutes before being immersed into liquid nitrogen (−196°C). The thickness of the platform can be adjusted to provide the cooling rate according to the selected protocol. As the volume of the container decreases the cooling rate increases. For example, a small container such as 250 µl straws cooled in liquid nitrogen vapors on a Styrofoam platform with a rate of 120°C/min (Kasai *et al.*, 1996). The large container used for blood products can be moved by a motor or by physically driving it through a temperature gradient inside a tall liquid nitrogen tank.

The specimen should be warmed slowly in air at room temperature or in a warm bath adjusted to different temperatures, not exceeding a physiological temperature. Based on selected protocols recommended in literature, the cooling and warming rates should be optimized by using a thermocouple. The thermocouple is usually placed in a container, filled with a cryopreservation solution, and positioned side by side with sample container.

13.2.1.2 *Controlled cooling*
Controlled cooling is achieved by means of a programmable rate freezer. During slow cooling,

biological specimens are usually pre-equilibrated in low concentrations of penetrating cryoprotectant. The concentration of cryoprotectants gradually increases inside and outside the tissue through several steps during a 2-hour period. By cooling sufficiently slowly, nearly all of the bulk water can be removed from the cells by osmosis. The advantages of using a programmed cooling device reside in the possibility to reduce the proportion of cells captured by ice, and to vary the cooling rate in order to better prepare cells to the temperature of cryostorage.

The supporting solution cooled at a constant rate remains unfrozen below 'true' freezing point. Ice formation occurs in a deeply supercooled solution. This leads to rapid propagation of ice throughout the solution. This may be detrimental to cells because they remain for the most part hydrated on the onset ice formation. However, some cells and tissues (Song *et al.*, 1995) can be cryopreserved using constant cooling rate varying from 0.5°C/min to 10°C/min.

Seeded ice allows a more regular formation of the ice front, i.e. prevention of spontaneous nucleation. Seeding of cells could be done manually or automatically in a machine. The color of the solution reflects the concentration of solute; the darker the color the more concentrated the solution (Figure 13.3B). Cells and tissue are usually loaded with low concentration of cryoprotectant (solute). The concentration of solute increases as the proportion of ice increases in the same sample during cooling. Finally, the concentration of solute becomes sufficiently high to accommodate vitrification (orange color on Figure 13.3B).

13.2.2 Rapid cooling technology

Rapid cooling of biological material involves the direct introduction of a specimen to a low temperature storage (Figure 13.3C).

The most common way used for rapid cooling is direct immersion into liquid nitrogen of cells/tissue by using a plastic straw following its special pretreatment. It employs the same tools and techniques as described for 'vitrification technology' (see later in

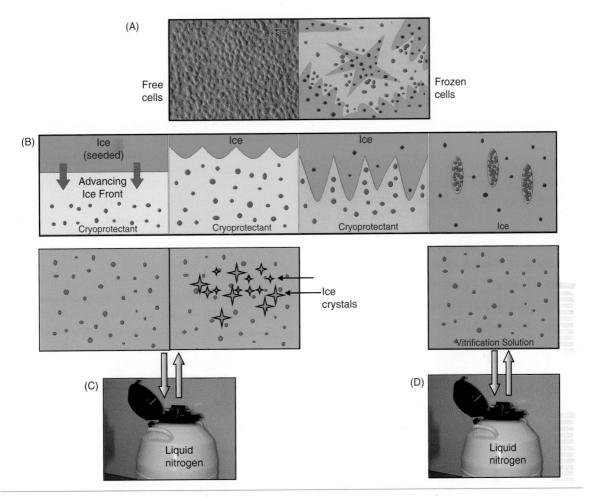

Figure 13.3 Schematic illustrations of cryopreservation: (A) Uncontrolled freezing; (B) Slow cooling procedure that involves seeding of ice; (C) Rapid cooling procedure; (D) Vitrification procedure.

this chapter) except for the solutions of intermediate concentrations solute that will vitrify on cooling but form ice during the warming step – this consequently could be damaging (for example, DMSO solutions used in rapid cooling contain 35–40 w/w% solute). The reduction of the solute concentration leads to an increase of the degree of crystallization. There is little potential in pursuing this approach since it is more effective to apply solutions which are ice-free during the whole cooling–warming cycle.

13.3 Vitrification technology

Vitrification is an alternative to customary freezing and this is the most promising methodology of cryopreservation. It was generally thought that supporting solutions for vitrification would be better for the preservation of living cells and tissues than solutions that crystallize, and hence damage cells during cooling and warming. Vitrification is of particular importance for TE research because the method is less time-consuming since it does not require the use of specialized equipment, making it easily applicable in most research laboratories and clinical settings. Current vitrification protocols are effectively optimized in such a way that a physiological temperature during the entire equilibration/dilution procedure is maintained. Cells can be observed and analyzed during almost the entire procedure; penetration and rehydration rates can be controlled.

These days there is a lot of interest in composing improved vitrification solutions. Vitrification can be achieved by partial replacement of water by penetrating agents which are easy glass formers and subsequent dehydration of biological material by non-penetrating cryoprotectant. The role of non-penetrating cryoprotectants is especially significant in vitrification protocols, since it requires fast dehydration before immersion into liquid nitrogen. In slow cooling protocols biological material might not survive prolonged dehydration and the damage occurred due to high concentrations of intercellular solutes.

It is believed that vitrification could substitute methods of low temperature preservation of cells and tissues based on the 'freezing concept' (Luyet, 1937; Luyet and Gehenio, 1940). Luyet and Gehenio were the first to convey a scientific and experimental approach to vitrification. Since the late 1930s, numerous studies on the behavior of cryoprotectant solutions during freezing (Amrhein and Luyet, 1966; Luyet, 1966) as well as the interaction between the living cells and ice have been conducted. Various methods have been developed or applied in broad research programs to investigate physical properties (mechanical, thermal, optical and electrical) of frozen solutions. In a series of investigations it was found that a vitrified state of samples was very hard to sustain since initially transparent, presumably vitrified samples, easily formed ice during re-warming.

Hence, Luyet and collaborators devoted a lot of their efforts to searching for conditions with which solutions make vitreous state. However, in the past, main technical issues concerning vitrification remained unsolved and this approach was neglected for three decades.

13.3.1 Determination of vitrification properties of solutions

In order to create conditions for vitrification, certain knowledge concerning the physical properties of the cryoprotective solutions employed is required. To achieve cryopreservation free from ice damage, it is necessary to cool rapidly through the temperature region of potential crystallization and reach the amorphous glassy state before ice crystals have the opportunity to form. Similarly, during the warming phase, the process needs to be rapid enough to avoid ice crystal formation. To achieve these conditions, it is necessary to determine the critical cooling and warming rates for each component of the vitrification solution. Temperature is not the only parameter that plays an important role in the determination of the critical warming and cooling rates. In fact, appropriate solute concentration is crucial in achieving the stability of the supercooled state, since it is known that dilute aqueous solutions are extremely difficult to vitrify. Systematic studies of non-equilibrium ice or hydrate crystallization on cooling and warming have been done on a variety of aqueous solutions, to find more efficient solutes and to determine conditions for vitrification (Koener and Luyet, 1966; Luyet and Rasmussen, 1967; Rapatz and Luyet, 1966, 1968b; Rasmussen and Mackenzie, 1968; Rasmussen amd Luyet, 1969; Sutton, 1992).

In area 1 of Figure 13.4 only cubic ice is crystallized. There is an absence of damage to cells due to the small size of cubic ice crystals. A high risk of damage to cells exists in area 3 because of hexagonal ice.

The cause and the events that take place on cooling/warming can be deduced from calorimetrical measurements. In addition to the measurement of the transformation that occurs in cryoprotective solutions, the structural states of samples can be observed by X-ray diffraction and by optical cryomicroscopy.

Characterizing the warming rate is particularly important, since in solutions of intermediate concentrations, ice crystallizes much easier on warming than on cooling. On warming a wholly amorphous solution, it first undergoes a glass transition and

Classical Experiment
(Luyet and Sager, 1967)

Luyet and his collaborators made important observations on how various aqueous solutions displayed different physical behaviors when warmed through different ranges of temperatures. There was a transition from the transparent, vitrified state to the opaque, crystallized state, in solutions of intermediate concentrations (25–50%). The passage from the transparent to opaque state corresponds to an increase of the radii of microscopic ice particles, spherulites that are present in frozen/vitrified solutions of intermediate concentration, into a size large enough to be observed under the microscope as ice crystals. These transformations have the characteristic of re-crystallization and the temperature at which this occurs is called 're-crystallization temperature'.

Luyet and Sager (1967) published a series of experiments performed on polyvinylpyrrolidone (PVP) solutions where the phenomenon of re-crystallization could be easily observed by direct visual observation. PVP, a high molecular weight polymer, is a popular non-penetrating agent used in cryopreservation. Their experiments, using eight PVP solutions with concentrations ranging from 28 to 48%, are described briefly as follows:

1. Droplets of sample PVP solutions were mounted between two coverslips (140 μm thick) to obtain a thin film of solution 12 μm in thickness. The cover slips were then held together by a ring of silicone grease around the edges.
2. The samples were frozen to a transparent state by abrupt immersion into an isopentane bath at −100°C for 30 seconds
3. Samples were re-warmed in another isopentane bath with the temperature preset and controlled by a pneumatic regulator.

The phenomenon of re-crystallization (opacity of solution) during re-warming was determined by direct visual observation. For each concentration of PVP, initial and final stages of re-crystallization (achieved within 2 minutes) were determined. Samples were observed against a strongly illuminated background; to determine the initial stages of re-crystallization, samples were observed with lateral illumination against a dark background. Opacity of the sample was an indicator of formation of large ice crystals. The sample would be considered as having reached its maximum level of crystallization when no further change in opacity could be observed during several minutes. It was found that the passage from one temperature range of instability to the other took place in rewarmed solutzions of PVP at certain total solute concentrations.

Based on this and other findings, the researchers gauged the range of temperatures for re-crystallization during re-warming for the PVP solutions of different concentrations, thus estimating the workable range of concentrations at which de-vitrification would not occur.

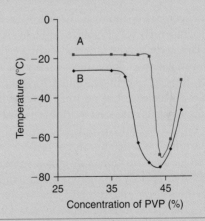

Temperatures at which rapidly cooled solutions of PVP of concentrations indicated in abscissa begin to turn opaque in two minutes (curve A) and become completely opaque in two minutes (curve B), upon being re-warmed (adapted from Luyet and Sager, 1967).

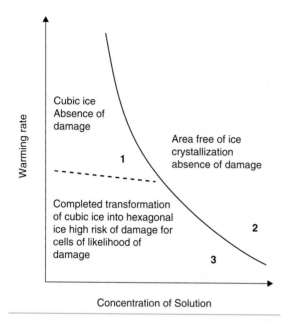

Figure 13.4 Schematic diagram demonstrating the various possibilities of ice crystallization versus concentration of aqueous solution on warming.

becomes a supercooled liquid. If the warming rate is not sufficiently high, ice crystallizes at higher temperatures. This event can be observed during calorimetrical investigations. Thus, the stability of the amorphous state can be defined by the critical warming rate above which ice has no time to crystallize on warming. Crystallization on warming is usually studied by optical cryomicroscopy. The samples are cooled and warmed at different rates.

In Figure 13.5, the formation of metastable cubic ice which forms from a wholly amorphous solution can be observed by optical cryomicroscopy. Cubic ice forms at as low as −100°C and then subsequently transforms into hexagonal ice at ~−90 to −78°C (Figure 13.5A). The kinetics of this transition have been studied since cubic ice seems to be innocuous to cells. Temperature range studied was −150°C (Figure 13.5B1), −80°C (Figure 13.5B2) and −63°C (Figure 13.5B3 and 13.5B4). Ice can also form on cracks and other free surfaces

(Figure 13.5B). X-ray diffraction is a more accurate method to detect cubical ice. Transformation into ordinary hexagonal ice occurs when the temperature rises. Cubic ice crystals are very small and cannot be detected by calorimetry. However, formation of hexagonal ice can be measured by differential scanning calorimetry (DSC). Phase and physical transitions of aqueous solutions can be seen on thermograms obtained by DSC or by related technique, namely differential thermal analysis. The differential scanning calorimeter measures the heat absorption of a sample as a function of temperature. These measurements can only be done on small samples. It can be estimated that the behavior of larger samples would be similar; however, geometry of sample and its size influences outcome.

Aqueous solutions of low concentrations undergo homogeneous nucleation during cooling. A spontaneous generation of new crystal nuclei in the liquid without the presence of free surfaces generally can only take place at a substantial supercooling (peak 0 on Figure 13.6A). Melting in the same solutions occurred at much higher temperatures (peak 4 on Figure 13.6A). The presence of the penetrating cryoprotectant suppresses equilibrium melting point of ice in the solution (T_m) to the extent indicated in Figure 13.6B. The temperature of homogeneous nucleation suppresses more markedly than T_m. As the concentration of solute increases, both temperatures decrease (10%, 20%, 30% PC in Figure 13.6B). Solutions of intermediate concentration form glass on cooling but form ice on warming (T_d) with subsequent melting (T_m) (40%, 58% PC). When the concentration of cryoprotectant increases, a process of devitrification and melting ice becomes less noticeable (58% PC). Thermograms representing vitrification solutions show no signs of devitrification and melting peak (60% PC).

Glass-forming ability of penetrating cryoprotectant has been investigated for decades (Boutron, 1990; Boutron and Kaufman, 1979; Luyet and Rasmussen, 1967). Finally, low molecular weight agents were characterized and placed in order with respect to their glass-forming ability which is deduced from diagram of phase and physical transitions: 2,3-butandiol

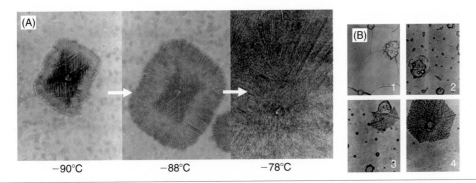

Figure 13.5 Dynamic studies: (A) on the transformation of cubic ice into hexagonal ice in propylene glycol-PVP-water system; (B) on the formation of ice in free surfaces in ethylene glycol-mono methyl ether-water system, dynamics study (1–4). Magnification × 50.

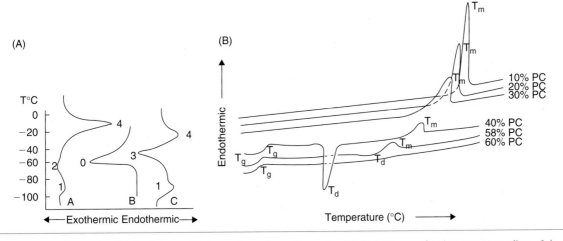

Figure 13.6 Schematic thermograms: (A) 0-exothermic peak corresponds to ice nucleation upon cooling; 1-heat consumption jump corresponds to glass transformation of the glass-like inclusions upon warming. The substance is transformed from the solid amorphous state into the supercooled liquid state; 2-an exothermic peak corresponds to completing ice crystallization process upon warming; 3-an exothermic effect denotes ice crystallization from the supercooled state of the system upon warming; 4-endothermic peak corresponds to melting of the whole system; (B) Schematic representation of processes that take place in aqueous solutions of cryoprotectants of different concentrations with the temperature changes, as indicated on thermograms: T_m-melting of ice in the solutions; T_d-temperature of devitrification; T_g-glass transition temperature; PC-hypothetical penetrating cryoprotectant.

(Boutron, 1990), propylene glycol (Boutron and Kaufman, 1979), ethylene glycol (Rasmussen and Luyet, 1969), dimethyl sulfoxide (Rasmussen amd Mackenzie, 1968), glycerol (Rasmussen and Luyet, 1969). Thus, 2,3-butandiol can form stable glass at the lowest solute concentration (35%) while another popular cryoprotectant, ethylene glycol, is able to form stable glass at 59% solute concentration only.

Table 13.2 Schematic illustration of toxicity of penetrating cryoprotectant and permeability of cell depending on temperature and duration of exposure

Duration of exposure	Cold room 4°C	Room temperature 20–25°C	Warming stage 37°C
1 minute		●	●
2 minute	●	●	●
3 minute	●	●	●
4 minute	●	●	●

Table 13.3 Evaluation of toxicity of penetrating cryoprotectants to cells and tissues

Any of PC Temperature	↓	↑	Survival/Viability Development potential Level of function Percentage of functioning cells
Concentration = constant			Differentiation
Any of PC Concentration	↑	↓	Survival/Viability Development potential Level of function Percentage of functioning cells
Temperature = constant			Differentiation

PC, penetrating cryoprotectant.

13.3.2 Determination of components toxicity

An important goal is to design cryopreservation solutions that are non-toxic to cells and tissues even after prolonged exposure. All penetrating cryoprotectants, which are low-molecular weight agents, are potentially toxic to cells. They permeate readily cell membranes and the level of toxicity primarily depends on duration of exposure and temperature of cells to one of these penetrating cryoprotectants (Tables 13.2 and 13.3).

Cryoprotectant toxicity could be assessed by the short-term exposure and long-term culture of cryoprotectant-candidates to cells and tissues. The

> ## Box 2 Example. Short-term exposure experiment
> ### (Valdez *et al.*, 1992)
>
> Valdez and co-workers extensively studied toxicity effects of penetrating cryoprotectants. They focused on the effects of type, concentration, temperature and time of exposure to cryoprotectants on the viability of cells. To accurately measure the contribution to solution toxicity for each of these parameters, the parameter in question was made variable, while the others remained constant. It is important that three out of four parameters remained unchanged. In their consistent study conducted on embryos the toxicity of solutions containing made up in a physiological solution and 10, 20, 30 and 40 v/v% of each of the following penetrating cryoprotectants: 2,3-butanediol, 1,3-butanediol, dimethyl sulfoxide, propylene glycol, ethylene glycol and glycerol were tested. Overall results revealed that at high concentration (30 v/v%) ethylene glycol was dramatically less toxic than its counterparts, this result was three fold higher than the second best cryoprotectant studied, which was dimethyl sulfoxide.
>
> The impact of the duration of exposure of cryoprotectant on cells was also studied. It was found that butandiols are significantly more toxic than ethylene glycol and dimethyl sulfoxide. At constant concentration (20 v/v%) it was found that exposure to 2,3-butandiol showed an adverse affect on cells after 20 minutes of exposure. As the time of exposure was doubled, the affects of exposure became more pronounced. Cells maintained high levels of viability after treatment with two cryoprotective agents, namely ethylene glycol and dimethyl sulfoxide solutions

classical way to assess the effects of cryoprotectant on cell viability is done by varying each of the following parameters one by one: type of cryoprotectant, cryoprotectant concentration, and time and temperature of exposure to cryoprotectant. It is important that other parameters remain invariable. The most consistent study on the short-term impact of penetrating cryoprotectants was conducted on embryos by Valdez and co-workers (see Box 2) (Valdez *et al.*, 1992).

Therefore, ethylene glycol was found to be the least toxic of the known cryoprotectants (Valdez *et al.*, 1992) and it very readily permeated cell membranes due to its lower molecular weight (Songsasen *et al.*, 1995).

Besides low molecular weight agents, sugars and high molecular weight polymers have demonstrated beneficial affects during cryopreservation. It seems that sugars and polymer cryoprotectants should have a low risk to toxicity since they could not penetrate cell membrane. However, adverse affects of polymers on cells could arise from low molecular weight impurities. Toxicity control is an essential part of vitrification studies. A systematic evaluation of polymers/sugars is especially important when sugars or polymers are present in designed vitrification solutions at high concentrations. Polymers such as PVP, Ficoll and Dextran are available in a wide range of molecular weights; the slightest variation in molecular weight or purity level can translate to dramatic differences in cell survival and viability (see for example, Box 3). Some of them could be used as supplied by manufacturers, others need to be purified by dialysis (Kuleshova *et al.*, 2001).

To comprehensively study the impact of sugars/polymers on sensitive cells, they also have to be cultured in the presence of low concentrations of these non-penetrating cryoprotectants (Kuleshova *et al.*, 1999a). The need to use polymer-based solutions arose through the understanding that traditional penetrating solutions have a high toxicity. Hence, these glass-like solidifying solutions with low concentrations of penetrating agents are used for vitrifying of mammalian embryos (Kuleshova *et al.*, 2001). Once toxicity information for a wide range

Box 4 Example. Determination of the lowest total solute concentration required for vitrification (Kuleshova *et al.*, 1999a)

How does the molecular weight and type of sugars influence vitrification properties of polyalcohols -based solutions?

In the absence of other additives 59% (w/w) EG formed a stable glass and when warmed at 10°C/min it showed no sign of devitrification. At lower concentrations of EG, clear evidence of ice crystal formation was found in the DSC thermograms. When EG was replaced by an equivalent amount of a monosaccharide, the total solute concentration required for vitrification did not depend on the amount of these monosaccharides in the solution (A, B, C).The wholly vitrified state occurred at a total solute concentration of 59% (w/w) the same as that of the EG solution without added carbohydrate. Thus, sugars

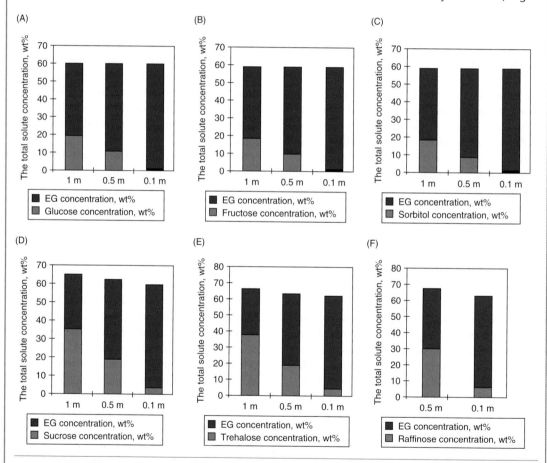

The lowest total solute concentration required for vitrification A. ethylene glycol-glucose-saline solution; B. ethylene glycol-fructose-saline solution; C. ethylene glycol-sorbitol-saline solution; D. ethylene glycol-sucrose-saline solution; E. ethylene glycol-trehalose-saline solution; F. ethylene glycol-raffinose-saline solution.

act to promote vitrification on a weight/weight basis which is equivalent to that of EG. Disaccharides appear to have a greater influence on the vitrification properties of EG solutions. The lowest total solute concentration required for vitrification solutions in which the EG was replaced by sucrose or trehalose became progressively higher as the proportion of sugar in the vitrification solution and the molecular weight increased (D, E). This presumably reflects the doubling in molecular weight, as compared to the monosaccharides. The effect becomes more prominent with increasing molecular weight of the disaccharides. Raffinose (*polysaccharides*) (F) had a significant effect on the vitrification properties of ethylene glycol solutions. When 0.5 molal or 0.1 molal raffinose was added to EG-solutions, vitrification was achieved at higher total concentrations. The raffinose significantly modified the Tg of these solutions.

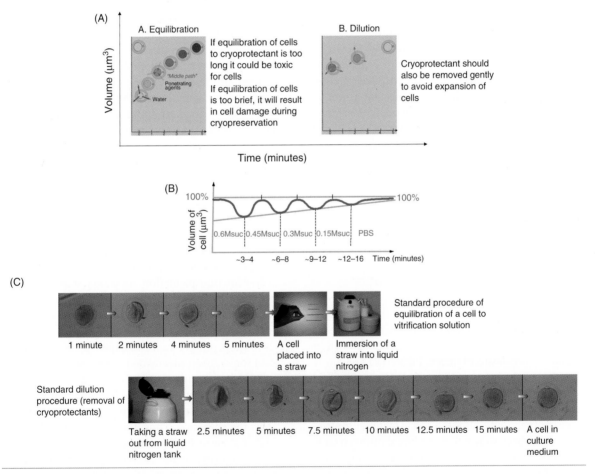

Figure 13.7 Schematic representations of: (A) equilibration and dilution procedures (A, equilibration; B, dilution); (B0 correct dilution procedure; (C) Illustration of step-wise equilibration of large cell (human oocyte) to vitrification solution following by removal of cryoprotectant.

brief equilibration in final VS, cells are prepared to be cryopreserved by direct immersion into liquid nitrogen. Removal of cryoprotectant is a rather more complicated event. Water enters the cells to dilute the cryoprotectant more rapidly than the cryoprotectant can leave the cells. Placing cells in solutions of progressively lower concentrations of cryoprotectant can partially reduce cell swelling.

These days a more effective method for removing cryoprotectant from cells is widespread (Figure 13.7A, B and C). The difference in osmolarity between the intra- and extracellular milieu can be reduced by establishing a balance between the high intracellular concentration of cryoprotectant and an extracellular hypertonic solution. As shown in Figure 13.7C cells containing cryoprotectant may even shrink when placed in a concentrated solution of sugar. This indicates that cryoprotectant as well as water is leaving the cell. The employment of sugars is effective and allows the cryoprotectant to be removed in a small number of steps. To demonstrate membrane excursion during equilibration/dilution procedure, human oocyte was examined. The large size of mammalian oocyte allows to distinctively observe changes in the cell (Figure 13.7B).

Figure 13.8 is a schematic representation using special manipulation skills.

13.4 Safety issues in cryopreservation

Liquid nitrogen is widely used for the storage of biological materials. Although the temperature of liquid nitrogen is low (−196°C), numerous viruses and micro organisms of infected materials previously stored in tanks can tolerate such temperatures and become a source of cross-contamination. There are several reported cases of contamination occurring via liquid nitrogen with some biological material applications. It has been shown that one liquid nitrogen tank, used to store blood products, could be unequivocally linked to the transmission of this disease to patients; the hepatitis virus taken from infected patients matched the virus isolated from the liquid nitrogen in the tank. Generally, many viruses, including papova virus, vesicular stomatitis virus, herpes simplex virus, adenovirus, may survive direct exposure to liquid nitrogen and could therefore potentially cause cross-contamination.

There are, however, several methods for reducing contamination. The risk of contamination by larger pathogens could be reduced by filtering the liquid nitrogen through a 0.2 μm filter, UV irradiation of the liquid nitrogen, or by inserting the cryopreserved specimen into an additional outer protective container before they are moved to a storage tank. As some, but not all, contaminating pathogens can be removed from cells and tissues by rigorous washing steps, it is best to adopt strategies that minimize the likelihood of either the straw or their contents from becoming contaminated. However, the methods outlined above have limited effectiveness; therefore, alternative strategies for reducing contamination should be pursued. Several strategies aimed at minimizing the likelihood of contamination during storage have been suggested and are described below.

13.4.1 Vapor storage

Storage of specimens in nitrogen vapor would reduce the likelihood of cross-contamination. Vapor storage (Figure 13.9A) is advantageous when large tissue samples need to be preserved. For example, heart valves for transplantation are widely stored in nitrogen vapor. Storage at the temperature of −145°C makes the samples less brittle and prevents them from cracking. The fluctuation of the temperature during opening of vapor storage tank is not significant for large samples (Wood, 1999). Small specimens warm significantly faster than large specimens – pulling canes out of a tank to check the information on a specimen before removing it could accidentally thaw the specimen. Thus, the storage temperature is higher than in liquid nitrogen and the risk of accidental thawing or the effects of greater fluctuations in storage temperatures may counteract the benefits of using vapor nitrogen storage for small samples. Hence, the use of a vapor storage system would be the best advantage for TECs.

Box 5   Example. Holders and manipulation skills

Right type of container or holder

A variety of new techniques and types of holders that enhance the cryopreservation of cells and tissues are under development. Sufficient amounts of biological material could be cryopreserved rapidly by immersion of containers into liquid nitrogen. Cooling of small living objects in straws in liquid nitrogen vapors was proven to be a gentle and effective method (Kasai *et al.*, 1990, 1996).

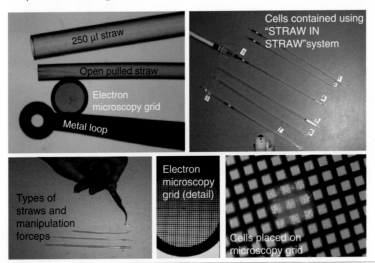

Different types of holders

A strategy for achieving very rapid cooling rates is to minimize the size of the sample, e.g. by placing the specimen on electron microscope grids (50,000–180,000°C/min) (Martino *et al.*, 1996), loops (<200,000°C/min) (Lane *et al.*, 1999) or narrow piece of aluminum foil, or inside heat-softened and pulled straws (20,000°C/min). The disadvantage of each of these approaches is that the vitrification solution comes in direct contact with liquid nitrogen during cooling or storage, consequently raising the risk of contamination (see section 13.4).

Cooling rates depending on type of container upon immersion in liquid nitrogen are shown below:

	Type of container/holder	Cooling rate
Cryopreservation by rapid cooling	Pouches	~100–400°C/min
	Cooling in 250 μl straw	2,500°C/min
	Cooling in 250 μl straw in liquid nitrogen vapors	~120°C/min
	Cooling in 250 μl straw inserted in 500 μl straw ('STRAW IN STRAW')	400°C/min
Cryopreservation by ultra rapid cooling	Between two metal surfaces	10,000°C/min
	Open pulled straw	20,000°C/min
	Cryotop	23,000°C/min
	Gold or copper grids	180,000°C/min
	Nylon or metal loop	<200,000°C/min
	Aluminum foil, sterile stripper tips	Not measurable

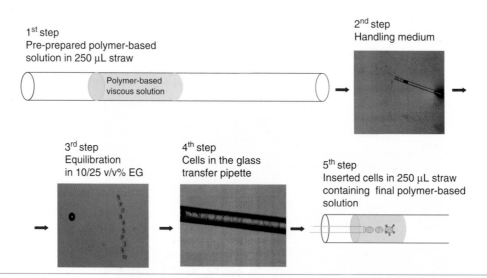

1st step
Pre-prepared polymer-based
solution in 250 µL straw

Polymer-based
viscous solution

2nd step
Handling medium

3rd step
Equilibration
in 10/25 v/v% EG

4th step
Cells in the glass
transfer pipette

5th step
Inserted cells in 250 µL straw
containing final polymer-based
solution

Figure 13.8 Schematic representation of the correct procedure for special manipulation skills.

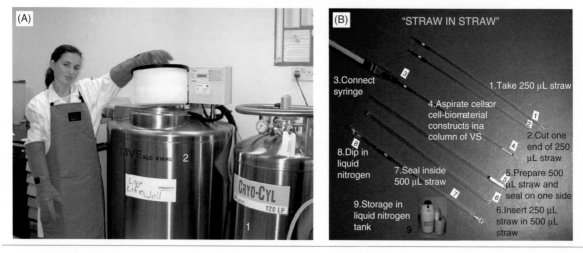

(A)

(B) "STRAW IN STRAW"

3. Connect
syringe

1. Take 250 µL straw

4. Aspirate cells or
cell-biomaterial
constructs in a
column of VS

8. Dip in
liquid
nitrogen

7. Seal inside
500 µL straw

2. Cut one
end of 250
µL straw

5. Prepare 500
µL straw and
seal on one side

9. Storage in
liquid nitrogen
tank

6. Insert 250 µL
straw in 500 µL
straw

Figure 13.9 Storage systems for cryopreservation: (A) Vapor storage system. Cylinder with liquid nitrogen (1) connected with vapor storage (2), it is designed to maintain the temperature of cryostorage automatically; (B) 'STRAW IN STRAW' system to eliminate the risk of cross-contamination. Our original 'STRAW IN STRAW' procedure (Kuleshova *et al.*, 2000) was developed for embryo cryopreservation and subsequently modified for the preservation of stem cells and cell-biomaterial constructs (Kuleshova, L., Wang, X.W., Wu, Y., *et al.* (2004). Vitrification of encapsulated hepatocytes with reduced cooling/warning rates. *Cryo Letters*, 25: 241–254).

13.4.2 Double packaging

It has been shown that properly sealed straws or bags do not leak and should, therefore, provide good protection. Some sealing strategies were better than others by providing extra protection, e.g. wrapping the straws in a plastic film or a double bag may be needed to further reduce the risk of leakage/contamination

(Russell *et al.*, 1997). Developing protocols to prevent contamination of very rapidly cooled or vitrified TECs may prove more difficult. One strategy to minimize the likelihood of contamination, thus allowing TECs and adult stem cells to be cooled by direct immersion in liquid nitrogen, and warmed by direct immersion into a water bath within a double straw arrangement was developed (Kuleshova and Shaw, 2000; Kuleshova *et al.*, 2004, 2005; Tan et al., 2007; Wu *et al.*, 2007) as a simple strategy for preventing cross-contamination (Figure 13.9B).

There is another issue needing to be addressed. Products of human and animal origin, such as serum and protein, are essential component parts of almost all cryopreservation solutions. Development of protein/serum-free solutions for entire vitrification/warming cycle and their application to TECs and stem cells, has been safe and successful and high viability achieved (Kuleshova *et al.*, 2004, 2005; Tan *et al.*, 2007; Wu *et al.*, 2007). This, in combination with placing the inner straw inside an outer sealed protective container to prevent the inner straw from ever coming into contact with liquid nitrogen and non-sterile surroundings, provides a simple, rapid and effective strategy for reducing or eliminating the risk of infection and biological contamination.

13.5 Cryopreservation: practical aspects

13.5.1 Cryopreservation of TECs (cell-matrix constructs)

The ability to successfully cryopreserve hepatocytes can be useful for bioartificial liver support systems. Typically, hepatocytes are cryopreserved as suspensions (Chesné *et al.*, 1991; Guillouzo *et al.*, 1999), in spheroids (Darr and Hubel, 2001; Hubel *et al.*, 2000), in collagen sandwich configuration (Koebe *et al.*, 1999) and on collagen monolayers (Stevenson *et al.*, 2004), on microcarriers (Demetriou *et al.*, 1986; Foy *et al.*, 1993) and microcapsules (Dixit *et al.*, 1993; Kuleshova *et al.*, 2004: Wu *et al.*, 2007). The most encouraging 'freezing' technique results are obtained for microencapsulated hepatocytes. For suspension,

some decrease in qualitative and quantitative results was observed, as in the past the slow cooling concept was explored (Guillouzo *et al.*, 1999). Although cryopreservation by slow cooling was studied thoroughly, literature often contains disagreeing conclusions as an application of the same protocol usually gives a variation in survival rates. Numerous reports have generally demonstrated that free hepatocytes survive slow cooling procedures only partially (Fautrel *et al.*, 1997; Guillouzo *et al.*, 1999) and few reports showed difficulties in obtaining good qualitative and quantitative results (Fautrel *et al.*, 1997; Guillouzo *et al.*, 1999). A decrease in the survival rate of rat hepatocytes after cryopreservation, such as 10–18%, 25% and 41%, has been reported (Fautrel *et al.*, 1997). This parameter, however, does not reflect the extent of hepatocyte functional capacity after preservation. Cell attachment is a more appropriate criterion. It may be assumed that hepatocytes which attach and spread on the substrate are viable and functional. Only about 50% of the parenchymal cells which survived after slow cooling was able to attach on plastic when plated (Diener *et al.*, 1993). A similar decrease in attachment abilities ranging from 27% to 62% has been observed for hepatocytes from most species (Guillouzo *et al.*, 1999). It should also be clarified that in the majority of studies, dead hepatocytes were removed by centrifugation through Percoll-gradient. Indeed, only 20% of originally isolated cells had attached to the matrix after cryopreservation (Hengstler *et al.*, 2000). Additional disadvantages of freezing hepatocyte suspensions are that some functions were reduced significantly compared to fresh culture (Diener *et al.*, 1993). Even if the efficacy of the procedure could be improved, there is limited success for the direct application of animal hepatocyte suspensions as a liver assistance device in modern practical medicine. Not surprisingly, the approach is far from satisfactory and hence the need to explore another concept becomes critical.

A method of vitreous cryopreservation of large quantities of small-sized cell-matrix constructs simultaneously was developed early at the turn of the century (Kuleshova *et al.*, 2004; Wu *et al.*, 2007) (Figure 13.10). It was established in principle that

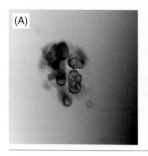 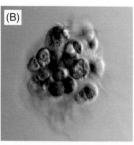

Figure 13.10 Confocal microscopy images of untreated (A) and vitrified/warmed (B) encapsulated hepatocytes. The fluorescence is from resorufin, formed as a result of EROD activity by cytochrome P450. Magnification is × 500.

pre-treated microencapsulated hepatocytes could be preserved by direct immersion into liquid nitrogen (Kuleshova *et al.*, 2004). This study determined that hepatocytes are capable of maintaining viability and higher levels of post-warming functions, in culture, for several days (Wu *et al.*, 2007). The hepatocytes exhibit a similar trend in the decline of their rate of cell deaths irrespective of treatment. It was also found that the developed novel vitrification procedure is applicable for a range of sizes of cell-containing constructs and could be applied for cryopreservation of microencapsulated hepatocytes with different cell density, morphology of cell aggregation and structure of biomaterial (Wu *et al.*, 2007). This comprehensive strategy provided excellent results and could be applied to other cell-biomaterial constructs.

13.5.2 Vitrification of precursor and stem cells

Studies on adult stem cells aiming at repairing the damage to the central nervous system, skin, muscle, bone and cornea are carried out worldwide. It is essential to establish cryopreservation protocols in the early stage of the research on adult stem cells for several reasons. Maintenance of the stem cells via

passaging may increase the chance of abnormalities due to excessive passaging and might not be cost-effective if the cells are not being used very often. Therefore, an efficient cryopreservation method is crucial for maintaining the cells at a particular passage number to reduce experimental variability.

Although survivability of certain adult SCs is not necessarily problematic (Bruder *et al.*, 1997; Kotobuki *et al.*, 2005), it was found that some of their functions could be affected by conventional freezing. Generally, the impact of freezing on adult SCs has not been investigated systematically; however, an adverse effect of freezing for human embryonic stem cells (ESCs) is widely reported in literature, as ESCs studies attracted broad attention. A comprehensive study on ESCs cryopreservation by conventional freezing and thawing methods is of particular interest as the assessment of cells was conducted shortly after thawing (Katkov *et al.*, 2006). This study demonstrated that more than half of these sensitive cells died in 3 days following slow freezing and thawing. The majority of ESCs expressed an apoptotic pathway shortly after freezing, while they were placed in incubator for recovery. Most cells were detached within 1.5 hours after thawing and cell viability was low. Earlier studies had assessed the pluripotency of ESCs much later after freezing and thawing; discarding dead and differentiated cells during passage control would eliminate crucial information on the amount of non-viable and differentiated cells after freezing. Conventional 'freezing' is particularly detrimental when used in combination with DMSO. This method can introduce specific changes that cause deleterious effects on SCs' viability and maintenance of pluripotency and mechanical stresses. DMSO may also promote differentiation and osmotic stresses in cells.

As cell-matrix interaction plays a crucial role in the development and regeneration of cartilage tissue, a cell carrier substance which closely mimics the natural environment in the cartilage specific extracellular matrix is a prerequisite for successful regeneration of *in vitro* cartilage tissue. Our group evaluated the feasibility of the vitrification approach for

---

Box 6   Example. Efficient vitrification procedure
(Kuleshova *et al.*, 2004)

Hepatocytes, entrapped at different cell densities in two types of engineered collagen matrices were used as models to evaluate efficacy and universality of our original vitrification method. The nature of collagens caused differences in capsule sizes. Microencapsulated hepatocytes were divided into three groups, namely: (1) control – untreated microencapsulated hepatocytes; (2) solution control – micro-encapsulated hepatocytes treated in cryopreservation solutions without undergoing cooling-warming procedure, whereby losses of cells may occur during treatment with non-physiological solutions; and (3) vitrification – microencapsulated hepatocytes treated with complete vitrification-warming cycle. All treatments, performed at room temperature (23 ± 2°C), were given on the next day (Day 1), approximately 24 hours after isolation. Encapsulated cells were placed in a conical tube, culture medium was removed and 10 v/v% EG was added followed by centrifugation at 10 × g for 2 min. Then 10 v/v% EG was replaced with 25 v/v% EG; the microcapsules were mixed evenly in the solution, and centrifugation was repeated. The centrifuge was maintained at +4°C during all steps of the procedure. Pellet was re-suspended in the final vitrification solution 40 v/v% EG 0.6 M sucrose and then a 250 μL plastic straw was filled up to a cotton plug without sealing. This straw was placed inside a 500 μL straw which was sealed and immersed into liquid nitrogen. This 'STRAW IN STRAW' system, developed by us, aimed to eliminate the risk of contamination (See Figure 13.9B). Total duration for these three steps did not exceed 11 min; the time spent in each solution was distributed evenly. The cells were warmed on the same day by immersion for 30 sec into water-bath adjusted at 38–39°C. The encapsulated hepatocytes were then expelled into 1 M sucrose. The concentration of sucrose was decreased to 0.7 M by dilution with 0.25 M sucrose and then to 0.2–0.15 M subsequently for each step with culture medium. Total time for the dilution procedure was 15 min (5 min for the first step and 2–2.5 min for each following step). All dilution steps were performed at room temperature (23 ± 2°C). The encapsulated hepatocytes were cultured routinely after the treatments.

---

cryopreservation of bone-marrow-derived MSCs cultured on modified alginate bead culture systems (ABCS) and the results are given in section 13.5.1 for vitrification protocol. We found that vitrification of MSCs cultured on ABCS resulted in a 30% higher viability rate than those achieved after applying the conventional freezing method. The cells had been grown through several passages with no evidence of any differences from non-cryopreserved cells. Modified ABCS efficiently support the proliferation as well as the capability of multi-linage differentiation of MSCs; more importantly, this spherical 3D culture system has been shown to enhance the protection of MSCs during vitrification procedure.

In addition to benefits of the above-mentioned SCs, the potential of non-embryonic SCs (particularly neural stem cells (NSCs)) to differentiate and repair damaged brain is very significant. The protocol for NSCs (as reported by Kuleshova *et al.*, 2005; Tan *et al.*, 2007) provides a sterile high viability method for cryopreservation of mammalian neural stem cells or precursor cells. A method comprising these characteristics will be critical for stem cell transplantation therapy. With complete avoidance of products of human or animal origin, we can believe that this protocol can serve as a starting point for the development of vitrification protocols for the cryopreservation of human neural stem cells, or other adult stem cells, and which may eventually be used in clinical settings.

The effectiveness of vitrification as a method of cryopreservation versus conventional slow and rapid

Box 7 Importance of employment of vitrification solutions

In general, the structure of living objects, cells/tissue or TEC which undergo glass–like solidification remains undamaged provided that the procedure is developed and performed correctly. As mentioned earlier in this chapter, the phase transition temperatures and heat flux characteristics of the samples can be measured specifically during warming by differential scanning calorimetry to determine lack of devitrification (see Figure 13.6). Primitively, massive ice crystals could be found by visual observation during warming (see Figure 13.5B). Our studies were specifically designed to demonstrate the importance of developing and utilizing final solutions with sufficient concentration to maintain an amorphous state during cooling and warming for maintenance of integrity of three dimensional cell aggregates (neuronal stem or precursor cells aggregated as neurospheres) as well as fragile TEC (Kuleshova *et al.*, 2004, 2005; Tan *et al.*, 2007; Wu *et al.*, 2007).

It has been established that the final solution with 40 v/v% EG 0.6M sucrose has sufficient concentration to achieve vitrification at the given cooling rate, with nearly 100% cell survival for both neurospheres and TEC after vitrification. Integrity of neurospheres was also preserved completely (Figure 13.11C). Although developed vitrification solution (40 v/v% EG 0.6M sucrose) alone did not significantly affect cell viability (Figure 13.11B), an attempt was made to further eliminate the risk of a toxic effect by reducing the concentration of EG. Rapid freezing with 37 v/v% EG 0.6M sucrose disrupted the structure of the neurospheres (Figure 13.11D). As 37 v/v% EG 0.6M sucrose appeared to protect against cell death (although not against loss of the structural integrity of the neurospheres) we also compared rapid freezing using a lower solute concentration, 30 v/v% EG 0.6M sucrose. The disruption of the structural integrity of the neurospheres was even more severe when 30 v/v% EG 0.6M sucrose was employed and there was markedly increased cell death (Figure 13.11E). Higher solute concentrations are used in vitrification to ensure the transition to a glass-like phase without formation of ice crystals on cooling or warming. The high cell viability on rapid freezing with 37 v/v% EG 0.6M sucrose could have implied either a near-vitrification or reduction in ice formation during cooling-thawing cycle. Therefore, we would not endorse rapid freezing as a viable cryopreservation procedure.

For TECs, the absence of free cells after warming, consistently over a hundred experiments, indicates that there has not been any ice formation during the cooling-warming cycle, and therefore the microcapsules remained undamaged (Kuleshova *et al.*, 2004). With the reduction of total solute concentration by 4%, the survival rate was reduced by 13%, which is not a significant difference. However, free cells were present, indicating the damage of the microcapsules due to ice-formation during the cooling-warming cycle. The viability of cells was further reduced significantly, when the concentration of the final solution was reduced by 7%, and broken microcapsules together with numerous free cells were observed, demonstrating necessity of application of vitrification.

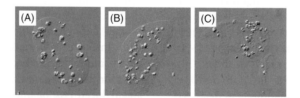

Confocal microscopy images of an intact microcapsule (A) a broken microcapsule which has ruptured on its right side in the picture (B) and a damaged microcapsule with the detached part seen on the below right of the main microcapsule (C). Magnification is × 100.

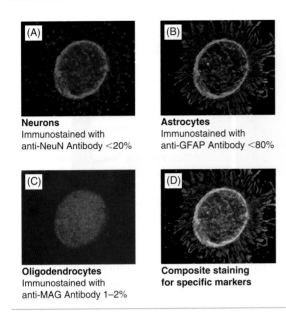

Neurons
Immunostained with
anti-NeuN Antibody <20%

Astrocytes
Immunostained with
anti-GFAP Antibody <80%

Oligodendrocytes
Immunostained with
anti-MAG Antibody 1–2%

**Composite staining
for specific markers**

Figure 13.12 Immunostaining for different lineages of differentiated neural stem cells following vitrification-warming cycle: (A) neuron; (B) astrocyte; (C) oigodendrocyte; (D) composite staining for specific markers.

immunological differences, operative technique and especially cryopreservation method.

The method employed in clinical settings for low temperature preservation of valves (below −135°C) uses the so-called 'freezing' approach and DMSO as a supporting chemical cryoprotectant. Following the discovery of cryoprotective properties imparted by glycerol in 1949 (Polge *et al.*, 1949) and the finding of another supporting chemical DMSO in 1959 (Lovelock and Bishop, 1959), many investigators have attempted the preservation of cells or tissues by adding those agents. Just three years later a successful replacement of the aortic valve using allograft was reported (Ross, 1962). It is not surprising that in the 1960s, investigators had focused on the development of cryopreservation protocols by the employment of newly discovered DMSO and, shortly after, cryopreservation of homograft valves using DMSO as a cryoprotectant implemented in practical medicine.

Since then the science of cryobiology has progressed significantly and usage of several other more advanced cryoprotectants has become widespread. In fact it has been shown that, in the majority of parameters, EG is the best one among common cryoprotectants such as glycerol, DMSO, propylene glycol, and EG; while DMSO is three times more toxic than EG when applied at high concentration (Valdez *et al.*, 1992). Cryopreservation exposes cells to major aniso-osmotic conditions, which results in potentially damaging volume excursion. EG has the ability to readily penetrate through cell membranes mainly due to lower molecular weight (EG has 62 MW versus 78 MW of DMSO and propylene glycol) (Woods *et al.*, 1999). EG also became more prevalent because of its low toxicity (Sakoju *et al.*, 1996; Valdez *et al.*, 1992). Armitage and Pegg (1979) showed that the rabbit heart tolerates EG better than DMSO. It has since been demonstrated that EG is more superior to DMSO for cryopreservation of mammalian pancreatic islets (Sakoju *et al.*, 1996) and reproductive cells (Mukaida *et al.*, 1998; Valdez *et al.*, 1992) in comparative studies.

Generally, the conventional methods used for blood vessel cryopreservation has produced unsatisfactory results, whereby viability was reduced to 50% (Gall *et al.*, 1998). In contrast, vitreous cryopreservation has been effectively applied for the preservation of veins (Song *et al.*, 2000a, 2000b). Using the vitrification approach to store vascular tissue results in a markedly improved tissue function compared to the standard method involving freezing. The maximum contractions achieved in vitrified vessels were >80% of fresh matched controls with similar drug sensitivities; whereas frozen vessels exhibited maximal contractions below 30% of controls and decreases in drug sensitivity (Song *et al.*, 2000b). *In vivo* studies of vitrified vessel segments in an autologous transplant model showed no adverse effect of vitrification compared to fresh tissue grafts. The popular vitrification solution employed for vascular grafts consists of DMSO, formamide, propylene glycol. Specification of VS 55 reflects that solution comprises 55% total cryoprotective solutes (Fahy *et al.*, 1995).

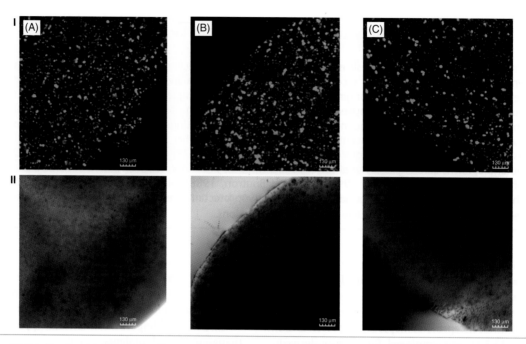

Figure 13.14 Confocal images of adipose derived stem cells cultured in alginate/fibrin glue beads: (A) control group; (B, C) Vitrification group. (Row I) Fluorescence images of cells stained with Calcein-AM (Green: live cells) and Ethidium Homodimer (Red: dead cells) show that the cell viability was maintained in the vitrification group. (Row II) Transmitted images of cells cultured on alginate beads. The transmitted images of the alginate beads indicate that the integrity of the beads has not been compromised.

State of the Art Experiment

The protocol in use is based on previously described protocols (Kuleshova *et al.*, 2004, 2005) with modifications.

In the current protocol, exposure to VS was done in two subsequent steps. A first exposure to VS consisting of 40 v/v% EG 0.6 M sucrose, followed by exposure to the same VS supplemented with 9% Ficoll, which has been shown to be free from impurities, thus having no adverse affect on biological material (Kasai *et al.*, 1992; Kuleshova *et al.*, 2001). Cell-scaffold constructs vitrified with this protocol had a surface area of 0.5 cm^2 × 1 mm thickness. This protocol assumes that usage of a small amount of VS will reduce the overall size of

a vitrifying sample and increase the cooling rate while exposed to cryogenic temperature. The protocol was devised so that cell-scaffolds constructs, almost free of external solution, will be subjected to cooling-warming cycle. The exposure to a second, more viscous vitrification solution was in order to reduce evaporation rate and protect cells on the surface of the nanofiber. An increase of total solute concentration of the second VS has enhanced vitrification properties and potentially plays a beneficial role in the protection of cell-scaffold construct during the cooling/warming cycle.

The cooling procedure, in liquid nitrogen vapors followed by plunging into liquid nitrogen,

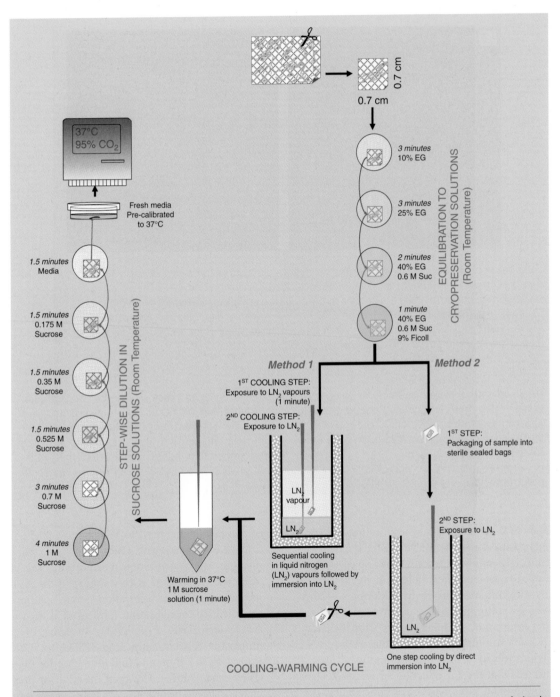

0.7 cm
0.7 cm

37°C
95% CO₂

EQUILIBRATION TO CRYOPRESERVATION SOLUTIONS (Room Temperature)

3 minutes
10% EG

3 minutes
25% EG

2 minutes
40% EG
0.6 M Suc

1 minute
40% EG
0.6 M Suc
9% Ficoll

Fresh media
Pre-calibrated
to 37°C

STEP-WISE DILUTION IN SUCROSE SOLUTIONS (Room Temperature)

1.5 minutes
Media

1.5 minutes
0.175 M
Sucrose

1.5 minutes
0.35 M
Sucrose

1.5 minutes
0.525 M
Sucrose

3 minutes
0.7 M
Sucrose

4 minutes
1 M
Sucrose

Method 1

1ST COOLING STEP:
Exposure to LN₂ vapours
(1 minute)

2ND COOLING STEP:
Exposure to LN₂

LN₂
vapour

LN₂

Sequential cooling
in liquid nitrogen
(LN₂) vapours followed by
immersion into LN₂

Warming in 37°C
1 M sucrose
solution (1 minute)

Method 2

1ST STEP:
Packaging of sample into
sterile sealed bags

2ND STEP:
Exposure to LN₂

LN₂

One step cooling by direct
immersion into LN₂

COOLING-WARMING CYCLE

Schematic representation of cryopreservation procedure of osteogenically induced bone marrow derived mesenchymal setm cells seeded on nanofiber mesh.

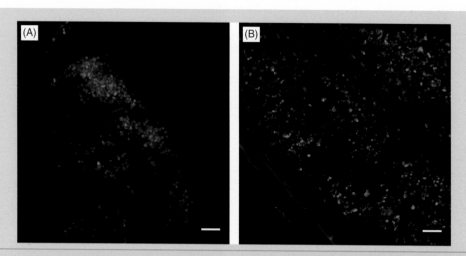

Comparison of the cell viability and structural integrity cell-scaffold construct undergoing vitrification versus control. Confocal micrographs of representative of bone marrow derived MSCs seeded on nanofiber mesh: (A) in untreated control culture; (B) after recovery from a complete vitrification-warming cycle. Vitrification solutions: in 40 v/v% EG, 0.6 M sucrose, 7% Ficoll in medium. Bone marrow derived MSCs labeled with a LIVE/DEAD kit. Green (Calcein-AM) fluorescence indicates live cells, while red fluorescence (Ethidium homodimer) indicates dead cells. Scale bar is 100 μm.

has been shown to reduce fractures in both cells (Kasai *et al.*, 1996) and tissue (Pegg *et al.*, 2006). The results of this developing study in vitreous preservation of cell-scaffold constructs have shown some encouraging results).

Confocal imaging of post-warm cell-scaffold constructs have shown cell viability and seeding density similar to those seen in control samples, indicating that vitrification protocol did not have a detrimental affect on either survivability of cells nor their attachment ability to scaffold. These promising results reinforce the belief that vitrification will play a pivotal role in tissue-engineered banking.

to organ such as rabbit kidney vitrification was conducted by Fahy *et al.* (2004b). During the last two decades of persistent investigations, Fahy and co-workers were able to overcome the problems of vitrification of organs with minimum injury by achieving a balance between the controlling of the cryoprotectant toxicity, chilling injury and warming organs with high rate, which is able to suppress devitrification.

Vitrification of organs may be improved in several ways. First, the introduction of cryoprotectant on a multimolar basis reduces osmotic injury; it is particularly important when high concentrations of cryoprotectants need to be used to achieve vitrification of whole organs (Pegg, 1977). Second, high pressure can also promote vitrification. The effect of increasing pressure is very similar to the effect of increasing solute concentration. Some solutions of intermediate concentration form glass at high pressure (MacFarlane *et al.*, 1981). A change of cryoprotectant concentration from 7.5 M to 8.4 M is equivalent to a thousand fold difference for pressure (Fahy *et al.*, 2004b). However, the limit of pressure increase should be properly controlled, since it might not be compatible with the survival or function of the organ. Perfusion of an organ at high pressure may in turn decrease the toxicity of a solution to some extend. The third keypoint is perfusion of vitrification solutions at reduced temperature such as −22°C. It should be clarified that perfusion of an organ is a step-wise process. The cryoprotectant is usually introduced gradually at $\frac{1}{8}$, $\frac{1}{4}$, $\frac{1}{2}$ and then 100% of full-strength solution. The initial steps of perfusion could be performed at only subzero temperature, while final concentration of solute allowed a flow at −22°C. All other steps, including the removal of cryoprotectant, should be done at 0°C. The vitrification of organs is a subject of importance and more success can be expected in the near future.

13.6 Future considerations

The development of an efficient method of cryopreservation by vitrification for SCs (Kuleshova *et al.*, 2005: Tan *et al.*, 2007), without the use of human and animal proteins, will be of great importance as TE progresses towards the realization of clinical applications.

It should be admitted that the currently available cryopreservation procedure uses a large amount of serum/protein, which is essential since cells should be cooled very slowly in a cooling machine or freezer. Serum/proteins play an important role in supporting cells during a gradual decrease of temperature in a cryopreservation solution before solidification occurs. While serum is vital in the slow cooling procedure, the presence or absence of serum does not influence solidified biological material at the temperatures of cryostorage. As culture conditions require the exclusion of the use of serum containing medium, using serum as part of the cryopreservation procedure may be detrimental in maintaining SCs as well as the progenitor cells which are widely used in TE in their undifferentiated state. In contrast, vitrification commonly employs direct immersion into liquid nitrogen of cells with high cooling rates and appears not to require the employment of serum/proteins.

Establishing an efficient method of cryopreservation by vitrification for these cells without the use of human and animal proteins will be of even greater importance as TE progresses toward a realization of clinical applications.

Cryopreservation of tissue is the most challenging since it requires preservation of overall structures as well as of living cells. The development of certain advanced concepts for effective cryopreservation of three-dimensional structures of cells, and a broad range of tissue grafts are the current aims. The development of low temperature preservation by vitirification of organs and even engineered organs in the future may be considered as long-term objectives of cryoscience.

13.7 Summary

1. Cryobiology is the science of living organisms, organs, biological tissue or biological cells at lower than physiological temperatures.
2. Low temperature preservation is a science/technology whereby cells and tissue are preserved at cryogenic temperatures and restored to original living state with a sufficient survival rate, viability and functionality.
3. The major directions of research and technology currently under investigation for cell, tissue and organ cryopreservation, can be classified as freezing and vitrification. Both approaches require cryoprotectants that are introduced with the aim to minimize cells, tissue or organs from being damaged during exposure to low temperature.
4. Cryoprotectants are classified into two main groups by their role in cryopreservation: (1) penetrating agents are introduced to replace water in cells and tissue; (2) non-penetrating agents aid in dehydration. This mitigates damage caused by ice formation and promotes the development of the amorphous state in cells rather than ice crystals during cooling cryostorage-warming cycle.
5. The freezing of cells is commonly achieved by a slow cooling technique, which is well understood as it was developed decades ago. Freezing is costly if it requires the use of a controlled rate freezer.

6. Slow cooling of cells/tissue breaches the principle that these should be exposed to non-physiological conditions as briefly as possible, before their storage at low temperatures.

7. Recently, the idea of replacing slow cooling protocols by the vitrification approach, initially developed for simple biological systems, is seen as an attractive prospect.

8. Vitrification is defined as glass-like solidification and/or complete avoidance of ice crystal formation during cooling and warming. Vitrification commonly employs direct immersion into liquid nitrogen of cells and TECs following brief introduction to cryoprotectants.

9. Cryopreservation of tissue, TECs and organs is the most challenging task since it requires the preservation of an overall tissue structure as well as of living cells. We have developed vitrification protocols that are highly effective for cryopreservation of TECs since cells and the constructs remain undamaged during the entire procedure.

10. In addition to considerations of performance based on a holistic vitrification strategy, practical considerations of manufacture arise. For realistic and practical clinical applications, it must be possible to vitrify in a reproducible and controlled fashion at an economic cost and speed. The manufacturing process must accommodate the presence of large number of cells (1 million cells per milliliter) in the construct; in the future it will be necessary to allow vitrification of large TECs (bigger than $5\,cm^3$).

References

Adem, C.G. and Harness, J.B. (1981). Computer control of a modified Langendorff perfusion apparatus for organ preservation using cryoprotective agents. *J Biomed Eng*, 3: 134–139.

Ali, J. and Shelton, J.N. (1992). Successful vitrification of day-6 sheep embryos. *J. Reprod. Fert*, 90: 63–70.

Amrhein, E.M. and Luyet, B.J. (1966). Evidence for an incipient melting at the 'recrystallization' temperatures of aqueous solutions. *Biodynamica*, 10: 61–67.

Armitage, W.J. and Pegg, D.E. (1979). The contribution of the cryoprotectant to total injury in rabbit hearts frozen with ethylene glycol. *Cryobiology*, 16: 152–160.

Boutron, P. and Kaufman, A. (1979). Stability of the amorphous state in the system water-1,2-propanediol. *Cryobiology*, 16: 557–568.

Boutron, P. (1990). Levo- and dextro-2,3-butanediol and their racemic mixtures: very efficient solutes for vitrification. *Cryobiology*, 27: 55–69.

Boutron, P. (1992). Cryoprotection of red blood cells by a 2,3-butanediol containing mainly the levo and dextro isomers. *Cryobiology*, 29: 347–358.

Bruder, S.P., Jaiswal, N. and Haynesworth, S.E. (1997). Growth kinetics, self-renewal, and the osteogenic potential of purified human mesenchymal stem cells during extensive subcultivation and following cryopreservation. *J Cell Biochem*, 64: 278–294.

Chesné, C., Guyomard, C., Grislain, L., *et al.* (1991). Use of cryopreserved animal and human hepatocytes for cytotoxicity studies. *Toxicol in vitro*, 5: 479–482.

Dahl, S.L., Chen, Z., Solan, A.K., *et al.* (2006). Feasibility of vitrification as a storage method for tissue-engineered blood vessels. *Tissue Eng*, 12: 291–300.

Darr, T.B. and Hubel, A. (2001). Postthaw viability of precultured hepatocytes. *Cryobiology*, 42: 11–20.

Demetriou, A.A., Levenson, S.M., Novikoff, P.M., *et al.* (1986). Survival, organization, and function of microcarrier-attached hepatocytes transplanted in rats. *Proc Natl Acad Sci*, 83: 7475–7479.

Diener, B., Utesch, D., Beer, N., *et al.* (1993). A method for the cryopreservation of liver parenchymal cells for studies of xenobiotics. *Cryobiology*, 30: 116–127.

Dixit, V., Darvasi, R., Arthur, M., *et al.* (1993). Cryopreserved microencapsulated hepatocytes-transplantation studies in Gunn rats. *Transplantation*, 55: 616–622.

Ekins, S. (1996). Vitrification of precision-cut rat liver slices. *Cryo-Letters*, 17: 7–14.

Fahy, G.M. (1981). Prospect for vitrification of whole organs. *Cryobiology*, 18: 617–625.

Fahy, G. M. (1989). Vitrification. In *Low Temperature Biotechnology: Emerging Applications and Engineering Contributions* (McGrath, J.J. and Diller, K.R., eds), BED-Vol 10/HTD Vol 98, pp. 313–146.

Fahy, G.M., da Mouta, C., Tsonev, L., *et al.* (1995). Cellular injury associated with organ cryopreservation: chemical toxicity and cooling injury. In *Cell Biology of Trauma* (Lemasters, J.J. and Oliver, C., eds). Boca Raton, FL, USA: CRC Press, pp. 333–356.

Fahy, G.M., Wowk, B., Wu, J. and Paynter, S. (2004a). Improved vitrification solutions based on the predictability of vitrification solution toxicity. *Cryobiology*, 48: 22–35.

Fahy, G.M., Wowk, B., Wu, J., *et al.* (2004b). Cryopreservation of organs by vitrification: perspectives and recent advances. *Cryobiology*, 48: 157–178.

Fautrel, A., Joly, B., Guyomard, C. and Guillouzo, A. (1997). Long-term maintenance of drug-metabolizing enzyme activities in rat hepatocytes after cryopreservation. *Toxicol Appl Pharmacol*, 147: 110–114.

Foy, B.D., Lee, J., Morgan, J., *et al.* (1993). Optimization of hepatocyte attachment to microcarriers: importance of oxygen. *Biotechnol Bioeng*, 42: 579–588.

Gall, K.L., Smith, S.E., Willmette, C.A. and O'Brien, M.F. (1998). Allograft heart valve viability and valve-processing variables. *Ann Thorac Surg*, 65: 1032–1038.

Guillouzo, A., Rialland, L., Fautrel, A. and Guyomard, A. (1999). Survival and function of isolated hepatocytes after cryopreservation. *Chem Biol Interac*, 121: 7–16.

Hengstler, J.G., Ringel, M., Biefang, K., *et al.* (2000). Cultures with cryopreserved hepatocytes: applicability for studies of enzyme induction. *Chem Biol Interact*, 125: 51–73.

Hotamisligil, S., Toner, M. and Powers, R.D. (1996). Changes in membrane integrity, cytoskeletal structure, and developmental potential of murine oocytes after vitrification in ethylene glycol. *Biol Reprod*, 55: 161–168.

Hubel, A., Conroy, M. and Darr, T.B. (2000). Influence of preculture on the prefreeze and postthaw characteristics of hepatocytes. *Biotechnol Bioeng*, 71: 173–183.

Kasai, M., Komi, J.H., Takakamo, A., *et al.* (1990). A simple method for mouse embryo cryopreservation in low toxicity vitrification solution, without appreciable loss of viability. *J Reprod Fertil*, 89: 91–97.

Kasai, M., Hamaguchi, Y., Zhu, S.E., *et al.* (1992). High survival of rabbit morulae after vitrification in ethylene glycol-based solution by a simple method. *Biol Reprod*, 46: 1042–1046.

Kasai, M., Zhu, S.E., Pedro, P.B., *et al.* (1996). Fracture damage of embryos and its prevention during vitrification and warming. *Cryobiology*, 33: 459–464.

Katkov, I., Isachenko, V., Isachenko, E., *et al.* (2006). Low- and high-temperature vitrification as a new approach to biostabilization of reproductive and progenitor cells. *Intl J Refrigeration*, 29: 346–357.

Knoener, C. and Luyet, B.J. (1966). Discontinuous change in expansion coefficient at the glass transition temperature in aqueous solutions of glycerol. *Biodynamica*, 10: 41–45.

Koebe, H.G., Mühling, B., Deglmann, C.J. and Schildberg, F.W. (1999). Cryopreserved porcine hepatocyte cultures. *Chem Biol Interact*, 121: 99–115.

Kotobuki, N., Hirose, M., Machida, H., *et al.* (2005). Viability and osteogenic potential of cryopreserved human bone marrow-derived mesenchymal cells. *Tissue Eng*, 11: 663–673.

Kuleshova, L.L., MacFarlane, D.R., Trounson, A.O. and Shaw, J.M. (1999a). Sugars exert a major influence on the vitrification properties of ethylene glycol-based solutions and have low toxicity to embryos and oocytes. *Cryobiology*, 38: 119–130.

Kuleshova, L., Gianoroli, L., Magli, C., *et al.* (1999b). Birth following vitrification of small number of human oocytes. *Hum Rep*, 14: 3077–3079.

Kuleshova, L.L. and Shaw, J.M. (2000). A strategy for rapid cooling of embryos within a double straw to eliminate the risk of contamination during cryopreservation and storage. *Hum Rep*, 15: 2604–2609.

Kuleshova, L.L., Shaw, J.M. and Trounson, A.O. (2001). Studies on replacing most of the penetrating cryoprotectant by polymers for embryo cryopreservation. *Cryobiology*, 43: 21–31.

Kuleshova, L.L. and Lopata, A. (2002). Vitrification can be more favorable than slow cooling. *Fertil Steril*, 78: 449–454.

Kuleshova, L., Wang, X.W., Wu, Y., *et al.* (2004). Vitrification of encapsulated hepatocytes with reduced cooling/warming rates. *Cryo-Letters*, 25: 241–254.

Kuleshova, L.L., Tan, C.K.F. and Dawe, G.S. (2005). Cryopreservation of adult stem cells cultured as neurospheres. *Cryobiology*, 50: 355.

Lane, M., Bavister, B.D., Lyons, E.A. and Forest, K.T. (1999). Containerless vitrification of mammalian oocytes and embryos. *Nat Biotechnol*, 17: 1234–1236.

Limaye, L.S. (1997). Bone marrow cryopreservation: improved recovery due to bioantioxidant additives in the freezing solution. *Stem Cells*, 15: 353–358.

Lovelock, J. and Bishop, M. (1959). Prevention of freezing damage to living cells by dimethyl sulfoxide. *Nature*, 183: 1394–1395.

Luyet, B. (1937). The vitrification of organic colloids and protoplasm. *Biodynamica*, 1: 1–14.

Luyet, B. J., and Gehenio, P. M. (1940). Life and death at low temperatures. Biodynamica, Normandy, Missouri, 341 pp.

Luyet, B.J. (1966). Anatomy of the freezing process in physical systems. In *Cryobiology* (Meryman, H.T., ed.). New York: Academic Press, pp. 115–138.

Chapter 14
Scaffold design and fabrication

Dietmar Hutmacher, Tim Woodfield, Paul Dalton and Jennifer Lewis

Chapter objectives:

- To outline the sequence of events in the formation of a tissue engineered construct, and the changes in important factors with time
- To understand the key rationale, characteristics and process parameters of the currently used scaffold fabrication techniques
- To understand which scaffold processing techniques allow the direct control over the design and fabrication of 3D pore architecture
- To provide a comprehensive list of various scaffold techniques, with the most relevant references

- To demonstrate the differences between pore size and pore interconnectivity and its physical and biological implications
- To learn how various types of scaffolds can be prepared; and how different configurations form various scaffolding materials
- To understand how a well considered design of 3D scaffold architecture can be used to influence the quality and quantity of engineered tissue *in vitro* and *in vivo*
- To assist research teams with their choice for a specific 3D scaffold processing technology which is able to function in the desired manner

"When one puts up a building one makes an elaborate scaffold to get everything into its proper place. But when one takes the scaffold down, the building must stand by itself with no trace of the means by which it was erected."

Andres Segovia (1893–1987), Spanish classical guitarist

14.1 Introduction

Scaffold-based tissue engineering concepts involve the combination of viable cells, biomolecules and a structural scaffold combined into a 'construct' to promote the repair and/or regeneration of tissues (Figure 14.1). The construct is intended to support cell migration, growth and differentiation, and guide tissue development and organization into a mature and healthy state. The science in the field is still in its infancy, and various approaches and strategies are under experimental investigation. It is by no means clear what defines ideal scaffold/cell or scaffold/neotissue constructs, even for a specific tissue type. The considerations are complex, and include architecture, structural mechanics, surface properties, degradation products and composition of biological components, and the changes of these factors with time *in vitro* and/or *in vivo* (Hollister, 2005; Hutmacher *et al.*, 2004; Woodfield *et al.*, 2004).

Scaffolds in tissue-engineered constructs will have certain minimum requirements for biochemical as well as chemical and physical properties. Any scaffold must provide sufficient initial mechanical strength and stiffness to substitute for the mechanical function of the diseased or damaged tissue that it aims at repairing

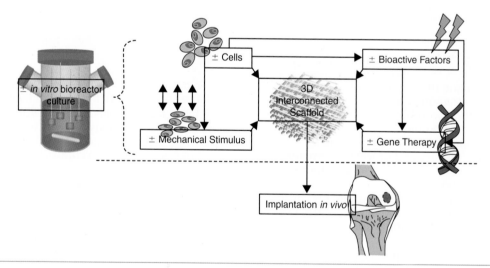

Figure 14.1 Scaffolds are central to tissue engineering strategies. Appropriate design of porous, biodegradable scaffold conduits in combination with the delivery of cells and/or growth factors to the defect site is the goal when fabricating a so-called TEC. Cells can be introduced *in vitro* and further subjected to mechanical stimuli or genetic enhancement prior to implantation. Alternatively, the scaffold alone can be manipulated with bioactive or gene factors to recruit reparative cells *in vivo* and enhance neotissue formation. Image reproduced with permission from Begell House Publishers: Woodfield, T.B., Bezemer, J.M., Pieper, J.S., *et al.* (2002). Scaffolds for tissue engineering of cartilage. *Crit Rev Eukaryot Gene Expr*, 12(3): 209–236.

or regenerating. Scaffolds may not necessarily be required to provide complete mechanical equivalence to healthy tissue, but stiffness and strength should be sufficient to at least support and transmit forces to the host tissue site in the context. For example, in skin tissue engineering, the construct should be able to withstand the wound contraction forces. In the case of bone engineering, external and internal fixation systems might be applied to support the majority load bearing forces until the bone has matured (Hutmacher, 2000).

Cell and tissue remodeling is important for achieving stable biomechanical conditions and vascularization at the host site. Hence, the three-dimensional (3D) scaffold/tissue construct should maintain sufficient structural integrity during the *in vitro* and/or *in vivo* growth and remodeling process. The degree of remodeling depends on the tissue itself (e.g. skin 4–6 weeks, bone 4–6 months), and its host anatomy and physiology. Scaffold architecture has to allow for initial cell attachment and subsequent migration into and through the matrix, mass transfer of nutrients and metabolites, and provision of sufficient space for development and later remodeling of organized tissue. The degradation and resorption kinetics of the scaffold need to be designed based on the relationships of mechanical properties, molecular weight (M_w/M_n), mass loss and tissue development described in Figure 14.2.

In addition to these essentials of mechanics and geometry, a suitable construct will possess surface properties that are optimized for the attachment and migration of cell types of interest (depending on the targeted tissue). The external size and shape of the construct must also be considered – especially if the construct is customized for an individual patient (Hutmacher, 2000; Sun and Lal, 2002).

In addition to considerations of scaffold performance based on a holistic tissue engineering strategy, practical considerations of manufacture arise. From a clinical point of view, it must be possible to manufacture scaffolds under GMP (Good Manufacturing Practice) conditions in a reproducible and quality controlled fashion at an economic cost and speed. To move the current tissue engineering practices to the next frontier, some manufacturing processes will accommodate the incorporation of cells and/or growth factors during the scaffold fabrication process. To address this, novel manufacturing processes such as robotic assembly and machine- and computer-controlled 3D cell encapsulation are under development so that the tissue-engineered construct not only has a controlled spatial distribution of cells and growth factors, but also a versatility of scaffold materials and microstructure within one construct.

14.2 Scaffold design

14.2.1 Introduction

It is important to emphasize, at the outset, that the field of scaffold-based tissue engineering is still in its infancy and many different approaches are under experimental investigation. Thus, it is by no means clear what defines an ideal scaffold/cell or scaffold/neotissue construct, even for a specific tissue type. Indeed, since some tissues perform multiple functional roles, it is unlikely that a single scaffold would serve as a universal foundation for the regeneration of even a single tissue. Hence, the considerations for scaffold design are complex, and they include material composition, porous architecture, structural mechanics, surface properties, degradation properties and products, together with the composition of any added biological components and, of course, the changes in all of these factors with time.

For each envisioned application, successful tissue engineering constructs (TECs) will have certain minimum requirements for biochemical as well as chemical and physical properties. Scaffolds may be required to provide sufficient initial mechanical strength and stiffness to substitute for the mechanical function of the diseased or damaged tissue which the TEC aims at repairing or regenerating. Scaffolds may not necessarily be required to provide complete mechanical equivalence to healthy tissue, indeed the variability in architecture for a single tissue type is so extensive that it is inconceivable that a single TEC would serve universal applications for even a single tissue. Nevertheless, stiffness and strength should be sufficient to at least either permit requisite cell seeding of the scaffold *in vitro* without compromising scaffold architecture or support and transmit forces in an *in vivo* healing site. Thus,

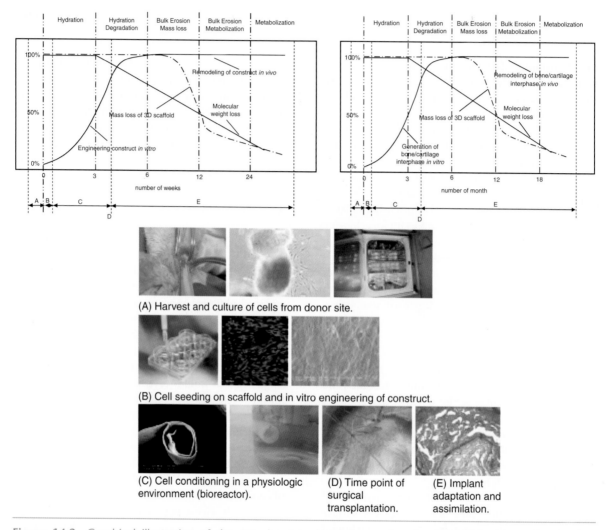

(A) Harvest and culture of cells from donor site.

(B) Cell seeding on scaffold and in vitro engineering of construct.

(C) Cell conditioning in a physiologic environment (bioreactor).

(D) Time point of surgical transplantation.

(E) Implant adaptation and assimilation.

Figure 14.2 Graphical illustration of the complex interdependence of molecular weight loss and mass loss of 3D scaffold matrix plotted against the time frame for tissue engineering a transplant. (A) Fabrication of bioresorbable scaffold. (B) Seeding of the cell populations into the polymeric scaffold in a static culture (Petri dish). (C) Growth of premature tissue in a dynamic environment (spinner flask) or growth of mature tissue in a physiologic environment (bioreactor). (D) Surgical transplantation. (E) Tissue-engineered transplant assimilation/ remodeling. In the first strategy (left), the physical scaffold structure supports the polymer/cell/tissue construct from the time of cell seeding up to the point where the hard tissue transplant is remodeled by the host tissue. In the case of load-bearing tissue such as articular cartilage and bone, the scaffold matrix must serve an additional function; it must provide sufficient temporary mechanical support to withstand *in vivo* stresses and loading. In Strategy I research programs, the material must be selected and/or designed with a degradation and resorption rate such that the strength of the scaffold is retained until the tissue-engineered transplant is fully remodeled by the host tissue and can assume its structural role. Mechanical stability of the scaffold carrying cells, genes or growth factors once deposited within the defect is a key requirement. If the mechanical integrity is compromised, cell viability may be jeopardized under loading and micromotion as well as fibrous encapsulation may result in failure of the construct. As a result, it is becoming more commonly accepted that the mechanical

in the context of skin tissue engineering the scaffold material should be sufficiently robust to not only resist a change in shape as a result of the introduction of cells into the scaffold, each of which would be capable of exerting tractional forces, but also the wound contraction forces that will be invoked during tissue healing *in vivo*. The same general rules would apply to bone engineering, although during the *in vivo* phase external and internal fixation systems, or other supports or restrictions on patient activity in early stage recovery, may reduce such mechanical roles of the intended scaffold.

Cell and tissue remodeling is important for achieving stable biomechanical conditions and vascularization at the host site. Hence, the 3D scaffold/tissue construct should maintain sufficient structural integrity during the *in vitro* and/or *in vivo* growth and remodeling process.

Scaffold architecture has to allow for initial cell attachment and subsequently migration into and through the matrix, mass transfer of nutrients and metabolites, and permit sufficient space for development and later remodeling of organized tissue. The porosity and internal space within a degradable scaffold will increase with time, allowing increased space for tissue to develop and/or remodeling (Figure 14.2).

14.2.2 Morphology/architecture

Scaffolds have numerous morphological architecture and examples are shown in Figure 14.3. The ASTM terminology (1999) for porous materials is similar and classified into three groups: interconnecting (open pores), nonconnecting (closed pores) or a combination of both. Ashby and Gibson (1997) describe that the mechanical properties of a porous solid depend mainly on its relative density, the properties of the material that make up the pore edges or walls and the anisotropic nature, if any, of the solid. In general, the stiffness (E^*) and yield strength (σ^*) in compression of porous solids are each related to the relative density by a power law relationship. Given that most constructs will require a high degree of porosity to accommodate mass transfer and tissue development, the volume fraction of the scaffold will necessarily be low. In all but the most biomechanically challenging applications, it is likely that the test for the scaffold engineer will be to achieve sufficient stiffness and strength in a highly porous structure to provide adequate mechanical integrity. One of the most demanding applications will be the repair and generation of musculoskeletal tissues, particularly bone, where scaffolds need to have a high elastic modulus in order to provide temporary mechanical support without showing symptoms of fatigue or failure, to be retained in the space they were designated for and to provide the tissue with adequate space for growth (Hutmacher, 2000). One of the fundamental challenges of scaffold design and materials selection is that to achieve sufficient strength and stiffness the scaffold material must have both a sufficiently high interatomic and intermolecular bonding and/or a physical and chemical structure which allows for hydrolytic attack and breakdown.

14.2.2.1 *Porosity definition*

A pore can be defined as a void space within a scaffold, whereas porosity can be considered as a collection of pores (Figure 14.3b). Pore size and porosity are important scaffold parameters (see Examples in Boxes 1 to 3). Macro-pores (i.e. above 50 μm) are of a scale to influence tissue function, e.g. pores greater then 300 μm in size are typically suggested for bone ingrowth in relation to vascularization of the construct. Micro-pores (i.e. below 50 μm) are of a scale to influence cell

---

Figure 14.2 (Continued) properties of the TEC should preferably match those of the surrounding host tissue. For Strategy II (right), the intrinsic mechanical properties of the scaffold architecture templates the cell proliferation and differentiation only up to the phases where the premature bone or cartilage is placed in a bioreactor. The degradation and resorption kinetics of the scaffold are designed to allow the seeded cells to proliferate and secrete their own extracellular matrix in the static and dynamic cell seeding phase (weeks 1–12), while the polymer scaffold gradually vanishes leaving sufficient space for new cell and tissue growth. The physical support by the 3D scaffold is maintained until the engineered tissue has sufficient mechanical integrity to support itself.

function (e.g. cell attachment) given that mammalian cells typically are 10–20 μm in size. Nano-porosity refers to pore architectures or surface textures on a nano-scale (i.e. 1–1000 nm). There is often a compromise between porosity and scaffold mechanical properties. High porosity (e.g. 90%) may provide a greater pore volume for cell infiltration and extracellular matrix (ECM) formation, but conversely decreases mechanical properties in accordance to a power-law relationship (Zein *et al.*, 2002). The compressive stiffness (E) and yield strength σ in compression are each related to the porosity of the scaffold by:

$$E = C_1 (100 - P)^n \qquad (1)$$
$$\sigma = C_2 (100 - P)^n \qquad (2)$$

where C_1, C_2 and n are constants, and $(100 - P)$ the relative density. For example, the theoretical exponent values are $n = 3$ and $n = 2$ for scaffolds with honeycomb and open-pore foam architectures, respectively.

14.2.2.2 Pore interconnectivity

Pore interconnectivity (Figure 14.3b) is a critical factor and is often overlooked in scaffold design and characterization. A scaffold may be porous, but unless the pores are interconnecting (i.e. voids linking one pore to another), they serve no purpose and become superfluous in a scaffold for tissue engineering. The interconnecting pores should be suitably large to support cell migration and proliferation in the initial stages, and consequently ECM infiltration of desired tissue. It is preferable that scaffolds for tissue engineering have 100% interconnecting pore volume, thereby also maximizing the diffusion and exchange of nutrients (e.g. oxygen, glucose) throughout the entire scaffold pore volume (see Chapter 12; Cell Nutrition).

14.2.2.3 Pore characterization

As a measure of pore interconnectivity, the accessible pore volume, or permeability, of a scaffold can be measured. Accessible pore volume can be defined as

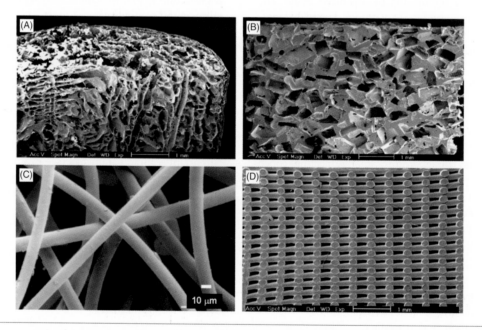

Figure 14.3a Scanning electron micrographs showing examples of scaffolds produced using various processing techniques. (A) liquid–solid phase separation and freeze-drying using 1,4-dioxane (PEGT/PBT), (B) compression molding and porogen leaching (PEGT/PBT), (C) nonwoven polylactic acid (PLA) mesh with 13 μm fiber diameter, (D) 3D fiber deposition (3DF) of PEGT/PBT scaffold with a 175-μm fiber diameter and 0.5-mm fiber spacing.

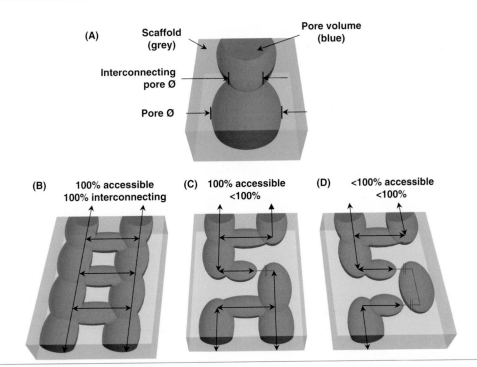

Figure 14.3b (A) Illustration of a section through a porous scaffold (grey) and corresponding pore geometries in relation to overall accessible pore volume (blue). (B) Pore volume 100% accessible and 100% interconnecting. (C) Pore volume 100% accessible but not 100% interconnecting. (D) Pore volume neither 100% accessible or 100% interconnecting.

Box 1 The influence of scaffold architecture on tissue formation in tissue-engineered constructs

Example 1: Optimizing scaffold architecture for bone tissue engineering applications

Recent SFF studies using scaffolds with designed pore architectures showed that pore size is less important for bone formation than first thought. The generally adopted dogma in bone tissue engineering suggests that optimal pore sizes of 200–600 μm are required to support bone growth into porous scaffolds, often in relation to infiltration of local supporting vasculature. To date, these studies have been performed on 3D scaffolds with a nondesigned, random collection of pores, varying in size and interconnectivity. Using designed RP scaffolds containing 100% interconnecting pores and homogeneous pore architecture, recent studies have demonstrated no significant difference in bone formation with varying pore size between 300 and 1200 μm as described below:

- RP PLGA scaffolds with Ø500 μm and Ø1600 μm interconnecting pores *in vitro* (Roy *et al.*, 2003b; Simon *et al.*, 2003).
- RP HA scaffolds with Ø300 μm and Ø800 μm interconnecting pores in mouse *in vivo*, and Ø400 μm and Ø1200 μm pores in mini-pig *in vivo* (Hollister, 2005).

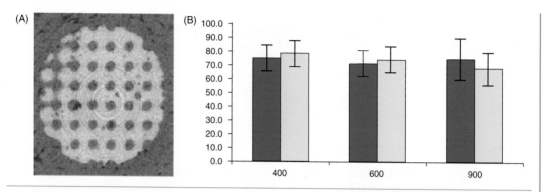

Example of bone growth in designed HA scaffolds in a mini-pig mandibular defect model. (A) μCT scan slice showing bone (grey) penetrating through entire scaffold (white). (B) Bar graph showing total bone ingrowth between 70 and 80% at 18 weeks for pore sizes between 400 and 900 μm where no significant difference was observed between pore sizes. Reproduced with permission from Blackwell Publishing: Hollister, S.J., Lin, C.Y., Saito, E. et al. (2005). Engineering craniofacial scaffolds. Orthod Craniofac Res, 8(3): 162–173.

Furthermore, using computational topology optimization techniques, it is possible to produce a porous 3D architecture optimized around key design criteria, such as porosity, permeability and mechanical stiffness (A and B). Topology-optimized scaffold designs can then be accurately reproduced using RP (C and D), containing appropriate architecture and mechanical properties suited to their final application (E).

Results from this work beg the question: is there an optimal biomaterial *and* architecture for regeneration of specific tissues? Although much progress has been made in the design of scaffolds with precise architecture, an answer to the above question still remains elusive.

Box 2 The influence of scaffold architecture on tissue formation in tissue-engineered constructs

Example 2: The effect of random versus defined pore architecture on tissue formation and nutrient gradients

For cartilage tissue engineering applications, recent studies have also investigated whether chondrocyte redifferentiation and cartilage tissue formation capacity can be regulated by controlled modifications in scaffold architecture and composition (Miot *et al.*, 2005). As a model system, copolymer scaffolds with different compositions [low or high poly(ethylene glycol) (PEG) content] and interconnecting pore architectures (random compression molding or designed 3DF, A and B) were produced as characterized previously in Section 14.2.2. 3DF scaffold architectures, with a more accessible pore volume (i.e. 90% or more at 98 μm for compression molding and at 380 μm for 3DF) and larger interconnecting pores (i.e. compression molding = 182 μm; 3DF = 380 μm), supported increased glycosaminoglycan (GAG)/DNA deposition (D) compared to compression molding scaffolds (C), but only if a high PEG composition was used.

Enhanced collagen type II mRNA was observed in 3DF compared with compression molding scaffolds irrespective of composition, suggesting that at the mRNA level, architecture alone is capable of instructing collagen type II synthesis pathways in chondrocytes and operates independently of GAG synthesis pathways.

By applying controlled and selective modifications of chemico-physical scaffold parameters, it is possible to demonstrate that both scaffold composition and architecture can regulate chondrocyte redifferentiation for cartilage tissue engineering. The observed effects of composition and architecture *in vitro* are likely mediated by differential surface protein adsorption and efficiency of nutrient/waste exchange, respectively (see Chapter 12).

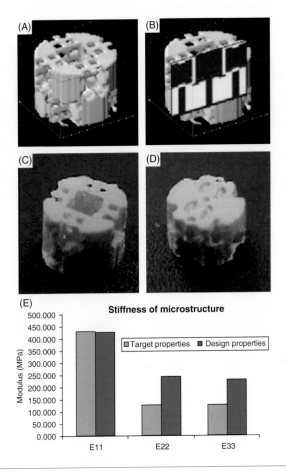

Topology optimization was used to design the scaffold (A) and its internal architecture (B). This design was accurately produced in a fabricated PPF/β-TCP scaffold [top (C) and bottom (D)]. Reproduced with permission from Mary Ann Liebert, publishers: Lin, C.Y., Schek, R.M., Mistry, A.S., et al. (2005). Functional bone engineering using ex vivo *gene therapy and topology-optimized, biodegradable polymer composite scaffolds. Tissue Eng, 11(9–10): 1589–1598. (E) Comparison of native bone anisotropic moduli (blue) with designed scaffold anisotropic moduli (red). Reproduced with permission from Blackwell Publishing: Hollister, S.J., Lin, C.Y., Saito, E., et al. (2005). Engineering craniofacial scaffolds. Orthod Craniofac Res, 8(3): 162–173.*

Box 3 The influence of scaffold architecture on tissue formation in tissue-engineered constructs

Example 3: Instructing or guiding cell attachment and ECM formation via scaffold design

The ability of designed pore architectures to instruct the zonal organization of chondrocytes and ECM components has also been investigated (Woodfield *et al.*, 2005). RP scaffolds were produced with either a homogeneous fiber spacing of 1 mm (A) or a pore-size gradient ranging between 0.5, 1 and 2 mm (B).

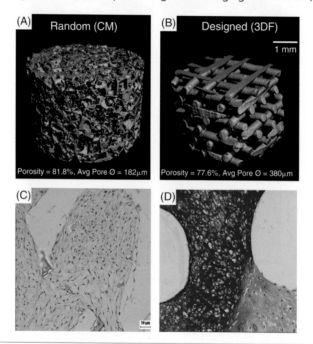

μCT images of (A) 'random' (compression molded) versus (B) 'designed' (3DF) architectures for cartilage tissue engineering. (C and D) Histological sections (safranin-O staining for GAG) showing that the quality and quantity of cartilage ECM components can be enhanced through careful design of scaffold architecture and biomaterial composition, as demonstrated for 3DF architectures (D). Reproduced with permission from Elsevier: Miot, S., Woodfield, T.B.F., Daniels, A.U., et al. (2005). Effects of scaffold composition and architecture on human nasal chondrocyte redifferentiation and cartilaginous matrix deposition. Biomaterials, 26: 2479–2489.

In vitro cell seeding showed that pore-size gradients promoted a similar anisotropic cell distribution to superficial (S), middle (M) and lower (L) zones in bovine articular cartilage, irrespective of seeding methods used. There was a direct correlation between zonal scaffold volume fraction (i.e. surface area to volume ratio) with DNA and GAG content. Prolonged tissue culture in vitro showed similar inhomogeneous distributions in zonal GAG and collagen type II synthesis, but not GAG/DNA, and levels were an order of magnitude less that in native cartilage.

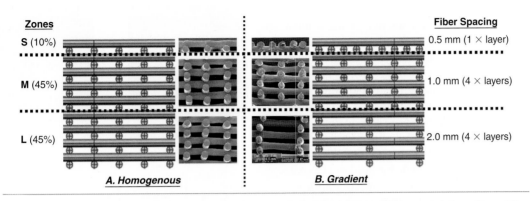

Zones

S (10%)

M (45%)

L (45%)

A. Homogenous

B. Gradient

Fiber Spacing

0.5 mm (1 × layer)

1.0 mm (4 × layers)

2.0 mm (4 × layers)

Illustration and scanning electron microscopy images of designed RP scaffolds containing either (A) a homogeneous 1-mm fiber spacing or (B) an anisotropic pore-size gradient. Regions corresponding to the superficial (S), middle (M) and lower zones (L) are indicated on the left, while the associated fiber spacing used in gradient scaffolds (0.5, 1.0 and 2.0 mm) are indicated on the right. Reproduced with permission from Mary Ann Liebert, Publishers: Woodfield, T.B., Van Blitterswijk, C.A., De Wijn, J., et al. (2005). Polymer scaffolds fabricated with pore-size gradients as a model for studying the zonal organization within tissue-engineered cartilage constructs. Tissue Eng, 11(9–10): 297–311.

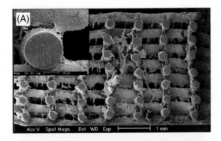

Scanning electron microscopy images showing chondrocytes attached to 3DF scaffolds containing (A) a homogeneous 1-mm fiber spacing or (B) an anisotropic pore-size gradient. By generating pore-size gradients within RP scaffolds. control over the distribution of cells and amount of ECM components, such as GAG and collagen, could be achieved. Reproduced with permission from Mary Ann Liebert, publishers: Woodfield, T.B.F., van Blitterswijk, C.A., de Wijn, J., et al. (2005). Polymer scaffolds fabricated with pore-size gradients as a model for studying the zonal organization within tissue-engineered cartilage constructs. Tissue Eng, 11: 1297–1311.

the total volume of pores which can be infiltrated from all peripheral borders to the interior of the scaffold (Malda *et al.*, 2005). Scaffold permeability can be measured by determining the flow rate of fluid flow through interconnecting pores (Li *et al.*, 2003). However, this technique is not suitable in scaffolds with large, 100% interconnected pore volumes (Figure 14.4B) as the scaffold provides no resistance to fluid flow. Alternatively, accessible pore volume, as well as vol% porosity, pore size distribution and scaffold surface area to volume ratio (i.e. volume fraction) can be characterized using techniques such as mercury intrusion porosimetry, micro-computed tomography (µCT) or image analysis (Ho and Hutmacher, 2006; Miot *et al.*, 2005; Woodfield *et al.*, 2004). Mercury porosimetry is a popular technique based on the principle that the pressure required to force a nonwetting liquid such as mercury into pores, against the resistance of liquid surface tension, is indicative of the pore size, assuming the pores are cylindrical in shape. However, the resolution of the technique is severely limited in scaffolds with large pore sizes (larger than 500 µm) where low mercury intrusion pressures are necessary and it has limitations when applied to materials that have irregular pore geometries. Alternative techniques such as µCT, which utilize 3D CT imaging to generate computer models of porous materials, have been developed for analyzing bone architecture and more recently scaffold architecture (Figure 14.4). Using 3D µCT techniques, considerably greater information can be obtained to characterize pore architectures containing features ranging from 6 to above 1500 µm, without the physical limitations associated with mercury porosimetry (Ho and Hutmacher, 2005 ; Woodfield *et al.*, 2004).

14.3 Scaffold fabrication

14.3.1 Introduction

A number of fabrication technologies have been applied to process biodegradable and bioresorbable materials into 3D polymeric scaffolds of high porosity and surface area (Table 14.1). From a scaffold

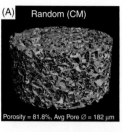
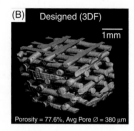
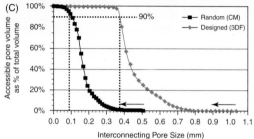
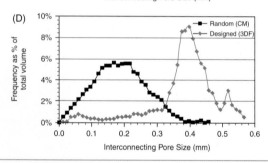

Figure 14.4 Example of µCT analysis comparing (A) 'random' versus (B) 'designed' scaffold architectures, produced using compression molding and particulate leaching or RP via 3D fiber deposition (3DF). (C) Accessible pore volume distribution and (D) pore size distribution for compression molding and particulate leaching and 3DF scaffolds represented as a percentage of total pore volume (PV). The dotted line in (C) represents the 90% threshold value indicating the interconnecting pore size at which 90% or more of the pore volume was accessible e.g. compression molding ≥98 µm; 3DF ≥380 µm. Arrows in (C) indicate direction of µCT pore accessibility measurement, from scaffold periphery towards scaffold interior. Reproduced with permission from Elsevier: Woodfield, T.B.F., van Blitterswijk, C.A., de Wijn, J., *et al.* (2005). Polymer scaffolds fabricated with pore-size gradients as a model for studying the zonal organization within tissue-engineered cartilage constructs. *Tissue Eng*, 11: 1297–1311.

Table 14.1 Summary of current literature on scaffold design and fabrication techniques and their application in tissue engineering

Technique	Porosity (%)	Pore size (μm)	Biomaterial	Comments	Cell type/tissue engineering application	Reference
Conventional techniques						
Foaming using blowing agents	–	20–1000	starch/SEVA-C	non porous outer layer (skin)	osteosarcoma – bone tissue engineering *in vitro*	Gomes et al., 2001a,b
	<80	<350	PLLA	pore interconnectivity >99%	–	Lin et al., 2003
Foaming using CaCO$_3$	<81	100–1000	chitin	difficult control	–	Chow and Khor, 2000
Foaming using NH$_4$HCO$_3$	<94	100–500	PLLA	interconnected pores; moldable shapes	hepatocyte – liver tissue engineering *in vitro*	Nam et al., 2000
Super-critical fluid technology (gas foaming)	<93	100	PLGA	only partially interconnected pores; formation of a skin layer	–	Mooney et al., 1996a
	10–30	<100	PLGA	high volume of noninterconnecting micro-pores	smooth muscle – muscle tissue engineering *in vitro*	Harris et al., 1998
Super-critical fluid technology (gas foaming) + particulate leaching	91–95	micro <35 macro <200	PLGA	low volume of noninterconnecting micro-pores low volume of interconnecting macro-pores	–	Whang et al., 1995
Sintered microspheres	32–39	67–300	PLGA	interconnected pores; reverse template of trabeculae	osteoblast/fibroblast – bone tissue engineering *in vitro*	Borden et al., 2002a,b, 2003
Fused salt particles + solvent casting + particulate leaching	<97	250–425	PLA, PEGT/PBT, PCL	improved interconnectivity compared to solvent casting + particulate leaching	–	Hou et al., 2003b

(continued)

Table 14.1 Continued

Technique	Porosity (%)	Pore size (µm)	Biomaterial	Comments	Cell type/tissue engineering application	Reference
Fused salt particles/gas foaming + particulate leaching	<94	<78	PLGA		–	Murphy et al., 2002
Coagulation + particulate leaching	<92	800–1500	PLGA	large interconnections (350 µm), irregular pore morphology	–	Holy et al., 1999
	<98	250–1180	PLA, PEGT/PBT, PCL	wide range of porosities and pore size	–	Hou et al., 2003a
Solvent casting + particulate leaching	82–92	<200	PLA/PVA	spherical pores, salt particles remain in matrix, only thin membranes or small devices	hepatocyte – liver tissue engineering in vitro	Mooney et al., 1995
Solvent merging + particulate leaching	<95	250–500	PLGA	good interconnectivity	–	Liao et al., 2002
Solvent casting + extrusion	<90	<30	PLGA, PLLA	highly porous; interconnected pores; severe salt breakdown during extrusion	–	Widmer et al., 1998
Solvent induced phase separation (immersion precipitation)	<95 <45	<10 <10	PU	only possible for films/tubes	–	Chen et al., 1999
TIPS	<97	<200	PLA, PLA/HA	high volume of interconnecting micro-pores	–	Schugens et al., 1996; Zhang and Ma, 1999a
Solid–liquid phase separation (freeze-drying)	<97	<500	PLA, PLLA/PLGA, PLA/HA	1,4-dioxane, phenol, benzene, naphthalene are toxic solvents	osteosarcoma, MC3T3 – cytotoxicity test in vitro	Liu et al., 1992; Lo et al., 1996; Zhang and Ma, 1999a,b; Ma and Zhang, 2001

Liquid–liquid phase separation	–	20	centrifugal casting produces porous inner layer, dense outer layer of tube	pHEMA/MMA	neural/spinal cord tissue engineering	Dalton et al., 2002
Polymerization-induced phase-separated scaffolds	<90	5–50	hydrogel scaffolds, mechanically weak; formed in water or acetone	pHEMA, pHPMA	ophthalmic – neural tissue engineering	Hicks et al., 1999; Loh et al., 2001; Hicks et al., 2003; Bakshi et al., 2004
Emulsion freeze drying	<97	<200	high volume of interconnecting micro-pores	PLGA, PEG	bone tissue engineering	Whang et al., 1995, 1998
Textile technologies						
Nonwoven fiber	<97	100–500	high porosity; insufficient mechanical properties	PGA, PLLA/PLGA	smooth muscle, endothelial, chondrocyte – muscle, blood vessel, cartilage tissue engineering in vitro and in vivo in nude mice, rabbit	Leidner et al., 1983; Mikos et al., 1993a; Freed et al., 1994; Mooney et al., 1996b
Woven fabrics	–	–	insufficient mechanical properties	LLDPE	–	Wintermantel et al., 1996; Shikinami et al., 1998
Fiber bonding	81	–	described process only possible for PGA-PLLA combination	PGA		Mikos et al., 1993a
	–	–	coating of fiber mesh with another polymer	PGA + PLLA/PLGA		Mooney et al., 1996b
Nanofiber electrospinning	–	–	interconnected pore structure with a high specific surface area	PCL, PU	MSC – bone tissue engineering in vitro	Yoshimoto et al., 2003; Kidoaki et al., 2005; Matsuda et al., 2005
	–	–	direct melt electrospinning technique; solvent-free process	PEO-PCL	fibroblast	Dalton et al., 2006

(continued)

Table 14.1 Continued

Technique	Porosity (%)	Pore size (µm)	Biomaterial	Comments	Cell type/tissue engineering application	Reference
Porous membrane lamination	<90	10–400	PLA, PLGA	irregular pore size, tedious procedure	–	Mikos et al., 1993b
Hydrocarbon templating	<88	300–1000	PLA	allows incorporation of proteins	chondrocyte – cartilage tissue engineering in vitro	Shastri et al., 2000
	<97	100–500	PLLA, PLGA	control over interconnectivity	–	Ma and Choi, 2001
SFF techniques *Laser and UV light sources*						
SLA	<90	>200	DEF/PPF + BAPO, P4HB/PHOH	100% interconnected macro-pore structure; polymers for bone, heart valve applications	–	Sodian et al., 2002; Cooke et al., 2003
	–	175–400	PEG-DMA	spatially patterned internal architectures; noninterconnecting dense pore/channel walls	MSC – bone tissue engineering in vitro	Mapili et al., 2005
Micro SLA	–	165–650	PEG-DA	allows fabrication of complex internal features along with precise spatial distribution of biological factors	BMSC – bone tissue engineering	Lu et al., 2006
SLS	–	<50	PEEK/HA	low porosity primarily for drug release; PEEK/HA composite scaffolds	drug release; bone tissue engineering	Cheah et al., 2002; Tan et al., 2003
	<25	>30	CPP	low porosity, basic shapes	Bone tissue engineering	Porter et al., 2001

	Porosity (%)	Pore size	Material	Characteristics	Application	Reference
	63–79	1750–2500	PCL	100% interconnected, topography optimized pore structure matching bone mechanical properties Anatomic shape	BMP-7-induced fibroblasts – bone tissue engineering *in vivo* in mouse and mini-pig	Williams *et al.*, 2005
SGC	3–10	–	vinylated polysaccharide/gelatine	noninterconnecting pores	tubular scaffolds for nerve, blood vessel tissue engineering	Matsuda and Magoshi, 2002
	–	100–500	PEG-DA	photo-patterning of single or multiple cell layers; photo-initiator concentration influences cell viability; photo-polymerizing hydrogels	hepatoma cell line – liver tissue engineering *in vitro*	Liu *et al.*, 2002
Printing Technology						
3D Printing (3DP)	<60	45–150	PLA	100% interconnected macro-pore structure; irregular pore architecture; limited resolution	fibroblast, vascular smooth muscle, epithelial cells – soft tissue engineering *in vitro*	Zeltinger *et al.*, 2001
	45	1000–1600	PLGA/β-TCP, HA	100% interconnected macro-pore structure containing 15-μm micro-pores; shrinkage postsintering	bone tissue engineering *in vivo* in rabbit calvarial defect	Roy *et al.*, 2003a,b
	–	>450	HA	100% interconnected macro-pore structure containing 10- to	bone tissue engineering	Seitz *et al.*, 2005

(continued)

Table 14.1 Continued

Technique	Porosity (%)	Pore size (μm)	Biomaterial	Comments	Cell type/tissue engineering application	Reference
				30-μm micro-pores; shrinkage postsintering; limited mechanical properties		
Cell Printing	–	<350	collagen I/soy-agar gel	cells embedded in gel; >90% cell survival; limited resolution	ovarian, endothelial, neuronal cells – cell viability in vitro	Boland et al., 2003; Wilson et al., 2003; Varghese et al., 2005; Xu et al., 2005
Extrusion/direct writing						
FDM	<80	150–700	PCL	100% interconnected macro-pore structure; > 200 μm diameter fiber; good mechanical properties	MSC – bone tissue engineering in vitro and in vivo in rabbits	Hutmacher et al., 2001; Zein et al., 2002; Schantz et al., 2003
	60	–	PLA	PLA degradation during processing	bone tissue engineering	Xiong et al., 2001
3DP	<75	100–1000	agar, fibrin/alginate	100% interconnected macro-pore structure; >500 μm diameter fiber; only hydrogel materials investigated; room temperature fabrication	fibroblast – soft tissue engineering in vitro	Landers et al., Landers and Mulhaupt, 2000; Landers et al., 2002a,b
	–	200–1000	chitosan/HA	100% interconnected macro-pore structure; >200 μm diameter fiber; room temperature fabrication	osteoblast – bone tissue engineering in vitro	Ang et al., 2002

Technique			Material	Features	Application	References
	<80	500	PCL, PEG/PCL	100% interconnected macro-pore structure; >400 μm diameter fiber	MSC – bone tissue engineering *in vitro*	Huang *et al.*, 2004
	<90	100–2000	PEGT/PBT	100% interconnected macro-pore structure; >175 μm diameter fiber; good dynamic mechanical properties; anisotropic pore architectures	chondrocyte – cartilage tissue engineering *in vitro* and *in vivo* in nude mice	Woodfield *et al.*, 2004, 2005; Malda *et al.*, 2005; Miot *et al.*, 2005; Moroni *et al.*, 2006a,b
	62–78	>100	PEGT/PBT, PBMA, PCL	100% interconnected macro-pore structure; core-shell or hollow fiber obtained via viscous encapsulation; growth factor/drug delivery	cartilage tissue engineering	Moroni *et al.*, 2006c
Direct writing	<90	5–100	PAA/PEI polyelectrolyte ink	100% interconnected macro-pore structure; 1 μm diameter fiber; fiber micro-pore structure with direct writing in low pH reservoir	–	Gratson and Lewis, 2005
Indirect SFF techniques Negative mould casting	–	50–800	PLA, PLA/HA, PPF/TCP	100% interconnected macro-pore structure; integration to form composite	biphasic scaffold for cartilage, bone tissue engineering	Taboas *et al.*, 2003; Schek *et al.*, 2004; Lin *et al.*, 2005

(continued)

Table 14.1 Continued

Technique	Porosity (%)	Pore size (µm)	Biomaterial	Comments	Cell type/tissue engineering application	Reference
				scaffolds; shrinkage after sintering; topography optimized pore structure matching bone mechanical properties; anatomic shape		
	50	50–400	HA (β-TCP)	100% interconnected macro-pore structure; shrinkage after sintering; preferential bone growth on rough, micro-pore surface	BMC – bone tissue engineering in vitro and in vivo	Wilson et al., 2004
	–	50–800	collagen I	100% interconnected vascular channels; incorporated into macro-porous, freeze-dried collagen type I	blood vessel tissue engineering	Sachlos et al., 2003
	polymer: 82, BCP: 66	polymer: 1300, BCP: 360	PEGT/PBT + BCP	100% interconnected biphasic composite combining 3D plotting of PEGT/PBT polymer and indirect SFF of calcium phosphate ceramic	cartilage/osteochondral tissue engineering	Moroni et al., 2006d
Micro-robotics/MEMS						
MEMS	–	45–3000	poly(glycerol sebacate)	interconnected channels in 2D with controlled architecture	endothelial cell – vascular tissue engineering in vitro	Fidkowski et al., 2005

design and function viewpoint each processing methodology has its pros and cons. It is beyond the scope of this chapter to cover all scaffold fabrication techniques available. Hence, the authors aim to provide the reader with a overview of the methods which currently are most relevant for scaffold-based tissue engineering.

It is the aim of this chapter to aggregate the compiled information and to present this data in a comprehensive form. The key rationale, characteristics and process parameters of the currently used scaffold fabrication techniques are presented. The aim of this part is to assist research teams with their choice for a specific 3D scaffold processing technology by providing the information for determining the critical issues. As discussed above, the current challenge in tissue engineering research is not only to design, but also to fabricate reproducible, bioresorbable 3D scaffolds, which are able to function as designed.

14.3.2 Conventional techniques

14.3.2.1 Porogen leaching

Porogen-leaching consists of dispersing a template (particles, etc.) within a polymeric or monomeric solution, gelling or fixing the structure, and removal of the template to result in a porous scaffold. The specific methods to achieve such scaffolds are numerous, and porogen-leaching is an inexpensive technique to manufacture cell-invasive scaffolds. An example of porogen leached scaffold with highly interconnected voids is shown in Figure 14.5. Scaffold manufacturing methods based on porogen leaching are capable of producing structures with locally porous internal architectures from a diverse array of materials. Local pores are voids characteristically defined by small struts and plates or spherical and tubular cavities generally less than 300 μm in diameter. Local pores are interconnected within local regions of the scaffold microstructure.

However, porogen leaching yields cavities with defined shape, while highly interconnected voids are desired for TECs. Solvent diffusion from emulsions

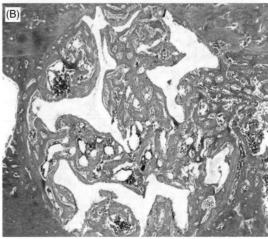

Figure 14.5 Use of a biomimetic strategy to engineer bone. PLGA/CAP scaffolds made by porogen leaching are applied in cancellous bone engineering strategies. Courtesy of Dr Jeff Karp, and Professor Jed Davies (MIT and University of Toronto).

can yield oriented pore structures. Although these methods yield interconnected pores that may comprise a continuous conduit throughout a scaffold, the pore connectivity is not an intentional result of a prior global design. Rather, the connectivity is a random product of variable, local void interconnections that are affected by polymer processing parameters. Such random connections may not provide optimal scaffold permeability for tissue ingrowth nor optimal

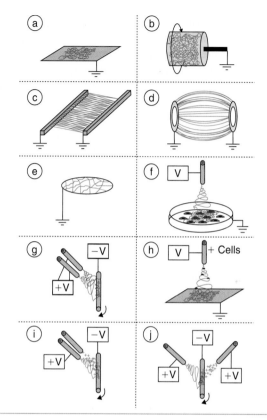

Figure 14.10 Selected collection schemes for electrospun fibers (red). Single ground (a), rotating single ground (b), dual bar (c), dual ring (d), single horizontal ring (e), electrospinning *in vitro* onto cells (f), dual spinneret electrospinning (with second green fibers) (g), electrospinning cells with polymer (h), and electrospinning/electrospraying with parallel (i) and perpendicular spinnerets (j).

14.4.3.5 *Cell/electrospun scaffold interactions*

Cells may adopt significantly different morphologies on electrospun fabrics, when compared to flat surfaces. Cardiomyocytes develop filopodia-like extensions along the electrospun fibers and form tissue constructs on ring supports (Figure 14.14) (Shin *et al.*, 2004), while smooth muscle cells orient their bodies along the length of the fiber (Figure 14.15) (Stankus *et al.*, 2006). Thin-layered TECs are therefore achievable and

demonstrate the potential of electrospinning in tissue engineering applications. Fibroblasts are particularly motile on electrospun fibers and stretch their cell body into 'spindle-like' morphologies (Dalton *et al.*, 2006). However, the surface properties of electrospun fibers are important for cell adhesion and spreading, demonstrated by different spreading of keratinocytes and fibroblasts on electrospun fabrics treated with different adsorbed proteins (Min *et al.*, 2004).

The literature covering 'direct *in vitro* electrospinning', i.e. the electrospinning of fibers directly onto cells, is currently sparse (Dalton *et al.*, 2006; Stankus *et al.*, 2006). Both solution and melt electrospinning onto cells is possible. Direct *in vitro* electrospinning is important for fabricating layered TECs, where cells and electrospun scaffolds are ordered into lamellar structures. Melt electrospinning may eventually be advantageous for directly electrospinning onto tissue for *in vivo* applications (direct *in vivo* electrospinning).

14.4.3.6 *Melt electrospinning*

Many tissue engineers aim to incorporate cells with electrospun material for 3D tissue constructs. The recent interest in electrospun fibers has generated a wealth of literature and yet includes only a handful of melt electrospinning approaches. Electrospinning polymers without solvents (via the melt), however, may be attractive for applications and systems where solvent accumulation or toxicity is a concern. Volatile solvents used for preparing electrospun fibers (chloroform, methanol, etc.) require removal before contact with cells, due to the cytotoxicity of residual solvent. Melt electrospinning may allow new approaches to such applications, overcoming technical restrictions governed by solvent accumulation and toxicity. Melt electrospinning will preclude the use of many biologically derived polymers of great interest to researchers; however, synthetic polymers are feasible.

14.4.3.7 *Solution electrospinning*

Many publications report various polymeric fibers electrospun from solution, both degradable and non-degradable, and from organic and aqueous solutions. Review articles of electrospinning include sources of solvent/polymer combinations for electrospinning

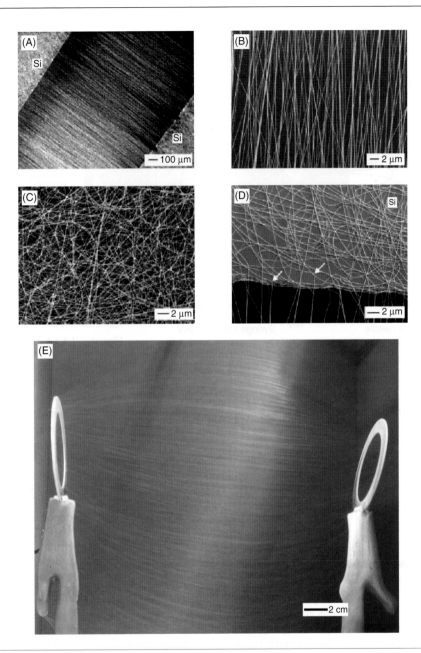

Figure 14.11 Electrospinning with dual collectors, an array of (aligned) fibers is collected between the gap (A and B). The random fibers are limited to the collector itself (C) while the aligned fibers are between the gap (D). (A–D) From Li *et al.* (2003) and reproduced with permission from the American Chemical Society. Dual rings also result in a suspended fiber array of significant length (E). (E) Reprinted from Dalton *et al.* (2005) with permission from Elsevier.

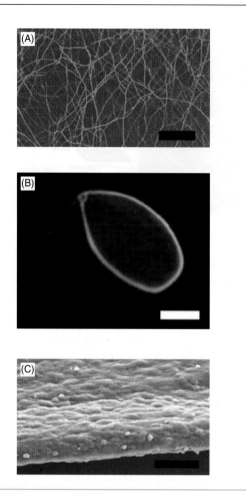

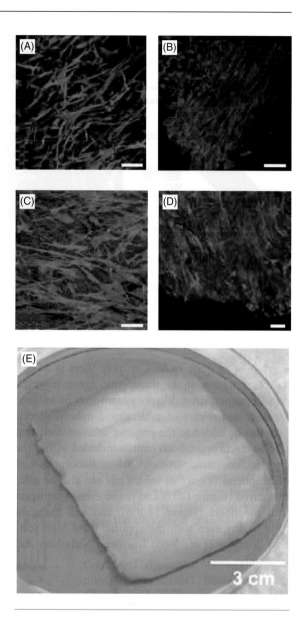

Figure 14.14 A highly split 'netting' of electrospun fibers (A) is collected onto a ring (B) and culturing with cardiomyocytes forms a thin tissue sheet (C). Scale bars for (A)–(C) = 100, 200 and 50 μm, respectively. Modified and reproduced with permission from Elsevier (2004). Contractile cardiac grafts using a novel nanofibrous mesh. *Biomaterials*, 25(17): 3717–3723.

the potential benefits offered by SFF technology is the ability to create parts with highly reproducible architecture and compositional variation across the entire matrix due to its computer controlled fabrication.

Figure 14.15 Smooth muscle cell constructs formed after electrospinning *and* electrospraying the cells. Culture in a bioreactor allows tissue constructs to be formed. Cells have greater viability when electrosprayed with gelatin. Reproduced with permission from Elsevier (2006). Microintegrating smooth muscle cells into a biodegradable elastomeric fiber matrix. *Biomaterials*, 27(5): 735–744.

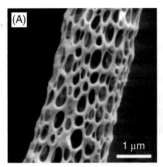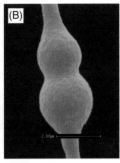

Figure 14.16 (A) Nanostructured morphologies may be observed within electrospun fibers, such as this PLLA example, electrospun from a dichloromethane solution. Reproduced with permission from Wiley: Dersch, D., Steinhart, M., Boudriot, U. *et al.* (2005). Nanoprocessing of polymers: applications in medicine, sensors, catalysis, photonics. *Polym Adv Technol*, 16(2–3): 276–282. (B) Image of a PLGA microspheres embedded in electrospun 2% PEO (500 K g/mol) from an ethanol/water (50:50 vol%) solution. Scale bar = 2 μm.

14.5.2 Systems based on laser and UV light sources

14.5.2.1 Stereolithography

The stereolithography (SLA) apparatus is often considered the pioneer of the RP industry with the first commercial system introduced in 1988 by 3D Systems Inc. Stereolithography is based on the use of a focused UV laser which is vector scanned over the top of a liquid bath of a photopolymerizable material. The UV laser causes the bath to polymerize where the laser beam strikes the surface of the bath, resulting in the creation of a first solid plastic layer at and just below the surface. The solid layer is then lowered into the bath and the laser generated polymerization process is repeated for the generation of the next layer, and so on, until a plurality of superimposed layers forming the desired scaffold architecture is obtained. The most recently created layer in each case is always lowered to a position for the creation of the next layer slightly below the surface of the liquid bath. Once the scaffold is complete, the platform rises out of the vat and the excess resin is drained. The scaffold is then removed from the platform, washed of excess resin and then placed in a UV oven for a final curing.

For industrial applications the photopolymer resins are mixtures of simple low-molecular-weight monomers capable of chain-reacting to form solid long-chain polymers when activated by radiant energy within specific wavelength range. The commercial materials used by SLA equipment are epoxy-based or acrylate-based resins that offer strong, durable and accurate parts/models. However, this material cannot be used as scaffold materials due to lack of biocompatibility and biodegradability. Hence, limited selection of photopolymerizable biomaterials is a major constraint for the use of the SLA technique in the design and fabrication of scaffolds for tissue engineering applications. However, biocompatible acrylic, anhydride and polyethylene oxide (PEO)-based polymers may be explored in future research, as they are already in the research or clinical stage, typically as curable bioadhesives or injectable materials. The variation of the laser intensity or traversal speed may be used to vary the cross-linking or polymer density within a layer so that the properties of the material can be varied from position to position within the scaffold. This would allow us to fabricate so-called biphasic or triphasic matrix systems. Microstereolithography (MSL), in particular, is thought to offer a great potential for the production of 3D polymeric structures with micrometer resolution.

14.5.2.2 Selective laser sintering

Selective laser sintering (SLS) also uses a focused laser beam, but to sinter areas of a loosely compacted powder. In this method, a thin layer of powder is spread evenly onto a flat surface with a roller mechanism. The powder is then raster-scanned with a high-power laser beam. The powder material that is struck by the laser beam is fused, while the other areas of powder remain dissociated. Successive layers of powder are deposited and raster-scanned, one on top of another, until an entire part is complete. Each layer is sintered deeply enough to bond it to the

the internal unbound powders to be removed if the part is designed to be porous, such as a scaffold for tissue engineering applications. The surface roughness and the aggregation of the powdered materials also affect the efficiency of removal of trapped materials. The resolution of the printer is limited by the specification of the nozzle size and position control of the position controller that defines the print head movement. Another factor is the particle size of the powder used, which simultaneously determines the layer thickness. A layer thickness between 100 and 400 µm can be achieved depending on the printer. The versatility of using a powdered material is both an advantage and a constraint of the 3DP process.

Most of the available biomaterials do not come in powder form and need special processing conditions to get a powder which fulfils the requirements for 3DP. Despite the restrictions discussed above, 3DP has been explored by several tissue engineering groups for more then a decade and a spin-off from MIT (Therics Inc.) has commercialized scaffolds made by 3D printing.

More recently, several groups (Mironov *et al.*, 2006; Xu *et al.*, 2005) were able to print cells in combination with hydrogels by simple modification of office inkjet printers showing the proof of principle to one day built tissue-engineered constructs in a fully automated system (see also Box 4).

Box 4 Organ printing

With the advantage of being both inexpensive and of high throughput, commercial thermal inkjet printers have been modified to print biomolecules onto target substrates with little or no reduction of their bioactivities, resulting in the creation of DNA chips, protein arrays and cell patterns. Most recently, by means of CAD, viable cells can be delivered to precise target positions within a matrix. Furthermore, using different cell types in combination with different hydrogels (bio-inks), which are then delivered to exact positions to mimic tissue structures of the original tissue, can be envisioned by using multiple nozzles. Thus, the printing of dissociated or aggregated cells based on specific patterns, and their subsequent fusion, may allow the development of replacement tissue or even whole organ substitutes (Jakab *et al.*, 2004; Mironov *et al.*, 2003).

Xu *et al.* (2005) and Mironov *et al.* (2005) proved the ability to print viable cells either by direct writing or the inkjet printing method. Since the physiological properties of mammalian cells strongly depend on the culture conditions, and are sensitive to heat and mechanical stress, a major concern was that the cells could be damaged or lysed by the conditions present during thermal printing. The temperature in the nozzle of the cartridge can be 300°C or higher. The study by Xu *et al.* (2005) indicates, however, that viable CHO cells and primary motor neurons can be delivered successfully by using a modified inkjet Hewlett-Packard (HP) printer and most of these cells (above 90%) were not lysed during printing. HP inkjet printer technology is based on vaporizing a micrometer-sized layer of liquid in contact with a thin film resistor. As the timescales involved in the drop ejection process are small, there is not enough time for heat to diffuse into the bulk liquid. While the surface in contact with the liquid can peak at 250–350°C, the bulk liquid does not rise more than a couple (approximately 4–10°C) degrees above ambient. This situation could change, however, depending on the liquid and thermophysical properties of the ink. Although mammalian cells are more sensitive to high heat and strong mechanical stress than bacteria, their volume is also relatively small compared to the total volume of the printed droplet. The fact that these mammalian cells are viable, can proliferate and differentiate indicates that damage from heat and mechanical stress during the very short timescale of printing is avoided. Moreover, the 3 × phosphate-buffered saline used for the cell print suspensions

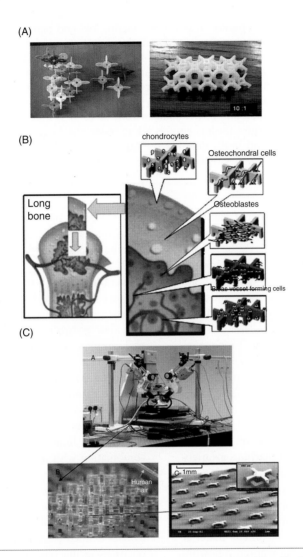

(A)

(B)

chondrocytes

Osteochondral cells

Long bone

Osteoblastes

Bleas vessel forming cells

(C)

Human hair

1mm

Figure reproduced with permission from: Zhang, H., Hutmacher, D.W., Chollet, F., et al. (2005). Microrobotics and MEMS-based fabrication techniques for scaffold-based tissue engineering. Macromol Biosci, 5(6): 477–489.

could have caused a further decrease in cell volume, because they are expected to shrink by osmosis in this hypertonic solution, effectively preventing clogging of the nozzle during printing. While these initial studies were mainly concerned with the survivability of cells, future studies will need to optimize the inkjet technology along with the hydrogels to be successfully used in tissue engineering.

Future studies will aim at modifying current inkjet printers based on piezo-electric technology for possible use in cell printing. However, there are some challenges in adapting commercial piezo-printers for organ printing. Commercial piezo-printers use a more viscous ink, and hence minimizing ink leakage and preventing formation of vapor is difficult. More viscous inks help to eliminate the need for complex fluid gates between the ink cartridge and print head to prevent the ink from backflow. However, this comes at the expense of requiring more power and higher vibration frequencies, both of which can break and damage the cell membranes. Typical commercial piezo-printers use frequencies up to 30 kHz and power sources ranging from 12 to 100 W. This becomes a problem because vibrating frequencies ranging from 15 to 25 kHz and power sources from 10 to 375 W are often used to disrupt cell membranes. Adapting piezo-printers for less viscous ink to lower the frequency and power would be challenging, since ink leakage and mist formation during printing could obscure the pattern. Future studies will show if these problems may be overcome.

14.5.4 Systems based on extrusion/direct writing

A number of groups (Ciardelli *et al.*, 2004; Hutmacher *et al.*, 2004; Landers *et al.*, 2002) have developed SFF machines that can perform extrusion of strands/filaments and/or plotting of dots in 3D with or without incorporation of cells (Figure 14.17). These systems are built to make use of a wide variety of polymer hot melts as well as pastes/slurries (i.e. solutions and dispersions of polymers and reactive oligomers). Techniques such as fused deposition modeling (FDM), 3D plotting, multiphase jet solidification (MJS) and precise extrusion manufacturing (PEM) all employ extrusion of a material in a layered fashion to build a scaffold. Depending on the type of machine, a variety of biomaterials can be used for scaffold fabrication.

A traditional FDM machine consists of a head-heated-liquefier attached to a carriage moving in the horizontal *x–y* plane. The function of the liquefier is to heat and pump the filament material through a nozzle to fabricate the scaffold following a programmed path which is based on CAD model and the slice parameters. Once a layer is built, the platform moves down one step in the *z*-direction to deposit the next layer. Parts are made layer by layer with the layer thickness varying in proportion to the nozzle diameter chosen. FDM is restricted to the use of thermoplastic materials with good melt viscosity properties; cells or other theromosensitive biological agents cannot be encapsulated into the scaffold matrix during the fabrication process.

A variation of the FDM process, the so-called 'precision extruding deposition (PED)' system, was developed and tested at Drexel University (Wang *et al.*, 2004). The major difference between PED and conventional FDM is that the scaffolding material can be directly deposited without filament preparation. Pellet-formed PCL is fused by a liquefier temperature provided by two heating bands and respective thermal couples, and is then extruded by the pressure created by a turning precision screw. One such technique, i.e. MJS, involves the extrusion of a melted material through a nozzle that has a jet-like design. The MJS process is usually used to produce metallic or ceramic parts via a lost-wax method.

A design limitation using extrusion systems in combination with thermoplastic polymers is that the pore openings for the scaffolds are not consistent in all three dimensions (see Box 5). The pore openings facing the *z*-direction are formed in between the intercrossing of material struts/bars and are determined by user-defined parameter settings. However, for pore openings facing both the *x*- and *y*-directions, these openings are formed from voids created by the stacking of material layers; hence, their sizes are restricted to the bar/strut thickness (diameter). As such, systems with a single extrusion head/liquefier do not show any variation in pore morphology in all three axes. One design variability exists by extruding one strut/bar directly on top of each other (see Box 5).

In the material science literature another term for extrusion-based systems is used, i.e. 'direct-write techniques' Gratson and Lewis (2005) showed that direct-writing techniques rely on the formulation of colloidal inks for a given deposition scheme (see 'State of the Art Experiment'). The techniques employed in direct writing are pertinent to many other fields next to scaffold fabrication such as the capability of controlling small volumes of liquid accurately. Direct-write techniques involving colloidal ink can be divided into two approaches: (i) droplet-based including direct inkjet printing and hot-melt printing, and (ii) continuous (or filamentary) techniques.

Box 5 Robot-assisted construct fabrication

The principle of micro-assembling a scaffold/cell constructs is based on the same concept as assembling a structure using small building block units like Lego®. Building blocks of different designs would be first fabricated via lithography or microfabrication technologies assembled by a dedicated precision robot with four degree-of-freedom micro-gripping capabilities, accomplishing a functional-sized scaffold with the required material, chemical and physical properties. A monolithic shape memory alloy micro-gripper was used to manipulate and assemble the unit micro-parts into a scaffold structure (Zhang et al., 2005). Even though preliminary data are exciting, future studies need to show if this technique can be translated into commercially viable concepts.

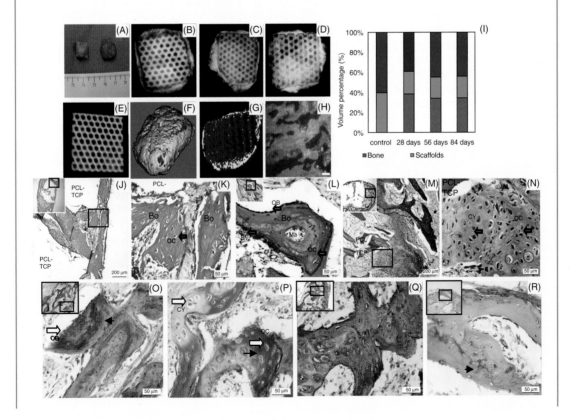

Constructs with bone marrow stromal cell (BMSC) sheet scaffolds [medical grade PCL–tricalcium phosphate (TCP)] were implanted into nude rat, harvested after 28, 56 and 84 days. (A) Gross appearance of BMSC sheet-scaffold constructs with osteogenic induction (right) and noninduction (control) (left) after 4 weeks. X-ray detected bone-like tissue formation in (B) 28-, (C) 56- and (D) 84-day implants. (E) X-ray image of constructs without cell seeding (Control). (F) μCT demonstrated the overall highly mineralized tissue similar to cortical (golden) and cancellous (red) bone in implanted constructs after 28 days. (G) μCT images disclosed the hard tissue formation within constructs. Mineralized tissue with similar density to cancellous bone was detected (darker areas), while the lighter color represented cortical bone. (H) Fluorescence was detected on the formed bone tissue after 28 days; the fluorescence came from the carboxy-3′,6′-diacetylfluorescein-labeled BMSC. (I) μCT quantification of implant tissue compositions depicted substantial bone formation in the induction group, accounting for 40% total volume for the 28-, 56- and 84-day implantations, while the control group formed only connective tissue. (J) Hematoxylin & eosin staining shows lamellar bone-like tissue formed in both the outer part and interior of constructs after 28 days. (K) High-magnification image shows well organized lamellar bone like tissue with distinct osteocytes located within bone tissue. (L) The typical osteoblasts (black arrow) located on the surface of neomineralized tissue with marrow cavities and blood vessels in 56-day implants. (M) Safranin-O staining demonstrated that hypertrophic chondrocytes (white arrow) were observed in 56-day implants at very low numbers. High magnification shows the chondrocytes (white arrow) with weak safranin-O staining surrounded by the osteocyte (black arrow). (N) OCN staining. Strong signals (arrow head) were detected on neomineralized tissue while very weak to no signals were detected on chondrocyte-like cells (O). (P) Collagen type I staining. Extensive staining was detected on the neomineralized tissue in constructs. (Q) Collagen type II staining. Limited signals (arrow head) were detected on chondrocyte-like cells, while no staining was seen for the mineralized tissues. BV: blood vessel + red blood cells; Ma: marrow; OB: osteoblast; Bo: bone; OC: osteocyte; CY: chondrocyte. Scale bar: (J) 200 μm; (H–Q) 50 μm. © 2007 with permission from Elsevier: Zhou, Y., Chen, F., Ho, S.T., et al. (2007). Combined marrow stromal cell-sheet techniques and high-strength biodegradable composite scaffolds for engineered functional bone grafts. Biomaterials, 28(5): 814–824.

State of the Art Experiment
Direct-Write Assembly of 3D Polyelectrolyte Scaffolds

Introduction

Direct-write techniques[1] that involve the layer-by-layer assembly of materials via 'ink' deposition offer a flexible, inexpensive route for creating complex 3D scaffolds for tissue engineering[2] applications. Recently, concentrated colloidal,[3,4] nanoparticle,[5] and polyelectrolyte inks[6,7] have been designed for direct writing in three dimensions. These inks all share the common feature of being able to flow through fine deposition nozzles and then rapidly 'set' to facilitate shape retention of the as-deposited features. Ink 'setting' involves a triggered change in rheological behavior (i.e. a fluid-to-gel transition) that arises either due to a change in shear rate or ink chemistry, rather than drying-induced solidification.

Both polyelectrolyte[6,7] and colloidal (e.g., concentrated hydroxyapatite inks[8]) inks can be utilized to produce 3D periodic scaffolds. Here, we describe the direct-write assembly of concentrated polyelectrolyte complexes[8] comprised of poly-anion-polycation mixtures that coagulate to form self-supporting filaments when deposited into an alcohol-rich reservoir. We first discuss the assembly process, and then describe the phase behavior and rheological properties of these polyelectrolyte inks. This is followed by a discussion of ink flow during deposition and reservoir effects on ink coagulation.

Direct-Write Assembly

3D periodic scaffolds are assembled by depositing a concentrated polyelectrolyte ink through a glass micro-capillary nozzle (diameter of $1\,\mu m$) into an alcohol-rich reservoir.[6] Representative images of 3D micro-periodic polyelectrolyte scaffolds created by this approach are shown in Figure 1a.[7] Each scaffold consists of spanning filaments patterned in serpentine sequence to yield parallel rods oriented orthogonally between layers and porous outer walls. These scaffolds, when patterned within the deposition reservoir, maintain their structural integrity during drying. Through focused ion beam milling, we can subsequently sculpt these scaffolds into various forms by removing defined sections, e.g. a triangular slice. By varying the deposition parameters, 3D polyelectrolyte scaffolds can be built with minimum and maximum filament diameters between ~0.5 and $10\,\mu m$.

Phase Behavior of Polyelectrolyte Complexes

Polyelectrolyte complexes exhibit a rich phase behavior that depends on several factors, including the polyelectrolyte type and architecture, their individual molecular weight and molecular weight ratio, the polymer concentration and mixing ratio, the ionic strength and pH of the solution, and the mixing conditions.[9] This vast array of experimental parameters makes it difficult to derive universal models for polyelectrolyte complex formation, but some trends have been reported.[10–13] Three types of phase behavior have been observed upon mixing oppositely charged polyelectrolytes together in solution: (1) a macroscopically homogeneous, soluble complex (single-phase), (2) a turbid suspension comprised of stable complex colloidal particles in a polymer-poor fluid (two-phase), or (3) a precipitated complex with a polymer-poor supernatant fluid (two-phase).[9] To create the desired ink fluidity for direct-write assembly, soluble complexes comprised of different molecular weight polyions must be mixed together at a non-stoichiometric ratio of ionizable groups, with the higher molecular weight (M_w) species in excess in solution. They must also be mixed together under ionic strength conditions that promote polyelectrolyte exchange reactions to yield a homogeneous fluid.[14]

The 3D micro-periodic scaffolds shown in Figure 1a were assembled from concentrated polyelectrolyte inks (40 wt% in water) comprised of a mixture of poly(acrylic acid) (PAA, Mw ~10,000 g/mol) and polyethylenimine (PEI, Mw ~600 g/mol). In these complexes, PAA served as the lyophilizing (host) polyelectrolyte and PEI served as the blocking (guest)

 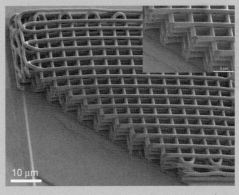

SEM images of 3D periodic structures with simple tetragonal symmetry (filament diameter = 1 μm, road width = 5 μm, 8 layers). (left) top view revealing registration between layers, (inset) angled view showing several layers of spanning filaments, (right) triangular section removed with focused ion beam milling, (inset) close-up of cut exposing the high integrity interfaces formed between layers. [Adapted from Ref. 7]

polyelectrolyte.[15,16] All PAA-PEI complexes were initially turbid (or aggregated) upon mixing and likely consist of aggregated colloidal particles (or a network at higher concentrations) with a nearly 1:1 charge ratio and an excess of 'free' lyophilizing species in solution.[14] These mixtures undergo structural rearrangement as they evolve towards a uniform distribution of the lower M_w component among chains of the higher M_w component. Soluble complexes emerge only when polyelectrolyte exchange reactions are favored and there is a sufficient excess of the lyophilizing species. The concentrated PAA-rich region is therefore most suitable for formulating inks for direct-write assembly.

Polyelectrolyte Ink Rheology

The rheological properties of concentrated polyelectrolyte inks both initially and after coagulation in the deposition reservoir must be optimized for 3D scaffold assembly. The initial viscosity of concentrated PAA-PEI inks (total polyelectrolyte concentration of ~40% by weight in solution) as a function of varying [COONa]/[NH$_x$] ratio, is shown here.[7]

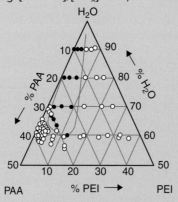

PAA-PEI-H$_2$O phase diagram showing (open circles) single-phase regions of macroscopically homogeneous, soluble complexes and (filled circles) two-phase region comprised of a turbid suspension of stable complex colloidal particles in a polymer-poor fluid (below c*) or a precipitated complex with a polymer-poor supernatant fluid (above c*). [Note: The solution pH ranged from pH ~3.5–4 in the PAA-rich to pH ~11 in the PEI-rich regions]. [Adapted from Ref. 7]

Under these conditions, a narrow two-phase region consisting of a polymer-rich aggregated network and a polymer-poor fluid is observed ranging from [COONa]/[NH$_x$] ratios of ~2.1 to 3.9. The soluble PAA-PEI complexes exhibit Newtonian flow behavior over the experimental conditions of interest. The ink viscosity increases from roughly 1 Pa•s for a pure PAA (40 wt%) solution to 10 Pa•s for PAA-PEI complexes (~40 wt%) with a [COONa]/[NH$_x$] ratio of five. On the opposite side of the two-phase region, PAA-PEI complexes with a [COONa]/[NH$_x$] of two exhibit an ink viscosity of 135 Pa•s, which is the maximum value observed. Upon further PEI addition, the ink viscosity decreases markedly as the PAA concentration was reduced below its c* value, where c* denotes the polymer concentration in solution at the dilute-to-semidilute transition.

Soluble PAA-PEI complexes used as inks for direct-write assembly coagulate when deposited into a reservoir comprised of two miscible liquids, isopropyl alcohol (IPA) and deionized water, mixed together at different volumetric ratios.[6,7] The observed rise in ink elasticity depends strongly on reservoir composition, as shown Figure 1d, for a representative PAA-PEI complex with a [COONa]/[NH$_x$] ratio of 5.7.[7] Prior to coagulation, this ink exhibits a shear elastic modulus (G') of ~1 Pa. However, in both pure water and alcohol-rich reservoirs (~83–88% alcohol), the ink experiences a nearly five-order of magnitude increase in elasticity (G'~5 × 10^4 Pa), which preserves the cylindrical (filament) shape during assembly.

The ink coagulation process is governed by many competing factors that depend strongly on reservoir composition. For example, ink coagulation occurs due to intensified electrostatic interactions between oppositely charged polyions when deposited in deionized water (pH ~6, low ionic strength). However, this mechanism results in the formation of a near-stoichiometric polyelectrolyte complex with excess species diffusion into the reservoir, which is not desired. In reservoirs ranging from pure water and 70% isopropyl alcohol (IPA), there is a systematic reduction in ink elasticity with increasing IPA content. This may arise due to a progressive screening of the intra-chain electrostatic interactions in response to the decreased dielectric constant of the medium. In addition, enhanced counterion condensation may also play a role suppressing the rise in ink elasticity observed for alcohol/water reservoirs (0–70% IPA).[17] If Na^+ counterions remain associated with the PAA backbone, the electrostatic interactions between oppositely charged PAA and PEI species in the ink are weakened. Unambiguously determining the role of electrostatic interactions in mixed solvent reservoirs is difficult, but it clearly has important implications for both direct-write assembly[6] and electrostatic driven layer-by-layer assembly[18] of polyelectrolyte-based structures.

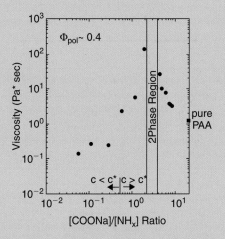

Log-log plot of initial ink viscosity for concentrated polyelectrolyte complexes (40% by weight) with varying [COONa:NH_x] ratio. As a benchmark, the viscosity of a concentrated PAA solution (40 wt%) is shown on the right axis. [Adapted from Ref. 7]

Finally, ink coagulation is driven by a change in solvent quality in an alcohol-rich reservoir (>70% IPA). Generally, in polyelectrolyte solutions, decreases in the electrostatic repulsions between the chain segments, either by increasing the ionic strength[19] or decreasing the dielectric constant[20] of the medium, cause transitions in the polymer conformation from extended coils to globules.[21] The extent of this collapse increases as the solvent quality decreases, eventually leading to precipitation and phase separation.[22] This mechanism does not yield a nearly stoichiometric complex with excess polyelectrolyte

species dissolved in the reservoir, but rather the entire complex aggregates together to form the patterned filaments shown here.

Ink Flow during Direct Writing

The laminar flow of viscous inks through a cylindrical tube can be described by the Hagen-Poiseuille equation:[23]

$$Q = \pi R^4 \, \Delta p/8 \, \eta L \qquad (1)$$

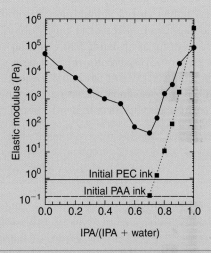

Semi-log plot of the elastic modulus of concentrated inks (40 wt% with a 5.7:1 [COONa:NH_x] ratio) coagulated in reservoirs of varying water/IPA ratio for polyelectrolyte complex ink (filled circles) and a concentrated PAA solution (40 wt%, filled squares). [Adapted from Ref. 7]

where Q = volumetric flow rate, R = tube radius, Δp = pressure drop, η = ink viscosity, and L = length of the tube. Our deposition nozzles consist of long tapered capillaries whose radius gradually decreases to the final nozzle size. Because the capillary radius far exceeds the nozzle radius, the pressure drop can be estimated using the following relationship for flow through a slowly-varying channel:[23]

$$\Delta p = \frac{8\eta Q}{\pi} \int_{x_1}^{a} R^{-4} dx \qquad (2)$$

where a = nozzle radius and x_1 = capillary radius. In this case R varies with x according to:

$$R = x\tan(\alpha) \tag{3}$$

where α = the angle of taper. Solving the integral yields:

$$\Delta p = \frac{8\eta Q}{3\pi(\tan \alpha)a^3} \tag{4}$$

where the pressure drop at x_1 can also be neglected because x_1 is much greater than a. This relation assumes a sufficiently small α and also neglects inertial forces, which is valid for the ink and deposition parameters described above:

$$\alpha\frac{2\rho Q}{\pi a\eta} \ll 1 \tag{5}$$

Using these expressions, the pressure required to maintain the desired volumetric flow rate ($Q = v\pi R^2$) at a constant deposition speed (v) can be estimated. As expected, this value increases with increasing ink viscosity or decreasing nozzle diameter. By mani pulating important variables, such as polyelectrolyte composition, molecular weight, and concentration, it should be possible to create 3D scaffolds with feature sizes ranging from 100 nm (desired ink viscosity of ~0.4 Pa•s) to 10 μm (desired ink viscosity of ~400 Pa•s) under modest applied pressures of ~20–200 kPa.

Summary

The design of concentrated polyelectrolyte inks comprised of viscous fluids that coagulate to yield highly elastic filaments upon contact with a deposition reservoir is central to our ability to create 3D polyelectrolyte scaffolds via direct-write assembly. These inks must possess the appropriate viscosity to facilitate flow under modest applied pressures. Several parameters can be manipulated to tailor the ink viscosity, including polyion molecular weight, concentration, and mixing ratio. However, the concentration of the lyophilizing polyelectrolyte should reside above c^* for the coagulation reaction to yield solid filaments rather than colloidal particles in suspension. Coagulation must also be rapid and produce filaments with the appropriate stiffness to maintain their shape and span unsupported regions during deposition and flexibility to adhere to the substrate or underlying layers within the scaffold. Looking towards the future, many ink and reservoir chemistries can be envisioned, including those based on physical (e.g. electrostatic, solvent quality) or chemical (e.g. polymerization) gelation. The inherent flexibility of direct-write assembly offers the potential to pursue a myriad of tissue engineering applications that either require or could benefit from 3D scaffolds with controlled composition and architecture.

References

1 Chrisey, D.B. *Science* **2000**, *289*, 879-881.
2 Griffith, L.G. and Naughton, G. *Science* **2002**, *295*, 1009–1014.
3 Smay, J.E., Cesarano, J. and Lewis, J.A. *Langmuir* **2002**, *18*, 5429–5437.
4 Smay, J.E., Gratson, G.M., Sheperd, R.F. *et al.* A. *Adv. Mater.* **2002**, *14*, 1279–1283.
5 Li, Q. and Lewis, J.A. *Adv. Mater.* **2003**, *15*, 1639–1643.
6 Gratson, G.M., Xu, M. and Lewis, J.A. *Nature* **2004**, *428*, 386.
7 Gratson, G.M. and Lewis, J.A. *Langmuir* 21, 457–464 (2005).
8 Michna, S., Wu, W. and Lewis, J.A. *Biomaterials* 26, 5632–5639 (2005).
9 Philipp, B., Dautzenberg, H., Linow, K.-J. *et al.* *Prog. Polym. Sci.* **1989**, *14*, 91–172.
10 Michaels, A.S. *Ind. Eng. Chem.* **1965**, *57*, 32–40.
11 Abe, K., Ohno, H. and Tsuchida, E. *Makromol. Chem.* **1977**, *178*, 2285–2293.
12 Zheleznova, I.V., Shulbaeva, G.B., Kalyuzhnaya, R.I. *et al.* *Dokl. Akad. Nauk SSSR* **1986**, *287*, 662–666.
13 Karibyants, N., Dautzenberg, H. and Colfen, H. *Macromolecules* **1997**, *30*, 7803–7809.
14 Zintchenko, A., Rother, G. and Dautzenberg, H. *Langmuir* **2003**, *19*, 2507–2513.

15 Zezin, A.B. and Kabanov, V.A. *Russ. Chem. Rev.* **1982**, *51*, 833–855.
16 Bakeev, I.H., Izumrudov, V.A., Kuchanov, S.I. *et al. Macromolecules* **1992**, *25*, 4249–4254.
17 Manning, G.S. *Acc. Chem. Res.* **1979**, *12*, 443–449.
18 Decher, G. *Science* **1997**, *277*, 1232–1237.
19 Dobrynin, A. V., Rubinstein, M. and Obukhov, S. *Macromolecules* **1996**, *29*, 2974–2979.
20 Micka, U., Holm, C. and Kremer, K. *Langmuir* **1999**, *15*, 4033–4044.
21 Kiriy, A., Gorodyska, G., Minko, S. *et al. J. Am. Chem. Soc.* **2002**, *124*, 13454–13462.
22 Poptoshev, E., Schoeler, B. and Caruso, F. *Langmuir* **2004**, *20*, 829–834.
23 Batchelor, G.K. *An Introduction to Fluid Dynamics*; Cambridge University Press: Cambridge, UK, 2000.

14.5.5 Indirect SFF

The key to versatile fabrication using SFF is its ability to literally build the model from its respective basic raw materials. However, the major limitations also lie in its methodology of building and bonding raw materials together. For the case of the SLA technique the raw material must be in a liquid form and must be photopolymerizable. The raw materials for SLS and laminated object manufacturing must be able to melt and be severed cleanly, respectively, and particles and layers must also be able to bond together based on the energy supplied. The 3DP powder and binder combination must be compatible and effectively adhere the bulk material. Finally, the FDM can only use a thermoplastic material. With these limitations on the building materials it further restricts the list and availability of biomaterials that can be used for forming scaffolds or devices using SFF technology. Some researchers have begun to explore other options to exploit the macroscopic geometry and internal intrinsic architecture attainable by the SFF, along with its convenience and accuracy of duplication from medical imaging sources. One emerging method is to fabricate a negative mold based on the scaffold design and cast the scaffold using the desired materials, which may not be usable in an SFF setting. Thus, the 'indirect' SFF techniques were born (Chu *et al.*, 2002).

There are two principal routes for the fabrication of scaffolds by RP: indirect and direct routes. Indirect routes rely on an additional molding step after fabricating the master pattern by RP. The term 'indirect' SFF was coined for scaffold fabrication by Hollister's group (Hollister, 2005; Taboas *et al.*, 2003). His group used the lost mold technique, combining the epoxy resin molds made by SLA (SL 5170 and SLA 250/40 from 3D Systems, respectively) based on 3D scaffold designs generated from CAD software or other imaging techniques, and a thermal-curable hydroxyapatite (HA)–acrylate suspensions as the raw or slurry material. After the molds were formed, the HA suspension was subsequently cast into the epoxy mold and cured at 85°C. The cured part was placed in a furnace at high temperature to simultaneously burn out the mold and the acrylate binder. Following the mold removal, the HA green body in the designed 3D structure was sintered at 1350°C into a 3D HA scaffold.

In their works published in 2003, Taboas *et al.* (2003) applied the indirect SFF method with conventional sponge scaffold fabrication procedures in creating a series of biomimetic scaffolds for multitissue and structural tissue interface engineering. They designed scaffolds of diameter 8 × 8 mm with a porosity of 50%. The first step was to create the molds for the scaffold, this was achieved using wax and polysulfonamide (PSA), which were commercial materials for the Model Maker II 3D printer (SolidScape Inc.). The versatility of this machine was that both wax and PSA can be used as the modeling material and/or the support material as there is a choice to remove either material by their respective means. With this set-up, four molds were made. PSA molds were made by melting and subsequently dissolving the wax portion in Bioact®. Wax molds were obtained by dissolving PSA in acetone, while cement and ceramic molds were created by casting cement paste in a wax mold and a HA–acrylic slurry in an inverse mold, respectively. The final ceramic (HA) mold was further burnt out and sintered.

With the molds made, four different casting routes were used to create different featured scaffolds. The first casting route used three sets of solvent casting. Porogen leaching was used to create a scaffold with local porosity. They are the traditional salt-leaching with salt (104–124 μm) and PLA (7.5%)–solvent combination and the emulsion-solvent diffusion method using PLA–THF and ethanol combination or snap-freezing technique. The second casting route was simply solvent casting PLA (25%) into the mold melt. The third casting route involved melt casting PGA and PLA into a top–bottom composite scaffold. Production of the fourth casting route produced a polymer–ceramic (HA–PLA) composite by a melt casting and an etching process. All of the final scaffolds were retrieved by the appropriate mold-removal processes involving melting, dissolving or etching.

Indirect SFF adds further versatility and detail in scaffold design and fabrication. The previous restriction on casting was the inability of molds to produce complex geometry and internal architecture. Now with indirect SFF, traditional casting processes with these SFF molds can meet the specific tissue engineering requirements, including mechanical integrity and customized shapes. Some highlighted advantages of indirect SFF include is cost savings as the materials required for mold casting are substantially less and need not be processed into a dedicated form for any particular SFF process, such as processing into a powder for SLS and 3DP. In addition, indirect SFF allows the use of a wide range of materials or a combination of materials (composites or copolymers). However, some drawbacks still revolve around this method including the resolution of the SFF method, as the cast model would inherit the errors and defects from the mold, such as cracks and dimensional changes.

Also, a mold removal method must be developed to remove the mold while preserving the as-cast scaffold intact and desired properties undisturbed.

14.5.6 Future directions

The main challenge in preparing a functional TEC is to obtain a homogenous distribution of cells, and hence new tissue, throughout the entire 3D scaffold volume. Furthermore, there exist two possibilities of incorporating cells into the scaffolds: (i) seeding of cells onto the surface of the scaffold subsequent to scaffold fabrication (Figure in Box 5) and (ii) the incorporation of cells into the scaffold fabrication process. This second approach is of interest especially when incorporating cells into the scaffold material; however, scaffold fabrication processes require cellular compatibility at the stage of inclusion. The concept of 'organ printing' (Mironov *et al.*, 2003) and robot-assisted construct fabrication (Zhang *et al.*, 2005) aims at making ideal TECs a reality (see 'Robot-assisted construct fabrication'). Electrospinning in combination with cells has much potential and has the benefit of being inexpensive. However, it is an untested process in the industrial tissue engineering world. Techniques that allow precise control over the location of cell and scaffold are sought after and offer new opportunities for TECs.

14.6 Conclusions

Scaffolds are of great importance for tissue engineering since they enable the fabrication of functional living implants out of cells obtained from cell culture. As the scaffolds for tissue engineering will eventually be implanted in the human body, the scaffold materials should be nonantigenic, noncarcinogenic, nontoxic, nonteratogenic and possess high cell/tissue biocompatibility so that they will not trigger pathological reactions after implantation. Apart from biomaterial issues, the macro- and micro-structural properties of the scaffold are also very important. In general, the scaffolds require individual external shape and well-defined internal structure with interconnected porosity to host most cell types. From a biological point of view, the designed matrix should serve various functions, including (i) acting as an immobilization site for transplanted cells, (ii) forming a protective space to prevent unwanted tissue growth into the wound bed and allow healing with differentiated tissue, (iii) directing migration or growth of cells via surface properties of the scaffold and (iv) directing migration or growth of cells via release of soluble molecules such as growth factors, hormones and/or cytokines. Future work has to provide further evidence that some of these techniques offer the right balance of capability and practicality to be suitable for fabrication of materials in sufficient quantity and quality to move holistic tissue engineering technology platforms into the clinical application.

References

Ang, T.H., Sultana, F.S.A., Hutmacher, D.W., *et al.* (2002). Fabrication of 3D chitosan–hydroxyapatite scaffolds using a robotic dispensing system. *Mater Sci Eng C*, 20: 35–42.

Bakshi, A., Fisher, O., Dagci, T., *et al.* (2004). Mechanically engineered hydrogel scaffolds for axonal growth and angiogenesis after transplantation in spinal cord injury. *J Neurosurg Spine*, 1: 322–329.

Boland, T., Mironov, V., Gutowska, A., *et al.* (2003). Cell and organ printing 2: fusion of cell aggregates in three-dimensional gels. *Anat Rec A Discov Mol Cell Evol Biol*, 272: 497–502.

Borden, M., Attawia, M., Khan, Y. and Laurencin, C.T. (2002a). Tissue engineered microsphere-based matrices for bone repair: design and evaluation. *Biomaterials*, 23: 551–559.

Borden, M., Attawia, M. and Laurencin, C.T. (2002b). The sintered microsphere matrix for bone tissue engineering: *in vitro* osteoconductivity studies. *J Biomed Mater Res*, 61: 421–429.

Borden, M., El-Amin, S.F., Attawia, M. and Laurencin, C.T. (2003). Structural and human cellular assessment of a novel microsphere-based tissue engineered scaffold for bone repair. *Biomaterials*, 24: 597–609.

Cheah, C.M., Leong, K.F., Chua, C.K., *et al.* (2002). Characterization of microfeatures in selective laser sintered drug delivery devices. *Proc Inst Mech Eng [H]*, 216: 369–383.

Chen, J.H., Laiw, R.F., Jiang, S.F. and Lee, Y.D. (1999). Microporous segmented polyetherurethane vascular graft: I. Dependency of graft morphology and mechanical properties on compositions and fabrication conditions. *J Biomed Mater Res B Appl Biomater*, 48: 235–245.

Chow, K.S. and Khor, E. (2000). Novel fabrication of open-pore chitin matrixes. *Biomacromolecules*, 1: 61–67.

Chu, T.M.G., Ortone, D.G., Hollister, S.J., *et al.* (2002). Mechanical and *in vivo* performance of hydroxyapatite implants with controlled architectures. *Biomaterials*, 23: 1283–1293.

Ciardelli, G., Chiono, V., Cristallini, C., *et al.* (2004). Innovative tissue engineering structures through advanced manufacturing technologies. *J Mater Sci Mater Med*, 15: 305–310.

Cima, M., Sachs, E., Fan, TL., Bredt, J.F., Michaels, S.P., *et al.* (1995) US Patent 5387380.

Cooke, M., Fisher, J., Dean, D., *et al.* (2003). Use of stereolithography to manufacture critical-sized 3D biodegradable scaffolds for bone ingrowth. *J Biomed Mater Res*, 64B: 65–69.

Dalton, P.D., Flynn, L. and Shoichet, M.S. (2002). Manufacture of poly(2-hydroxyethyl methacrylate-co-methyl methacrylate) hydrogel tubes for use as nerve guidance channels. *Biomaterials*, 23: 3843–3851.

Dalton, P.D., Klee, D. and Möller, M. (2005). Electrospinning with dual collection rings. *Polymer*, 46: 611–614.

Dalton, P.D., Klinkhammer, K., Salber, J., *et al.* (2006). Direct *in vitro* electrospinning with polymer melts. *Biomacromolecules*, 7: 686–690.

Deitzel, J.M., Kleinmeyer, J.D., Hirvonen, J.K. and Tan, N.C.B. (2001). Controlled deposition of electrospun poly(ethylene oxide) fibers. *Polymer*, 42: 8163–8170.

Dersch, D., Steinhart, M., Boudriot, U., *et al.* (2005). Nanoprocessing of polymers: applications in medicine, sensors, catalysis, photonics. *Polym Adv Technol*, 16: 276–282.

Fidkowski, C., Kaazempur-Mofrad, M.R., Borenstein, J., *et al.* (2005). Endothelialized microvasculature based on a biodegradable elastomer. *Tissue Eng*, 11: 302–309.

Freed, L.E., Vunjak-Novakovic, G., Biron, R.J., *et al.* (1994). Biodegradable polymer scaffolds for tissue engineering. *Biotechnology (NY)*, 12: 689–693.

Giordano, R.A., Wu, B.M., Borland, S.W., *et al.* (1996). Mechanical properties of dense polylactic acid structures fabricated by three dimensional printing. *J Biomater Sci Polym Ed*, 8: 63–75.

Gomes, M.E., Reis, R.L., Cunha, A.M., *et al.* (2001a). Cytocompatibility and response of osteoblastic-like cells to starch-based polymers: effect of several additives and processing conditions. *Biomaterials*, 22: 1911–1917.

Gomes, M.E., Ribeiro, A.S., Malafaya, P.B., *et al.* (2001b). A new approach based on injection molding to produce biodegradable starch-based polymeric scaffolds: morphology, mechanical and degradation behavior. *Biomaterials*, 22: 883–889.

Gratson, G.M. and Lewis, J.A. (2005). Phase behavior and rheological properties of polyelectrolyte inks for direct-write assembly. *Langmuir*, 21: 457–464.

Harris, L.D., Kim, B.S. and Mooney, D.J. (1998). Open pore biodegradable matrices formed with gas foaming. *J Biomed Mater Res*, 42: 396–402.

Hicks, C.R., Clayton, A.B., Vijayasekaran, S., *et al.* (1999). Development of a poly(2-hydroxyethyl methacrylate) orbital implant allowing direct muscle attachment and tissue ingrowth. *Ophthal Plast Reconstr Surg*, 15: 326–332.

Hicks, C.R., Crawford, G.J., Lou, X., *et al.* (2003). Corneal replacement using a synthetic hydrogel cornea, AlphaCor: device, preliminary outcomes and complications. *Eye*, 17: 385–392.

Ho, S.T. and Hutmacher, D.W. (2006). A comparison of micro CT with other techniques used in the characterization of scaffolds. *Biomaterials*, 27(1): 362–1376.

Hollister, S.J. (2005). Porous scaffold design for tissue engineering. *Nat Mater*, 4: 518–524.

Hollister, S.J., Lin, C.Y., Saito, E. *et al.* (2005). Engineering craniofacial scaffolds. *Orthod Craniofac Res*, 8: 162–173.

Holy, C.E., Dang, S.M., Davies, J.E. and Shoichet, M.S. (1999). *In vitro* degradation of a novel poly(lactide-*co*-glycolide) 75/25 foam. *Biomaterials*, 20: 1177–1185.

Hou, Q., Grijpma, D.W. and Feijen, J. (2003a). Porous polymeric structures for tissue engineering prepared by a coagulation, compression molding and salt leaching technique. *Biomaterials*, 24: 1937–1947.

Hou, Q., Grijpma, D.W. and Feijen, J. (2003b). Preparation of interconnected highly porous polymeric structures by a replication and freeze-drying process. *J Biomed Mater Res*, 67B: 732–740.

Huang, M.H., Li, S., Hutmacher, D.W., *et al.* (2004). Degradation and cell culture studies on block copolymers prepared by ring opening polymerization of epsilon-caprolactone in the presence of poly(ethylene glycol). *J Biomed Mater Res*, 69A: 417–427.

Huang, Z.M., Zhang, Y.Z., Kotaki, M. and Ramakrishna, S. (2003). A review on polymer nanofibers by electrospinning and their applications in nanocomposites. *Composites Sci Technol*, 63: 2223–2253.

Hutmacher, D.W. (2000). Scaffolds in tissue engineering bone and cartilage. *Biomaterials*, 21: 2529–2543.

Hutmacher, D.W., Schantz, T., Zein, I., *et al.* (2001). Mechanical properties and cell cultural response of polycaprolactone scaffolds designed and fabricated via fused deposition modeling. *J Biomed Mater Res*, 55: 203–216.

Hutmacher, D.W., Sittinger, M. and Risbud, M.V. (2004). Scaffold-based tissue engineering: rationale for computer-aided design and solid free-form fabrication systems. *Trends Biotechnol*, 22: 354–362.

Jakab, K., Neagu, A., Mironov, V. and Forgacs, G. (2004). Organ printing: fiction or science. *Biorheology*, 41: 371–375.

Karp, J.M., Schoichet, M.S. and Davies, J.E. (2003). Bone formation on two-dimensional poly(DL-lactide-*co*-glycolide) (PLGA) films and three-dimensional PLGA tissue engineering scaffolds *in vitro*. *J Biomed Mater Res A*, 64: 388–396.

Kidoaki, S., Kwon, I.K. and Matsuda, T. (2005). Mesoscopic spatial designs of nano- and microfiber meshes for tissue-engineering matrix and scaffold based on newly devised multilayering and mixing electrospinning techniques. *Biomaterials*, 26: 37–46.

Landers, R. and Mulhaupt, R. (2000). Desktop manufacturing of complex objects, prototypes and biomedical scaffolds by means of computer-assisted design combined with computer-guided 3D plotting of polymers and reactive oligomers. *Macromolec Mater Eng*, 282: 17–21.

Landers, R., Hubner, U., Schmelzeisen, R. and Mulhaupt, R. (2002a). Rapid prototyping of scaffolds derived from thermoreversible hydrogels and tailored for applications in tissue engineering. *Biomaterials*, 23: 4437–4447.

Landers, R., Pfister, A., Hubner, U., *et al.* (2002b). Fabrication of soft tissue engineering scaffolds by means of rapid prototyping techniques. *J Mater Sci*, 37: 3107–3116.

Langer, R. and Vacanti, J.P. (1993). Tissue engineering. *Science*, 260: 920–926.

Lavik, E.B., Klassen, H., Warfvinge, K., *et al.* (2005). Fabrication of degradable polymer scaffolds to direct the integration and differentiation of retinal progenitors. *Biomaterials*, 26: 3187–3196.

Li, D., Wang, Y.L. and Xia, Y.N. (2003). Electrospinning of polymeric and ceramic nanofibers as uniaxially aligned arrays. *Nano Letters*, 3: 1167–1171.

Li, S., de Wijn, J., Li, J., *et al.* (2003). Macroporous biphasic calcium phosphate scaffold with high permeability/porosity ratio. *Tissue Eng*, 9: 535–548.

Liao, C.J., Chen, C.F., Chen, J.H., *et al.* (2002). Fabrication of porous biodegradable polymer scaffolds using a solvent merging/particulate leaching method. *J Biomed Mater Res*, 59: 676–681.

Lin, A.S., Barrows, T.H., Cartmell, S.H. and Guldberg, R.E. (2003). Microarchitectural and mechanical characterization of oriented porous polymer scaffolds. *Biomaterials*, 24: 481–489.

Lin, C.Y., Schek, R.M., Mistry, A.S., *et al.* (2005). Functional bone engineering using *ex vivo* gene therapy and topology-optimized, biodegradable polymer composite scaffolds. *Tissue Eng*, 11: 1589–1598.

Liu, S.Q. and Kodama, M. (1992). Porous polyurethane vascular prostheses with variable compliances. *J Biomed Mater Res*, 26: 1489–1502.

Liu, V.A. and Bhatia, S.N. (2002). Three-dimensional photopatterning of hydrogels containing living cells. *Biomed Microdevices*, 4: 257–266.

Liu, X. and Ma, P.X. (2004). Polymeric scaffolds for bone tissue engineering. *Ann Biomed Eng*, 32: 477–486.

Lo, H., Kadiyala, S., Guggino, S. and Leong, K. (1996). Poly(L-lactic acid) foams with cell seeding and controlled-release capacity. *J Biomed Mater Res*, 30: 475–484.

Loh, N.K., Woerly, S., Bunt, S.M., *et al.* (2001). The regrowth of axons within tissue defects in the CNS is promoted by implanted hydrogel matrices that contain BDNF and CNTF producing fibroblasts. *Exp Neurol*, 170: 72–84.

Loscertales, I.G., Barrero, A., Márquez, M. and Spretz, R. (2004). Electrically forced coaxial nanojets for one-step hollow nanofiber design. *J Am Chem Soc*, 126: 5376–5377.

Lu, Y., Mapili, G., Suhali, G., *et al.* (2006). A digital micromirror device-based system for the microfabrication of complex, spatially patterned tissue engineering scaffolds. *J Biomed Mater Res A*, 77: 396–405.

Ma, P.X. (2004). Scaffolds for tissue fabrication. *Mater Today*, 7: 30–40.

Ma, P.X. and Choi, J.W. (2001). Biodegradable polymer scaffolds with well-defined interconnected spherical pore network. *Tissue Eng*, 7: 23–33.

Ma, P.X. and Zhang, R. (2001). Microtubular architecture of biodegradable polymer scaffolds. *J Biomed Mater Res*, 56: 469–477.

Malda, J., Woodfield, T.B.F., van der Vloodt, F., *et al.* (2005). The effect of PEGT/PBT scaffold architecture on the composition of tissue engineered cartilage. *Biomaterials*, 26: 63–72.

Mapili, G., Lu, Y., Chen, S. and Roy, K. (2005). Laser-layered microfabrication of spatially patterned functionalized tissue-engineering scaffolds. *J Biomed Mater Res B Appl Biomater*, 75: 414–424.

Matsuda, T. and Magoshi, T. (2002). Preparation of vinylated polysaccharides and photofabrication of tubular scaffolds as potential use in tissue engineering. *Biomacromolecules*, 3: 942–950.

Matsuda, T., Ihara, M., Inoguchi, H., *et al.* (2005). Mechano-active scaffold design of small-diameter artificial graft made of electrospun segmented polyurethane fabrics. *J Biomed Mater Res A*, 73A: 125–131.

Mikos, A., Bao, Y., Cima, L., *et al.* (1993a). Preparation of poly(glycolic acid) bonded fiber structures for cell attachment and transplantation. *J Biomed Mater Res*, 27: 183–189.

Mikos, A., Sarakinos, G., Leite, S., *et al.* (1993b). Laminated three-dimensional biodegradable foams for use in tissue engineering. *Biomaterials*, 14: 323–330.

Min, B.M., Lee, G., Kim, S.H., *et al.* (2004). Electrospinning of silk fibroin nanofibers and its effect on the adhesion and spreading of normal human keratinocytes and fibroblasts *in vitro*. *Biomaterials*, 25: 1289–1297.

Miot, S., Woodfield, T.B.F., Daniels, A.U., *et al.* (2005). Effects of scaffold composition and architecture on human nasal chondrocyte redifferentiation and cartilaginous matrix deposition. *Biomaterials*, 26: 2479–2489.

Mironov, V., Boland, T., Trusk, T., *et al.* (2003). Organ printing: computer-aided jet-based 3D tissue engineering. *Trends Biotechnol*, 21: 157–161.

Mironov, V., Reis, N. and Derby, B. (2006). Bioprinting: a beginning. *Tissue Eng*, 12: 631–634.

Mooney, D.J., Park, S., Kaufmann, P.M., *et al.* (1995). Biodegradable sponges for hepatocyte transplantation. *J Biomed Mater Res*, 29: 959–965.

Mooney, D., Baldwin, D., Suh, N., *et al.* (1996a). Novel approach to fabricate porous sponges of poly(d,l-lactic-*co*-glycolic acid) without the use of organic solvents. *Biomaterials*, 17: 1417–1422.

Mooney, D.J., Mazzoni, C.L., Breuer, C., *et al.* (1996b). Stabilized polyglycolic acid fiber-based tubes for tissue engineering. *Biomaterials*, 17: 115–124.

Moroni, L., de Wijn, J.R. and van Blitterswijk, C.A. (2006a). 3D fiber-deposited scaffolds for tissue engineering: influence of pores geometry and architecture on dynamic mechanical properties. *Biomaterials*, 27: 974–985.

Moroni, L., Poort, G., Van Keulen, F., *et al.* (2006b). Dynamic mechanical properties of 3D fiber-deposited PEOT/PBT scaffolds: an experimental and numerical analysis. *J Biomed Mater Res A*, 78: 605–614.

Moroni, L., Schotel, R., Sohier, J., *et al.* (2006c). Polymer hollow fiber three-dimensional matrices with controllable cavity and shell thickness. *Biomaterials*, 27: 5918–5926.

Moroni, L., Paoluzzi, L., Pieper, J.S., *et al.* (2006d) Incorporation of designed ceramic particles and demineralized bone matrix in 3D fiber deposited polymeric scaffolds for bone tissue engineering. *Proceedings of the European Tissue Engineering and Regenerative Medicine International Society Conference*, Rotterdam, pp. 108–189.

Moroni, L., Paoluzzi, L., Pieper, J.S. *et al.* (2008) Development of tailor-made hybrid scaffolds for bone and osteochondral tissue engineering. Submitted.

Murphy, W.L., Dennis, R.G., Kileny, J.L. and Mooney, D.J. (2002). Salt fusion: an approach to improve pore interconnectivity within tissue engineering scaffolds. *Tissue Eng*, 8: 43–52.

Nam, Y.S. and Park, T.G. (1999). Porous biodegradable polymeric scaffolds prepared by thermally induced phase separation. *J Biomed Mater Res*, 47: 8–17.

Nam, Y., Yoon, J. and Park, T. (2000). A novel fabrication method of macroporous biodegradable polymer scaffolds using gas foaming salt as a porogen additive. *J Biomed Mater Res*, 53: 1–7.

Nguyen, K.T. and West, J.L. (2002). Photopolymerizable hydrogels for tissue engineering applications. *Biomaterials*, 23: 4307–4314.

Nimni, M.E. and Harkness, R.D. (1988) Molecular structures and functions of collagen. *Collagen Vol. I – Biochemistry*, Nimni ME (ed.), CRC Press, Boca Raton, FL, pp. 1–79.

Porter, N., Pilliar, R. and Grynpas, M. (2001). Fabrication of porous calcium polyphosphate implants by solid free-form fabrication: a study of processing parameters and *in vitro* degradation characteristics. *J Biomed Mater Res*, 56: 504–515.

Reneker, D.H., Yarin, A.L., Fong, H. and Koombhongse, S. (2000). Bending instability of electrically charged liquid jets of polymer solutions in electrospinning. *J Appl Phys*, 87: 4531–4547.

Roy, T.D., Simon, J.L., Ricci, J.L., *et al.* (2003a). Performance of degradable composite bone repair products made via three-dimensional fabrication techniques. *J Biomed Mater Res A*, 66: 283–291.

Roy, T.D., Simon, J.L., Ricci, J.L., *et al.* (2003b). Performance of hydroxyapatite bone repair scaffolds created via three-dimensional fabrication techniques. *J Biomed Mater Res A*, 67: 1228–1237.

Sachlos, E., Reis, N., Ainsley, C., *et al.* (2003). Novel collagen scaffolds with predefined internal morphology made by solid freeform fabrication. *Biomaterials*, 24: 1487–1497.

Sachs, E.M., Haggerty, J.S., Cima, M.J. and Williams, P.A. (1993) US Patent 5204055.

Schantz, J.T., Lim, T.C., Ning, C., *et al.* (2003). Cranioplasty after trephination using a novel biodegradable burr hole cover: Technical case report. *Neurosurgery*, 58 (ONS Suppl 1): ONS-176

Schek, R.M., Taboas, J.M., Segvich, S.J., *et al.* (2004). Engineered osteochondral grafts using biphasic composite solid free-form fabricated scaffolds. *Tissue Eng*, 10: 1376–1385.

Schugens, C., Maquet, V., Grandfils, C., *et al.* (1996). Polylactide macroporous biodegradable implants for cell transplantation. II. Preparation of polylactide foams by liquid–liquid phase separation. *J Biomed Mater Res*, 30: 449–461.

Seitz, H., Rieder, W., Irsen, S., *et al.* (2005). Three-dimensional printing of porous ceramic scaffolds for bone tissue engineering. *J Biomed Mater Res B Appl Biomater*, 74B: 782–788.

Shastri, V., Martin, I. and Langer, R. (2000). Macroporous polymer foams by hydrocarbon templating. *Proc Natl Acad Sci USA*, 97: 1970–1975.

Schmelzeisen, R., Schimming, R. and Sittinger, M. (2003). Making bone: implant insertion into tissue-engineered bone for maxillary sinus floor augmentation – a preliminary report. *J Craniomaxillofac Surg*, 31: 34–39.

Shikinami, Y. and Kawarada, H. (1998). Potential application of a triaxial three-dimensional fabric (3-DF) as an implant. *Biomaterials*, 19: 617–635.

Shin, M., Ishii, O., Sueda, T. and Vacanti, J.P. (2004). Contractile cardiac grafts using a novel nanofibrous mesh. *Biomaterials*, 25: 3717–3723.

Simon, J.L., Roy, T.D., Parsons, J.R., *et al.* (2003). Engineered cellular response to scaffold architecture in a rabbit trephine defect. *J Biomed Mater Res A*, 66: 275–282.

Smeal, R.M., Rabbitt, R., Biran, R. and Tresco, P.A. (2005). Substrate curvature influences the direction of nerve outgrowth. *Ann Biomed Eng*, 33: 376–382.

Sodian, R., Loebe, M., Hein, A., *et al.* (2002). Application of stereolithography for scaffold fabrication for tissue engineered heart valves. *ASAIO J*, 48: 12–16.

Stankus, J.J., Guan, J., Fujimoto, K. and Wagner, W.R. (2006). Microintegrating smooth muscle cells into a biodegradable elastomeric fiber matrix. *Biomaterials*, 27: 735–744.

Taboas, J., Maddox, R., Krebsbach, P. and Hollister, S. (2003). Indirect solid free form fabrication of local and global porous, biomimetic and composite 3D polymer–ceramic scaffolds. *Biomaterials*, 24: 181–194.

Tan, K.H., Chua, C.K., Leong, K.F., *et al.* (2003). Scaffold development using selective laser sintering of polyetheretherketone–hydroxyapatite biocomposite blends. *Biomaterials*, 24: 3115–3123.

Teng, Y.D., Lavik, E.B., Qu, X., *et al.* (2002). Functional recovery following traumatic spinal cord injury mediated by a unique polymer scaffold seeded with neural stem cells. *Proc Natl Acad Sci USA*, 99: 3024–3029.

Townsend-Nicholson, A. and Jayasinghe, S.N. (2006). Cell electrospinning: a unique biotechnique for encapsulating living organisms for generating active biological microthreads/scaffolds. *Biomacromolecules*, 7: 3364–3369.

van de Witte, P., Dijkstra, P.J., van den Berg, J.W.A. and Feijen, J. (1996). Phase separation processes in polymer solutions in relation to membrane formation. *J Memb Sci*, 117: 1–31.

Varghese, D., Deshpande, M., Xu, T., *et al.* (2005). Advances in tissue engineering: cell printing. *J Thorac Cardiovasc Surg*, 129: 470–472.

Wang, F., Shor, L., Darling, A., *et al.* (2004). Precision extruding deposition and characterization of cellular poly-ε-caprolactone tissue scaffolds. *Rapid Prototyping J*, 10: 42–49.

Weiss, L.E., Amon, C.H., Finger, S., *et al.* (2005). Bayesian computer-aided experimental design of heterogeneous scaffolds for tissue engineering. *Comp-Aided Des*, 37: 1127–1139.

Whang, K., Thomas, C.H., Healy, K.E. and Nuber, G. (1995). A novel method to fabricate bioabsorbable scaffolds. *Polymer*, 36: 837–842.

Whang, K., Tsai, D.C., Nam, E.K., *et al.* (1998). Ectopic bone formation via rhBMP-2 delivery from porous bioabsorbable polymer scaffolds. *J Biomed Mater Res*, 42: 491–499.

Widmer, M., Gupta, P., Lu, L., *et al.* (1998). Manufacture of porous biodegradable polymer conduits by an extrusion process for guided tissue regeneration. *Biomaterials*, 19: 1945–1955.

Williams, J.M., Adewunmi, A., Schek, R.M., *et al.* (2005). Bone tissue engineering using polycaprolactone scaffolds fabricated via selective laser sintering. *Biomaterials*, 26: 4817–4827.

Wilson, C.E., de Bruijn, J.D., van Blitterswijk, C.A., *et al.* (2004). Design and fabrication of standardized hydroxyapatite scaffolds with a defined macro-architecture by rapid prototyping for bone-tissue-engineering research. *J Biomed Mater Res*, 68A: 123–132.

Wilson, W.C. Jr. and Boland, T. (2003). Cell and organ printing 1: protein and cell printers. *Anat Rec A Discov Mol Cell Evol Biol*, 272: 491–496.

Wintermantel, E., Mayer, J., Blum, J., *et al.* (1996). Tissue engineering scaffolds using superstructures. *Biomaterials*, 17: 83–91.

Woodfield, T.B., Bezemer, J.M., Pieper, J.S., *et al.* (2002). Scaffolds for tissue engineering of cartilage. *Crit Rev Eukaryot Gene Expr*, 12: 209–236.

Woodfield, T.B.F., Malda, J., de Wijn, J., *et al.* (2004). Design of porous scaffolds for cartilage tissue engineering using a three-dimensional fiber-deposition technique. *Biomaterials*, 25: 4149–4161.

Woodfield, T.B.F., van Blitterswijk, C.A., de Wijn, J., *et al.* (2005). Polymer scaffolds fabricated with pore-size gradients as a model for studying the zonal organization within tissue-engineered cartilage constructs. *Tissue Eng*, 11: 1297–1311.

Xiong, Z., Yan, Y.N., Zhang, R.J. and Sun, L. (2001). Fabrication of porous poly(L-lactic acid) scaffolds for bone tissue engineering via precise extrusion. *Scr Mater*, 45: 773–779.

Xu, T., Jin, J., Gregory, C., *et al.* (2005). Inkjet printing of viable mammalian cells. *Biomaterials*, 26: 93–99.

Yoshimoto, H., Shin, Y.M., Terai, H. and Vacanti, J.P. (2003). A biodegradable nanofiber scaffold by electrospinning and its potential for bone tissue engineering. *Biomaterials*, 24: 2077–2082.

Yoon, J.J. and Park, T.J. (2001). Degradation behaviours of biodegradable macroporous scaffolds prepared by gas foaming of effervescent salts. *J Biomed Mater Res*, 55: 401–408.

Zein, I., Hutmacher, D.W., Tan, K.C. and Teoh, S.H. (2002). Fused deposition modeling of novel scaffold architectures for tissue engineering applications. *Biomaterials*, 23: 1169–1185.

Zeltinger, J., Sherwood, J., Graham, D., *et al.* (2001). Effect of pore size and void fraction on cellular adhesion, proliferation, and matrix deposition. *Tissue Eng*, 7: 557–572.

Zhang, H., Hutmacher, D.W., Chollet, F., *et al.* (2005). Microrobotics and MEMS-based fabrication techniques for scaffold-based tissue engineering. *Macromol Biosci*, 5: 477–489.

Zhang, R. and Ma, P.X. (1999a). Poly(alpha-hydroxyl acids)/hydroxyapatite porous composites for bone-tissue engineering. I. Preparation and morphology. *J Biomed Mater Res*, 44: 446–455.

Zhang, R. and Ma, P.X. (1999b). Porous poly(L-lactic acid)/apatite composites created by biomimetic process. *J Biomed Mater Res*, 45: 285–293.

Zhou, Y.F., Chou, A.M., Li, Z.M., *et al.* (2007). Combined marrow stromal cell sheet techniques and high strength biodegradable composite scaffolds for engineered functional bone grafts. *Biomaterials*, 28: 814–824.

Chapter 15
Controlled release strategies in tissue engineering

Jeffrey Hubbell

Chapter objectives:

- To understand the methods employed in controlled morphogen release
- To understand the influence of the physicochemical properties of the incorporated bioactive on its controlled release
- To be able to explain the classes of morphogens used in tissue engineering, and the implications of their mode of action on appropriate controlled release strategies
- To be able to relate materials physicochemical and molecular

characteristics with their resorption behavior and controlled release characteristics
- To appreciate examples of current successes and future challenges of controlled morphogen release in tissue engineering therapies
- To be able to list important morphogens used in tissue engineering and relate their biological function to therapeutic targets

"In commenting on the remarkable passion with which tissue engineering scientists from different disciplines work together to develop controlled release strategies for tissue repair scaffolds, 'Kids are dying', Linda Griffith from MIT said, 'so people tend to work together'."

Quoted in the *New York Times*, 10 September 1996

15.1 Introduction

Tissue engineering strategies often depend heavily on the provision of bioactive signals, be they substrate-bound adhesion signals, or bound or diffusible growth factor signals. As such, controlled release technologies play an important role in the overall task of tissue engineering. In tissue engineering, one often uses biomaterial scaffolds or matrices, here using the word *scaffold* to refer to a macro- or microporous material used to provide a structural support (e.g. a polymer sponge) and the word *matrix* to refer to a nanoporous continuous material for the same purpose (e.g. a hydrogel). Cells can be transplanted within such scaffolds and matrices, e.g. the cells having been cultured and placed there *in vitro*. In many cases, one must coax the cells to behave in desired ways by providing them with bioactive signals in a controlled manner. It is sometimes even hoped that by providing bioactive signals within a matrix, we may succeed at coaxing cells, including progenitor and stem cells, to migrate into the matrix from the surrounding tissues, to obviate the need for cell transplantation and the logistical complexities that are associated with it. Thus, controlled release strategies are needed to provide signals to cells within scaffolds and matrices to differentiate along desired pathways.

In vitro, when cells are cultured within a scaffold or matrix, controlled release may not be necessary: the differentiation signals may simply be added to the medium in the bioreactor. Controlled release may be beneficial, even if not necessary, to reduce the amounts of very expensive signals, such as growth factors, by providing them only locally within the scaffold or matrix. *In vivo*, however, it makes little sense to administer such factors to the whole system, i.e. systemically, since very little and possibly even no growth factor would actually find its way to the implant, being cleared by other pathways in the body before it reaches that site. Thus, *in vitro*, controlled release may be beneficial; *in vivo*, it is usually essential.

The term *controlled release* is often taken to refer to release of a bioactive molecule from the bound state to a freely diffusible state, whereupon it can exert its biological activity. In tissue engineering, however, many of the bioactive molecules of interest are active in their immobilized state and some, such as growth factors, can be active either immobilized or freely diffusible. As such, in this chapter we shall use the term controlled release to refer to the provision of a bioactive molecule over time in a manner such that its biological activity can be productively harnessed. Thus, controlled release more precisely refers to the controlled provision of a biologically active molecule in its most appropriate state. For a molecule that induces adhesion and thereby cell differentiation, controlled release would refer to immobilization of that ligand and presentation on a surface for a prolonged duration. For a growth factor that signals intracellularly, it would refer to release of the factor into the diffusible state, so that it could be bound by cells and internalized.

The overall goal of controlled release it to present a biologically active molecule to the tissues at a concentration that is optimal – not so low as to be ineffective and not so high as to display toxic side-effects, i.e. in the so-called *therapeutic window*, as illustrated in Figure 15.1. If simply injected into the tissue site as a bolus, without incorporation into a biomaterial scaffold or matrix, release will usually occur as a sudden burst, perhaps well above the upper limit of the therapeutic window, and then rapidly decline to concentrations that are too low to be effective. If released

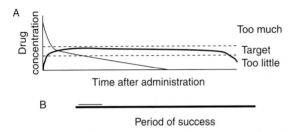

Figure 15.1 The goal of controlled release is to provide effective concentrations of a drug to a tissue space for an effective period of time. Most drugs are active in a so-called therapeutic window, above which concentration the drug demonstrates toxic side-effects and below which concentration the drug is ineffective. In the absence of controlled release (thin line in A), the drug is injected at too high a concentration, rapidly disperses in the tissue, passes through the therapeutic window at effective concentrations for a brief period of time (thin line in B) and then rapidly declines to too low a concentration to be effective. With controlled release (bold line in A), the drug increases in tissue concentration after implantation up to the therapeutic window, is maintained there for a desired duration and eventually is depleted from the implant. This places the drug within the therapeutic window for a much longer period of time (bold line in B).

slowly from a scaffold, a typical case in tissue engineering, the concentration of the drug in the vicinity of the implant would rise to within the therapeutic window, remain there for a prolonged duration while morphogenesis occurs and then gradually decline to zero as the factor is depleted from the scaffold.

A number of strategies have been explored and can be envisioned for the controlled release of biological signals in tissue engineering (Table 15.1). These include (i) simple admixing of factors within matrices with which they may have some biological affinity, (ii) entrapping factors within gel networks, (iii) entrapping factors within hydrophobic degradable polymers that either make up the matrix or that are used to form microparticles that are themselves entrapped within the matrix, (iv) binding factors to affinity sites that are conjugated to the matrix and (v) binding the factors directly to the matrix material.

15.1.1 Bioactive molecules of interest in tissue engineering

Before considering the approaches to controlled release that are likely to be successful in tissue engineering, it is important to consider what types of molecules might be important to release. For example, release of low-molecular-weight hydrophobic drugs, as might be used in contraception or cancer, requires vastly different materials types than might be used in release of proteins drugs, as might be used in diabetes or growth hormone deficiency. In tissue engineering, it is these latter biomolecules that are more of

Table 15.1 Drug delivery strategies commonly encountered in tissue engineering

Case	Binding or affinity	Example
Physical mixtures	none or physicochemical	growth factors in collagen
Gel networks	physical entrapment	growth factors in hydrogels
Polymer fibers and particles	physical entrapment	growth factors in microparticles
Specific affinity mixtures	biochemical	growth factors in heparin-containing matrices
Chemical immobilization	covalent	adhesion factors on scaffolds

interest – bioactive proteins. These proteins include growth factor morphogens, such as bone morphogenetic proteins (BMPs), transforming growth factor (TGF)-β and vascular endothelial growth factors (VEGFs), and adhesion factor morphogens, such as whole and domains of fibronectin, vitronectin and laminin, and peptide mimics of these proteins. These molecules are large (molecular weight 5000–2,000,000 Da), are water-soluble and are very sensitive to their environment. Even small changes in their environment, such as exposure to hydrophobic surfaces, can result in protein denaturation and loss of biological activity. Perhaps even more importantly, some of these denaturative changes in protein structure can create sites that are immunogenic and may lead to the formation of antibodies against the protein that is being therapeutically delivered. While these antibodies may not have clinical consequences, they may also be neutralizing, blocking the favorable biological interactions between the protein and its receptors, and thus render the implant less effective or ineffective altogether. As such, one must pay a great deal of attention to the details of the released molecule, to ensure not only that it is released, but that it is released in a native form and with the desired kinetics.

Growth factors are by far the drugs of highest current interest in tissue engineering. These molecules are ligands for cell-surface receptors that turn on complex cascades of signal transduction within the cell. Many, but not all, of the growth factors are dimers and bind to receptor tyrosine kinases, typically a receptor that is bound to one subunit of the growth factor phosphorylating the receptor that is bound to the other, and *vice versa*, to result in receptors that are phosphorylated in the active states and not in their inactive states. Phosphorylation results in different binding of other intracellular proteins, which then triggers the signal transduction cascade.

Examples of growth factors that are important in tissue engineering are BMPs, TGF-βs, platelet-derived growth factors (PDGFs), fibroblast growth factors (FGFs), the neurotrophins, the VEGFs and the insulin-like growth factors – a nonexhaustive list. The plural is used in the list for the names, because the names represent families of proteins, e.g. the angiogenic growth factors of the VEGFs consist of VEGF-A, VEGF-B, VEGF-C and VEGF-D, and moreover several isoforms of these growth factors exists, such as VEGF-A$_{110}$, VEGF-A$_{121}$, VEGF-A$_{165}$ and VEGF-A$_{189}$. The different family members are related by partial sequence homology and sometimes by overlap in receptor specificity, but may have vastly different function. For example, VEGF-A induces primarily hemangiogenesis (i.e. the formation of new blood vessels from existing ones), whereas VEGF-C induces primarily lymphangiogenesis (i.e. the formation of new lymphatic vessels). The different isoforms are typically different splice variants of the same parent sequence and may also have different signaling capacity. For example, VEGF-A$_{165}$ binds to both VEGF receptors 1 and 2 (among other receptors) as well as to cell-surface heparan sulfate, whereas VEGF-A$_{121}$ binds to VEGF receptors 1 and 2, but not cell-surface heparan sulfate, which leads to very different biological activities between the two isoforms.

Some of the growth factors mentioned above have already been turned into commercial tissue engineering products, e.g. PDGF-BB for healing chronic wounds in the skin (Regranex®; Johnson & Johnson) and BMP-2 for healing surgical and traumatic defects in bone (e.g. INFUSE®; Medtronic Sofamor Danek). In both cases, prolonged delivery of the growth factor is the key to efficacy. In the case of Regranex®, this is achieved by frequent readministration; in the case of INFUSE®, this is achieved by affinity between the growth factor and a biomaterial matrix, as will be discussed further below.

Adhesion factors are also important candidates for drug delivery in tissue engineering. These molecules are normally bound irreversibly to the extracellular matrix and serve a role of stress transmission between the cell's cytoskeleton and the extracellular matrix. The extracellular domains of transmembrane receptors bind to the adhesion factors and the intracellular domains interact, with the help of several other intermediary proteins, with the cytoskeleton. Since stress transmission is the ultimate function of receptor ligation, clearly the adhesion factor must be immobilized in the extracellular milieu. As such, these factors function as agonists (i.e. inducing a function in its desired form) in the immobilized state and as competitive antagonists in their diffusible state. Thus, in the case of these molecules controlled release means controlled immobilized presentation.

Adhesion factors of interest include fibronectin, vitronectin, fibrinogen, the laminins and thrombospondin – also a very nonexhaustive list. In addition to the whole proteins, subdomains within the proteins that contain the receptor-binding domains have been identified for most of these proteins, and these smaller domains can be expressed independently and employed. Moreover, still smaller receptor-binding domains within those protein domains have been identified and have been mimicked with small synthetic peptides. Among these, the most widely studied in tissue engineering is the tripeptide sequence Arg–Gly–Asp, or RGD in the one-letter code, from fibronectin and a number of other proteins, which bind to the integrin family of adhesion receptors. While the very small RGD sequence binds to a number of integrins without a great deal of specificity between integrins, extending the sequence with flanking amino acids or conformationally constraining it by cyclizing the sequence can both increase its affinity for the receptor and recreate some of the specificity between different integrin receptors that exists in the proteins themselves.

From the perspective of current tissue engineering, fibrinogen and fibronectin are by far the most important adhesion factors. Fibrinogen is the basis for fibrin – a material widely commercially used as a surgical sealant and adhesive, as well as experimentally in a number of cell transplantation applications. Here, fibrinogen forms the entire basis for the biomaterial scaffold (the fibrin scaffold is formed by polymerization of fibrinogen after its cleavage by the coagulation protease thrombin) and then is also one of the key adhesion factors in the scaffold. This fibrinogen is purified from blood plasma by a cryoprecipitation procedure, which also copurifies substantial amounts of fibronectin. This is why it is also listed as a commercially important protein at this date – the protein is not purified and added back to

fibrinogen preparations to make fibrin. Rather, it is copurified to produce a fibrin that contains substantial amounts of fibronectin, which is then one of the important active agents in promoting cell adhesion and migration within the fibrin scaffold. A number of other adhesion factors are under commercial development in tissue engineering, including members of the laminin family, vitronectin and peptides based on the RGD sequence.

15.1.2 Modes of controlled release

Drugs may be incorporated within biomaterials in a homogeneous or a heterogeneous manner, as shown in Figure 15.2(A). When incorporated in a homogeneous manner, an implicit requirement is that the drug be somehow soluble in the matrix material or, at least, soluble in its precursors. In this way, the drug can be homogeneously dissolved in the biomaterial and, although a solid, the matrix acts as a solvent for the drug; the entire complex of drug and matrix are referred to as a *solid solution* of drug within the matrix biomaterial. Given that the drugs of interest in tissue engineering are typically water-soluble and macromolecular, this situation usually occurs only when the biomaterial is very hydrophilic. Such biomaterials then absorb large amounts of water and are referred to as hydrogels. Hydrogels may either be microporous, e.g. composed of hydrophilic nanoscopic fibrils or cellular structures, such as a fibrin hydrogel, or they may be nanoporous, composed of a homogeneous network of water-soluble polymer chains cross-linked into a solid by covalent or physicochemical bonds.

When hydrophobic polymers are used, i.e. those not forming hydrogels, then the drugs that are of most interest in tissue engineering are not soluble within them and they form a heterogeneous structure. Domains of drug are dispersed throughout the biomaterial scaffold heterogeneously as nano- or microdomains. This situation would exist with polymers such as the degradable polyesters poly(lactic acid), poly(glycolic acid) (PEG) and poly(ε-caprolactone),

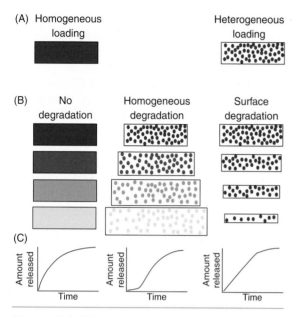

Figure 15.2 (A) Protein drugs can be loaded within hydrophilic polymer matrices in a homogeneous manner (left), but within a hydrophobic polymer matrix they are loaded in a heterogeneous manner of protein particles or highly concentrated protein in pools of buffer in the polymer (right). (B; left) Protein may be released from polymer matrices by simple diffusion in the absence of polymer degradation, which can occur with hydrophilic gel matrices. In this manner, the rate of diffusion of protein controls the rate of release. (B; center) Alternatively, the polymer may degrade homogeneously, swelling as it degrades, releasing drug either homogeneously (not shown) or heterogeneously (shown) dispersed throughout the matrix. In this manner, the rate of degradation controls the rate of release. Homogeneous degradation often occurs with hydrophilic gel matrices. (B; right) With hydrophobic degradable polymers, surface degradation can occur, which releases the heterogeneous particles or pools of protein as they are exposed to the surface of the polymer implant. (C) With diffusion-controlled release, release rates are not linear (left); however, with homogeneous degradation (center) and surface degradation (right), near-linear release profiles can be obtained.

and their copolymers. If the protein drug were incorporated within the polymer scaffold as a dry powder, such as produced by spray-drying or by lyophilization, then pools of dry or highly concentrated protein would exist dispersed throughout the scaffold. It is also possible to incorporate the protein drug as an aqueous solution dispersed throughout the polymer scaffold, using an emulsion process, resulting in a polymer scaffold with pools of dissolved protein drug dispersed heterogeneously throughout (see Section 15.4 for more detail on this method and others).

Just as there is more than one mode of drug incorporation (homogeneous, heterogeneous), there is also more than one mode of drug release, depending upon the mode of drug interaction with the biomaterial and the mode of material degradation, as shown in Figure 15.2(B). A first mode of release may be found in the case of hydrogel matrices that have mesh sizes that are sufficiently large as to not present a substantial resistance to drug diffusion, e.g. mesh sizes larger than around 100 nm. In the case of highly porous hydrogels, the drug is not entrapped within the matrix and drug release thus proceeds independently of biomaterial degradation; the biomaterial may degrade quickly, or slowly, or not at all, but drug release proceeds more-or-less independently of these considerations. Such a situation is referred to as *diffusion-controlled release*. In this case, the diffusion coefficient of the drug in the hydrogel matrix is constant over time, since the hydrogel structure does not change during the process of release. Since the concentration of drug remaining in the hydrogel matrix decreases over time, due to the release process, the rate of release decreases over time and a constant rate of release is unachievable. When relatively short-term release characteristics are desired, this approach may be appropriate. The biomaterial plays its main role as a vehicle for local, if transient, administration of the drug. The biomaterial may also play a second role, as a matrix for cell transplantation or cell invasion, independent of its role in drug release.

A second mode of release may be found in the case of hydrogel matrices with smaller mesh sizes. The polymer network that composes the gel may efficiently entrap the protein drug and hinder its diffusion from the material. In this case, the drug cannot escape from the matrix until the matrix begins to degrade. As hydrolysis occurs in the material, the architecture of the polymer network is cleaved, resulting in increase in the mesh size and an increase in the overall swelling of the material. As this occurs, the drug becomes less entrapped and may begin to diffuse from the matrix. Given that hydrolysis occurs throughout the material and drug release then ensues from the entire material, this situation is referred to as *homogeneous degradation*. Such situations would exist in a hydrolytically degradable hydrogel, for example. Wherever water is in the hydrogel structure, hydrolysis occurs. Since the hydrogel is highly swollen with water, water is equally present throughout the hydrogel matrix, and as such degradation proceeds equally fast throughout the matrix. This leads to porosity of the hydrogel network that increases over time and then diffusive release of drug that correspondingly increases over time. In this case, the diffusion coefficient of the drug in the hydrogel matrix starts out at a very low value – too low to demonstrate substantial release. As the gel degrades, the diffusion coefficient increases over time, leading to increasing release rates. However, as release occurs, the concentration of drug remaining in the hydrogel matrix decreases over time, somewhat offsetting this increasing release rate. When these two effects are balanced, release characteristics can be obtained that are nearly *zero-order*, i.e. that are nearly constant over time. The extent to which this is true depends upon the extent to which the two opposing effects (diffusion coefficient increasing over time and concentration in the matrix decreasing over time) are appropriately counter-balanced. When prolonged release is needed, e.g. for slow morphogenetic processes, this mechanism of controlled release can be very appropriate.

A third mode of release may be found with the use of more hydrophobic materials. In the case of such hydrophobic polymers used as biomaterial scaffolds, water will naturally have much better access to the surface of the scaffold than the internal domains of the scaffold, leading to polymer hydrolysis in a heterogeneous manner, first on the surface, and then deeper in as the surface erodes and exposes new polymer to the surface, much like a bar of hand soap dissolves. Such degradation is referred to as *surface degradation*. With surface degradation, drug is likewise released from the biomaterial scaffold layer by layer, and the rate of drug release depends sensitively on the rate of polymer degradation and also the morphology of the implant. For example, the surface area of a polymer film does not change very much as the film degrades and thus the rate of release in such morphologies is close to constant over time, which is referred to as *zero-order release*. In cases of release by surface degradation, it is easer to achieve zero-order release than with other release mechanisms.

Finally, a fourth mode of controlled drug release may be found in the situation in which the bioactive molecule is not desired to be released at all, i.e. *biomolecule immobilization*. The principle example of when this is desired is with adhesion proteins and peptides that are active in their bound, rather than free, state. Here, both surface and bulk immobilization must be considered. If an adhesion molecule is being immobilized on the surface of a surface-degrading scaffold, to support cell attachment or to trigger cell differentiation, clearly it must remain on the surface for the desired duration. If the surface layer is removed from the scaffold too rapidly, then the biomolecule will not be present sufficiently long to carry out is desired signaling function. In this case, it would have to be immobilized throughout the bulk of the material, so that when one polymer surface layer is removed, new adhesion molecule is exposed. If the scaffold degrades only slowly, compared with the timescale of cell differentiation and tissue formation within the scaffold, then these considerations are less important. Likewise, if the scaffold degrades by bulk degradation, then the rate of removal of the adhesion molecule from the polymer surface will likewise be slow, even if it is not immobilized throughout the bulk. Thus, the necessity of immobilization throughout the bulk, or the satisfactory nature of surface immobilization, depends on the details of the application and the material.

Having now introduced the topic to consider what types of drugs are interesting to release (mostly proteins and peptides) and the different modes of controlled release that may be encountered (diffusive release, release with bulk or surface degradation, no release at all), we can now turn our attention to specific examples, as indicated in Table 15.1.

15.2 Bioactive factors admixed with matrices

In some cases, the biomolecules that form biomaterial matrices and scaffolds may serve directly as a carrier for controlled delivery of a bioactive molecule. A number of possibilities exist by which the scaffolds may prolong release of diffusible protein drugs like growth factors. The scaffold or matrix may merely serve to localize the placement of the drug as a diffusion source, as described above in regard to diffusion-controlled release. Alternatively, even a relatively small affinity between the biomaterial matrix and the factor may results in a useful prolongation of release, as has been observed with collagen and gelatin matrices. Finally, for drugs that display relatively poor solubility, the matrix or scaffold can contain a continuous source of slowly dissolving factor, also providing sustained release for an effective duration.

Type I collagen makes up a large fraction of connective tissues such as skin and tendon, and bovine skin from closed and controlled herds makes up a useful supply of type I collagen for use as biomaterial matrices in tissue engineering (Clark *et al.*, 1989; Koide, 2005; Lee *et al.*, 2001). The material may be used in granular form, as is used in the commercial product OP-1® Implant in the US and in Europe (Stryker Biotechnology), which consists of osteogenic protein-1 (also known as BMP-7) dispersed in granules of type I collagen. This product has been approved for clinical use in tibial nonunions after failed autograft. Bovine type I collagen may also be used in the form of a macroporous sponge as a drug delivery and cell ingrowth matrix. Such sponges may be placed in bony defects, as well as within medical devices such as spinal fusion cages – hollow structural devices used to maintain the spacing

between the vertebral bodies when the intervertebral disc is removed and bony fusion between the two vertebrae is sought. BMP-2 has been developed as a product for such bone growth applications, in the product INFUSE® in the US and InductOs® in Europe (Medtronic Sofomor Danek). BMP-2 in such collagen sponges has been shown to remain at the implantation site *in vivo* for at least 2 weeks, using radiolabeled protein (Seeherman and Wozney, 2005).

BMPs in collagen matrices have gone forward to clinical testing in man, after substantial testing in preclinical animal models (De Biase and Capanna, 2005; Seeherman *et al.*, 2002; Seeherman and Wozney, 2005). For example, BMP-2 in bovine type I collagen sponges was evaluated in two clinical studies (Boden *et al.*, 2000; Burkus *et al.*, 2003) applied within cylindrical interbody spinal fusion cages, comparing the cage plus the BMP-2 in a collagen sponge versus the standard of care at the time, which was the spinal fusion cage with bone autograft taken from the iliac crest. Bone formation in and around the cage is illustrated in Figure 15.3. The BMP-2-treated group performed as well in terms of fusion between the two vertebrae as the group treated with standard of care. This demonstrated a substantial advantage to treatment with BMP-2, in that it was not necessary to harvest bone from the iliac crest – a procedure that is associated with significant morbidity. BMP-2 in collagen sponges was also evaluated in treatment of open tibial fractures with an intermedulary nail, compared to treatment with the nail alone and soft tissue management in both cases (Govender *et al.*, 2002). The group treated with BMP-2 in a collagen sponge was demonstrated to heal faster and with fewer complications than the group treated with the standard of care. BMP-2 in granular collagen matrices has been compared to bone autograft in both spinal fusion indications with interbody fusion cages (Johnsson *et al.*, 2002) and in treatment of tibial nonunions (i.e. fractures that failed to heal and resulted in a chronic lack of union between the proximal and distal tibial portions) (Friedlaender *et al.*, 2001), and in both cases the BMP-7 in collagen granules demonstrated efficacy that was statistically

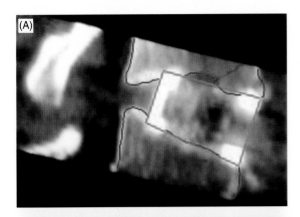

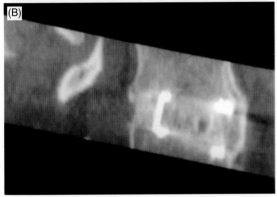

Figure 15.3 Radiographic image of vertebrae and the intervertebral space in a human patient after treatment with a spinal fusion cage filled with BMP-2 contained within an absorbable collagen sponge. Thin-cut computed tomography (1-mm) sagittal reconstructions are shown. Mineralized tissue is white and the metal cage is dense white. (A) A section 2 days after implantation of the cage. For clarity, the edges of the two vertebrae are outlined in red and the edges of the fusion cage are outlined in green. No bone is present within the cage or on either end of it. (B) A section 2 years after implantation of the cage. Bone is present both within the cage, as well as at both ends. Reproduced with permission from Burkus, J.K., Dorchak, J.D. and Sanders, D.L. (2003). Radiographic assessment of interbody fusion using recombinant human bone morphogenetic protein type 2. *Spine*, 28: 372–377.

equivalent to bone autograft. These clinical studies have led to approval for use of BMP-2 in treatment of open tibial fractures and in spinal fusion, and of BMP-7 in treatment of tibial nonunions after failed autograft, products that are among the most successful controlled release products in the field of tissue engineering.

In the above examples, we see that bovine type I collagen can be used successfully as a carrier of growth factors in tissue engineering, here BMPs in bone tissue repair. This is true even though the factor is not retained in the matrix very long – a fact that requires the use of relatively high doses of the growth factors in the successful therapies. In addition to the lack of highly prolonged delivery, there may be immunological consequences of the use of bovine proteins. For example, in the clinical evaluation of BMP-7 in bovine type I collagen granules, 5% of the patients developed antibodies for the bovine collagen and 10% of the patients developed antibodies against the recombinant human BMP-7 (Friedlaender *et al.*, 2001). One could hypothesize that the bovine collagen acted as an adjuvant to enhance immune reactions against the recombinant human protein and these worries create a driving force for the development of either human-derived or synthetic matrices that would function as well. Nevertheless, the success of the currently existing bovine type I collagen matrices should not be underestimated.

Human-derived proteins are being developed as an alternative to bovine-derived proteins. While human collagen is available for clinical use (e.g. purified from placentas), its expense is such that it is not widely used for implants that require high masses of protein. Rather, human fibrin is widely available and broadly utilized, purified from human blood plasma donations (Albala and Lawson, 2006). Human fibrin has the advantage that it can be injected as a liquid: double-barrel syringes are usually used, in which a fibrinogen stream from one syringe is mixed via a static mixer with a thrombin stream from the second syringe and coagulation into a gel occurs over a period of seconds to minutes, depending on the concentration of thrombin in the final mixture.

As an example of another sustained release strategy for these admixtures in which the gel does not provide a substantial permeability barrier (the network structure of collagen and fibrin gels is very open compared to the size of the proteins being released), proteins of low solubility can be used. We present another example with BMP-2 to illustrate this point. Recombinant BMP-2 when produced in mammalian cells is glycosylated, whereas BMP-2 produced in engineered bacterial cells is not glycosylated. The gylcosylation of BMP-2 (and other proteins) has an effect to increase its solubility and nonglycosylated bacterially produced BMP-2 is much less soluble than mammalian cell-produced BMP-2. Delayed release due to this limited solubility is illustrated in Figure 15.4. This has been utilized to produce BMP-2 that is retained in fibrin matrices for prolonged periods relative to mammalian cell-produced BMP-2, which demonstrated in clinical studies in dogs (veterinary patients) healing in bone fusions that was statistically equivalent to that obtained with bone autografts (Schmoekel *et al.*, 2004). A typical example of healing in these animals is illustrated in Figure 15.5. This principle of controlled release would presumably also work in other highly porous biological matrices, such as type I collagen matrices.

As an alternative to native protein matrices being used as admixtures with protein drugs such as growth factors, denatured protein matrices can also be used. The main example of this is bovine-derived gelatin. Gelatin is produced by denaturation of type I collagen, producing an unfolded protein ensemble that self-assembles into a gel by hydrophobic interactions. The gel that so forms, identical to that in edible gelatin, is not terribly stable at physiological temperature and one can achieve much more stable gels by chemical cross-linking with bifunctional agents such as glutaraldehyde, which reacts with amine functionalities in the denatured protein chains coming mostly from lysine residues. Cross-linked gelatin matrices have been widely utilized in animal studies with proteins such as BMP-2, both with and without inorganic particles of tricalcium phosphate to assist in stability during bone formation (Takahashi *et al.*, 2005; Yamamoto *et al.*, 2006). In this case, it is likely that retention of

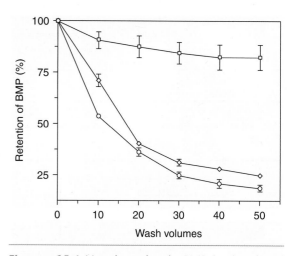

Figure 15.4 Nonglycosylated BMP-2 (produced in bacteria) demonstrates much less solubility than glycosylated BMP-2 (e.g. native or produced in mammalian cells). This limited solubility has been used to prolong release of protein from fibrin surgical matrices. Release *in vitro* is shown. Release of the nonglycosylated BMP-2 (squares) is very slow compared to release of the glycosylated protein (diamonds). When the nonglycosylated protein was complexed with heparin (circles), its release was accelerated and became as fast as the glycosylated protein. The limited solubility of certain drugs can be used as a tool to prolong release from matrices. Reproduced with permission from Schmoekel, H., Schense, J.C., Weber, F.E., Gratz, K.W., Gnagi, D., Muller, R., and Hubbell, J.A. (2004). Bone healing in the rat and dog with nonglycosylated BMP-2 demonstrating low solubility in fibrin matrices. *J Orthoped Res*, 22: 376-381.

the growth factor within the gelatin matrix is due to many low-affinity interactions due to electrostatic attraction with the gelatin matrix. This has been more clearly demonstrated in a study with two hepatocyte growth factor isoforms interacting with gelatin. In this study, native hepatocyte growth factor was demonstrated to be released very slowly from cross-linked gelatin matrices, almost as slowly as material degradation. By contrast, when a highly charged region of the protein was deleted to make a less charged variant protein, that variant was released by simple diffusion

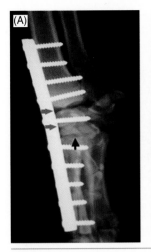
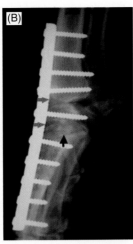

Figure 15.5 Nonglycosylated BMP-2 was used in fibrin surgical matrices to fuse the carpal joint complex in canine patients in pancarpal arthrodesis procedures. Normally, the joint is stabilized with a surgical plate (shown on the radiographs) and autologous bone is implanted within the joint space to induce joint fusion. Here, fibrin with nonglycosylated BMP-2 was injected in the joint space, rather than bone autograft. (A) Immediately after the procedure, no bone is visible within the three joints of the complex (arrowheads). (B) After healing, the joints were completely fused, as demonstrated by radio-opacity within the joint space (arrowheads). In a study comparing nonglycosylated BMP-2 in fibrin, arthrodesis was as effective as in a control cohort treated with both autograft. Reproduced with permission from Schmoekel, H., Schense, J.C., Weber, F.E., Gratz, K.W., Gnagi, D., Muller, R., and Hubbell, J.A. (2004). Bone healing in the rat and dog with nonglycosylated BMP-2 demonstrating low solubility in fibrin matrices. *J Orthoped Res,* 22: 376–381.

from the matrix, much faster than the native isoform (Ozeki and Tabata, 2006). This demonstrates an additional principle for controlled release in highly porous biopolymer matrices, the use of numerous but low-affinity electrostatic interactions between the protein drug and the matrix to obtain sustained release.

In addition to low-affinity interactions, some biopolymeric matrices have the potential to exhibit high-affinity biophysical interactions for some protein drugs, especially some growth factors. This principle can be illustrated with fibrin. Fibrin forms in the body at sites of injury and accordingly some growth factors have evolved to be immobilized in such fibrin clots (Mosesson, 2005). Cells such as blood platelets are naturally present within fibrin clots and these cells release large quantities of stored growth factors to induce healing. The best studied of these proteins is FGF-2, which contains a very high-affinity binding site for fibrinogen and fibrin, with a dissociation constant lower than 10 nM (Peng *et al.*, 2004; Sahni *et al.*, 2003). This affinity immobilizes the FGF-2 within fibrin matrices, protects it from proteolytic degradation (Sahni and Francis, 2000) and may potentiate its activity in biological processes as complicated as induction of angiogenesis (Sahni *et al.*, 1999). Other growth factors also bind to fibrin, such as VEGF-A$_{165}$ (Sahni and Francis, 2000). Fibrin is not the only biological matrix molecule with activity to bind certain growth factors: FGF-2 has also been reported to bind to type I collagen (Kanematsu *et al.*, 2004), and BMP-7 has been reported to bind to type II and type IV collagen (Scheufler *et al.*, 1999; Sieron *et al.*, 2002).

In this section, we have seen that relatively simple approaches to controlled release of protein growth factors as mixtures in biopolymer matrices can be very useful. Examples of such biopolymers are type I collagen granules and sponges, cross-linked gelatin matrices, and fibrin. The scaffold or matrix may merely localize the drug or it may serve to delay release of the drug by subtle electrostatic affinity. In the case of particular matrix–growth factor pairs, such as fibrin and FGF-2, naturally evolved high-affinity interactions may result in prolonged retention of the drug within the matrix. Admixtures of growth factors in biopolymer matrices have already resulted in commercial products in tissue engineering, such as those used in bone repair.

15.3 Bioactive factors entrapped within gel matrices

It is possible to make hydrogel matrices have characteristics and porosities such that protein permeation is directly inhibited by entrapment. This may be done in two ways: either the network structure of the hydrogel matrix may be sufficiently tight as to resist protein diffusion or the polymer chains in the network may drive the protein out of solution. Both concepts are described in this section.

Homogeneous hydrogels have a structure not unlike a three-dimensional (3D) fishing net, with polymer chains connecting cross-links within the network, which may be covalent or physical (Anseth *et al.*, 2002; Kissel *et al.*, 2002; Martens *et al.*, 2004). The molecular weight of the chains that connect the cross-links and the basic structure of the polymer chain determine the distance between the cross-links, and thus the average pore size within the material (Lustig and Peppas, 1988). These pores should not be thought of as static pores with rigid walls, rather they are formed dynamically and represent a time-average, like the porosity created by the strings of a fishing net under water, sometimes looser and sometimes tighter as the polymer chains self-diffuse. When the average distance between cross-links within the hydrogel network becomes of the same order of magnitude as the diameter of the protein drug, diffusion of the drug becomes significantly restricted and the drug becomes entrapped within the hydrogel. This situation can be useful in controlled release in tissue engineering.

Protein can be entrapped within small hydrogel particles and these particles can be distributed throughout a matrix or scaffold to provide a sustained source of a diffusible bioactive factor such as a growth factor. One such hydrogel structure is obtained when water-soluble polymer chains are attached to a polymerizable species at their termini and then this macromonomer is further polymerized. An example of this is with PEG, a polymer that is highly soluble in water. This chain can be reacted with ester monomers, so as to make a still water-soluble ABA block copolymer, where the A block is an oligomer (e.g. lactic acid) and the B block is the central PEG chain, as illustrated in Figure 15.6. This polymer is still water-soluble and as such it is not useful in sustained release (unless the A blocks are sufficiently large to drive the entire polymer out of solution). If the chain termini

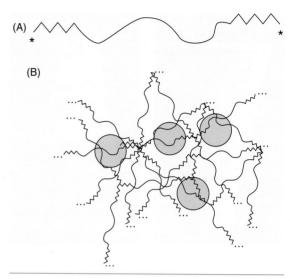

Figure 15.6 A number of hydrogel systems have been devised to create nanoporous networks for entrapment of protein drugs. One material is formed by polymerization of a water-soluble macromonomer (A), consisting of a central large chain of PEG, flanked by short degradable regions of oligo(lactic acid), which are then end-caped with polymerizable acrylate groups. The central chain provides solubility and creates the strings in the meshwork, the flanking regions make the strings degradable, and the end-capping groups create knots in the 3D network to tie the strings together. Such a network is shown in (B). When the distance between cross-links is of the same order as the diameter of the protein, the protein drug may be entrapped. As the strings in the network degrade, this distance between cross-links increases and the protein may escape.

are attached to polymerizable acrylate groups (Ac), to form Ac–oligo(lactic acid)–PEG–oligo(lactic acid)–Ac, then the resulting macromonomer can be polymerized to form a cross-linked hydrogel (the cross-links, or knots in the fishing net, are chains of Ac groups attached to each other) connected by PEG chains (the strings in the fishing net), with a hydrolytically sensitive link between the cross-links [the oligo(lactic acid) blocks]. These polymers have been created (Sawhney *et al.*, 1993) and the polymerization can be carried out in water under sufficiently mild conditions to not degrade proteins incorporated in the hydrogels. If the molecular weight of the PEG is sufficiently low, then sustained protein release can be obtained by entrapment of the protein within the hydrogel network (Mason *et al.*, 2001).

The incorporation of proteins within a 3D network is not without its tricks. With a 2D net, like a fishing net, the net can be wrapped around the objects within – the net can be preformed and then deformed around the objects. However, with a 3D net, somehow the objects must be placed within the net; if they are too large (relative to the distance between cross-links) to get out, however, then they are also too large to get in! In the above example, this dilemma was solved by constructing the net around the objects – the protein was dissolved in the macromonomer solution and the macromonomer was polymerized to form the net *in situ*. Other chemistries have also been developed to accomplish these ends. For example, PEG has been derivatized to form a chemically reactive polymer; PEG can be branched, like a four- or eight-arm starburst, and when these branched reactive polymers are cross-linked by addition of a corresponding chemically reactive group, a cross-linked hydrogel will form (Elbert *et al.*, 2001; Metters and Hubbell, 2005; Schoenmakers *et al.*, 2004; van de Wetering *et al.*, 2005). Here, the knots in the network are the central portions of the starburst polymers, and the strings between the knots consist of one PEG chain, the cross-linking molecule and the second PEG chain. In the example cited, the polyethylene chains are reacted with acrylate groups and these are cross-linked to small-molecular-weight dithiols like dithiotreitol or a low-molecular-weight PEG diacrylate. When the networks are very loose, incorporated protein is released by simple diffusion; however, when the meshwork is tighter, because the molecular weight of the arms on the PEG chains is less, degradation-controlled release can be obtained. Control over release rate via meshwork architecture is illustrated in Figure 15.7.

In the above examples, sustained release was obtained when the network side was of the same

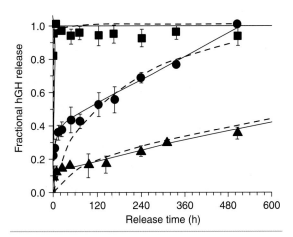

Figure 15.7 The network architecture of cross-linked PEG hydrogels can be adjusted by using different molecular weight PEGs. Here, release of human growth hormone (hGH) from PEG gels formed from an eight-arm, 2-kDa PEG octaacrylate (triangles – a very tight network), an eight-arm 10-kDa PEG octaacrylate (squares – a very open network) and 50:50% blend of the two (circles – an intermediate network). The PEG multiacrylates were cross-linked by reaction with equiequivalent dithiothreitol. The solid lines represent fitting of the experimental data and the dashed lines represent results of a predictive model based on measured network characteristics. It is seen that the protein release rate can be controlled by careful selection of the network architecture. Reproduced with permission from Wetering, P. van de, Metters, A.T., Schoenmakers, R.G. and Hubbell, J.A. (2005). Poly(ethylene glycol) hydrogels formed by conjugate addition with controllable swelling, degradation, and release of pharmaceutically active proteins. *J Control Rel*, 102: 619–627.

order of magnitude as the protein diameter. If this is the case, of course the network porosity is very, very small relative to the size of a cell and cells are unable to penetrate the hydrogel network. This is why these sorts of delivery vehicles would commonly be presented as microparticles dispersed through another scaffold or matrix. An alternative would be to make a microporous scaffold out of these nanoporous

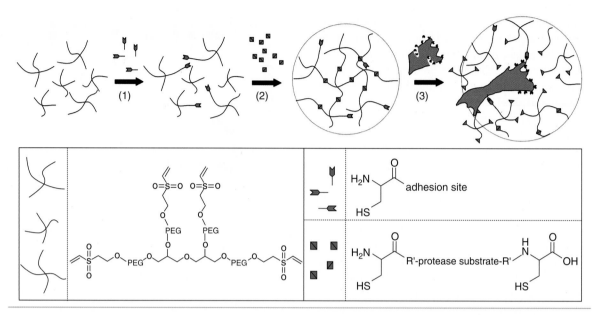

Figure 15.8 Biomimetic gel matrices can be constructed by cross-linking of reactive branched PEGs, e.g. terminated with vinylsulfone moieties, and dithiol peptides (red squares), with a cysteine residue on each end of the peptide. As the PEG chain contains more than two reactive groups and the peptide contains two, a cross-linked network forms in step (2) of the reaction. In step (1), an adhesion peptide, such as the RGD peptide, can be incorporated into the network as a peptide that contains a single cysteine residue. If the cross-linking peptide is designed as a substrate for cell-expressed enzymes, such as matrix metalloproteinases, then cells can locally degrade the matrix and migrate through it, as in step (3) of the reaction. Given that the enzymes are localized near the surfaces of the cells, matrix degradation occurs only at the surface of the cells as well, degrading the matrix as the cells migrate though it. Reproduced with permission from Lutolf, M.P., Raeber, G.P., Zisch, A.H., Tirelli, N. and Hubbell, J.A. (2003). Cell-responsive synthetic hydrogels. *Adv Mater,* 15: 888–892.

materials, such that cells could migrate through the micropores (say, of diameter 10–100 μm) while the proteins diffused through the nanopores. A second alternative would be to somehow endow the nanoporous network with an ability to be degraded by cells as they migrate. When cells migrate, they activate a number of proteolytic cascades at their surface (Lutolf and Hubbell, 2005). If one could somehow construct the hydrogel network to be sensitive to these enzymes, then cells could create their own micropores in the matrix as they migrate, allowing the protein to be released from the nanoporous material or to be released from it as the matrix is degraded by the cell-associated enzymes.

This concept of a cell-degradable, nanoporous matrix has been achieved (Gobin and West, 2002;

Lutolf *et al.*, 2003a–c; West and Hubbell, 1999). For example, peptides can be used as cross-linkers for correspondingly reactive PEG chains, the peptides being designed to be enzyme substrates for cell-activated enzymes such as plasmin (Pratt *et al.*, 2004) or matrix metalloproteinases (Lutolf *et al.*, 2003a), as illustrated in Figure 15.8 (Lutolf *et al.*, 2003b). These hydrogel matrices offer a great deal of design flexibility: if the peptides are designed to be fast-degrading substrates, then cell migration through the materials can be very fast; if slower-degrading substrates are used, then cell migration and material degradation (and thus release) slower. In one example, BMP-2 was incorporated into such matrices and they were implanted in bone defects in rats, then bone repair was induced (Lutolf *et al.*, 2003c), as shown in Figure 15.9. In the above example

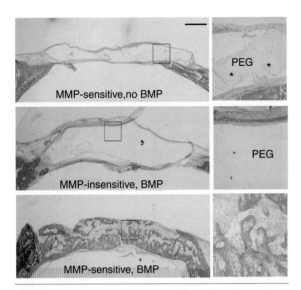

Figure 15.9 Matrix metalloproteinase-sensitive biomimetic gels, as illustrated in Figure 15.8, were used to release BMP-2 in calvarial defects in rats. (Upper panels) A negative control in which a matrix metalloproteinase-sensitive gel was implanted, however without BMP-2, demonstrated no bone formation. (Middle panels) A negative control in which BMP-2 was implanted, in a gel that was insensitive to MMPs, demonstrated no bone formation. The gel was not degraded or infiltrated by cells. (Lower panels) The positive, in which a gel was used that was sensitive to MMPs and that contained BMP-2, demonstrated matrix infiltration and good bone formation (green in the histology). All gels contained the RGD sequences as an adhesion moiety. All images are at 3 weeks postoperatively. Reproduced with permission from Lutolf, M.R. (2003). Repair of bone defects using synthetic mimetics of collagenous extracellular matrices. *Nature Biotechnol,* 21: 513–518.

of BMP-2 release from an enzymatically sensitive gel, the protein was incorporated throughout the gel in an entrapped state. As cells migrated into the gel, degrading it enzymatically as they proceeded, they also liberated the growth factor into the diffusible state, micrometer by micrometer (Lutolf *et al.*, 2003). Such a release concept can be very advantageous for morphogenesis into a matrix, in that the rates of release of

the bioactive are automatically tied to the rate of cell infiltration into the matrix.

In this section, we have seen that proteins that are essentially hydrophilic in nature can be entrapped within hydrophilic networks. The network structure, if sufficiently small, can resist the diffusion of the protein, and as such, as the network degrades and becomes thus more permeable, controlled release from the network can be obtained. Network degradation can be hydrolytic, or it can be proteolytic.

15.4 Bioactive factors entrapped within hydrophobic scaffolds or microparticles

Degradable hydrophobic polymers such as poly(lactic acid), poly(glycolic acid), poly(ε-caprolactone) and their copolymers have been widely explored for controlled release of proteins, for applications that include tissue engineering. These polymers can display distinct advantages for many applications – these advantages include the long duration over which controlled release may be sustained and the ability to integrate drug delivery directly into a structural scaffold built from the same polymer. Since the polymer is hydrophobic and the drug usually hydrophilic, the drug is distributed heterogeneously throughout the polymer and the drug escapes as the polymer degrades. In the paragraphs below, we describe several means by which protein drugs have been incorporated into hydrophobic polymers, whether as microparticles distributed throughout a scaffold or within a scaffold itself directly.

One of the easiest ways to incorporate protein drug within a polymeric scaffold is by simple adsorption to the scaffold surface. While this does not allow great control over the release characteristics of the drug, useful sustained release and efficacy can sometimes be obtained when the therapeutic window of the drug is relatively wide. A good example has been provided in scaffolds intended for bone regeneration (Borden *et al.*, 2002, 2004; Kofron *et al.*, 2004). Polyester microparticles, 600–700 μm in diameter, were compressed within a metal mold, which was then heated to a temperature in excess of the polymer's glass transition temperature. At such temperatures, molecular mobility within the polymer is enabled, allowing chains from one microparticle to diffuse into the surface of neighboring microparticles. When the system is cooled, this gives a sintered solid, with porosity remaining from the interstitial space between the microparticles, which creates a very high surface area within the implant per unit volume. This surface can be used to adsorb protein, at surface concentrations up to 100–500 ng/cm². Given that many growth factors are active at amounts from 10 to 100 ng, this can represent a large quantity of the growth factor. One favorable feature of such porous solids is that they can be very strong and thus can be used as structural implants in repair of load-bearing tissues, such as bone. When BMP-7 was incorporated into such porous sintered solids and used for bone repair in the rabbit, very effective bone healing was observed. When cells from the bone marrow were also seeded within the scaffolds, even more effective bone healing was observed. This example demonstrates that growth factor can be incorporated by simple means such as adsorption (because very little of the growth factor is in principle required), that the drug delivery vehicle can be the same as the scaffold and that such approaches can be useful for cells invading a matrix as well as in differentiation of cells that are transplanted into a matrix.

More sophisticated approaches for drug incorporation can also be undertaken, with correspondingly better control of the controlled release characteristics of the construct. For example, the protein drug can be formulated as a dry powder in a finely distributed state, such as by lyophilization into a fine cake and then micronizing that cake into small particles in a ball mill. Alternatively, the protein can be spray-dried into fine particles directly. Processes have been described for incorporation of such powders directly into polymer scaffolds (Lu *et al.*, 2003) – see also 'Classical Experiment: Controlled protein release from hydrophobic polymers' describing the seminal work of Robert Langer on the topic. BMP-7 was suspended in degradable polyester polymer in an organic solvent, films were cast and the solvent was evaporated to result in polymer films containing the polymer fine particles of the protein drug distributed throughout. When these films with BMP-7 where seeded with cells isolated from muscle, the growth factor enhanced their osteogenic differentiation and these cell-seeded polymer films were useful in repair of bone in rabbit models. This example demonstrates that protein drug can be loaded as a particulate discontinuous phase within the degradable hydrophobic polymers, and that these constructs can be useful in cell differentiation and tissue repair.

Classical Experiment 1
Controlled protein release from hydrophobic polymers

In the 1970s, many investigators, both academic and commercial, were working on the delivery of hydrophobic drugs from hydrophobic degradable and nondegradable materials; however, very few investigators were working on controlled release of protein drugs. In a seminal paper in 1976 in *Nature* (Langer and Folkman, 1976), this limitation was decisively broken. Robert Langer, now a professor and world-renowned researcher in drug delivery at the Massachusetts Institute of Technology and then a postdoctoral fellow working with Professor Judah Folkman at Harvard Medical School, wrote of his results on controlled release of proteins from hydrophobic polymers,

both *in vitro* and *in vivo*. The method that Langer and Folkman used was simple and elegant, and has been used by countless researchers around the world and in numerous clinical products to date. The method takes advantage of the fact that many polymers are soluble in organic solvents, but that proteins are generally not. If a protein is dissolved in a buffer and then a water-miscible organic solvent (such as ethanol) is mixed into the water, denaturation of the protein typically occurs, where it partially unfolds and perhaps permanently loses some or all of its biological activity. However, if the protein is mixed as a solid particle directly into an organic solvent, it cannot be denatured, because it has no molecular mobility in the solid state. In the method developed by Langer and Folkman, protein particles are mixed into an organic solution of a hydrophobic polymer, such as ethylene–vinylacetate copolymer in CH_2Cl_2. The solvent is selected to be highly volatile, to facilitate subsequent evaporation of the polymer to a film or some other form. This was done with a number of proteins to understand how the film fabrication influences release. If the protein particles are very few in number and are completely surrounded by polymer, then little protein release occurs. However, if the protein particles are denser and make contact with each other, then a slowly dissolving tortuous path of protein may exist within the polymer, allowing access of water outside the sample to the protein-rich interstices within the polymer, this in turn allowing slow release. In the experiments reported in 1976, release of up to 100 days was observed and the rate of release depended on the molecular weight, and thus the diffusion coefficient, of the test protein, as illustrated in the figure. To test this concept *in vivo*, Langer encapsulated an angiogenic protein within ethylene–vinylacetate pellets and implanted these in the corneas of rabbits – a tissue in which vessels can be easily visualized. The protein released from these pellets induced angiogenesis from the sclera into the cornea in a manner that was dependent on the protein being released. One of the reasons

that this method has been so successful is that it so easy to carry out and to scale into an industrial process.

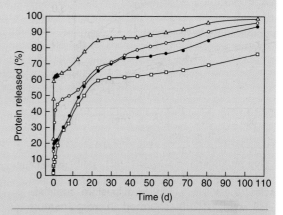

Proteins released from ethylene–vinylacetate copolymer demonstrated that long-term release could be obtained and that the rate of release depended upon the molecular weight, and thus the diffusion coefficient, of the probe protein. Lysozyme (14 kDa, triangle), soybean trypsin inhibitor (21 kDa, open circle), alkaline phosphatase (88 kDa, closed circle) and catalase (250 kDa, square) were used. Reproduced with permission from Langer, R. and Folkman, J. (1976). Polymers for sustained-release of proteins and other macromolecules. Nature, 263: 797–800.

Methods such as those described above can be used with a variety of polymeric scaffolds in tissue repair, even in rather complicated conditions. For example, angiogenic growth factors such as VEGF-A can be incorporated into polymeric scaffolds formed by a gas foaming method (Sun *et al.*, 2005). When the VEGF-A was incorporated into the surface of the polymer, it was released fairly quickly from the implant, whereas if it was pre-encapsulated within degradable hydrophobic polyester microparticles which were then incorporated into the scaffold, release was substantially delayed (Ennett *et al.*, 2006). This difference in sustained release rates has been used to strategic advantage in angiogenesis (Richardson *et al.*, 2001). The growth factor VEGF-A induces angiogenesis; however, the

newly formed blood vessels are often immature, being hyperpermeable and unusually branched in architecture. Other growth factors, such as PDGF, are known to induce maturation of these vessels, by associating them with other cells that stabilize the new vessels and reduced permeability. This two-stage release method was used to release VEGF-A quickly, being incorporated into the surface of the scaffold, and PDGF more slowly, being pre-incorporated into microparticles that were then incorporated into the scaffold. When angiogenesis was studied in animal models, substantially more mature angiogenesis was obtained with the dual growth factor release system than with the single growth factor release system releasing VEGF-A (Richardson *et al.*, 2001). The difference in the maturity of newly formed vessels with dual release is illustrated in Figure 15.10. These examples demonstrate that multiple release rates can be achieved from growth factors in hydrophobic degradable polymer scaffolds and that this can be used to obtain high-level biological function in tissue responses.

It is instructive to think about the details of how protein drugs can be incorporated into polymer microparticles. The most common method is referred to as the water-in-oil-in-water method (Cohen *et al.*, 1991). The final structure consists of a polymer microparticle with dispersed domains of protein-containing water

State of the Art Experiment
Matrix immobilization of VEGF-A in therapeutic angiogenesis

Many scientists are searching for new growth factors for use as drugs in solving tissue engineering problems. Often it is hoped that a new growth factor lies just around the corner, waiting to be discovered to solve important problems. However, not all the magic lies in the growth factor itself, rather the details of how the growth factor is delivered – the controlled release strategy – may matter as much as the growth factor itself in determining efficacy. An example is found in controlled release strategies with VEGF-A. VEGF-A induces very potent angiogenesis, i.e. growth of new blood vessels from existing ones, yet it also induces blood vessels to form in a relatively immature manner. Controlled release scientists have sought to corelease vessel-stabilizing growth factors to correct this tendency (Richardson *et al.*, 2001), as outlined elsewhere in this chapter. Here, we show also that scientists thinking about the details of controlled release of VEGF-A can also solve the problem, without any additional factors. Research led by graduate student Martin Ehrbar and senior scientist Andreas Zisch, then in the author's laboratory at the Swiss Federal Institute of Technology in Zurich, produced a recombinant VEGF-A variant, i.e. $\alpha_2 PI_{1-8}$–pl–VEGF-A (Ehrbar *et al.*, 2004). This tripartite fusion protein consisted of an 8-amino-acid sequence from the protein α_2 plasmin inhibitor (PI) that served as a substrate for the coagulation transglutaminase factor XIIIa, connected to a substrate for plasmin (pl), connected to the growth factor VEGF-A. Factor XIIIa is an enzyme that cross-links fibrin to itself and also as a number of nonfibrin proteins to fibrin during coagulation. In this case, the fusion protein has incorporated at its terminus a sequence that is linked into fibrin during coagulation under

the influence of this enzyme. In this way, the fusion protein, if merely mixed into fibrinogen and spiked with thrombin and factor XIII, is spontaneously grafted into a fibrin matrix. In order to allow release of the growth factor from the matrix, Ehrbar, Zisch and their team engineered a cleavage domain between the growth factor and the fibrin-binding site that would be sensitive to enzymes produced by cells as they migrate into the fibrin matrix and remodel it, here engineering an intervening substrate for the enzyme plasmin. Thus, as cells migrate into the fibrin matrix, they encounter a field of immobilized VEGF-A, which they can then liberate by local proteolytic remodeling. When Ehrbar *et al.* in Valentin Djonov's laboratory at the University of Bern quantified the angiogenic response in animal models, they discovered that this cell-demanded concept of controlled release yielded an angiogeneic response that was much more mature that occurred with the free, wild-type VEGF-A: vessels were less diffuse and leaky than the problematic hyperpermeable vessels that form with the native growth factor (Ehrbar *et al.*, 2004, 2005). Typical morphologies are illustrated in the figure. It may be that constant exposure to exceedingly low doses of VEGF-A afforded by the cell-demanded release strategy solved the problems of unusual angiogenic morphology and pathological function that can be observed with native VEGF-A. Interestingly, the fibrin-bound VEGF-A was more potent in induction of angiogenesis in the animal models – one normally thinks of biological effects being minimal at very low doses, but in this case steady exposure to very low quantities yielded a very desirable response, both in terms of quantity and quality.

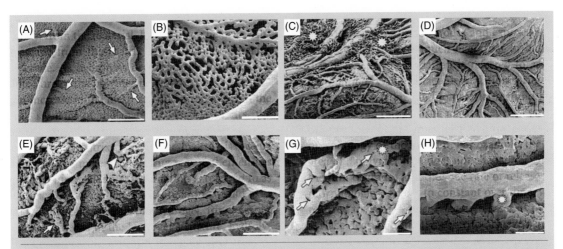

Under most conditions of release, VEGF induces a pathological morphology of angiogenesis, with vessels being tortuous and hyperpermeable. When VEGF was covalently bound to a fibrin cell infiltration matrix via a linker that cells could locally cleave via cell-activated matrix metalloproteinases, a very low and continuous supply of VEGF was generated. This resulted in a morphologically normal angiogenic response in the chick chorioallantoic membrane (CAM) assay, as well as in mice. Scanning electron micrographs (SEM) are shown of vascular casts obtained from the CAM assay. (A and B) SEM images are shown of angiogenesis in unmodified fibrin. These demonstrate physiologically normal morphology: (A) adjacent to the fibrin implant and (B) beneath the implant, showing a somewhat reduced vascular network. (C, E and G) Images of angiogenesis obtained with VEGF freely diffusible in fibrin matrices. White stars in (C) demonstrate a pathological morphology, much like a hemangioma. Arrowheads in (E) denote abrupt changes in vessel diameter as they transition to capillaries. Arrows in (G) denote regions of vessel splitting and remodeling. (D, F and H) Images of angiogenesis obtained with fibrin-bound VEGF are shown. These resulted in a physiologically more normal morphology. Connections of arterioles and venules to the capillary network are normal (star in H). Bars: A, B, E and F = 200 μm, C and D = 500 μm or G and H = 50 μm. Reproduced with permission from Ehrbar, M. Djonov, (2004). Cell-demanded liberation of VEGF(121) from fibrin implants induces local and controlled blood vessel growth. Circ Res, 94: 1124–1132.

droplets within it and it is possible in some circumstances to further dry the watery domains to consist of merely dry protein dispersed throughout the polymer microparticle. The construct is made as follows. In one preparation, the protein is dissolved in a buffer, which is usually optimized for long-term storage stability of the protein to be encapsulated. In another preparation, the polymer is dissolved in a water-immiscible organic solvent (the 'oil') – the solvent is selected to be highly volatile, to facilitate its removal, and CH_2Cl_2 is a common choice. The protein solution and the polymer solution are mixed; since the organic solvent is not miscible in water, an emulsion forms and the size of the watery droplets in the polymer continuous phase may be controlled via the energy that is used

in preparing the mixture. Now what exists is a pot of polymer solution in organic solvent, with microdroplets of protein solution in buffer dispersed throughout it. How does one make microparticles from this? This emulsion is further mixed with water that contains an emulsion stabilizer, such as poly(vinyl alcohol) and again energy is added, although at a lower level so as to obtain larger droplets. Now what exists is a watery continuous phase, with microdroplets of polymer solution in organic solvent, which further have smaller droplets within them of protein solution in buffer – a so-called water-in-oil-in-water double emulsion, i.e. an emulsion within an emulsion. This double emulsion can be subjected to low pressures to evaporate off the organic solvent, during which the polymer

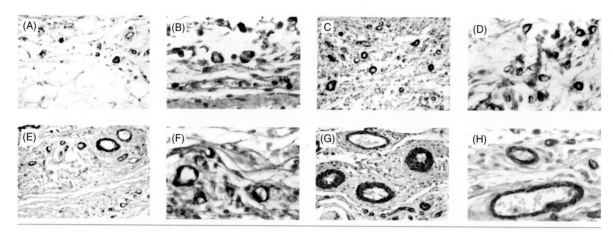

Figure 15.10 Physiological angiogenesis involves more than one growth factor. Dual growth factor release systems have been explored as a means by which to induce mature angiogenesis, using VEGF as an angiogenesis-inducing factor and PDGF as a vascular maturation factor, to induce cladding of the new vessels with stabilizing smooth muscle cells. Here, VEGF was incorporated within a scaffold and was released quickly, whereas PDGF was incorporated within microparticles contained in the scaffold and was released slowly. α-Smooth muscle actin was stained, as an indication of vessel stabilization by smooth muscle cells. (A and B) Angiogenesis within scaffolds containing no growth factor was minimal. (C and D) Angiogenesis within scaffolds containing only VEGF. (E and F) Angiogenesis within scaffolds containing only microencapsulated PDGF. (G and H) Angiogenesis within scaffolds with dual release of VEGF and PDGF demonstrated a vascular response that was clearly more mature than with either growth factor alone, with statistically more vessels that were positive for both endothelial cell and smooth muscle cell markers. Magnification: A, C, E and G = ×400; B, D, F and H = ×1000. Reproduced with permission from Richardson, T.P. (2001). Polymeric system for dual growth factor delivery. *Nature Biotechnol,* 19: 1029–1034.

solution then merely becomes a solid polymer, containing the watery droplets of protein throughout – microparticles of polymer with smaller micro- and nano-droplets of protein in buffer. This method has been used to develop a number of commercial products in protein drug delivery.

One difficulty with the use of hydrophobic polyesters is that they degrade to yield acidic products and the pH of the interior of the microparticle may become very low (Ding *et al.*, 2006; Li and Schwendeman, 2005). This can lead to chemical changes in the protein, such as deamidation of glutamine and asparagine residues to glutamic acid and aspartic acid residues, respectively. Since this unfavorable side reaction is due to the acidic microenvironment within the particle, approaches have been developed to coencapsulate bases, such as $Mg(OH)_2$, a common antacid (Zhu

et al., 2000). This has been demonstrated to dramatically improve stability of the encapsulated protein, and has been shown to enhance the delivered activity of protein growth factors such as FGF-2 and BMP-2.

The examples in this section demonstrate that protein drugs can be incorporated throughout hydrophobic degradable polymers, both in forms of microparticles and finished scaffolds, to serve multiple roles. Both single and multiple protein drugs can be released.

15.5 Bioactive factors bound to affinity sites within matrices

In synthetic and biomolecular scaffolds that are highly permeable, it is difficult to retain protein drugs for prolonged periods. It was mentioned above, e.g. in the context of gelatin matrices, that low-affinity electrostatic interactions can partially remedy this. It was further mentioned that some biomolecular pairs, such as fibrin and FGF-2, have an evolved high-affinity interaction. This evolved affinity can be further engineered to prolong sustained release and this forms the topic of the present section.

Most growth factors of interest in tissue engineering are referred to as heparin-binding growth factors. They contain on their surfaces domains that are positively charged (Maxwell *et al.*, 2005) that bind with relatively high affinity to heparan sulfate-containing proteoglycans in the extracellular matrix. This affinity for heparin has been utilized in engineering systems for controlled release of growth factors. Early work utilized heparin bound into hydrogel matrices, where the heparin then served to bind the growth factors. Microparticles of such biofunctional gels have been used in controlled release of FGF-2 (Laham *et al.*, 1999; Sellke *et al.*, 1998). Chemically modified heparin has been bound within PEG hydrogels to bind FGF-2 for use in inducing differentiation of mesenchymal stem cells (Benoit and Anseth, 2005a,b).

Quite sophisticated delivery matrices have been developed based on this principle of heparin affinity. For example, fibrin derivatives have been constructed that contain heparin-binding peptides (Maxwell *et al.*, 2005; Sakiyama-Elbert and Hubbell, 2000a,b; Taylor *et al.*, 2004). These peptides then bind heparin, which can then bind heparin-binding growth factors. The higher the ratio of bound heparin to incorporated growth factor, the more delayed is release of the heparin-binding growth factor, as illustrated in Figure 15.11. Although the overall interactions are a bit indirect (to fibrin is coupled a peptide, which then binds heparin, which then binds the growth factor), the approach has been demonstrated to lead to dramatically prolonged release of growth factors such as FGF-2 and nerve growth factor.

The examples in this section demonstrate that biologically evolved affinities between protein drugs and biomolecules can be exploited in engineering systems for sustained release, and that these affinities can be

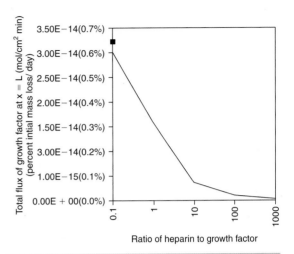

Figure 15.11 Many growth factors used in tissue engineering bind heparin. Here, heparin was bound within fibrin gels into which a heparin-binding peptide was covalently cross-linked during coagulation. Results from a computational model of release from these gels are shown. Release of the growth factor in the absence of heparin if shown by the square, and release at various ratios of heparin-to-growth factor is shown by the line, using the affinity for basic FGF for heparin. At a ratio of 10 and higher, very substantial prolongation of growth factor release was obtained. Reproduced with permission from Sakiyama-Elbert, S.E. and Hubbell, J.A. (2000). Development of fibrin derivatives for controlled release of heparin-binding growth factors. *J Control Rel*, 65: 389–402.

endowed upon both synthetic and biological matrices for tissue engineering.

15.6 Bioactive factors covalently bound to matrices

As mentioned in the Introduction, controlled release in tissue engineering does not always imply that the bioactive agent is actually released from the support; a desired controlled release outcome may mean that the bioactive is retained on the surface of a polymer matrix or scaffold. This is true especially for adhesion factors as drugs, but may be true as well for growth factors. We begin with the obvious case of the adhesion factor and then proceed to the less obvious case of the growth factor to make some cases-in-point.

In order for an adhesion factor to carry out its biological function of transmitting stress between a substrate outside the cell and the cytoskeleton within the cell, the adhesion factor must be immobilized to the substrate. In the case of biopolymeric scaffolds like collagen and fibrin, there exist many adhesion ligands that are built into the scaffold material by evolution. In the case of synthetic matrices, it is necessary to incorporate exogenous ligands into them to enable biospecific cell adhesion.

A number of chemical approaches have been developed by which to endow materials with exogenous cell adhesion sites (Hubbell, 1995; Lutolf and Hubbell, 2005) and a detailed coverage of these methods is outside the scope of this chapter. For example, with hydrogels, adhesion peptides may be covalently coupled to an immobilization site though a linking polymer, such as a PEG chain (Hern and Hubbell, 1998). Such approaches have been used to direct cell migration on hydrogel surfaces (DeLong et al., 2005), e.g. by incorporating gradients of the adhesion peptide on the biomaterial. By carefully selecting the peptide that is incorporated within the overall nonadhesive hydrogel platform, one may select between different adhesion receptors on the surface of the cell to induce different functions, for example between different β_1- and β_3-integrins dependent upon the identity of the peptide and the protein that it mimics (Gonzalez et al., 2004). Such approaches can be used to create biomaterials that demonstrate selectivity to specific cell types, such as endothelial cells (Hubbell et al., 1991; Jun and West, 2005). One can also trigger specific cell functions with immobilized adhesion ligands, such as differentiation of precursor cells (Patel et al., 2005) and osteoblasts (Benoit and Anseth, 2005a,b).

Even biopolymeric matrices that already have cell adhesion characteristics can benefit from incorporation of additional cell adhesion domains. This is because the cell adhesion peptides and proteins do more than just provide adhesion – they also provide specific cell differentiation cues much like growth factors do. Interesting examples can be taken in the context of engineering of fibrin scaffolds. Peptides and proteins can be incorporated into fibrin during coagulation using the enzymology of coagulation. During coagulation, an enzyme, the transglutaminase factor XIIIa, links glutamine residues to lysine residues to cross-link the fibrin gel in to a strong and stable matrix. It is possible to synthesize exogenous peptides or to express recombinant proteins so as to incorporate a substrate site for factor XIIIa (Schense and Hubbel, 1999). The net effect of this is that the exogenous peptide or protein will be covalently incorporated into the fibrin network during coagulation. This has been explored with adhesion peptides from the protein laminin – a protein that is involved in a number of morphogenetic processes, including nerve growth. When four laminin peptides were covalently coupled into fibrin matrices and the derivatized fibrin was used to repair resected nerves in the rat, the number of myelinated axons bridging the resection site was nearly doubled compared to unmodified fibrin (Schense et al., 2000), as illustrated in Figure 15.12. Fibrin was already adhesive for nerve cells, but the laminin-derived peptides bound to other adhesion receptors and signaled greater regeneration than with the native fibrin alone. Similar effects have been seen with other adhesion proteins linked into fibrin and inducing complicated morphogenetic processes such as angiogenesis (Hall and Hubbell, 2004).

Growth factors are also interesting drugs to consider for immobilization within biomaterial matrices and on the surfaces of biomaterial scaffolds. One of the earliest demonstrations that substrate-bound growth factors could be biologically active was carried out with epidermal growth factor, as highlighted in 'Classical Experiment: Surface immobilization of growth factors', with work on cell differentiation *in vitro* (Kuhl and Griffith-Cima, 1996, 1997). One of the earliest schemes to construct biomaterials for implantation as matrices was by derivatization of type I collagen with the growth factor transforming growth factor-β2. The growth factor was linked to collagen via a reactive bifunctional linker and was

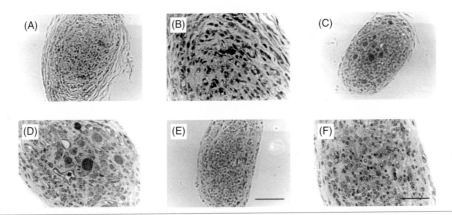

Figure 15.12 Exogenous adhesion ligands can be covalently incorporated within fibrin matrices during coagulation by constructing them as substrates for the coagulation transglutaminase factor XIIIa. This was explored in regeneration of peripheral nerves in rat models, using resections of the dorsal root. (A and B) Transmission electron microscopy (TEM) showing nerve structure in the middle of gaps left empty within a polymer tube. Only a small number of myelinated axons were visible. (C and D) TEM in the middle of a gap treated with unmodified fibrin. Here, also, only a small number of myelinated axons was visible. (E and F) TEM in the middle of a gap treated with fibrin containing four peptides derived from laminin, i.e. peptides containing the RGD, YIGSR, IKVAV and RNIAEIIKDI sequences. The number of myelinated axons in these samples was increased by 85% compared to nerves treated with unmodified fibrin. Length bar: A, C and E = 50 μm; B, D and F = 25 μm. Reproduced with permission from Schense, J.C., Bloch, J., Aebischer, P., and Hubbell, J.A. (2000). Enzymatic incorporation of bioactive peptides into fibrin matrices enhances neurite extension. *Nature Biotechnol,* 18: 415–419.

Classical Experiment 2
Surface immobilization of growth factors

Growth factors are typically thought to be active in their freely diffusible state. It is certainly the case that many growth factors bind to components of the extracellular matrix; however, the extracellular matrix is enzymatically remodeled, resulting in mobilization and release of the growth factors. A very early study on permanently bound growth factors was performed by Philip Kuhl, then a graduate student at Massachusetts Institute of Technology (MIT), and Professor Linda Griffith, also at MIT. The study was published in *Nature Medicine* in 1996 (Kuhl and Griffith-Cima, 1996, 1997). The experiment was testing whether epidermal growth factor retained and could exert its biological activity when permanently immobilized upon the surface of a biomaterial. Kuhn and Griffith worked with mouse epidermal growth factor, which contains no lysine residues; this means that the only amine on the protein is at the N-terminus, and if an amine-reactive chemistry is used to immobilize the protein, the protein can only be immobilized in one conformation and orientation, by the N-terminus. Glass surfaces were derivatized to graft a multiarmed star PEG, to which the mouse epidermal growth factor was immobilized, as illustrated in the figure. The surface-bound growth factor was found to be competent in inducing cell proliferation and morphological changes as effectively as controls with soluble growth factor, demonstrating that the factor merely needed to ligate its receptors on the cell surface, i.e. that internalization of the growth factor and recycling was not needed to induce the positive signaling of the growth factor. This concept set the stage for a number of other investigators to consider surface immobilization of growth factors as an approach to controlled release, i.e. an approach to providing the sustained activity of a growth factor in the vicinity of a biomaterial matrix or scaffold.

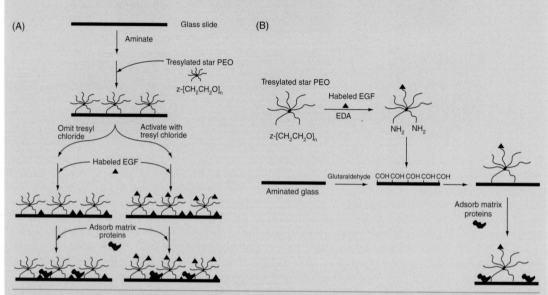

Surface-immobilized epidermal growth factor (EGF) was demonstrated to be biologically active in hepatocyte cultures. Epidermal growth factor was bound to a branched star-PEG, either after grafting the PEG to the surface (A) or prior to grafting to the surface (B). The PEG served to tether the growth factor to the surface in a manner that retained its biological activity and ability to ligate the EGF receptor. Reproduced with permission from Kuhl, P.R. and Griffith-Cima, L.G. (1996). Tethered epidermal growth factor as a paradigm for growth factor-induced stimulation from the solid phase. Nature Med, 2: 1022–1027.

useful as an injectable biomaterial scaffold for cell infiltration (Bentz *et al.*, 1998).

The scheme described above for functionalization of fibrin using the transglutaminase factor XIIIa has also been used for chemical linking of growth factors within fibrin. Growth factors are recombinantly expressed as a fusion protein to include a sequence that is coupled by factor XIIIa to fibrin, with an intervening amino acid sequence that is cleaved by cell-activated enzymes. This scheme has been carried out with VEGF-A, with BMP-2 and with nerve growth factor, demonstrating high activity *in vitro* and *in vivo* (Ehrbar *et al.*, 2004; Sakiyama-Elbert *et al.*, 2001; Schmoekel *et al.*, 2005). In the case of

VEGF-A, it was possible to obtain surprising activities from the growth factor, as highlighted in the text box. These substrate-bound growth factors can interact with the physical aspects of their environment, such as flow near the surface of cells, in special ways (Helm *et al.*, 2005).

15.7 Summary

- Most drugs of interest for controlled release in tissue engineering are proteins and peptides. These include adhesion factors and growth factors. The hydrophilic nature of these drugs imposes many constraints on the controlled release strategies that can be employed.

- In tissue engineering, controlled release sometimes means that the drug is released from the surface or bulk of an implant slowly over long periods of time, and sometimes it means that the drug is immobilized on the surface of or throughout the material and is not released at all. The difference depends on the mode of action of the drug, e.g. adhesion factors are only active when immobilized.
- Hydrophilic drugs may be incorporated homogeneously through hydrophilic matrices, but only heterogeneously through hydrophobic matrices. In hydrophobic matrices, they may be incorporated as solid drug particles through solvent processes, or as aqueous pools rich in protein, for example through water-in-oil-in-water double emulsion processes.
- Protein drugs may be released from hydrophilic matrices through diffusive release or through bulk degradation, or from hydrophobic matrices through surface degradation. With diffusive release, it is not possible to obtain constant releases rates over time. With bulk degradation, this can be obtained if certain features of network structure are appropriately balanced. It is easier to achieve with surface degradation.
- Many growth factors bind to other biological molecules, such as heparin and heparan sulfate proteoglycan. This affinity can be exploited in heparin-containing systems for controlled release. Some growth factors have evolved affinities for some collagen types and for other biological matrices such as fibrin. This can also be exploited in controlled release.
- Hydrogels resemble 3D fishing nets and protein drugs can be efficiently entrapped within hydrogels if the distance between cross-links within the network is sufficiently close to the diameter of the protein being released.
- Degradable polyesters, such as poly(lactic acid), poly(glycolic acid), poly(ε-caprolactone) and their copolymers, are useful in controlled release from tissue engineering scaffolds.

- The internal pH in degradable polyesters is often very acidic, which may cause problems in the long-term stability of protein drugs being released. This can be corrected by coencapsulating pharmaceutically acceptable bases, such as $Mg(OH)_2$.

References

Albala, D.M. and Lawson, J.H. (2006). Recent clinical and investigational applications of fibrin sealant in selected surgical specialties. *J Am Coll Surg*, 202: 685–697.

Anseth, K.S., Metters, A.T., Bryant, S.J., *et al.* (2002). *In situ* forming degradable networks and their application in tissue engineering and drug delivery. *J Control Release*, 78: 199–209.

Benoit, D.S.W. and Anseth, K.S. (2005a). Heparin functionalized peg gels that modulate protein adsorption for HMSC adhesion and differentiation. *Acta Biomater*, 1: 461–470.

Benoit, D.S.W. and Anseth, K.S. (2005b). The effect on osteoblast function of colocalized RGD and PHSRN epitopes on peg surfaces. *Biomaterials*, 26: 5209–5220.

Bentz, H., Schroeder, J.A. and Estridge, T.D. (1998). Improved local delivery of TGF-beta 2 by binding to injectable fibrillar collagen via difunctional polyethylene glycol. *J Biomed Mater Res*, 39: 539–548.

Boden, S.D., Zdeblick, T.A., Sandhu, H.S. and Heim, S.E. (2000). The use of rhBMP-2 in interbody fusion cages – definitive evidence of osteoinduction in humans: a preliminary report. *Spine*, 25: 376–381.

Borden, M., Attawia, M., Khan, Y., *et al.* (2004). Tissue-engineered bone formation *in vivo* using a novel sintered polymeric microsphere matrix. *J Bone Joint Surg Br*, 86B: 1200–1208.

Borden, M., Attawia, M. and Laurencin, C.T. (2002). The sintered microsphere matrix for bone tissue engineering: *In vitro* osteoconductivity studies. *J Biomed Mater Res*, 61: 421–429.

Burkus, J.K., Dorchak, J.D. and Sanders, D.L. (2003). Radiographic assessment of interbody fusion using recombinant human bone morphogenetic protein type 2. *Spine*, 28: 372–377.

Clark, D.P., Hanke, C.W. and Swanson, N.A. (1989). Dermal implants – safety of products injected for soft-tissue augmentation. *J Am Acad Dermatol*, 21: 992–998.

Cohen, S., Yoshioka, T., Lucarelli, M., *et al.* (1991). Controlled delivery systems for proteins based on poly(lactic glycolic acid) microspheres. *Pharm Res*, 8: 713–720.

De Biase, P. and Capanna, R. (2005). Clinical applications of BMPs. *Injury*, 36: 43–46.

DeLong, S.A., Gobin, A.S. and West, J.L. (2005). Covalent immobilization of RGDS on hydrogel surfaces to direct cell alignment and migration. *J Control Release*, 109: 139–148.

Ding, A.G., Shenderova, A. and Schwendeman, S.P. (2006). Prediction of microclimate pH in poly(lactic-*co*-glycolic acid) films. *J Am Chem Soc*, 128: 5384–5390.

Ehrbar, M., Djonov, V.G., Schnell, C., *et al.* (2004). Cell-demanded liberation of VEGF$_{121}$ from fibrin implants induces local and controlled blood vessel growth. *Circ Res*, 94: 1124–1132.

Ehrbar, M., Metters, A., Zammaretti, P., *et al.* (2005). Endothelial cell proliferation and progenitor maturation by fibrin-bound VEGF variants with differential susceptibilities to local cellular activity. *J Control Release*, 101: 93–109.

Elbert, D.L., Pratt, A.B., Lutolf, M.P., *et al.* (2001). Protein delivery from materials formed by self-selective conjugate addition reactions. *J Control Release*, 76: 11–25.

Ennett, A.B., Kaigler, D. and Mooney, D.J. (2006). Temporally regulated delivery of VEGF *in vitro* and *in vivo*. *J Biomed Mater Res*, 79: 176–184.

Friedlaender, G.E., Perry, C.R., Cole, J.D., *et al.* (2001). Osteogenic protein-1 (bone morphogenetic protein-7) in the treatment of tibial nonunions – a prospective, randomized clinical trial comparing rhOP-1 with fresh bone autograft. *J Bone Joint Surg Am*, 83A: S151–S158.

Gobin, A.S. and West, J.L. (2002). Cell migration through defined, synthetic extracellular matrix analogues. *FASEB J*, 16: 751–753.

Gonzalez, A.L., Gobin, A.S., West, J.L., *et al.* (2004). Integrin interactions with immobilized peptides in polyethylene glycol diacrylate hydrogels. *Tissue Eng*, 10: 1775–1786.

Govender, S., Csimma, C., Genant, H.K. and Valentin-Opran, A. (2002). Recombinant human bone morphogenetic protein-2 for treatment of open tibial fractures – a prospective, controlled, randomized study of four hundred and fifty patients. *J Bone Joint Surg Am*, 84A: 2123–2134.

Hall, H. and Hubbell, J.A. (2004). Matrix-bound sixth Ig-like domain of cell adhesion molecule II acts as an angiogenic factor by ligating alpha v beta 3-integrin and activating VEGF-r2. *Microvasc Res*, 68: 169–178.

Helm, C.L.E., Fleury, M.E., Zisch, A.H., *et al.* (2005). Synergy between interstitial flow and VEGF directs capillary morphogenesis *in vitro* through a gradient amplification mechanism. *Proc Natl Acad Sci USA*, 102: 15779–15784.

Hern, D.L. and Hubbell, J.A. (1998). Incorporation of adhesion peptides into nonadhesive hydrogels useful for tissue resurfacing. *J Biomed Mater Res*, 39: 266–276.

Hubbell, J.A. (1995). Biomaterials in tissue engineering. *Bio-Technology*, 13: 565–576.

Hubbell, J.A., Massia, S.P., Desai, N.P. and Drumheller, P.D. (1991). Endothelial cell-selective materials for tissue engineering in the vascular graft via a new receptor. *Bio-Technology*, 9: 568–572.

Johnsson, R., Stromqvist, B. and Aspenberg, P. (2002). Randomized radio stereometric study comparing osteogenic protein-1 (BMP-7) and autograft bone in human noninstrumented posterolateral lumbar fusion – 2002 Volvo Award in Clinical Studies. *Spine*, 27: 2654–2661.

Jun, H.W. and West, J.L. (2005). Endothelialization of microporous YIGSR/PEG-modified polyurethaneurea. *Tissue Eng*, 11: 1133–1140.

Kanematsu, A., Marui, A., Yamamoto, S., *et al.* (2004). Type I collagen can function as a reservoir of basic fibroblast growth factor. *J Control Release*, 99: 281–292.

Kissel, T., Li, Y.X. and Unger, F. (2002). ABA-triblock copolymers from biodegradable polyester A-blocks and hydrophilic poly(ethylene oxide) B-blocks as a candidate for in situ forming hydrogel delivery systems for proteins. *Adv Drug Del Res*, 54: 99–134.

Kofron, M.D., Li, X.D. and Laurencin, C.T. (2004). Protein- and gene-based tissue engineering in bone repair. *Curr Opin Biotechnol*, 15: 399–405.

Koide, T. (2005). Triple helical collagen-like peptides: engineering and applications in matrix biology. *Connect Tiss Res*, 46: 131–141.

Kuhl, P.R. and Griffith-Cima, L.G. (1996). Tethered epidermal growth factor as a paradigm for growth factor-induced stimulation from the solid phase. *Nat Med*, 2: 1022–1027.

Kuhl, P.R. and Griffith-Cima, L.G. (1997). Tethered epidermal growth factor as a paradigm for growth factor-induced stimulation from the solid phase [erratum]. *Nat Med*, 3: 93.

Laham, R.J., Sellke, F.W., Edelman, E.R., *et al.* (1999). Local perivascular delivery of basic fibroblast growth factor in patients undergoing coronary bypass surgery – results of a phase I randomized, double-blind, placebo-controlled trial. *Circulation*, 100: 1865–1871.

Langer, R. and Folkman, J. (1976). Polymers for sustained-release of proteins and other macromolecules. *Nature*, 263: 797–800.

Lee, C.H., Singla, A. and Lee, Y. (2001). Biomedical applications of collagen. *Int J Pharmaceut*, 221: 1–22.

Li, L. and Schwendeman, S.P. (2005). Mapping neutral microclimate pH in PLGA microspheres. *J Control Release*, 101: 163–173.

Lu, H.H., Kofron, M.D., El-Amin, S.F., *et al.* (2003). *In vitro* bone formation using muscle-derived cells: a new paradigm for bone tissue engineering using polymer–bone morphogenetic protein matrices. *Biochem Biophys Res Commun*, 305: 882–889.

Lustig, S.R. and Peppas, N.A. (1988). Solute diffusion in swollen membranes. 9. Scaling laws for solute diffusion in gels. *J Appl Polym Sci*, 36: 735–747.

Lutolf, M.P. and Hubbell, J.A. (2005). Synthetic biomaterials as instructive extracellular microenvironments for morphogenesis in tissue engineering. *Nat Biotechnol*, 23: 47–55.

Lutolf, M.P., Lauer-Fields, J.L., Schmoekel, H.G., *et al.* (2003a). Synthetic matrix metalloproteinase-sensitive hydrogels for the conduction of tissue regeneration: engineering cell-invasion characteristics. *Proc Nat Acad Sci USA*, 100: 5413–5418.

Lutolf, M.P., Raeber, G.P., Zisch, A.H., *et al.* (2003b). Cell-responsive synthetic hydrogels. *Adv Mater*, 15: 888–892.

Lutolf, M.R., Weber, F.E., Schmoekel, H.G., *et al.* (2003c). Repair of bone defects using synthetic mimetics of collagenous extracellular matrices. *Nat Biotechnol*, 21: 513–518.

Martens, P.J., Bowman, C.N. and Anseth, K.S. (2004). Degradable networks formed from multi-functional poly(vinyl alcohol) macromers: comparison of results from a generalized bulk-degradation model for polymer networks and experimental data. *Polymer*, 45: 3377–3387.

Mason, M.N., Metters, A.T., Bowman, C.N. and Anseth, K.S. (2001). Predicting controlled-release behavior of degradable PLA-b-PEG-b-PLA hydrogels. *Macromolecules*, 34: 4630–4635.

Maxwell, D.J., Hicks, B.C., Parsons, S. and Sakiyama-Elbert, S.E. (2005). Development of rationally designed affinity-based drug delivery systems. *Acta Biomater*, 1: 101–113.

Metters, A. and Hubbell, J. (2005). Network formation and degradation behavior of hydrogels formed by Michael-type addition reactions. *Biomacromolecules*, 6: 290–301.

Mosesson, M.W. (2005). Fibrinogen and fibrin structure and functions. *J Thromb Hemost*, 3: 1894–1904.

Ozeki, M. and Tabata, Y. (2006). Affinity evaluation of gelatin for hepatocyte growth factor of different types to design the release carrier. *J Biomater Sci Polym Edn*, 17: 139–150.

Patel, P.N., Gobin, A.S., West, J.L. and Patrick, C.W. (2005). Poly(ethylene glycol) hydrogel system supports preadipocyte viability, adhesion, and proliferation. *Tissue Eng*, 11: 1498–1505.

Peng, H., Sahni, A., Fay, P., *et al.* (2004). Identification of a binding site on human FGF-2 for fibrinogen. *Blood*, 103: 2114–2120.

Pratt, A.B., Weber, F.E., Schmoekel, H.G., *et al.* (2004). Synthetic extracellular matrices for in situ tissue engineering. *Biotechnol Bioeng*, 86: 27–36.

Richardson, T.P., Peters, M C., Ennett, A.B. and Mooney, D.J. (2001). Polymeric system for dual growth factor delivery. *Nat Biotechnol*, 19: 1029–1034.

Sahni, A., Altland, O.D. and Francis, C.W. (2003). FGF-2 but not FGF-1 binds fibrin and supports prolonged endothelial cell growth. *J Thromb Hemost*, 1: 1304–1310.

Sahni, A., Baker, C.A., Sporn, L.A. and Francis, C.W. (2000). Fibrinogen and fibrin protect fibroblast growth factor-2 from proteolytic degradation. *Thromb Hemostas*, 83: 736–741.

Sahni, A. and Francis, C.W. (2000). Vascular endothelial growth factor binds to fibrinogen and fibrin and stimulates endothelial cell proliferation. *Blood*, 96: 3772–3778.

Sahni, A., Sporn, L.A. and Francis, C.W. (1999). Potentiation of endothelial cell proliferation by fibrin(ogen)-bound fibroblast growth factor-2. *J Biol Chem*, 274: 14936–14941.

Sakiyama-Elbert, S.E. and Hubbell, J.A. (2000a). Controlled release of nerve growth factor from a heparin-containing fibrin-based cell ingrowth matrix. *J Control Release*, 69: 149–158.

Sakiyama-Elbert, S.E. and Hubbell, J.A. (2000b). Development of fibrin derivatives for controlled release of heparin-binding growth factors. *J Control Release*, 65: 389–402.

Sakiyama-Elbert, S.E., Panitch, A. and Hubbell, J.A. (2001). Development of growth factor fusion proteins for cell-triggered drug delivery. *FASEB J*, 15: 1300–1302.

Sawhney, A.S., Pathak, C.P. and Hubbell, J.A. (1993). Bioerodible hydrogels based on photopolymerized poly(ethylene glycol)-*co*-poly(alpha-hydroxy acid) diacrylate macromers. *Macromolecules*, 26: 581–587.

Schense, J.C., Bloch, J., Aebischer, P. and Hubbell, J.A. (2000). Enzymatic incorporation of bioactive peptides into fibrin matrices enhances neurite extension. *Nat Biotechnol*, 18: 415–419.

Schense, J.C. and Hubbell, J.A. (1999). Cross-linking exogenous bifunctional peptides into fibrin gels with factor XIIIa. *Bioconj Chem*, 10: 75–81.

Scheufler, C., Sebald, W. and Hulsmeyer, M. (1999). Crystal structure of human bone morphogenetic protein-2 at 2.7 angstrom resolution. *J Mol Biol*, 287: 103–115.

Schmoekel, H., Schense, J.C., Weber, F.E., *et al.* (2004). Bone healing in the rat and dog with nonglycosylated BMP-2 demonstrating low solubility in fibrin matrices. *J Orthoped Res*, 22: 376–381.

Schmoekel, H.G., Weber, F.E., Schense, J.C., *et al.* (2005). Bone repair with a form of BMP-2 engineered for incorporation into fibrin cell ingrowth matrices. *Biotechnol Bioeng*, 89: 253–262.

Schoenmakers, R.G., van de Wetering, P., Elbert, D.L. and Hubbell, J.A. (2004). The effect of the linker on the hydrolysis rate of drug-linked ester bonds. *J Control Release*, 95: 291–300.

Seeherman, H., Wozney, J. and Li, R. (2002). Bone morphogenetic protein delivery systems. *Spine*, 27: S16–S23.

Seeherman, H. and Wozney, J.M. (2005). Delivery of bone morphogenetic proteins for orthopedic tissue regeneration. *Cytokine Growth Factor Rev*, 16: 329–345.

Sellke, F.W., Laham, R.J., Edelman, E.R., *et al.* (1998). Therapeutic angiogenesis with basic fibroblast growth factor: technique and early results. *Ann Thorac Surg*, 65: 1540–1544.

Sieron, A.L., Louneva, N. and Fertala, A. (2002). Site-specific interaction of bone morphogenetic protein 2 with procollagen II. *Cytokine*, 18: 214–221.

Sun, Q.H., Chen, R.R., Shen, Y.C., *et al.* (2005). Sustained vascular endothelial growth factor delivery enhances angiogenesis and perfusion in ischemic hind limb. *Pharmaceut Res*, 22: 1110–1116.

Takahashi, Y., Yamamoto, M. and Tabata, Y. (2005). Enhanced osteoinduction by controlled release of bone morphogenetic protein-2 from biodegradable sponge composed of gelatin and beta-tricalcium phosphate. *Biomaterials*, 26: 4856–4865.

Taylor, S.J., McDonald, J.W. and Sakiyama-Elbert, S.E. (2004). Controlled release of neurotrophin-3 from fibrin gels for spinal cord injury. *J Control Release*, 98: 281–294.

van de Wetering, P., Metters, A.T., Schoenmakers, R.G. and Hubbell, J.A. (2005). Poly(ethylene glycol) hydrogels formed by conjugate addition with controllable swelling, degradation, and release of pharmaceutically active proteins. *J Control Release*, 102: 619–627.

West, J.L. and Hubbell, J.A. (1999). Polymeric biomaterials with degradation sites for proteases involved in cell migration. *Macromolecules*, 32: 241–244.

Yamamoto, M., Takahashi, Y. and Tabata, Y. (2006). Enhanced bone regeneration at a segmental bone defect by controlled release of bone morphogenetic protein-2 from a biodegradable hydrogel. *Tissue Eng*, 12: 1305–1311.

Zhu, G.Z., Mallery, S.R. and Schwendeman, S.P. (2000). Stabilization of proteins encapsulated in injectable poly (lactide-*co*-glycolide). *Nat Biotechnol*, 18: 52–57.

Chapter 16
Bioreactors for tissue engineering

David Wendt, Nicholas Timmins, Jos Malda, Frank Janssen, Anthony Ratcliffe, Gordana Vunjak-Novakovic and Ivan Martin

Chapter objectives:

- To define the concept of a bioreactor
- To discuss the key roles of bioreactors in tissue engineering applications: from basic research through tissue manufacturing
- To discuss the main functions and benefits of bioreactors in tissue engineering processes
- To present an engineering approach to bioreactor design

- To illustrate how bioreactors, which recapitulate aspects of the cellular microenvironment, can be used as 3D *in vitro* model systems
- To discuss the impact of varying levels of automation and control on various manufacturing strategies and the resulting engineered products

"Scientists investigate that which already is; Engineers create that which has never been"

Albert Einstein

"Simplicity is the ultimate sophistication"

Leonardo da Vinci

16.1 Introduction

The term 'bioreactor' may initially conjure up images of a fermentation tank mixing a suspension of bacteria, possibly for the production of an antibiotic. In this classical application, the main functions of the bioreactor are to control the environmental conditions (e.g. pH, temperature, pressure) and the nutrient/product concentrations during the bioprocess. The level of control, reproducibility and automation that an optimized bioreactor system enables is essential to manufacture products that must meet specific regulations and criteria regarding efficacy, safety and quality, in addition to being cost-effective.

In the context of tissue engineering, the key functions of a bioreactor are essentially the same, i.e. to provide control and standardization (Figure 16.1): (i) by establishing control over the physicochemical culture parameters during cell/tissue culture, bioreactors offer much potential for improving the quality of engineered tissues; (ii) by standardizing, automating and possibly scaling the manufacture of tissue grafts for clinical applications, bioreactors have a key role to play in facilitating the economically viable and reproducible production of tissue-engineered products. However, in contrast to established industrial fermentation processes, tissue engineering is still an area of ongoing research, and thus bioreactors have the additional role of providing well-defined model systems supporting controlled investigations on cell function and tissue development in three-dimensional (3D) environments. Specifically, tissue engineering bioreactors should be designed to enable the application of multiple regulatory signals (e.g. growth factors, hydrodynamic, mechanical or electrical stimuli), to accommodate replicates via modular design and to provide biosensor or imaging compatibility.

In this chapter, we will discuss the role of bioreactors in cell-based tissue engineering approaches, focusing primarily on their applications to 3D culture systems. We will begin by discussing the functions of bioreactors in three key processes of tissue engineering: (i) cell seeding of porous scaffolds, (ii) maintaining adequate mass transport in the seeded constructs and (iii) physical conditioning the developing tissues. We will then present a general strategy for the design and development of a bioreactor system, exemplifying practical considerations in choosing specific bioreactor components. Finally, we will discuss the implementation of bioreactors in the context of *in vitro* 3D model systems and the manufacture of tissue-engineered products for clinical applications.

16.2 Key functions of bioreactors in tissue engineering

This section relates to the role of bioreactors in establishing and maintaining a 3D cell culture. Special focus is given to a typical approach in tissue engineering, whereby the development of a tissue is initiated by seeding cells into porous 3D scaffolds.

16.2.1 Bioreactors for cell seeding

Considering that the initial cell density and cell distribution within a 3D scaffold can have a significant impact on the ultimate structure, composition and

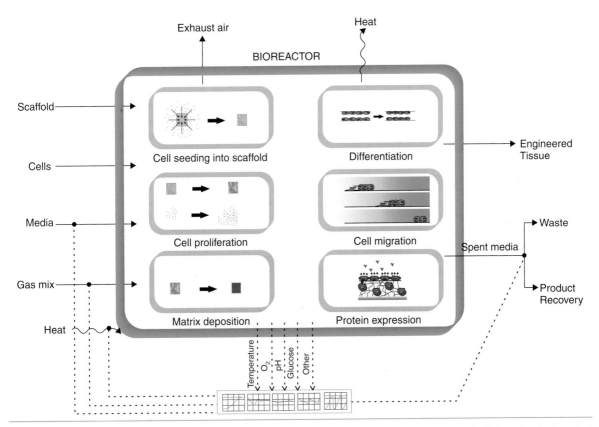

Figure 16.1 Tissue engineering bioreactors. Bioreactors can be used to establish control of the physiochemical parameters during cultivation of cells and tissues. In the context of tissue engineering this controlled environment is supportive of processes such cell seeding into 3D scaffolds, cell proliferation, ECM deposition, cellular differentiation, cell migration and protein expression. Control over the environment in which these processes take place, in turn, leads to higher quality, greater reproducibility and improved scalability.

function of an engineered tissue, cell seeding can be one of the critical steps in the generation of functional tissues. For many applications, including the production of autologous grafts for clinical applications, where the availability of the cell source (e.g. tissue biopsies) is often limited, cells should be seeded with the highest possible efficiency and viability. While cell seeding of hydrogels is a relatively straightforward process, distributing cells into porous 3D scaffolds effectively and reproducibly can be a major challenge, particularly for large scaffolds or those with complex pore architectures.

The most commonly used seeding technique, termed 'static seeding', consists in simply pipetting a concentrated cell suspension into a porous scaffold. This manual, user-dependent process clearly lacks control and standardization. Stirred-flask 'bioreactors' can improve the quality and reproducibility of the seeding process, in particular for thin and highly porous scaffolds (Vunjak-Novakovic *et al.*, 1999). However, due to insufficient convection of cells into thick or less porous scaffolds, stirred-flask systems can result in low seeding efficiencies and generate nonuniform cell distributions, with a high density of cells lining the scaffold surface (Wendt *et al.*, 2003). Perfusing a cell suspension directly through the pores of a 3D scaffold in a bioreactor can result in a more efficient and effective cell seeding, with more uniformly distributed cells than the above techniques, particularly when seeding thick scaffolds of low porosity (Wendt *et al.*, 2003). The use of a perfusion seeding technique in combination with scaffolds having anisotropic architectures could also allow to control the distribution of cells within large porous scaffolds according to specific, nonuniform patterns (Figure 16.2).

Seeding techniques that involve agitation or convective flow may, however, have adverse consequences on cellular viability and phenotype, necessitating the development of new seeding techniques for particularly shear sensitive cells. For instance, a recently described method to seed cardiac myocytes, which are sensitive to both shear and low oxygen levels, consisted of two steps: cells were first inoculated into the scaffold using a thermally polymerizing gel as a delivery vehicle (for low mechanical stress), then medium perfusion initiated immediately following gelation (for immediate oxygen supply) (Radisic *et al.*, 2003). In the future, the design of bioreactors for cell seeding into scaffolds will benefit from experimental and/or theoretical analysis of flow parameters within specific scaffold types.

16.2.2 Bioreactors for enhanced mass transport

After distributing cells throughout the volume of a porous scaffold, a key challenge is maintenance of this distribution and cell viability within the interior of the construct during prolonged culture. This requires sufficient mass transport of nutrients and oxygen to the cells, along with adequate removal of their metabolic waste products. (A more fundamental discussion on cell nutrition and mass transfer can be found in Chapter 12.) The implications of inadequate mass transfer can often be observed following the culture of 3D constructs under conventional static conditions (i.e. with unmixed culture media).

Due to diffusional limitations, statically cultured constructs are frequently inhomogeneous in structure and composition, containing a necrotic central region and dense layers of viable cells encapsulating the construct periphery. Although the limiting species are not decisively known, insufficient oxygen transport has been associated with inhomogeneous development of both engineered cardiac (Radisic *et al.*, 2003, 2004, 2005) and cartilage tissues (Lewis *et al.*, 2005; Malda *et al.*, 2004).

Convective transport of media within stirred-flask and rotating-vessel bioreactor systems can enhance mass transport to and from the construct surface and, to a certain extent, within the construct pores. Initially, convection within the pores is dependent upon scaffold geometry and permeability, and may only be effective to a limited distance into the scaffold. Subsequently, as the construct develops, the scaffold pores may occlude with cells and the extracellular matrix (ECM), further decreasing the efficacy of convection. Bioreactors that perfuse culture medium directly through the pores of the scaffold not only enhance transport at the construct periphery, but also within the internal pores, potentially eliminating mass transport limitations (Figure 16.3). Perfusion bioreactors have been shown to enhance the growth, differentiation and mineralized matrix deposition by bone cells (Bancroft *et al.*, 2002; Goldstein *et al.*, 2001; Sikavitsas *et al.*, 2005), expression of cardiac-specific markers by cardiomyocytes (Radisic *et al.*, 2003, 2004), and ECM synthesis, accumulation and distribution uniformity by chondrocytes (Davisson *et al.*, 2002b; Wendt *et al.*, 2006). Another advantage of perfusion bioreactors over convective systems is the possibility to easily monitor the metabolite consumption of the cells (such as oxygen and glucose) by the use of online biosensors (Wendt *et al.*, 2006; Janssen *et al.*, 2006a). This provides an important tool during the development and set-up of bioreactor systems, e.g. allowing us to monitor the cellular proliferation online (Figure 16.4) (Janssen *et al.*, 2006b). However, while perfusion bioreactors can offer greater control of mass transport than other convective systems, there still remains the potential for flow to follow a preferential path through the construct

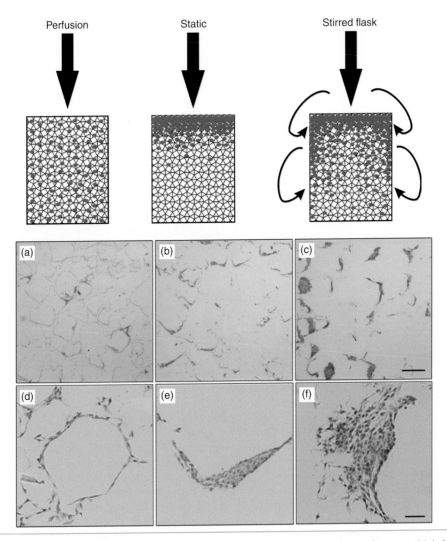

Figure 16.2 Cell seeding distributions. Hematoxylin & eosin-stained cross-sections of 4-mm thick foam scaffolds following cell seeding by (a and d) perfusion, (b and e) static loading and (c and f) stirred-flask. Foams seeded by perfusion were uniformly seeded with cells outlining the foam pores. Foams seeded statically contained large cell clusters nonuniformly distributed within the scaffold. Foams seeded by stirred-flask were highly nonuniform and contained very large clusters in the scaffold region directly exposed to fluid flow. Scale bar: a–c = 500 μm; d–f = 100 μm. (a to f, from Wendt, D., Marsano, A., Jakob, M. *et al.* (2003). Oscillating perfusion of cell suspensions through three-dimensional scaffolds enhances cell seeding efficiency and uniformity. *Biotechnol Bioeng*, 84(2): 205–214).

(particularly for scaffolds with a wide pore size distribution or if the tissue develops nonuniformly), leaving other regions poorly nourished. Furthermore, optimizing a perfusion system may require a balance between mass transport of nutrients and waste products to and from cells, retention of newly synthesized *ECM* components within the construct, and fluid-induced shear stresses within the scaffold pores.

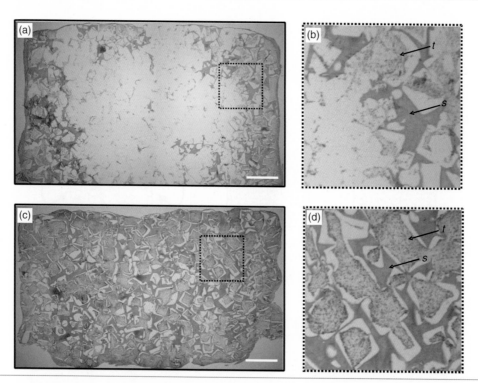

Figure 16.3 Implications of mass transfer on cell and matrix distributions. Human chondrocyte–foam constructs following perfusion cell seeding and 2 weeks of culture. (a and b) Statically cultured constructs; (c and d) perfusion-cultured constructs; (a and c) low magnification images show the tissue distribution throughout the entire cross-section (scale bar = 1 mm); (b and d) higher magnification images identify the tissue '*t*' and scaffold '*s*' within the cross-sections. Statically cultured constructs contained cells and matrix only at the construct surface, reaching a depth of approximately 1 mm into the scaffold (note: the central region of foam scaffold did not adhere to the histology slide due to the absence of cells). In contrast, perfusion-cultured constructs were highly homogeneous, containing a uniform distribution of cells and matrix throughout the cross-section. (From Wendt, D., Stroebel, S., Jakob, M. *et al.* (2006). Uniform tissues engineered by seeding and culturing cells in 3D scaffolds under perfusion at defined oxygen tensions. *Biorheology*, 43(3–4): 481–488.)

Optimization of the operating conditions should ideally be supported by computational fluid dynamics (CFD) modeling, possibly in conjunction with flow visualization techniques (Sucosky *et al.*, 2004). Models developed to date have been valuable for estimation of fluid velocity and shear profiles within the pores of 3D scaffolds. As the fluid dynamics will be dependent on the architecture of a scaffold's porous network, highly relevant CFD models could be based on a reconstruction of the actual scaffold microarchitecture, perhaps generated from micro-computed

tomography (μCT) imaging of the porous scaffold (Cioffi *et al.*, 2005; Porter *et al.*, 2005) (Figure 16.5). A major challenge that still remains is to extend CFD models based on empty scaffolds to later times of construct development when the porous network is filled with cells and ECM.

16.2.3 Bioreactors for physical conditioning

Our body's tissues and organs are subjected to a highly complex biomechanical environment of dynamic

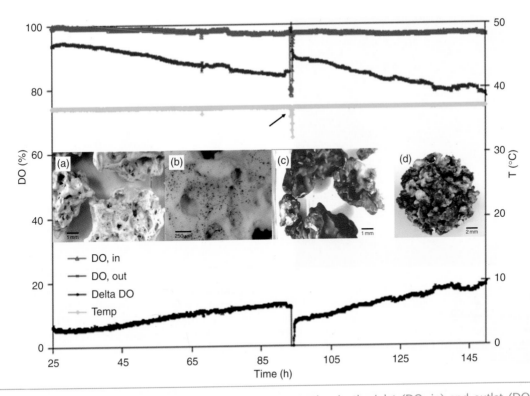

Figure 16.4 On-line monitoring of dissolved oxygen concentration in the inlet (DO, in) and outlet (DO, out) medium during dynamic proliferation of BMSCs on 10 cm³ of ceramic granular scaffolds in a perfusion bioreactor. Oxygen concentrations in the inlet media remained near saturation levels (100% DO) throughout the culture period, whereas oxygen levels measured in the outlet media decreased throughout the 6 days. This increase in oxygen consumption (delta DO) was correlated to the higher number of BMSC observed at day 6 (c and d) as compared to immediately after cell seeding (a and b) (indicated by more intense methylene blue staining). Black arrow indicates temperature and dissolved oxygen disturbance of the system by opening the incubator. Scale bars: b = 250 μm; a and c = 1 mm; top view of all scaffolds and d = 2 mm. (Reproduced with permission from Frank Janssen.)

stresses, strains, fluid flow and hydrostatic pressure. It is widely accepted that physiological forces not only play an important role in cell physiology *in vivo*, but can also modulate the activity of cells in 3D scaffolds *in vitro*. Innovative bioreactors have been developed to apply one or more regimes of controlled physical stimuli to 3D engineered constructs in an attempt to improve or accelerate the generation of a functional tissue (Figure 16.6). These bioreactor-based model systems have provided compelling evidence that mechanical conditioning of 3D constructs can

(i) stimulate ECM production [e.g. dynamic compression to engineered cartilage (Davisson *et al.*, 2002a; Demarteau *et al.*, 2003) or bone], (ii) improve cell/tissue structural organization [e.g. fluid flow through engineered blood vessels (Niklason *et al.*, 1999)], (iii) direct cell differentiation [e.g. translational and rotational strain to induce mesenchymal progenitor cell differentiation toward the ligament lineage (Altman *et al.*, 2002a)] and/or (iv) enhance a specific tissue function [e.g. surface motion to engineered cartilage to enhance lubrication capacity (Grad *et al.*, 2005)].

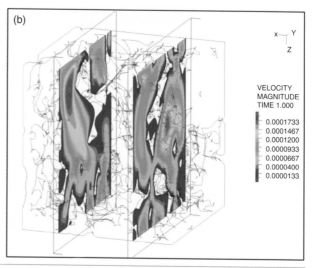

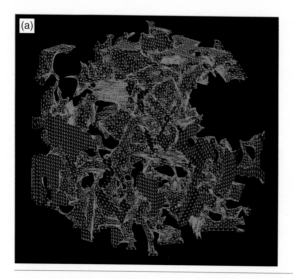

Figure 16.5 CFD modeling of perfusion through a foam scaffold. (a) μCT was used to precisely reconstruct the actual pore microstructure of a foam scaffold. (b) While models based on assumptions with a simplified geometry can provide order-of-magnitude estimates of velocities and shear stresses, the μCT-based 3D reconstruction provides the basis for a more realistic simulation of the profiles within the tortuous pores. (Images provided by Professor Jürg Küffer, Fachhochschule Beider Basel, Basel, Switzerland.)

Figure 16.6 Bioreactors for mechanical conditioning of engineered constructs. Illustration (examples) of bioreactor systems designed to apply (a) dynamic compression (Demarteau, O., Wendt, D., Braccini, A., *et al.* (2003). Dynamic compression of cartilage constructs engineered from expanded human articular chondrocytes. *Biochem Biophys Res Commun*,

Figure 16.6 (Continued) 310(2): 580–588); (b) simulated articular motion (Wimmer, M.A., Grad, S., Kaup, T., *et al.* (2004). Tribology approach to the engineering and study of articular cartilage. *Tissue Eng*, 10(9–10): 1436–1445); (c) torsion and tension (Altman, G.H., Lu, H. *et al.* (2002). Advanced bioreactor with controlled application of multi-dimensional strain for tissue engineering. *J Biomech Eng*, 124(6): 742–749).

While mechanical conditioning has the potential to improve the structural and functional properties of engineered tissues, little is known about which specific mechanical force(s), or regimes of application (magnitude, frequency, duty cycle), are stimulatory for a particular tissue. At this time, the selection of optimal physical conditioning parameters is greatly complicated by the wide variety of model systems that have been used, such as varying cell types, scaffolds, forces and applied regimes, and culture times. Moreover, since cell–scaffold and cell–ECM interactions play a key role in mechanotransduction

Classical Experiment

Cardiovascular disease is a major cause of mortality in the developed world. Bypass grafting with autologous veins and arteries is limited by the availability of functional tissue that can be taken from the patient and the biomaterial grafts that are smaller than 6 mm in diameter fail at a rate of 40% over 6 months. An approach taken to overcome this problem and grow functional

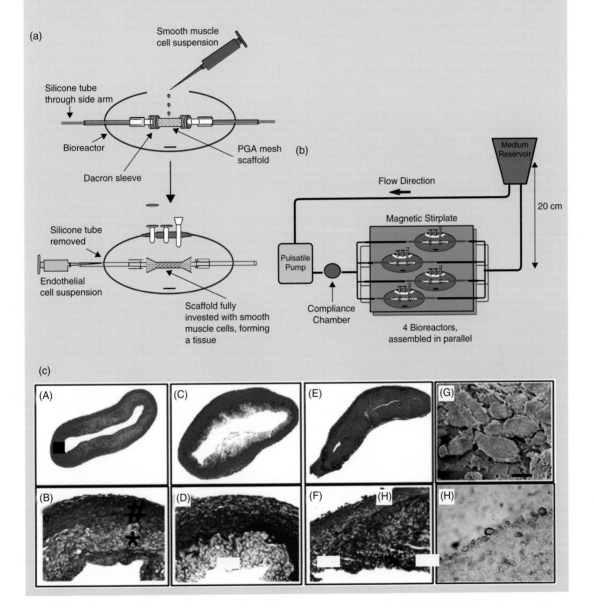

arteries by using cells, biomaterial scaffolds and bioreactors is one of the classical examples of the use of bioreactors for functional tissue engineering. This is also one of the first demonstrations that regulatory factors that play a role in tissue development and remodeling *in vivo* can enhance functional tissue assembly *in vitro*. (a) Tubular biodegradable scaffolds made from fibrous polyglycolic acid placed over silicone tubing were seeded with aortic smooth muscle cells derived from a biopsy of bovine vascular tissue. (b) After seeding, bioreactors were filled with culture medium that was supplemented by factors regulating vascular development and a pulsatile radial stress was applied to mimic the flow of blood through developing blood vessels. After 8 weeks, the silicone tubing was removed, suspension of endothelial cells was injected into the lumen and the flow rate was then gradually increased over 3 days of culture. (c) Pulsatile flow enhanced the structure and mechanical integrity of engineered vessels (A and B) as compared to nonpulsed vessels (C and D). The combination of medium supplements and pulsatile flow gave best results (E and F) as shown by dense cells ( ), accumulation of collagen (blue), presence of endothelial cells (scanning electron micrograph in G) and PECAM antigen (H), and rupture strengths greater than 2000 mmHg. Engineered vessels remained patent for more than 3 weeks following implantation in miniature swine, but only if precultured in the presence of pulsatile radial stress. The result highlights the importance of the bioreactor component in the engineering process. (From Niklason, L.E., Gao, J., Abbott, W.M., *et al.* (1999). Functional arteries grown *in vitro*. *Science*, 284(5413): 489–493.)

(reviewed in van der Meulen and Huiskes, 2002), engineered tissues at different stages of development, which contain different types and amounts of ECM components, may require different regimes of stimuli. To date, operating parameters for bioreactors applying physical conditioning have generally been determined and refined by a largely trial and error approach. To gain a more comprehensive understanding of how physical factors modulate tissue development, it will be necessary to integrate controlled bioreactor studies with quantitative analyses and computational modeling of mechanically induced fluid flows, changes in mass transport and physical forces experienced by the cells.

16.3 Bioreactor design and development

As our understanding of the aforementioned aspects of tissue engineering deepens, we are entering a time when the rational design of tissue engineering bioreactor systems is possible and, indeed, essential. Depending on its application, a tissue engineering bioreactor might be as simple as a spinner flask in an incubator or as complex as a fully self-contained, clinically deployed unit for the generation of implantable materials in humans. Regardless of the level of complexity, design of a tissue engineering bioreactor system can be approached as for classical engineering problems, beginning with a definition of the problem.

16.3.1 Problem definition

At its most basic level, the problem definition is a simple statement of the task at hand, e.g. 'to develop a tissue engineering bioreactor for the cultivation of chondrocytes'. However, a more precise definition of the objectives, such as 'to develop a tissue engineering bioreactor for the cultivation and mechanical stimulation of autologous human articular chondrocyte constructs in the clinic, for implantation into humans', can dramatically influence the design requirements. This detailed definition identifies the basic features of the reactor system (a mechanism for mechanical stimulation is required) and immediately introduces numerous constraints, including regulatory requirements. Additional restrictions regarding operation, performance and economics should also be identified (e.g. the client requires that cuboidal constructs of $1\,cm^3$ be produced in $6\text{-}cm^3$ lots, total system cost less than $50,000), as these will directly influence design and construction.

16.3.2 System design and components

With the problem defined, it is possible to conceptualize a solution for the bioreactor design (see the Example in Box 1). A basic concept of how to implement the desired seeding, cultivation and conditioning regimes evolves into a more in-depth system design and specification detailing operational parameters such as maximum/minimum temperature, pH, humidity, flow rates and pressure. Based on fundamental engineering principles, the operating parameters can in turn be used to calculate specific component requirements, such as tube diameter, pump speed and heating/cooling requirements. Basic calculations of this type are provided in the following example of bioreactor design. Details of these, and more in-depth calculations applicable to bioreactor design, can be found in any good text on process engineering or transport phenomena (e.g. Doran, 1995; Geankoplis, 2003; Perry & Green, 1997) and it is strongly recommended that for anyone

---

Box 1 Example

The following example illustrates basic design features and considerations for a simple perfusion system.

Description   To develop a bioreactor system for the seeding and culture of an osteoprogenitor cell suspension on porous ceramic scaffolds to generate osteoinductive grafts.

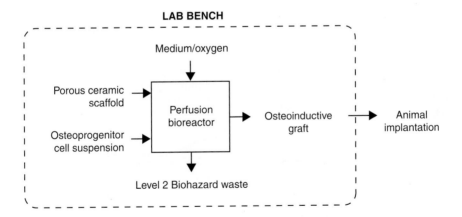

System concept   As the resulting construct should be uniform, both seeding and cultivation will utilize a perfusion-based approach. The reactor itself will be independent of incubators. Resulting grafts are for implantation into animals for research purposes. Scaffolds will be disks (8 mm diameter × 4 mm thick). Reusable components (reactor vessel and tubing) will be steam sterilized. An online biosensor can be added to monitor nutrient consumption online or, more simply, a port can be provided for offline medium sampling. In order to minimize medium usage (and cost), three-quarters of the medium is recycled.

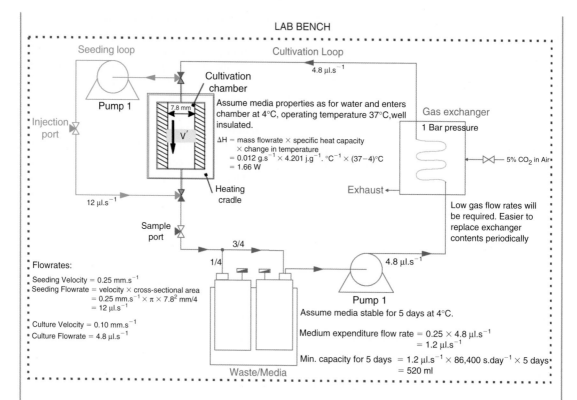

LAB BENCH

Seeding loop
Cultivation Loop

4.8 μl.s^{-1}

Pump 1

Cultivation chamber

7.8 mm

V′

Assume media properties as for water and enters chamber at 4°C, operating temperature 37°C, well insulated.

ΔH = mass flowrate × specific heat capacity × change in temperature
= 0.012 g.s^{-1} × 4.201 j.g^{-1}. °C^{-1} × (37−4)°C
= 1.66 W

Gas exchanger

1 Bar pressure

5% CO_2 in Air

Injection port

Heating cradle

Exhaust

Low gas flow rates will be required. Easier to replace exchanger contents periodically

12 μl.s^{-1}

Sample port

3/4

1/4

Pump 1

4.8 μl.s^{-1}

Flowrates:

Seeding Velocity = 0.25 mm.s^{-1}
Seeding Flowrate = velocity × cross-sectional area
= 0.25 mm.s^{-1} × π × 7.8^2 mm/4
= 12 μl.s^{-1}

Culture Velocity = 0.10 mm.s^{-1}
Culture Flowrate = 4.8 μl.s^{-1}

Waste/Media

Assume media stable for 5 days at 4°C.

Medium expenditure flow rate = 0.25 × 4.8 μl.s^{-1}
= 1.2 μl.s^{-1}

Min. capacity for 5 days = 1.2 μl.s^{-1} × 86,400 s.day^{-1} × 5 days
= 520 ml

System components

- *Pumps.* Peristaltic pumps provide reliable continuous flow and are ideal for aseptic operations. Pump 1: cells may be damaged during transit of the pump head. Convex rollers should be tested for suitability. Pump 2: flat rollers will prove more controlled flow at later stages of the culture as the scaffold pores become occluded, increasing the required operating pressure.

Flat rollers   Convex rollers

- *Tubing.* Tubing shown in red and green should exhibit low permeability to gas, be opaque (to prevent photodegradation of medium components) and have sufficient mechanical properties to withstand continuous operation in the selected pumps. Tubing shown in aqua acts as a gas exchange surface and must therefore be gas-permeable. Platinum-cured silicon tubing is ideal (high mechanical strength and opacity are not required).
- *Connectors and valves.* A wide variety of medical-grade items are readily available for purchase and routinely used in the clinic, and these are therefore a natural choice. Other major considerations are reliability and compatibility.

- *Heating cradle/medium reservoir.* The calculated minimum heating capacity of 1.66 W is small and easily within the capabilities of a Peltier unit. Potentially the cooling half of the unit might be coupled to the refrigeration requirements for the media reservoirs. The medium containers should be protected from light, and cooled/insulated to prevent degradation and prevent growth in the waste bottle.
- *Culture chamber.* The culture chamber must be constructed from biocompatible materials that exhibit minimal cell adhesion properties. As heat must be transferred from the cradle through the chamber and to the medium, it should be constructed from a good thermal conductor. In this case fluoropolymer-coated (internal) anodized aluminum provides a good combination of biocompatibility, corrosion resistance, mechanical strength and anti-adhesive properties, and is able to withstand repeated autoclaving.

not already familiar with the underlying theory that these be referred to.

In many cases, suitable products are available 'off the shelf', and can be selected based on the calculated requirements and materials considerations. Key points to consider are that bioreactor materials should be biocompatible (i.e. all materials in contact with the culture environment should be tested for cytotoxicity), nonsupportive of cell adhesion (unless specifically required) and durable (for the required operational life time). For clinical applications the reactor must be designed and developed in compliance with relevant GMP (Good Manufacturing Practice) regulations and materials.

Although often overlooked, features such as handling and ease of cleaning should also be given attention throughout the design process. Likewise, integration of various components and unit operations in a possibly modular design deserves major consideration.

16.3.3 Development and implementation

With the basic bioreactor established, ongoing development and design refinements lead towards an optimized system. During this time it is critical that effective means of quantifying reactor performance are implemented. Such measures might include quantification of cell seeding efficiency, extent of cell differentiation or metabolic parameters and may well have been specified within the context of the problem definition (e.g. must achieve a cell seeding efficiency of not less than 70%).

Ultimately, at the time of implementation the bioreactor system as a whole must meet the specified design criteria (an optimized component does not always correspond to an optimized system). The bioreactor system should also provide the advantages identified earlier in the chapter, i.e. to provide control and standardization or to scale and automate tissue manufacture.

16.4 Bioreactors as 3D *in vitro* model systems

There is an increasing recognition that 3D culture of cells on scaffolds has significantly more relevance for fundamental biological research than standard Petri dish cultures (Abbott, 2003). To serve as a biologically sound *in vitro* system, a 3D culture needs to recapitulate some aspects of the actual cellular microenvironment that exists *in vivo*. *In vivo*, the processes of cell differentiation and tissue assembly are directed by multiple factors acting in concert, and according to specific spatial and temporal sequences (Figure 16.7). It is thought that the cell function *in vitro* can be mediated by the same factors known to play a role *in vivo*. The factors of interest for tissue engineering include cytokines, growth and transcription factors, hydrodynamic shear and pressure, and mechanical and electrical signals. Biophysical regulation of cells cultured on scaffolds by combinations of these factors can be achieved by using bioreactors that provide the necessary environmental control and the application of regulatory factors (Figure 16.8).

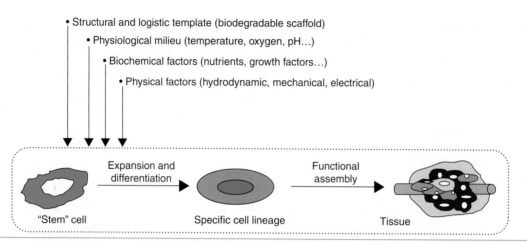

- Structural and logistic template (biodegradable scaffold)
  - Physiological milieu (temperature, oxygen, pH…)
    - Biochemical factors (nutrients, growth factors…)
      - Physical factors (hydrodynamic, mechanical, electrical)

"Stem" cell → Expansion and differentiation → Specific cell lineage → Functional assembly → Tissue

Figure 16.7 Developmental paradigm. Tissue development and remodeling, *in vivo* and *in vitro*, involves the proliferation and differentiation of stem/progenitor cells and their subsequent assembly into tissues. Cell function and tissue assembly depend on (a) the availability of a scaffold for cell attachment and tissue formation, (b) the maintenance of physiological conditions in cell/tissue environment, (c) supply of nutrients, oxygen, metabolites and growth factors, and (d) presence of physical regulatory factors.

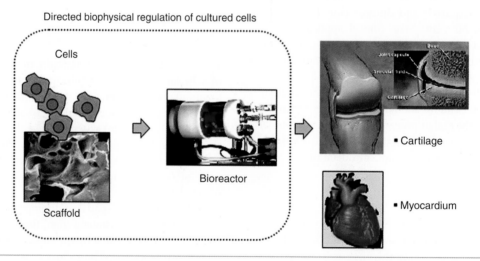

Directed biophysical regulation of cultured cells

Cells

Scaffold

Bioreactor

- Cartilage

- Myocardium

Figure 16.8 Tissue engineering paradigm. The regulatory factors of cell function and tissue assembly depicted in Figure 16.7 can be utilized *in vitro* to engineer functional tissues. The cells themselves (either differentiated or progenitor/stem cells seeded onto a scaffold and cultured in a bioreactor) carry out the process of tissue formation in response to regulatory signals. The scaffold provides a structural, mechanical and logistic template for cell attachment and tissue formation. Bioreactors provide the environmental conditions necessary for cell growth and differentiation, via control of flow patterns and mass transport, and the application of regulatory signals (biochemical and physical) according to specific spatial and temporal regimes. Clearly, the regulatory cascades associated with specific cell functions are still largely unknown. In this respect, bioreactors help determine the individual and interactive effects of specific factors and the underlying mechanisms of their action.

State of the Art Experiment

Bone marrow stromal cells (BMSCs) represent a very small percentage of the total number of nucleated cells in a bone marrow aspirate (approximately 0.01%). To obtain a sufficient number of cells for bone tissue engineering applications, BMSCs are therefore typically first selected and expanded in monolayer (2D) prior to loading into 3D scaffolds. However, 2D-expanded BMSC have a dramatically reduced differentiation capacity in comparison with those found in fresh bone marrow (Banfi *et al.*, 2000; Mendes *et al.*, 2002), placing potential limits on their clinical utility. To bypass the process of 2D expansion and its associated limitations, Braccini *et al.* used an innovative bioreactor-based approach to seed, expand and differentiate BMSC directly in a 3D ceramic scaffold (Braccini *et al.*, 2005). Nucleated cells, freshly isolated from a bone marrow aspirate, were introduced into direct perfusion bioreactor system and perfused through the pores of 3D ceramics for 5 days, then further cultured under perfusion for an additional 2 weeks. Using the developed procedure, (a) BMSC could be seeded and extensively expanded (eight doublings) within the 3D environment of the ceramic pores. Interestingly, (b) the system supported the formation of stromal-like tissues within the pores, where BMSC could be cocultured with hematopoietic progenitor cells. (c) When constructs were implanted ectopically in nude mice, those engineered in the bioreactor reproducibly generated bone tissue that was uniformly distributed throughout the scaffold volume and filled an average of 52.1 ± 7.7% of the ceramic pores. (d) In marked contrast, when similar numbers of 2D-expanded BMSC were loaded into ceramic scaffolds and implanted, bone was infrequently generated and even in the most osteoinductive constructs was localized to peripheral regions, filling only 9.6 ± 2.7% of the ceramic pore volume. This paradigm represents a promising approach towards establishing a 3D *in vitro* model system simulating bone marrow, which could be used to investigate interactions between different populations of bone marrow cells in a more physiological environment than previously established systems (e.g. Petri dishes or spinner flasks). Furthermore, this study has significant implications for clinical applications of osteoinductive grafts considering that the elimination of a 2D expansion phase would facilitate the development of a more streamlined, effective, reproducible and economical manufacturing process for generating autologous BMSC-based bone grafts.

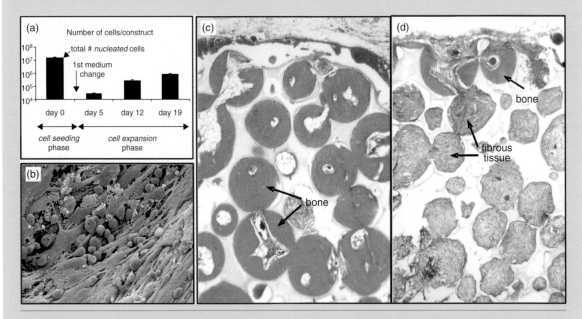

(From Braccini, A., Wendt, D., Jaquiery, C., et al. (2005). Three-dimensional perfusion culture of human bone marrow cells. Stem Cells, 23(8): 1066–1072.)

We will now discuss two examples in which the use of bioreactors was key to establish biologically sound yet controllable *in vitro* models and to obtain quantitative experimental data, which were then rationalized by mathematical models: (i) the progression of cartilage development and (ii) oxygen transport in engineered cardiac muscle.

16.4.1 Bioreactor studies of tissue-engineered cartilage

Engineered cartilages cultured in bioreactors with hydrodynamically active environments (involving mixing of culture medium) are generally structurally and functionally superior to those grown in static cultures. A system that gave particularly good results is the rotating bioreactor. One configuration comprises two concentric cylinders (the inner one serving as a gas exchange membrane) that are rotated around their horizontal axis at the rate of 15–40 r.p.m. The annular space between the cylinders is 110 ml in volume, and can hold up to 12 tissue constructs that are 1 cm in diameter and 5 mm thick. The rotating rate is adjusted such that the settling constructs are suspended in culture medium without external fixation. Importantly, settling of tissue constructs in rotating flow is associated with dynamic changes in the velocity, shear and pressure at construct surfaces. Although the mechanisms underlying the observed enhancement of cartilage development (chondrogenesis) under these conditions are yet to be determined, the effects were attributed to the convective flow at construct surfaces and enhanced mass transport at tissue surfaces.

Based on the high quality of tissue-engineered cartilage, rotating bioreactors were used to study the spatial and temporal patterns of cartilage development by bovine calf chondrocytes cultured on fibrous polyglycolic acid scaffolds. The distributions of cells and cartilaginous tissue matrix are shown in Figure 16.9. Cells at the construct periphery proliferated more rapidly during the first 4 days of culture and initiated matrix deposition in this same region (Figure 16.9d). Over time, chondrogenesis progressed both inward towards the construct center and outward from its surface. Cell density gradually decreased and became more uniform as the cells separated themselves by newly synthesized matrix and the construct size increased (Figure 16.9d–f). After 10 days of culture, cartilaginous tissue was formed at the construct periphery (Figure 16.9b). By 6 weeks of culture, self-regulated cell proliferation and deposition of cartilaginous matrix yielded constructs that had physiological cell density and spatially uniform distributions of matrix components (Figure 16.9c).

The development of engineered cartilage was analyzed using a spatially varying, deterministic continuum model (Obradovic *et al.*, 2000). The model accounted for the diffusion of oxygen and its utilization by the cells, and diffusion of newly synthesized glycosaminoglycan (GAG) and its deposition within the tissue constructs. Model predictions of GAG concentration profiles were consistent with those measured via high-resolution (40 μm) image processing of histological sections of tissue samples harvested at timed intervals (Martin *et al.*, 1999). Mathematical models of this kind helped rationalize experimental observations, identify some of the mechanisms affecting tissue regeneration and design advanced systems for cartilage tissue engineering. For example, the model confirmed the first-order dependence of GAG synthesis on local concentration of oxygen. The model also helped identify that there is product inhibition of GAG synthesis via a feedback mechanism by which the cells control their immediate environment. Together, these two effects helped explain the experimentally measured gradients in GAG concentrations within cultured tissues. Notably, these relationships could not be identified without the mathematical model by simply correlating the measured data. Model development and verification would not be possible without the utilization of bioreactors capable of providing controlled conditions for tissue growth.

Further studies are needed to extend the experimentation and modeling to functional parameters of engineered cartilage and to the bioreactors that more closely mimic the environment of a joint, and include perfusion and mechanical loading during cultivation.

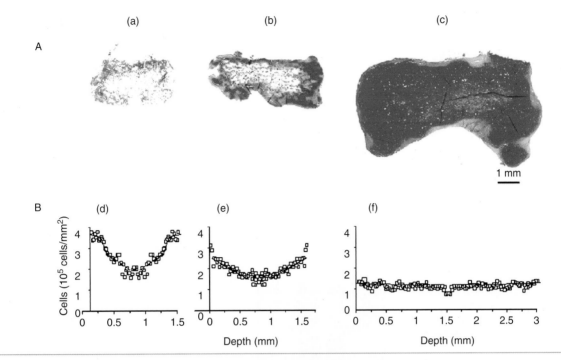

Figure 16.9 Bioreactor studies of *in vitro* chondrogenesis. Bovine chondrocytes were seeded into fibrous scaffolds made of polyglycolic acid and transferred into rotating bioreactors. At timed intervals, tissue constructs were sampled and analyzed to assess the amounts and distributions of cells and tissue matrix. (A) Full cross-sections of tissue constructs after (a) 3 days, (b) 10 days and (c) 6 weeks of culture. Stain: safranin-O/fast green. Scale bar: 1 mm.The intensity of stain correlates with the concentration of glycosaminoglycan (GAG), one of the two main components of cartilaginous tissue matrix. (B) Spatial profiles of cell distribution after (d) 3 days, (e) 10 days and (f) 6 weeks of culture (measured by image processing). (From Obradovic, B., Meldon, G., et al. (2000). Glycosaminoglycan deposition in engineered cartilage: experiments and mathematical model. *AICHE J*, 46(9): 1860–1871.)

16.4.2 Bioreactor studies of engineered cardiac muscle

In native cardiac tissue, a high cell density (around 10^8 cells/cm^3) is supported by the flow of oxygen-rich blood through a dense capillary network. Oxygen diffuses from the blood into the tissue space surrounding each capillary. Since the solubility of oxygen in plasma at 37°C is very low ($130\,\mu M$ in arterial blood), the presence of a natural oxygen carrier, hemoglobin, increases the total oxygen content of blood by carrying 65 times more oxygen than the blood plasma alone ($8630\,\mu M$) (Fournier, 1998). Under physiologic conditions, only a fraction of oxygen is depleted from the blood hemoglobin in a single pass through a capillary network.

Some aspects of the native environment in cardiac muscle were recapitulated using bioreactors, with the goal to study factors that determine the development of thick, synchronously contracting cardiac constructs consisting of functionally coupled viable cells. To mimic the capillary network, cardiomyocytes and fibroblasts isolated from neonatal rat hearts were cultured on elastomer scaffolds with a parallel array of channels that were perfused with culture medium (Figure 16.10). To mimic oxygen supply

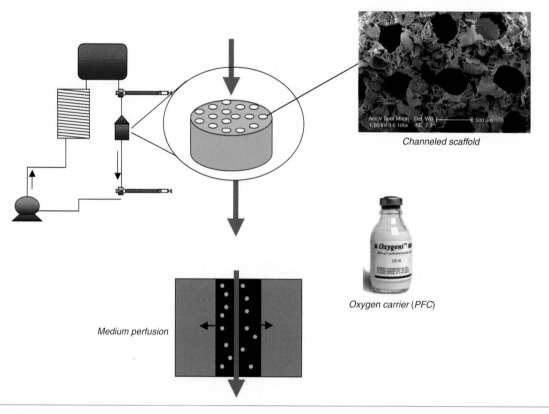

Channeled scaffold

Oxygen carrier (PFC)

Medium perfusion

Figure 16.10 Biomimetic system for studies of oxygen transport in engineered cardiac tissue. Cell populations obtained from neonatal rat hearts were seeded into highly porous elastomer scaffolds with an array of channels (to mimic the capillary network) and perfused with culture medium containing an oxygen carrier (to mimic to role of hemoglobin). (From Radisic, M., Deen, W., Langer, R. and Vunjak-Novakovic, G. (2005). Mathematical model of oxygen distribution in engineered cardiac tissue with parallel channel array perfused with culture medium containing oxygen carriers. *Am J Physiol Heart Circ Physiol*, 288: H1278–H1289.)

by hemoglobin, culture medium was supplemented with a perfluorocarbon (PFC) emulsion; constructs perfused with unsupplemented culture medium served as controls. In PFC-supplemented medium, the decrease in the partial pressure of oxygen in the aqueous phase was only 50% of that in control medium (28 versus 45 mmHg between the construct inlet and outlet at the flow rate of 0.1 ml/min). Consistently, constructs cultivated in the presence of PFC had higher amounts of DNA, troponin I and Cx-43, and significantly better contractile properties as compared to control constructs. Improved construct

properties were correlated with the enhanced supply of oxygen to the cells within constructs.

A mathematical model of oxygen distribution in cardiac constructs with an array of channels was developed. Concentration profiles of oxygen and cells within the constructs were obtained by numerical simulation of the diffusive-convective oxygen transport and its utilization by the cells. The model was used to evaluate the effects of medium perfusion rate, oxygen carrier and scaffold geometry on viable cell density. The model was first implemented for the set of parameters relevant for the work proposed in this application: construct thickness 2 mm, channel diameter 330 μm, channel wall-to-wall spacing 370 μm and medium perfusion velocity of 0.049 cm/s. Subsequently, the model was used to define scaffold geometry and flow conditions necessary to cultivate cardiac constructs with clinically relevant thicknesses (5 mm). In future, the model can be used as a tool for optimization of scaffold geometry and flow conditions.

Taken together, these studies suggest that bioreactor-based 3D model systems, designed to mimic specific aspects of the native cell/tissue milieu, can be a powerful *in vitro* tool for quantitative biological research. Optimization and standardization of these culture systems offers the additional possibility for controlled studies of cell–cell interactions, enzyme induction and cell metabolism, with the potential for automation and high-throughput drug screening (Gebhardt *et al.*, 2003).

16.5 Bioreactors in clinical applications

A major challenge to bring a tissue-engineered product into routine clinical practice is to translate research-scale production models into clinically applicable manufacturing designs that are reproducible, clinically effective, economically acceptable and compliant with GMP requirements. While production processes for tissue engineering products currently rely on unautomated manual techniques, it appears inevitable that innovative and low-cost bioreactor systems which automate, standardize and scale the production process will be *central* to future manufacturing strategies, and will play a *key* role in the successful exploitation of a tissue-engineered product for widespread clinical use (Figure 16.11).

Design of an effective manufacturing process, which fits the long-term goals and expectations of the company, must be initiated early in the product development strategy. Altering an established process which down the road proves inefficient or cannot meet product demand will require considerable resources and

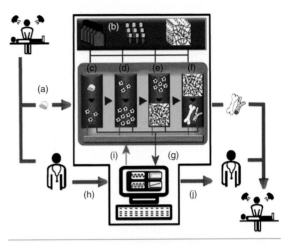

Figure 16.11 Vision for a closed-system bioreactor for the automated production of tissue-engineered grafts. (a) The surgeon would take a biopsy from the patient and introduce it into the bioreactor located on-site at the hospital. (b) All reagents (e.g. culture medium, medium supplements and scaffolds) would be stored in compartments under appropriate conditions (i.e. temperature, humidity). The bioreactor system could then (c) automatically isolate the cells, (d) expand the cells, (e) seed the cells onto a scaffold and (f) culture the construct until a suitably developed graft is produced. (g) Environmental culture parameters and tissue development would be monitored and inputs fed into a microprocessor unit for analysis. In conjunction with data derived from clinical records of the patient (h), the inputs would be used to control culture parameters at predefined optimum levels automatically (i) and provide the surgical team with data on the development of the tissue, enabling timely planning of the implantation (j) Figure generated by M. Moretti.

incur substantial costs, requiring revalidation, and new regulatory review and approval. While a simple manufacturing design may minimize initial development costs, it may compromise the capacity for scale-up and increase product cost at later stages when a company is attempting to move to profitability. Alternatively, a manufacturing design that addresses scale-up and efficiency early in the product development process maximizes the potential to manufacture large numbers in a uniform and reproducible manner in the long-term, but incurs high initial costs, possibly at a time when a company may be trying to conserve funds and minimize costs. Below we describe several manufacturing strategies for engineering tissue grafts, ranging from current labor-intensive approaches through the concept of a fully automated and controlled closed bioreactor system.

Carticel®, produced by Genzyme Tissue Repair, is an autologous cell transplantation product for the repair of articular cartilage defects currently used in the clinic. To produce the Carticel® product, a cartilage biopsy is harvested by a surgeon and sent to a centralized facility where the chondrocytes are isolated and expanded to generate a sufficient number of cells (Mayhew *et al.*, 1998) using routine culture systems (i.e. manually by a lab technician, inside a biological safety cabinet, housed in a bio-class 10,000 clean room). Hyalograft C™, marketed by Fidia Advanced Biopolymers, is an alternative autologous cell-based product for the treatment of articular cartilage defects. Chondrocytes, obtained from a biopsy and expanded at a centralized facility, are seeded onto a 3D scaffold and cultured for 14 days using routine tissue culture techniques (Scapinelli *et al.*, 2002). For both of these products, the simple production systems kept initial product development costs down as the products were established within the marketplace. However, considering that these manufacturing designs are processes requiring a large number of manual and labor-intensive manipulations, in addition to the autologous nature of the product (each cell preparation is treated individually), the costs of the products are rather high and the ability to serve an increasing number of patients per year may be challenging.

Dermagraft, developed by Advanced Tissue Sciences and currently manufactured by Smith & Nephew, is an allogeneic product manufactured with dermal fibroblasts grown on a scaffold for the treatment of chronic wounds such as diabetic foot ulcers (Marston *et al.*, 2003). Dermagraft is produced using a closed manufacturing system (Figure 16.12), consisting of a set of

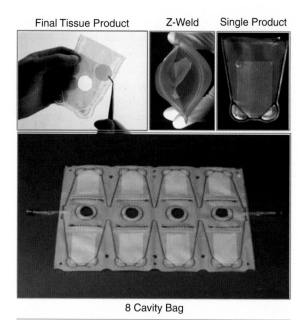

Final Tissue Product   Z-Weld   Single Product

8 Cavity Bag

Figure   16.12 Dermagraft   production   system and final product unit. One 2 × 3-inch piece of Vicryl scaffold is Z-welded into an ethylene vinyl acetate (EVA) compartment into which cells are seeded to grow the dermal tissue. The scaffold is held in place along the top and the bottom with welds. The single product unit is manufactured from an eight-cavity multiple processing bag, which is grown vertically in a corrugated manifold system (manifolding 12 of the eight-cavity bags, resulting in a lot of 96 units), designed for even mass transfer and exposure to environmental conditions. The EVA bioreactor is designed to be translucent, enabling placement on the wound, tracing of the wound and cutting of a piece of tissue to the desired size for implantation. (Naughton, G.K. (2002). From lab bench to market: critical issues in tissue engineering. *Ann NY Acad Sci*, 961: 372–385.)

cultivation bags with eight cavities, in which cells are seeded onto a scaffold, the growth process occurs, cryopreservation is performed and the cells are finally transported to the clinic. Twelve sets of eight-cavity bags are manifolded together to result in a lot of 96. This simple system allows for incremental scale-up, multiple lots to be manufactured together and the uniform generation of large numbers of units at any one time. Although not applicable to autologous cell-based tissue engineering approaches, this scaleable manufacturing system has significant implications on costs of production for allogeneic products. Drawbacks to this strategy include the high level of technical difficulty in product development, long product development time, resultant high product development costs and high initial costs of manufacture (until large enough numbers of units are manufactured).

As an alternative to manufacturing engineered products within *centralized* production facilities (Figure 16.13a) as in the examples described above, a *decentralized* production system, such as a fully automated closed-bioreactor system, could be located on-site within the confines of a hospital. This strategy is exemplified by the concept of ACTES™ (Autologous Clinical Tissue Engineering System), currently under development by Millenium Biologix (Figure 16.13b). As a fully automated, closed bioreactor system, ACTES™ would digest a patient's cartilage biopsy, expand the chondrocytes, and provide (i) an autologous cell suspension or (ii) an osteochondral graft (CartiGraft™), generated by seeding and culturing the cells onto the surface of an osteoconductive porous scaffold. Clearly, a decentralized strategy such as this possesses the greatest risks upfront, requiring an extensive development time, and significant investments in costs in a highly technical and complex bioreactor system. In the long-term, however, systems like ACTES™ would eliminate complicated logistical issues of transferring biopsies and engineered products between locations, eliminate the need for large and expensive GMP tissue engineering facilities, and facilitate scale-up and minimize labor-intensive operator handling, likely reducing the cost of the engineered product.

The broad spectrum of manufacturing designs described show a range of technologies, level of difficulty, time for development and expense. Low-level technology allows for more rapid entry to clinical trials and to market with reduced cost, but downstream in the commercialization efforts this can result in major production challenges as demand for product increases. Higher-level technology involves longer development time and increased costs, and prepares the company for a higher level of sales. However, the higher costs may be incurred for a substantial period of time before the level of sales is able to provide a return of the initial substantial investment. Ultimately, the most appropriate strategy should be determined based on the amount of scientific and clinical data already available to support commercialization of the envisaged cell-based product.

16.6 Future perspectives for bioreactors in tissue engineering

The generation of 3D tissues *ex vivo* not only requires the development of new biological models rather than those already established for traditional monolayer or micromass cell cultures (Cukierman *et al.*, 2001), but also poses new technical challenges owing to the physicochemical requirements of large cell masses. In this context, bioreactors represent a key tool for all processes involved in the engineering of 3D tissues based on cells and scaffolds, including monitoring and control of the relevant culture parameters.

These features are essential for the streamlined and possibly more cost-effective manufacture of engineered grafts, and could thus facilitate a broader commercialization and clinical implementation of tissue-engineered products. The possibility of reproducible and controlled 3D cultures would also support fundamental studies aiming at investigating mechanisms of cell differentiation and tissue development in biologically relevant models. Understanding the role of specific biochemical or physical factors in tissue development will in turn not only support a more efficient engineering of grafts, but also set the stage for the application of those same factors directly at the site where tissue has to be regenerated. In this regard, bioreactors for *in vitro* culture might be instrumental to identify the requirements for '*in vivo* bioreactors' (Stevens *et al.*, 2005), thus promoting the shift from tissue engineering approaches to the more challenging field of regenerative medicine.

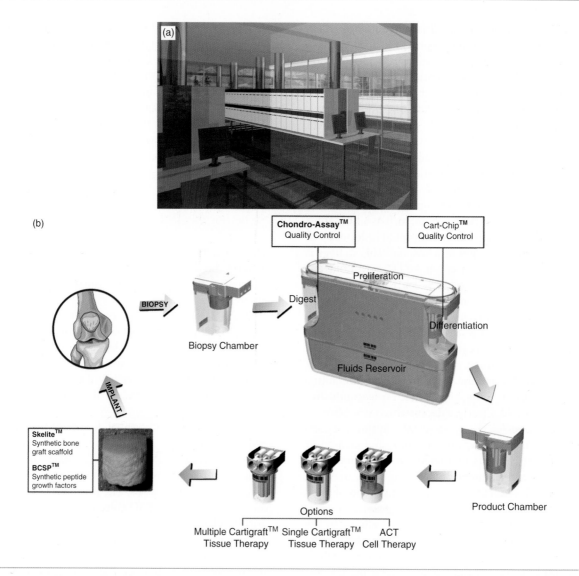

Figure 16.13 Centralized and decentralized production units. (a) Computer controlled culture system in a clean room for the *centralized* production of autologous tissue-engineered grafts (EGM Architects/IsoTis Orthobiologics). (b) A single unit of the decentralized ACTES™ production system (Millenium Biologix).

16.7 Summary

- Similar to bioreactors in classical applications, the key functions of bioreactors in tissue engineering are to provide control and standardization of physiochemical culture parameters during cell/tissue culture.

- Bioreactors can improve the quality (i.e. cell distribution and cell utilization) and reproducibility of the process of seeding cells into 3D porous scaffolds.

- Mass transport of nutrients and waste products to and from cells within engineered constructs can be enhanced by convective bioreactor systems. Bioreactors which perfuse media directly through the scaffold have the greatest potential to eliminate mass transport limitations and maintain cell viability within large 3D constructs.
- Mechanical conditioning within controlled bioreactor systems has the potential to improve the structural and functional properties of engineered tissues. However, optimizing the operating parameters (i.e. which specific mechanical force(s) and regimes of application) for a particular tissue will require significant quantitative analysis and computational modeling.
- By recapitulating aspects of the actual cellular microenvironment that exists *in vivo*, bioreactors can provide *in vitro* model systems to investigate cell function and tissue development in 3D environments.
- Design and development of a tissue engineering bioreactor system should be approached as for classical engineering problems: define the problem, conceptualize the solution, develop a prototype, quantify reactor performance, refine the design, validate reactor performance.
- Innovative and low-cost bioreactor systems that automate, standardize and scale the production of a tissue-engineered product will be *central* to future manufacturing strategies, and will play a *key* role in the successful exploitation of an engineered product for widespread clinical use.

References

Abbott, A. (2003). Cell culture: biology's new dimension. *Nature*, 424: 870–872.

Altman, G.H., Horan, R.L., Martin, I., *et al.* (2002a). Cell differentiation by mechanical stress. *FASEB J*, 16: 270–272.

Altman, G.H., Lu, H.H., Horan, R.L., *et al.* (2002b). Advanced bioreactor with controlled application of multi-dimensional strain for tissue engineering. *J Biomech Eng*, 124: 742–749.

Bancroft, G.N., Sikavitsas, V.I., van den Dolder, J., *et al.* (2002). Fluid flow increases mineralized matrix deposition in 3D perfusion culture of marrow stromal osteoblasts in a dose-dependent manner. *Proc Natl Acad Sci USA*, 99: 12600–12605.

Banfi, A., Muraglia, A., Dozin, B., *et al.* (2000). Proliferation kinetics and differentiation potential of *ex vivo* expanded human bone marrow stromal cells: implications for their use in cell therapy. *Exp Hematol*, 28: 707–715.

Braccini, A., Wendt, D., Jaquiery, C., *et al.* (2005). Three-dimensional perfusion culture of human bone marrow cells. *Stem Cells*, 23: 1066–1072.

Cioffi, M., Boschetti, F., Raimondi, M.T. and Dubini, G. (2006). Modeling evaluation of the fluid-dynamic microenvironment in tissue-engineered constructs: a micro-CT based model. *Biotechnol Bioeng*, 93: 500–510.

Cukierman, E., Pankov, R., Stevens, D.R. and Yamada, K.M. (2001). Taking cell–matrix adhesions to the third dimension. *Science*, 294: 1708–1712.

Davisson, T., Kunig, S., Chen, A., *et al.* (2002a). Static and dynamic compression modulate matrix metabolism in tissue engineered cartilage. *J Orthop Res*, 20: 842–848.

Davisson, T., Sah, R.L. and Ratcliffe, A. (2002b). Perfusion increases cell content and matrix synthesis in chondrocyte three-dimensional cultures. *Tissue Eng*, 8: 807–816.

Demarteau, O., Wendt, D., Braccini, A., *et al.* (2003). Dynamic compression of cartilage constructs engineered from expanded human articular chondrocytes. *Biochem Biophys Res Commun*, 310: 580–588.

Doran, P.M. (1995). *Bioprocess Engineering Principles*. London: Academic Press.

Fournier, R.L. (1998). *Basic Transport Phenomena in Biomedical Engineering*. Philadelphia, PA: Taylor & Francis.

Gebhardt, R., Hengstler, J.G., Muller, D., *et al.* (2003). New hepatocyte *in vitro* systems for drug metabolism: metabolic capacity and recommendations for application in basic research and drug development, standard operation procedures. *Drug Metab Rev*, 35: 145–213.

Geankoplis, C.J. (2003). *Transport Processes and Separation Process Principles*. (4th edn.). Upper Saddle River, NJ: Prentice Hall.

Goldstein, A.S., Juarez, T.M., Helmke, C.D., *et al.* (2001). Effect of convection on osteoblastic cell growth and function in biodegradable polymer foam scaffolds. *Biomaterials*, 22: 1279–1288.

Grad, S., Lee, C.R., Gorna, K., *et al.* (2005). Surface motion upregulates superficial zone protein and hyaluronan production in chondrocyte-seeded three-dimensional scaffolds. *Tissue Eng*, 11: 249–256.

Janssen, F.W., Oostra, J., van Oorschot, A., *et al.* (2006a). A perfusion bioreactor system capable of producing clinically relevant volumes of tissue engineered bone: *In vivo* bone formation showing proof of concept. *Biomaterials*, 27: 315–323.

Janssen, F.W., Hofland, I., Van Oorschot, A., *et al.* (2006b). Online measurement of oxygen consumption by bone marrow stromal cells in a combined cell-seeding and proliferation bioreactor. *J Biomater Res A*, 79: 338–348.

Lewis, M.C., Macarthur, B. D., Malda, J., *et al.* (2005). Heterogeneous proliferation within engineered cartilaginous tissue: the role of oxygen tension. *Biotechnol Bioeng*, 91: 607–615.

Malda, J., Rouwkema, J., Martens, D.E., *et al.* (2004). Oxygen gradients in tissue-engineered PEGT/PBT cartilaginous constructs: measurement and modeling. *Biotechnol Bioeng*, 86: 9–18.

Marston, W.A., Hanft, J., Norwood, P. and Pollak, R. (2003). The efficacy and safety of Dermagraft in improving the healing of chronic diabetic foot ulcers: results of a prospective randomized trial. *Diabetes Care*, 26: 1701–1705.

Martin, I., Obradovic, B., Freed, L.E. and Vunjak-Novakovic, G. (1999). Method for quantitative analysis of glycosaminoglycan distribution in cultured natural and engineered cartilage. *Ann Biomed Eng*, 27: 656–662.

Mayhew, T.A., Williams, G.R., Senica, M.A., *et al.* (1998). Validation of a quality assurance program for autologous cultured chondrocyte implantation. *Tissue Eng*, 4: 325–334.

Mendes, S.C., Tibbe, J.M., Veenhof, M., *et al.* (2002). Bone tissue-engineered implants using human bone marrow stromal cells: effect of culture conditions and donor age. *Tissue Eng*, 8: 911–920.

Naughton, G.K. (2002). From lab bench to market: critical issues in tissue engineering. *Ann NY Acad Sci*, 961: 372–385.

Niklason, L.E., Gao, J., Abbott, W.M., *et al.* (1999). Functional arteries grown *in vitro*. *Science*, 284: 489–493.

Obradovic, B., Meldon, J.H., Freed, L.E. and Vunjak-Novakovic, G. (2000). Glycosaminoglycan deposition in engineered cartilage: experiments and mathematical model. *AICHE J*, 46: 1860–1871.

Perry, R.H. and Green, D.W. (1997). *Perry's Chemical Engineers' Handbook*. (7th edn.). New York: McGraw-Hill.

Porter, B., Zauel, R., Stockman, H., *et al.* (2005). 3-D computational modeling of media flow through scaffolds in a perfusion bioreactor. *J Biomech*, 38: 543–549.

Radisic, M., Euloth, M., Yang, L., *et al.* (2003). High-density seeding of myocyte cells for cardiac tissue engineering. *Biotechnol Bioeng*, 82: 403–414.

Radisic, M., Yang, L., Boublik, J., *et al.* (2004). Medium perfusion enables engineering of compact and contractile cardiac tissue. *Am J Physiol Heart Circ Physiol*, 286: H507–H516.

Radisic, M., Deen, W., Langer, R. and Vunjak-Novakovic, G. (2005). Mathematical model of oxygen distribution in engineered cardiac tissue with parallel channel array perfused with culture medium containing oxygen carriers. *Am J Physiol Heart Circ Physiol*, 288: H1278–H1289.

Scapinelli, R., Aglietti, P., Baldovin, M., *et al.* (2002). Biologic resurfacing of the patella: current status. *Clin Sports Med*, 21: 547–573.

Sikavitsas, V.I., Bancroft, G.N., Lemoine, J.J., *et al.* (2005). Flow perfusion enhances the calcified matrix deposition of marrow stromal cells in biodegradable nonwoven fiber mesh scaffolds. *Ann Biomed Eng*, 33: 63–70.

Stevens, M.M., Marini, R.P., Schaefer, D., *et al.* (2005). *In vivo* engineering of organs: the bone bioreactor. *Proc Natl Acad Sci USA*, 102: 11450–11455.

Sucosky, P., Osorio, D.F., Brown, J.B. and Neitzel, G.P. (2004). Fluid mechanics of a spinner-flask bioreactor. *Biotechnol Bioeng*, 85: 34–46.

van der Meulen, M.C. and Huiskes, R. (2002). Why mechanobiology? A survey article. *J Biomech*, 35: 401–414.

Vunjak-Novakovic, G., Martin, I., Obradovic, B., *et al.* (1999). Bioreactor cultivation conditions modulate the composition and mechanical properties of tissue-engineered cartilage. *J Orthop Res*, 17: 130–138.

Wendt, D., Marsano, A., Jakob, M., *et al.* (2003). Oscillating perfusion of cell suspensions through three-dimensional scaffolds enhances cell seeding efficiency and uniformity. *Biotechnol Bioeng*, 84: 205–214.

Wendt, D., Stroebel, S., Jakob, M., *et al.* (2006). Uniform tissues engineered by seeding and culturing cells in 3D scaffolds under perfusion at defined oxygen tensions. *Biorheology*, 43: 481–488.

Wimmer, M.A., Grad, S., Kaup, T., *et al.* (2004). Tribology approach to the engineering and study of articular cartilage. *Tissue Eng*, 10: 1436–1445.

Chapter 17
Tissue engineering for skin transplantation

Richard Price, Edwin Anthony, Simon Myers and Harshad Navsaria

Chapter objectives:

- To understand the structure and function of skin
- To know the function of differentiation markers and role of basement membrane
- To understand the concept of keratinocyte cell expansion in vitro
- To know different delivery systems for keratinocyte application

- To understand the significance of dermis for keratinocyte grafting
- To distinguish different types of tissue engineered dermal templates/products
- To understand the role of keratinocyte stem cells for clinical application

"The need for artificial material capable of substituting for skin has long been recognized. . ."

Burke *et al.* (1981)

17.1 Introduction

There are a number of pathologies in which skin loss or damage is a salient feature. Most notable amongst these are burns, in particular large-area injuries, and chronic ulcers. Problems common to both pathologies include the loss of superficial epidermis and dermis, abnormal wound healing, and the failure of (or delay associated with) wound healing. Both pathologies present enormous economical and medical challenges, and have received a large amount of interest with the advent of tissue engineering.

The scale of the problem of burns injuries has recently been outlined (Wedler *et al.*, 1999). In 1995, almost 25,000 patients (of a population of 512 million) were hospitalized throughout Europe; McManus and Pruitt (1996) have given a figure of 1.2 million in the US (population 268 million). House fires account for only 4% of admissions, but 70% of fatalities, mainly (75%) because of the inhalation injury associated with fires in enclosed spaces; 20% of such patients die of the burn injury itself and the remaining 5% of metabolic or infectious complications (Hasselgren, 1999). Between 5 and 10% of admitted patients require admission to a specialized burn unit and at least 1% die (Nguyen *et al.*, 1996). Ten years ago the financial cost of the treatment of major burns was calculated to be NZ$927 (€451) per percentage burn surface area (Lofts, 1991).

When contemplating reconstruction in burns injuries, the most important feature is the depth of injury; this determines the innate ability of the epidermis to undergo self-regeneration and, therefore, to achieve wound closure. The first efforts to classify the depth of the burn injury were by Jean de Vigo (1460–1525) of Italy in 1483, who recognized superficial and deep wounds, but it was not until 1953 that Jackson clearly defined the current method of burn classification (Jackson, 1953). Jackson's description includes superficial partial skin loss, deep partial skin loss and full-thickness skin loss, and went on to note the poor and delayed healing associated with the latter, recommending skin grafting in these cases (Wallace, 1982).

Chronic leg ulcers represent another spectrum of epithelial pathology; in these wounds there is altered vascular pathophysiology and other microscopic features such as cell senescence (Bruce and Deamond, 1991; Regan *et al.*, 1991; Hehenberger *et al.*, 1998; Van de Berg *et al.*, 1998) and altered cytokine profiles or responsiveness (Unemori *et al.*, 1994; Hasan *et al.*, 1997; Stanley *et al.*, 1997; He *et al.*, 1999). The cost of such ulcers is difficult to assess and estimates vary from less than £20 (€29) per week (Morrell *et al.*, 1998) to more than $250 (€194) per week (Marston *et al.*, 1999); given the chronic nature of the pathology the total cost per patient may be enormous.

17.2 Structure of the epidermis

The normal epidermis consists of a multilayered cellular structure with little or no extracellular matrix (ECM). The major cell, making up 95% of the total, is the keratinocyte, a terminally differentiated cell that moves progressively from deep to superficial before being shed from the outer surface as a keratin scale. On the basis of changes in the cell structure, the epidermis may be divided (from deep to superficial) into four layers: the strata basale, spinosum (prickle), granulosum and corneum (Figure 17.1).

The stratum basale consists of cuboidal cells, usually one cell deep, in a continuous layer and attached to the basement membrane. The epibasal layer, the stratum spinosum, is distinguished by numerous desmosomes interconnecting the cells to produce a stable network of cells. The desmosomes appear to provide not only

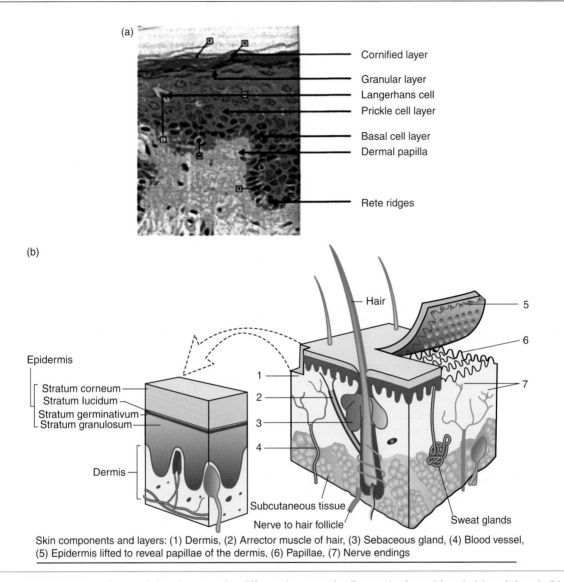

Skin components and layers: (1) Dermis, (2) Arrector muscle of hair, (3) Sebaceous gland, (4) Blood vessel, (5) Epidermis lifted to reveal papillae of the dermis, (6) Papillae, (7) Nerve endings

Figure 17.1 Structure of normal skin showing the different layers and cell types in the epidermis (a) and dermis (b).

a means of intercellular adhesion, but also to form a focus for the internal organization of cytoskeletal components, particularly tonofilaments. The next layer, the stratum granulosum, is characterized by the intracellular deposition of keratohyalin granules, particles of irregular shape and approximately 2 nm in diameter which are arranged in rows or lattices. Another feature

of this layer is Odland bodies – lamellated granules of 100–300 nm in size containing lipid for discharge into the extracellular space. It is this lipid that binds cells in the stratum corneum. The cells in this layer (corneocytes) become flattened as they lose their nuclei and organelles, and as the keratin filaments within them polymerize by the formation of disulfide bonds.

This is carried out under the influence of filaggrin, the protein component of the keratohyalin granule, and results in the formation of a flattened, highly insoluble keratin shell bound to the cell membrane by involucrin. The shedding of such a cell involves the degradation of the intercellular lipid component and the desmosomes before groups of cells are lost as a flakes of skin.

The cell cycle within the epidermis begins with a putative stem cell (thought to reside in either the bulge region of the lower outer root sheath of the hair follicle or, less likely, in the deeper rête ridges of glabrous skin). This cell gives rise, by definition, to all cells in the epidermis which are then terminally differentiated, although the transient amplifying population may undergo further limited proliferation and may form up to 32% of all cell reproduction (Penneys *et al.*, 1970). Whilst the description given above presents a linear progression from strata basale to corneum, there are at least two other patterns recognized from work in animal models. The first shows a vertical arrangement of groups, with, for example, the stratum spinosum thickly layered beneath a corneocyte and apparently differentiating in a horizontal plane, whilst another theory proposes proliferative units consisting of a stem cell surrounded by transit-amplifying and post-mitotic cells. This latter theory extends to include other cell types such as Langerhans cells and melanocytes within the unit, and would explain the apparently patchy distribution of these cells within the epidermis. The latter organization has yet to be described in humans and as such cannot be universally adopted.

The kinetics of epidermal proliferation are complicated by the observation that post-mitotic cells may be very active metabolically and may increase tissue mass without an increase in cell numbers. The turnover time is the amount of time taken for the whole cell population to replace itself and depends on the length of the cell cycle (the time between successive mitoses) and the growth fraction, i.e. that proportion of cells undergoing mitosis. In normal mouse skin this is estimated to be 60% of cells; higher proliferative rates may be achieved by either shortening the cell cycle or increasing the number of cells undergoing mitosis. In normal human skin the net effect of both parameters is to generate an epidermal turnover time of 52–75 days.

17.3 Keratins

The cytoskeleton of the keratinocyte (and all mammalian cells) is formed from three groups of filaments: actin (7 nm diameter), tubulin (20–25 nm) and intermediate filaments (7–10 nm). The latter group contains several recognized components, including vimentin, desmin, the nuclear laminins and the keratins, the polypeptide molecules characteristic of all epithelial cells and which account for 65% of the final mass of a corneocyte. The keratin family consists of over 30 individual members who may be separated by electrophoresis into the basic keratins (numbered 1–8) and the acidic keratins (numbered 9–12), and whose corresponding gene families are termed, respectively, types I and II. The molecules pair into acidic/basic dimers that are specifically expressed according to cell physiology and morphology. For example, simple epithelia are characterized by the keratin pair K8/K18 and stratified epithelia by K5/K14. Further, palmoplantar skin expresses K9, differentiated skin K1/K10 and hyperproliferative or regenerating epidermis K6/K16/K17. This expression provides a convenient way with which to investigate morphology and function in the epidermis with the use of immunohistochemistry (Table 17.1).

17.4 Structure of the dermo-epidermal junction

The dermis and epidermis meet at the dermo-epidermal junction (DEJ), which forms an intimate and complex attachment that allows not only the physical attachment of the epidermis to deeper structures, but also a physiological link between the two components that is integral to the normal function of both (Figure 17.2). It extends in convoluted manner throughout its length and extends in continuity to the epidermal

Table 17.1 Epithelial keratin expression

Keratin	Epidermal expression
K5/K14	stratified epithelia (mature)
K1/K10	differentiated epidermis
K6/K16	proliferative/regenerating epithelia
K9	skin of palmoplantar origin

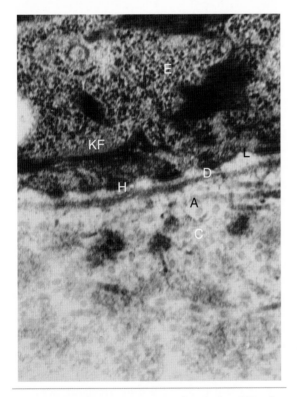

Figure 17.2 Electron micrograph of the DEJ after treatment with hyaluronic acid and application of cultured keratinocytes on Laserskin® showing a well-differentiated basement membrane and abundant collagen fibers in the newly formed dermis (×50,000).

appendages. By significantly increasing the surface area of contact the interdigitations not only allow more intimate communication, but also prevent shear forces from separating the two layers. The DEJ consists of the cell membranes of those cells superficial to it (i.e. keratinocytes, melanocytes and Langerhans cells) and those deep to it (fibroblasts, mast cells, etc.) bound to a common membrane – the basement membrane. It and the surrounding complexes are referred to as the basement membrane zone (BMZ).

The basal surface of the keratinocyte attaches to the lamina lucida via hemidesmosomes. These consist primarily of inner plaques on the inner surface of the cell membrane (allowing continuity with tonofilaments within the cell), outer plaques (closely associated with the cell membrane itself) and the sub-basal plate, external to the cell and immediately adjacent (and embedded into) the basement membrane. The basement membrane itself may be resolved into three laminae by electron microscopy: the outer lucida, intermediate densa and deep fibroreticularis. Within the lamina lucida are numerous anchoring filaments related to the hemidesmosomes and orientated perpendicular to the cell and basement membranes. The lamina densa appears relatively amorphous, and the fibroreticular layer appears to demonstrate anchoring fibrils, short, curving structures that extend into the dermis either singly or in relation to amorphous bodies and then termed anchoring plaques.

In addition to the above electron microscopic description, a number of specific molecular components can be identified within the basement membrane; these allow assessment of maturity by specific immunohistochemical staining (Table 17.2).

The basement membrane is intrinsic to the stability of healing wounds. Rigal *et al.* (1991) delineated the pattern of basement membrane component deposition with healing over a period of 10 days and showed a near-linear production of most components. Initially, collagen VII is absent, whilst bullous pemphigoid antigen (BPA) and collagen IV are the predominant components. Collagen VII production only begins after day 3, by which time BPA, laminin, fibronectin and collagen IV are demonstrated in appreciable quantities. By day 7, the levels of all these components are relatively equal and the basement membrane may be described as fully formed. Similarly, Carver *et al.* (1993) showed that *in vivo* the

Table 17.2 Molecular components of the basement membrane

Component	Location
Bullous pemphigoid antigen-1	hemidesmosome plaque
Bullous pemphigoid antigen-2	hemidesmosome plaque and anchoring fibrils
Laminins 5, 6 and 7	anchoring filaments
Collagen IV	lamina densa
Chondroitin sulfate	lamina densa
Collagen VII	anchoring fibrils

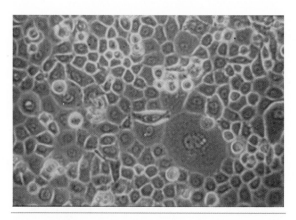

Figure 17.3 Cultured keratinocytes on a plastic substrate (×200).

lack of stability of keratinocyte grafts may be due to the inherent delays associated with full maturation of the basement membrane and, therefore, of the cells to the wound bed.

17.5 *In vitro* keratinocyte culture

Replacement of whole skin relies upon the provision of both the dermis and epidermis. Whilst split-thickness skin grafting satisfies these criteria, it is not always appropriate, particularly after a burn injury involving a large total burn surface area. In this instance two obvious problems present themselves. The first is that there is, by definition, insufficient healthy skin available to resurface the burn area, even with mesh-expansion of those grafts. Second, inflicting a further injury to the integument only adds to the physiological challenge facing a patient who may already be critically ill. With this in mind, tissue engineering of skin has evolved in the hope of developing an artificial skin substitute that may be made available for grafting and that might, in the future, negate the need for a split-thickness autograft.

The epidermal component of skin was the first to be investigated using tissue engineering techniques. The concept of *in vitro* culture of epidermal cells to provide large expansion of numbers was first proposed by Earle *et al.* (1954), who advocated its use in

burns. The earliest methods of keratinocyte culture involved the use of either organ or explant cultures, where the keratinocyte population grew along the side of the material and onto the flask (Cruickshank *et al.*, 1960; Karasek, 1968). These methods were fraught with fibroblast overgrowth and showed limited proliferative capacity.

In 1956, Puck *et al.* (1956) demonstrated that a layer of lethally irradiated epithelial cells could be used as a source for mitogens, without the cells themselves proliferating. This allowed a second population of cells to proliferate within the same chamber. In 1975, Rheinwald and Green published their chance discovery that human epidermal cells proliferated under the same conditions as mouse teratoma cells (Rheinwald and Green, 1975). In this instance, it was the provision of a feeder layer of lethally irradiated Swiss 3T3 mouse fibroblasts and of mitogens within the culture media that enabled human epithelial cells to proliferate rapidly, and proved one of the most important breakthroughs in the field (Figure 17.3). Keratinocyte culture methods have since been modified in a number of ways. Further work (Rheinwald and Green, 1977) optimized the growth conditions resulting in a mean cell cycle time of 22–24 hours, much greater than that seen in either normal or hyperproliferative (psoriatic) skin (Bauer and Grood 1975). Since then a number of minor modifications have been made to the culture medium resulting in the conditions used today (Fahmy *et al.*, 1993). A 1-cm^2 biopsy yields approximately 3×10^6 cells (Tenchini *et al.*, 1992) and, in optimal conditions, a 3-cm^2 piece of skin can be expanded more than 5000-fold within 3–4 weeks, yielding sufficient sheets of keratinocytes to cover the entire surface of an adult human (Green *et al.*, 1979). Furthermore, cells in culture have not only a shorter cell cycle time, but also a higher growth fraction than either normal or hyperproliferative skin (Carver and Leigh, 1991).

Other factors are also clearly important, particularly the age of the donor. Neonatal foreskin is capable of some 50–60 population doublings, but this decreases with age (Gilchrest, 1983), and both fibroblasts and keratinocytes have a finite number of population doublings, after which they are unsuitable for culture (Green *et al.*, 1977). In addition, donor systemic illness may decrease the proliferative capacity of keratinocytes (Stoner and Wood, 1996).

Within the culture flask, proliferating cells form islands which spread laterally and, to a lesser extent, stratify; growth is predominantly in a horizontal plane, but the daughter cells of those in the center of the island start to stratify. If the entire flask is allowed to stratify, a so-called Green Sheet is formed, consisting of up to 12 cell layers. The number of layers is determined by the calcium ion concentration in the medium; low calcium solutions promote the formation of a monolayer and high that of a stratified sheet (Boyce and Ham, 1983). After removal, this material is very fragile and is so thin as to be transparent to the naked eye; in order to minimize trauma due to handling, backing materials are needed, one example

Classical Experiment
Discovery of skin tissue engineering

In 1977, Rheinwald and Green performed the classic experiments that founded tissue-engineered skin as we know it today (Rheinwald and Green, 1977). They grew keratinocytes in colonies that eventually merged into a sheet. Keratinocytes were made to proliferate on a plastic substrate with the support of the combination of growth factors and irradiated mouse 3T3 fibroblasts. This opened the doors for therapeutic applications of the keratinocyte grafts (Gallico *et al.*, 1984). Since that time further advances have been made in the nutritional requirements of the keratinocytes. The need for calcium was discovered by Henning's research group (Hennings *et al.*, 1980). Time subsequently led to the knowledge that trace elements (Barnes and Sato, 1980) and hormones such as transferrin, insulin and hydrocortisone (Tsao *et al.*, 1982) were also important.

These studies led to the formulation of a medium which was specific to keratinocytes and did not require the need for feeders for cell proliferation leading to the growth of single-layered nondifferentiated keratinocytes (Tsao *et al.*, 1982). The addition of fetal bovine serum to the cells allowed the cells to stratify rather than to remain as a single-layered sheet. Cells grown by these two different methods made the application of keratinocyte grafts more practical for clinical application (Pittelkow and Scott, 1986). Another important step was the demonstration that a dermis or a dermal equivalent is required for optimal graft take. The grafting of keratinocyte sheets alone led to poor take rates and graft failure. The significance of dermis was demonstrated by Kangesu *et al.* (1993) in the porcine model and by Cuono *et al.* (1986) in humans. This initiated the development of artificial dermal substitutes and dermal templates. Some of the landmark observations in the field are: allogenic keratinocytes do not survive transplantation, the concept of preconfluent delivery systems for keratinocyte grafting, enrichment of KSCs for permanent graft take, micrografting of hair follicles in tissue-engineered templates and identification of signaling pathways that are involved in epithelial regeneration.

of which is the polyurethane backing material described by Rennekampff *et al.* (1996).

In order to circumvent handling problems associated with transfer materials, alternative delivery systems have been developed. Recent advances in biotechnology have allowed the development of transfer systems where cells are primarily grown (*in vitro*) on a substrate, which is then used to transfer the cells to the wound bed without the need for enzymatic dissociation (Myers *et al.*, 1995, 1997; Wright *et al.*, 1998). This carries two important advantages: the handling of the cells by clinical staff is minimized and there exists the possibility that cells could be transplanted prior to forming actual sheets (preconfluent grafting). In this method, cells are plated onto a transfer material (e.g. Laserskin®; Fidia Advanced Biopolymers) *in vitro*, but it is not necessary for them to form a confluent sheet – they are transferred whilst still in individual colonies (Figure 17.4). The lack of confluence in this instance is beneficial since there is minimal stratification and, therefore, differentiation of the cells. When the material is placed, inverted, onto the wound bed, the vast majority of cells presented are of a proliferative phenotype. This method has been shown to be at least as effective as other methods and has the potential benefits of decreasing the time required for culture or increasing the area that can be covered at a given time point (Harris *et al.*, 1998).

More recently, an aerosol method has been described (Horch *et al.*, 1998; Kaiser *et al.*, 1994) for keratinocyte delivery. In this technique, a mix of fibrin and keratinocytes is sprayed directly onto the wound bed, obviating the need for either sheets or a handling system and allowing for full use of preconfluent cells. The fibrin binds the cells to the wound bed upon contact, but does not provide any barrier function until the cells stratify and cornify. The fibrin glue serves an important intermediary function; cells are attached to the wound bed mechanically by a mesh of material. Cells cultured *in vitro* are devoid of hemidesmosomes, even if grown as a composite graft, and are therefore unable to attach directly to a basement membrane. By binding cells individually or in clusters to the wound surface, a temporary method of preventing the shear forces associated with both delivery systems and confluent sheet grafting exists. At the time of writing the technique is still in its infancy, but early reports are promising.

An alternative technique for enhancing take rates has been recently described by Takeda *et al.* (1999). These authors demonstrated improved epithelial cover and basement membrane formation after 1 week *in vivo* following coculture with laminin V.

The cellular composition of the material returned to the wound after culture differs from that initially harvested from the patient. Melanocyte numbers decrease rapidly in culture and eventually stabilize at approximately 10% of their initial levels (Staiano-Coico *et al.*, 1990). This means that the end-product obtained by current tissue engineering technologies is inferior in terms not only of skin adnexae, but also in terms of the color match of the original and the potential for the skin to darken in reaction to UV exposure.

Histologically, the grafts prepared consist of sheets of cells with much simpler organization than that of native epidermis. The basal layer is relatively well formed, but the suprabasal layers are irregular in depth, consisting of two to 10 layers, with poor organization. Although desmosomes and gap junctions are present (Franzi *et al.*, 1992), clinically the skin is extremely fragile.

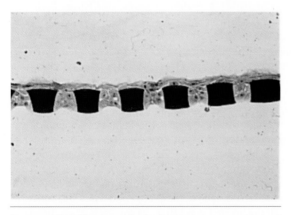

Figure 17.4 Cross-section of keratinocytes cultured on a Laserskin® membrane (×250).

17.6 Decreasing immunogenicity within cultured keratinocytes

Researchers are currently modulating the way in which a host reacts against grafted donor cells in order to improve keratinocyte culture survival. The need for this immune modulation was realized after published evidence by Brain *et al.* (1989) and Carver *et al.* (1991). They proved that grafted keratinocytes do not persist in an immunocompetent host and that an immune reaction could be demonstrated.

Langerhans cells are epidermal dendritic cells, part of the body's immune system, that perform as antigen-presenting cells, pick up antigens and deliver them to the lymphatic system. It is possible to deplete Langerhans cell number by *in vitro* UV irradiation and this may attenuate the immune response against allograft (Granstein *et al.*, 1987). It is also known that Langerhans cells proliferate poorly if at all *in vitro*; cell cultures do not stain for HLA-DR after 7 days in culture (Morhenn *et al.*, 1982) and functional assays fail to demonstrate such cellular activity (Hefton *et al.*, 1984). Using this information it forms one of several methods proposed for inducing clinical tolerance of allogeneic epidermis. Aübock *et al.* (1988) demonstrated prolonged but not indefinite survival of grafts with Langerhans cell depletion, whilst postgrafting application of UV appears to prolong allograft survival (Granstein *et al.*, 1987). Unfortunately this method cannot be employed indefinitely and may adversely affect wound healing (Das *et al.*, 1991; Ozcan *et al.*, 1993). Other, systemic, methods of preventing allograft loss include cyclosporin A administration, alone (Petri *et al.*, 1998) or in conjunction with antibody administration (Ossevoort *et al.*, 1999), as part of a standard immunosuppression protocol used for renal transplantation (Wendt *et al.*, 1994), or as a combination of UV irradiation and 8-methoxypsoralen (Granstein *et al.*, 1987). Ultimately, however, there appears little doubt that allogeneic cells are lost from the wound (Brain *et al.*, 1989; Roseeuw *et al.*, 1990) and that they initiate a degree of rejection (Carver *et al.*, 1991).

17.7 Development of *in vivo* grafting

17.7.1 Progression of epidermal replacements

Early keratinocyte culturing developed from the 1930s onwards; however, cultured keratinocyte characterization was not performed made until 1956 (Perry *et al.*, 1956; Wheeler *et al.*, 1957). The finding that the epidermal cells were able to proliferate in culture and that they had mitotic activity was made in 1960 (Cruickshank *et al.*, 1960). It was also discovered that early differentiation occurred in keratinocyte culture in the absence of fibroblasts (Pruniéras *et al.*, 1965). The classic Rheinwald–Green technique was not published until the mid 1970s (Rheinwald and Green, 1977); keratinocyte colonies proliferated and merged to become a confluent sheet when grown on a plastic substrate in growth factor media with irradiated mouse 3T3 fibroblasts. This knowledge then allowed for the beginning of cultured keratinocyte grafts as real therapeutic applications (Briggaman and Wheeler, 1968; Gallico *et al.*, 1984).

Although keratinocyte expansion *in vitro* has many potential clinical applications (Gallico *et al.*, 1984), keratinocyte attachment, proliferation and differentiation depend on interaction with the dermis. The key to the development of epidermal replacement has been the study of the interactions between the epithelial and mesenchymal layers, and skin injuries such as full-thickness burns result in the need for both epidermal and dermal layers to be repaired too. Understanding that the presence of a basement membrane encourages rapid re-epithelialization has been vital to the development of epidermal replacements.

The initial excitement surrounding the discovery was dampened by early reports of poor 'take', fragile epidermis and even a late blistering phenomenon. Autologous grafting came to clinical fruition in 1981 with the grafting of two patients suffering 40 and 80% total body surface area (TBSA) burns (O'Connor *et al.*, 1981). Only a small portion of their wounds were grafted and it was noted that graft 'take' was superior on fresh rather than granulating wounds. Since that time a number of authors have reported similar experiences (Cuono *et al.*, 1986; Faure *et al.*, 1987) up to a maximum TBSA of 98% (Gallico *et al.*, 1984) in two children. Unfortunately, successful engraftment is highly variable, and reports have varied between 0 and 95% mean 'take'. Even experienced practitioners have been unable to attain similar engraftment rates as split-thickness autografting; De Luca *et al.* (1992) reported a case series of 14 burns with a mean take of only 63% (range 10–100%) and although the Shriners Burn Institute (Galveston, USA) quotes rates as high as 90%, they note that the long-term outcome is much poorer because of the blistering seen with minimal shearing forces, as late as 6 months (Rose and Herndon, 1997).

17.7.2 Progression of dermal replacements

The dermal layer has similarly been found to be essential for the quality and 'take' of cultured keratinocyte grafts (Bell *et al.*, 1981; Boyce and Hansbrough, 1988; Briggaman and Wheeler, 1968; Burke *et al.*, 1981a). Improvement of functional and cosmetic outcome occurs when split-thickness or full-thickness skin grafting involves more dermis due to less scarring and wound contracture (Burke *et al.*, 1981a). During the care of massive burns whole allograft is used where there is a limited amount of donor skin available for use. In fact, the European Skin Bank uses a glycerol-preserved version of caderveric allografts and the American Association of Tissue Banks uses a more expensive cryopreserved caderveric allograft (Myers *et al.*, 1996). Ultimately the antigenic epidermal component of the allograft skin caused a rejection even in view of the fact that there was an initial 'take' and angiogenesis into the graft which lasted for several weeks. Later removal of the epidermis from the allograft and its replacement with cultured autologous keratinocyte sheets were techniques which were developed to counteract the rejection. An early dermal replacement was developed that was made up of fibroblasts incorporated into a glycosaminoglycan (GAG) membrane (Hansbrough *et al.*, 1989). The concept was that the intercellular signaling from the dermis would contribute to keratinocyte growth. Over the years allograft dermis has been advanced to reduce the risk of rejection and disease transmission, and improve long-term 'take'.

17.8 Failure of keratinocyte 'take'

The reasons for graft failure are elusive, but possibly the most important is the state of the wound bed onto which sheets are grafted. Both the nature of the tissue forming the bed and the bacterial load upon it have been identified as significant factors. Bacterial load adversely affects cultured keratinocyte take; Nuzzo *et al.* (2000) have demonstrated bacterial-induced apoptosis *in vitro*, whilst several authors (e.g. Blight *et al.*, 1991; Odessey 1992; Paddle-Ledinek *et al.*, 1997) have documented high bacterial loads in association with poor take rates. Teepe *et al.* (1990) demonstrated 15% take on granulating wounds, whilst De Luca *et al.* (1989) have reported rates varying between 24 and 47% on freshly excised or early granulating wounds. Furthermore, superior grafting has been demonstrated by a number of authors in conjunction with dermal analogues (Compton *et al.*, 1989; Gallico 1990; Kangesu *et al.*, 1993; Teepe *et al.*, 1990).

Early excision of the burn wound reduces infection rates and enhances keratinocyte take (Odessey, 1992). Paradoxically, wound bed preparation may also play a part; topical antiseptics are toxic to keratinocytes (Tatnall *et al.*, 1990) and this may have played a part in the poor results reported in some studies (Eldad *et al.*, 1987). Even in those circumstances in which the wound bed was sterile and culture conditions optimal, e.g. the elective excision of congenital nevi, Donati *et al.* (1992) have reported mean take rates of only 56%.

As early as 1979, Prunerias and coworkers described a method for producing what is now referred to as de-epidermalized dermis (DED) (Pruniéras *et al.*, 1983). The cellular content of whole skin was removed using the same method as for epithelial cell culture, but the remnant was utilized as a dermal analogue.

This sparked a wave of research, led by Bell and coworkers into the possibility of creating a bio-engineered dermis based upon a simple molecular scaffold. In the same year, Bell and coworkers published their first description of a contracted collagen lattice suitable for *in vitro* research (Bell *et al.*, 1979) and which was eventually grafted *in vivo* by Burke *et al.* (1981). The initial composition has been modified to include chondroitin-6-sulfate for biochemical stability and is now marketed commercially as Integra® (Figure 17.5). Alternative strategies have included the use of polyglycolic acid matrices (Cooper *et al.*, 1991) and of hyaluronic acid. The latter is if particular note since the degradation products of the macromolecule seem to exert a marked influence on the phenotype of the healing wound (reviewed in Chen and Abatangelo, 1999).

The common link to all such dermal replacement techniques is to provide a system whereby granulation tissue may regenerate the dermis in a controlled fashion, leading to the eventual production of a wound bed that more closely mimics the normal dermis. Although it is unclear whether the technologies merely provide a suitable scaffold or whether a more involved response occurs, the net effect is the same.

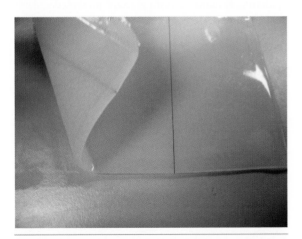

Figure 17.5 Integra with the silicon upper layer *in situ*. From Anthony, E.T. *et al.* (2006). The development of novel dermal matrices for cutaneous wound repair. *Drug Discovery Today: Therapeutic Strategies,* 3(1): 81–86.

17.9 Enhanced dermal grafting

Recent innovation has led to the introduction of two important methods of enhancing the grafted material. The first is a period of *in vitro* coculture with dermal fibroblasts, and the second is composite grafting. In 1992, Landeen and coworkers described the changes associated with a period of *in vitro* co-culture of a collagen gel and dermal fibroblasts (Landeen *et al.*, 1992). They demonstrated a rise in ECM components such as hydroxyproline, hyaluronic acid and chondroitin-4,6-disulfate. These findings suggested that such a period of culture might have profound effects on the dermal analogue eventually grafted; after 4 weeks in culture the material was clearly much more complex and, in some ways, more closely resembled normal dermal constitution. Since that time a number of authors have described benefits associated with *in vitro* co-culture, summarized in Table 17.3.

This has, most recently, led to the production of dermal replacement technologies containing living cells, in the hope that this will in some way enhance dermal regeneration, and improve both keratinocyte adherence and wound healing. Nevertheless, a

Table 17.3 Reported benefits associated with *in vitro* seeding of dermal analogues

Proposed benefit	References
Improved collagen deposition and remodeling	De Vries *et al.*, 1995; Lamme *et al.*, 1996; Berry *et al.*, 1998
Improved DEJ formation	Okamoto and Kitano 1993; Coulomb *et al.*, 1998
Enhanced survival of keratinocytes	Maruguchi *et al.*, 1994; Tseng *et al.*, 1996
Decreased myofibroblast formation	De Vries *et al.*, 1995; Lamme *et al.*, 1998
Decreased wound contraction	Desmouliere 1995; Berry *et al.*, 1998; Coleman *et al.*, 1998; Coulomb *et al.*, 1998

number of milestones have been achieved with the current keratinocyte technology, treating a range of diseases in a variety of anatomical sites (Table 17.4).

Another important innovation to be given serious consideration of late is that of composite grafting. This method utilizes *in vitro* techniques to grow a bilaminar skin equivalent consisting of dermis (with of without fibroblasts) and epidermis. The material may then be grafted as a whole-skin equivalent. At the time of writing only one such technology has been marketed (Apligraf®), but it is likely that this number will grow as the techniques involved are mastered.

17.9.1 Dermal component

Despite mounting evidence that allogenic grafting was not a permanent solution and seemed to be rejected, efforts continued to develop more complex composites with mesenchymal elements. These would approximate

more closely with normal skin *in vitro* and also could possibly allow adnexal structures to be incorporated. It had been noted as far back as 1952 that pure epidermal grafts were less durable than those containing dermis during work with explants (Billingham and Reynolds 1952). Later investigators identified that initially covering burns with cryopreserved allogenic skin, followed by later removal of the epidermis provided an incorporated allogenic dermal bed, but eliminated the much more antigenic epithelium (Cuono *et al.*, 1987; Heck *et al.*, 1985). When harvested blister epithelia (Heck *et al.*, 1985) or cultured keratinocytes (Cuono *et al.*, 1987) were applied in these conditions, full wound healing was reported and the resulting skin matured rapidly, as observed by histological analysis. Seemingly, allogenic fibroblasts elicit much less of a rejection response compared to allogenic keratinocytes. The importance of a dermal element for successful grafting has been further demonstrated (Kangesu *et al.*, 1993). Currently, Integra provides the best demonstration of the concept of a two-stage procedure that reconstructs dermis (Heimbach *et al.*, 2003) and will be discussed in the next section. DED has also been used as a dermal substitute prior to application of keratinocytes. The epidermis is removed from glycerol-preserved cadaver skin (Figure 17.6).

Table 17.4 Applications of cultured keratinocyte technology

Site of use	References
Autologous	
intra-oral	Langdon *et al.*, 1990
mastoid cavities	Premachandra *et al.*, 1990
superficial burn	Soeda *et al.*, 1993
full-thickness burn injury	Blight *et al.*, 1991
facial scar revision	Tsai *et al.*, 1997
chronic ulcers	Hollander *et al.*, 1999
giant congenital nevi	Gallico *et al.*, 1989
hypospadias	Romagnoli *et al.*, 1990
separation of Siamese twins	Higgins *et al.*, 1994
Recessive epidermolysis bullosa	McGrath *et al.*, 1993
Allogenic	
chronic venous ulcers	Leigh *et al.*, 1987
diabetic leg ulcers	Harvima *et al.*, 1999
skin graft donor sites	Fratianne *et al.*, 1993
burn injuries	Burt *et al.*, 1989

Figure 17.6 Glycerol-preserved caderveric skin used for preparing DED.

17.9.2 Tissue-engineered constructs

The composition of the range of constructs available can be divided up into whether they are synthetic or natural, with cells or without. Synthetic devices are easier to handle, store and are cheaper, whereas nonsynthetics may have a much shorter shelf-life, but may bio-interact with the wound of choice. See Tables 17.5 and 17.6.

17.9.3 Biobrane and transcyte

Biobrane is a thin bilaminar dressing with a porous silicone outer layer bonded to an inert nylon fabric that is coated with peptides derived from porcine type I collagen in order to promote adherence. It is a very useful dressing for partial-thickness wounds that adheres to the raw surface, removing the need for painful dressing changes, and is then pushed off by the re-epithelialization below over the following days. It also reduces the length of stay in hospital for

Table 17.5 Types of replacement matrices

Dermal replacements	Epidermal and dermal combination replacements
Cryopreserved allograft	Apligraf™
Glycerol preserved allograft	
Alloderm™	
Integra™	
Dermagraft™	
Dermagraft-TC™	
Hyalograft 3D™	

Table 17.6 Products for wound repair

Product	Epidermis	Dermis	Characteristics
Dermagraft™	N/A	neonatal allogeneic fibroblasts, absorbable scaffold (polyglycolic acid or polyglactic-910)	remains metabolically active after cryopreservation and the dermal matrix is reabsorbable; similar to Alloderm™
Integra™	silicone epidermis	bovine tendon collagen and shark GAGs	allows for immediate coverage of large acute wounds reducing bacterial ingress and water egress
Apligraf™	neonatal foreskin allogeneic human keratinocytes	human neonatal fibroblasts and bovine type 1 collagen matrix	requires less donor site so that the split skin graft size is decreased
Biobrane™	porous silicone outer layer	nylon fabric coated with cross-linked peptides derived from porcine type I collagen	the initial adherence is promoted by the collagen derived peptides on the nylon which bind to the fibrin and collagen on the surface of the wound
Trancyte™	polymer membrane	allogeneic human fibroblasts	the tissue matrix and the growth factors are left intact post-freezing
Hyalograft 3D™	NA	human fibroblasts and microperforated hyaluronic acid membrane (Figure 17.7)	provides a hyaluronic acid scaffold
Laserskin™	autologous human keratinocytes	N/A	used in combination with Hyalograft 3D™

scalded children (Hankins *et al.*, 2006). When seeded with allogenic fibroblasts, an immature ECM of fibronectin, type I collagen and growth factors is generated before the cells are suspended at −70°C after 17 days. The resultant product, formerly known as Dermagraft-TC, now Transcyte (Figure 17.7), requires storage at −20°C and is 16 times more expensive than Biobrane.

17.9.4 AlloDerm

AlloDerm is an acellular human allogenic dermal matrix which has been preserved by freeze-drying. It is claimed that the basement membrane is preserved and that it can be used to prepare the wound bed before application of thin split-skin graft in full-thickness wounds from burn excision (Lattari *et al.*, 1997; Wainwright 1995). [Using an *in vivo* method such as the Cuono technique (Cuono *et al.*, 1987), the physical removal of the epidermis will tend to remove the basement membrane proteins such as collagen IV and laminin.] Clinical results were reasonable with no evidence of rejection and the former

small, pilot study was extended into a larger, nonrandomized trial. The clinical behavior was similar, but also the patients were tested after for hypersensitivity reactions to homogenized aliquots of the acellular dermis and no problems were observed (Reagan *et al.*, 1997). The obvious drawback is that it has no epidermal elements so requires immediate coverage with a split-skin graft.

17.9.5 Dermagraft

This is a similar construct to AlloDerm, but arrived at using an absorbable polymer scaffold (polyglycolic acid or polyglactin-910) seeded with neonatal allogenic fibroblasts (Figure 17.8) (Hansbrough *et al.*, 1992a). Screened neonatal fibroblasts are enzymatically isolated and placed into tissue culture. After number expansion the cells are placed on the polymer scaffold and, as with Transcyte, an ECM is laid down consisting of growth factors and collagens,

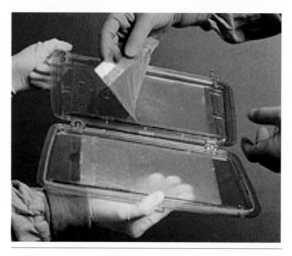

Figure 17.7 Commercially available product 'Transcyte'. Reproduced with permission from Advanced Biohealing.

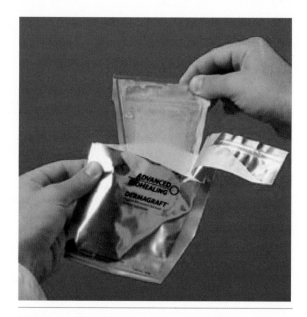

Figure 17.8 Commercially available product 'Dermagraft'. Reproduced with permission from Advanced Biohealing.

tenascin, vitronectin, and GAGs, but unlike Transcyte this assembly remains metabolically active after cryopreservation and the dermal matrix is resorbable (Cooper *et al.*, 1991). As with AlloDerm there is no epidermal element, but unlike AlloDerm there is no exogenous collagen. Wound closure is not achieved without a split-thickness graft, but it has been used more to stimulate wound healing in chronic wounds such as diabetic foot ulcers (Naughton *et al.*, 1997) and was as successful as split-thickness grafts on the treatment of burns wounds excised to varying depths (Hansbrough *et al.*, 1992b). As with other allogenic devices, long-term survival of the grafted cells wound not be expected.

17.9.6 Integra

Dermal templates have been translated to clinical practice. The most prevalent construct is derived from bovine collagen and shark chondroitin sulfate, covered by a silicone epidermis (Figure 17.5). It allows immediate coverage of large, acute wounds, reducing bacterial ingress and water egress (Burke *et al.*, 1981b; Yannas and Burke, 1980). A pore score allocated to 70–200 μm allows migration of endogenous fibroblasts and endothelial cells into the dermal matrix producing a vascularized neodermis (Moiemen *et al.*, 2001), and Integra is currently the most widely spread and proven 'artificial skin' of any kind (Heimbach *et al.*, 1988). However, there is still a delay before the definitive epidermal cover is provided in the form of an ultra-thin split-skin graft (around 0.15 mm), but the donor sites heal more rapidly than a conventional split-thickness graft (around 0.33 mm) and can be reharvested more frequently. Long-term results over an 11-year period are good in experienced hands, with take rates of around 80%, minimal hypertrophic scarring and cosmetic results superior to standard split-thickness graft. Areas grafted onto children were able to grow with the child and histological examination showed remodeling of the dermal component over the first month.

The 3-week delay while the neodermis vascularized is very convenient in allowing time for cell culture of autologous keratinocytes. However, a case report of this was not very successful (Pandya *et al.*, 1998). More recently, one case report details using widely spaced thin split-thickness grafts 1:6 in conjunction with cultured keratinocytes and fibroblasts with better results (Wisser and Steffes, 2003) or another alternative shown in animals has been to apply autologous keratinocytes directly to the underside of the Integra at the time of application, as the cells will then migrate through to the silicone/collagen interface (Jones *et al.*, 2003).

17.9.7 Cultured skin substitutes

A variation on the dermal template idea outlined above with Dermagraft and Integra has been in use for some time and warrants discussion. Boyce and Hansbrough (1988) used a dermal collagen–GAG

State of the Art Experiment
Tissue-engineered skin

Owing to the significant costs involved, tissue-engineered skin's most appropriate place remains in the treatment of life-threatening burn injuries and even here the role is adjuvant, e.g. filling the interstices of a wide mesh split-skin autograft on a prepared wound bed. In this clinical situation and for various reasons the approach is a staged approach, even when subconfluent systems are used. The combination of keratinocyte cultures with Integra in relatively small, clean full-thickness wounds probably represents the state of the art in terms of cosmetic and functional closure, but this experience is very limited at present. The single case of hair follicle micrografting to an Integra bed is an exciting step forward that pre-empts the hope of amplifying follicle numbers in organ culture.

membrane and seeded it with autologous keratinocytes. After maturation and differentiation of the construct these were successfully applied to full-thickness defects. The down side of application to clinical burns is the delay of at least 2 weeks while the composites were grown, which is very labor intensive and almost requires a dedicated laboratory per patient to avoid cross-contamination. The benefits are that much less of a donor site is required so the split-thickness graft requirements can be decreased. This procedure advanced to the logical next step of incorporating autologous fibroblasts in the dermal layer and applied them to four patients; 13 grafts were applied and nine took successfully (Hansbrough et al., 1989). Later papers showed similar results and were less successful than the conventional autograft (Boyce et al., 1995b). Of note, histological differences included lack of a vascular plexus in the dermis of the cultured skin substitute (CSS) when compared to conventional split-thickness graft and (or possibly subsequent to) a less keratinized epithelium. Grossly, the CSS was more hyperpigmented but also smoother, lacking any meshed pattern. Fully 11 out 17 wounds grafted with CSS later needed regrafting, but in those that did heal the epidermis was more like normal skin than the blister-prone epithelium from grafting keratinocyte sheets alone. A variation of their technique would enable a CSS with allogeneic keratinocytes and fibroblasts to be made. Obviously this would only differ from Apligraf by the type of dermal scaffold. Application to a small number of patients had mixed results from complete healing to limb amputation and it was felt that the allogeneic cells were ultimately replaced by 'circumferential ingrowth of autologous cells from the wound perimeter' (Boyce et al., 1995a).

17.9.8 Apligraf

The first method of recombining epidermal and mesenchymal tissue, in an attempt to mimic normal skin, was surprisingly early on in the history of cell culture (Freeman, Igel et al., 1976), and not long after Bell and coworkers performed the pioneering work that laid down the basic template for Apligraf (Bell et al., 1979) and published the proof of concept in vivo 2 years later in *Science* (Bell et al., 1981), with further studies with autogenic and allogenic tissue later in rats (Bell et al., 1983). The complete technique was devised some time later (Parenteau et al., 1991), albeit on the somewhat dubious principle that 'parenchymal and stromal cells such as the fibroblast and keratinocyte, may lack the antigenicity necessary to elicit an immune response' despite evidence to the contrary from more than one study (Aubock et al., 1988; Brain et al., 1989; Burt et al., 1989). Nevertheless, an organotypic model had been devised with an impressive range of properties resembling that of normal skin. Morphological analysis revealed seeded, neonatal fibroblasts aligned normally in a contracted bovine type I collagen matrix, overlayed with an epidermis with distinct basal, spinous, granular and cornified layers (Figure 17.9). While the DEJ was conspicuously flat with no epidermal ridges nor rête pegs, transmission electron microscopy showed numerous desmosomes between keratinocytes that contain abundant keratins, and evidence of hemidesmosomes, organized basement membrane and anchoring filaments at the DEJ (Parenteau et al., 1992). There was a normal keratinocyte transit time of 11–14 days (Parenteau et al., 1992), but the trans-epidermal water loss was 30-fold that of normal skin (Ernesti et al., 1992). Apligraf seemed to be immunologically inert and this was ascribed to the lack of Langerhan's cells (Demarchez et al., 1992).

Proof of concept began initially with nude mouse studies where integration and persistence was reported (Parenteau et al., 1992). Further animal studies suggested vascularization (Hansbrough et al., 1994). In humans it is important to note that a biological skin equivalent, based on the work published by Bell et al., was tried clinically in the late 1980s, but with poor results – the graft was lysed with 48 hours in all but one patient in a burns scenario (Wassermann et al., 1988). However, the authors developing Apligraf do not refer to this very relevant early work. In the acute wounds of immunocompetent humans, 15 patients received Apligraf. Although no frank rejection was

Figure 17.9 Apligraf in Transwell (a) and Apligraf being freed from agarose media prior to clinical application (b).

by the grafted keratinocytes or fibroblasts, nor to any of the xenogenic collagen used. When used on split-skin graft donor sites, Apligraf had similar healing advantages as meshed autograft on the donor site compared to conventional dressings and again no clinical rejection was observed in these partial-thickness wounds (Kirsner, 1998). For chronic wounds, in a study of 293 patients successful healing was increased from about half to nearly two-thirds of wounds, with a more significant benefit seen in long-term ulceration. Clinically, rejection was not observed (Falanga *et al.*, 1998). In diabetic foot ulceration a similar picture was seen on the clinical course. More ulcers were healed successfully in the group treated with Apligraf (90 versus 60) and no clinical rejection was mentioned (Veves *et al.*, 2001). Another uncontrolled trial in 23 patients gave a comparable story (Brem *et al.*, 2000). In summary, it was felt that Apligraf was easy to handle and showed little evidence of rejection. Observation revealed that it took on a very different stage compared to conventional skin grafts where upon it became translucent. When it was mentioned that the Apligraf did not remain, then the consensus seemed to be that it did help speed up the healing. Keen to look ahead, some hypothesized about the possibility of gene therapy and one paper looked at a model of wounding Apligraf *in vitro*, then culturing it in conditions that would 'mimic those of a chronic wound' before application close to a transduced cell line where it was observed that the fibroblasts at the edge of the mock wound picked up the retroviral vector (Badiavas *et al.*, 1996). This is some steps away from a possibly more useful approach of transforming unwounded Apligraf and then using that as a fairly robust vehicle to treat congenital disorders with known gene defects such has already been successfully shown *in vivo* with junctional epidermolysis bullosa patients (Robbins *et al.*, 2001).

17.9.9 Keratinocyte stem cells

Stem cells are receiving a lot of attention at the moment in many fields because of their transgermal

seen, the Apligraf took on an unusual appearance compared to the normal process of skin grafts, whereupon it became translucent and jelly-like below the allogenic epidermis. Another significant finding was that a biopsy of this area showed marked dermal fibrosis, i.e. scarring (Eaglstein *et al.*, 1995). Wounds were noted to contract more than full-thickness skin grafts. With regard to immunogenicity, tests showed that no antibodies could be detected to any antigens expressed

and pluripotential nature. Some populations, e.g. hematogenous stem cells, have been well described and clinical therapies developed. Less is known with regard to the keratinocyte stem cell (KSC) and controversies remain, but it may well be as useful and is more easily accessible. Also, there are already methods developed that can deliver cultured keratinocytes to cutaneous pathologies.

The KSC (holoclone) cycles slowly, minimizing potential DNA mutations, and probably as a consequence is long lived. Each stem is capable of dividing many times (more than 140) into another copy and a separate daughter cell. The resulting transient amplifying cell (meroclone) may divide a number of times providing an explosion in cell numbers, vital in a high turnover environment, before further differentiation into a terminal cell (paraclone) which migrates away from the basal layer.

The location of the KSC seems to vary. Typically, they are found in patches or clusters related to the DEJ along the basal lamina. There is probably a segmental distribution of progeny related to a stem cell, resulting in their patchy distribution along the DEJ. In the glabrous (hairless) skin of the soles and palms, the clusters are found deeper, at the tips of the deep rête ridges (Michel *et al.*, 1996). Estimates vary as to how bountiful they are, but probably 5–10% of the basal population is stem cells. However, the majority of the stem cells in the integument arise in the 'bulge region' of the hair follicle – a morphologically distinct region in mice, but not humans. This area may represent the ultimate source of stem cells, in hair-bearing skin, but is probably fed during anagen by the lower dermal papilla – a vital structure for follicle development.

In vivo identification of stem cells relies on their ability to retain labels for long periods because they cycle slowly. More recently, p63, a p53 homologue required for epithelial development, has been shown to be 'abundantly expressed by epidermal and limbal (cornea) holoclones', *in vitro* (Pellegrini *et al.*, 2001). The level of p63 expressed by meroclones drops dramatically as they leave the stem cell niche.

A population of cells that may remain for the whole life of the animal and has the potential to

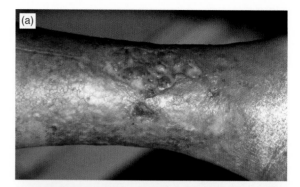
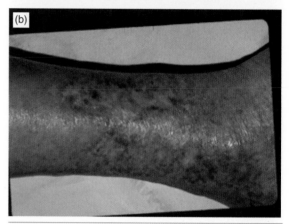

Figure 17.10 Cultured keratinocyte grafts used to cover a leg ulcer before (a) and after 3 weeks (b).

reconstitute the whole epidermis offers a convenient vehicle for genetic manipulation. Keratinocytes have been cultured *in vitro* for 26 years and can be grafted onto wounds via carrier dressings or on skin constructs. See Figures 17.10–17.12. One *in vivo* study (Robbins *et al.*, 2001) has demonstrated the transfection of keratinocytes from patients suffering from junctional epidermolysis bullosa, a lethal blistering disorder, and successful creation of phenotypically normal skin on severe combined immune deficient (SCID) mice. If an adequate population of stem cells has been transfected and carried over through the *in vitro* cultivation then longevity should be assured (Bianco and Robey, 2001). If the correct signals can

Figure 17.11 Hyalomatrix 3D consisting of esterified hyaluronic acid with a silicone membrane.

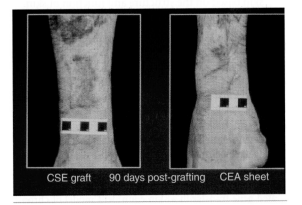

CSE graft   90 days post-grafting   CEA sheet

Figure 17.12 Cultured skin equivalent (CSE) (keratinocytes grown on DED) and cultured epithelial autograft (CEA) grafted on a patient with tattoo excision 90 days post-grafting. The clinical appearance is superior with CSE.

be identified and the right environments provided then could we also progress from just stem cells producing regeneration of skin to complex skin constructs with adnexal structures as well?

Lastly, are there any notes of caution ringing in the euphoria surrounding 'stemness'? How many steps are there between a pluripotential, slow cycling cell capable of ordered multiple divisions and the unregulated propagation of malignancy? It is felt that follicular and epidermal stem cells are pivotal in the pathogenesis of skin tumors, being targeted by chemical and viral attack. As with all treatments, potential or otherwise, we must first strive to do no harm.

17.10 The future of tissue-engineered skin

- At present it is possible to reliably extract and cultivate keratinocytes from a small skin biopsy and then grow up the body surface area of an adult in 2–3 weeks. Unfortunately, this period is at odds with the current management of acute burns and is affected by the age of the patient.
- The resultant sheets are difficult to handle, have variable take rates and are susceptible to shear forces due to the unsophisticated DEJ. Depletion of slow-cycling stem cells in current culture systems could contribute to late failure of grafts after apparently successful grafting.
- To date, no one ideal delivery method has emerged as the technique of choice as evidenced by the number of different systems described in the literature. Keratinocytes have been delivered on a multitude of backing substrates and as confluent or preconfluent cells.
- However, while epidermal cover will save a life, a substantial dermis is required for quality of life and is lacking if cultured autologous keratinocytes only are used on full-thickness wounds.
- This is an effective technique for large burns victims and the healed skin is tough and supple, a testament to dermal reconstruction. Unfortunately, a delay of 3–4 weeks is required while the neodermis develops. If this process were more prompt dermal formation would allow quicker grafting and definitive cover.
- The now classic process of keratinocyte culture, at least in Europe, is going to have to change. Imminent legislation will mean that culturing keratinocytes on a mouse 3T3 feeder layer will class the cells as a xenotransplant. Similarly, recent

lesions induced by allogeneic cultured epidermis: a multi-centre study in the treatment of children. *Burns*, 18(Suppl 1): S16–S19.

De Vries, H.J., Middelkoop, E., *et al.* (1995a). Stromal cells from subcutaneous adipose tissue seeded in a native collagen/elastin dermal substitute reduce wound contraction in full thickness skin defects. *Lab Invest*, 73: 532–540.

De Vries, H.J., Zeegelaar, J.E., *et al.* (1995b). Reduced wound contraction and scar formation in punch biopsy wounds. Native collagen dermal substitutes. A clinical study. *Br J Dermatol*, 132: 690–697.

Demarchez, M., Asselineau, D., *et al.* (1992). Migration of Langerhans cells into the epidermis of human skin grafted into nude mice. *J Invest Dermatol*, 99: 54S–55S.

Desmouliere, A. (1995). Factors influencing myofibroblast differentiation during wound healing and fibrosis. *Cell Biol Int*, 19: 471–476.

Donati, L., Magliacani, G., *et al.* (1992). Clinical experiences with keratinocyte grafts. *Burns*, 18(Suppl 1): S19–S26.

Eaglstein, W.H., Iriondo, M., *et al.* (1995). A composite skin substitute (graftskin) for surgical wounds. A clinical experience. *Dermatol Surg*, 21: 839–843.

Earle, W.R., Bryant, J.C., *et al.* (1954). Certain factors limiting the size of the tissue culture and the development of massive cultures. *Ann NY Acad Sci*, 58: 1000–1011.

Eldad, A., Burt, A., *et al.* (1987). Cultured epithelium as a skin substitute. *Burns Incl Therm Inj*, 13: 173–180.

Ernesti, A.M., Swiderek, M., *et al.* (1992). Absorption and metabolism of topically applied testosterone in an organotypic skin culture. *Skin Pharmacol*, 5: 146–153.

Fahmy, F.S., Navsaria, H.A., *et al.* (1993). Skin graft storage and keratinocyte viability. *Br J Plast Surg*, 46: 292–295.

Falanga, V., Margolis, D., *et al.* (1998). Rapid healing of venous ulcers and lack of clinical rejection with an allogeneic cultured human skin equivalent. Human Skin Equivalent Investigators Group. *Arch Dermatol*, 134: 293–300.

Faure, M., Mauduit, G., *et al.* (1987). Growth and differentiation of human epidermal cultures used as auto- and allografts in humans. *Br J Dermatol*, 116: 161–170.

Franzi, A.T., D'Anna, F., *et al.* (1992). Histological evaluation of human cultured epithelium before and after grafting. *Burns*, 18(Suppl 1): S26–S31.

Fratianne, R., Papay, F., *et al.* (1993). Keratinocyte allografts accelerate healing of split-thickness donor sites: applications for improved treatment of burns. *J Burn Care Rehabil*, 14(2 Pt 1): 148–154.

Freeman, A.E., Igel, H.J., *et al.* (1976). Growth and characterization of human skin epithelial cell cultures. *In vitro*, 12: 352–362.

Gallico, G.G. 3rd (1990). Biologic skin substitutes. *Clin Plast Surg*, 17: 519–526.

Gallico, G.G. 3rd, O'Connor, N.E., *et al.* (1984). Permanent coverage of large burn wounds with autologous cultured human epithelium. *N Engl J Med*, 311: 448–451.

Gallico, G.G. 3rd, O'Connor, N.E., *et al.* (1989). Cultured epithelial autografts for giant congenital nevi. *Plast Reconstr Surg*, 84: 1–9.

Gilchrest, B.A. (1983). *In vitro* assessment of keratinocyte aging. *J Invest Dermatol*, 81(1 Suppl): 184s–189s.

Granstein, R.D., Smith, L., *et al.* (1987). Prolongation of murine skin allograft survival by the systemic effects of 8-methoxypsoralen and long-wave ultraviolet radiation (PUVA). *J Invest Dermatol*, 88: 424–429.

Green, H., Kehinde, O., *et al.* (1979). Growth of cultured human epidermal cells into multiple epithelia suitable for grafting. *Proc Natl Acad Sci USA*, 76: 5665–5668.

Green, H., Rheinwald, J.G., *et al.* (1977). Properties of an epithelial cell type in culture: the epidermal keratinocyte and its dependence on products of the fibroblast. *Prog Clin Biol Res*, 17: 493–500.

Hankins, C.L., Tang, X.Q., *et al.* (2006). Hot beverage burns: an 11-year experience of the Yorkshire Regional Burns Centre. *Burns*, 32: 87–91.

Hansbrough, J.F., Boyce, S.T., *et al.* (1989). Burn wound closure with cultured autologous keratinocytes and fibroblasts attached to a collagen–glycosaminoglycan substrate. *J Am Med Ass*, 262: 2125–2130.

Hansbrough, J.F., Cooper, M.L., *et al.* (1992a). Evaluation of a biodegradable matrix containing cultured human fibroblasts as a dermal replacement beneath meshed skin grafts on athymic mice. *Surgery*, 111: 438–446.

Hansbrough, J.F., Dore, C., *et al.* (1992b). Clinical trials of a living dermal tissue replacement placed beneath meshed, split-thickness skin grafts on excised burn wounds. *J Burn Care Rehabil*, 13: 519–529.

Hansbrough, J.F., Morgan, J., *et al.* (1994). Evaluation of Graftskin composite grafts on full-thickness wounds on athymic mice. *J Burn Care Rehabil*, 15: 346–353.

Harris, P.A., Leigh, I.M., *et al.* (1998). Pre-confluent keratinocyte grafting: the future for cultured skin replacements? *Burns*, 24: 591–593.

Harvima, I.T., Virnes, S., *et al.* (1999). Cultured allogeneic skin cells are effective in the treatment of chronic diabetic leg and foot ulcers. *Acta Derm Venereol*, 79: 217–220.

Hasan, A., Murata, H., *et al.* (1997). Dermal fibroblasts from venous ulcers are unresponsive to the action of transforming growth factor-beta 1. *J Dermatol Sci*, 16: 59–66.

Hasselgren, P.O. (1999). Burns and metabolism. *J Am Coll Surg*, 188: 98–103.

He, C., Hughes, M.A., *et al.* (1999). Effects of chronic wound fluid on the bioactivity of platelet-derived growth factor in serum-free medium and its direct effect on fibroblast growth. *Wound Repair Regen*, 7: 97–105.

Heck, E.L., Bergstresser, P.R., *et al.* (1985). Composite skin graft: frozen dermal allografts support the engraftment and expansion of autologous epidermis. *J Trauma*, 25: 106–112.

Hefton, J.M., Amberson, J.B., *et al.* (1984). Loss of HLA-DR expression by human epidermal cells after growth in culture. *J Invest Dermatol*, 83: 48–50.

Hehenberger, K., Heilborn, J.D., *et al.* (1998). Inhibited proliferation of fibroblasts derived from chronic diabetic wounds and normal dermal fibroblasts treated with high glucose is associated with increased formation of L-lactate. *Wound Repair Regen*, 6: 135–141.

Heimbach, D., Luterman, A., *et al.* (1988). Artificial dermis for major burns. A multi-center randomized clinical trial. *Ann Surg*, 208: 313–320.

Heimbach, D.M., Warden, G.D., *et al.* (2003). Multicenter postapproval clinical trial of Integra dermal regeneration template for burn treatment. *J Burn Care Rehabil*, 24: 42–48.

Hennings, H., Michael, D., *et al.* (1980). Calcium regulation of growth and differentiation of mouse epidermal cells in culture. *Cell*, 19: 245–254.

Higgins, C.R., Navsaria, H., *et al.* (1994). Use of two stage keratinocyte-dermal grafting to treat the separation site in conjoined twins. *J R Soc Med*, 87: 108–109.

Hollander, D., Stein, M., *et al.* (1999). Autologous keratinocytes cultured on benzylester hyaluronic acid membranes in the treatment of chronic full-thickness ulcers. *J Wound Care*, 8: 351–355.

Horch, R.E., Bannasch, H., *et al.* (1998). Single-cell suspensions of cultured human keratinocytes in fibrin-glue reconstitute the epidermis. *Cell Transplant*, 7: 309–317.

Jackson, D.M. (1953). The diagnosis of the depth of burning. *Br J Surg*, 40: 588–596.

Jones, I., James, S.E., *et al.* (2003). Upward migration of cultured autologous keratinocytes in Integra artificial skin: a preliminary report. *Wound Repair Regen*, 11: 132–138.

Kaiser, H.W., Stark, G.B., *et al.* (1994). Cultured autologous keratinocytes in fibrin glue suspension, exclusively and combined with STS-allograft (preliminary clinical and histological report of a new technique). *Burns*, 20: 23–29.

Kangesu, T., Navsaria, H.A., *et al.* (1993). Kerato-dermal grafts: the importance of dermis for the *in vivo* growth of cultured keratinocytes. *Br J Plast Surg*, 46: 401–409.

Karasek, M.A. (1968). Growth and differentiation of transplanted epithelial cell cultures. *J Invest Dermatol*, 51: 247–252.

Kirsner, R.S. (1998). The use of Apligraf in acute wounds. *J Dermatol*, 25: 805–811.

Lamme, E.N., de Vries, H.J., *et al.* (1996). Extracellular matrix characterization during healing of full-thickness wounds treated with a collagen/elastin dermal substitute shows improved skin regeneration in pigs. *J Histochem Cytochem*, 44: 1311–1322.

Lamme, E.N., van Leeuwen, R.T., *et al.* (1998). Living skin substitutes: survival and function of fibroblasts seeded in a dermal substitute in experimental wounds. *J Invest Dermatol*, 111: 989–995.

Landeen, L.K., Zeigler, F.C., *et al.* (1992). Characterization of a human dermal replacement. *Wounds*, 4: 167–175.

Langdon, J.D., Leigh, I.M., *et al.* (1990). Autologous oral keratinocyte grafts in the mouth. *Lancet*, 335: 1472–1473.

Lattari, V., Jones, L.M., *et al.* (1997). The use of a permanent dermal allograft in full-thickness burns of the hand and foot: a report of three cases. *J Burn Care Rehabil*, 18: 147–155.

Leigh, I.M., Purkis, P.E., *et al.* (1987). Treatment of chronic venous ulcers with sheets of cultured allogenic keratinocytes. *Br J Dermatol*, 117: 591–597.

Lofts, J.A. (1991). Cost analysis of a major burn. *NZ Med J*, 104: 488–490.

Marston, W.A., Carlin, R.E., *et al.* (1999). Healing rates and cost efficacy of outpatient compression treatment for leg ulcers associated with venous insufficiency. *J Vasc Surg*, 30: 491–498.

Maruguchi, T., Maruguchi, Y., *et al.* (1994). A new skin equivalent: keratinocytes proliferated and differentiated on collagen sponge containing fibroblasts. *Plast Reconstr Surg*, 93: 537–544. discussion 545–546.

McGrath, J.A., Schofield, O.M., *et al.* (1993). Cultured keratinocyte allografts and wound healing in severe recessive dystrophic epidermolysis bullosa. *J Am Acad Dermatol*, 29: 407–419.

McManus, W.F. and Pruitt, B.A. (1996). Thermal injuries. In *Trauma* (Feliciano, D.V., Moore, E.E. and Mattox, K.L., eds). Stamford, CT: Appleton & Lange, pp. 937–949.

Michel, M., Torok, N., *et al.* (1996). Keratin 19 as a biochemical marker of skin stem cells *in vivo* and *in vitro*: keratin 19 expressing cells are differentially localized in function of anatomic sites, and their number varies with donor age and culture stage. *J Cell Sci*, 109: 1017–1028.

Moiemen, N.S., Staiano, J.J., *et al.* (2001). Reconstructive surgery with a dermal regeneration template: clinical and histologic study. *Plast Reconstr Surg*, 108: 93–103.

Morhenn, V.B., Benike, C.J., *et al.* (1982). Cultured human epidermal cells do not synthesize HLA-DR. *J Investig Dermatol*, 78: 32–37.

Morrell, C.J., Walters, S.J., *et al.* (1998). Cost effectiveness of community leg ulcer clinics: randomized controlled trial. *Br Med J*, 316: 1487–1491.

Myers, S., Navsaria, H., *et al.* (1995). Transplantation of keratinocytes in the treatment of wounds. *Am J Surg*, 170: 75–83.

Myers, S.R., Grady, J., *et al.* (1997). A hyaluronic acid membrane delivery system for cultured keratinocytes: clinical take rates in the porcine kerato-dermal model. *J Burn Care Rehabil*, 18: 214–222.

Myers, S.R., Machesney, M.R., *et al.* (1996). Skin storage. *Br Med J*, 313: 439.

Naughton, G., Mansbridge, J., *et al.* (1997). A metabolically active human dermal replacement for the treatment of diabetic foot ulcers. *Artif Organs*, 21: 1203–1210.

Nguyen, T.T., Gilpin, D.A., *et al.* (1996). Current treatment of severely burned patients. *Ann Surg*, 223: 14–25.

Nuzzo, I., Sanges, M.R., *et al.* (2000). Apoptosis of human keratinocytes after bacterial invasion. *FEMS Immunol Med Microbiol*, 27: 235–240.

O'Connor, N., Mulliken, J., *et al.* (1981). Grafting of burns with cultured epithelium prepared from autologous epidermal cells. *Lancet*, 317: 75–78.

Odessey, R. (1992). Addendum: multicenter experience with cultured epidermal autograft for treatment of burns. *J Burn Care Rehabil*, 13: 174–180.

Okamoto, E. and Kitano, Y. (1993). Expression of basement membrane components in skin equivalents – influence of dermal fibroblasts. *J Dermatol Sci*, 5: 81–88.

Ossevoort, M.A., Lorre, K., *et al.* (1999). Prolonged skin graft survival by administration of anti-CD80 monoclonal antibody with cyclosporin A. *J Immunother*, 22: 381–389.

Ozcan, G., Shenaq, S., *et al.* (1993). Ultraviolet-A induced delayed wound contraction and decreased collagen content in healing wounds and implant capsules. *Plast Reconstr Surg*, 92: 480–484.

Paddle-Ledinek, J.E., Cruickshank, D.G., *et al.* (1997). Skin replacement by cultured keratinocyte grafts: an Australian experience. *Burns*, 23: 204–211.

Pandya, A.N., Woodward, B., *et al.* (1998). The use of cultured autologous keratinocytes with integra in the resurfacing of acute burns. *Plast Reconstr Surg*, 102: 825–858. discussion 829–930

Parenteau, N.L., Nolte, C.M., *et al.* (1991). Epidermis generated *in vitro*: practical considerations and applications. *J Cell Biochem*, 45: 245–251.

Parenteau, N.L., Bilbo, P., *et al.* (1992). The organotypic culture of human skin keratinocytes and fibroblasts to achieve form and function. *Cytotechnology*, 9: 163–171.

Pellegrini, G., Dellambra, E., *et al.* (2001). p63 identifies keratinocyte stem cells. *Proc Natl Acad Sci USA*, 98: 3156–3161.

Penneys, N.S., Fulton J.E., Jr. *et al.* (1970). Location of proliferating cells in human epidermis. *Arch Dermatol*, 101: 323–327.

Perry, V.P., Evans, V.J., *et al.* (1956). Long-term tissue culture of human skin. *Am J Hyg*, 63: 52–58.

Petri, J.B., Schurk, S., *et al.* (1998). Cyclosporine A delays wound healing and apoptosis and suppresses activin beta-A expression in rats. *Eur J Dermatol*, 8: 104–113.

Pittelkow, M.R. and Scott, R.E. (1986). New techniques for the *in vitro* culture of human skin keratinocytes and perspectives on their use for grafting of patients with extensive burns. *Mayo Clin Proc*, 61: 771–777.

Premachandra, D.J., Woodward, B.M., *et al.* (1990). Treatment of postoperative otorrhoea by grafting of mastoid cavities with cultured autologous epidermal cells. *Lancet*, 335: 365–367.

Pruniéras, M., Mathivon, M.F., *et al.* (1965). Euploid culture of adult epidermal cells in monocellular layers. *Ann Inst Pasteur (Paris)*, 108: 149–165.

Pruniéras, M., Régnier, M., *et al.* (1983). Nouveau procédé de culture des cellules épidermiques humaines sur derme homologue ou hétérologue: préparation de greffons recombinés. *Ann Chir Plast* (Suppl): 28s–33s.

Puck, T.T., Marcus, P.I., *et al.* (1956). Clonal growth of mammalian cells *in vitro*; growth characteristics of colonies from single HeLa cells with and without a feeder layer. *J Exp Med*, 103: 273–283.

Reagan, B.J., Madden, M.R., *et al.* (1997). Analysis of cellular and decellular allogeneic dermal grafts for the treatment of full-thickness wounds in a porcine model. *J Trauma*, 43: 458–466.

Regan, M.C., Kirk, S.J., *et al.* (1991). The wound environment as a regulator of fibroblast phenotype. *J Surg Res*, 50: 442–448.

Rennekampff, H.O., Hansbrough, J.F., *et al.* (1996). Wound closure with human keratinocytes cultured on a polyurethane dressing overlaid on a cultured human dermal replacement. *Surgery*, 120: 16–22.

Rheinwald, J.G. and Green, H. (1975). Formation of a keratinizing epithelium in culture by a cloned cell line derived from a teratoma. *Cell*, 6: 317–330.

Rheinwald, J.G. and Green, H. (1977). Epidermal growth factor and the multiplication of cultured human epidermal keratinocytes. *Nature*, 265: 421–424.

Rigal, C., Pieraggi, M.-T., *et al.* (1991). Healing of full-thickness cutaneous wounds in the pig. I. Immunohistochemical study of epidermo-dermal junction regeneration. *J Invest Dermatal*, 96: 777–785.

Robbins, P.B., Lin, Q., *et al.* (2001). *In vivo* restoration of laminin 5 beta 3 expression and function in junctional epidermolysis bullosa. *Proc Natl Acad Sci USA*, 98: 5193–5198.

Romagnoli, G., De Luca, M., *et al.* (1990). Treatment of posterior hypospadias by the autologous graft of cultured urethral epithelium. *N Engl J Med*, 323: 527–530.

Rose, J.K. and Herndon, D.N. (1997). Advances in the treatment of burn patients. *Burns*, 23(Suppl 1): S19–S26.

Roseeuw, D.I., De Coninck, A., *et al.* (1990). Allogeneic cultured epidermal grafts heal chronic ulcers although they do not remain as proved by DNA analysis. *J Dermatol Sci*, 1: 245–252.

Soeda, J., Inokuchi, S., *et al.* (1993). Use of cultured human epidermal allografts for the treatment of extensive partial thickness scald burn in children. *Tokai J Exp Clin Med*, 18: 65–70.

Staiano-Coico, L., Hefton, J.M., *et al.* (1990). Growth of melanocytes in human epidermal cell cultures. *J Trauma*, 30: 1037–1042. discussion 1043.

Stanley, A.C., Park, H.Y., *et al.* (1997). Reduced growth of dermal fibroblasts from chronic venous ulcers can be stimulated with growth factors. *J Vasc Surg*, 26: 994–999. discussion 999–1001.

Stoner, M.L. and Wood, F.M. (1996). Systemic factors influencing the growth of cultured epithelial autograft. *Burns*, 22: 197–199.

Takeda, A., Kadoya, K., *et al.* (1999). Pretreatment of human keratinocyte sheets with laminin 5 improves their grafting efficiency. *J Invest Dermatol*, 113: 38–42.

Tatnall, F.M., Leigh, I.M., *et al.* (1990). Comparative study of antiseptic toxicity on basal keratinocytes, transformed human keratinocytes and fibroblasts. *Skin Pharmacol*, 3: 157–163.

Teepe, R.G., Kreis, R.W., *et al.* (1990). The use of cultured autologous epidermis in the treatment of extensive burn wounds. *J Trauma*, 30: 269–275.

Tenchini, M.L., Ranzati, C., *et al.* (1992). Culture techniques for human keratinocytes. *Burns*, 18(Suppl 1): S11–S16.

Tsai, C.Y., Ueda, M., *et al.* (1997). Clinical results of cultured epithelial cell grafting in the oral and maxillofacial region. *J Craniomaxillofac Surg*, 25: 4–8.

Tsao, M.C., Walthall, B.J., *et al.* (1982). Clonal growth of normal human epidermal keratinocytes in a defined medium. *J Cell Physiol*, 110: 219–229.

Tseng, S.C., Kruse, F.E., *et al.* (1996). Comparison between serum-free and fibroblast-cocultured single-cell clonal culture systems: evidence showing that epithelial anti-apoptotic activity is present in 3T3 fibroblast-conditioned media. *Curr Eye Res*, 15: 973–984.

Unemori, E.N., Ehsani, N., *et al.* (1994). Interleukin-1 and transforming growth factor-alpha: synergistic stimulation of metalloproteinases, PGE2, and proliferation in human fibroblasts. *Exp Cell Res*, 210: 166–171.

Van de Berg, J.S., Rudolph, R., *et al.* (1998). Fibroblast senescence in pressure ulcers. *Wound Repair Regen*, 6: 38–49.

Veves, A., Falanga, V., *et al.* (2001). Graftskin, a human skin equivalent, is effective in the management of non-infected neuropathic diabetic foot ulcers: a prospective randomized multicenter clinical trial. *Diabetes Care*, 24: 290–295.

Wainwright, D.J. (1995). Use of an acellular allograft dermal matrix (AlloDerm) in the management of full-thickness burns. *Burns*, 21: 243–248.

Wallace, A.F. (1982). *The Progress of Plastic Surgery*. Oxford: Oxford University Press.

Wassermann, D., Schlotterer, M., *et al.* (1988). Preliminary clinical studies of a biological skin equivalent in burned patients. *Burns Incl Therm Inj*, 14: 326–330.

Wedler, V., Kunzi, W., *et al.* (1999). Care of burns victims in Europe. *Burns*, 25: 152–157.

Wendt, J.R., Ulich, T.R., *et al.* (1994). Indefinite survival of human skin allografts in patients with long-term immunosuppression. *Ann Plast Surg*, 32: 411–417.

Wheeler, C.E., Canby, C.M., *et al.* (1957). Long-term tissue culture of epithelial-like cells from human skin. *J Invest Dermatol*, 29: 383–391. discussion 391–392.

Wisser, D. and Steffes, J. (2003). Skin replacement with a collagen based dermal substitute, autologous keratinocytes and fibroblasts in burn trauma. *Burns*, 29: 375–380.

Wright, K.A., Nadire, K.B., *et al.* (1998). Alternative delivery of keratinocytes using a polyurethane membrane and the implications for its use in the treatment of full-thickness burn injury. *Burns*, 24: 7–17.

Yannas, I.V. and Burke, J.F. (1980). Design of an artificial skin. I. Basic design principles. *J Biomed Mater Res*, 14: 65–81.

Chapter 18
Tissue engineering of cartilage

Mats Brittberg and Anders Lindahl

Chapter objectives:

- To know the location and differences between hyaline, fibrous and elastic cartilage
- To recognize the composition of adult hyaline cartilage with emphasis on cells, type of collagen and proteoglycans
- To understand that cartilage has a limited repair capacity, and the differences in repair capacity between fetal and adult cartilage
- To know that articular chondrocytes can be cultured and used for induction of cartilage repair
- To understand the rationale for cell implantation, and the advantages and

- limitations of other cell sources including stem cells for cartilage repair
- To understand what part of the cartilage is suitable for cell isolation, and what growth factors are important for cartilage growth and differentiation
- To recognize that bioreactors are central to the development of cartilage implants, and what type of scaffold materials are used for cartilage repair
- To understand that the extracellular matrix and cells are part of the signaling system

"If we consult the standard Chirugical Writers from Hippocrates down to the present Age, we shall find, that an ulcerated Cartilage is universally allowed to be a very troublesome disease . . . when destroyed, it is never recovered."

William Hunter (1718–1783)

"There are, I believe, no instances in which a lost portion of cartilage has been restored, or a wounded portion repaired, with new and well-formed permanent cartilage."

James Paget (1814–1899)

18.1 Introduction

Chondrocytes are cells of mesodermal origin forming three different types of cartilage where the properties are dependent on the extracellular matrix (ECM) composition: (i) hyaline cartilage is found in the joints (Figure 18.1), rib cartilage, nose, trachea and larynx, (ii) elastic cartilage is found in the ear, epiglottis and larynx, and (iii) fibrous cartilage is found in the intervertebratal discs. Hyaline cartilage is also transiently involved in skeletal development through the process of endochondral ossification.

(A) (B)

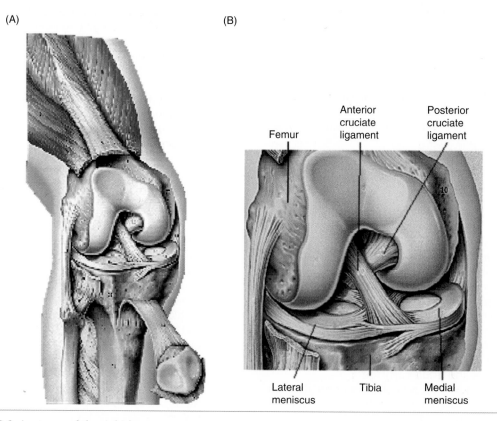

Figure 18.1 Anatomy of the right knee.

18.2 Composition of adult hyaline human articular cartilage

The cartilage formed in the embryo is gradually changed during development to meet the needs and demands of the adult organism. In the embryo there is initially a high cell to matrix ratio, whereas the adult cartilage consists only of 2% cells (the chondrocytes). The major part of the matrix consists of water (65–80%) and the rest of solid material (20–35%). Of the solid part, about 60% is collagen, 30% proteoglycans and 10% other proteins. The mechanical function of the proteoglycans is to be shock absorbers, whereas the collagens are resistant to shear stress. The exact composition of the cartilage matrix and the biomechanical properties vary between different locations and are also dependent on age. The average thickness in loaded areas of the cartilage is 2.4 mm.

In young animals there are only two zones that can be identified in articular cartilage: the superficial and growth zones. Four structurally different zones with different functions and protein composition can be identified in mature cartilage: the superficial, transitional, radial and calcified zones (Figure 18.2) (Poole *et al.*, 2001). The superficial zone lines the synovial joint and its major components are lubricin or superficial zonal protein (Swann *et al.*, 1985). This protein functions as a lubricant and is normally only found in this area of the articular cartilage. The elongated and flat cells in this area synthesize mainly type I collagen.

In the transitional zone the cells start to become more rounded and pericellularly type VI collagen can be found. The major proteoglycans in this area is decorin. The collagen fibers in these two first regions are horizontally oriented while in the radial zone,

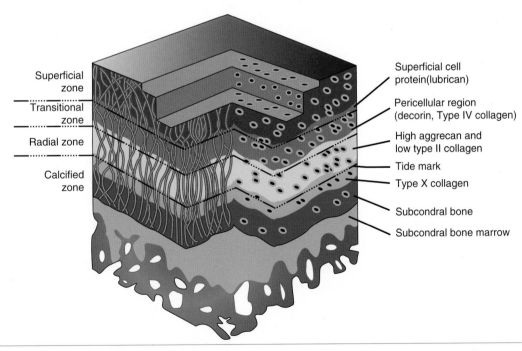

Figure 18.2 Structural and functional zones of articular cartilage. Adapted from Poole, A.R., Kojima, T. *et al.* (2001). Composition and structure of articular cartilage: a template for tissue repair. *Clin Orthop* (391 Suppl): S26–S33.

where the major parts of type II collagen is found, the fibers are vertically oriented (ap Gwynn *et al.*, 2002). In the radial zone the chondrocytes are more sparsely distributed, and the major matrix proteins are the typical cartilage type II collagen and the proteoglycan, aggrecan. Next to the subchondral bone the cartilage becomes mineralized and type X collagen is found in this area.

18.3 Cartilage components

18.3.1 Collagen

The major part of the solid phase of adult articular cartilage is collagen. The main functions of the collagens, e.g. collagen fibrils, are to give structure and tensile strength to the articular cartilage. The fibrils

are composed of aggregates of five tropocollagens (three identical α_{II} polypeptide chains assembled in a super-helix) that have a diameter of 20 nm. This unique structure of the fibrils makes collagen insoluble and resistant to attack by degrading enzymes (Figure 18.3).

To create maximal strength to the cartilage the fibrils in the surface are oriented horizontally and are thereby able absorb the shear forces created by the movement of the joint. In the radial zone the fibers are vertically oriented in a hexameric structure. The chondrocytes are trapped within the fiber hexamers and this could be an explanation for the columnar appearance of chondrocytes typically seen in the radial zone.

The main type of collagen in articular cartilage is type II collagen, but small amounts of types I, V, VI, IX, X and XI can be detected. The different types

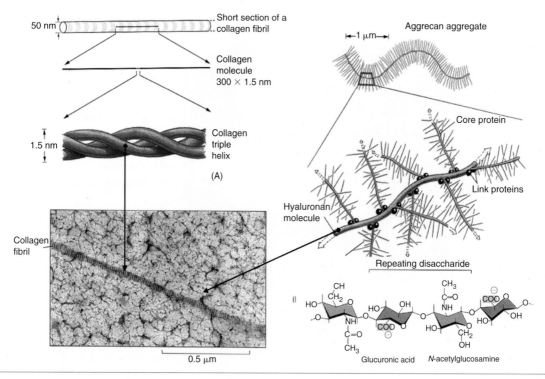

Figure 18.3 Cartilage matrix components: a unique molecular structure.

of collagen are typically found in certain zones and areas of the cartilage as described above. Type II collagen can be found in two different splicing variants, IIa and IIb. Of these two variants, IIb, is considered to represent a highly differentiated form specific for articular cartilage. Collagen is defined as a structural protein of the ECM which contains one or more domains having the conformation of a triple helix (van der Rest and Garrone, 1991). Collagens participating in quarter-staggered fibrils are type II and XI, in collagen-forming sheets are type X, in collagen-beaded filaments are type VI, and in fibrin-associated collagens are type IX. The principal differences between collagens are outlined in Figure 18.4. The turnover of collagens in adult normal cartilage is considered to be very slow [e.g. the turnover of the whole collagenous component of human femoral head cartilage matrix is estimated to be 400 years (Eyre, 2002)]. However, in certain disease states the synthesis can be increased, indicating an imbalance in tissue homeostasis.

18.3.2 Proteoglycans

Two different types of proteoglycans are present in the articular cartilage: nonaggregating and aggregating. The smaller nonaggregating proteoglycans consist of a core protein to which a large number of sulfated glycosaminoglycans (GAGs) are attached laterally (Figure 18.3). The GAGs normally found in articular cartilage are keratin sulfate, chondroitin sulfate and dermatan sulfate. The distribution of these molecules varies between different proteoglycans, but their internal function is relatively unknown.

Larger aggregated proteoglycans are formed by the binding of smaller nonaggregating proteoglycans to hyaluronic acid, which has binding sites for the core proteins. The hyaluronic acid is further bound to the chondrocytes through the surface receptor CD44. The concentration of hyaluronic acid is higher during the embryonic formation of cartilage and it has been shown that hyaluronic acids of certain molecular weight are chondroinductive.

Within the proteoglycans the negatively charged GAGs are packed around the core protein at a distance of 20–50 nm. Water molecules are bound to the GAGs and when the cartilage is compressed, there is a resistance for the water to leave the matrix, thereby limiting the forces of compression. In compression studies with cartilage explants constructed *in vitro*, it has been shown that dynamic compression selectively stimulates the synthesis of GAGs, which indicates an adaptation capacity of the tissue to mechanical force.

The main fraction of the proteoglycans, especially aggrecan, is found in the radial zone, but smaller proteoglycans like decorin and biglycan are found in the superficial zone. Decorin and biglycan both interact with collagens through the leucine-rich structures of their core proteins (Figure 18.5).

One matrix protein, which was initially identified in association with aggrecan, is cartilage-derived protein or Matrilin-1. The amount of this protein has been shown to change during maturation of cartilage and is increased in osteoarthritis (OA). Other proteins involved in macrofibril organization and stabilization of type II collagen include the five-armed member of the thrombospondin family, cartilage oligomeric protein (COMP). This protein has been used as a diagnostic marker in serum for the progress of matrix degradation in OA.

18.3.3 The chondrocyte

The cells of the cartilage are called chondrocytes. These cells are characterized by their special ability to produce matrix and their highly developed cytoskeleton of actin filaments. The actin filaments contribute to the biomechanical properties of the chondrocytes as well as connecting the chondrocyte to the articular matrix. The interactions with the ECM are critical for both the development and maintenance of the cartilage (Guilak, 1995). The immediate microenvironment of the chondrocyte is called a chondron and contains especially of type VI collagen. The chondron functions in insulating and physically separating the

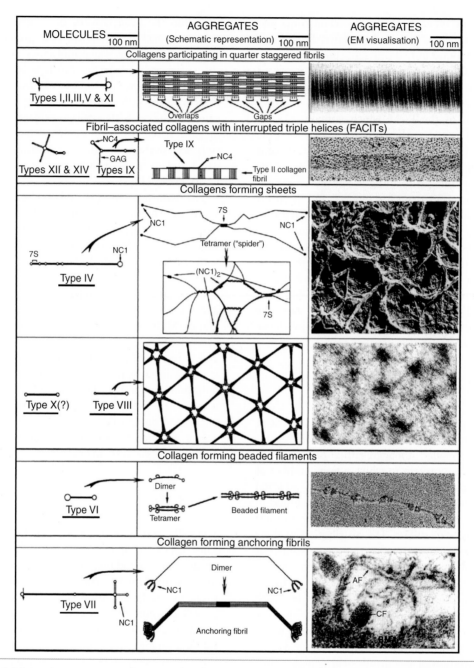

Figure 18.4 Molecular structure and supramolecular assemblies of collagens. From van der Rest, M. and Garrone, R. (1991). Collagen family of proteins. *Faseb J*, 5(13): 2814–2823.

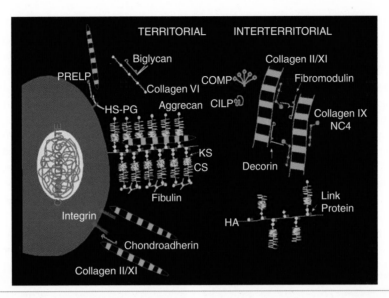

Figure 18.5 Proteins and proteoglycans in cartilage ECM. Source: Dick Heinegaard, Connective Tissue Biology, University of Lund, Sweden.

chondrocyte from direct interaction with the bulk of the load-bearing matrix. The number of chondrocytes per chondron varies from one to eight, with the lowest number in the superficial zone and the highest in the radial zone (Hunziker *et al.*, 2002). The distribution of chondrocytes varies in the different zones from approximately 7000 to 24,000 cells/mm³. From these figures, the number of chondrocytes beneath a certain area of an articular joint surface can be calculated to be approximately 24,000 cells/mm².

18.3.3.1 Proliferation of chondrocytes

Although one of the characteristics of articular cartilage is low cell turnover, there is some evidence in the literature showing that proliferation may occur in healthy cartilage, OA cartilage or in experimentally induced trauma. Generally, the proliferating cells have been found either in the superficial layer or in the deep calcified layer.

Due to the fact that proliferative cells are normally associated with some primitive features and stem cell properties, the findings of proliferative cells in these layers also raises the question from which area is the cartilage growing, i.e. intrinsic or appositional. Appositional growth has recently been detected by injection of the thymidine analogue bromodeoxyuridine (BrdU) into the developing joints of a marsupial (*Monodelphis domestica*) every other day for 2 weeks. BrdU integrates into the cells during the S phase and can be detected by immunohistochemistry. The conclusion of the study was that the progenitor cells of cartilage are located in the superficial layer and one explanatory hypothesis is that articular cartilage is growing appositionally (Hayes *et al.*, 2001).

This hypothesis is supported by additional studies in which the cells of the superficial zone were isolated by their affinity to fibronectin. These cells of the superficial zone were carefully characterized and shown to express the cell fate selector gene *Notch-1*, which is involved in the embryonic development of articular cartilage, especially in the condensation process (Watanabe *et al.*, 2003).

18.4 Pathophysiology of cartilage lesion development

Cartilage has a very low self-repair ability and it was already in the middle of the 18th century that Hunter "If we consult the standard Chirurgical Writers from Hippocrates down to the present Age, we shall find, that an ulcerated Cartilage is universally allowed to be a very troublesome disease … when destroyed, it is never recovered." This statement is still valid 200 years later even if spontaneous repair of small defects has been reported.

The major explanation for this limited capacity of repair is that cartilage is a unique tissue in its total lack of innervations and vascular supply. These two components normally work together upon an injury to obtain an inflammatory response attracting phagocytotic and multipotent stem cells. Another component limiting the ability of repair is that chondrocytes, even if they could respond to an injury, are trapped within the tight cartilage matrix.

In contrast to the adult cartilage, fetal cartilage has the ability to repair isolated lesions and ulcers (Figure 18.6). This specific self-repairing ability could possibly be explained by the structural and compositional differences between adult and embryonic cartilage (Namba *et al.*, 1998). Fetal cartilage contains higher amounts of hyaluronic acid and has

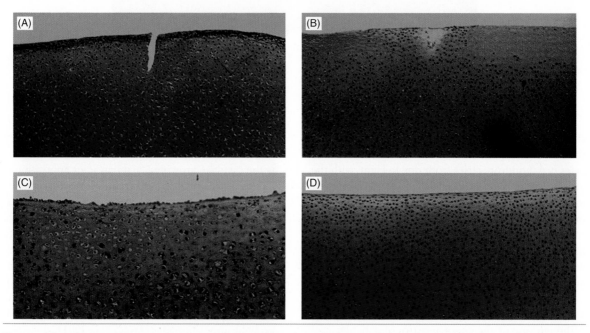

Figure 18.6 Spontaneous repair of superficial defects in articular cartilage in a fetal lamb model. (A) Acute injury consisting of a relatively parallel incisional defect that did not violate any blood vessels (×10). (B) Photomicrograph made 3 days after the injury, showing dead chondrocytes adjacent to the incisional defect and beneath sites where the footpad of the ophthalmic knife came into contact with the cartilage (×10). (C) Seven days after the injury, the defect is completely filled. The site of the defect and the areas that came into contact with the footpads were hypocellular and contained sparsely populated chondrocytes. Increased mitotic figures were visible adjacent to the zone of injury (×40). (D) Twenty-eight days after the injury, the cellularity, matrix staining and architectural arrangement of the cells matched those of the sham controls (×10). Reproduced from Namba, R.S., Meuli, M., *et al.* (1998). Spontaneous repair of superficial defects in articular cartilage in a fetal lamb model. *J Bone Joint Surg Am*, 80(1): 4–10.

a higher matrix to cell ratio. At the matrix level the embryonic cartilage contains type I collagen and the GAG chondroitin-6 sulfate, whereas while the adult cartilage mainly contains type II collagen and chondroitin-4 sulfate.

Acute or repetitive blunt joint trauma can damage articular cartilage and result in the formation of isolated defects. Due to the fact that the biomechanical properties surrounding the defect are altered, the remaining cells will be subjected to new mechanical forces that may result in cell death or apoptosis. The cellular processes ongoing on the rim of isolated defects have been compared to the process of the development of idiopathic OA. It is therefore thought that even if smaller defects may self-repair, larger isolated lesions caused by trauma that are left untreated may progress into OA – one of the most common forms of musculoskeletal diseases affecting millions of people throughout the world. The clinical symptoms of OA are pain and functional impairment that induce joint stiffness and dysfunction with subsequent impaired performance in daily life. As many as 25% of patients cannot cope with daily activities, often resulting in depression and social isolation (Kean *et al.*, 2004). In the European Union and USA combined, over 1 million joint replacements are performed each year as a consequence of OA. The main risk factors for OA are age, obesity and any form of joint trauma where the limited repair capacity of articular cartilage is a confounding factor. OA is often thought of as a disease of the elderly; however, OA is evident in younger patients; with approximately 5% of the population between 35 and 54 years having radiographic signs of OA (Petersson *et al.*, 1997). In a long-term follow-up study from the John Hopkins Precursor Study knee injuries increased the risk of subsequent knee OA over 5 times (Gelber *et al.*, 2000) corresponding to a similar risk as being overweight by 24 kg at the same age.

18.5 Artificial induction of cartilage repair

The limited repair capacity of cartilage has forced researchers and surgeons to find methods to induce cartilage repair. Several methods have been used where the general concept has been to use cells or tissues with chondrogenic potential to repair cartilage defects. The most frequently used methods are drilling of the subchondral bone in order to attract bone marrow cells (Insall, 1974), and resurfacing with periosteum (O'Driscoll and Salter, 1984) and perichondrium (Homminga *et al.*, 1989). However, the repair tissue formed is usually of the fibrous cartilage type unable to withstand the joint load over time.

The use of *in vitro* expanded chondrocytes for transplantation has been used as an alternative method for cartilage repair and the first cellular approach in humans for cartilage tissue engineering was initiated in the late 1980s. The treatment was named autologous chondrocyte transplantation (ACT) (Brittberg *et al.*, 1994) and has since become a clinical cell-based therapy for the repair of cartilage lesions worldwide under the alternative name of autologous chondrocyte implantation (ACI). In this procedure cartilage pieces are harvested arthroscopically from a healthy part of the joint. The cells are isolated by enzymatic digestion, subsequently culture-expanded *in vitro* to 10–30 times the initial amount and finally implanted into the defect site in high density (Figure 18.7).

18.5.1 Rationale for cell implantation

Articular chondrocytes are responsible for the unique features of articular cartilage; therefore, it seems rational to use true committed chondrocytes to engineer *in vitro* or *in vivo* cartilage in order to repair a cartilaginous defect. Human articular chondrocytes derived from articular cartilage biopsies have a limited proliferative potential and the number of cell divisions they undergo *in vitro* decreases with age (Dozin *et al.*, 2002). Chondrocytes grown as a monolayer dedifferentiate and loose both their chondrogenic phenotype with type II collagen production as well as their redifferentiation potential (Figure 18.8). The use of specific growth factors, such as basic fibroblast growth factor (FGF-2) and transforming

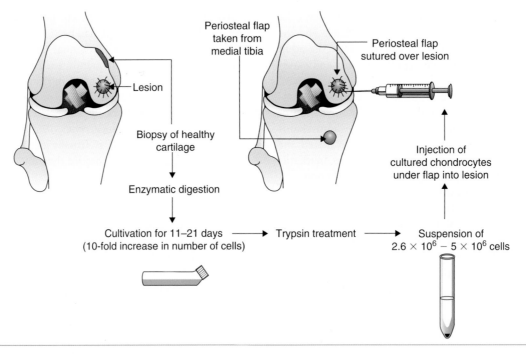

Figure 18.7 ACT procedure.

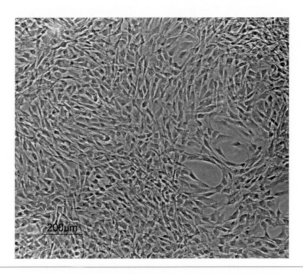

Figure 18.8 Human articular chondrocytes in monolayer cultures prior to implantation in ACT procedure.

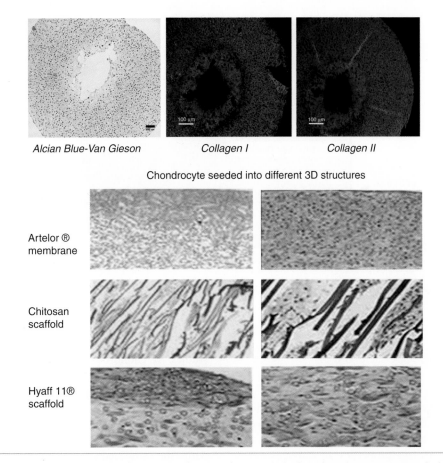

| Alcian Blue-Van Gieson | Collagen I | Collagen II |

Chondrocyte seeded into different 3D structures

Artelor ® membrane

Chitosan scaffold

Hyaff 11® scaffold

Figure 18.9 Cartilage regeneration from monolayer cultures. Articular chondrocytes expanded in monolayer redifferentiate when grown in high-density cultures (top panel) or when grown in 3D scaffolds (lower panel).

growth factor (TGF)-β1, stimulates cell proliferation and maintains the cell's ability to redifferentiate upon transfer into a three-dimensional (3D) environment (Jakob *et al.*, 2001). Expansion of human articular chondrocytes to be used for cell therapy is presently performed in the presence of 10% fetal calf serum (FCS) or autologous serum (Brittberg *et al.*, 1994), and their redifferentiation *in vitro* and *in vivo* remains a challenge. However, if dedifferentiated chondrocytes are cultured in 3D cultures in gels as well as in porous scaffolds in the presence of chondrogenic inducers

(TGF-β1) the cells re-express phenotypic markers of the articular chondrocyte (Figure 18.9).

In a mesenchymal tissue defect site the initial number of multipotential cells that can take part in the repair process is critical (Caplan *et al.*, 1997). The dedifferentiated chondrocytes have similarities with primitive mesenchymal cells, and a high-density cell implantation imitates the prechondrogenic cell condensation seen during limb formation and results in subsequent cartilage formation (Peterson *et al.*, 2002) (Figure 18.10).

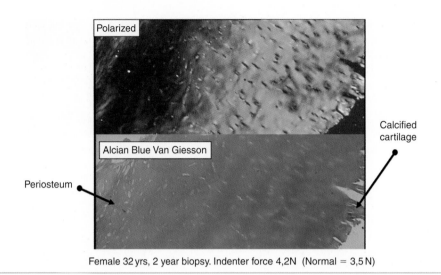

Female 32 yrs, 2 year biopsy. Indenter force 4,2N (Normal = 3,5 N)

Figure 18.10 Regenerated cartilage from the ACT procedure 2 years after cell implantation.

18.5.2 Cartilage specimens for implantation

Articular cartilage is composed of zones with different mechanical and functional properties (see above); furthermore, different locations in the joint have different cartilage properties. In a tissue engineering approach, cartilage has to be obtained as autologous grafts or allogeneic grafts, but does it matter from what location? Several studies have addressed this question. Using a surface abrasion technique the superficial zone was separated from the underlying growth zone in young articular cartilage. The superficial cells comprise approximately 4% of the total cells obtained from cartilage and the zone has a higher production capacity of superficial zone proteins (SZPs) (Darling *et al.*, 2004) while the growth zone produces more collagen type II; no difference in aggrecan production is seen between the zones. There was no regional variation seen between the femoral groove and medial and lateral condyles, which makes the pooling of zonal cells from different regions an acceptable option for tissue engineering

studies. If chondrocytes from the superficial and middle zones of immature bovine cartilage are cultured separately or sequentially, cartilaginous tissue formed from superficial zone chondrocytes exhibits less matrix growth and lower compressive properties than constructs from middle zone chondrocytes, with the stratified superficial/middle constructs exhibiting intermediate properties. Expression of SZPs was highest at the construct surfaces, with the localization of SZPs in superficial/middle constructs being concentrated at the superficial surface (Klein *et al.*, 2003).

Full-thickness, mid- and deep-zone, and deep-zone chondrocytes isolated from bovine cartilage cultured on ceramic substrates formed cartilage tissues with different structures. Tissue developed from deep-zone chondrocytes was thicker and had accumulated larger amounts of ECM than the tissues formed by the full-thickness and mid- and deep-zone chondrocytes. The tissue formed by the full-thickness chondrocytes accumulated the greatest amount of collagen, whereas the tissue formed by the mid- and deep-zone chondrocytes accumulated significantly more proteoglycans. The mid- and deep-zone chondrocytes produced tissue that had compressive mechanical properties up to 4 times greater than the cartilaginous tissues formed by cells from either the full-thickness or deep-zone part of the cartilage (Waldman *et al.*, 2003). A conclusion from these studies is that full-thickness cartilage biopsies are needed for the tissue engineering of articular cartilage.

18.5.3 Cell seeding density

Hunziker *et al.* (2002) have estimated that the mean number of cells of human cartilage tissue is 9626 cells/mm^3. With a mean height of the hyaline articular cartilage layer of 2.4 mm, a 1-cm^2 defect would have a volume of 24,000 mm^3 and need approximately 2.4×10^7 cells. *In vitro* experiments with chondrocytes directly seeded to cartilage explants show a linear relationship between biosynthetic activity and the number of seeded chondrocytes (Chen *et al.*, 1997). However, the number of cells needed for the implantation either as a suspension or in a scaffold has been studied extensively, but without reaching a consensus of seeding density. Seeding of polylactide–polyglycolide scaffolds with a density of less than 10×10^6 cells/ml resulted in very little cartilaginous material (LeBaron and Athanasiou, 2000), while seeding scaffolds at a density ranging from 20 to 100×10^6 cells/ml resulted in formation of clinically appropriate cartilage when implanted subcutaneously into nude mice (Puelacher *et al.*, 1994). Chondrocytes seeded in high-density cultures (10^7 cells/ml) in Atelocollagen gel generated a cartilage-like tissue (Iwasa *et al.*, 2003). All the publications that explore the appropriate cell seeding density have used using different cell sources, scaffolds cell seeding methods and growth media. There are no published studies that evaluate the influence of cell seeding concentration using human articular chondrocytes and clinically validated methods of culture in 3D. However, it is accepted that chondrogenesis is enhanced by increasing the cell seeding density and cell–cell interactions. However, the ideal number of

Table 18.1 Cells with chondrogenic capacities

Chondrocytes	Mesenchymal stem cells
Articular	Bone marrow
Auricular	Adipose cells
Septal nasal	Periosteum
Costal	Synovial membrane

cells needed to start and retain secure chondrogenesis is not known, although many commercial companies have as their goal to deliver 1×10^6 cells/cm^2.

18.6 What type of chondrogeneic cells are ideal for cartilage engineering?

Pure chondrocytes, epiphyseal or mature chondrocytes and allogeneic or autologous chondrocytes have been used in similar processes, as well as ear and nasal chondrocytes and autologous mesenchymal bone marrow stromal cells, as summarized in Table 18.1.

18.6.1 Allogeneic versus autologous cells

Chondrocytes express transplantation antigens and can theoretically participate in immunological reactions although the matrix produced by the chondrocytes protects the cells from rejection. Experiments have demonstrated that allogeneic growth plate chondrocyte implantation in cartilage defects of adult rabbits yields neocartilage; however, it is degenerated after a few weeks not only due to the humoral immune response, but also because of cell-mediated cytotoxicity (Kawabe and Yoshinao, 1991). When comparing long-term follow-up of allografted chondrocytes with autograft cells, both treatments gave cartilage regeneration. In the human situation, however, autologous chondrocytes seem to be preferable since there may be a risk of disease transmission with allografted cells.

18.6.2 Articular chondrocytes versus other cells

Articular chondrocytes have already been determined towards the cartilage lineage, but other cartilage sources could be considered e.g. nose, rib ear and growth plate chondrocytes. However, the behaviors of these cells are not similar. When porcine auricular, costal and articular chondrocytes were isolated and mixed with a fibrin polymer, and subsequently implanted into severe combined immunodeficiency (SCID) mice, new cartilaginous matrix was synthesized by the transplanted chondrocytes in all experimental groups. However, the ratios of dimension and mass for auricular chondrocyte constructs increased by 20–30%, the ratios for costal chondrocyte constructs were equal to the initial values and the ratios for articular chondrocyte constructs decreased by 40–50%. Furthermore, the biomechanical properties of the engineered cartilage made with auricular or costal chondrocytes were superior to those of cartilage made with articular chondrocytes (Xu *et al.*, 2004). The difference might be explained by the fact that auricular and costal cartilages normally exist in a vascularized area, while the articular chondrocytes exist in a nonvascularized area.

Human ear, nasal and rib chondrocytes are equally suited to the generation of autologous cartilage grafts for nonarticular reconstructive surgery, and their regeneration potential could be enhanced with a combination of growth factors (Tay *et al.*, 2004).

Within the mesenchymal tissue, hyaline cartilage chondrocytes are usually considered tissue restricted and without broader differentiation potential (Friedenstein and Chailakhjan, 1970). Culture-expanded mesenchymal stem cells have multipotency, giving rise to bone, cartilage, muscle, fat and stromal cells under defined culture conditions. Hyaline chondrocytes are generally considered to lack multipotency and be restricted to the cartilage lineage. However, when articular chondrocytes are subjected to the same culture techniques and analysis used in the study of phenotypic plasticity of marrow-derived mesenchymal stem cells, chondrocytes have the potential to form cartilage in pellet mass cultures, adipose cells in dense monolayer cultures and a calcium-rich matrix in an osteogenic assay. In contrast to mesenchymal stem cells, chondrocytes formed only cartilage, and not bone, in an *in vivo* osteochondrogenic ceramic implantation assay (Tallheden *et al.*, 2003), indicating a different *in vivo* default pathway. However, additional work will be necessary to assess whether cells of nonarticular origin can be successfully used for articular cartilage repair.

18.6.3 Embryonic stem (ES) cells

In the early 1980s several laboratories were able to culture cells derived from the inner cell mass of 3.5-day-old mouse embryos. If cells are cultured on feeders or in conditioned media containing inhibitors of differentiation the ES cells are able to maintain their undifferentiated stage in cell cultures over numerous cell doublings. Since then ES cells have been derived from several species including the human blastocyst (22).

ES cells exhibit all the characteristics of stem cells with properties including self-renewal, pluripotency in differentiation, high levels of telomerase activity, short G_1 cycle checkpoint and ability to initiate DNA replication without external stimulation. When injected into muscles or testis of SCID mice, the cells form teratomas with cells representing all three germ cell layers (Figure 18.11).

ES cells cultured in confluent cultures that are devoid of their factors of inhibiting differentiation (leukemia inhibitory factor or continuous expansion of cells) demonstrate aggregate formation. After several days *in vitro* the cells form large aggregates or embryoid bodies (EBs) due to the presence of many early embryoid derivatives. When subjecting the EBs to differentiation factors, the differentiation fate can be modified. Bone morphogenetic protein-2 and -4 have been demonstrated to induce differentiation of chondrogenesis in mouse ES cell (Kramer *et al.*, 2003). In human ES cells, induced differentiation to the hyaline cartilage phenotype has not been described, but potentially such cells could be developed as an 'of-the-shelf' product.

18.6.4 Xenograft cells

Another alternative cell source is xenograft cartilage, although the immunologic problem of cross-species transplantation as well as the problem of cross-species pathogen infectivity have to be solved if those cell type are to be used instead of true committed chondrocytes (Deschamps *et al.*, 2005).

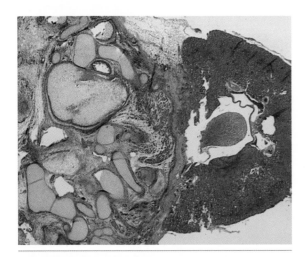

Figure 18.11 Teratoma formation of human ES cells in SCID mice.

18.6.5 Direct isolation of tissue

Cell isolation and direct implantation has been described. The method involves a tissue fragmentation step to mince the cartilage tissue mechanically into small tissue fragments before reimplantation. The purpose is to promote outgrowth of embedded chondrocytes through the increased tissue surface area. Using this methodology the researchers established that chondrocytes can effectively grow into adjacently placed scaffold materials and produce neocartilage (Lu *et al.*, 2006). Using such a technique it might be possible to engineer cartilage by a one-stage procedure.

18.7 Scaffolds in cartilage tissue engineering

The primary goal of the tissue engineering approach to full-thickness cartilage defects is to provide engineered cartilage with the emphasis on regeneration rather than repair. The ideal scaffold should provide an immediate support to cells and have mechanical properties matching those of the tissue being repaired. Gradually the material then has to degrade, as the cells begin secreting their own ECM, thus allowing for an optimal integration between newly formed and existing tissue (Freed *et al.*, 1994) (see Table 18.2).

Table 18.2 Criteria for scaffolds for tissue engineering (Freed *et al.*, 1994)

A	The surface should permit cell adhesion and growth
B	Neither the polymer nor its degradation should provoke inflammation or toxicity when implanted *in vivo*
C	The material should be reproducible in 3D structures
D	The porosity should be at least 90% in order to provide a high surface area for cell–polymer interactions and sufficient space for ECM regeneration; minimal diffusional constraints during *in vitro* cultures
E	The scaffolds should be absorbed once it has served the template function
F	The scaffold degradation should match the rate of tissue regeneration

Table 18.3 Different scaffolding materials used for cartilage repair

Absorbable fabricated matrices		
Polygalactin	Vicryl, multifilament yarn	Vacanti *et al.*, 1990
poly-l-Lactic acid		Wedge *et al.*, 1986
polyglycolic acid	Dexon	Freed *et al.*, 1994
Polyurethane		Klompmaker *et al.*, 1992
Non-absorbable fabricated matrices		
polyvinyl alcohol	Ivalon, porous sponge	Cobey, 1967
nylon		Kuhn, 1953
polyester	Dacron	Messner and Gillquist, 1993
polytetrafluoroethylene	Teflon	Messner and Gillquist, 1993
carbon fiber meshworks		Muckle and Minns, 1989
polyethylene		Mahmood *et al.*, 2005
Matrices from animals or humans		
collagen sponges (mainly collagen type I)		Wakitani *et al.*, 1989
meniscus		Heatley and Revell, 1985
decalcified bone		Dahlberg and Kreicberg, 1991
fibrin polymers		Nixon *et al.*, 1992
hyaluronic acid		Robinson *et al.*, 1989

In cartilage tissue engineering, the scaffold must promote chondrocyte attachment, *in vitro* proliferation and cell–cell interactions, as well as favor the expression and maintenance of a cell-differentiated phenotype. For the development of a suitable polymer scaffold, it is essential to consider that chondrocytes must organize in 3D to stimulate the synthesis of ECM molecules and prevent the loss of cell phenotype for a proper cartilage function once implanted. Ideally the scaffold should provide a temporary framework until the repaired cells have produced their own matrix. Both biologic and synthetic matrices [collagen and fibrin gels, decalcified bone, polyglycolic acids (PGAs), ceramics, and carbon fibers] have been used for the induced repair of cartilage. The different types of scaffolds can be preseeded by cultured cells before the introduction into the cartilage defect or the cells may migrate into the scaffold after implantation. Cartilage repair with allogenic or autologous cells of mesenchymal origin with different types of scaffolds have been tried in different models (see Table 18.3).

The successful molding of a human ear in PGA fibers subsequently seeded with calf articular chondrocytes and subcutaneously implanted in severe combined immunodeficiency (SCID) mice demonstrated the potential of combination of artificial scaffolds and cells (Cao *et al.*, 1997) (see Classical Experiment).

The use of a matrix for chondrocyte grafting started in the 1980s when embryonic chondrocytes in fibrin gel were used to maintain transferred chondrocytes within prefabricated cartilage defects in roosters (Itay *et al.*, 1987). Biodegradable PGA polymer seeded with chondrocytes was an alternative material used to resurface joint cartilage defects in the rabbit knee (Vacanti *et al.*, 1994). Type I collagen gel seeded with adherent

Classical Experiment

Total reconstruction of distinct anatomical structures is one of the most difficult problems in the field of plastic surgery. In the case of a total external ear reconstruction the surgeon's goal is to create an ear that is similar in appearance to the contralateral auricle. The surgeon has two paths to follow in this respect: ear cartilage formed from autogenous cartilage grafts or alloplastic implants.

The latter suffer from problems associated with any implanted prosthetic device, and they have an increased susceptibility to infection and uncertain long-term durability. Sculpted autogenous costal cartilage grafts are able to overcome implant disadvantages and can even grow with the patients. However, the esthetic outcome is highly reliant on the skill of the surgeon, extensive operation time

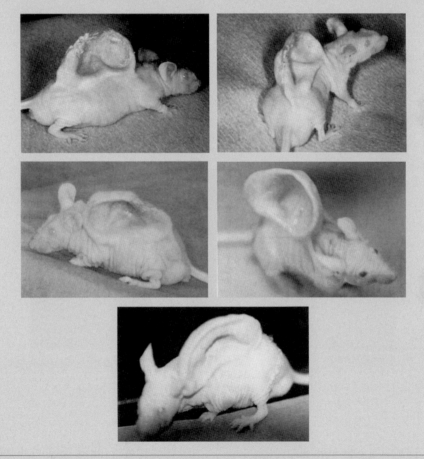

Gross appearance of human ear construct of PGA fibers seeded with calf articular chondrocytes and subcutaneously implanted into a nude mouse. Pictures taken after 12 weeks. Note the 3D shape that is almost identical to that of the human ear. From Cao, Y., Vacanti, J.P. et al. (1997). Transplantation of chondrocytes utilizing a polymer-cell construct to produce tissue-engineered cartilage in the shape of a human ear. Plastic and Reconstructive Surgery, *100(2): 297–302.*

is required and the harvest of cartilage from the rib results in donor-site morbidity.

The tissue engineering challenge is thus to produce a flexible framework in the shape of the complex 3D structure of the auricle with the potential to grow autogenous cartilage in a precisely predetermined shape and require minimal operative time.

A polymer template was formed in the shape of a human auricle using a nonwoven mesh of PGA molded after being immersed in a 1% solution of polylactic acid (PLA). Each PGA–PLA template was seeded with chondrocytes isolated from bovine articular cartilage and then implanted into subcutaneous pockets on the dorsa of 10 athymic mice. The 3D structure was cultured for 12 weeks, and subjected to gross morphologic and histologic analysis.

The overall geometry of the cultured tissue-engineered constructs closely resembled the 3D structure of the external auricle. Histological analysis demonstrated the formation of new cartilage. The human auricle is a rather simple anatomical structure of skin and cartilage, but nevertheless a formidable challenge to reconstruct for the surgeon. Although skin coverage is a critical element of any ear reconstruction, the techniques reported in this classical experiment address the formation of a cartilage scaffold. The authors were able to demonstrate that PGA–PLA constructs can be fabricated in a very intricate configuration and, if seeded with chondrocytes, the tissue-engineered construct is able to generate new cartilage that would be useful in plastic and reconstructive surgery.

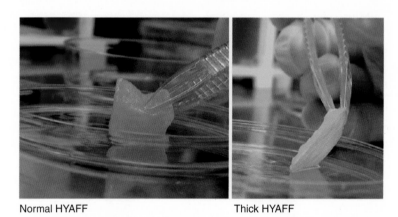

Normal HYAFF   Thick HYAFF

Figure 18.12 Cartilage graft construct based on chondrocytes cultured in hyaluronic acid.

cells from bone marrow or cells from the periosteum grown in culture were both able to regenerate cartilaginous tissue, although with mechanical properties similar to hyaline cartilage (Wakitani *et al.*, 1994). Despite numerous experiments (Table 18.2), the combinations of scaffolds and cells have shown variable results, and the clinical use of these scaffolds has been lacking. In the search for the ideal scaffold material, hyaluronic acid has proven to be an ideal molecule for tissue engineering strategies in cartilage repair, given its impressive multifunctional activity in cartilage homeostasis (Chen and Abatangelo, 1999) (Figure 18.12).

Nature often gives clues for the researcher in the tissue engineering field and before starting with a new tissue engineering concept it is necessary to gather knowledge of the biology of tissue formation during embryogenesis, to possess scientific knowledge to mimic the consistency and architecture of tissues in the adult organism, and to understand tissue mechanical properties and their alterations due to the physiological activity. Furthermore, the interactions between cells and the ECM have to be studied since the ECM not only acts as a structural support, but is also involved in informational exchange with the cellular component. Cells produce many kinds of molecules that form a particular environment providing both a scaffold and guidance during development and throughout adult life. The recapitulation of embryonic events represents a major goal for tissue engineering, but it is important to emphasize that the adult environment is quite different from the one present during the embryonic stage and tissues respond differently to specific information (Caplan, 2003).

All this is performed through complicated multistep differentiation cascades implying fine cross-talk between cells and ECM with the synthesis of site-specific specialized molecules (Toole, 2001). Therefore, the appropriate scaffold should be designed to play this multifunctional role and not be used as a mere vehicle for reparative cells or growth factors. The scaffold has to change its features as the regenerative process takes place – starting with tissue induction in the early events and then supporting tissue formation during differentiation. With this notion in mind the materials used in tissue engineering have to be biocompatible and biodegradable. Physical properties have to match those of the tissue being replaced, promoting cell attachment and ultimately possessing the capability of being remodeled by tissue-specific cells (Frenkel and Di Cesare, 2004).

Hyaluronan (Figure 18.13) is present in all soft tissues of higher organisms and, in particularly high concentrations, in the ECM of articular cartilage and in the mesenchyme of the developing embryo (Toole, 2001). It is a linear polymer consisting of a regular

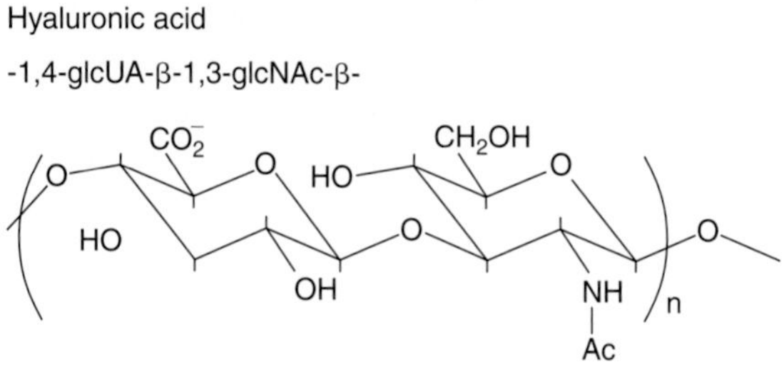

Hyaluronic acid

-1,4-glcUA-β-1,3-glcNAc-β-

Figure 18.13 Hyaluronic acid.

repeating sequence of a nonsulfated disaccharide units, glucuronic acid and N-acetyl-glucosamine with a molecular weight varying from 4000 to 8×10^6 Da. Hyaluronan is basically unmodified by evolution, underlying its importance in the physiological environment. Hyaluronan plays different biological roles depending on its physical and chemical properties. It interacts with binding proteins, proteoglycans and active molecules, such as growth factors which influence matrix structure, water balance lubrication and cell interaction. Hyaluronan can influence cell movement by its ability to alter the osmotic pressure leading to the formation of hydrated pathways in the tissue.

In the last few years the use of hyaluronan, in its highly purified form, has become common practice in medicine. Intra-articular administration of hyaluronan is widely being used to relieve pain and improve joint mobility in the nonsurgical treatment of OA. Hyaluronan is a major component of the ECM of embryonic mesenchymal tissues, thus the use of hyaluronan-based scaffolds will help create an embryonic-like milieu.

The greatest barrier to the successful use of this polymer in tissue engineering is the fact that hyaluronan, in its purified form, has such characteristics that limit its use as a biomaterial. Water solubility, rapid resorption and short residence time in the tissue along with its poor ductility hamper its possible applications. Cross-linking and coupling reactions are two ways to obtain a modified, stable form of hyaluronan (Campoccia et al., 1998). With the right application, hyaluronan derivatives can be extruded to produce fibers or membranes, lyophilized to obtain sponges, or processed to obtain microspheres. The combination of hyaluronan-based scaffolds and fibers has found its way into the clinic, where several hundred patients have now been treated (Dickinson *et al.*, 2005; Pavesio *et al.*, 2003).

18.8 Bioreactors in cartilage tissue engineering

Despite the good results of the first-generation technology of ACT, some of the concerns regarding the transplantation of immature chondrocytes in suspension led to the second generation of ACT. This technique involves the transplantation of chondrocytes in a 3D scaffold. This technique has several potential advantages such as the possibility of arthroscopic implantation, predifferentiation of cells *in vitro* and implant stability, among others. The nutritional requirements of cells that synthesize ECM increase along the differentiation process. The mass transfer must be increased according to the tissue properties. Bioreactors therefore represent a great tool to accelerate the biochemical and mechanical properties of the engineered tissues, providing adequate mass transfer and physical stimuli. The tissue engineering challenges in combining scaffolds and cells include the seeding properties where cells usually do not penetrate deep into the scaffold in static cultures. However, if biomaterials and bioreactor culture vessels are successfully combined, a basis for systematic and controlled *in vitro* tissue growth can be obtained. Such model systems have been created based on chondrocytes, biodegradable PGA scaffolds, and bioreactors where the scaffold induces cell differentiation and degrades at a defined rate, whereas the bioreactor maintains controlled *in vitro* culture conditions that permit tissue growth and development. The bioreactor has enabled tests for cartilage formation in different gravitations where Earth gravitation was superior to low gravitation in space (Freed *et al.*, 1997) (see State of the Art Experiment).

State of the Art Experiment

Tissue engineering was a new emerging scientific field in the early 1990s. Basically, the concept was to combine cells with biomaterials with the aim to create artificial tissue equivalent to human body tissue in case of injury or damage. The constructs were initially cultured in static culture systems but bioreactors were developed to control the cultures more rigorously as well as introducing a mechanical component by rotation (see Experimental design). The classical experiment was initiated to address questions regarding effects on cells and tissue from exposure to microgravity. The experimental setting was spectacular since it involved bioreactor culture both on Earth and in space, and thus gave the field of tissue engineering important publicity. The objectives were to examine chondrocyte viability and differentiation in a long-term space flight study and assess the effect of microgravity on long-term cartilage growth and function.

Chondrocytes were obtained from the femoropatellar grooves of young bovine calves and seeded (5×10^6 cells) on 5-mm diameter $\times$2-mm thick disks made of PGA formed as a 97% porous mesh of 13-μm diameter fibers. Cell/scaffold constructs were subsequently cultured in rotating bioreactors. Culture medium was replaced at a rate of 50% every 3–4 days and the vessel rotation speed was gradually increased from 15 to 28 r.p.m. over 3 months to induce mixing by gravitational construct settling. After 3 months, constructs were transferred into each of two flight-qualified rotating, perfused bioreactors (BTS) for an additional 4 months of cultivation on either the Mir Space Station or on Earth. Specifically, one BTS containing 10 constructs was

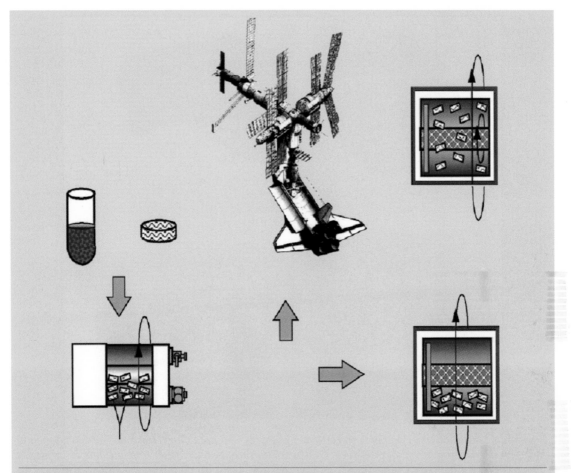

Experimental design: cartilage cells were seeded onto polymer scaffolds and the resulting constructs were cultivated in rotating bioreactors first for 3 months on Earth and then for an additional 4 months either on the Mir Space Station (10^{-4} to 10^{-6} g) or on Earth (1 g).

transferred to Mir via the US Space Shuttle and brought back to Earth 4 months later. A second BTS with 10 constructs served as an otherwise identical study conducted on Earth. Constructs were analyzed after 3 months (prior to launch) and after additional 4 months on Earth or in space.

Both environments yielded cartilaginous constructs, each weighing between 0.3 and 0.4 g, and consisting of viable, differentiated cells that synthesized proteoglycan and type II collagen. Compared with the Earth group, Mir-grown constructs were more spherical, smaller and mechanically inferior (see Result from bioreactor-grown cartilage). No differences in cell viability were demonstrated, but the differences in mechanical properties and inferior GAG production as well as more spherical form was attributed to the weightless environment.

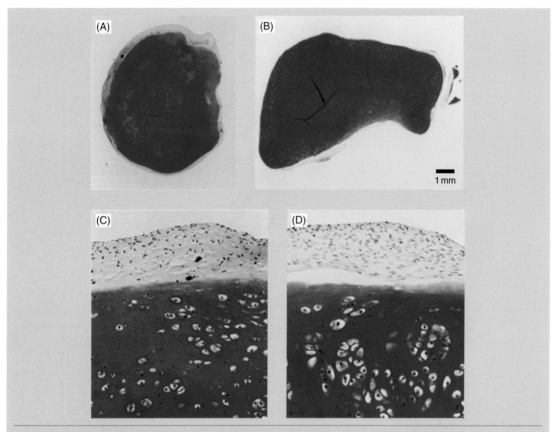

Result from bioreactor-grown cartilage on Earth and in space. The space-grown cartilage was more spherical and with inferior biomechanical properties. Construct structure. (A and B) Full cross-sections of constructs from Mir and Earth groups (×10). (C and D) Representative areas at the construct surfaces (×200). GAG is stained red with safranin-O. From Freed, L.E., Langer, R., et al. (1997). Tissue engineering of cartilage in space. Proc Natl Acad Sci USA, *94(25): 13885–13890.*

The experiment demonstrated the robustness of a bioreactor system and also demonstrated the importance of gravity, i.e. mechanical environment, on cartilage differentiation and function. The bioreactor system was demonstrated as a suitable system for controlled microgravity studies of cartilage and other tissues. Furthermore, the results gained had implications for long-term human spaceflight as well as managing clinical conditions of pseudo-weightlessness in prolonged immobilization, hydrotherapy and intrauterine development.

18.9 Growth factors that stimulate chondrogenesis

The cartilage matrix contains several growth factors that are released at the time of injury from the damaged cartilage matrix as well as from the damaged subchondral bone. Such growth factors include TGF-β, insulin-like growth factor-I (IGF-I), bone morphogenetic proteins (BMPs) and basic FGF (FGF-2) (Morales and Hascall, 1989). The factors orchestrate different effects. bFGF stimulates chondrocyte cell proliferation, stabilizes phenotypic expression and inhibits terminal chondrocyte differentiation. TGF-β controls chondrocyte proliferation and differentiation as well as inducing chondrogenic differentiation of mesenchymal cells *in vitro* and chondrogenesis *in vivo* (Joyce *et al.*, 1990). It also inhibits chondrocyte terminal differentiation and stimulates chondrocyte proliferation in cooperation with FGF, and it has been shown to promote repair of damaged cartilage. BMPs initiate cartilage differentiation and maintain the chondrocytic phenotype. IGFs have important anabolic effects on cartilage metabolism with increased matrix production.

Little is known about the interactions among these growth factors, and knowledge of their effect and role in chondrogenesis is limited. However, these findings indicate a potential therapeutic way to use exogenous growth factors to enhance the repair of cartilage defects.

18.10 Future developments

Research in regenerative medicine involves cell and molecular biology, developmental cell biology, immunology, and polymer chemistry. This new direction in medicine will use three strategies: transplantation of cells to form new tissue in the transplant site, implantation of bioartificial tissues constructed *in vitro* and induction of regeneration *in vivo* from healthy tissues next to an injury (Stocum, 2004).

However, regarding the future for cartilage repair in the new century, the idea is to transplant stem/progenitor cells, or their differentiated products, into a cartilage lesion site where they may form new tissue, or the cells could be used to construct a bioartificial tissue *in vitro* to replace the original tissue or organ. Bioartificial tissues are made by seeding stem or differentiated cells into a natural or artificial biomaterial scaffold shaped in the appropriate form and then implanting the construct in place of the damaged tissue or organ. Theoretically, the use of stem cells is preferable to the use of differentiated cells harvested directly from a donor because stem cells

have the potential for unlimited growth and thus supply. Such so-called uncommitted cells are capable of a broad range of chondrogeneic expression and could provide a regenerative tissue that recreates the embryonic lineage transitions originally involved in joint tissue formation. However, the recent research described in this chapter shows that the use of true committed chondrocytes is still reasonable, but more research is needed to know how to make use of them in a more efficient way. We may also see semibiological approaches where cells are used in combination with artificial implants. Such techniques could be suitable for resurfacing of large osteochondral defects and may be one way to develop semibiological articular prostheses.

18.11 Summary

1. Adult articular chondrocytes exhibit a level of phenotypic plasticity that is comparable with that of mesenchymal stem cells.
2. The number of chondrocytes needed to start chondrogenesis in a scaffold is more than 20×10^6 cells or at least 1×10^6 cells/cm^2.
3. There is a need to study the use of freshly isolated chondrocytes for direct implantation.
4. Pooling of zonal chondrocytes from different regions is an acceptable source for chondrocytes to be used for cartilage tissue engineering.
5. There is, as yet, no ideal scaffold for cartilage engineering. In small defects *in vitro* developed mature osteochondral constructs may be used in the future instead of autologous osteochondral plugs with or without cells. In larger defects, hyaluronan-based scaffolds in combination with chondrocytes could be an option.
6. Small- to medium-sized traumatic defects need to be treated differently compared to large osteoarthritic defects where the cartilage engineering process has to be directed not only to fill a defect, but also to withstand a diseased joint's negative influence. The tissue engineering approach should address several phenomena including blocking the production of pro-inflammatory factors and

suppressing the progression of the degenerative process affecting both cartilage and the bone compartment.

References

ap Gwynn, I., Wade, S., *et al.* (2002). Novel aspects to the structure of rabbit articular cartilage. *Eur Cell Mater*, 4: 18–29.

Brittberg, M., Lindahl, A., *et al.* (1994). Treatment of deep cartilage defects in the knee with autologous chondrocyte transplantation. *N Engl J Med*, 331: 889–995.

Campoccia, D., Doherty, P., *et al.* (1998). Semisynthetic resorbable materials from hyaluronan esterification. *Biomaterials*, 19: 2101–2127.

Cao, Y., Vacanti, J.P., *et al.* (1997). Transplantation of chondrocytes utilizing a polymer-cell construct to produce tissue-engineered cartilage in the shape of a human ear. *Plast Reconstr Surg*, 100: 297–302. discussion 303–304.

Caplan, A.I. (2003). Embryonic development and the principles of tissue engineering. *Novartis Found Symp*, 249: 17–25. discussion 25–33, 170–174, 239–241.

Caplan, A.I., Elyaderani, M., *et al.* (1997). Principles of cartilage repair and regeneration. *Clin Orthop Relat Res*, 342: 254–269.

Centers for Disease Control and Prevention (2002). Prevalence of self-reported arthritis or chronic joint symptoms among adults – United States. *MMWR Morb Mortal Wkly Rep*, 51(42): 948–950.

Chen, A.C., Nagrampa, J.P., *et al.* (1997). Chondrocyte transplantation to articular cartilage explants in vitro. *J Orthop Res*, 15: 791–802.

Chen, W.Y. and Abatangelo, G. (1999). Functions of hyaluronan in wound repair. *Wound Repair Regen*, 7: 79–89.

Darling, E.M., Hu, J.C., *et al.* (2004). Zonal and topographical differences in articular cartilage gene expression. *J Orthop Res*, 22: 1182–1187.

Deschamps, J.Y., Roux, F.A., *et al.* (2005). History of xenotransplantation. *Xenotransplantation*, 12: 91–109.

Dickinson, S.C., Sims, T.J., *et al.* (2005). Quantitative outcome measures of cartilage repair in patients treated by tissue engineering. *Tissue Eng*, 11: 277–287.

Dozin, B., Malpeli, M., *et al.* (2002). Response of young, aged and osteoarthritic human articular chondrocytes to inflammatory cytokines: molecular and cellular aspects. *Matrix Biol*, 21: 449–459.

Eyre, D. (2002). Collagen of articular cartilage. *Arthritis Res*, 4: 30–35.

Freed, L.E., Vunjak-Novakovic, G., *et al.* (1994). Biodegradable polymer scaffolds for tissue engineering. *Biotechnology (NY)*, 12: 689–693.

Freed, L.E., Langer, R., *et al.* (1997). Tissue engineering of cartilage in space. *Proc Natl Acad Sci USA*, 94: 13885–13890.

Frenkel, S.R. and Di Cesare, P.E. (2004). Scaffolds for articular cartilage repair. *Ann Biomed Eng*, 32: 26–34.

Friedenstein, A.J., Chailakhjan, R.K., *et al.* (1970). The development of fibroblast colonies in monolayer cultures of guinea-pig bone marrow and spleen cells. *Cell Tissue Kinet*, 3: 393–403.

Gelber, A.C., Hochberg, M.C., *et al.* (2000). Joint injury in young adults and risk for subsequent knee and hip osteoarthritis. *Ann Intern Med*, 133: 321–328.

Guilak, F. (1995). Compression-induced changes in the shape and volume of the chondrocyte nucleus. *J Biomech*, 28: 1529–1541.

Hayes, A.J., MacPherson, S., *et al.* (2001). The development of articular cartilage: evidence for an appositional growth mechanism. *Anat Embryol (Berl)*, 203: 469–479.

Homminga, G.N., van der Linden, T.J., *et al.* (1989). Repair of articular defects by perichondral grafts. Experiments in the rabbit. *Acta Orthop Scand*, 60: 326–329.

Hunziker, E.B., Quinn, T.M., *et al.* (2002). Quantitative structural organization of normal adult human articular cartilage. *Osteoarthritis Cartilage*, 10: 564–572.

Insall, J. (1974). The Pridie debridement operation for osteoarthritis of the knee. *Clin Orthop*, 101: 61–67.

Itay, S., Abramovici, A., *et al.* (1987). Use of cultured embryonal chick epiphyseal chondrocytes as grafts for defects in chick articular cartilage. *Clin Orthop Relat Res*, 220: 284–303.

Iwasa, J., Ochi, M., *et al.* (2003). Effects of cell density on proliferation and matrix synthesis of chondrocytes embedded in Atelocollagen gel. *Artif Organs*, 27: 249–255.

Jakob, M., Demarteau, O., *et al.* (2001). Specific growth factors during the expansion and redifferentiation of adult human articular chondrocytes enhance chondrogenesis and cartilaginous tissue formation in vitro. *J Cell Biochem*, 81: 368–377.

Joyce, M.E., Roberts, A.B., *et al.* (1990). Transforming growth factor-beta and the initiation of chondrogenesis and osteogenesis in the rat femur. *J Cell Biol*, 110: 2195–2207.

Kawabe, N. and Yoshinao, M. (1991). The repair of full-thickness articular cartilage defects. Immune responses to reparative tissue formed by allogeneic growth plate chondrocyte implants. *Clin Orthop Relat Res*, 268: 279–293.

Kean, W.F., Kean, R., *et al.* (2004). Osteoarthritis: symptoms, signs and source of pain. *Inflammopharmacology*, 12: 3–31.

Klein, T.J., Schumacher, B.L., *et al.* (2003). Tissue engineering of stratified articular cartilage from chondrocyte subpopulations. *Osteoarthritis Cartilage*, 11: 595–602.

Kramer, J., Hegert, C., *et al.* (2003). In vitro differentiation of mouse ES cells: bone and cartilage. *Methods Enzymol*, 365: 251–268.

LeBaron, R.G. and Athanasiou, K.A. (2000). *Ex vivo* synthesis of articular cartilage. *Biomaterials*, 21: 2575–2587.

Lu, Y., Dhanaraj, S., *et al.* (2006). Minced cartilage without cell culture serves as an effective intraoperative cell source for cartilage repair. *J Orthop Res*, 24: 1261–1270.

Morales, T.I. and Hascall, V.C. (1989). Factors involved in the regulation of proteoglycan metabolism in articular cartilage. *Arthritis Rheum*, 32: 1197–1201.

Namba, R.S., Meuli, M., *et al.* (1998). Spontaneous repair of superficial defects in articular cartilage in a fetal lamb model. *J Bone Joint Surg Am*, 80: 4–10.

O'Driscoll, S.W. and Salter, R.B. (1984). The induction of neochondrogenesis in free intra-articular periosteal autografts under the influence of continuous passive motion. An experimental investigation in the rabbit. *J Bone Joint Surg Am*, 66: 1248–1257.

Pavesio, A., Abatangelo, G., *et al.* (2003). Hyaluronan-based scaffolds (Hyalograft C) in the treatment of knee cartilage defects: preliminary clinical findings. *Novartis Found Symp*, 249: 203–217. discussion 229–233, 234–238, 239–241.

Peterson, L., Brittberg, M., *et al.* (2002). Autologous chondrocyte transplantation. Biomechanics and long-term durability. *Am J Sports Med*, 30: 2–12.

Petersson, I.F., Boegard, T., *et al.* (1997). Radiographic osteoarthritis of the knee classified by the Ahlback and Kellgren & Lawrence systems for the tibiofemoral joint in people aged 35–54 years with chronic knee pain. *Ann Rheum Dis*, 56: 493–496.

Poole, A.R., Kojima, T., *et al.* (2001). Composition and structure of articular cartilage: a template for tissue repair. *Clin Orthop*, 391(Suppl): S26–S33.

Puelacher, W.C., Kim, S.W., *et al.* (1994). Tissue-engineered growth of cartilage: the effect of varying the concentration of chondrocytes seeded onto synthetic polymer matrices. *Int J Oral Maxillofac Surg*, 23: 49–53.

Stocum, D.L. (2004). Tissue restoration through regenerative biology and medicine. *Adv Anat Embryol Cell Biol*, 176: III–VIII. 1–101, back cover.

Swann, D.A., Silver, F.H., *et al.* (1985). The molecular structure and lubricating activity of lubricin isolated from bovine and human synovial fluids. *Biochem J*, 225: 195–201.

Tallheden, T., Dennis, J.E., *et al.* (2003). Phenotypic plasticity of human articular chondrocytes. *J Bone Joint Surg Am*, 85A(Suppl 2): 93–100.

Tay, A.G., Farhadi, J., *et al.* (2004). Cell yield, proliferation, and postexpansion differentiation capacity of human ear, nasal, and rib chondrocytes. *Tissue Eng*, 10: 762–770.

Toole, B.P. (2001). Hyaluronan in morphogenesis. *Semin Cell Dev Biol*, 12: 79–87.

Vacanti, C.A., Kim, W., *et al.* (1994). Joint resurfacing with cartilage grown *in situ* from cell–polymer structures. *Am J Sports Med*, 22: 485–488.

van der Rest, M. and Garrone, R. (1991). Collagen family of proteins. *FASEB J*, 5: 2814–2823.

Wakitani, S., Goto, T., *et al.* (1994). Mesenchymal cell-based repair of large, full-thickness defects of articular cartilage. *J Bone Joint Surg Am*, 76: 579–592.

Waldman, S.D., Grynpas, M.D., *et al.* (2003). The use of specific chondrocyte populations to modulate the properties of tissue-engineered cartilage. *J Orthop Res*, 21: 128–132.

Watanabe, N., Tezuka, Y., *et al.* (2003). Suppression of differentiation and proliferation of early chondrogenic cells by Notch. *J Bone Miner Metab*, 21: 344–352.

World Health Organization (2003). The burden of musculoskeletal conditions at the start of the new millennium. *World Health Organ Tech Rep Ser.*, 919: 1–218.

Xu, J.W., Zaporojan, V., *et al.* (2004). Injectable tissue-engineered cartilage with different chondrocyte sources. *Plast Reconstr Surg*, 113: 1361–1371.

Chapter 19
Tissue engineering of bone

Steven van Gaalen, Moyo Kruyt, Gert Meijer, Amit Mistry, Antonios Mikos, Jeroen van den Beucken, John Jansen, Klaas de Groot, Ranieri Cancedda, Christina Olivo, Michael Yaszemski and Wouter Dhert

Chapter contents

Chapter objectives:

- To understand the different types and functions of bone
- To identify the benefits and drawbacks of autologous bone as a graft
- To indicate the different components of a bone tissue engineering graft
- To point out the host-related issues that are important for a graft to succeed
- To summarize the different steps in bone tissue engineering
- To know the different levels of evidence needed prior to performing clinical studies
- To realize that in bone tissue engineering research some of these levels are not yet completed

"It is anticipated that the orthopedic surgeon will be required to transform from a hardware expert into a 'mesenchemist' to ensure that the proper bioactive factor is placed in the proper location at the correct time and in the optimal amount to facilitate the body's self-repair by controlling its intrinsic repair-regeneration capacity."

Arnold Caplan (2005)

19.1 Introduction: bone

19.1.1 General description of bone (Figure 19.1)

Bone is the main supporting system in the human body. It is a unique combination of minerals and tissue that provides excellent tensile and loading strength. Most prominent is the inorganic mineral phase that is responsible for its stiffness. Cells and tissue constitute the organic phase, responsible for maintenance, tensile strength and elasticity. The characteristic bone cell is the osteocyte that resides inside little chambers (lacunae) surrounded by mineralized bone matrix. This surrounding bone matrix has been deposited by the osteoblasts as an uncalcified matrix (i.e. osteoid) that is calcified by calcium salt deposition. The osteoblasts are present where new bone is formed and border these areas with so-called osteoblast zones. The osteoblasts that are incorporated in the lacunae become osteocytes; others enter a resting state as flat-bone-lining cells. Active osteoblasts are derived of these lining cells or are descendants of osteoprogenitor cells that reside in bone marrow stroma and the well-vascularized membranes that line the outer and inner surface of bone (periosteum and endosteum, respectively). In addition to a supportive function, bone has an important function in the regulation of calcium and phosphate blood levels. It is believed that besides the osteoclasts that can resorb bone, the osteocytes and osteoblasts are also crucial in this process. Finally, bone harbors the bone marrow that is the main breeding place for white and red blood cells.

19.1.1.1 Types of bone (Figure 19.1)
Macroscopically, the skeleton consists of long bones, like the femur, flat bones like the skull, and cuboid bones like vertebrae and carpals. On a structural level, two types of bone are identified: cancellous (trabecular, spongy) and compact (dense, cortical) bone (Figure 19.1A–C). The cancellous bone is the most changing and active part, like the crown of a tree. It is active in growth, calcium homeostasis and hematopoiesis, and its supportive function is mainly in locations with a predominantly compression type of loading, such as in vertebral bodies and adjacent to articulating joints (e.g. the knee joint). Compact bone is more static and the strongest, like the stem of a tree. Its main locations are the shafts of long bones and peripheral lining of flat bones. Microscopically, two major forms of bone can be identified: woven and lamellar bone (Figure 19.1D–E). Woven bone is the immature unorganized type of bone. It is present abundantly in newborns and in locations where fast bone formation takes place, like in the growth plates (physes) and in fracture repair. After woven bone is laid down, it is organized by remodeling to become lamellar bone. Haversian bone is the most complex of compact lamellar bone, it contains vascular Haversian channels circumferentially surrounded by lamellar bone, and this complex is called an osteon.

19.1.1.2 Extracellular bone matrix
In this context, the difference between calcification and ossification (or osteogenesis) should be emphasized (see Definitions). Calcification can take place in extracellular bone matrix, but also in artery walls (atherosclerosis). Ossification is exclusively the mineralization/calcification of osteoid. Osteoid is

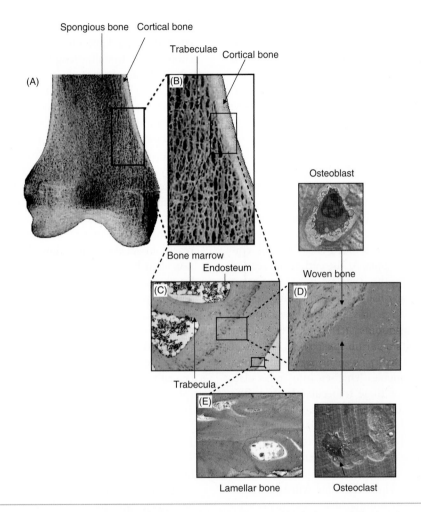

Figure 19.1 An overview of a macroscopic aspect of a right femur (thigh bone, (A)) showing the mechanically strong cortical bone on the outside and the inner spongious bone in this metaphysic end of the bone; (B) is a detail of the cortical bone and the inner spongious bone showing the trabeculae of which it is composed; (C) is a detail of spongeous bone where a single trabecula is pictured with bone marrow cavities, lined with endosteum, on each side. The cavities make this type of bone rather porous as compared to the more dense cortical bone; (D) shows a woven type of bone that is typically young and unorganised; (E) is an example of lamellar bone that has been remodelled into a more mechanically strong type of bone. On top of (D) an example of an osteoblast is portrayed that can typically be found in areas of active bone formation. Similar to this, an osteoclast is found in the same areas for further remodelling of the bone (insert below (D)). Note the channel of resorbed bone behind the osteoclast.

predominantly constituted of collagen type 1 fibers. It is generally accepted that osteoid calcification is initiated by the formation of little buds inside the matrix, so-called matrix vesicles that act as nucleation sites for the first mineral deposition. Alkaline phosphatase (ALP) is an important enzyme in this process, as it can hydrolyze organic phosphate and calcium-containing substrates to create supersaturated concentrations to enhance mineral deposition. Inappropriate ectopic (at extraskeletal sites) calcification is thought to occur due to the absence of naturally occurring inhibiting substances. Specific processes such as necrosis (cell death) and inflammation are associated with reducing the activity of these inhibitors, often resulting in local calcification.

19.1.1.3 Bone formation and repair

During the formation of the skeleton and with maintenance and repair of bone, two distinct mechanisms can be observed separately or combined. These are intramembranous and endochondral bone formation.

Intramembranous bone formation is the mechanism for the formation of flat bones like the cranial bones and the scapula. At the end of the second month of gestation, this type of bone formation begins in condensates of loose mesenchymal tissue that contain osteogenic cells. In association with adequate vascularization, ossification starts and progresses without an intermediate. The first small mass of bone with its irregular shape is called a spicule. Several spicules lengthen into trabeculae. Continued growth of these trabeculae leads to the formation of trabecular bone (also known as spongy or cancellous bone). On the trabecular surfaces, the osteoblast zones will continue osteoid formation layer upon layer which is called appositional bone formation. This process continues until sufficient bone density is obtained. Then, the bone will be remodeled to gain the most optimal shape and density by simultaneous bone resorption by osteoclasts and bone formation by osteoblasts. This process will lead to either compact (dense) bone present in the cortex or the more spongeous cancellous bone in the interior of bones. Despite the fact that intramembranous bone formation is highly efficient, it is inappropriate for the fast longitudinal growth of the appendicular skeleton (of arms and legs) in childhood.

Endochondral bone formation is the mechanism for long bone formation and lengthening. Characteristic of this mechanism is that bone is preceded by the formation of cartilage. In the development of the initial limb bud, mesenchymal cells condense and differentiate in an avascular environment into chondroblasts

that produce cartilage matrix (anlage). This anlage is covered with a dense fibrous layer known as the perichondrium. Initially, lengthening is established by repeated chondrocyte divisions and matrix production, and widening of the limb bud is established by appositional growth from the perichondrium. Then, ossification centers appear in each bone, first centrally (diaphyseal) and later at the extremities (epiphysial). Before ossification, the chondrocytes enlarge (hypertrophy) and further mature, which results in increased intracellular calcium concentration. The thin cartilage matrix that surrounds the hypertrophied cells serves as a substrate for calcification that coincides with death (apoptosis) of the chondrocytes. This results in calcified scaffolding for later bone apposition. Synchronously, the perichondrium is invaded by capillaries and differentiates into periosteum that delivers pre-osteoblasts. From the periosteum capillaries and osteogenic cells invade the calcified cartilage and start to deposit osteoid, which is the start of ossification. Almost the entire bone is ossified except for to the central canal and two transverse plates just beneath the epiphysis. These are the growth plates (physes) responsible for bone elongation until adolescence (Figure 19.2).

19.1.1.3 Fracture healing

The fracture healing process only aims at rapid stabilization of broken bone parts with little commitment to anatomy. Both of the above mechanisms of bone formation are utilized when appropriate. When a fracture occurs, an area of cell death (necrosis) will result. Before any restoration, this area has to be cleaned, which is called the inflammatory phase. Then, the repair phase follows with initial stabilization by cartilage (soft callus) that is replaced by bone (hard callus) as in endochondral bone formation. Simultaneously (direct) intramembranous bone formation can be found depending on the local oxygen supply. Only when a fracture is stable and with unchanged anatomy (like a crack) intramembranous repair alone will be sufficient. After the repair phase, the remodeling phase follows. Comparable to nonfractured bone, this phase will result in the best achievable morphology (Figure 19.3).

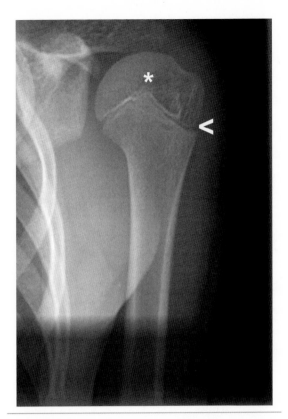

Figure 19.2 X-ray of the left shoulder of a 10-year-old girl showing the open physis (arrow) and ossification in the epiphysis forming the humeral head (*).

19.1.2 Bone grafting (Figure 19.4)

When the bone repair mechanism fails as a result of magnitude, infection or other causes, bone grafting has been shown to be a highly successful therapy. Bone grafting means that bone from somewhere else is applied to stimulate bone formation. Preferably, this bone is the patient's own (autologous) and harvested from locations of relative excess, e.g. from the pelvis. Bone of other humans (allogenic) or animals (xenogenic) is also applied. In 1881, Macewen was one of the first to describe the procedure for allograft reconstruction of the humerus in a young child (Macewan, 1881). Nowadays, bone grafting is still used as a therapy, e.g. in complicated fractures that fail to heal or severely atrophied regions (like the mandibula after

teeth extraction) that need augmentation. In addition, it is often applied in the case of fractures with a high risk of nonunion, when extensive regions of bone have been removed (e.g. in case of malignancies or infections), and to facilitate fusion of bones after extensive reconstructive procedures that are performed in orthopedic and maxillofacial surgery.

The exact mechanism of bone graft incorporation is not completely understood, although several distinct mechanisms of bone formation have been determined. Probably most important is the provision of scaffolding that allows deposition of bone and subsequent integration by the surrounding bone. The feature of a material to do so is called osteoconductivity. Another property of grafted bone is that the extracellular matrix (ECM) contains factors that can induce surrounding nonbone cells to differentiate towards osteoblasts. This property is called osteoinduction. Finally, although the vast majority of osteogenic cells will not survive after transplantation, some may contribute to bone formation after the graft is sufficiently vascularized, this obviously is osteogenesis. When massive osteogenesis is crucial, e.g. in very large defects, vascularized grafts (where the cells can survive) are an option with the current microsurgical possibilities.

Another important aspect for the success of a bone graft is that it can be completely resorbed and remodeled, and therefore does not interfere with physiologic bone adaptation. The autologous cancellous bone graft is considered most optimal, because it possesses all of the above qualities. A disadvantage is that it carries limited mechanical strength that is sometimes required. In that case more dense cortical bone can be applied at the expense of the other qualities.

Estimates on the number of bone graft procedures each year approximate 500,000 procedures in the US alone. The vast majority of these procedures are spinal fusions (approximately 50%) followed by general orthopedic procedures and craniomaxillofacial procedures (Boden, 2002).

19.1.2.1 Orthopedic applications of bone grafts
Spinal fusion is necessary to stabilize fractured vertebrae, deformities or to treat low back pain as a result of disk degeneration or intervertebral instability. It is

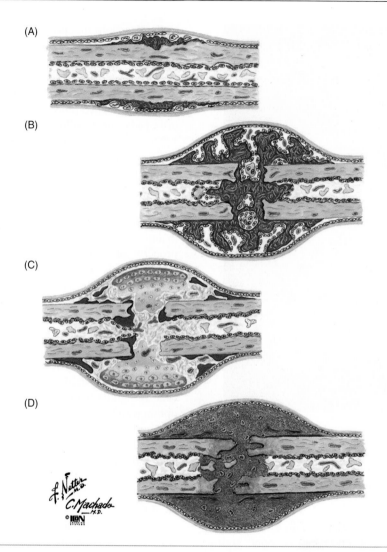

Figure 19.3 Schematic representation of the sequential steps in fracture healing: (A) a fresh hematoma is formed, (B) inflammatory and mesenchymal cells arrive after small vessels have invaded the area, (C) callus is formed by both endosteal and periosteal reaction and (D) the callus is remodeled. Reprinted with permission from Elsevier: http://www.netterimages.com.

achieved by an anterior approach to create fusion of the invertebral space and/or a posterolateral approach to create a bone bridge between the vertebral arches and transverse processes of neighboring vertebrae. In most cases, internal fixation with cages (anterior) or rods and screws (on the lateral side of the vertebral bodies) will be performed in addition to bone grafting.

Augmentation of osteolytic bone deformities or defects also requires grafting. Osteolysis is often seen after joints have been replaced by prostheses and the mechanical loading of the bone has changed. This

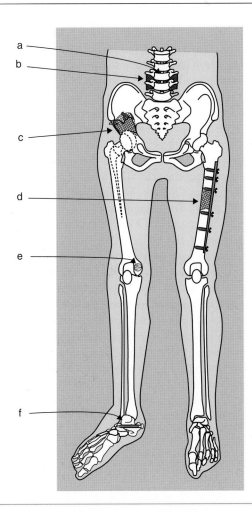

Figure 19.4 Potential clinical orthopedic applications for the use of bone grafts: (a) anterior spine fusion, (b) posterolateral spine fusion, (c) osteolytic defect (acetabular defect from a prosthetic implant), (d) traumatic bone defect/pseudoarthrosis or tumor defect in femur diaphysis, (e) cystic lesion femur condyle metaphysis and (f) subtalar joint arthrodesis.

results in loss (osteolysis) of bone where natural biomechanics is overtaken by the implant biomechanics (stress shielding). Also defects as a result of cysts, infection or (tumor) resection can be augmented with bone grafts. Small-sized bone chips are ideal if mechanical strength can be obtained with instrumentation (i.e.

with the use of nails, plates and screws). For nonhealing fractures (nonunions), grafting may be indicated especially in so-called atrophic nonunions (when the stability may be sufficient but biology is lacking).

19.1.2.2 Oral/maxillofacial applications of bone grafts
Augmentation is often required because continuous resorption of the alveolar ridge after extraction of all teeth eventually results in an unfavorable denture-bearing area. As a consequence, denture support as well as retention and stability decreases (Cawood and Howell, 1991). For the extremely atrophic mandible, various ridge augmentation techniques are available that use cancellous rib or corticocancellous iliac grafts (Fazili *et al.*, 1986). For single tooth replacement situations, augmentation can be realized by means of a local bone graft taken from the symphysis area of the chin. In trauma, forehead defects can be covered by cortical plates harvested from the outer table of the cranial vault (calvarium) or by rib grafts. Orbital fractures can be restored by struts of rib, iliac wing or calvarial bone. Continuity defects in the mandible can be reconstructed using rib or iliac wing bone and stabilization with bridging plates. An alternative approach was introduced by Boyne (1992): a titanium scaffold, modeled after the original mandible, was filled with autologous bone paste and attached with screws to the original bone surface.

19.1.3 Alternatives for the autologous bone graft

For decades, the autologous bone graft has been the golden standard for the vast majority of graft applications. This means it is most optimal and alternative operations and/or grafts should be compared to it. The need for alternative grafts has been recognized for the following reasons:

(1) *Autograft harvesting prolongs the operation*: especially for relatively small operations, harvesting from the iliac wing or resection of a rib considerably lengthens the procedure;
(2) *Availability is limited*: this is a serious drawback for those procedures that require large amounts,

like posterior spinal fusion or situations where additional operations are anticipated;

(3) *Surgical complications*: these include chronic donor site pain in over 10% of patients (Banwart *et al.*, 1995; Silber *et al.*, 2003) and (wound) infections.

These drawbacks and the estimated economic market value of bone graft substitutes of up to $2 billion per year (US only) have boosted research in this field exponentially (Boden, 2002). Another (future) reason for an alternative is that on some occasions the autologous graft fails and requires improvement. However, contemporary research aims at equal performance, which has appeared to be very challenging. Currently there are some well-established alternatives that will be discussed briefly. Remember, in general, none of these is as good as the autologous bone graft. Therefore, application is limited to the use as an extender to the autologous graft volume (Gupta and Maitra, 2002) or to less challenging locations. Several graft-related qualities are summarized in Table 19.1.

19.1.3.1 Allografts

Allografts by nature are derived from different human donors. The major concerns are the transmission of diseases and the immunogenicity of the graft. Therefore allografts undergo several treatments. The grafts are devitalized by freezing ($-80°C$), and often additional procedures like sterilization with ethylene oxide and/or high-dosage γ irradiation are applied. These sterilization steps, however, can significantly reduce the osteoinductivity of the graft (Friedlaender, 1987). Prior to these steps, physical debridement and ultrasonic baths are applied to remove most soft tissue, including the cells. The most common methods of preservation are freezing and freeze-drying (lyophilization). The latter obviously changes material properties substantially. Another treatment is demineralization of the allograft bone which results in demineralized bone matrix (DBM). By using a special protocol, the osteoinductive proteins are retained in this material that can be applied as putty. With allografts, similar stages of incorporation by the host are observed as in the autologous bone graft; however, the whole process is significantly delayed. This may be explained by the fact that allograft contains no living cells which may be osteogenic and osteoinductive. Furthermore, it has been observed that the greater the genetic difference between donor and recipient, the more delayed and/or incomplete the repair process.

Table 19.1 Relative qualities of different grafts [based on data from Goldberg and Stevenson (1990) and Giannoudis *et al.* (2005)]

Bone graft	Immunogenicity	Mechanical properties	Osteoconduction	Osteoinduction	Osteogenesis
Autograft					
cancellous	0	0	3	3	3
cortical	0	3	2	2	2
Allograft					
cancellous					
frozen	2	0	2	1	0
freeze-dried	1	0	2	1	0
cortical					
frozen	2	3	1	0	0
freeze-dried	1	1	1	0	0
DBM	1	0	1	2	0

0 = no effect, 1 = minimal effect, 2 = large effect, 3 = very large effect.

This implies that despite the preparation procedure, some immunogenicity remains (Bos *et al.*, 1983). Despite these disadvantages, bone allografting has increased considerably, especially in the US (Friedlaender, 1983; Mankin *et al.*, 1983; Prolo and Rodrigo, 1985).

19.1.3.2 Xenografts

Bovine DBM is also used as a bone graft substitute. Its osteoinductive potential is attributed to the remaining ECM that contains bone morphogenetic proteins (Pietrzak *et al.*, 2005). Clinical outcomes are variable, with moderate to good clinical results in some oral/maxillofacial applications (Maiorana *et al.*, 2006; Nevins *et al.*, 2003). In contrast, its use in hip revision surgery showed poor results with graft rejection (pseudo-infection) (Charalambides *et al.*, 2005). Immunological responses *in vivo* can be related to the origin of the donor and processing steps, such as the use of photooxidation (Hetherington *et al.*, 2005).

19.1.3.3 Collagen

Collagen is a major component of the ECM in bone. It plays an important role in vascular ingrowth, mineral deposition and binding of growth factors (Cornell, 1999). Clinical applications of collagen as a bone substitute show disappointing results due to minimal structural support and its potential immunogenicity (Fleming *et al.*, 2000). As a composite material with other bone substitutes, it may function as an effective autograft extender (Muschler *et al.*, 1996) or stand-alone graft (Boden *et al.*, 2000).

19.1.3.4 Porous ceramics

Ceramics are synthetic materials often made from calcium and phosphate that have been used extensively in dentistry and orthopedics during the past three decades (Hollinger *et al.*, 1996; Bohner, 2000). Coralline hydroxyapatite (HA) is a hydrothermal conversion of the carbonate matrix of natural porous sea coral. Its successful use in clinical applications has been reported (Elsinger and Leal, 1996), although its structural support is very limited. Some ceramics showed osteoinductive properties (Habibovic

et al., 2006) (see Section 19.2.1). Tricalcium phosphate (TCP) is a ceramic that will quickly resorb *in vivo*; HA, on the other hand, will hardly be resorbed. Both materials are biocompatible but due to the slow resorption of HA, the stress shielding in time is a matter of concern. Therefore combined implants are made such as biphasic calcium phosphate (BCP; a combination of HA and TCP) (Daculsi *et al.*, 1989). High-porosity constructs provide good surface areas for vascular ingrowth and osteoconduction.

From the above it will be clear that a good alternative that has all the advantages of autologous bone without its disadvantages does not exist. Only recently has the scientific world realized that the solution to this challenge requires the input and collaboration of both life scientists and engineers. The interdisciplinary research field that has evolved from this insight, striving to create biological substitutes to restore or reinforce tissue, is referred to as tissue engineering (Caplan and Goldberg, 1999; Langer and Vacanti, 1993).

19.2 Strategies for bone tissue engineering

In this chapter bone tissue engineering will be regarded as any attempt to stimulate bone formation with implants that are designed according to principles of life sciences and engineering. Ideally, tissue engineering can substitute the autologous bone graft, which necessitates that the product is at least as good as the autologous bone graft. Also, tissue-engineered constructs that cannot completely substitute for the use of autologous bone may be added to it in order to extend the graft volume and to diminish the amount of autologous bone needed (graft extenders). In general, tissue-engineered implants are constructs of a carrier (scaffold) and biologically active factors. These biological factors can be (a combination of) cells and proteins that stimulate host cells. As mentioned previously, constructs are designed to act for one or more of the following qualities as mentioned previously: osteoconduction, osteoinduction and osteogenesis. Therefore, the ingredients for bone tissue engineering can basically be divided into scaffolds, growth factors and cells. This simple division, of course, is not as distinct as it seems and will be outlined in the following sections.

19.2.1 Scaffolds

A well-designed scaffold for bone tissue engineering serves multiple purposes. One of these is to provide a delivery vehicle for osteoinductive molecules and/or osteogenic cells. Furthermore, this component of a tissue engineering therapy must fill the gap in a bone defect and should facilitate healing. To do so effectively, several qualities of an effective scaffold material have been identified and will be discussed, including biocompatibility, osteoconductivity, porosity, biodegradability, mechanical properties and intrinsic osteoinductivity (Habibovic *et al.*, 2006). Additionally, one must bear in mind the storage, handling and sterilization of the material prior to its final use. Two classes of materials that are widely studied for tissue engineering of bone are ceramics and polymers; each of these will be discussed in the following sections on the scaffold material qualities.

19.2.1.1 Biocompatibility

Perhaps the most important quality of a bone tissue engineering scaffold is biocompatibility. Once implanted, a scaffold must not instigate an inflammatory or toxic response that will result in cell death and further aggravation of the injury. All components and products of the material must also be biocompatible, including unreacted monomers, initiators, stabilizers, cross-linking agents, emulsifiers, solvents, and released degradation products (Temenoff *et al.*, 2000).

Calcium phosphate (CaP) ceramics, such as HA and TCP (see Section 19.1.3), are very similar to the inorganic component of natural bone tissue. Thus, it is no surprise that CaP ceramics demonstrate exceptional biocompatibility. In fact, direct bonding with bone tissue is very possible (Ruhe *et al.*, 2006). While biocompatibility is more of a concern with synthetic polymers as compared to CaP ceramics, degradable polymers such as poly(L-lactic acid) (PLLA), poly(glycolic acid) (PGA), poly(DL-lactic-*co*-glycolic acid) (PLGA) and

poly(ε-caprolactone) (PCL) are already widely used for certain biological applications. Additionally, various other polymers such as polyorthoester (POE), polyanhydrides and poly(propylene fumarate) (PPF) are being investigated for tissue engineering, and have shown favorable biocompatibility in animal models (Hutmacher, 2000; Mistry and Mikos, 2005).

19.2.1.2 Osteoconductivity

Osteoconductivity is essential for successful bone substitution. A scaffold must provide a substrate upon which bone cells can adhere, proliferate, migrate and deposit bone. CaP ceramics offer excellent osteoconductivity, likely because they are chemically similar to the inorganic phase of bone. An important characteristic of ceramics that has shown to considerably improve the osteoconductivity considerably is a large surface area that is achieved by increased microporosity (Wilson *et al.*, 2006). Figure 19.5(A) shows osteoconduction into the pores of a BCP scaffold implanted in a rabbit radius defect implanted for 6 weeks in an adult rabbit (van Gaalen, unpublished data).

Some biodegradable polymers have shown to exhibit osteoconductivity as well, albeit to a lesser extent than ceramics. Figure 19.5(B) shows a PPF/PLGA scaffold implanted in a rabbit radius (Hedberg *et al.*, 2005a). The osteoconductivity of many polymers may be enhanced by the attachment of certain peptides derived from ECM proteins. One such peptide, Arg–Gly–Asp (RGD), is derived from the ECM proteins fibronectin and laminin, and is widely studied for promoting biomolecular recognition, cell attachment and cell function (Shin *et al.*, 2003). Biomaterials modified with specific peptides such as RGD have been shown to promote growth and migration of osteoprogenitor cells as well as osteoblasts (Shin *et al.*, 2003). A different approach towards osteoconductive scaffolds is the fabrication of collagen fibrils which mimic the nanostructure of ECM in natural bone tissue (Hartgerink *et al.*, 2001). These materials are prepared by the synthesis of collagen peptide chains, which are subsequently polymerized and self-assembled into nanofibrils only a few nanometers in diameter and many microns in length (Hartgerink *et al.*, 2001).

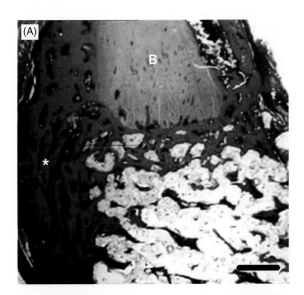

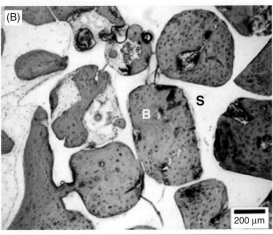

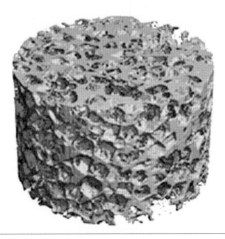

Figure 19.6 Micro-computed tomography image of a PPF/PLGA scaffold with interconnected porous architecture prepared by salt leaching. Bar = 1 mm. Reproduced with permission from Hedberg, E.L., Shih, C.K., Lemoine, J.J., *et al.* (2005). Mikos, *in vitro* degradation of porous poly(propylene fumarate)/poly(DL-lactic-co-glycolic acid) composite scaffolds. *Biomaterials*, 26(16): 3215–3225.

Figure 19.5 (A) Undecalcified histology of a BCP scaffold implanted in a rabbit radius segmental defect. Newly formed bone is indicated with *. Scale bar = 1 mm. (B) PPF/PLGA scaffold implanted in a rabbit radius. S = scaffold structure; B = bone.

19.2.1.3 Porosity

Another crucial requirement for facilitating bone growth inside a material is an interconnected porous architecture. Void space is required to allow blood vessels and surrounding bone to grow into the scaffold, and to allow implanted and/or migrating cells to deposit bone within it. Additionally, porosity increases the available surface area for cell attachment, growth and function throughout a defect (Temenoff *et al.*, 2000). Various methods may be used to create porous scaffolds depending on material properties and desired pore architecture. One such example is the method of salt leaching as shown for the PPF polymer in Figure 19.6 (Hedberg *et al.*, 2005b). Other methods for creating porous scaffolds include hydrocarbon templating (an anhydrous method of particulate leaching in polymers), gas foaming and phase separation (Hacker *et al.*, 2003). More recently, electrospinning (Yoshimoto *et al.*, 2003) and stereolithography [three-dimensional (3D) printing] have been added to this list. The latter allowing more specific control of scaffold geometry and pore architecture as well as optimization of mechanical properties (Cook *et al.*, 1995; Lin *et al.*, 2004; Wilson *et al.*, 2004).

19.2.1.4 Biodegradability

Currently used permanent implant materials, such as metals, are associated with infection, corrosion, fatigue or failure, any of which may require implant replacement (Temenoff *et al.*, 2000). Additionally, nondegradable materials impede the stimuli for growth of the bone around it (stress shielding). In contrast, biodegradable materials used as bone tissue engineering scaffolds temporarily fill a defect and gradually degrade as bone is regenerated. While the material degrades, space is created for developing tissue and load is gradually transferred from scaffold to bone. Ultimately, the defect is replaced by natural bone tissue, and the biomaterial is safely broken down and removed from the body (Timmer *et al.*, 2003a).

Ceramics and polymers vary widely in degradation behavior depending on composition. As mentioned previously, HA hardly resorbs and is therefore considered practically nondegradable, whereas TCP degrades quite fast (Ruhe *et al.*, 2007). Poly(methylmethacrylate) (PMMA) polymer is considered nondegradable, while polymers such as PLGA, PCL and PPF degrade by ester hydrolysis. The degradation rate of PPF-based systems, for example, can be varied based on composition or synthesis conditions, allowing one to tailor design a scaffold towards a specific application (Peter *et al.*, 1998). Furthermore, many materials can be modified with certain peptide sequences enabling degradation by protease enzymes (Shin *et al.*, 2003).

The degradation products of these biomaterials must also demonstrate biocompatibility. For example, PPF-based polymer networks degrade into fumaric acid and propylene glycol, as well as small quantities of acrylic acid and poly(acrylic acid-*co*-fumaric acid). Timmer *et al.* (2003b) demonstrated that these individual degradation components induced *in vitro* cytotoxicity only at high concentrations. Thus, the expected slow release of these products from the slowly degrading polymer is not expected to adversely affect biocompatibility.

19.2.1.5 *Mechanical properties*
The interaction of hard HA crystals and elastic collagen fibers in the nanoscale structure of bone contribute to its unique mechanical properties, thus making bone difficult to replace using synthetic materials (Taton, 2001). Metals, such as titanium and stainless steel, are often used in orthopedics, because of superior strength compared to bone. However, bone tissue relies on mechanical stimulation for regeneration and remodeling. Permanent metal implants absorb the majority of these stimulating forces. This phenomenon is called a stress shielding effect and results in bone resorption around the implant (Bobyn *et al.*, 1992). Ideally, a tissue engineering scaffold should be designed to match the mechanical properties of the specific bone it will replace. One novel strategy to improve the mechanical properties of biodegradable polymers is the incorporation of nanoparticle fillers into bulk polymers. For example, the incorporation of very low concentrations of surface-modified alumoxane nanoparticles or functionalized carbon nanotubes into biodegradable PPF-based systems resulted in significant increases in flexural and compressive mechanical properties (Horch *et al.*, 2004; Shi *et al.*, 2006) (Figure 19.7).

19.2.1.6 *Intrinsic osteoinductivity*
Bone induction is a phenomenon that is well known for growth factors (see Section 2.2). However, also without the administration of growth factors, intrinsic bone induction is a quality of specifically prepared scaffolds. Recently, it has been shown that this quality significantly enhances bone healing (Habibovic *et al.*, 2006). In 1969, Winter reported on ectopic bone induction by a sponge made of a porous polymer, polyhydroxyethylmethacrylate (pHEMA), which showed that not only factors that reside in bone matrix, but also physical structures themselves could have bone-inductive properties. Several decades later, others (Ripamonti, 1991; Zhang, 1991) discovered that porous CaPs also could induce bone formation when implanted extraskeletally. It was demonstrated later that bone induction by biomaterials requires both a specific 3D porous structure of the implanted biomaterial and an *in vivo* remineralization of pore surfaces (Habibovic *et al.*, 2005). The mechanism of this phenomenon has not been elucidated yet, although there are indications that CaP dissolution

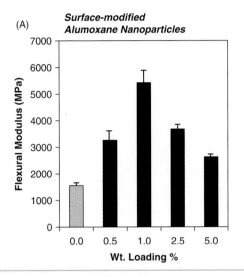

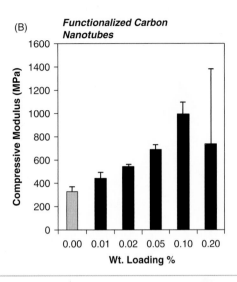

Figure 19.7 Mechanical properties of PPF-based nanocomposites. (A) Flexural modulus of PPF-based polymer infused with surface-modified alumoxane nanoparticles as a function of nanoparticle loading percentage (1% = 1 g of nanoparticles per 100 g of polymer/nanoparticle scaffold material). Reproduced with permission from Horch, R.A., *et al.* (2004). Nanoreinforcement of poly(propylene fumarate)-based networks with surface modified alumoxane nanoparticles for bone tissue engineering. *Biomacromolecules*, 5(5): 1990–1998. (B) Compressive modulus of PPF-based polymer with incorporated functionalized carbon nanotubes as a function of nanotube loading concentration. Reproduced with permission from Shi, X., *et al.* (2006). Injectable nanocomposites of single-walled carbon nanotubes and biodegradable polymers for bone tissue engineering. *Biomacromolecules*, 7(7): 2237–2242.

and subsequent reprecipitation possibly with the incorporation of inductive factors plays a role (Yuan, 2001).

19.2.1.7 Other considerations
Other considerations for scaffold design involve the ease with which the scaffold can be used. Injectable materials can be molded into irregular shapes or directly injected into a defect and cross-linked *in situ*, thus enhancing implant–tissue contact and minimizing the invasiveness of surgery (Temenoff and Mikos, 2000). Both ceramic pastes and polymers are available in injectable formulations and have demonstrated these advantages. Additionally, the scaffold should be capable of sterilization in a manner such that its bioactivity and chemical composition are not modified during sterilization and processing.

19.2.2 Growth factors

Growth factors are signaling molecules that can influence certain cellular functions through their binding to specific cell membrane receptors (Figure 19.8). Most illustrative for this function is osteoinduction, a phenomenon described more than 40 years ago by Friedenstein (1962) and extensively investigated by Urist (1965), who discovered the presence of BMPs inside the extracellular bone matrix (see Classical Experiment). The precise nature of these BMPs remained enigmatic until about 15 years ago, when it was discovered that not only each protein separately (especially BMP-2 and BMP-7) but a whole class of BMPs can induce bone (Wozney *et al.*, 1988; Wozney, 1992).

Apart from the osteoinductive factors, other factors have also been associated with the stimulation of bone formation and fracture healing, such

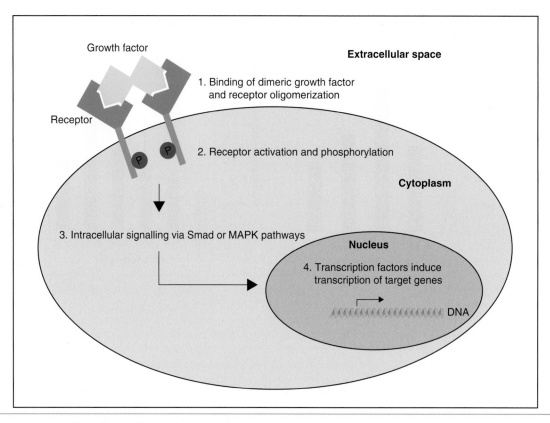

Figure 19.8 Binding of growth factors to membrane-bound receptors (1) results in receptor oligomerization and activation. Phosphorylation of the intracellular parts of the receptors (2) activates intracellular signaling pathways (3) and subsequently results in the transcription of target genes (4). These transcription products can influence cellular behavior (such as cell proliferation and cell differentiation).

as transforming growth factor (TGF)-β, vascular endothelial growth factor (VEGF), insulin-like growth factor (IGF), fibroblastic growth factor (FGF) and platelet-derived growth factor (PDGF).

19.2.2.1 Osteoinductive growth factors (Table 19.2)

The use of growth factors in bone tissue engineering is currently mainly focused on the osteoinductive factors. Structurally, BMPs appear as dimeric proteins that have multiple signaling epitopes. For these ligands, several membrane-bound serine/threonine kinase receptors have been identified. Their working mechanism is described in Figure 19.8. The effect

of BMPs depends on many factors like the type and combination of factors, the concentration, exposure time, cell responsiveness and carrier type. Although currently over 20 different BMPs have been identified, mainly BMP-2 and BMP-7 (OP-1) have been used in bone tissue engineering (Borden *et al.*, 2004; Jansen *et al.*, 2005; Kroese-Deutman *et al.*, 2005; Riley *et al.*, 1996; Ripamonti *et al.*, 2001; Ruhe *et al.*, 2004; Vehof *et al.*, 2002; Wan and Cao, 2005). The concentration is a crucial factor and is species dependent. For example, in humans, unfortunately, much higher concentrations are needed than in rodents and dogs (Boden, 1999). The cells that are likely most sensitive

Classical Experiment
Bone morphogenetic proteins and the work of Urist

In the 1960s Marshall R. Urist (Urist, 1965) investigated decalcified bone that was implanted at various locations, both hetero- and homotopic, in several types of animals and in humans. These acellular, decalcified bones (i.e. only the organic part of bone tissue) yielded new bone in amounts proportional to the volume of the implant, with over 90% positive experimental results. Evaluation using light microscopy showed that initially (3 weeks postimplantation) these implants were enveloped in highly vascular, inflammatory and fibrous connective tissue. Subsequently, invading cells of the host started the process of resorption of the collagenous matrix, after which new bone deposits became apparent after 4–6 weeks. These early deposits were always found in the interior of well-vascularized excavation chambers. Also, cartilage formation was observed in the closed ends of old, vascular channels, which was gradually replaced by bone tissue between 8 and 16 weeks via the process of endochondral ossification. At this point, Urist called this process 'osteogenesis by autoinduction'. In later experiments (Urist et al., 1976), Urist introduced the term BMP for a proteinaceous compound that was obtained by chromatographic separation of a collagenase digest of bone matrix gelatin. This compound was loaded into diffusion chambers, which were implanted in muscle pouches in adult rats, and demonstrated to induce osteogenesis in the direct vicinity of the diffusion chamber.

The purification and characterization of BMPs became established in the late 1980s, as did their amino acid sequence and DNA clones (Luyten et al., 1989; Wozney et al., 1988; Wozney, 1992). Currently, many different types of BMPs have been identified, which differ in function (Table 19.2) (Rengachary, 2002). Although the yield of fresh BMPs out of bone is extremely low (10 kg of bovine bone powder yields 10 μg BMP), recombinant DNA technologies have allowed now the production of large quantities of so-called recombinant BMP (rhBMP) for commercial exploitation and clinical use. An example of the *in vivo* potential of rhBMP-2 on porous CaP implants in rabbits is shown below. Several clinical reports on nonunions in tibia, femur and the spine have also shown beneficial results to the use of rhBMPs. Spine fusion models (see also Figures 19.4 and 19.12), such as an anterior (cage) or posterolateral approach (both with or without instrumentation), have shown additional positive results for the use of both rhBMP-2 and OP-1 (rhBMP-7) as an alternative to the use of autograft bone. Several studies have measured both clinical signs of fusion (such as the OSWERTY pain and disability scoring system) and radiographic evidence for successful fusion (Table 19.5). Unfortunately, large quantities of BMPs of up to 20 mg per side (posterolateral fusion model) are needed with subsequent considerable financial consequences.

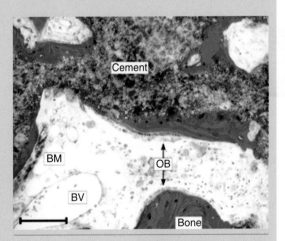

Detailed light microscopic aspect of bone formation in rhBMP-2-loaded porous CaP cement implants after 10 weeks of subcutaneous implantation in a rabbit. OB = osteoblasts, BM = bone marrow, BV = blood vessel. Bar = 100 μm.

Table 19.2 BMPs that are members of the large TGF-β family (their individual functions *in vivo* are indicated)

BMP	Function
BMP-2	osteoinductive, osteoblast differentiation, apoptosis
BMP-3 (osteogenin)	most abundant BMP in bone, inhibits osteogenesis
BMP-4	osteoinductive, lung and eye development
BMP-5	chondrogenesis
BMP-6	osteoblast differentiation, chondrogenesis
BMP-7 (OP-1)	osteoinductive, development of kidney and eye
BMP-8 (OP-1)	osteoinductive
BMP-9	nervous system, hepatic reticuloendothelial system, hepatogenesis
BMP-10	cardiac development
BMP-11 (GDF-8, myostatin)	patterning mesodermal and neuronal tissues
BMP-12 (GDF-7)	induces tendon-like tissue formation
BMP-13 (GDF-6)	induces tendon and ligament-like tissue formation
BMP-14 (GDF-5)	chondrogenesis, enhances tendon healing and bone formation
BMP-15	modifies follicle-stimulating hormone activity

GDF = growth/differentiation factor. BMP-1 is not a member of the TGF-β family; it has protease activity and is therefore very different from the other BMPs.

to induction are close to the osteogenic lineage, like mesenchymal progenitor cells, although osteoinduction of lymphocytes has been reported (Friedenstein *et al.*, 1970). The carrier material is crucial for an optimal release profile, but also the surface characteristics appear to be important. For example, CaP-coated titanium scaffolds showed better results with BMP than similar noncoated scaffolds (Vehof *et al.*, 2002).

19.2.2.2 Osteostimulating growth factors (Table 19.3)
In mammals, three isoforms of TGF-β have been identified: TGF-β1, TGF-β2 and TGF-β3. Of these, TGF-β1 has been used most frequently to enhance bone formation in bone tissue engineering approaches. In contrast to BMPs, which can induce osteogenesis, these are not osteoinductive (Borden *et al.*, 2004; Sampath *et al.*, 1987). Nevertheless, TGF-β1 has been investigated frequently and shown to stimulate bone formation in titanium or degradable polyester scaffolds at orthotopic sites (Jansen *et al.*, 2005; Ripamonti *et al.*, 2001; Sampath *et al.*, 1987). FGF exists in acidic (FGF-1)

and basic (FGF-2) isoforms. Both stimulate proliferation of endothelial cells, fibroblasts, chondroblasts and osteoblasts, and are present in and acting on bone. *In vivo* animal studies have shown the more powerful effects of FGF-2 (Chiou *et al.*, 2006; Fujita *et al.*, 2005) in comparison to FGF-1 (Nakajima *et al.*, 2001). PDGF stimulates replication in osteoblast cell cultures *in vitro* and it increases type I collagen synthesis. PDGF also stimulates bone resorption *in vitro* by a mechanism that involves prostaglandin synthesis (an inflammatory factor). The main functions of PDGF are chemotactic: it attracts inflammatory cells to the fracture site and stimulates mesenchymal cell proliferation (Nash *et al.*, 1994). IGF-1 promotes the proliferation and differentiation of osteoprogenitor cells (Thaller *et al.*, 1993), but has not (yet) convincingly shown its effect in large bone healing studies. Recombinant human growth hormone plays a similar role as IGF-1, but seems to show a more potent and dose-dependent effect on rat tibia strength (ultimate load to failure) in both intact and healed bones after fracture (Carpenter *et al.*, 1992).

Table 19.3 Overview of several growth factors (GF) and their potential roles in bone healing

GF	Source	Receptor type	Function	Literature							
				GF	Animal	Model	Dose	Time	Methods	Results	Reference
TGF-β	platelets, ECM bone, cartilage matrix	STS	mesenchymal cell proliferative factor	TGF-β1	rabbit	Tibia	1–10 µg/day control GF	6 weeks	H+M	1 µg increased callus 1 µg increased strength	Lind et al., 1993
				TGF-β1	rat	tibia	4/40 ng every other day	40 days	M	40 ng optimal strength	Nielsen et al., 1994
				TGF-β2	rabbit	tibia	60/600 ng single shot	5/7/10/14 days	H	600 ng increased callus	Critchlow et al., 1995
BMP	osteoprogenitors, osteoblasts, ECM bone	STS	mesenchymal cell and progenitor cell differentiation factor	BMP-2	dog	radius	0/150/600/2400 µg control AG	12/24 weeks	H+M+X	H: unions AG = BMP X: unions AG = BMP M: superior 150 µg	Sciadini and Johnson, 2000
				BMP-2	rabbit	ulna	0/20 µg control empty	2/3/4/6 weeks	M+X	increased stiffness and strength in BMP	Bostrom and Camacho, 1998
				BMP-2	monkey	spine (cage)	0/0.3/0.6 mg control GF	24 weeks	H+M+X	all BMP cages fused non unions in GF	Boden et al., 1998
				BMP-7	monkey	ulna tibia	0/250/500/ 1000/2000 µg control AG	20 weeks	H+M+X	increased torsion strength in BMP versus untreated limb	Cook et al., 1995
				BMP-7	human	fibula	0 or 2.5 mg control empty control DBM	12 months	X+dexa	DBM and BMP similar results 0mg nonunions	Geesink et al., 1999

(Continued)

Table 19.3 Continued

GF	Source	Receptor type	Function	Literature							
				GF	Animal	Model	Dose	Time	Methods	Results	Reference
				BMP-7	human	tibia	3.5 mg or autograft bone	9 months	X	autograft = BMP	Friedlaender et al., 2001
FGF	macrophages, mesenchymal cells, chondrocytes, osteoblasts	TK	mitogenic stimulus	FGF-2	rabbit	tibia	0/50/100/200/400 µg single dose	5 weeks	H + X	≥100 µg more bone mineral content	Kato et al., 1998
				FGF-2	dog	tibia	0/200 µg single dose	2/4/8/16/32 weeks	H + M	FGF accelerates bone repair and strength no dose response	Nakamura et al., 1998
				FGF-2	monkey	fibula	0.4/1/3 mg single dose control graft	0/3/5/10 weeks	H + M + X	increase in strength no effect on stiffness	Radomsky et al., 1999
IGF	bone matrix, osteoblasts, chondrocytes	TK	proliferation/ differentiation osteoprogenitor cells	IGF-1	rat	skull	0 or 2 mg continuous 14 days	1/2/3/4/5/ 6/8 weeks	H	complete fill IGF no healing controls	Thaller et al., 1993
GH	pituitary gland cells		regulation of skeletal growth	GH	rat	tibia	0/0.08/0.4/2 10 mg/kg 2dd control no inj.	40 days	M	2 and 10 mg more optimal mechanical performance	Bak et al., 1990
				GH	rabbit	tibia	0/150 µg/kg 5 times per week	4, 6, 8 weeks	M + X	no added effect GH	Carpenter et al., 1992

Factor	Source cells	Receptor	Function		Animal	Site	Dose	Time	Testing	Result	Reference
PDGF	platelets, osteoblasts	TK	proliferation mesenchymal cells and osteoblasts macrophage chemotaxis	PDGF	rabbit	tibia	0 or 80 µg	28 days	H+M+X	more advanced callus in PDGF no mechanical difference	Nash et al., 1994
EGF	(deep) epidermal cells, chondrocytes, osteoprogenitor cells, osteoblasts, osteocytes	TK	proliferation chondrocytes, promotes formation and function of osteoclasts and osteoblasts	EGF	mice	subcutis	2 × 80 µl (1–20 ng/ml)	4,8 w	H	successful bone formation in 10 ng/ml EGF + 1% FCS	Elabd et al., 2007
VEGF	endothelial cells, hypertrophic chondrocytes, fibroblasts, smooth muscle, cells, osteoblasts	TK	bone angiogenesis, chondrocyte differentiation, osteoblast differentiation, osteoclast recruitment	VEGF-A	rat	femur	single dose adenovirus	1,2,4 w	H	shortened endochondral phase and enhanced angio-/osteogenesis	Tarkka et al., 2003
				VEGF-A	rabbit	femur	single dose adenovirus	1,3 w	H	increased osteoblast number, increased osteoid volume, increased bone volume	Hiltunen et al., 2003

TK = tyrosine kinase, STS = serine threonine sulfate, TGF = transforming growth factor, BMP = bone morphogenetic protein, FGF = fibroblast growth factor, IGF = insulin-like growth factor, GH = growth hormone, PDGF = platelet-derived growth factor, EGF = epidermal growth factor, VEGF = vascular endothelial growth factor, H = histology, M = mechanical testing, X = X-rays, dexa = bone densitometry, AG = autograft, DBM = demineralized bone matrix, FCS = fetal calf serum.

The availability and costs of growth factors remain an encumbrance for their structural application in clinical settings. Although prevailing molecular techniques are capable of producing growth factors as an alternative to their isolation from bone (BMP-2 and TGF-β1) or platelets (TGF-β1), the costs remain extremely high and these recombinant growth factors appear less potent (Osborn, 1980) in comparison to their bone matrix-derived counterparts.

Despite these costs, the results of bone tissue engineering experiments with BMPs indicated their potential as a stand-alone graft in a femoral defect (Takagi and Urist, 1982) or with recombinant BMP-2 in a spine fusion model (Boden *et al.*, 1998) even in a (percutaneous) injectable form (Alden *et al.*, 1999). Recently, the combination of allograft and BMPs was shown to be as good as autograft for recalcitrant tibia fractures in a randomized controlled trial (Jones *et al.*, 2006). Fortunately, despite the high potency of BMP-based bone tissue engineering, such implants have never been shown to grow new bone at undesired sites such as the spinal canal (Boden *et al.*, 1999) after the removal of a vertebral arch (laminectomy).

19.2.3 Cells

Cell-based bone tissue engineering dates back to the discovery that ectopically implanted bone marrow could form bone (de Bruyn, 1955). Interestingly, the purpose of this ectopic bone appeared to be a new shelter for bone marrow so that it could continue hematopoiesis. It was the Russian scientist Friedenstein who discovered which cells were responsible for this phenomenon. These cells adhered to culture plastic and, where present, in fresh bone marrow at a ratio of 1–10 per 100,000 nucleated cells. As they formed distinct colonies and resembled fibroblasts, they were originally named colony-forming unit fibroblasts (CFU-Fs) (Friedenstein, 1968; Friedenstein and Kuralesova, 1971; Friedenstein *et al.*, 1978). Although the initial number is quite limited, they can easily be expanded in culture, which allows sheer infinite amounts of bone to develop. Since their discovery,

many researchers have given many names to (specific subsets of) these CFU-Fs. Some regarded them as (mesenchymal) stem cells because the CFU-F populations obviously showed stem cell characteristics, i.e. self-renewal, clonogenicity and the potential to form many different mesenchymal tissues. However, they are not homogeneous (of one single cell type). This is reflected in the differences in growth potential (colony size) and spontaneous differentiation pathways within and between colonies (Ashton *et al.*, 1980; Aubin, 1998; Bianco *et al.*, 2001; Mendes, 2002; Phinney *et al.*, 1999; Prockop, 1997). Another deviation from the definition of mesenchymal stem cell is that not all types of mesenchymal tissues can be generated from (all) the CFU-Fs. Therefore, these cells are better referred to as what they are essentially: bone marrow-derived stromal cells (BMSCs).

19.2.3.1 Cell origin

With sophisticated sublethal irradiation experiments it was proven that the BMSCs constitute a separate mesenchymal lineage within the predominantly hematopoietic bone marrow (Friedenstein, 1968). In fact, ectopically transplanted donor BMSCs first create the appropriate environment for subsequent homing of host hematopoietic cells. Within the bone, the BMSCs are believed to be continuous with the cells lining the endothelium of the vascular network, so called pericytes (Bianco *et al.*, 2001). In this context they are referred to as stromal fibroblasts or reticular cells [Westen–Bainton (WB) cells (Bianco and Boyde, 1993; Westen and Bainton, 1979)]. The WB cells actively produce a stock of pre-osteogenic cells which make them important contributors in bone development. Later in development, the WB cells maintain the hematopoietic micro-environment (Krebsbach *et al.*, 1999). When the bone marrow is disturbed these cells are easily released into the blood. Other sources of progenitor cells can be found in the periosteum that covers the bone. Cell cultures obtained from this source show similar fibroblast-like morphology, and have proven similar osteogenic potential and colony regeneration capacity in comparison to BMSCs (Nakahara *et al.*, 1990; Takushima *et al.*, 1998). Recent studies also showed the potential of muscle-derived (Peng *et al.*, 2004; Sun *et al.*, 2005) and adipose tissue-derived progenitor cells (Ogawa *et al.*, 2004), as well as dermal multipotent cells (Chunmeng and Tianmin, 2004). Apparently, pluripotent mesenchymal cells are present in several human tissues that contain regenerative capacities.

19.2.3.2 Bone marrow retrieval

Two different approaches in bone marrow retrieval can be applied: *in vivo* bone marrow retrieval (usually from the iliac crest) and postmortem femoral or tibial marrow cavity flushing, as used frequently for research in syngeneic mouse and rat models.

With *in vivo* bone marrow retrieval, the sample will predominantly be populated with blood cells and only a minor fraction of CFU-Fs with the ability to adhere to culture plastic (Krebsbach *et al.*, 1999). As an example, only an average of 36 CFU-Fs that stained positive for ALP (a marker of the osteogenic lineage) were reported per 1 million nucleated cells in aspirates of 32 middle-aged healthy donors (Muschler *et al.*, 1997). Interestingly, 70% variation in ALP-positive CFU-F content was donor related and 20% resulted from the aspiration technique. It was also observed that multiple low-volume (less than 2 ml) bone aspirates yielded higher concentrations of BMSCs than single larger-volume aspirates (Muschler *et al.*, 2004). One can imagine this is because single bone marrow aspirates only locally disturb the marrow stroma. Another method that is frequently used to increase the BMSC content of the aspirate is density gradient centrifugation – his method allows separation from erythrocytes and other relatively dense blood contents and increases the density about 5 times (Aubin, 1998).

With respect to postmortem marrow cavity flushing, the complete stroma of the marrow cavity is obtained. This stroma is primarily the scaffold for hematopoietic cells, and contains many cell types such as marrow adipocytes, bone-lining cells (inactive osteoblasts), osteoblastic cells and of course the WB cells (Bianco and Boyde, 1993; Westen and Bainton, 1979).

Although it has been reported that the structural orderliness of stromal tissue does not need to be preserved for the bone marrow microenvironment to be transferred (Friedenstein *et al.*, 1982), mechanical dissociation, by either centrifugation or resuspension through a needle, decreased the amount of ALP-positive CFU-Fs, as well as their osteogenic potential in mice. The gentle use of the trypsin, an enzyme used for cell segregation, as an alternative for these methods, successfully counteracted this phenomenon (Friedenstein *et al.*, 1992).

19.2.3.3 BMSC selection

Already in the early work of Friedenstein on the culture of BMSCs, it can be seen that the strong adherence of these cells to tissue culture dishes was the main selection criterion for the retrieval of mesenchymal cells. After the majority of the CFU-Fs have attached within 1 day, the contaminating cells (mostly of hematopoietic origin) can be easily removed by washing (Friedenstein *et al.*, 1976, 1987). This results, however, in relatively heterogenic BMSC cultures. With the development of new cell-specific markers, insight into subgroups within the heterogeneous pool of CFU-Fs is obtained. One such marker, that specifies early and very immature osteoblast progenitors, is the monoclonal antibody called STRO-1 (Gronthos and Simmons, 1995; Stewart *et al.*, 1999). Other markers of osteogenecity can be found among the noncollagenous bone matrix proteins, such as osteocalcin, which accounts for 10–20% of all noncollagenous proteins in bone and is closely associated with bone mineralization. Osteonectin is a protein found in bone that has calcium and mineral-to-collagen binding properties, it regulates cell shape and cell migration. Osteopontin, a bone matrix protein with integrin (cell-binding) activity, is related to the regulation of mineral proliferation. In addition, an important role in angiogenesis, callus formation and bone remodeling was recently shown in an osteopontin-deficient mouse model (Duvall *et al.*, 2006). By using immunolabeling it is possible to isolate the BMSCs by either positive or negative selection. In a negative selection a subset of cells can be obtained by

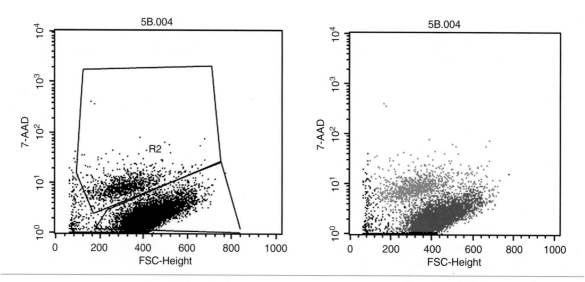

Figure 19.9 Two panels showing the result of FACS analysis on cell vitality. On the *y*-axis, the 7-amino-actinomycin D viability staining solution indicates the number of positively stained cells (indicating dead cells); on the *x*-axis, the forward scatter height (FSC) indicates the size of the cells. In the left panel, the separate 'clouds' of cells are detected with R2 indicating the 'cloud' of dead cells. In the right panel, this cloud is colored green, whereas the vital cells are indicated in red. Black cells are too small and were left out of this analysis. The group of vital cells (red cloud) can subsequently be selected for further testing.

removing target cells (such as CD45$^+$ and glycophorin A$^+$, hematopoietic cells) from the whole bone marrow cell population to obtain the mesenchymal progenitor cells (Reyes *et al.*, 2001). In a positive selection BMSCs are sorted out after labeling with antibodies directed against the membrane molecules specific for primitive mesenchymal cells like STRO-1 (Simmons and Torok-Storb, 1991). However, STRO-1-positive cells are not a uniform population of cells, and other cells of both stromal and hematopoietic descent can also be STRO-1 positive (Gronthos *et al.*, 1994). Therefore, HOP-26 (Joyner *et al.*, 1997), SB-10 (CD 166) (Bruder *et al.*, 1997, 1998a) and CD49a (Deschaseaux and Charbord, 2000) have further defined the subgroups of cell markers specific to primitive human marrow stromal cells. *In vivo*, the majority of cells are HOP-26-negative (osteoblasts) except for some cells located close to the bone surface. *In vitro*, HOP-26$^+$/ALP$^+$ cells are found in BMSC cultures with an increase of this

fraction upon treatment with dexamethasone [a cell differentiation factor for osteoprogenitor cells (Walsh *et al.*, 2001)]. Recently, a novel method of phage display (Griffiths *et al.*, 1994) (using the filaments of bacteriophages to display diverse libraries (up to 10^{10}) of antibody fragments) has uncovered C15 as a new antibody to detect immature and mature osteogenic cells in bone marrow cultures (Letchford *et al.*, 2006).

The use of fluorescence-activated cell sorting (FACS) seems essential in these developments. In this FACS technique, fluorescently labeled antibodies are attached to target cells that are sent one by one through a tube at a speed of thousands of cells per second. Per cell, the presence of the label can be determined which allows very precise measurements and even sorting of the cells (Figure 19.9). Less complicated is the addition of small magnetic beads to the antibodies that allow cell sorting by magnetic attraction (Oyajobi *et al.*, 1999).

19.2.3.4 Culture methods

In the last 40 years, generally accepted methods for selection, culture and replating of especially human BMSCs have been developed (Colter et al., 2000, 2001). Often whole marrow is separated by density gradient [Ficoll (Baksh et al., 2003) or Percoll (Kadiyala et al., 1997) gradients] to select the mononuclear cells with centrifugation. These cells are then usually cultured in medium containing 10–20% fetal calf serum (FCS), L-glutamine and antibiotics. Some prefer the use of a standardized synthetic serum such as 2% Ultroser (de Bruijn, 1998) instead of FCS; others prefer the addition of FGF-2 to stimulate the proliferation of BMSCs (Kruyt et al., 2003a). Cells are usually cultured in plastic culture dishes (often coated with fibronectin for optimal cell adhesion) of varying surface areas with the addition of roughly $1 cm^3$ of culture medium for each $5 cm^2$. Medium is replaced every 3–4 days until cells reach confluency in 7–14 days depending on the initial plating density and species of BMSCs used. When medium is removed and cells have been washed with PBS, Trypsin 0.25% in EDTA is used as a thin film and incubated for 5 min at 37°C to detach the cells. Cells can now be counted (see Section 3.1.1) and replated at the desired density using standardized medium or stored in liquid nitrogen using 5% dimethylsulfoxide and 30% FCS.

In an effort to produce clinically useful volumes of tissue-engineered bone products, a direct perfusion bioreactor system was developed as depicted in Figure 19.10 (Janssen et al., 2006). Perfusion flow rate, flow direction and stance position of the bioreactor are factors which influence the amounts and homogeneity of the cells seeded on the scaffold surface. The successful seeding of goat bone marrow cells into a bioreactor system with volumes of up to $10 cm^3$ of BCP scaffolds (2–6 mm) was shown. Unfortunately the osteogenicity of these constructs was only shown in an ectopic mouse model where $0.03 cm^3$ granules were implanted (Janssen et al., 2006). For more information on bioreactors, see also Chapter 16.

19.2.3.5 BMSC differentiation

Although most of the BMSCs can differentiate into the osteogenic lineage in vitro, in vivo bone formation remains quite unpredictable especially in constructs with human BMSCs (Kuznetsov and Gehron Robey, 1996). This may result from a relatively low commitment towards the osteogenic lineage. A method that may overcome this problem is to stimulate the cells by either dexamethasone (10 nM) (Maniatopoulos et al., 1988; Walsh et al., 2001) or more persuading factors like BMPs. Even more radical is to genetically modify the cells by transducing the cells with growth factor genes (Bianco et al., 2001; Caplan, 2000). This method can be regarded as a development in cell-based tissue engineering, but also as an improvement of growth factor-based tissue engineering in that it finally found the ideal vehicle for regulated factor release in these (mesenchymal) cells.

19.3 Steps in bone tissue engineering research – from idea to patient

19.3.1 Mechanistic studies

Bone tissue engineering is a promising technology that may obviate the need for the autologous bone graft in the future. At this very moment, thousands of scientists are investigating many aspects of it on a worldwide scale, of which biomaterial engineering and cell biology are most prominent. Biomaterial technology can nowadays produce scaffolds shaped precisely according to a predesigned 3D virtual model, which can carry cells, growth factors or both (Warnke et al., 2006; Wilson et al., 2004). However, the optimal material configuration for the scaffolds still has to be defined. Cell biology has made great advancements in the isolation and functional differentiation of adult BMSCs (Jiang et al., 2002). However, the exact role of the cells, the optimal cell concentration and the optimal level of differentiation are also far from clear. Despite these uncertainties, the separate disciplines have come a far way in cell-based bone tissue engineering.

Preclinical tissue engineering research is typically done at different levels of resemblance to the clinical situation. At the first level, in vitro studies are

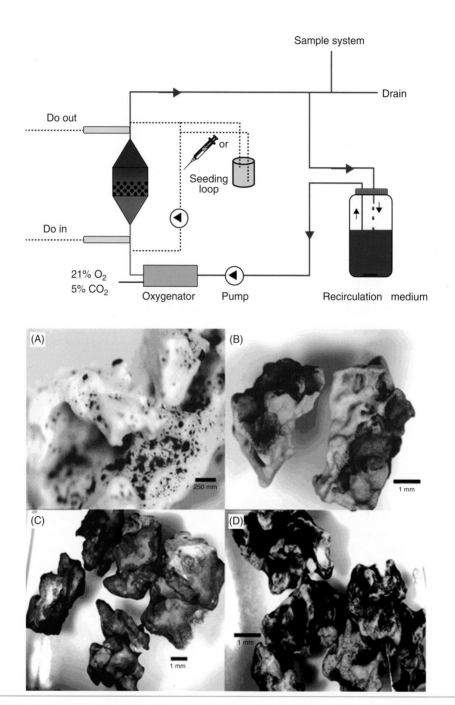

Figure 19.10 (Top) Process scheme of a bioreactor system. Two loops can be distinguished: a seeding loop (dashed line) and a proliferation loop (solid line). Red lines indicate oxygen-rich and blue lines indicate oxygen-depleted nutrient flows. (Bottom) Methylene blue (A–C) and MTT indicating viable cells (D) staining of BMSCs on BCP scaffolds after dynamic culture (A, scale bar = 0.25 mm) and proliferation after 6 (B), 19 (C) and 25 (D) days. Scale bars in (B–D) = 1 mm.

performed and then the proof of concept is evaluated in small animal models, followed by feasibility studies with more clinically relevant sizes and locations. Finally, efficacy studies may be required that mimic the clinical situation as much as possible (e.g. primate studies). It is important to recognize these different levels in order to use the appropriate models. For example, to investigate the differentiation and proliferation of cells in combination with the scaffolds, *in vitro* studies will be informative and preferable. On the other hand, to asses the feasibility of tissue-engineered bone as a graft substitute in spinal fusion, models that mimic the biomechanics and size of the clinical situation are needed. In the following sections, different subsets of studies are discussed that will be required to bring tissue engineering to the patient.

19.3.1.1 In vitro *models*

In vitro models are the cornerstone of all developments in tissue engineering. Every idea should first be tested thoroughly *in vitro* before even thinking of *in vivo* tests. With respect to growth factors, this involves the purification, screening of release systems and effect on BMSCs or characterized cell lines like 3T3 fibroblasts. With respect to BMSCs, extensive *in vitro* research has been performed to identify these cells and to assess the interaction between these cells, scaffolds and (growth) factors. Such studies also give insight in the osteogenic potential of the cells themselves and the benefits of additional treatments, i.e. drug or gene therapy. Some essential *in vitro* techniques to study the cells in culture are summarized.

Cell counting Obviously this is essential for standardizing and optimizing culture and cell seeding procedures. It seems easy, but it is in fact rather complicated. Until now, it is even impossible to accurately quantify the cells in a 3D environment such as a hybrid construct. Moreover, most of the methods are end-stage investigations involving cell lysis. In the 2D culture environment counting by using an inverted phase contrast microscope is most widely used when the density is not too high. Detached cells are counted by

a cytometer chamber, with a predefined volume and grid. When cells are in suspension, automated systems like FACS, as discussed previously for cell selection, and cytometers are also convenient. DNA quantification is applied when individual cells cannot be observed. This is usually the case in tissue-engineered constructs where information on cell attachment, proliferation, survival and differentiation is essential. The cells are lysed to release the DNA that is subsequently colored and quantified by photospectrometry. Many commercial kits are available. To relate the DNA content to the amount of cells, it is necessary to make a standard curve of reference, and to calibrate the assay according to the cell type and scaffold that is used. Alternatively, one can use metabolic assays, where a substrate is converted by viable cells into a product of a specific color or fluorescence. Subsequently the reaction is stopped and the product is measured by spectrometry (Wilson *et al.*, 2002). A pitfall of this technique is that cell metabolism changes with cell density. Most assays require cell lysis, e.g. tetrazolium salt assays, although, nonlethal assays based on REDOX indicators have been developed (Nikolaychik *et al.*, 1996).

Colony-forming unit efficiency assay To determine the quality of bone marrow aspirates, the colony-forming unit efficiency (CFU-E) is determined. This is the fraction of nucleated cells that adheres to plastic and forms a colony (CFU-F). After the bone marrow is harvested, the nucleated cells are plated at different densities, from about 10^3 to 10^5 cells/cm^2, the medium is refreshed after 3–5 days, and at 7–10 days relatively small colonies can be stained and counted. The number of colonies per initially seeded number of nucleated cells appears to vary with species, technique and age and lies between 1 and 10 per 100,000 seeded cells (Bianco *et al.*, 2001; Friedenstein *et al.*, 1970).

Differentiation assays The most specific markers for cell differentiation towards the osteogenic lineage obviously are mRNA probes or specific antigens as discussed in the previous section on cells. For a fair indication of osteogenicity, easier *in vitro* assays that also allow quantification exist. Most well known is coloring the characteristic product of osteogenesis i.e. calcium

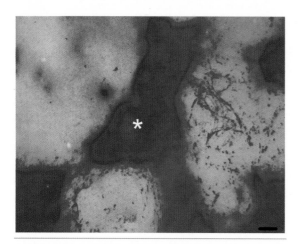

Figure 19.11 Low-magnification image of a culture disk containing rabbit cells within mineralized matrix (*) after several days of culture. Staining performed with Alizarin red to indicate mineralization of the ECM. Bar = 2 mm.

deposition. This can be done with Alizarin red (Figure 19.11) or Calcein green. More specific is to quantify the activated form of the characteristic enzyme ALP. Many commercial ALP kits exist. By division with the simultaneously obtained amount of DNA, the ALP/cell number gives a reasonable indication for cell differentiation. Another assay focuses on hydroxyproline, an important component of the triple helix of collagen and therefore a collagen quantifier (Acil *et al.*, 2000). However, hydroxyproline is not collagen type I specific (the predominant form of collagen in bone) and can therefore not be used as a single indicator for the presence of bone matrix-producing cells.

19.3.1.2 In vivo *models: proof of concept*
This type of study is designed to prove the hypothetical working mechanism (concept) of a technique. In case of cell-based bone tissue engineering, most researchers believe the cells are responsible for new bone formation in the sense that these cells create the new bone. This concept is straightforward and seems easy to prove. However, despite the fact that many studies indeed show an increased amount of

bone when BMSCs are added, only few researchers really managed to prove the direct relation between the cells and new bone (Bruder *et al.*, 1998b; Goshima *et al.*, 1991). The reason why this is rather difficult is because the observed bone formation is not by definition derived from the implanted cells. As we know, it can be derived by at least two other mechanisms: osteoconduction or osteoinduction. Therefore, the proof of concept of cell-based bone tissue engineering needs to exclude these mechanisms as much as possible. In practice, implantations are made preferably in a non-bony environment to exclude osteoconduction. This is referred to as an ectopic location, contrary to the orthotopic location (inside the bone) that should be used for functional studies. A control condition without cells or, even better, with devitalized cells should be incorporated to serve as a control for osteoinduction (Kruyt *et al.*, 2003a). Even more effective to investigate the contribution of the implanted cells is to place the constructs in an environment shielded from the influx of surrounding cells, so-called diffusion chambers. A disadvantage of these, however, is that there can be no vascularization inside the chambers. Well-established ectopic models are subcutaneous and intramuscular implantations in small animals, such as mice and rats. Immunodeficient strains of these animals are available, which is advantageous to overcome the problems linked to implantation of genetically different cells. Another advantage of small-animal models is the almost complete absence of osteoinduction by the biomaterial itself which largely excludes this confounder (Yuan *et al.*, 2006).

With simple *in vivo* models, however, the proof of osteogenesis promoted by BMSCs remains indirect. Even when both requirements of an ectopic location and absence of bone induction in the controls have been fulfilled, the only way to directly address the conceptual mechanism of osteogenesis is to identify the original BMSCs inside the new bone. An elegant method consists in identifying genetically different cells in immunodeficient animals (Goshima *et al.*, 1991), e.g. human cells in mice or male cells in a female (gender mismatch). However, when immunocompetent models are used, labeling of the cells is the method of choice. Such a label should neither affect nor select the cells in any way. It should not fade with cell divisions and it must be compatible with bone histology. Obviously, the label should be specific and not transfer to neighboring cells. Currently there are only a few potential off-the-shelf labels available, which are still far from ideal. Most well-known labels are fluorescent dyes that can be incorporated inside the membrane bilayer or inside the cytosol (Horan and Slezak, 1989). At each division, these labels are divided equally over the daughter cells, which allows tracing of the cell progeny, but decreases the label intensity over time. Unfortunately, these labels are not reliable in the long term because of label transfer (Emans et al., 2006; Kruyt et al., 2003b). In recent years, genetic marking has become available and was shown to be compatible with bone histology. This technique applies the ability of (retro)viruses to introduce marker genes into the cells without apparently disturbing these cells (Verma and Somia, 1997). The marker genes can encode for short amino acid sequences that result in proteins which can be revealed by immunohistochemistry. Alternatively, the specific proteins are released in the circulation and can be measured in the blood (e.g. interleukin-3). Recent techniques use proteins that constitute enzymes such as firefly luciferase, which produces long wavelength light that penetrates through the bone and can be measured outside the animal (Contag et al., 1998; Feeley et al., 2006) by noninvasive bioluminescence imaging (see State of the Art Experiment).

19.3.1.3. In vivo *models: feasibility studies* (*Figure 19.12*)

In a next step, a proof of feasibility is needed to determine whether the technique results in functional bone. Many models have been developed, ranging from radius defects in mice to equine knee models (Table 19.4 and Figure 19.12). In general, the feasibility models can be divided into several groups: bone augmentation models (e.g. posterior spinal fusion and mandibula onlays), nonweight-bearing defect models, segmental defect models and functional joint models.

The segmental defect appears to be most popular to study bone tissue engineering in clinically sized models. In 1998, Bruder et al. were the first to show an effect of BMSCs in a dog segmental femur defect; this study was later repeated with allogenic cells that yielded a similar effect (Arinzeh et al., 2003; Brodke et al., 2006; Bruder et al., 1998a; Kraus et al., 1999). Currently, sheep and goat segmental defect studies represent the majority of successful cell-based bone tissue engineering publications. These involve defects in the cranium (Shang et al., 2001), mandibula (Schliephake et al., 2001; Yuan et al., 2007), femur (Zhu et al., 2006), tibia (Kon et al., 2000; Mastrogiacomo et al., 2006) and metatarsals (Bensaid et al., 2005; Petite et al., 2000; Viateau et al., 2004).

An important advantage for any model is the potential to evaluate multiple (different) samples in the same animal. This not only fractionates the number of animals needed by the number of implants per animal, but also reduces the sample size (n) because the inter-animal variance is excluded with paired comparisons. Recently, a model that allows simultaneous assessment of many conditions was introduced for bone augmentation on the transverse processes in the spine as mentioned in the previous paragraph (Kruyt et al., 2006; Wilson et al., 2006). Naturally, in the case of paired comparisons, the samples should not influence each other in any way. Therefore, functional spinal fusion models should not have different conditions on each side of the spine, as fusion on one side gives stability and therefore influences fusion on the other side. For similar reasons, some authors advocate the use of only one-level spinal fusion models. With respect to segmental or non-load-bearing defect models, it is preferable to use critical size defects, meaning that the defect will never heal spontaneously. This is a fundamental quality that permits the conclusion that, in case of success, the therapy healed a defect instead of only enhanced its spontaneous healing. A rule of thumb (which should never be applied as a law and needs verification in case of new defect models) is that segmental defects in the mature skeleton are critically sized when they are 2.5 times the diaphyseal shaft width. Advantages of

State of the Art Experiment
Stem cell labeling with luciferase and *in vivo* imaging

In cell-based tissue engineering, understanding the response of the implanted cells is important to determine the viability of the implants and the cellular reaction to new drugs or biomaterials. Marker genes for *in vivo* real-time imaging provide a way to trace implanted cells in living animals. The firefly luciferase and *Aequora* Enhanced Green Fluorescent Protein (EGFP) are the paradigms for these types of markers. Firefly luciferase is an enzyme that oxidates its substrate (D-luciferin) in the presence of ATP and oxygen. During the oxidation, low-energy photons are emitted (bioluminescence). Owing to their long wavelength, these photons can be detected through body tissues by using a CCD camera. This can be quantified. EGFP, instead, belongs to a family of proteins which have fluorescent properties. Upon excitation with a certain wavelength, the emitted light of a longer wavelength (fluorescence) is traceable both *in vitro* and *in vivo*, although the *in vivo*

applications are limited due to its low penetration energy. Double labeling of BMSCs with luciferase and EGFP, e.g. in a fusion protein, provides a nondisruptive way to verify the labeling of the cells prior to implantation (EGFP by FACS) and to allow follow-up after surgery (luciferase by bioluminescence; see below). The gene for the marker protein is inserted in the BMSCs by genetic labeling, using defective viral particles (viral vectors), which retain the infective but not the replicative property. When added to a culture of BMSCs (cell transduction), the viral particles enter the cell and stably integrate the marker genes into the cell genome (see next page).

Transduced BMSCs can be seeded on CaP scaffolds and implanted subcutaneously into small animals (mice or rats). Since the amount of bioluminescence is directly dependent on the number of viable BMSCs expressing the luciferase, periodic imaging of the cells provides quantitative

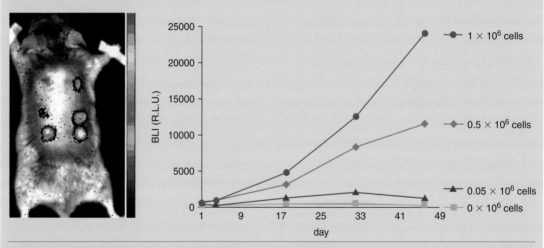

In vivo bioluminescence imaging. (Left) Localization and intensity of light emitted from subcutaneous implants: the bioluminescence is quantified and converted into a color scale that ranges from dark blue (for low light intensities) to white (for high light intensity). (Right) kinetic curve of bioluminescence. The variation of the bioluminescence intensity over time and the trend of the curve are indicative of survival and growth of the cells in vivo.

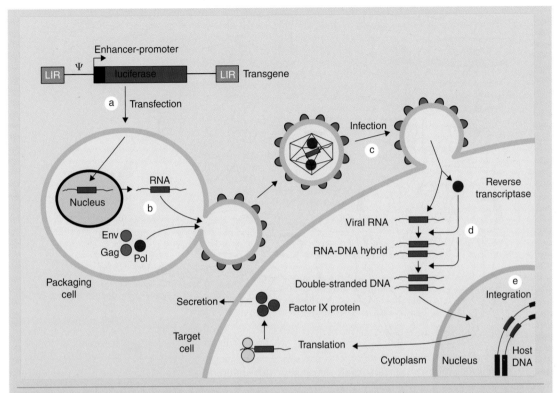

Production of retroviral vectors for cell transduction. (a) The transgene (in this case the gene for luciferase) in a vector backbone is put into a packaging cell which expresses the genes that are required for viral integration (gag, pol and env). (b) The transgene is incorporated into the nucleus where it is transcribed to RNA. This is then packaged into the retroviral particle (vector), which cannot replicate itself, which is shed from the packaging cell. (c) The vector is delivered to the target cell by infection. (d) The virally encoded enzyme reverse transcriptase converts the vector RNA into an RNA–DNA hybrid and then into double-stranded DNA. (e) The vector DNA is integrated into the host genome. The host-cell machinery will transcribe and translate it to make RNA and, in this case, the luciferase protein. LTR = long terminal repeat; Ψ = packaging sequence. Reprinted with permission from Macmillan Publishers Ltd: Verma, I.M. and Somia, N. Nature, 389(6648): 239–242, copyright 1997.

information about cell survival, proliferation and location. The samples can be retrieved for bone quantification and visualization of the implanted BMSCs by immunostainining EGFP. Compared analysis of bioluminescence and histology have demonstrated that in ectopic implants with transduced goat BMSCs, bone deposition occurs when bioluminescence increases during the weeks postsurgery, likely as a consequence of the BMSC proliferation. On the contrary, bone formation did not occur when bioluminescence decreased, likely because of the death of the implanted BMSCs.

Application of bioluminescence *in vivo* imaging to regenerative medicine is now expanding and might become a powerful tool to gain more bright insights into the development of cell-engineered constructs.

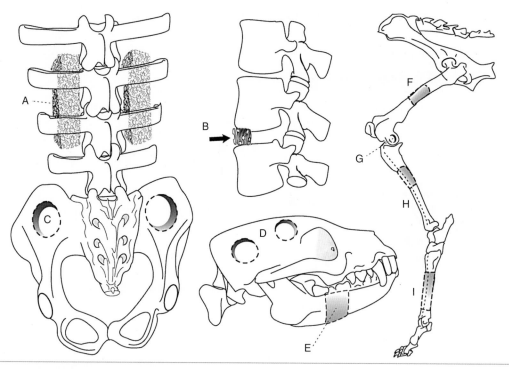

Figure 19.12 Animal models to test graft feasibility (for references, see Table 19.4): (A) posterolateral intertransverse spine fusion, (B) anterior interbody fusion (using a cage), (C) os ilium defect, (D) skull defect, (E) mandibular segmental defect, (F) femoral segmental defect, (G) intracondylar metaphyseal defect, (H) tibial segmental defect and (I) metatarsal defect.

segmental defects are the good possibilities for radiological evaluation in time and to finally evaluate the bone functionality by mechanical tests. Despite hardware support, diaphyseal defects are usually unilateral. However, ulnar and radial defects may be tolerated bilaterally. Non-weight-bearing defects obviously are less complicated and usually can be paired. Well-known examples are skull defects, which allow up to quadruple evaluations (four defects in one skull). More functional are the cage-assisted anterior spinal fusion models. Within the cage the graft is shielded, but the intervertebral body fusion allows for mechanical testing. Relatively new are total joint models, which combine bone and cartilage tissue engineering to create functional joints. The potential of this was shown in a clinical case study (Ohgushi *et al.*,

2005). Efficacy studies are preferably done in non-human primates. So far none have been reported for cell-based bone tissue engineering. BMP-based bone tissue engineering, however, has already been investigated in non-human primates (Boden *et al.*, 1998).

19.3.2 Upscaling to clinical application

The final proof of the technology will be by randomized clinical trials (RCTs). Currently, pilot studies and RCTs have been published regarding BMP-based tissue engineering (Table 19.5). With regard to cell-based bone tissue engineering, no RCTs have been reported yet, although some investigators believe that the technology can be applied in humans (Cancedda

Table 19.4 Overview of feasibility models used in bone tissue engineering including their locations, types of grafts used and animal models

Type	Location	Graft type	Intra-animal control	Animals	More reading
Posterolateral spinal fusion	thoracal	augmentation of cancellous bone	uni- and multi-segmental	mice, rat, sheep, dog	Reviewed in Kruyt et al., 2004b
	lumbar			rat, rabbit, sheep, goat,	Reviewed in Kruyt et al., 2004b
Segmental critical-sized defect	ulna/radius	corticocancellous cylinder	yes	mouse, rat	Gazit et al., 1999
	metatarsus diaphysis		no	rabbit, dog, monkey sheep	Reviewed in An et al., 2000 Petite et al., 2000; Bensaid et al., 2005
	femur diaphysis		no	dog, goat	Reviewed in An et al., 2000
	tibia diaphysis		no	rabbit, dog, sheep	Reviewed in An et al., 2000
	mandible		no	mini-pig, goat	Reviewed in An et al., 2000
Nonweight-bearing critical-sized defect	cervical/thoracal	cancellous in spinal cages	uni- and multi-segmental	dog, sheep, goat	Reviewed in Kruyt et al., 2004b
	lumbar			dog, sheep	Reviewed in Kruyt et al., 2004b
	frontal or parietal bone	cancellous disks	yes	rat, rabbit, dog, monkey	Reviewed in Schmitz and Hollinger, 1986
			yes	sheep	Shang et al., 2001
	iliac wing	cancellous disks	yes	goat, dog	Anderson et al., 1999
	femur condyle	cancellous rods	yes/no	dog, goat, rabbit	Overgaard et al., 2000

Table 19.5 Chronological summary of clinical studies on BMPs

Model	Treatment	No. of patients	Reference	Outcome
Femur nonunion	BMP ± auto- or allograft	12	Johnson et al., 1988a,b	11/12 healed
Tibia nonunion	BMP + autograft	6	Johnson et al., 1988a,b	6/6 healed
Spine nonunion	BMP + autograft	5	Sandhu et al., 1997	5/5 solid fusion (lumbar spine)
Food and Drug Administration approval for the use of rhBMP-2 in degenerative disk disease (INFUSE)				
Lumbar interbody cage model	INFUSE laparoscopic (2–4mg BMP)	4	Boden et al., 2000	11/11 INFUSE solid fusion
	INFUSE retroperitoneal (2–4mg BMP)	7		2/3 autograft solid fusion
	autograft retroperitoneal	3		
Lumbar interbody cage model	INFUSE transperitoneal	27	Gornet et al., 2001	operative time and blood loss less in BMP group
	INFUSE retroperitoneal	116		2 patients antibodies against rhBMP-2
	autograft transperitoneal	26		34 patients antibodies against bovine collagen
	autograft retroperitoneal	110		no differences in outcome technique or graft
				>95% fusion at 6 and 12 months
Lumbar interbody cage model	INFUSE laparoscopic	136	Zdeblick et al., 2001	>90% fusion at 6 and 12 months
				historical control group
				1 patient antibodies against rhBMP-2
				32 patients antibodies against bovine collagen
Lumbar interbody dowel model	allograft + INFUSE transperitoneal	24	Burkus et al., 2002	blood loss less in allograft/BMP group (12–18mg BMP)
	autograft transperitoneal	23		90% fusion in BMP group versus 65% in AG (6 months)
				100% fusion in BMP versus 90% in AG (12 months)
				39% persistent donor site pain at 24 months in AG

Model	Treatment	n	Author	Results
Posterior interbody cage model	INFUSE posterior laminectomy	34	Alexander et al., 2002	higher fusion rate in INFUSE (92 versus 78%)
	autograft posterior laminectomy	33		
Anterior cervical ring allograft	INFUSE filled ring allograft	18	Baskin et al., 2002	single- or two-level anterior cervical fusion model (1:1)
	autograft filled ring allograft	15		less blood loss in INFUSE, all patients fused at 6 months
				negligible pain at donor sites in autograft after 6 months
				less improvement in pain scores in two-level autograft
Posterolateral lumbar fusion	(1) rhBMP-2 (15 mg) + 7 cm³ BCP	7	Luque, 2002	group (1) 6/7 BMP sides fused versus 4/7 autografts at 12 months
	unilateral + autograft contralateral, (2) rhBMP-2 (20 mg) + 10 cm³ BCP/side	8		group (2) 8/8 fused at 12 months
Posterolateral lumbar fusion	autograft + instrumentation	5	Boden et al., 2002	2/5 autograft fused versus 20/20 BMP ± instrumentation
	rhBMP-2/BCP (20 mg/10 cm³/side) + instr.	11		
	rhBMP-2/BCP (20 mg/10 cm³/side) − instr.	9		
Degenerative spondylolisthesis model with instrumentation				
Posterolateral lumbar fusion	OP-1 (BMP-7) putty (3.5 mg/side)	24	Vaccaro et al., 2004	74% fusion in OP-1 versus 60% in autograft group at 1 year
	autograft	12		
Posterolateral lumbar fusion	OP-1 (BMP-7) putty (3.5 mg/side)	24	Vaccaro et al., 2005	55% fusion in OP-1 versus 40% in autograft group at 2 year
	autograft	12		

The majority of the publications were based on spine studies with either the laparoscopic retroperitoneal or transperitoneal approach. Fusions are based on radiological data. INFUSE = rhBMP-2 (1.5 mg/ml) + bovine collagen sponge; inst. = instrumentation in the spine (such as pedicle screws and rods for mechanical stability).

et al., 2003; Kitoh *et al.*, 2004; Marcacci *et al.*, 2007; Nocini *et al.*, 2006; Ohgushi *et al.*, 2005; Quarto *et al.*, 2001; Schimming and Schmelzeisen, 2004; Vacanti *et al.*, 2001). It has become clear that upscaling tissue engineering to the clinical situation presents several challenges that can be related to species, the size of the construct and the location of the implants.

Species-related difficulties include the inferior quality of the human bone marrow aspirates and the BMSCs *per se* when compared to those of other species, e.g. murine BMSCs (Krebsbach *et al.*, 1997; Majors *et al.*, 1997). Furthermore, a limitation of using human BMSCs is the relatively high age at which the treatment will be performed (Bab *et al.*, 1984; Bianco *et al.*, 2001; Mendes, 2002). The potential of BMSCs might be diminished as a result of the replicative senescence (the total number of potential cell divisions is restricted as a result of breaking down of the telomeres that cap the chromosomes) (Bianco *et al.*, 2001). To overcome this problem, transducing BMSCs with telomerase (which partially restores the telomeres) inhibits the replicative senescence of the BMSCs and allows up to 10 times more population doublings (Shi *et al.*, 2002; Simonsen *et al.*, 2002).

Increasing the size from the cubic millimeter range (in mice) to the clinical, cubic centimeter range (1000-fold) has dramatic repercussions for both the implant and the wound bed. Large tissue engineering implants, until now, are nonvascularized constructs and therefore a poor setting for cell survival. It is generally accepted that organisms that exceed the volume of several cubic millimeters need vascularization for oxygen and nutritional supply – an axioma that exists in many life sciences. Typical examples are the limited size of nonvascularized animals such as insects and the early development of active perfusion in the human embryo already during the fifth week of gestation, when it is about 3 mm long. In addition, when applying larger tissue-engineered constructs, a more extensive injury with a more profound wound healing reaction will ensue. Inside the wound hematoma, cell survival will be compromised, due to the high potassium concentration as a result of erythrocyte lysis (Street *et al.*, 2000).

19.3.2.1 *Graft viability and vasculogenesis*

The intimate connection between osteogenesis and blood vessels was first proposed in the second half of the 18th century. In 1963, Trueta predicted the existence of a vascular-stimulating factor, acting at the sites of bone damage. Recently, knockout animals (animals lacking genes, e.g. those involved in the angiogenic process such as VEGF, see also Table 19.3) and the treatment of fractures with angiogenic inhibitors have given new insights in the interaction of bone formation and angiogenesis. Accordingly, the inhibition of VEGF was shown to prevent BMP-2-mediated bone formation (Peng *et al.*, 2005). In line with this, a variety of other growth factors present within the fracture hematoma, like PDGF and FGF, have high angiogenic potential. A few of them have also been shown to significantly stimulate and modulate bone remodeling and fracture healing (Gerber *et al.*, 1999; Hausman *et al.*, 2001) (see also Table 19.3). It is now clear that the vascular network contributes to the development of bone in several ways. First, systemically by acting as a transport system for oxygen, hormones, nutrients and waste products, but also for the host stem cells and the immune system. Second, by releasing factors locally, produced by the endothelial cells, e.g. TGF-β1 and VEGF (see also Table 19.3) and BMPs, which stimulate osteoblast proliferation and differentiation, respectively. Third, by stimulating bone marrow stromal cell proliferation by direct contact with endothelial cells (Villars *et al.*, 2000). Closely related growth factors, like epidermal growth factor (EGF) and PDGF, can have antagonistic effects and steer the BMSCs fate in opposite directions. By binding to their respective receptors, which are both tyrosine kinase receptors, EGF elicits the osteoblastic differentiation of human BMSCs, while PDGF will inhibit their osteoblastic differentiation (Kratchmarova *et al.*, 2005).

In cell-based tissue-engineered bone constructs, indications for vascularization-dependent bone formation exist. For example, in the ectopic mice model (Kruyt *et al.*, 2004a), bone formation starts at the periphery of the implants and appears in the core of the scaffolds at later time points. In parallel, vascularization of implants proceeds slowly and the first vessels reach the centre of the scaffolds exceeding the millimeter scale only after several weeks of implantation (Pacheco *et al.*, 1997; Pelissier *et al.*, 2003). Apparently, neo-osteogenesis requires neovascularization; therefore accelerating vascularization may enhance bone formation considerably.

In vitro models to study bone vascularization In regenerative medicine, endothelial cell-derived tube formation in culture is a major tool to understand the mechanisms behind neoangiogenesis in vascular-engineered constructs. Such models were used to study the reciprocal effects of mature human endothelial cells (HUVECs), osteoblasts or BMSCs (Meury *et al.*, 2006). In cocultures, cells will influence each other by releasing soluble factors with a paracrine effect, by producing ECM for cellular adhesion and by establishing physical contacts. These interactions can be analyzed separately using simple 2D culture systems, or simultaneously, using 3D systems (Wenger *et al.*, 2004). To study the effects of soluble factors, BMSCs in culture are exposed to medium that is derived from endothelial cell cultures. This conditioned medium will contain an undetermined cocktail of growth factors produced by the endothelium (Villars *et al.*, 2000). A better method to study paracrine effects, however, is the coculture in transwell filtered chambers. Different cell types can be seeded separately, one on the bottom of the well and, the other on the filter above it. The filter permits the diffusion of nutrients and cell-secreted factors, while retaining the cells in their chambers. The advantage of the transwell system is that the different cell types can interact dynamically, which is not possible in a system with conditioned medium.

In vivo models for vascularized bone tissue-engineered constructs The majority of *in vivo* models on improved vascularization of bone tissue-engineered implants investigates the effect of angiogenic growth factors that can be combined with the scaffolds, e.g. as a coating. After implantation the factors are locally released, providing a stimulus for endothelial cells, precursor cells and infiltrating osteoblasts. For example, a VEGF-coating on PLG (polylactide-*co*-glycolide) scaffolds enhanced bone formation when compared to uncoated scaffolds (Sun *et al.*, 2005). Angiogenic factors can also be delivered at the desired site *in vivo* by injection: locally administered into a necrotic bone implant, FGF-2 increased the formation and length of new blood vessels by about 2-fold (Nakamae *et al.*, 2004). In delivering angiogenic growth factors at the implant site, the concentration and combinations of such factors are of critical importance, since this might interfere with the osteogenic differentiation of BMSCs (Kratchmarova *et al.*, 2005). In addition, many angiogenic factors with a similar mode of action might be very cell-specific in reality. For instance, VEGF acts selectively on the proliferation and differentiation of endothelial cells, while FGF-2 stimulates migration and proliferation of smooth muscle cells and fibroblasts. Finally, the timing of delivery is important, since different growth factors will be effective and stimulate osteogenesis only at defined moments during this process. When applied *in vivo*, the angiogenic potential of the treatment might be limited by the degradation of the growth factors and by the fraction of active factor which is released. Moreover, it might still take a long while before the scaffolds will be fully vascularized, during which BMSCs will likely not survive (Rouwkema *et al.*, 2006).

Another recent development in the field of tissue engineering is *in vitro* prevascularization. In this case, a prevascular network is grown inside the tissue-engineered construct *in vitro*. After implantation, the construct can be vascularized by a connection of the host vasculature with these pre-existing vessels (Levenberg *et al.*, 2005).

19.3.2.2 Host-related issues

The far from ideal host conditions that are encountered in the human patient are a major concern for upscaling the technique. Two important aspects are the local wound repair response and the fracture healing cascade. Both phenomena in general constitute an inflammatory response with removal of debris followed by an influx of blood vessels and cells

Table 19.6 Flow chart representing the sequential steps in wound healing (left panel) in comparison to fracture healing (right panel)

Wound healing (Anderson 2000; Dvorak *et al.*, 1987)	Fracture repair (Bolander 1992; Jingushi and Bolander 1991)
Excudative phase (immediately)	*Excudative phase* (immediately)
Hematoma forms with release of signaling factors	Hematoma in bone and soft tissue
Provisional fibrin matrix formation	Release of factors, e.g. IL-1, IL-6, TGF-β, PDGF
	Provisional fibrin matrix formation
Acute inflammatory phase	*Immediate inflammatory phase*
More excudation	Proliferation of adjacent undifferentiated cells
Influx of inflammatory cells like macrophages	Influx of inflammatory cells like macrophages
Chronic inflammation phase	*Organizing phase*
Ingrowth of blood vessels	Ingrowth of blood vessels
Fibroblast proliferation and repair with provisional fibrous tissue formation	Clot is transformed into a fibrous reparative granuloma, called external callus
Reorganizing phase	*Initial hard tissue phase*
Reorganization of the scar tissue	Formation of a stabilizing cartilage (soft callus) by differentiated, migrated mesenchymal cells
A scar will always remain	Membranous (direct) bone formation is initiated from the periost (hard callus)
	Remodeling phase
	Transformation of cartilage to bone and complete remodeling of preliminary bone to normal bone

There is an obvious resemblance with the exception of the final product: a scar is formed in wound healing and bone is formed in bone healing.

for repair (Table 19.6). The fracture (bone) healing response will depend on the mechanical situation. If sufficiently stable and with a minimal gap, bone will bridge the defect by so-called creeping substitution of the damaged fracture sides (intramembranous or direct bone formation). This is the equivalent of primary wound closure in surgical wounds. If the local situation is not absolutely stable (which is usually the case with implants), the inflammatory phase is followed by a cartilaginous intermediate before transformation into bone (endochondral bone formation). Cell-based tissue engineering, by directly delivering osteoprogenitor cells, therefore actually deviates from Mother Nature's strategy of wound repair and may not support, or even interfere with, this practice. On the other hand, in a later stage the cells are obviously needed and supplementation may complement the process, especially when the physiologic response is inadequate. Irrespective of their function, cell survival will be difficult and may be impossible due to the following generally accepted compromising circumstances:

- The larger implant necessitates a larger wound bed with an increased diffusion distance.
- The cells will be subjected to an initial inflammatory response that is designed to clean the area of nonfunctional substances.
- A toxic potassium concentration will be present in the hematoma as a result of lysed erythrocytes (Street *et al.*, 2000).

Thus, when aiming at cell-derived bone formation in cell-based bone tissue engineering, we will need a radical change in design in order to improve cell survival. This may be achieved by stimulating rapid vascularization of the constructs by providing endothelial cells and/or factors (Levenberg *et al.*, 2005). Alternatively, the concept of the implanted cells as the main source of restoration tissue should be revised. With respect to bone tissue engineering, it would mean that the cells will predominantly have a stimulating and inducing role, a role that is already well accepted for the cells in the autologous bone graft (Burchardt, 1983; Yaszemski *et al.*, 1996).

19.4 Current status of bone tissue engineering

19.4.1 Clinical experience in orthopedics

During the last decade, several groups have made the step towards clinical application of cell-based bone tissue engineering, published in recognized journals as case reports (Quarto *et al.*, 2001; Vacanti *et al.*, 2001). Despite sometimes promising results, it should be realized that, with current knowledge, none of these studies was designed to investigate the actual contribution of the added BMSCs.

Quarto *et al.* (2001) were the first to report the treatment of large (4–7 cm) bone defects of the tibia,

ulna and humerus in three patients from 16 to 41 years old, where the more 'conventional' surgical therapies had failed. They implanted a custom-made unresorbable porous HA scaffold seeded with *in vitro* expanded autologous BMSCs. An external fixation device was applied for stabilization for 6–13 months after surgery. No major complications occurred in any of these patients. Using plain radiographs and computed tomography (CT), integration at the implant–host bone interface was observed by the presence of callus formation and bone bridging in the second month after surgery. Limb function was recovered between 6 and 12 months. No complete fractures in the implant region were observed during the follow-up of about 6.5 years (Figure 19.13), although with time the implants revealed a progressive appearance of cracks and fissures indicative of ceramic disintegration (Marcacci *et al.*, 2007).

Vacanti *et al.* (2001) reported the case of a man who had a traumatic avulsion of the distal phalanx of the thumb. The phalanx was replaced with a porous coral implant that was seeded with *in vitro* expanded autologous periosteal cells. Plain radiographs and magnetic resonance imaging were performed in addition to histological biopsies. The radiographs were claimed to show remodeling of the implant at 28 months postsurgery. However, histomorphometry at 10 months after implantation showed only 5% of lamellar bone and ossified endochondral tissue filling the implant.

Morishita *et al.* (2006) treated a defect resulting from surgery of benign bone tumors in three patients using HA scaffolds seeded with *in vitro* expanded autologous BMSCs after osteogenic differentiation of the cells. Two bone defects in a tibia and one defect in a femur were treated. Although ectopic implants in nude mice were mentioned to show the osteogenicity of the cells, details such as the percentage of the implants containing bone and at what quantities were not reported. The same research group also performed ankle arthroplasty in three patients with an ankle prosthesis composed of alumina coated with HA on which BMSCs were seeded and cultured prior to implantation (Kitamura *et al.*, 2004; Ohgushi *et al.*, 2005).

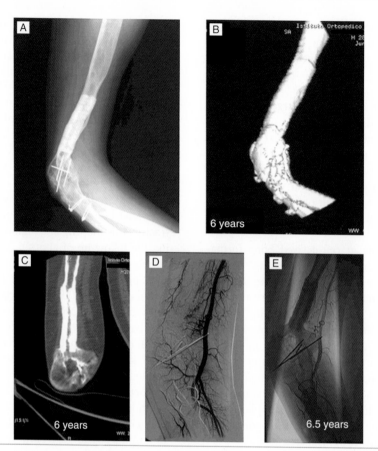

Figure 19.13 Long-term follow-up data (6–6.5 years) published by Marcacci *et al.* (2007, Stem cells associated with macroporous bioceramics for long bone repair: 6 to 7 year outcome of a pilot clinical study. *Tissue Eng,* 13(5): 947–955) of the same 22-year-old man who lost a 7-cm segment of the right humerus as a result of a multi-fragmented fracture as shown in Figure 1(G–I) in the report by Quarto *et al.* in 2001. (A) Plain radiograph. (B) The 3D reconstructed CT image based on the 2D data as shown in (C). Additional angiography was performed at 6.5 years of follow-up and shown in (D) and (E).

Another technique where BMSCs may be advantageous is distraction osteogenesis. This technique comprises of very slow distraction of bone (1 mm/day) to allow the formation of a repair callus in the defect. Distraction osteogenesis of the femur and tibia with contemporary transplantation of autologous BMSCs and platelet-rich plasma (PRP) was performed by Kitoh *et al.* (2004) in three patients. Two patients with achondroplasia (Dwarphism) and one

patient with congenital pseudoarthrosis of the tibia were treated. Autologous culture expanded BMSCs (on average $14–50 \times 10^6$ cells) were injected into the distraction callus together with autologous PRP (on average 1.7×10^{10} platelets per PRP preparation). Results were compared between two of the three experimental cases (achondroplasts) and 30 achondroplasts undergoing similar lengthening procedures with the same monolateral external fixation device in the control group not receiving injections. Average healing index seems in favor of the two experimental cases (22.0 days/cm of distraction as compared to 37.8 days/cm in the controls). However, the group size of only two patients in the experimental group, as admitted by the authors, is too small for any statistically sound conclusions.

Obviously the proof of the technique is still missing in the clinical setting. Since osteoconduction is an important contributor in the incorporation of the implants, the effect of BMSCs can only be analyzed when controls without cells are available. Additionally, in none of these reports could solid proof for cell survival in the implanted grafts be given. This is because the only reliable cell tracing methods are based on retroviral transfection of these cells, which is incompatible with human studies. The best method to assess cell functionality in the clinical setting probably is the analysis of biopsies. Unfortunately, this is seldom an opportunity with the exception of dental applications as described in Section 19.4.2. Since mechanical testing of the implants in the living individual is not an option, high-resolution CT scans might be the most powerful tool to analyze the success in terms of bone formation in clinical studies. Most valuable of course will be the use of randomized clinical trials where the outcome of the tissue engineering therapy should be compared to adequate control conditions.

19.4.2 Clinical experience in oral/maxillofacial surgery

Only a few clinical studies have been reported on maxillofacial surgery using a combination of a scaffold and seeded cells. The outcomes of these studies are disappointing and indicate that cell-based bone tissue engineering still has a long way to go. Schimming and Schmelzeisen (2004) published a clinical study in which augmentation of the posterior maxilla was carried out using a matrix derived from mandibular periosteal cells on a polymer fleece. In total, 15 patients were treated according to a two-step procedure. The first stage was to reconstruct the host area with tissue-engineered bone; 3 months later, the second stage was performed with dental implant placement that allowed retrieval of a histological sample. In eight of these 15 patients an unsuccessful outcome was observed; connective tissue was found in the histological biopsies. In the case of a positive outcome (seven patients), the authors did not mention whether the observed bone formation was derived from osteoinduction or osteoconduction. Meijer et al. (2007) conducted a human study to investigate a potential benefit of BMSCs seeded on a scaffold in restoring an intra-oral osseous defect. Seven out of 10 cultured constructs that were additionally evaluated in the subcutaneous mouse model did show bone, which indicated the osteogenicity of these constructs. However, in only one patient was bone formation observed that could have been derived from the implanted cells.

19.4.3 Limitations of clinical applicability

The cell-based bone tissue engineering approach, although successful in the reported cases, obviously requires further investigation before it can be widely applied to clinical practice. During the last decade, many aspects of the technique have been improved with regard to scaffold development and cell culture techniques. Currently, exciting possibilities are under investigation, like the potential of using allogenic cells (Le Blanc et al., 2003; Di Nicola et al., 2002) and cells derived of other tissues like fat tissue (Liu et al., 2007) or even peripheral blood (Wan et al., 2006). However, some aspects of the technique remain a formidable challenge that will have to be solved in the future. First, the supply of oxygen and nutrients to the cells when clinically sized constructs are implanted is a concern for all cell-based engineered tissues, including bone. Research on the nature and mechanism of action of growth factors promoting angiogenesis is one way to tackle this problem. Another strategy is to prevascularize the constructs by adding endothelial cells. This will also require new technologies to allow the connection of the implant to the host vasculature.

A second concern is the continuous search for a better scaffold, which should be biocompatible, and have a suitable 3D porous structure and suitable biomechanical properties, while remaining degradable and osteoconductive. Like the (autologous) bone graft, it should act as a temporary framework for the cells until new bone is formed. Bioceramics composed of 100% HA have high strength to mechanical loading, but are not resorbed. On the contrary, scaffolds made of natural coral or tricalcium phosphate are resorbed, but the resorption is occurring too rapidly and they are too fragile to sustain the weight load. Recently, ceramics have been developed that can be resorbed by osteoclasts as in physiologic bone remodeling. Possibly this is a step in the direction of the ideal scaffold (Langstaff *et al.*, 1999, 2001; Mastrogiacomo *et al.*, 2006, 2007).

19.4.4 Future perspectives

As it is clear from this chapter, bone tissue engineering is promising, but many hurdles will have to be overcome in order to consider it a true alternative for the autologous bone graft. From a cellular perspective, in order to achieve accelerated vascularization to enhance bone formation and to better understand the reciprocal interaction of osteoblasts and endothelium in bone tissue engineering, extensive preclinical research is still required. *In vitro* studies will need to constitute a base of selection for factors and criteria to develop new strategies to accelerate the perfusion of the scaffold and its connection to the vascular network of the host.

Adding growth factors to stimulate vessel ingrowth could bring the solution. Another approach to bypass the aspect of cell nutrition diffusion depths and potassium cell toxicity, induced by the hematoma, may be to first implant the scaffold, followed by the injection of the BMSCs in a second procedure. At this time point, new sprouts of blood vessels are present in the hematoma, and a sufficient supply of oxygen and nutrients can be delivered after the BMSC injection, increasing the chances for survival of the injected cells. Finally, it may turn out that the host cells can be recruited efficiently after vascularization has been established, which would obviate the need for exogenously cultured progenitors.

In the meantime a cautious approach towards clinical application should be maintained as some conflicting and often worrying results of pioneering clinical work have already been reported, such as a patient case in an attempt to reconstruct a thumb (Hentz and Chang, 2001; Vacanti *et al.*, 2001) or in a series of 27 patients receiving jaw augmentation (Schimming and Schmelzeisen, 2004). These two publications clearly lack evidence for survival of the transplanted cells. The mere presence of bone in the constructs could simply indicate successful osteoconduction from the adjacent host bone in the transplants used. This concept was expressed in a similar, but more profound study by Meijer *et al.* (2007) who did add an ectopic mouse model to show the osteogenic potency of the cells, but again stressed the limitations of the biopsies taken from the tissue-engineered constructs where cell origin could not be deduced. These studies indicate that cell-based bone tissue engineering is still far from approval by regulatory health care authorities for clinical application at this time (Logeart-Avramoglou *et al.*, 2005; Muschler *et al.*, 2004). Therefore, it should be realized that an important long-term challenge will be the further development of strategies that utilize the endogenous regenerative capacity of the local tissues, including host cells. This could lead to a situation where *ex vivo* expansion and, sometimes, differentiation of host cells will not be necessary any more, which can be more cost-effective and will be less complex in terms of regulatory issues.

The major challenge facing bone tissue engineers in the realm of scaffold development is achieving a proper balance between mechanical properties, a porous architecture and degradability, while maintaining osteoconductivity and biocompatibility. The most promising methods of scaffold development are focused on understanding biological and material properties in the nanoscale. The study of biomolecule (growth factors incorporated inside the scaffold structure)–cell interactions and novel scaffold synthesis methods, such as electrospinning and the use of nanofillers, may ultimately lead to better, more effective scaffolds. An interesting development with great potential, because no limitations related to growth factors and cell survival exist, is intrinsically bone-inductive implants. Recently, this quality of the scaffold was proven effective in enhancing bone formation in animal models (Habibovic et al., 2006).

As a researcher, one has to be aware that the responsibility for the clinical implementation of a competitive alternative to the use of autograft bone begins with our cell cultures in the laboratory setting. As a clinician, one has the responsibility to know the exact benefits and limitations of this alternative and, when it reaches the clinical setting, inform our patients about this. As a student, one has the responsibility to critically study the many reports that have been published by a diversity of research groups worldwide since we are the scientists of tomorrow. Hopefully, we may now all benefit from this collective work that aims to be a guide for the interested scientists showing the diversity of both the biological and material science aspects in bone tissue engineering.

19.5 Summary

- Bone serves a main biomechanical function with excellent tensile and loading strength, and is involved in calcium metabolism and hematopoiesis.
- The currently most optimal bone graft is cancellous autograft bone with osteoconductive, osteoinductive and, to some extent, osteogenic qualities. Some of the drawbacks are limited supply, longer surgical procedures to obtain the graft and concomitant complications. These drawbacks and financial reasons have boosted research into alternative bone grafts.

- A scaffold for bone tissue engineering should ideally match the mechanical properties of the bone it will replace and participate in the natural remodeling process, which means that it should be resorbable.
- Several osteoinductive and osteostimulating growth factors have been identified of which the osteoinductive BMP-2 and BMP-7 are the most potent.
- Characterization and identification of BMSCs has been enhanced by new insights and developments in harvesting from different tissue sources in vivo, improved sample acquisition techniques, more refined in vitro cell handling and newly discovered specific antibodies.
- For solid preclinical testing, in vitro cell studies followed by small animal 'proof-of-concept' studies and subsequent large animal 'feasibility studies' should be performed. Finally, efficacy studies (e.g. in nonhuman primates) may be needed prior to clinical testing.
- One of the limitations for solid proof-of-concept studies in bone tissue engineering is the lack of a cell marker that will not fade with subsequent cell divisions, that is cell specific, does not interfere with cell metabolism and is easy to detect in a noninvasive manner.
- Upscaling towards clinically relevant defect sizes for bone tissue engineering, moving from the cubic millimeter towards the cubic centimeter scale, constitutes a poor setting for cell survival without adequate solutions for early vascularization. Neo-osteogenesis requires neovascularization. Without this solution, cell-free concepts for tissue engineering should be considered such as the use of osteoinductive biomaterials.
- Several clinical applications of cell-based bone tissue engineering have been reported in the literature without any comparisons to control groups and with no evidence for the actual working mechanism. Despite all efforts, the level of evidence for any benefit related to the use of bone tissue engineering in literature has not reached beyond the level of proof-of-concept studies.

Definitions

Allograft	Tissue transplanted between genetically different individuals of the same species.
Autograft	Tissue transplanted within one individual.
Bioactive	Characterized by bonding osteogenesis. This means that bone formation starts at the material surface, and has a very tight and direct bond with the material. This typically occurs when an apatite layer can precipitate on the material surface (de Groot, 1980; Geesink *et al.*, 1988).
Biocompatibility	Ability of a material to perform with an appropriate host response in a specific application (Williams, 1999, 2003).
Biomaterial	Material that can be used for any period of time as a whole or as part of a system that treats, augments or replaces any tissue, organ or function in the body (Sakkers, 1999).
Calcification in tissues	Calcification is the deposition of insoluble calcium salts, which can be in extracellular bone matrix (a normal, physiological condition), but also in artery walls (a pathological condition called atherosclerosis (Cormack, 1987a) when calcifications occurs without ossification (Cormack, 1987b).
Hybrid construct	Combination of a synthetic porous scaffold and (biologic) cells.
Ossification	Production of organic bone matrix and its subsequent mineralization or calcification. Ossification can also occur without calcification in abnormal circumstances such as with rickets in children or osteomalacia in adults (Cormack, 1987b).
Osteoconduction	Ability of a material/graft to allow ingrowth of vessels and osteoprogenitor cells from the recipient bed (Damien and Parsons, 1991) or spreading of bone over the surface proceeded by ordered migration of differentiating osteogenic cells (Davies, 1998). An important aspect is the direct bonding of bone to the material surface without fibrous tissue interposition, so-called 'contact osteogenesis' or 'bonding osteogenesis' (Davies, 1998).
Osteogenesis	Bone formation by determined osteoprogenitor cells (Friedenstein, 1966; Friedenstein *et al.*, 1966). The generation of bone by a certain tissue/cell by itself, i.e. in the absence of host cells (e.g. in diffusion chambers) (Bab *et al.*, 1984).
Osteoinduction	Urist *et al.* (1967): The mechanism of cellular differentiation towards bone of one tissue due to the physicochemical effect or contact with another tissue, at that time only tissues (like demineralized bone matrix and uroepithelium) were known to have this ability. In the last decades this ability has also been recognized for some biomaterials (Ripamonti, 1991; Yuan, 1999).
	Friedenstein (1968): The induction of undifferentiated inducible osteoprogenitor cells that are not yet committed to the osteogenic lineage to form osteoprogenitor cells.
Stem cell	Stem cells are defined as resting cells capable of asymmetric cell division to allow both self-renewal (to prevent depletion of the stem cell pool) and the production of progeny cells that start proliferation and differentiation (to generate one or more tissue types) (Aubin, 1998; Bianco *et al.*, 2001; Muschler and Midura, 2002; Owen, 1998).
Tissue engineering	National Science Foundation (US) (1988): Tissue engineering is the application of the principles and methods of engineering and the life sciences toward the fundamental understanding of structure/function relationships in normal and pathological mammalian tissues and the development of biological substitutes to restore, maintain or improve functions.

Langer and Vacanti (1993) refer to tissue engineering as: 'an interdisciplinary field that applies the principles of engineering and life sciences toward the development of biological substitutes that restore, maintain or improve tissue function'.

Caplan and Goldberg (Caplan and Goldberg, 1999) described this 'interdisciplinary application of basic biological principles' as biomimetics. '[Tissue engineering] is the application of biomimetics toward the development of biological substitutes that restore, maintain or improve tissue function'.

Xenograft	Tissue transplanted between different species.

References

Acil, Y., Terheyden, H., *et al.* (2000). Three-dimensional cultivation of human osteoblast-like cells on highly porous natural bone mineral. *J Biomed Mater Res*, 51: 703–710.

Alden, T.D., Pittman, D.D., *et al.* (1999). Percutaneous spinal fusion using bone morphogenetic protein-2 gene therapy. *J Neurosurg*, 90(1 Suppl): 109–114.

Alexander, J.T., Branch, C.L.J., *et al.* (2002). An analysis of the use of rhBMP-2 in PLIF constructs: clinical and radiolographic outcomes. Presented at *AANS/CNS Section on Disorders of the Spine and Peripheral Nerves*, Orlando, FL.

An, Y.H., Woolf, S.K., *et al.* (2000). Pre-clinical *in vivo* evaluation of orthopaedic bioabsorbable devices. *Biomaterials*, 21: 2635–2652.

Anderson, J.M. (2000). *The Cellular Cascades of Wound Healing*. Toronto: EM Squared.

Anderson, M.L., Dhert, W.J., *et al.* (1999). Critical size defect in the goat's os ilium. A model to evaluate bone grafts and substitutes. *Clin Orthop Relat Res*, 364: 231–239.

Arinzeh, T.L., Peter, S.J., *et al.* (2003). Allogeneic mesenchymal stem cells regenerate bone in a critical-sized canine segmental defect. *J Bone Joint Surg Am*, 85A: 1927–1935.

Ashton, B.A., Allen, T.D., *et al.* (1980). Formation of bone and cartilage by marrow stromal cells in diffusion chambers *in vivo*. *Clin Orthop Relat Res*, 151: 294–307.

Aubin, J.E. (1998). Bone stem cells. *J Cell Biochem*(Suppl 31): 73–82.

Bab, I., Ashton, B.A., *et al.* (1984). Assessment of an in vivo diffusion chamber method as a quantitative assay for osteogenesis. *Calcif Tissue Int*, 36: 77–82.

Bak, B., Jorgensen, P.H., *et al.* (1990). Dose response of growth hormone on fracture healing in the rat. *Acta Orthop Scand*, 61: 54–57.

Baksh, D., Davies, J.E., *et al.* (2003). Adult human bone marrow-derived mesenchymal progenitor cells are capable of adhesion-independent survival and expansion. *Exp Hematol*, 31: 723–732.

Banwart, J.C., Asher, M.A., *et al.* (1995). Iliac crest bone graft harvest donor site morbidity. A statistical evaluation. *Spine*, 20: 1055–1060.

Baskin, D.S., Ryan, P., *et al.* (2002). ACDFP with cornerstone-SR allograft and plate: rhBMP-2 vs autograft. Presented at *AANS/CNS Section on Disorders of the Spine and Peripheral Nerves*, Orlando, FL.

Bensaid, W., Oudina, K., *et al.* (2005). *De novo* reconstruction of functional bone by tissue engineering in the metatarsal sheep model. *Tissue Eng*, 11: 814–824.

Bianco, P. and Boyde, A. (1993). Confocal images of marrow stromal (Westen–Bainton) cells. *Histochemistry*, 100: 93–99.

Bianco, P., Riminucci, M., *et al.* (2001). Bone marrow stromal stem cells: nature, biology, and potential applications. *Stem Cells*, 19: 180–192.

Bobyn, J.D., Mortimer, E.S., *et al.* (1992). Producing and avoiding stress shielding. Laboratory and clinical observations of noncemented total hip arthroplasty. *Clin Orthop Relat Res*, 274: 79–96.

Boden, S.D. (1999). Bioactive factors for bone tissue engineering. *Clin Orthop*, 367(Suppl): S84–S94.

Boden, S.D. (2002). Overview of the biology of lumbar spine fusion and principles for selecting a bone graft substitute. *Spine*, 27(16 Suppl 1): S26–S31.

Boden, S.D., Martin, G.J., Jr., *et al.* (1998). Laparoscopic anterior spinal arthrodesis with rhBMP-2 in a titanium interbody threaded cage. *J Spinal Disord*, 11: 95–101.

Boden, S.D., Martin, G.J., Jr., *et al.* (1999). Posterolateral lumbar intertransverse process spine arthrodesis with recombinant human bone morphogenetic protein 2/ hydroxyapatite–tricalcium phosphate after laminectomy in the nonhuman primate. *Spine*, 24: 1179–1185.

Boden, S.D., Zdeblick, T.A., *et al.* (2000). The use of rhBMP-2 in interbody fusion cages. Definitive evidence of osteoinduction in humans: a preliminary report. *Spine*, 25: 376–381.

Boden, S.D., Kang, J., *et al.* (2002). Use of recombinant human bone morphogenetic protein-2 to achieve posterolateral lumbar spine fusion in humans: a prospective, randomized clinical pilot trial: 2002 Volvo Award in clinical studies. *Spine*, 27: 2662–2673.

Bohner, M. (2000). Calcium orthophosphates in medicine: from ceramics to calcium phosphate cements. *Injury*, 31(Suppl 4): 37–47.

Bolander, M.E. (1992). Regulation of fracture repair by growth factors. *Proc Soc Exp Biol Med*, 200: 165–170.

Borden, M., Attawia, M., *et al.* (2004). Tissue-engineered bone formation *in vivo* using a novel sintered polymeric microsphere matrix. *J Bone Joint Surg Br*, 86: 1200–1208.

Bos, G.D., Goldberg, V.M., *et al.* (1983). Immune responses of rats to frozen bone allografts. *J Bone Joint Surg Am*, 65: 239–246.

Bostrom, M.P. and Camacho, N.P. (1998). Potential role of bone morphogenetic proteins in fracture healing. *Clin Orthop Relat Res*, 355(Suppl): S274–S282.

Boyne, P.J. (1992). *Maxillofacial Surgery*. Philadelphia, PA: Saunders.

Brodke, D., Pedrozo, H.A., *et al.* (2006). Bone grafts prepared with selective cell retention technology heal canine segmental defects as effectively as autograft. *J Orthop Res*, 24: 857–866.

Bruder, S.P., Horowitz, M.C., *et al.* (1997). Monoclonal antibodies reactive with human osteogenic cell surface antigens. *Bone*, 21: 225–235.

Bruder, S.P., Kraus, K.H., *et al.* (1998a). The effect of implants loaded with autologous mesenchymal stem cells on the healing of canine segmental bone defects. *J Bone Joint Surg Am*, 80: 985–996.

Bruder, S.P., Kurth, A.A., *et al.* (1998b). Bone regeneration by implantation of purified, culture-expanded human mesenchymal stem cells. *J Orthop Res*, 16: 155–162.

Burchardt, H. (1983). The biology of bone graft repair. *Clin Orthop Relat Res*, 174: 28–42.

Burkus, J.K., Transfeldt, E.E., *et al.* (2002). Clinical and radiographic outcomes of anterior lumbar interbody fusion using recombinant human bone morphogenetic protein-2. *Spine*, 27: 2396–2408.

Cancedda, R., Mastrogiacomo, M., *et al.* (2003). Bone marrow stromal cells and their use in regenerating bone. *Novartis Found Symp*, 249: 133–143. discussion 143–147, 170–174, 239–241.

Caplan, A.I. (2000). Mesenchymal stem cells and gene therapy. *Clin Orthop Relat Res*, 379(Suppl): S67–S70.

Caplan, A.I. (2005). Review: mesenchymal stem cells: cell-based reconstructive therapy in orthopedics. *Tissue Eng*, 11: 1198–1211.

Caplan, A.I. and Goldberg, V.M. (1999). Principles of tissue engineered regeneration of skeletal tissues. *Clin Orthop Relat Res*, 367(Suppl): S12–S16.

Carpenter, J.E., Hipp, J.A., *et al.* (1992). Failure of growth hormone to alter the biomechanics of fracture-healing in a rabbit model. *J Bone Joint Surg Am*, 74: 359–367.

Cawood, J.I. and Howell, R.A. (1991). Reconstructive pre-prosthetic surgery. I. Anatomical considerations. *Int J Oral Maxillofac Surg*, 20: 75–82.

Charalambides, C., Beer, M., *et al.* (2005). Poor results after augmenting autograft with xenograft (Surgibone) in hip revision surgery: a report of 27 cases. *Acta Orthop*, 76: 544–549.

Chiou, M., Xu, Y., *et al.* (2006). Mitogenic and chondrogenic effects of fibroblast growth factor-2 in adipose-derived mesenchymal cells. *Biochem Biophys Res Commun*, 343: 562–644.

Chunmeng, S. and Tianmin, C. (2004). Effects of plastic-adherent dermal multipotent cells on peripheral blood leukocytes and CFU-GM in rats. *Transplant Proc*, 36: 1578–1581.

Colter, D.C., Class, R., *et al.* (2000). Rapid expansion of recycling stem cells in cultures of plastic-adherent cells from human bone marrow. *Proc Natl Acad Sci USA*, 97: 3213–3218.

Colter, D.C., Sekiya, I., *et al.* (2001). Identification of a subpopulation of rapidly self-renewing and multipotential adult stem cells in colonies of human marrow stromal cells. *Proc Natl Acad Sci USA*, 98: 7841–7845.

Contag, P.R., Olomu, I.N., *et al.* (1998). Bioluminescent indicators in living mammals. *Nat Med*, 4: 245–247.

Cook, S.D., Wolfe, M.W., *et al.* (1995). Effect of recombinant human osteogenic protein-1 on healing of segmental defects in non-human primates. *J Bone Joint Surg Am*, 77: 734–750.

Cormack, D.H. (1987a). *The Circulatory System. Part Three: The Tissues of the Body*. Philadelphia, PA: Lippincott.

Cormack, D.H. (1987b). *Bone. Part Three: The Tissues of the Body*. Philadelphia, PA: Lippincott.

Cornell, C.N. (1999). Osteoconductive materials and their role as substitutes for autogenous bone grafts. *Orthop Clin North Am*, 30: 591–598.

Critchlow, M.A., Bland, Y.S., et al. (1995). The effect of exogenous transforming growth factor-beta 2 on healing fractures in the rabbit. *Bone*, 16: 521–527.

Daculsi, G., LeGeros, R.Z., et al. (1989). Transformation of biphasic calcium phosphate ceramics *in vivo*: ultrastructural and physicochemical characterization. *J Biomed Mater Res*, 23: 883–894.

Damien, C.J. and Parsons, J.R. (1991). Bone graft and bone graft substitutes: a review of current technology and applications. *J Appl Biomater*, 2: 187–208.

Davies, J.E. (1998). Mechanisms of endosseous integration. *Int J Prosthodont*, 11: 391–401.

de Bruijn, J.D. (1998). Tissue engineering of goat bone: osteogenic potential of goat bone marrow cells. *Bioceramics*: 497–500.

de Bruyn, P.H. (1955). Bone formation by fresh and frozen autogenous and homogenous transplants of bone, bone marrow and periosteum. *Am J Anat*, 96: 375–417.

de Groot, K. (1980). Bioceramics consisting of calcium phosphate salts. *Biomaterials*, 1: 47–50.

Deschaseaux, F. and Charbord, P. (2000). Human marrow stromal precursors are alpha 1 integrin subunit-positive. *J Cell Physiol*, 184: 319–325.

Di Nicola, M., Carlo-Stella, C., et al. (2002). Human bone marrow stromal cells suppress T-lymphocyte proliferation induced by cellular or nonspecific mitogenic stimuli. *Blood*, 99: 3838–3843.

Duvall, C.L., Taylor, W.R., et al. (2006). Impaired angiogenesis, early callus formation, and late stage remodeling in fracture healing of osteopontin deficient mice. *J Bone Miner Res*, 22: 286–297.

Dvorak, H.F., Harvey, V.S., et al. (1987). Fibrin containing gels induce angiogenesis. Implications for tumor stroma generation and wound healing. *Lab Invest*, 57: 673–686.

Elabd, C., et al. (2007). Human adipose tissue-derived multipotent stem cells differentiate *in vitro* and *in vivo* into osteocyte-like cells. *Biochem Biophys Res Commun*, 361(2): 342–348.

Elsinger, E.C. and Leal, L. (1996). Coralline hydroxyapatite bone graft substitutes. *J Foot Ankle Surg*, 35: 396–399.

Emans, P.J., Pieper, J., et al. (2006). Differential cell viability of chondrocytes and progenitor cells in tissue-engineered constructs following implantation into osteochondral defects. *Tissue Eng*, 12: 1699–1709.

Fazili, M., van der Dussen, F.M., et al. (1986). Long-term results of augmentation of the atrophic mandible. *Int J Oral Maxillofac Surg*, 15: 513–520.

Feeley, B.T., Conduah, A.H., et al. (2006). *In vivo* molecular imaging of adenoviral versus lentiviral gene therapy in two bone formation models. *J Orthop Res*, 24: 1709–1721.

Fleming, J.E. Jr., Cornell, C.N., et al. (2000). Bone cells and matrices in orthopedic tissue engineering. *Orthop Clin North Am*, 31: 357–374.

Friedenstein, A.J. (1962). Humoral nature of osteogenic activity of transitional epithelium. *Nature*, 194: 698–699.

Friedenstein, A.Y. (1968). Induction of bone tissue by transitional epithelium. *Clin Orthop Relat Res*, 59: 21–37.

Friedenstein, A. and Kuralesova, A.I. (1971). Osteogenic precursor cells of bone marrow in radiation chimeras. *Transplantation*, 12: 99–108.

Friedenstein, A.J., Piatetzky, S., I, et al. (1966). Osteogenesis in transplants of bone marrow cells. *J Embryol Exp Morphol*, 16: 381–390.

Friedenstein, A.J., Chailakhjan, R.K., et al. (1970). The development of fibroblast colonies in monolayer cultures of guinea-pig bone marrow and spleen cells. *Cell Tissue Kinet*, 3: 393–403.

Friedenstein, A.J., Gorskaja, J.F., et al. (1976). Fibroblast precursors in normal and irradiated mouse hematopoietic organs. *Exp Hematol*, 4: 267–274.

Friedenstein, A.J., Ivanov-Smolenski, A.A., et al. (1978). Origin of bone marrow stromal mechanocytes in radiochimeras and heterotopic transplants. *Exp Hematol*, 6: 440–444.

Friedenstein, A.J., Latzinik, N.W., et al. (1982). Marrow microenvironment transfer by heterotopic transplantation of freshly isolated and cultured cells in porous sponges. *Exp Hematol*, 10: 217–227.

Friedenstein, A.J., Chailakhyan, R.K., et al. (1987). Bone marrow osteogenic stem cells: *in vitro* cultivation and transplantation in diffusion chambers. *Cell Tissue Kinet*, 20: 263–272.

Friedenstein, A.J., Latzinik, N.V., et al. (1992). Bone marrow stromal colony formation requires stimulation by haemopoietic cells. *Bone Miner*, 18: 199–213.

Friedlaender, G.E. (1983). Immune responses to osteochondral allografts. Current knowledge and future directions. *Clin Orthop Relat Res*, 174: 58–68.

Friedlaender, G.E. (1987). Bone banking. In support of reconstructive surgery of the hip. *Clin Orthop Relat Res*, 225: 17–21.

Friedlaender, G.E., Perry, C.R., *et al.* (2001). Osteogenic protein-1 (bone morphogenetic protein-7) in the treatment of tibial nonunions. *J Bone Joint Surg Am*, 83A(Suppl 1; Pt 2): S151–S158.

Fujita, M., Kinoshita, Y., *et al.* (2005). Proliferation and differentiation of rat bone marrow stromal cells on poly(glycolic acid)–collagen sponge. *Tissue Eng*, 11: 1346–1355.

Gazit, D., Turgeman, G., *et al.* (1999). Engineered pluripotent mesenchymal cells integrate and differentiate in regenerating bone: a novel cell-mediated gene therapy. *J Gene Med*, 1: 121–133.

Geesink, R.G., de Groot, K., *et al.* (1988). Bonding of bone to apatite-coated implants. *J Bone Joint Surg Br*, 70: 17–22.

Geesink, R.G., Hoefnagels, N.H., *et al.* (1999). Osteogenic activity of OP-1 bone morphogenetic protein (BMP-7) in a human fibular defect. *J Bone Joint Surg Br*, 81: 710–718.

Gerber, H.P., Vu, T.H., *et al.* (1999). VEGF couples hypertrophic cartilage remodeling, ossification and angiogenesis during endochondral bone formation. *Nat Med*, 5: 623–628.

Giannoudis, P.V., Dinopoulos, H., *et al.* (2005). Bone substitutes: an update. *Injury*, 36(Suppl 3): S20–S27.

Goldberg, V.M. and Stevenson, S. (1990). *Bone Transplantation*. New York: Churchill Livingstone.

Gornet, M.F., Burkus, K., *et al.* (2001). rh-BMP-2 with tapered cages: A prospective randomized lumbar fusion study. Presented at the *Annual Meeting of the North American Spine Society*, Seattle, WA.

Goshima, J., Goldberg, V.M., *et al.* (1991). The origin of bone formed in composite grafts of porous calcium phosphate ceramic loaded with marrow cells. *Clin Orthop*, 269: 274–183.

Griffiths, A.D., Williams, S.C., *et al.* (1994). Isolation of high affinity human antibodies directly from large synthetic repertoires. *EMBO J*, 13: 3245–3260.

Gronthos, S., Graves, S.E., *et al.* (1994). The STRO-1$^+$ fraction of adult human bone marrow contains the osteogenic precursors. *Blood*, 84: 4164–4173.

Gronthos, S. and Simmons, P.J. (1995). The growth factor requirements of STRO-1-positive human bone marrow stromal precursors under serum-deprived conditions *in vitro*. *Blood*, 85: 929–940.

Gupta, M.C. and Maitra, S. (2002). Bone grafts and bone morphogenetic proteins in spine fusion. *Cell Tissue Bank*, 3: 255–267.

Habibovic, P., Yuan, H., *et al.* (2005). 3D microenvironment as essential element for osteoinduction by biomaterials. *Biomaterials*, 26: 3565–3575.

Habibovic, P., Yuan, H., *et al.* (2006). Relevance of osteoinductive biomaterials in critical-sized orthotopic defect. *J Orthop Res*, 24: 867–876.

Hacker, M., Tessmar, J., *et al.* (2003). Towards biomimetic scaffolds: anhydrous scaffold fabrication from biodegradable amine-reactive diblock copolymers. *Biomaterials*, 24: 4459–4473.

Hartgerink, J.D., Beniash, E., *et al.* (2001). Self-assembly and mineralization of peptide-amphiphile nanofibers. *Science*, 294: 1684–1688.

Hausman, M.R., Schaffler, M.B., *et al.* (2001). Prevention of fracture healing in rats by an inhibitor of angiogenesis. *Bone*, 29: 560–564.

Hedberg, E.L., Kroese-Deutman, H.C., *et al.* (2005a). Effect of varied release kinetics of the osteogenic thrombin peptide TP508 from biodegradable, polymeric scaffolds on bone formation *in vivo*. *J Biomed Mater Res A*, 72: 343–353.

Hedberg, E.L., Shih, C.K., *et al.* (2005b). *In vitro* degradation of porous poly(propylene fumarate)/poly(DL-lactic-co-glycolic acid) composite scaffolds. *Biomaterials*, 26: 3215–3225.

Hentz, V.R. and Chang, J. (2001). Tissue engineering for reconstruction of the thumb. *N Engl J Med*, 344: 1547–1548.

Hetherington, V.J., Kawalec, J.S., *et al.* (2005). Immunologic testing of xeno-derived osteochondral grafts using peripheral blood mononuclear cells from healthy human donors. *BMC Musculoskelet Disord*, 6: 36.

Hiltunen, M.O., *et al.* (2003). Adenovirus-mediated VEGF-A gene transfer induces bone formation *in vivo*. *Faseb J*, 17(9): 1147–1149.

Hollinger, J.O., Brekke, J., *et al.* (1996). Role of bone substitutes. *Clin Orthop Relat Res*, 324: 55–65.

Horan, P.K. and Slezak, S.E. (1989). Stable cell membrane labeling. *Nature*, 340: 167–168.

Horch, R.A., Shahid, N., *et al.* (2004). Nanoreinforcement of poly(propylene fumarate)-based networks with surface modified alumoxane nanoparticles for bone tissue engineering. *Biomacromolecules*, 5: 1990–1998.

Hutmacher, D.W. (2000). Scaffolds in tissue engineering bone and cartilage. *Biomaterials*, 21: 2529–2543.

Jansen, J.A., Vehof, J.W., *et al.* (2005). Growth factor-loaded scaffolds for bone engineering. *J Control Release*, 101(1–3): 127–136.

Janssen, F.W., Oostra, J., *et al.* (2006). A perfusion bioreactor system capable of producing clinically relevant volumes of tissue-engineered bone: *in vivo* bone formation showing proof of concept. *Biomaterials*, 27: 315–323.

Jiang, Y., Jahagirdar, B.N., *et al.* (2002). Pluripotency of mesenchymal stem cells derived from adult marrow. *Nature*, 418(6893): 41–49.

Jingushi, S. and Bolander, M.E. (1991). Biological cascades of fracture healing as models for bone–biomaterial interfacial reactions. In *The Bone–Biomaterial Interface* (Davies, J.E. and Albrektsson, T., eds). Toronto: University of Toronto Press.

Johnson, E.E., Urist, M.R., *et al.* (1988a). Bone morphogenetic protein augmentation grafting of resistant femoral nonunions. A preliminary report. *Clin Orthop Relat Res*, 230: 257–265.

Johnson, E.E., Urist, M.R., *et al.* (1988b). Repair of segmental defects of the tibia with cancellous bone grafts augmented with human bone morphogenetic protein. A preliminary report. *Clin Orthop Relat Res*, 236: 249–257.

Jones, A.L., Bucholz, R.W., *et al.* (2006). Recombinant human BMP-2 and allograft compared with autogenous bone graft for reconstruction of diaphyseal tibial fractures with cortical defects. A randomized, controlled trial. *J Bone Joint Surg Am*, 88: 1431–1441.

Joyner, C.J., Bennett, A., *et al.* (1997). Identification and enrichment of human osteoprogenitor cells by using differentiation stage-specific monoclonal antibodies. *Bone*, 21: 1–6.

Kadiyala, S., Young, R.G., *et al.* (1997). Culture expanded canine mesenchymal stem cells possess osteochondrogenic potential *in vivo* and *in vitro*. *Cell Transplant*, 6: 125–134.

Kato, T., Kawaguchi, H., *et al.* (1998). Single local injection of recombinant fibroblast growth factor-2 stimulates healing of segmental bone defects in rabbits. *J Orthop Res*, 16: 654–659.

Kitamura, S., Ohgushi, H., *et al.* (2004). Osteogenic differentiation of human bone marrow-derived mesenchymal cells cultured on alumina ceramics. *Artif Organs*, 28: 72–82.

Kitoh, H., Kitakoji, T., *et al.* (2004). Transplantation of marrow-derived mesenchymal stem cells and platelet-rich plasma during distraction osteogenesis – a preliminary result of three cases. *Bone*, 35: 892–898.

Kon, E., Muraglia, A., *et al.* (2000). Autologous bone marrow stromal cells loaded onto porous hydroxyapatite ceramic accelerate bone repair in critical-size defects of sheep long bones. *J Biomed Mater Res*, 49: 328–337.

Kratchmarova, I., Blagoev, B., *et al.* (2005). Mechanism of divergent growth factor effects in mesenchymal stem cell differentiation. *Science*, 308: 1472–1477.

Kraus, K.H., Kadiyala, S., *et al.* (1999). Critically sized osteo-periosteal femoral defects: a dog model. *J Invest Surg*, 12: 115–124.

Krebsbach, P.H., Kuznetsov, S.A., *et al.* (1997). Bone formation *in vivo*: comparison of osteogenesis by transplanted mouse and human marrow stromal fibroblasts. *Transplantation*, 63: 1059–1069.

Krebsbach, P.H., Kuznetsov, S.A., *et al.* (1999). Bone marrow stromal cells: characterization and clinical application. *Crit Rev Oral Biol Med*, 10: 165–181.

Kroese-Deutman, H.C., Ruhe, P.Q., *et al.* (2005). Bone inductive properties of rhBMP-2 loaded porous calcium phosphate cement implants inserted at an ectopic site in rabbits. *Biomaterials*, 26: 1131–1138.

Kruyt, M.C., De Bruijn, J., *et al.* (2003a). Application and limitations of chloromethyl-benzamidodialkylcarbocyanine for tracing cells used in bone Tissue engineering. *Tissue Eng*, 9: 105–115.

Kruyt, M.C., de Bruijn, J.D., *et al.* (2003b). Viable osteogenic cells are obligatory for tissue-engineered ectopic bone formation in goats. *Tissue Eng*, 9: 327–336.

Kruyt, M.C., Stijns, M.M., *et al.* (2004a). Genetic marking with the DeltaLNGFR-gene for tracing goat cells in bone tissue engineering. *J Orthop Res*, 22: 697–702.

Kruyt, M.C., van Gaalen, S.M., *et al.* (2004b). Bone tissue engineering and spinal fusion: the potential of hybrid constructs by combining osteoprogenitor cells and scaffolds. *Biomaterials*, 25: 1463–1473.

Kruyt, M.C., Wilson, C.E., *et al.* (2006). The effect of cell-based bone tissue engineering in a goat transverse process model. *Biomaterials*, 27: 5099–5106.

Kuznetsov, S. and Gehron Robey, P. (1996). Species differences in growth requirements for bone marrow stromal fibroblast colony formation *in vitro*. *Calcif Tissue Int*, 59: 265–270.

Langer, R. and Vacanti, J.P. (1993). Tissue engineering. *Science*, 260: 920–926.

Langstaff, S., Sayer, M., *et al.* (1999). Resorbable bioceramics based on stabilized calcium phosphates. Part I: rational design, sample preparation and material characterization. *Biomaterials*, 20: 1727–1741.

Langstaff, S., Sayer, M., *et al.* (2001). Resorbable bioceramics based on stabilized calcium phosphates. Part II: evaluation of biological response. *Biomaterials*, 22: 135–150.

Le Blanc, K., Tammik, L., *et al.* (2003). Mesenchymal stem cells inhibit and stimulate mixed lymphocyte cultures and mitogenic responses independently of the major histocompatibility complex. *Scand J Immunol*, 57: 11–20.

Letchford, J., Cardwell, A.M., *et al.* (2006). Isolation of C15: a novel antibody generated by phage display against mesenchymal stem cell-enriched fractions of adult human marrow. *J Immunol Methods*, 308: 124–137.

Levenberg, S., Rouwkema, J., *et al.* (2005). Engineering vascularized skeletal muscle tissue. *Nat Biotechnol*, 23: 879–884.

Lieberman, J.R., Daluiski, A., *et al.* (2002). The role of growth factors in the repair of bone. Biology and clinical applications. *J Bone Joint Surg Am*, 84-A: 1032–1044.

Lin, C.Y., Kikuchi, N., *et al.* (2004). A novel method for biomaterial scaffold internal architecture design to match bone elastic properties with desired porosity. *J Biomech*, 37: 623–636.

Lind, M., Schumacker, B., *et al.* (1993). Transforming growth factor-beta enhances fracture healing in rabbit tibiae. *Acta Orthop Scand*, 64: 553–556.

Liu, T.M., Martina, M., *et al.* (2007). Identification of common pathways mediating differentiation of bone marrow and adipose tissues derived human mesenchymal stem cells (MSCs) into three mesenchymal lineages. *Stem Cells*, 25: 750–760.

Logeart-Avramoglou, D., Anagnostou, F., *et al.* (2005). Engineering bone: challenges and obstacles. *J Cell Mol Med*, 9: 72–84.

Luque, E. (2002) Latest clinical results using demineralized bone materials and rhBMP-2: the Mexican experience. Presented at *Total Spine: Advanced Concepts and Constructs*, Cancun, Mexico.

Luyten, F.P., Cunningham, N.S., *et al.* (1989). Purification and partial amino acid sequence of osteogenin, a protein initiating bone differentiation. *J Biol Chem*, 264: 13377–13380.

Macewan, W. (1881). Observations concerning transplantation of bone. Illustrated by a case of inter-human osseous transplantation, whereby over two-thirds of the shaft of a humerus was restored. *Proc Roy Soc Lond*, 32: 232–247.

Maiorana, C., Sigurta, D., *et al.* (2006). Sinus elevation with alloplasts or xenogenic materials and implants: an up-to-4-year clinical and radiologic follow-up. *Int J Oral Maxillofac Implants*, 21: 426–432.

Majors, A.K., Boehm, C.A., *et al.* (1997). Characterization of human bone marrow stromal cells with respect to osteoblastic differentiation. *J Orthop Res*, 15: 546–557.

Maniatopoulos, C., Sodek, J., *et al.* (1988). Bone formation *in vitro* by stromal cells obtained from bone marrow of young adult rats. *Cell Tissue Res*, 254: 317–330.

Mankin, H.J., Doppelt, S., *et al.* (1983). Clinical experience with allograft implantation, The first ten years. *Clin Orthop Relat Res*, 174: 69–86.

Marcacci, M., Kon, K., *et al.* (2007). Stem cells associated with macroporous bioceramics for long bone repair: 6 to 7 year outcome of a pilot clinical study. *Tissue Eng*, 13: 947–955.

Mastrogiacomo, M., Corsi, A., *et al.* (2006). Reconstruction of extensive long bone defects in sheep using resorbable bioceramics based on silicon stabilized tricalcium phosphate. *Tissue Eng*, 12: 1261–1273.

Mastrogiacomo, M., Papadimitropoulos, A., *et al.* (2007). Engineering of bone using bone marrow stromal cells and a silicon-stabilized tricalcium phosphate bioceramic: evidence for a coupling between bone formation and scaffold resorption. *Biomaterials*, 28: 1376–1384.

Meijer, G.J., de Bruijn, J.D., *et al.* (2007). Cell-based bone tissue engineering. *PLoS Med*, 4: e9.

Mendes, S.C., *et al.* (2002). Bone tissue-engineered implants using human bone marrow stromal cells: effect of culture conditions and donor age. *Tissue Eng*, 8(6): 911–920.

Meury, T., Verrier, S., *et al.* (2006). Human endothelial cells inhibit BMSC differentiation into mature osteoblasts *in vitro* by interfering with osterix expression. *J Cell Biochem*, 98: 992–1006.

Mistry, A.S. and Mikos, A.G. (2005). Tissue engineering strategies for bone regeneration. *Adv Biochem Eng Biotechnol*, 94: 1–22.

Morishita, T., Honoki, K., *et al.* (2006). Tissue engineering approach to the treatment of bone tumors: three cases of cultured bone grafts derived from patients' mesenchymal stem cells. *Artif Organs*, 30: 115–118.

Muschler, G.F., Negami, S., *et al.* (1996). Evaluation of collagen ceramic composite graft materials in a spinal fusion model. *Clin Orthop Relat Res*, 328: 250–260.

Muschler, G.F., Boehm, C., *et al.* (1997). Aspiration to obtain osteoblast progenitor cells from human bone marrow: the influence of aspiration volume [published erratum appears in *J Bone Joint Surg Am* 1998; 80: 302]. *J Bone Joint Surg Am*, 79: 1699–1709.

Muschler, G.F. and Midura, R.J. (2002). Connective tissue progenitors: practical concepts for clinical applications. *Clin Orthop Relat Res*, 395: 66–80.

Muschler, G.F., Nakamoto, C., *et al.* (2004). Engineering principles of clinical cell-based tissue engineering. *J Bone Joint Surg Am*, 86A: 1541–1558.

Nakahara, H., Bruder, S.P., *et al.* (1990). *In vivo* osteochondrogenic potential of cultured cells derived from the periosteum. *Clin Orthop*, 259: 223–232.

Nakajima, F., Ogasawara, A., *et al.* (2001). Spatial and temporal gene expression in chondrogenesis during fracture healing and the effects of basic fibroblast growth factor. *J Orthop Res*, 19: 935–944.

Nakamae, A., Sunagawa, T., *et al.* (2004). Acceleration of surgical angiogenesis in necrotic bone with a single injection of fibroblast growth factor-2 (FGF-2). *J Orthop Res*, 22: 509–513.

Nakamura, T., Hara, Y., *et al.* (1998). Recombinant human basic fibroblast growth factor accelerates fracture healing by enhancing callus remodeling in experimental dog tibial fracture. *J Bone Miner Res*, 13: 942–949.

Nash, T.J., Howlett, C.R., *et al.* (1994). Effect of platelet-derived growth factor on tibial osteotomies in rabbits. *Bone*, 15: 203–208.

Nevins, M.L., Camelo, M., *et al.* (2003). Evaluation of periodontal regeneration following grafting intrabony defects with bio-oss collagen: a human histologic report. *Int J Periodontics Restorative Dent*, 23: 9–17.

Nielsen, H.M., Andreassen, T.T., *et al.* (1994). Local injection of TGF-beta increases the strength of tibial fractures in the rat. *Acta Orthop Scand*, 65: 37–41.

Nikolaychik, V.V., Samet, M.M., *et al.* (1996). A new method for continual quantitation of viable cells on endothelialized polyurethanes. *J Biomater Sci Polym Ed*, 7: 881–891.

Nocini, P.F., Schlegel, K.A., *et al.* (2006). Two techniques for the preparation of cell-scaffold constructs suitable for sinus augmentation: steps into clinical application. *Tissue Eng*, 12: 2649–2656.

Ogawa, R., Mizuno, H., *et al.* (2004). Chondrogenic and osteogenic differentiation of adipose-derived stem cells isolated from GFP transgenic mice. *J Nippon Med Sch*, 71: 240–241.

Ohgushi, H., Kotobuki, N., *et al.* (2005). Tissue engineered ceramic artificial joint – *ex vivo* osteogenic differentiation of patient mesenchymal cells on total ankle joints for treatment of osteoarthritis. *Biomaterials*, 26: 4654–4661.

Osborn, J.F. (1980). *Dynamic Aspects of the Implant–Bone Interface*. Munchen: Carl Hansen Verlach.

Overgaard, S., Soballe, K., *et al.* (2000). Efficiency of systematic sampling in histomorphometric bone research illustrated by hydroxyapatite-coated implants: optimizing the stereological vertical-section design. *J Orthop Res*, 18: 313–321.

Owen, M.E. (1998). *The Marrow Stromal System*. Cambridge: Cambridge University Press.

Oyajobi, B.O., Lomri, A., *et al.* (1999). Isolation and characterization of human clonogenic osteoblast progenitors immunoselected from fetal bone marrow stroma using STRO-1 monoclonal antibody. *J Bone Miner Res*, 14: 351–361.

Pacheco, E.M., Civelek, A.C., *et al.* (1997). Clinicopathological correlation of technetium bone scan in vascularization of hydroxyapatite implants, A primate model. *Arch Ophthalmol*, 115: 1173–1177.

Pelissier, P., Villars, F., *et al.* (2003). Influences of vascularization and osteogenic cells on heterotopic bone formation within a madreporic ceramic in rats. *Plast Reconstr Surg*, 111: 1932–1941.

Peng, H., Usas, A., *et al.* (2004). Converse relationship between *in vitro* osteogenic differentiation and *in vivo* bone healing elicited by different populations of muscle-derived cells genetically engineered to express BMP4. *J Bone Miner Res*, 19: 630–641.

Peng, H., Usas, A., *et al.* (2005). VEGF improves, whereas sFlt1 inhibits, BMP2-induced bone formation and bone healing through modulation of angiogenesis. *J Bone Miner Res*, 20: 2017–2027.

Peter, S.J., Miller, S.T., *et al.* (1998). *In vivo* degradation of a poly(propylene fumarate)/beta-tricalcium phosphate injectable composite scaffold. *J Biomed Mater Res*, 41: 1–7.

Petite, H., Viateau, V., *et al.* (2000). Tissue-engineered bone regeneration. *Nat Biotechnol*, 18: 959–963.

Phinney, D.G., Kopen, G., *et al.* (1999). Donor variation in the growth properties and osteogenic potential of human marrow stromal cells. *J Cell Biochem*, 75: 424–436.

Pietrzak, W.S., Perns, S.V., *et al.* (2005). Demineralized bone matrix graft: a scientific and clinical case study assessment. *J Foot Ankle Surg*, 44: 345–353.

Prockop, D.J. (1997). Marrow stromal cells as stem cells for nonhematopoietic tissues. *Science*, 276: 71–74.

Prolo, D.J. and Rodrigo, J.J. (1985). Contemporary bone graft physiology and surgery. *Clin Orthop*, 200: 322–342.

Quarto, R., Mastrogiacomo, M., *et al.* (2001). Repair of large bone defects with the use of autologous bone marrow stromal cells. *N Engl J Med*, 344: 385–386.

Radomsky, M.L., Aufdemorte, T.B., *et al.* (1999). Novel formulation of fibroblast growth factor-2 in a hyaluronan gel accelerates fracture healing in nonhuman primates. *J Orthop Res*, 17: 607–614.

Rengachary, S.S. (2002). Bone morphogenetic proteins: basic concepts. *Neurosurg Focus*, 13: e2.

Reyes, M., Lund, T., *et al.* (2001). Purification and *ex vivo* expansion of postnatal human marrow mesodermal progenitor cells. *Blood*, 98: 2615–2625.

Riley, E.H., Lane, J.M., *et al.* (1996). Bone morphogenetic protein-2: biology and applications. *Clin Orthop Relat Res*, 324: 39–46.

Ripamonti, U. (1991). Bone induction in nonhuman primates. An experimental study on the baboon. *Clin Orthop Relat Res*, 269: 284–294.

Ripamonti, U., Crooks, J., *et al.* (2001). Induction of bone formation by recombinant human osteogenic protein-1 and sintered porous hydroxyapatite in adult primates. *Plast Reconstr Surg*, 107: 977–988.

Rouwkema, J., Boer, J.D., *et al.* (2006). Endothelial cells assemble into a 3-dimensional prevascular network in a bone tissue engineering construct. *Tissue Eng*, 12: 2685–2693.

Ruhe, P.Q., Kroese-Deutman, H.C., *et al.* (2004). Bone inductive properties of rhBMP-2 loaded porous calcium phosphate cement implants in cranial defects in rabbits. *Biomaterials*, 25: 2123–2132.

Ruhe, P.Q., Wolke, J.G., *et al.* (2006). *Calcium Phosphate Ceramics for Bone Tissue Engineering*. Boca Raton, FL: CRC Press.

Ruhe, P.Q., Wolke, J.G., *et al.* (2007). *Calcium Phosphate Ceramics for Bone Tissue Engineering*. Boca Raton, FL: CRC Press.

Sakkers, R.J. (1999). *Bone and Bone Bonding Copolymers: A Useful Combination?* Utrecht: UMC.

Sampath, T.K., Muthukumaran, N., *et al.* (1987). Isolation of osteogenin, an extracellular matrix-associated, bone-inductive protein, by heparin affinity chromatography. *Proc Natl Acad Sci USA*, 84: 7109–7113.

Sandhu, H.S., Kanim, L.E.A., *et al.* (1997). The safety and efficacy of purified native human bone morphogenetic protein for spinal fusion. A ten-year followup study. Presented at *Annual Meeting of the American Academy of Orthopedic Surgeons*, San Francisco, CA.

Schimming, R. and Schmelzeisen, R. (2004). Tissue-engineered bone for maxillary sinus augmentation. *J Oral Maxillofac Surg*, 62: 724–729.

Schliephake, H., Knebel, J.W., *et al.* (2001). Use of cultivated osteoprogenitor cells to increase bone formation in segmental mandibular defects: an experimental pilot study in sheep. *Int J Oral Maxillofac Surg*, 30: 531–537.

Schmitz, J.P. and Hollinger, J.O. (1986). The critical size defect as an experimental model for craniomandibulofacial nonunions. *Clin Orthop Relat Res*, 205: 299–308.

Sciadini, M.F. and Johnson, K.D. (2000). Evaluation of recombinant human bone morphogenetic protein-2 as a bone-graft substitute in a canine segmental defect model. *J Orthop Res*, 18: 289–302.

Shang, Q., Wang, Z., *et al.* (2001). Tissue-engineered bone repair of sheep cranial defects with autologous bone marrow stromal cells. *J Craniofac Surg*, 12: 586–593. discussion 594–595.

Shi, S., Gronthos, S., *et al.* (2002). Bone formation by human postnatal bone marrow stromal stem cells is enhanced by telomerase expression. *Nat Biotechnol*, 20: 587–591.

Shi, X., Hudson, J.L., *et al.* (2006). Injectable nanocomposites of single-walled carbon nanotubes and biodegradable polymers for bone tissue engineering. *Biomacromolecules*, 7: 2237–2242.

Shin, H., Jo, S., *et al.* (2003). Biomimetic materials for tissue engineering. *Biomaterials*, 24: 4353–4364.

Silber, J.S., Anderson, D.G., *et al.* (2003). Donor site morbidity after anterior iliac crest bone harvest for single-level anterior cervical discectomy and fusion. *Spine*, 28: 134–139.

Simmons, P.J. and Torok-Storb, B. (1991). Identification of stromal cell precursors in human bone marrow by a novel monoclonal antibody, STRO-1. *Blood*, 78: 55–62.

Simonsen, J.L., Rosada, C., *et al.* (2002). Telomerase expression extends the proliferative life-span and maintains the osteogenic potential of human bone marrow stromal cells. *Nat Biotechnol*, 20: 592–596.

Stewart, K., Walsh, S., *et al.* (1999). Further characterization of cells expressing STRO-1 in cultures of adult human bone marrow stromal cells. *J Bone Miner Res*, 14: 1345–1356.

Street, J., Winter, D., *et al.* (2000). Is human fracture hematoma inherently angiogenic? *Clin Orthop Relat Res*, 378: 224–237.

Sun, J.S., Wu, S.Y., *et al.* (2005). The role of muscle-derived stem cells in bone tissue engineering. *Biomaterials*, 26: 3953–3960.

Sun, Q., Chen, R.R., *et al.* (2005). Sustained vascular endothelial growth factor delivery enhances angiogenesis and perfusion in ischemic hind limb. *Pharm Res*, 22: 1110–1116.

Takagi, K. and Urist, M.R. (1982). The role of bone marrow in bone morphogenetic protein-induced repair of femoral massive diaphyseal defects. *Clin Orthop*, 171: 224–231.

Takushima, A., Kitano, Y., *et al.* (1998). Osteogenic potential of cultured periosteal cells in a distracted bone gap in rabbits. *J Surg Res*, 78: 68–77.

Tarkka, T., *et al.* (2003). Adenoviral VEGF-A gene transfer induces angiogenesis and promotes bone formation in healing osseous tissues. *J Gene Med*, 5(7): 560–566.

Taton, T.A. (2001). Nanotechnology. Boning up on biology. *Nature*, 412: 491–492.

Temenoff, J.S. and Mikos, A.G. (2000). Injectable biodegradable materials for orthopedic tissue engineering. *Biomaterials*, 21: 2405–2412.

Temenoff, J.S., Lu, L., *et al.* (2000). *Bone-Tissue Engineering Using Synthetic Biodegradable Polymer Scaffolds.* Toronto: EM Squared.

Thaller, S.R., Dart, A., *et al.* (1993). The effects of insulin-like growth factor-1 on critical-size calvarial defects in Sprague-Dawley rats. *Ann Plast Surg*, 31: 429–433.

Timmer, M.D., Ambrose, C.G., *et al.* (2003a). *In vitro* degradation of polymeric networks of poly(propylene fumarate) and the crosslinking macromer poly(propylene fumarate)-diacrylate. *Biomaterials*, 24: 571–577.

Timmer, M.D., Shin, H., *et al.* (2003b). *In vitro* cytotoxicity of injectable and biodegradable poly(propylene fumarate)-based networks: unreacted macromers, cross-linked networks, and degradation products. *Biomacromolecules*, 4: 1026–1033.

Trueta, J. and Buhr A.J. (1963). The vascular contribution to osteogenesis. V. the vasculature supplying the epiphysial cartilage in rachitic rats. *J Bone Joint Surg Br*, 45: 572–581.

Urist, M.R. (1965). Bone: formation by autoinduction. *Science*, 150: 893–899.

Urist, M.R., Silverman, B.F., *et al.* (1967). The bone induction principle. *Clin Orthop Relat Res*, 53: 243–283.

Urist, M.R., Nogami, H., *et al.* (1976). A bone morphogenetic polypeptide. *Calcif Tissue Res*, 21(Suppl): 81–87.

Vacanti, C.A., Bonassar, L.J., *et al.* (2001). Replacement of an avulsed phalanx with tissue-engineered bone. *N Engl J Med*, 344: 1511–1514.

Vaccaro, A.R., Patel, T., *et al.* (2004). A pilot study evaluating the safety and efficacy of OP-1 Putty (rhBMP-7) as a replacement for iliac crest autograft in posterolateral lumbar arthrodesis for degenerative spondylolisthesis. *Spine*, 29: 1885–1892.

Vaccaro, A.R., Anderson, D.G., *et al.* (2005). Comparison of OP-1 Putty (rhBMP-7) to iliac crest autograft for posterolateral lumbar arthrodesis: a minimum 2-year follow-up pilot study. *Spine*, 30: 2709–2716.

Vehof, J.W., Takita, H., *et al.* (2002). Histological characterization of the early stages of bone morphogenetic protein-induced osteogenesis. *J Biomed Mater Res*, 61: 440–449.

Verma, I.M. and Somia, N. (1997). Gene therapy – promises, problems and prospects. *Nature*, 389: 239–242.

Viateau, V., Guillemin, G., *et al.* (2004). A technique for creating critical-size defects in the metatarsus of sheep for use in investigation of healing of long-bone defects. *Am J Vet Res*, 65: 1653–1657.

Villars, F., Bordenave, L., *et al.* (2000). Effect of human endothelial cells on human bone marrow stromal cell phenotype: role of VEGF? *J Cell Biochem*, 79: 672–685.

Walsh, S., Jordan, G.R., *et al.* (2001). High concentrations of dexamethasone suppress the proliferation but not the differentiation or further maturation of osteoblast precursors *in vitro*: relevance to glucocorticoid-induced osteoporosis. *Rheumatology*, 40: 74–83.

Wan, C., He, Q., *et al.* (2006). Allogenic peripheral blood derived mesenchymal stem cells (MSCs) enhance bone regeneration in rabbit ulna critical-sized bone defect model. *J Orthop Res*, 24: 610–618.

Wan, M. and Cao, X. (2005). BMP signaling in skeletal development. *Biochem Biophys Res Commun*, 328: 651–657.

Warnke, P.H., Wiltfang, J., *et al.* (2006). Man as living bioreactor: fate of an exogenously prepared customized tissue-engineered mandible. *Biomaterials*, 27: 3163–3167.

Wenger, A., Stahl, A., *et al.* (2004). Modulation of *in vitro* angiogenesis in a three-dimensional spheroidal coculture model for bone tissue engineering. *Tissue Eng*, 10(9–10): 1536–1547.

Westen, H. and Bainton, D.F. (1979). Association of alkaline-phosphatase-positive reticulum cells in bone marrow with granulocytic precursors. *J Exp Med*, 150: 919–937.

Williams, D. (2003). Revisiting the definition of biocompatibility. *Med Device Technol*, 14: 10–13.

Williams, D.F. (1999). *The Williams' Dictionary of Biomaterials.* Liverpool: Liverpool University Press.

Wilson, C., Dhert, W., *et al.* (2002). Evaluating 3D bone tissue engineered constructs with different seeding densities using the alamar blue assay and the effect on *in vivo* bone formation. *J Mater Sci Mater Med*, 13: 1265–1269.

Wilson, C.E., de Bruijn, J.D., *et al.* (2004). Design and fabrication of standardized hydroxyapatite scaffolds with a defined macro-architecture by rapid prototyping for bone-tissue-engineering research. *J Biomed Mater Res A*, 68: 123–132.

Wilson, C.E., Kruyt, M.C., *et al.* (2006). A new *in vivo* screening model for posterior spinal bone formation: comparison of ten calcium phosphate ceramic material treatments. *Biomaterials*, 27: 302–314.

Winter, G.D. and Simpson B.J. (1969). Heterotopic bone formed in a synthetic sponge in the skin of young pigs. *Nature*, 223(201): 88–90.

Wozney, J.M. (1992). The bone morphogenetic protein family and osteogenesis. *Mol Reprod Dev*, 32: 160–167.

Wozney, J.M., Rosen, V., *et al.* (1988). Novel regulators of bone formation: molecular clones and activities. *Science*, 242: 1528–1534.

Yaszemski, M.J., Payne, R.G., *et al.* (1996). Evolution of bone transplantation: molecular, cellular and tissue strategies to engineer human bone. *Biomaterials*, 17: 175–185.

Yoshimoto, H., Shin, Y.M., *et al.* (2003). A biodegradable nanofiber scaffold by electrospinning and its potential for bone tissue engineering. *Biomaterials*, 24: 2077–2082.

Yuan, H. (2001). *Osteoinduction of calcium phosphates*, Faculty of Medicine: Biomaterials Research Group, 2001, Leiden, The Netherlands, p. 156.

Yuan, H., *et al.* (1999). A preliminary study on osteoinduction of two kinds of calcium phosphate ceramics. *Biomaterials*, 20(19): 1799–1806.

Yuan, H., van Blitterswijk, C.A., *et al.* (2006). A comparison of bone formation in biphasic calcium phosphate (BCP) and hydroxyapatite (HA) implanted in muscle and bone of dogs at different time periods. *J Biomed Mater Res A*, 78: 139–147.

Yuan, J., Cui, L., *et al.* (2007). Repair of canine mandibular bone defects with bone marrow stromal cells and porous beta-tricalcium phosphate. *Biomaterials*, 28: 1005–1013.

Zdeblick, T.A., Helm, S.E., *et al.* (2001) Laparoscopic approach with tapered metal cages: rhBMP-2 vs. autograft. Presented at *Annual Meeting of the North American Spine Society*, Seattle, WA.

Zhang, X. (1991). *A Study of Prorous Block HA Ceramics and its Osteogenesis*. Amsterdam: Elsevier Science.

Zhu, L., Liu, W., *et al.* (2006). Tissue-engineered bone repair of goat-femur defects with osteogenically induced bone marrow stromal cells. *Tissue Eng*, 12: 423–433.

Chapter 20
Tissue engineering of the nervous system

Paul Dalton, Alan Harvey, Martin Oudega and Giles Plant

Chapter objectives:

- To understand the different regenerative capacities of the PNS and CNS, and identify the pros and cons of using different cell types for transplantation into the injured CNS
- To appreciate the critical gap length in peripheral nerves, and how therapies can increase this distance
- To gain knowledge of different bioengineering strategies being used to promote peripheral nerve repair
- To recognize the importance of growth factors in regenerative events in the nervous system, and the importance of modulating both growth inhibitory and growth promoting events after brain and spinal cord injury

- To understand the potential role of genetic modification of cells for use in biohybrid implants in the nervous system, and the limitations of animal models for human neurotrauma and neurodegenerative disease
- To recognize the different types of spinal cord injury and how this affects potential treatment strategies
- To understand the importance of using biologically relevant peptide sequences in polymer scaffolds
- To gain a knowledge of the potential role of gene therapy and transplantation in the treatment of retinal diseases

"When peripheral nerve segments were used as 'bridges' between the medulla and spinal cord, axons from neurons at both these levels grew approximately 30 millimeters."

From the abstract of David and Aguayo (1981), an article that altered the widely held view that regeneration in the spinal cord was not possible.

20.1 Introduction

The loss of touch, sight, hearing or movement often results from diseases or injuries to the nervous system. These injuries are not only devastating to the individual, but have substantial societal implications and costs. The retina, cochlear, spinal cord, brain and peripheral nerve have very different molecular and cellular environments, and tissue engineering within each requires different strategies. The limited regeneration in the central nervous system (CNS) is primarily due to the different cellular and molecular environment, and less to the internal properties of the corresponding neurons. Scaffolds and matrices in the nervous system, therefore, provide a substrate for axonal growth that has efficacy even in the inhibitory CNS environment. The focus of tissue engineering in the peripheral nerve has been the reconnection of severed nerves separated from their targets or from the distal nerve segments. Here, the therapeutic approach demands the implantation of artificial structures that support and guide spontaneously regenerating axons.

In this chapter, the authors will outline the various issues and approaches associated with tissue engineering in the nervous system. The molecular and cellular environment demands particular emphasis, as does the natural anatomical structure and relevant animal models for repair. However, given the complex interactions between neurons, glia and the immune system *in vivo* it is important to note that tissue engineering to date has only encompassed limited areas of neuroscience. The induction of neurite outgrowth and nerve regeneration with neurotrophic factors and the use of ECM proteins/peptides have attracted the most attention, while materials for neuroprotection strategies has so far been more limited.

20.2 Peripheral nerve

Peripheral nerve injuries (PNI) can lead to lifetime loss of function and disfigurement. Several hundred thousand such traumatic injuries occur each year in Europe and the US alone. The peripheral nerve, when severed, is capable of a substantial amount of regeneration. The peripheral nerve contains only the axon part of the neuron and one could consider the peripheral nerve trunk as a protective structure for axons (Figure 20.1). The cell bodies of sensory neurons are located in structures just next to the spinal cord (dorsal root ganglia (DRG)), or in cranial ganglia, while the cell bodies of motor neurons are within the CNS (spinal cord or brainstem) (Figure 20.2). Regenerating axons are accurately guided for long distances along naturally occurring Bands of Büngner once the nerve defect is bridged. The most popular approach in peripheral nerve tissue engineering involves *in vivo* implantation of artificial scaffolds and substrates that will guide naturally regenerating axons to the distal segment.

20.2.1 Peripheral nerve anatomy

Peripheral nerves are discrete trunks filled with sensory and motor axons, and support cells such as Schwann cells and fibroblasts. Due to limb movements and the resulting tensile and compressive stresses, the epineurium provides a protective structure to the axons. The epineurium is a sheath of loose fibro-collagenous tissue that binds individual fascicles into one nerve trunk (Figure 20.1). Inside these fascicles are the axons, myelinated by Schwann cells. Peripheral spinal nerves originate at the dorsal or

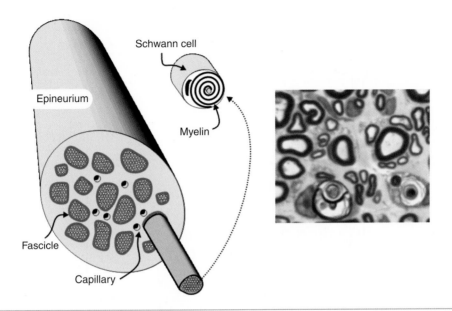

Figure 20.1 Anatomy of the peripheral nerve. The axons from the selected fascicle are indicated in yellow. The Schwann cell myelin sheath is drawn to demonstrate the wrapping of the cell body around the axon; however, the myelin is compact and only a single dark ring is typically seen with histology (right). This image is courtesy of Dr Rajiv Midha (University of Calgary) and shows myelinated axons regenerated within a pHEMA-*co*-methyl methacrylate nerve guide implanted in a rat sciatic nerve.

ventral roots of the spinal cord while cranial nerves originate from the brainstem. Dorsal roots contain sensory axons, carrying signals into the CNS; ventral roots carry motor signals from CNS-originating neurons to muscles and glands (Figure 20.2). Cranial nerves can be purely sensory or motor, or may contain both types of axons.

20.2.2 PNI

Regeneration of severed axons requires that the parent cell bodies survive the initial trauma. Injuries closer to the spinal cord or brainstem are more likely to cause neuronal death and are therefore more complex than an injury at the distal periphery. After initial PNI, the proximal nerve stump will swell, but undergoes minimal damage compared to the distal end. After a nerve is severed, the distal portion degenerates due to protease activity termed 'Wallerian

Degeneration' (reviewed by Koeppen, 2004). In the distal end, the axonal cytoskeleton breaks down and Schwann cells shed their myelin lipids. Macrophages clear myelin and other debris – a critical event that increases the potential for nerve regeneration. This is because myelin-associated glycoprotein (MAG) present in peripheral nervous system (PNS) myelin inhibits peripheral axon regeneration (Shen *et al.*, 1998; Torigoe and Lundborg, 1998), and successful PNS regeneration occurs only after myelin is cleared and myelin-specific proteins are downregulated by Schwann cells. Delay of myelin clearance impedes peripheral nerve regeneration unless the MAG gene is disrupted and the myelin is therefore MAG-free (Schafer *et al.*, 1996). Importantly, Schwann cells in the distal portion of the severed nerve also begin to secrete specific factors that stimulate axon regrowth. After debris clearance, regeneration begins at the proximal end and continues toward the distal stump.

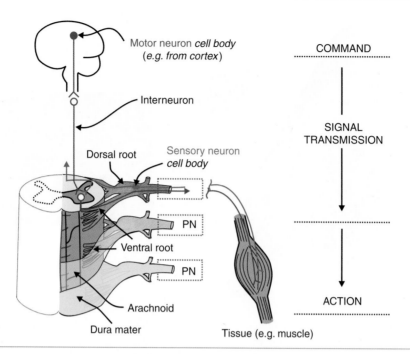

Figure 20.2 Origins of PNS axons. The junction between the spinal cord and the PNS, with schematically drawn cell bodies of sensory neurons (green) and motor neurons (red), and their corresponding neurites. For the length of the peripheral nerve (PN), ony axons are present in the tissue. The motor neurons are typically in the brain; however, other intermediate neurons located within the grey matter can play a role in transferring the electrical impulse. The right-hand side text and arrows indicate the direction of the signal for a motor command – a sensory signal is not shown. Dr Ian Galea's assistance with this image is appreciated.

Accurate alignment of the original fascicles is critical to microsurgical repair (Figure 20.3). In the distal nerve, form inside tubular structures and Bands of Büngner (see boxed text, page 621) play a crucial role in successful peripheral nerve regeneration (Scherer and Salzer, 2001). Functional reinnervation in humans occurs at a rate of about 2–5 mm/day – significant injuries can take many months to heal and it is uncommon for full restoration of sensory or motor function to result.

20.2.3 Autologous nerve grafts (autograft)

PNI most commonly results from blunt trauma or from penetrating missiles such as bullets or other objects, but it is also associated with fractures and fracture-dislocations. Crush injuries are therefore more common than nerve transections. When nerve endings are unable to be rejoined without tension, then a bridging section of nerve is used and two end-to-end sutures are performed. The crushed section of nerve is cut, removed and replaced by a nerve taken from another (less important) site, typically the sural nerve (from the back of the leg) (Evans, 2001). The autograft works relatively well in practice and is the gold standard upon which all alternative therapies are judged. However, a second surgery is required to obtain the bridging nerve and there is loss of function at the donor site, often leading to detrimental changes such as scarring and the possible formation

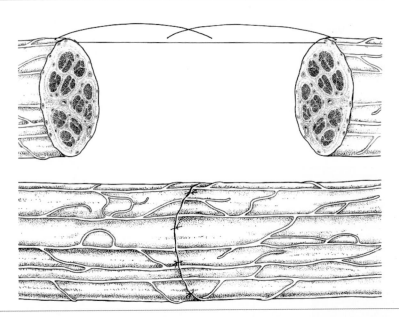

Figure 20.3 End-to-end suturing of peripheral nerves. Alignment of the fascicles is critical in successful regeneration and microsurgical techniques have been developed to optimize this surgery. Reprinted from Lundborg, G. (2000). *J Hand Surg*, 25A: 391–414, with permission from the American Society for Surgery of the Hand.

of painful neuromas. Furthermore, donor nerves are often of small caliber and limited in number. These problems drive the search for a tissue engineering alternative to this treatment.

20.2.4 Use of tubes (nerve guides) in the PNS

Off-the-shelf alternatives for small injury gaps (around 3–10 mm) are nerve guides which function by enclosing the therapy area. For nerve gaps up to 10 mm, hollow tubes (nerve guides or nerve guidance channels) have successful outcomes to the level of the autograft in various animal and comparable clinical situations. Many different nerve guides have been implanted and are reviewed elsewhere (Heath & Rutkowski, 1998; Lundborg, 2004; Meek & Coert, 2002); however, the silicone nerve guide has been the most frequently implanted. Degradable collagen nerve guides (Neuragen™; Integra Life Sciences) have been

approved for use by the Food and Drug Administration and are being clinically used to treat paralysis due to PNI. Nerve guides have followed many tissue engineering-based strategies, such as matrices, scaffolds, cell transplants and/or drug delivering therapies.

The general sequence of regeneration within the most-studied silicone nerve guides is schematically drawn in Figure 20.4. After implantation, fluid and cytokines (including neurotrophic factors) primarily generated by the Schwann cells build up within the nerve guide (Figure 20.4a). Within 7 days an oriented fibrin scaffold is formed and cells – primarily fibroblasts – penetrate into the nerve guide while axon debris is removed by Schwann cells and macrophages in the distal end (Figure 20.4b). From 7 to 14 days, regenerating axons and Schwann cells penetrate into the nerve guide using the newly formed fibrin scaffold as a substrate (Figure 20.4c). Schwann cells in the distal component begin to form part of the Bands of Büngner – long cellular tracts from the severed distal nerve ending to the final muscular targets. Between

(a)

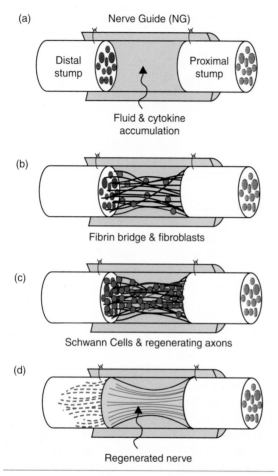

Figure 20.4 Sequence of events in empty silicone nerve guides. Initially, fluid and cytokines fill up the nerve guide (a), followed by the formation of a fibrin matrix inside the lumen which initially supports invading fibroblasts (blue) (b), followed by Schwann cells (green) and axons (red) (c), which eventually penetrate to the distal stump. The regenerated nerve usually is thinner in the middle of the bridge (yellow).

14 and 56 days, Schwann cells myelinate the regenerating axons that exit the nerve guide, directed by the Bands of Büngner in the distal nerve stump to their targets.

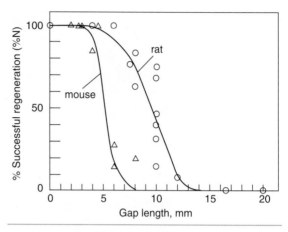

Figure 20.5 The chance of successful reinnervation with nerve guides is dramatically reduced once an injury gap reaches a certain value. This length, termed the critical gap length, L_C, is where successful regeneration occurs 50% of the time. Reprinted from Zhang, M. and Yannas (2005). *Adv Biochem Eng Biotechnol*, 94: 67–89, with permission from Springer-Verlag.

20.2.5 Critical gap length

A critical gap length (L_C) (Zhang & Yannas, 2005) limits successful peripheral nerve regeneration using nerve guides. When the transected peripheral nerve approaches this experimentally determined length the chances of successful regeneration within nerve guides are diminished. Figure 20.5 shows the frequency of successful regeneration (%N) within rat and murine models using silicone nerve guides, and L_C (when %$N = 50$) has been determined to be 9.7 ± 1.8 mm for rats and 5.4 ± 1.0 mm for mice. The L_C is most likely greater in primates with gaps of 30 mm able to be bridged (Mackinnon *et al.*, 1985). The introduction of matrices, fibers and cells within nerve guides increases the L_C to varying degrees, and it is this parameter that is important in assessing therapeutic success. The concept of a shift length (ΔL) has been proposed, which is the difference between the L_C of the experimental device and the L_C of the standard tube. The introduction of a therapy will generally increase the L_C; Figure 20.6 has been drawn based upon this approach of assessing additional therapies.

20.2.6 Matrices and scaffolds

Both matrices and scaffolds are supporting structures for tissue engineering – full definitions are given in Chapter 9. Cells must penetrate and migrate through the matrix, while scaffolds have distinct interconnecting open pores. Peripheral nerve tissue engineering matrices have either oriented or random structures, while scaffolds predominantly have an oriented morphology.

20.2.6.1 Matrices

The L_C generally increases for nerve guides when they contain matrices within the lumen (Figure 20.6b). Effective matrices have similar mechanical properties; they are all weak, viscoelastic hydrogels with high water content. Collagen matrices cannot be too concentrated as they obstruct cell and axonal penetration; between 1.28 and 2.0 mg/ml has widely been used both *in vivo* and *in vitro* (Ceballos *et al.*, 1999; Labrador *et al.*, 1998; Midha *et al.*, 2003). Matrigel™ is a solubilized basement membrane extracted from mouse sarcoma; while the introduction of Matrigel™ into nerve guides regenerates the nerve more than collagen, it is not a suitable matrix for clinical use (Labrador *et al.*, 1998). Fibrin formed with peptide sequences from laminin demonstrates increased axon counts in nerve guide experiments (Schense *et al.*, 2000).

20.2.6.2 Oriented matrices

In vivo experiments with oriented matrices have a higher L_C value than isotropic matrices of the same material (Figure 20.6c). The magnetic alignment of positive or negative diamagnetic anisotropic molecules such as fibrin and collagen, respectively, elicits guided regeneration *in vitro* and improved *in vivo* regeneration (Ceballos *et al.*, 1999; Dubey *et al.*, 2001; Verdu *et al.*, 2002). Figure 20.7(a) shows fibrin that has been oriented by the application of a magnetic field during gelation. Oriented substratum and techniques to form such structures on a relevant

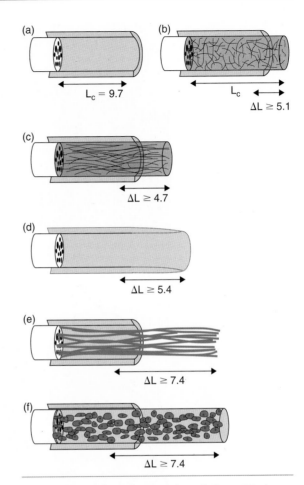

Figure 20.6 Selected examples of the critical gap length gap, L_C, adapted from Yannas and Hill (2004) for the rat sciatic nerve. Empty silicone nerve guides (a) have a L_C of 9.7 mm and the introduction of a collagen–GAG matrix (b) increases the L_C to at least 14.8 mm with a ΔL of 5.1 mm. When oriented fibrin matrices are used (c), this results in a ΔL of at least 4.7 mm *when compared to the same, nonoriented matrices*. Degradable collagen nerve guides (d) have a ΔL of at least 5.4 mm compared to silicone tubes. When polyamide or collagen fibers are introduced (e), the ΔL is at least 7.4 mm compared to empty nerve guides, while Schwann cells (f) result in a similar ΔL value of 7.4 mm, compared to nerve guides filled with phosphate-buffered saline.
Note: the shift length (ΔL) is the result of the therapy – *not necessarily compared to empty nerve guides*.

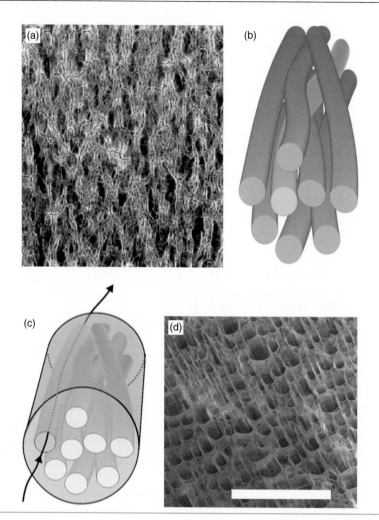

Figure 20.7 Selected approaches for oriented scaffolds/matrices for peripheral nerve repair. Magnetically aligned structures have so far been demonstrated with fibrin (a), collagen and Matrigel™ matrices. Both fibers (b) and channels (c) have guidance properties, while scaffolds formed via uniaxial freezing result in many oriented channels, as shown by a PEG hydrogel (d); scale bar = 100 μm. Image (a) is reprinted from Dubey, N., Letourneau, P.C. and Tranquillo, R.T. (2001). *Biomaterials,* 22: 1065–1075, with permission from Elsevier. Dr Jürgen Groll and Professor Martin Möller (RWTH-Aachen, Germany) provided image (d).

scale are therefore important issues in neural tissue engineering. The overall orientation of the scaffold/matrix should be similar to the gap, i.e. the orientation effect should be of centimeter-length distances.

20.2.6.3 Scaffolds
Scaffolds of collagen–glycosaminoglycan (GAG), when included into the lumen of degradable

Neuragen™ collagen nerve guides, have demonstrated excellent regenerative properties (Chamberlain *et al.*, 1998), while a scaffold/nerve guide based on collagen only has improved on these results (Harley *et al.*, 2004). Fibers will guide axons along their length when their diameter is approximately 250 μm or less (Smeal *et al.*, 2005). Fibers are attractive guidance substrates as they can be bundled together and used as oriented scaffolds (Figure 20.7b). *In vivo* results of nerve guides using fibers show their ability to guide axons over a large injury gap (Yoshii and Oka, 2001). Shaped fibers with specific cross-sections also demonstrate efficacy in guiding nerves, while the stacking ability of shaped fibers is of interest (Lietz *et al.*, 2006). The accurate alignment of the fascicles is imperative for successful repair, however the tortuous path of individual fibres, when bundled together, may not provide an accurate path for nerve regeneration.

The use of channels also results in the directed growth of neurites (Figure 20.7c and d). One simple technique involves the suspension of linear polymeric fibers into a liquid that is subsequently crosslinked, followed by the dissolution of the fiber with a suitable solvent (Flynn *et al.*, 2003). Another approach involves the low-pressure injection molding of poly(DL-lactic-*co*-glycolic acid) (PLGA) foams, resulting in a scaffold that supports neurite outgrowth when delivered with Schwann cells *in vivo* (Hadlock *et al.*, 2000). Such approaches are applicable to a range of materials, essentially converting isotropic materials into scaffolds with guidance properties. The surface modification of such channels significantly increases neurite growth *in vitro* (Yu and Shoichet, 2005). Channels can also be manufactured through isothermal crystallization processing, which have morphology similar to the natural tubules (Figures 20.7d and 20.8 and Box 1). Although these scaffolds have been investigated for use in the spinal cord (Stokols and Tuszynski, 2006), their use in the peripheral nerve is limited.

20.2.7 Schwann cell grafts

When Schwann cells are included within nerve guides, large shift lengths (ΔL) result and significant gap lengths are bridged (Figures 20.6f). Schwann cells can be simply suspended within a matrix inside a nerve guide or cultured on surfaces that are then rolled up and inserted into a nerve guide for regeneration (Hadlock *et al.*, 2001). The use of chimeric grafts in which donor freeze-thawed (cell-free) peripheral nerve sheaths are reconstituted with autologous cultured Schwann cells also shows promise (Cui *et al.*, 2003a). Schwann cell transplantation has great potential as a tool in PNS tissue engineering; however, any eventual clinical solution for bridging injury gaps will have to compete with the relatively low-cost alternative of the autograft.

20.2.8 Neurotrophic factors

After implantation, the lumen of nerve guides rapidly fills up with cytokines and trophic factors, among them the neurotrophic factors. These soluble molecules assist in neuron survival after axotomy, through complex molecular interactions. Neurotrophic factors include the neurotrophins (NTs), neuropoetic cytokines and fibroblast growth factors (FGFs) (Lundborg, 2000; Scherer and Salzer, 2001). The NTs signal through tropomyosin-related kinase (Trk) receptors in the cell membrane in conjunction with low-affinity receptors such as p75. The NTs are nerve growth factor (NGF), NT-3, NT-4/5 and brain-derived neurotrophic factor (BDNF). After injury, NGF is elevated in the distal nerve, and acting through TrkA assists in the survival and outgrowth of a proportion of sensory neurons. Other sensory neurons express different receptors and are supported by NT-3 or by another factor – glial-derived neurotrophic factor (GDNF). Motor neurons have TrkB and TrkC receptors, and NGF therefore has no effect on neuron survival or axon outgrowth of these neurons. BDNF and NT-3, respectively, signal through TrkB and TrkC receptors that are present on motor neurons and increase neuron viability after axotomy. Increased brainstem and spinal cord motor neuron survival has also been reported using GDNF.

The neuropoetic cytokines include ciliary neurotrophic factor (CNTF) and interleukin-6. Studies on neurotrophic factor production after nerve guide implantation reveal that CNTF and NGF are rapidly elevated to maximum levels only 3–6 hours post-implantation. Application of CNTF increases motor neuron viability and increases the number

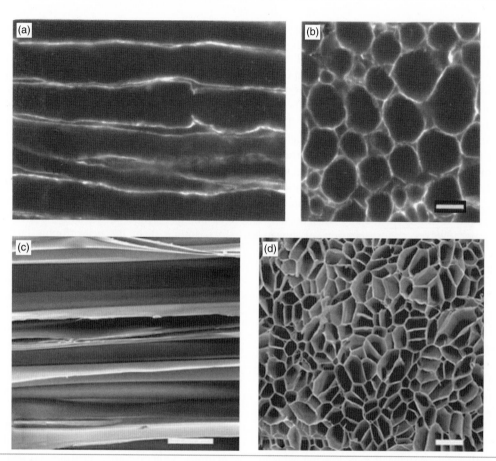

Figure 20.8 Natural lamina tubules that guide regenerating axons in the distal nerve to their targets (a and b; scale bar = 10 μm). Although the pore sizes are different, morphological similarities between these tubules and scaffolds formed with uniaxial freezing (c and d; scale bar = 200 μm) are apparent. The images of the tubules are provided by Dr Steve Scherer (University of Pennsylvania), and modified from Scherer, S.S. and Salzer, J.L. (2001). *Glial Cell Development*. Oxford Press. pp. 299–330. reprinted with permission from Oxford University Press. The images of oriented agarose scaffolds are reprinted from Stokols, S. and Tuszynski, M. (2004). *Biomaterials*, 25: 5839–5846, with permission from Elsevier.

of myelinated axons in the distal nerve (Sahenk *et al.*, 1994), while delivery of both CNTF and BDNF through pump systems to the sciatic nerve results in improved functional recovery following end-to-end repair (Lewin *et al.*, 1997).

The addition of acidic FGF to collagen-filled hydrogel nerve guides also increases the quantity and thickness of myelinated axons (Midha *et al.*, 2003). In cell/biomatrix hybrid structures the long-term supply of

specific neurotrophic factors can be obtained by the targeted genetic modification of Schwann cells using viral vectors *in vivo* (Blits *et al.*, 1999) or *ex vivo* (Hu *et al.*, 2005, 2007) (see also 'Genetically modified cells').

Selected methods of delivering neurotrophic factors in combination with acellular nerve guides are schematically drawn in Figure 20.9. The growth factor can be included within the nerve guide,

---

Bands of Büngner

Although the critical gap length for humans is approximately 30 mm, once the regenerating axons bridge this defect there is a natural scaffold prepared in the distal segment of the peripheral nerve. Axons will travel distances well above the critical gap length along Schwann cell-containing tubular basal lamina, which are termed 'Bands of Büngner' after the German neurologist Otto von Büngner (1858–1905) and are often described as 'Schwann cell columns'. After PNI, the cytoskeleton of the severed axon is degraded, and Schwann cells, previously wrapped around the axons, dispense with their myelin sheath and clear debris, adopting a nonmyelinating phenotype. The Schwann cells partly clear the myelin while increasing numbers of macrophages in the distal stump complete the task. Throughout such Wallerian degeneration the microtubules of basal lamina that support these cells remain and are specific tracts that channel the regenerating axons to the final targets.

With the development and optimization of microsurgical techniques for peripheral nerve repair, the success of end-to-end suturing greatly depends on the alignment of the fascicles. Therefore, the penetration of a sensory axon down a tubule previously filled with an axon from a motor neuron (and *vice versa*) is an ineffective (but frequent) result. Whether the regenerating axons from a transection selectively enter their original Schwann cell tubes is problematic. Careful microsurgery, binding the individual fascicles together, will maximize the functional recovery; however, axonal misdirection is frequent. Nevertheless, the Bands of Büngner are an excellent substrate for regeneration over long distances.

The tubules in which the Bands of Büngner form have morphologies similar to oriented tissue engineering scaffolds, particularly those formed with uniaxial freezing. As demonstrated in Figure 20.8, the channels of both structures are long, continuous and relatively noninterconnected. While the application of such scaffolds for bridging nerve defects is still in its developmental stages, the natural basal lamina tubules, or Bands of Büngner, offer an insight into scaffolds associated with guided regeneration of axons over long distances.

---

homogeneously, or specifically at the lumen to inwardly direct the release (Fine *et al.*, 2002; Piotrowicz and Shoichet, 2006). The matrix in the lumen of the nerve guide is also a source of growth factors – either simply incorporated (Midha *et al.*, 2003) or as part of a noncovalent controlled release system (Lee *et al.*, 2003). Figure 20.9(e) shows heparin-binding growth factors (e.g. NGF, NT-3) are locally delivered from matrices as the axons regenerate. The heparin-incorporated fibrin matrix significantly retards the release rate of a growth factor such as NGF. Axonal regeneration in such delivery systems is dose-dependent, while histomorphometric analysis demonstrates no statistical difference between such matrix-filled nerve guides and autograft controls (Lee *et al.*, 2003).

The potential permutations and concentrations of neurotrophic factors are many and knowledge of their clinical toxicity or carcinogenic nature is still limited. Cocktails of different neurotrophic factors may result in enhanced neuron survival or increased axon regrowth. Although neurotrophic factors have been restricted in their clinical use so far, their application for neuron survival and promoting axonal outgrowth for the PNS (and CNS) will likely increase.

20.2.9 PNS summary

The clinical outcome after peripheral nerve repair has not significantly improved in recent decades (Lundborg, 2000). While it is rare for all sensory function to return to a patient, the availability and success of an autograft provides an economic challenge for tissue engineering strategies – the ideal therapy needs

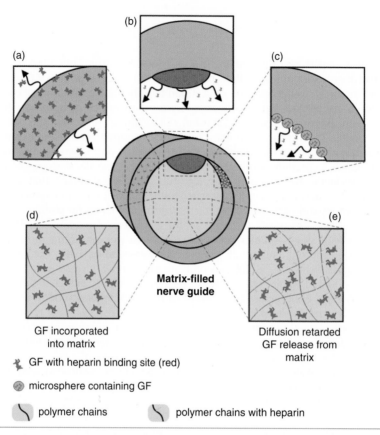

Figure 20.9 A selection of approaches for delivering growth factors into peripheral nerve guides. Growth factors can be incorporated into the nerve guide (a) or introduced closer to the lumen with a rod (b) or as growth factor-containing microspheres (c). The growth factors may be incorporated into the matrix as nonbound (d) or as a noncovalent controlled release system (e).

a relatively low-cost solution. The most successful tissue engineering alternatives to the autograft to date include strategies involving Schwann cell grafts, bundled fibers or channels, or oriented scaffolds/matrices placed within nerve guides.

20.3 CNS: spinal cord

The spinal cord has considerable impediments to regeneration. Damage to the spinal cord results in the immediate death of neural cells and disrupts the blood supply to the injury area. A major difference with the PNS is that in the spinal cord the neuronal

cell bodies are damaged, whereas in the PNS only axons are injured. Another difference is that in the spinal cord a growth-inhibitory glial scar develops that impedes any effective axon growth activity. This scar consists mainly of reactive astrocytes and fibroblasts, which express growth-inhibitory molecules such as chondroitin sulfate proteoglycans and semaphorins, and thus forms a physical and chemical barrier for regenerating axons. There is also secondary neuronal and glial degeneration as a delayed consequence of the initial injury. Additionally, with or without injury, white matter, which contains myelin and myelin-associated axonal growth inhibitors, prevents significant axonal regenerative response in the adult spinal cord and brain. Tissue engineering in the spinal cord has focused on cellular substrates that replace lost spinal cells and/or provides a terrain for severed axons to grow across the site of injury. Acellular substrates (scaffolds/matrices) also elicit an axonal regeneration response, but appear to be less effective than cellular transplants. The mechanisms of reduced cyst formation and astrocytic scarring associated with scaffold implantation are unknown.

20.3.1 Summary of anatomy and injury response

The middle of the spinal cord consists of a butterfly-shaped grey matter containing the cell bodies of neurons and glial cells and numerous blood vessels. The white matter surrounding this structure contains mainly axons and support cells (termed glia) such as oligodendrocytes (providing myelin insulation to the axons), astrocytes and microglia (immune cells). Traumatic spinal cord injury (SCI) results in disruption of both sensory and motor axons, and the lack of continuity in the spinal cord have devastating consequences for the injured patients. The initial injury response following bruising or transection of the spinal cord involves a primary event in which cells and tissue are crushed or cut by the injury and cells undergo necrosis. Following this primary injury there are secondary events that involve influx of Ca^{2+}, release of inflammatory cytokines, oxidative stress and fluid build-up within the lesion site. Inflammatory, or 'secondary events', cause neurons in the grey matter, and oligodendrocytes in the white matter, to die from apoptosis. Death of oligodendrocytes and subsequent loss of the myelin sheath interferes with the speed and reliability of electrical conduction in any surviving axons, significantly contributing to a sustained and long-lasting loss of function. Successful spinal cord repair requires all these issues to be tackled, which is a complex and difficult task.

20.3.2 SCI injury models

There are a number of experimental animal models of SCI, such as complete transection, hemisection and contusion/compression injuries. Different models are best suited for the measurement of tissue sparing (contusion), axon regeneration (complete transection) or sprouting of intact fibers (hemisection). Most studies to date involve rodents (rats or mice) (see also 'Animal models').

20.3.2.1 Contusion animal model
Most (73%) human SCI is the result of a compression or contusion injury (Bunge et al., 1993; Kakulas, 1999), often resulting in the formation of a cavity or cyst as schematically shown in Figure 20.10. Experimentally, such injuries are modeled either by transient compression of the spinal cord using a weight drop method or by displacement of the spinal cord. If performed appropriately, both approaches result in reproducible injuries of graded severity. With a contusion injury, analysis of the axon regeneration response is complex due to the presence of spared axons and their collateral sprouts. To repair the contused spinal cord, different cell types have been introduced directly into the contusion cavity such as Schwann cells (Figure 20.11; Hill et al., 2006; Pearse et al., 2004; Takami et al., 2002), olfactory ensheathing glia (OEG; Barakat et al., 2005; Feron et al., 2005; Plant et al., 2003), fibroblasts (McTigue et al., 1998; Mitsui et al., 2005) and stem cells (Keirstead et al., 2005; Liu et al., 2000; Mitsui et al., 2005). In general, cell grafting into the contusion cavity likely elicits a neuroprotective effect as well as an axon regeneration response, often resulting in some degree of functional recovery (Plant et al., 2003; Takami et al., 2002).

20.3.2.2 Hemisection experiments
Approximately 27% of human SCI results from a laceration with disruption of the dura mater (Bunge et al., 1993; Kakulas, 1999). This particular type of injury can be mimicked in the laboratory using a surgical microknife or microscissors as shown in Figure 20.10. Although with a partial (hemi) section

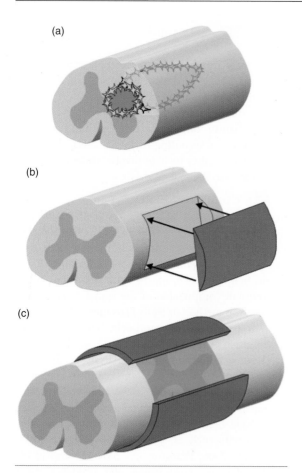

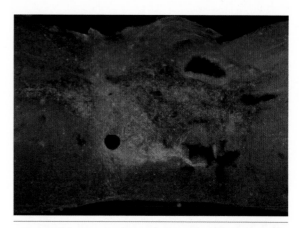

Figure 20.11 Transplantation of Schwann cells (red) after a contusion injury (moderate injury with New York University device), with the green labeling the p75 receptor. The Schwann cells could be visualized by *ex vivo* gene therapy using a lentivirus-encoding DSRED-2. The cells were introduced into a 2-month-old spinal cord, 7 days post-injury. The rostral end of the nerve is to the left of the image.

Figure 20.10 Contusion (a), hemisection (b) and full-transection (c) models for SCI. With contusion injuries, the formation of an astrocytic scar (often termed glial scar) presents a barrier to the axons which lines a fluid-filled cyst. The reactive astrocytes (yellow) lining the cyst are a formidable barrier to regenerating axons. In the surgically created hemisection model, the space can be filled with a matrix/scaffold (dark blue), either injected to fill the lesion or preformed and implanted (shown). With full-transection models, the cord is severed and a scaffold is positioned between or the stumps are inserted into a nerve guide (green and sectioned to expose lumen) filled with a therapeutic agent (light blue). In some species, the spinal cord will retract after full transection, creating a gap to fill with cells/matrix/scaffold.

the loss of tissue and function is limited, over time injury-induced secondary tissue loss may add to the overall destruction. With a hemisection SCI model, as with a contusion injury, the presence of spared axons and the formation of axon collaterals make a proper analysis of any transplant-induced regeneration response difficult. To repair the spinal cord following a hemisection, cells/scaffolds can be transplanted as a preformed cable that bridges the injury site (Bamber *et al.*, 2001) or as a suspension/*in situ* hydrogel that fills the lesion site (Lu *et al.*, 2004; Jain *et al.*, 2006). An alternative approach is the transplantation of pieces of peripheral nerve (Oudega and Hagg, 1996). As with other types of SCI, the grafted cells or tissue elicit axonal growth in a hemisection; however, the responding axons do not exit the transplant without additional interventions such as an increase in the levels of neurotrophic factors in the spinal tissue beyond the transplant using osmotic minipumps or direct injection of neurotrophic factors (Bamber *et al.*, 2001; Lu *et al.*, 2004; Oudega and Hagg, 1996).

20.3.2.3 Full-transection experiments

Full-transection experiments are not common due to surgical complications such as instability of the spinal column and the difficulty in animal care. However, this type of lesion does provide absolute evidence, when performed correctly, of true axon regeneration. A number of studies have been performed utilizing multiple peripheral nerve bridges contained in fibrin with acidic FGF (Cheng *et al.*, 1996). Cheng *et al.* reported long-distance regeneration of important motor tracts, but unfortunately this study has not been replicated with the same success. While poly(acrylonitrile-*co*-vinylchloride) [P(AN-VC)] nerve guides have been used for full transection (see Classical Experiment), more recent investigations using soft, flexible hydrogel nerve guides in the fully transected cord demonstrate improved regenerative capacity (Spilker *et al.*, 2001; Tsai *et al.* 2004).

20.3.2.4 Gender, Genetic Background and Age

Accumulating evidence suggests that there are gender differences in the cascade of degenerative processes and functional outcomes after traumatic and ischemic CNS injuries (Bramlett, 2005; Roof and Hall, 2000). The presence of estrogens and progestins likely plays an important role in the greater neuroprotection seen in females versus males after injury, but the underlying biological mechanisms are not fully understood. In rodent studies, there are also important strain-specific differences in degenerative and regenerative responses after CNS injury (Cui *et al.*, 2007; Dimou *et al.*, 2006; Kigerl *et al.*, 2006). Finally, age is also an important factor in the pathophysiology of CNS and PNS injuries, and in the ability to recover. It is important to realize that animal models do not always reflect these various differences, and it may well be that for a particular type of injury different animal models should be developed that represent such differences.

20.3.3 Cell transplantation

Cell transplants into the spinal cord are used as a 'bridge' to encourage injured axons to grow through

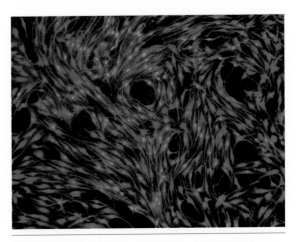

Figure 20.12 Schwann cells in culture. Schwann cells are good candidates to use as bridging grafts in the CNS, producing (naturally or after genetic modification) a variety of growth-promoting molecules. The development of culture techniques has resulted in the isolation and purification of Schwann cell populations from the peripheral nerve.

areas of damage towards and into their target areas. Various cell types have been tested; most recently stem cells derived from either embryonic or adult tissue have been investigated.

20.3.3.1 Schwann cells

Schwann cells are the myelin-forming cells of the PNS (see 'Peripheral nerve anatomy'). After PNI, Schwann cells facilitate repair by expressing axonal growth-promoting molecules, including neurotrophic factors, surface cell adhesion molecules and extracellular matrix. The ability to produce a variety of growth-promoting molecules makes the Schwann cell a potential candidate for providing a growth substrate for injured CNS axons (Figure 20.12). In the injured spinal cord, Schwann cell transplants elicit axon regeneration (see Classical Experiment) (Bunge and Pearse, 2003; Pearse *et al.*, 2004; Oudega *et al.*, 2005, 2007; Raisman, 1997; Xu *et al.*, 1997, 1999). Schwann cells also form myelin sheaths around CNS axons allowing signal conduction in regenerated CNS axons (Imaizumi *et al.*, 2000). In addition to axon

Classical Experiment
Schwann cell and OEG transplantation in P(AN-VC) nerve guides

The use of Schwann cell transplants within nerve guides was pioneered in the Bunge laboratory at The Miami Project to Cure Paralysis during the 1990s. Schwann cells can be isolated and cultured from peripheral nerves (and are therefore autologous), and elicit an axonal regeneration and myelination response in the CNS. This was elegantly demonstrated by Xu *et al.* (1997), who transplanted a P(AN-VC) nerve guide seeded with Schwann cells and Matrigel™ into the completely transected adult rat thoracic spinal cord. Several weeks after transplantation, myelinated and unmyelinated axons were present within the Schwann cell cable that connected the opposing cord stumps (Figure 1). These and other experiments demonstrated the potential of the Schwann cell for spinal cord repair approaches. However, the experiments also emphasized a drawback of these types of transplants. The responding axons grew across the Schwann cell graft, but did not exit the graft to grow into the spinal cord tissue beyond, thereby preventing the formation of new synaptic connections, which are necessary to re-establish axonal circuits involved in controlled motor function. Figure 2 visually depicts the axon penetration

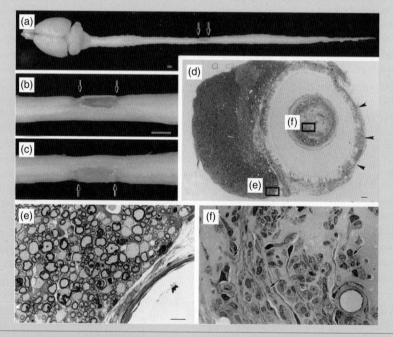

Images of Schwann cell/Matrigel™ transplants in P(AN-VC) nerve guides implanted in hemisection. Dorsal views of brain and spinal cord (a), and closer magnification of the hemisection area and implant (b and c) are shown. A transverse section depicts the nerve guide lumen filled with a tissue bridge (d), and the intact ventral and dorsal horn regions of the spinal cord. Magnifications of (d) show myelinated axons at the outer interface of the nerve guide (e) and smaller caliber axons present within the lumen (f). Modified from Xu, X.M., Zhang, S.-X., Li, H. et al. (1999). Eur J Neurosci, 11: 1723–1740, with permission from the European Neuroscience Association.

and exit of these nerve guides implanted within hemisection models. When Schwann cells are not included within the nerve guides (Matrigel™ only) there is no penetration of axons. While Schwann cell inclusion with the Matrigel™ demonstrates axonal penetration, axons are guided out of the nerve guide via the delivery of BDNF and/or NT-3 into the caudal end of the distal nerve.

OEG share some similarities with both Schwann cells and astrocytes. Most likely OEG enfold the elongating olfactory axons so that they are prevented from recognizing the growth inhibitory cues in the CNS environment. This unique feature of OEG was thought to be beneficial for eliciting axonal growth across an injury site such as scar tissue. However, while OEG will ensheath axons they may not myelinate them; Schwann cells, however, successfully and consistently remyelinate in the spinal cord. In a pivotal experiment by Ramon-Cueto, Plant and Bunge at The Miami Project to Cure Paralysis in 1998, OEG and Schwann cells were combined. A P(AN-VC) nerve guide was filled with Schwann cells mixed in Matrigel™. The Matrigel™ shrunk following overnight gelation at 37°C, during which the Schwann cells aligned longitudinally, and the tubing with the Schwann cell/Matrigel™ cable inside was cut in 10-mm pieces. Next, the adult thoracic spinal cord was exposed and an 8-mm segment (thoracic level 7–9) removed and immediately replaced by the nerve guide with the Schwann cell/Matrigel™ cable inside. OEG were then injected into the rostral and caudal stumps. Animals survived for 8 weeks and anterograde tracing from the cervical spinal cord level was performed before termination of the experiment.

The results showed that the additional injections of OEG either end of the Schwann cell channels allowed greater numbers of axons to exit the Schwann cell bridging grafts (Xu et al., 1997; Plant et al., 2001). Descending axons such as those of the raphe system (located in the brainstem) grew in areas devoid of many Schwann cells, but were seen in areas containing fibroblasts and OEG. Sensory axons also regenerated for long distances – up to 18mm. Such an experiment demonstrates the utility of multiple treatments for a successful spinal cord repair outcome.

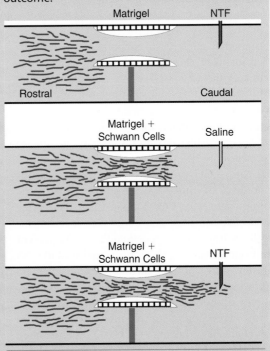

Illustration of the effect of P(AN-VC) nerve guides implanted within hemisections. When Schwann cells are not included within the nerve guides (Matrigel™ only) there is no penetration of axons (red) into the graft. While Schwann cell inclusion with the Matrigel™ demonstrates axonal penetration, the axons can be guided out of the nerve guide via the delivery of BDNF and/or NT-3 into the caudal end of the distal nerve a few millimeters from the implant. Without elevated levels of neurotrophic factor, then axons do not exit the implant. The blue line is a lesion made just prior to tracing.

regeneration, an intraspinal Schwann cell graft has a neuroprotective effect, reducing the secondary injury (Takami *et al.*, 2002). Pearse *et al.* (2004) injected Schwann cells into the cystic cavity while maintaining an increased level of cyclic adenosine monophosphate (cAMP). Such a combinatory approach resulted in improved functional recovery.

20.3.3.2 Olfactory ensheathing glia
The uninjured adult CNS environment is nonpermissive for axon elongation. However, there is one structure in the adult CNS where axon growth does occur – the olfactory bulb. Throughout adulthood, olfactory neurons generate axons that grow into the olfactory bulb to make their appropriate connections. This particular growth response occurs due to the presence of OEG. When grafted into the injured spinal cord, OEG promote the regrowth of axons beyond areas of injury and into distal spinal tissue (Ramon-Cueto *et al.*, 1998; Ramon-Cueto, 2000) (see Classical Experiment). The underlying mechanisms are still unclear, but it is possible that OEG prevent axons from recognizing the nonpermissive adult CNS environment by enfolding them, as they normally do for growing olfactory axons. In addition, OEG may elicit only a moderate astrocytic reaction, resulting in only low expression of growth-inhibitory molecules (Plant *et al.*, 2003; Takami *et al.*, 2002). As described in the Classical Experiment, OEG and Schwann cells have been used in combination, resulting in growth of proprio- and supraspinal axons beyond the lesion gap (Ramon-Cueto *et al.*, 1998), which was not achieved with Schwann cell transplants alone (Xu *et al.*, 1997).

20.3.3.3 Stem cells
The application of stem cells for spinal cord repair has to date been limited by technical and ethical concerns and, sometimes, political restrictions. However, different types of stem cells such as embryonic or adult neural stem cells (McDonald *et al.*, 2004; Picard-Riera *et al.*, 2004) or bone marrow stromal cells (Kamada *et al.*, 2005; Lu *et al.*, 2005; Nandoe Tewarie *et al.*, 2006) have been transplanted into

the injured spinal cord eliciting some neuroprotective and/or regenerative responses. Stem cells generally differentiate into astrocytes, which restrict their effects on axon regeneration and myelination. It may be necessary to differentiate stem cells before/during cell transplantation, as the injury environment does not favor differentiation into neurons or oligodendrocytes. Differentiation of the stem cells prior to implantation may be accomplished by a variety of factors, although the most efficient combination has yet to be determined experimentally. Promising results have been reported with grafts of embryonic stem cell-derived oligodendrocyte precursor cells (Keirstead *et al.*, 2005) and adult brain-derived neural precursor cells (Karimi-Abdolrezaee *et al.*, 2006). It will also be essential to ensure that stem cells do not continue to multiply after transplantation, potentially forming tumor-like structures. Although there is some enthusiasm for treatment of SCI with stem cells, at present there is little understanding as to how their presence aids in the repair of the spinal cord.

20.3.3.4 Genetically modified cells
Gene therapy holds great potential for the treatment of neurotrauma and many types of neurodegenerative CNS disease. Delivery of therapeutic molecules by gene therapy has been shown to stop or delay the progression of neurodegeneration (Tuszynski, 2002) and stimulate the regenerative capacity of CNS neurons using different cell types, including Schwann cells. This involves fibroblasts, olfactory ensheathing glia and neural progenitors (Blits *et al.*, 2003, 2004; Cao *et al.*, 2004; Grill *et al.*, 1997; Hu *et al.*, 2005; Kobayashi *et al.*, 1997; Ruitenberg *et al.*, 2002, 2003, 2005; Weidner *et al.*, 1999).

In addition, appropriate engineering of cells allows neurotrophic factors to be selectively delivered to specific CNS locations (Jakobsson and Lundberg, 2006). For example, OEG genetically modified to secrete NTs such as NT-3, BDNF and GDNF have stimulated long-distance growth in certain spinal cord tracts (Cao *et al.*, 2004; Ruitenberg *et al.*, 2003, 2005). In addition, NT-secreting OEG improves spinal tissue sparing (Ruitenberg *et al.*, 2005), which is believed to be correlated with improved behavioral outcomes (Ruitenberg *et al.*, 2003).

Genetic engineering of cells can involve the use of different viral systems. Viral vector-mediated delivery systems that have been developed involve the use of adenovirus, adeno-associated virus or lentivirus. These vectors allow the delivery of foreign genes into neural tissue either by direct injection or by indirect *ex vivo* techniques (Dijkhuizen *et al.*, 1998; Hermens and Verhaagen, 1998; Huber *et al.*, 2000; Ruitenberg *et al.*, 2002, 2003, 2005). Transfer of genes to transplanted cells enables better monitoring of the cells within the spinal cord or CNS when using reporter genes, e.g. Green Fluorescent Protein (GFP), β-Gal and DSRED-2. Gene delivery using carriers/vectors such as nanoparticles is another approach currently in the developmental stage (see Box 2).

20.3.4 Matrices and scaffolds

The matrix/substrate must allow axons to penetrate, grow through and exit so that they can re-enter their target regions. After SCI, both descending and ascending axons tracts are disrupted, each of these tracts having specific locations in spinal cord white matter (Figure 20.13). Since various axonal populations have different growth requirements, a future graft may consist of composite bridging materials, each tailored to a particular axon population. Whether matrices or scaffolds are effective due to regenerative properties, or because of other factors such as neuroprotection, is a matter for further investigation.

20.3.4.1 *Relevant peptide sequences*
Many peptide sequences investigated for neuron/material interactions are derived from various bioactive

---

Delivering nanoparticles to the brain

To reach specific areas of the brain, noninvasive (or minimally invasive) delivery of the therapy is imperative. Nanoparticles have advantages in this respect in that they can contain a therapy which can be internalized by the cell and have a high permeability. Loaded therapies may also be released in a time-dependent manner. Hydrophobic molecules can also be incorporated into micelles, which permeate freely through water. The use of nanoparticles as delivery systems for difficult-to-access parts of the body is an exciting and soon to be investigated aspect of medical science. An example of their applicability is the delivery of gene therapy to the brain, demonstrating efficacy that is comparable to viral methodologies. Research by the group of Prasad has established the feasibility of using amino-functionalized organically modified silica nanoparticles as nonviral vectors for *in vivo* gene transfection (Bharali *et al.*, 2005). The nanoparticles were delivered to the midbrain region, which is involved in controlling voluntary movement and regulating mood, and contains dopaminergic neurons.

The nanoparticles are prepared as micelles from oil-in-water microemulsions, and modification of silica results in improved micelle structures and stability. Cationic charges are introduced to the outer surface, while the DNA is loaded by simple incubation. The DNA-loaded nanoparticles are suspended in phosphate-buffered saline and then sterilized by filtration. The nanoparticle diameter of 30 nm results in good permeability and a high incidence of transfection is observed after injection. Side-effects typically observed with gene transfer, such as carrier toxicity, immunological side-effects and conversion to a pathogenic form, were not observed with these nanoparticles. The DNA-loaded nanoparticles induced the expression of Enhanced GFP into dopamine-producing cells, but could be used for Parkinson's disease as well as for other disorders. While the delivery of genes with nanoparticles is a promising methodology, ideally protocols need to be developed that allow specificity of action and targeting of specific cell populations in the nervous system.

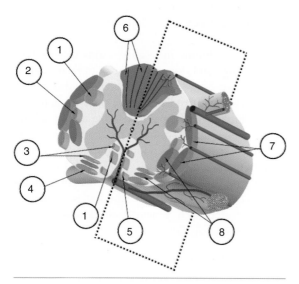

Figure 20.13 Selected tracts within the human spinal cord leading to/from the brain. Disruption of tracts will affect the corresponding sensory or motor signals. The tracts are symmetrical along the saggital plane (shown). Motor tracts: (1) corticospinal (voluntary, discrete movement, particularly in distal part of limbs), (2) rubrospinal (facilitates activity of flexor muscles), (3) reticulospinal (facilitates or inhibits voluntary movement or reflex), (4) vestibulospinal (facilitates postural activity associated with balance) and (5) tectospinal (reflex postural movements in response to vision). Sensory tracts: (6) fasciculus gracilis/cuneatus (discriminative touch and conscious perception of body and muscle position), (7) spinocerebellar (subconscious perception of body and muscle position) and (8) spinothalamic (pathway for crude touch, pain, pressure, temperature).

domains of laminin (Aumailley and Gayraud, 1998). The peptide sequence of YIGSR is generally preferred as a ligand for *in vitro* applications and is widely proposed as an alternative to RGD for cell adhesion (Borkenhagen *et al.*, 1998; Yu and Shoichet, 2005). Typically, such laminin-derived peptide sequences are grafted onto a scaffold surface or incorporated as part of a matrix (Plant *et al.*, 1997; Schense *et al.*, 2000). Another important peptide sequence of laminin is IKVAV, promoting cell adhesion, neurite outgrowth,

experimental metastasis, collagenase IV activity, angiogenesis, plasminogen activator activation, cell growth, tumor growth and differentiation of neural progenitor cells (Silva *et al.*, 2004; Tashiro *et al.*, 1989). Although the aforementioned active peptide sequences are effective at promoting neuron adhesion and axon growth alone, the use of both YIGSR and IKVAV has a synergistic effect on neurite outgrowth. The extended peptides CDPGYIGSR and/or CQAASIKVAV further increase neurite outgrowth (Yu and Shoichet, 2005). Other investigated laminin-derived peptides include RNIAEIIKDI and RGD, the latter being a cryptic peptide that is naturally only accessible after proteolytic cleavage of the adjacent domain (Aumailley and Gayraud, 1998).

The peptide VFDNFVLK from tenascin C is also a growth-promoting substrate (Meiners *et al.*, 2001). In addition to these molecules, one can design structures that actively impair other cells from entering the structure. Neural cell adhesion molecule (NCAM) is one such molecule that allows interaction of neurons and astrocytes (Crossin and Krushel, 2000). The L1 adhesion molecule, for instance, strongly promotes axon growth (Lemmon *et al.*, 1989) and its Fc portion is also active (Doherty *et al.*, 1995).

20.3.4.2 Matrices

In situ gelling matrices are particularly interesting for noninvasive approaches to neural tissue engineering. Fibrin is commonly used as a matrix and can be modified with peptide sequences from laminin (Schense *et al.*, 2000) or used to deliver neurotrophic factors with a heparin-binding site (Taylor *et al.*, 2004). Fibrin also resulted in the greatest number of neurons traced with retrograde-labeling, compared to collagen I, methylcellulose and Matrigel™ (Tsai *et al.*, 2006). The latter matrix has also been used in many CNS tissue engineering investigations, particularly with Schwann cell or OEG transplantation into the spinal cord (see Classical Experiment).

Zhang *et al.* have developed one particular peptide system culminating in a commercial matrix named Puramatrix™ that supports axonal growth (Holmes *et al.*, 2000). Exploiting the properties of charged peptide sequences, Puramatrix™ consists of amino acids that self-assemble into fibrillar structures, and can encapsulate neurons and other cells. Another self-assembling matrix consists of amphiphilic molecules with the bioactive sequence of IKVAV at the terminus. Aggregation of the peptide solution results after mixing the peptide solution with cell suspensions (Silva *et al.*, 2004). Self-aggregating amphiphilic molecules result in nanofibrillar structures that are 5–8 nm in diameter and up to micrometers in length. The IKVAV sequence on the end of the amphiphile induces the differentiation of neural progenitor cells into neurons more effectively than soluble peptides or bound to substrates. Significantly fewer neural progenitor cells differentiated into astrocytes when cultured in the amphiphilic matrix. Use of self-assembled peptide nanofibers *in vivo* to bridge tissue defects in the spinal cord and visual system has recently been reported (Ellis Behnke *et al.*, 2006; Guo *et al.*, 2007).

20.3.4.3 Scaffolds

Guidance scaffolds/polymers based upon poly(α-hydroxy acids) demonstrate the ability to stimulate axon growth (Hurtado *et al.*, 2006; Patist *et al.*, 2004; Teng *et al.*, 2002). The acidic degradation products of these polymers do not have a negative impact on spinal cord tissue (Gautier *et al.*, 1998), although this is likely concentration dependent. Controlled uniaxial freezing can generate channels within a polymer solution or hydrogels by using the forces associated with crystal growth (Figure 20.7d). The resulting oriented scaffolds demonstrate neurite and cell penetration into the channels (Patist *et al.*, 2004). Such scaffolds have been used as a substrate for neural stem cell and genetically modified Schwann cell transplantation into the spinal cord (Hurtado *et al.*, 2006; Stokols and Tuszynski, 2006; Teng *et al.*, 2002).

Hydrogel scaffolds also demonstrate robust neurite and cell penetration. It should be mentioned that the implantation of such scaffold materials into the injured spinal cord typically results in a significantly different injury response to that commonly observed with a traumatic lesion alone. Cystic cavity formation can be completely absent when scaffolds are implanted

and astrocytic scarring is often reduced in density. A notable lack of cystic cavity formation compared to lesions only has been observed with a range of scaffolds, such as Neurogel™, poly(2-hydroxyethyl methacrylate) (pHEMA), agarose, collagen or fibronectin (Giannetti *et al.*, 2001; Jain *et al.*, 2006; King *et al.*, 2006; Moellers *et al.*, 2002; Stokols and Tuszynski, 2006; Woerly *et al.*, 1999, 2004). In addition, axons will penetrate into the scaffold, in a highly oriented and ordered manner if the scaffold itself has oriented pores. Some axons are myelinated; the likely sources of Schwann cells are the dorsal and ventral roots near the injury site. Endogenous Schwann cells are known to infiltrate Neurogel™, agarose scaffolds and pHEMA hydrogel nerve guides (Stokols and Tuszynski, 2006; Tsai *et al.*, 2004; Woerly *et al.*, 2004).

Bundled fibers have also shown promise within the injured spinal cord (Novikov *et al.*, 2002; Yoshii *et al.*, 2004). Collagen fibers have been implanted in full-transection models, which were created 10 days after an initial contusion injury. In this second operation, the collagen fibers only (without a nerve guide) were inserted. After 12 weeks the fibers were well integrated with the spinal cord tissue and functional improvements observed (Yoshii *et al.*, 2004). Alginate- and fibronectin-coated poly-β-hydroxybutyrate (PHB) fibers have also been inserted after a second injury to the spinal cord (Novikov *et al.*, 2002). Initially the rubrospinal tract only was disrupted, followed 7 days later by the insertion of the filaments in a hemisection experiment. The number of rubrospinal neurons lost decreases by 50% when the filaments were inserted, while the absence of PHB fibers (i.e. alginate and fibronectin only) resulted in a similar neuron survival to the controls (Novikov *et al.*, 2002).

20.3.5 Nerve guides and transplantation

The classical experiment of cell transplantation with nerve guides in the fully transected spinal cord was pioneered in the Bunge group (see Classical Experiment) with P(AN-VC) nerve guides. Collagen nerve guides and pHEMA-based conduits can also be used to promote regeneration in the spinal cord (Spilker *et al.*, 2001; Tsai *et al.*, 2004). Hydrogel guides result in the presence of axons within the lumen even when containing no other therapeutic factor. Regeneration into P(AN-VC) nerve guides occurs with both Matrigel™ and Schwann cells present in the lumen.

As is the case with peripheral nerve repair, the inclusion of a matrix into a hydrogel nerve guide increases the number of penetrating axons (Tsai *et al.*, 2006). Collagen type I matrices (1.28 mg/ml), while supporting robust regeneration in the peripheral nerve (Labrador *et al.*, 1998; Midha *et al.*, 2004),

promote less penetrating axons in the hydrogel nerve guides as fibrin, methylcellulose and Matrigel™ (Tsai *et al.*, 2006). Such results demonstrate the different nature of the PNS and CNS, and materials used in one system are not necessarily effective in another.

20.3.6 Neuronal and axon tracing

Most axon pathways in the brain and spinal cord do not have unique markers that enable easy identification. Also, if present, expression of any such markers may be affected by injury-related processes and/or during

---

State of the Art Experiment
Sequences of scaffold degradation in the spinal cord

Many materials encourage nerve repair after SCI. The implantation of such scaffolds into the injured spinal cord often results in a significantly different injury response to that commonly observed with a traumatic lesion alone. Cystic cavity formation can be completely absent when scaffolds are implanted and scarring is often reduced in density. In addition, axons will penetrate into the scaffold in a highly oriented and ordered manner if the scaffold itself has oriented pores. Such observations have been reported from various laboratories, *with a range of materials used for the scaffold.* Reduced cystic cavity formation is reported with cell-invasive scaffolds of Neurogel™, pHEMA, agarose or fibronectin (Woerly *et al.*, 1999, 2004; Giannetti *et al.*, 2001; Jain *et al.*, 2006; King *et al.*, 2006; Stokols and Tuszynski, 2006) while the control lesions typically were accompanied by cyst formation and tissue loss.

Our understanding of the sequence of events after scaffold or matrix implantation into the spinal cord is limited. Of the investigations that report the effect of implanted materials in the spinal cord, almost all are single timepoints with survival times of 4–8 weeks and there is little knowledge (other than behavioral) outlining the initial stages after implantation. A notable exception to this observation is an excellent study of fibronectin scaffolds

implanted into a hemisection lesion (King *et al.*, 2006). Immunohistochemistry at the post-implantation times of 3 days, 1 week, 2 weeks, 4 weeks and 12–16 weeks permits some understanding to the interaction between the implant and spinal cord tissue.

As outlined in Figure 1, the immediate response is the recruitment of macrophages to a fibronectin scaffold and vascularization with blood vessels. After 1 week, axons begin their penetration into the scaffold, as do Schwann cells, while macrophages remain present. At week 3 axonal growth continued and the axons were myelinated. Reactive astrocytes begin to surround the fibronectin scaffold and macrophages migrate away from the implant site. The deposition of laminin into tubules within the scaffold is first detected after 3 weeks. At week 4, which is a common time point for other investigations, the axons remain within the scaffold, are myelinated with Schwann cells and astrocytes have migrated into the implant site, but are not activated. The laminin tubules that were observed in weeks 3 and 4 have diminished. The fibronectin scaffold also degrades during implantation and is fully resorbed after 4 months. While a lesion will often result in a cystic cavity, there is no cyst formation at the implant site of a fibronectin scaffold.

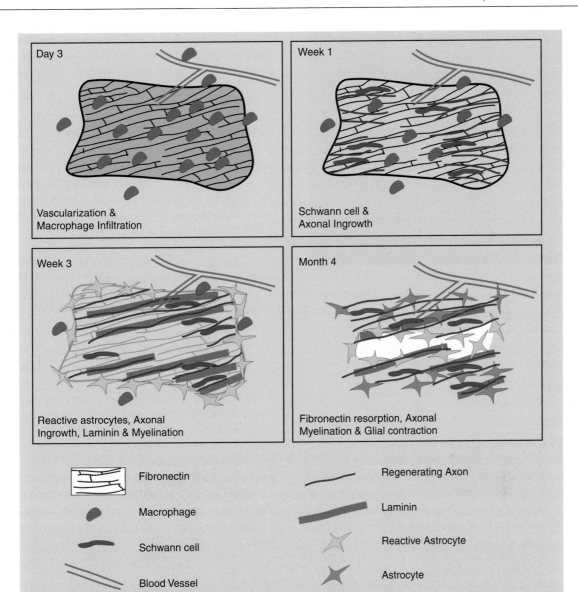

The sequence of events after implantation of a fibronectin scaffold in the hemisection lesion of the spinal cord. Image modified from King, V.R., Phillips, J.B., Hunt-Grubbe, H. et al. (2006). Biomaterials, 27: 485–496, (2006), and gratefully permitted by Dr Von King and Professor John Priestley (Queen Mary University of London).

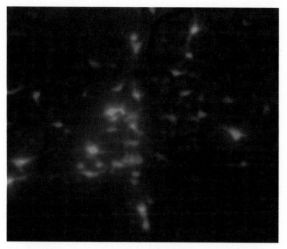

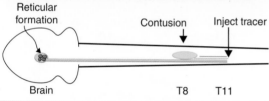

Figure 20.14 Fast blue-positive neurons in the reticular formation. Retrograde labeled from T11 spinal cord, with Schwann cell graft at a T8 contusion.

the process of regeneration. Thus, neuronal and axon tracers are used to label neurons and axons, respectively, to identify and evaluate the response of fiber tracts to intraspinal transplants. Administration of a tracer near the axon endings allows uptake and retrograde labeling of the neuronal cell bodies of the responding systems. Often used fluorescent tracers are Fluorogold and Fast blue, which in the rat require anywhere between 5 and 14 days for retrograde transport from lower spinal cord levels to the cell bodies in the brain (the transport time depends on the distance as well as the properties of the tracer). These tracers are usually injected a few millimeters caudal to, or into, a transplant. When used appropriately and combined with a thorough analysis of the location and number of back-filled neurons in the spinal cord and brain, the response of individual fiber tracts can be assessed (Figure 20.14). A drawback of this approach is that it is difficult to evaluate how far axons have

regenerated. Injection of a tracer at the level of the cell bodies in the brain allows for uptake and anterograde transport towards the axonal endings. Three often-used anterograde tracers are wheat germ-agglutinated horseradish peroxidase, biotinylated dextran amine and the B subunit of cholera toxin. The first tracer has a relatively fast transport time, but the axons are often difficult to quantify after identification using enzyme histochemistry. Biotinylated dextran amine has a relatively slow transport time, but its identification using immunocytochemistry produces easily recognizable axons. This is also true for the B subunit of cholera toxin, which in addition has a relative short transport time. The use of tracers is crucial for quantitative analysis of the effect of therapeutic strategies on axon regeneration after SCI, thereby relating anatomical plasticity to any observed behavioral recovery.

20.3.7 Approaches to overcome the inhibitory nature of the injury-induced scar

After SCI a physical and chemical barrier (glial scar) forms at the impact site which prevents the growth of new axon sprouts. This scar consists mainly of reactive astrocytes and meningeal fibroblasts, which express extracellular axonal growth-inhibitory molecules such as chondroitin sulfate proteoglycans and semaphorins. It has become clear that manipulation of this barrier will be an essential component of a successful repair strategy for the injured spinal cord. Such manipulation of the scar may include: (i) decreasing the inhibitory nature of the scar, (ii) preventing axons from recognizing inhibitory molecules and (iii) enhancing the intrinsic growth ability of axons. The permissiveness of the scar for axons can be increased by preventing receptor–ligand (inhibitor) binding (Schnell and Schwab, 1990), by obstructing the synthesis of inhibitors (Grimpe and Silver, 2004) or by degrading biologically active components of the inhibitors (Moon *et al.*, 2001; Bradbury *et al.*, 2002). The intrinsic growth ability of axons can be increased by targeting molecules downstream in the intracellular pathways that promote neurite extension or prevent growth cone collapse (Spencer and Filbin, 2004). In the laboratory, several approaches have resulted in axon regrowth beyond the scar (Silver and Miller, 2004), but success was usually obtained using injuries that could be considered as 'smaller size' injuries, i.e. with less formation of scar tissue. With larger injuries the same approaches have had only minimal success. The barrier of scarring can be overcome by implantation of OEG into the scar environment (Ramon-Cueto et al., 1998; Ramon-Cueto, 2000) or by increasing the levels of neurotrophic factors in the nervous tissue using osmotic mini-pumps (Bamber et al., 2001; Oudega and Hagg, 1996).

20.3.8 Summary

SCI initiates a plethora of destructive events. Some of the cells and molecules present at the site of injury may contribute to protection and repair, but the ultimate outcome is loss of tissue and loss of function in motor and sensory systems. One of the main obstacles for regenerating axons in the injured spinal cord is the presence of a glial scar at the site of injury or, in the case of an intraspinal transplant, at the transplant–host spinal cord interface. Unlike the PNS and the use of peripheral nerve autografts and nerve bridges, there is currently no reproducible, successful clinical therapy for SCI. Cell transplantation, combined with supportive matrices and scaffolds or neurotrophic factors, has been the predominant focus of tissue engineering strategies. The emphasis has been on neuroprotection, enhanced tissue sparing and re-establishment of continuity across an injury site. Future objectives include increasing the number of damaged ascending and descending axons that regenerate, improving the remyelination of these regenerating axons, and the guidance of regenerating axons back to their appropriate target regions in the brain or spinal cord. The development of new techniques involving gene therapy and ex vivo manipulation of cells prior to transplantation will hopefully provide additional, and economically sustainable, tools that will improve therapeutic outcomes in injured patients.

20.4 CNS: optic nerve injury model

20.4.1 Advantages of the model

Damage to fiber tracts interconnecting different CNS regions causes disruption of neural circuits and often leads to the death of associated neurons. Therapeutic strategies are therefore needed to ensure the viability of affected neurons, promote the regrowth of damaged axons across lesion sites, maintain growth within CNS tissue distal to the injury, and finally ensure that regrowing axons reinnervate appropriate target areas and reconnect with appropriate target neurons.

The visual system is for many a preferred model in which to study regenerative success or failure after nerve fiber damage in the mature CNS:

(i) Retinal ganglion cells (RGCs), the neurons that project axons out of the retina and through the optic nerve to central targets in the brain, can be directly targeted via eye injections.
(ii) There is detailed knowledge of intra-retinal and visual brain circuitry.
(iii) The optic nerve is a discrete, centrally derived tract that is surgically accessible and can be manipulated without affecting other fiber tracts and structures.
(iv) The extent of RGC survival and axonal regrowth can readily be quantified.

Normally, adult mammalian RGCs do not regrow axons beyond an optic nerve injury and, if the axotomy occurs within the orbit, about 90% of RGCs die within 2 weeks (Villegas-Perez et al., 1993). However, regenerative failure is not inevitable. Some adult RGCs will survive axotomy and regrow their axons when appropriately stimulated and/or when provided with an appropriate local environment.

20.4.2 Regenerative therapies

After optic nerve crush, adult RGC survival can be enhanced by intraocular injections of recombinant neurotrophic factors, such as BDNF, NT-4/5, CNTF or GDNF (Yip and So, 2000), or by injections of factors that block intracellular signaling pathways associated with cell death (Bähr, 2000). However, these effects are transient and viability is generally enhanced for only a few weeks. Use of viral vectors, particularly adeno-associated viral vectors, encoding appropriate growth-promoting genes to transduce and genetically modify RGCs provides more long-term protection (Dinculescu *et al.*, 2005; Harvey *et al.*, 2006; Martin and Quigley, 2004).

Limited regrowth of RGC axons across an optic nerve crush can be elicited using a number of approaches including the implantation of peripheral nerve fragments into the vitreous, injuring the lens (Leon *et al.*, 2000), or activating macrophages within the eye (Yin *et al.*, 2003). Some regeneration is also seen after blockade of growth-inhibitory molecules or after inactivation of Rho, a GTPase that mediates growth cone inhibition (Bertrand *et al.*, 2005). Overall, however, adult RGC axons fail to grow more than just a few millimeters past the injury site and additional tissue engineering strategies are required to stimulate more sustained regenerative growth.

The most successful approach has been to auto-graft peripheral nerve segments from the same animal onto the cut optic nerve (Bray *et al.*, 1987). Under these conditions, regrown axons reinnervate target areas in the brain and some limited function is restored. However, as with the use of such grafts in peripheral nerve repair (see 'Use of tubes (nerve guides) in the PNS') there is a functional cost involved in removing the nerves for transplantation. Alternatives are therefore being sought, including the use of gene-activated collagen matrices (Berry *et al.*, 2001), matrices containing cultured olfactory ensheathing cells (Li *et al.*, 2003) and, most recently, chimeric peripheral nerve grafts containing genetically modified Schwann cells (Hu *et al.*, 2005) (Figures 20.15 and 20.16). Various types of hydrogel matrices, some containing peptide and amino sugar sequences, have also been used in attempts to promote the regeneration of RGC axons after more central optic tract injures in the brain (Harvey, 1999). In particular, the use of polymers

containing genetically modified cells that secrete RGC growth factors shows promise (Loh *et al.*, 2001).

Finally, there is increasing evidence for synergistic interactions in CNS repair (Lu *et al.*, 2004; Schense *et al.*, 2000; Yu and Shoichet, 2005). These interactions can be assessed at a single location, using different combinations of factors, or at multiple locations, e.g. testing the effect of factors applied to the cell body and supplying cells/appropriate factors near the growing tips of the axons (Cui *et al.*, 2003b; Hu *et al.*, 2005, 2007). The optic nerve injury model is an ideal system in which to examine such combinatorial effects; proteins or viral vectors can be injected into the eye to directly influence RGCs, while tissue and genetic engineering can be used to modify the environment within which RGC axons regenerate.

20.5 CNS: retina

The retina is a thin, layered neural structure consisting of RGCs, bipolar and amacrine cells, photoreceptors (cones and rods), Müller cells and retinal pigment epithelial (RPE) cells, the latter attached to a thin membrane called Bruch's membrane. Bruch's membrane is partially synthesized by the RPE cells, and contains laminin, collagen type IV, fibronectin and heparin sulfate proteoglycans. RPE cells maintain the vitality of the photoreceptors. When the RPE and photoreceptors are detached from Bruch's membrane, photoreceptor cell death (and blindness) ensues.

20.5.1 Diseases of the retina

Diabetic retinopathy, retinitis pigmentosa and macular degeneration are three major diseases of the eye, while glaucoma – often associated with raised intraocular pressure – results in damage to the optic cup and retina. Age-related macular degeneration (ARMD) is the leading cause of blindness in the US for people aged above 55 years. Current effective treatments for retinal disorders are limited and tissue

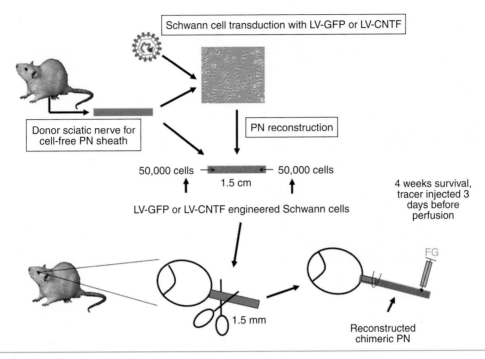

Schwann cell transduction with LV-GFP or LV-CNTF

Donor sciatic nerve for cell-free PN sheath

PN reconstruction

50,000 cells — 1.5 cm — 50,000 cells

LV-GFP or LV-CNTF engineered Schwann cells

4 weeks survival, tracer injected 3 days before perfusion

FG

1.5 mm

Reconstructed chimeric PN

Figure 20.15 Chimeric, genetically engineered peripheral nerve grafts act as bridges for axon regeneration. Donor peripheral nerves (e.g. the sciatic nerve) are freeze-thawed to kill all cells, but retain organized extracellular matrix. Sheaths are then reconstituted with host adult Schwann cells that have been expanded in culture and genetically modified using lentiviral (LV) vectors to express appropriate neurotrophic factors. These engineered, chimeric nerves are sutured onto the cut end of adult rat optic nerves. After about 4 weeks a tracer (Fluorogold, FG) is injected into the distal end of the grafts to retrogradely label the RGCs in the operated eye that have regenerated axons through the grafts. Animals are perfused 3 days later, and the number of surviving and regenerating RGCs is quantified in each eye (further details, see Hu *et al.*, 2005). Picture courtesy of Ying Hu, School of Anatomy and Human Biology, University of Western Australia.

engineering strategies are currently emerging. One nontissue engineering approach of restoring vision is with neuroprosthetics (see Box 3).

20.5.2 Cell transplantation and gene therapy

Intraocular cell transplantation is being assessed as a potential therapeutic strategy for the treatment of degenerative diseases that primarily affect photoreceptors. Different transplantation methods (retinal sheets, fragments, dissociated mixed or purified cell suspensions) have been trialed and different

sources of cells have been used, including neural precursors, embryonic and postnatal retinal progenitor cells, neural stem cell lines, and bone marrow stem cells (Aramant and Seiler, 2005; Das *et al.*, 2005; MacLaren *et al.*, 2006). One of the leading experimental strategies for ARMD is the transplantation of RPE cells to the subretinal space (Lund *et al.*, 2003). RPEs injected into the subretinal space as a suspension maintain some photoreceptors for a limited period; however, the injected cells are unable to form ordered monolayers and eventually die. Iris pigment epithelium (IPE), obtained as an iridectomy specimen, has also been investigated as an autologous

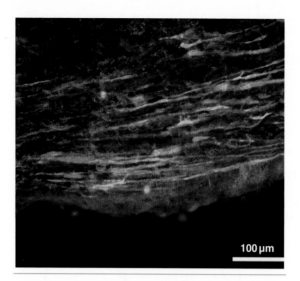

100 μm

Figure 20.16 Engineered Schwann cells are aligned along the axis of the reconstructed peripheral nerves – transplanted onto the cut optic nerve – and express green fluorescent protein (GFP). The red staining shows location of S-100 protein in the tissue, which is located primarily in the cytoplasm of glial cells.

source of cells for transplantation into the retina (Figure 20.17). IPE and RPE cells originate from the neuroectoderm and develop distinct properties when they migrate into different embryogenic microenvironments. *In vitro* investigations have demonstrated that IPE cells can acquire properties of RPE cells (Thumann, 2001).

Cell transplantation substrates have been used as carriers to deliver cells, such as RPEs, IPEs and retinal progenitor cells (RPCs) to the retina. The seeded substrate is typically inserted subretinally through a slit in the sclera and choroid. Injected cell suspensions are exposed to high shear forces as they pass through the injecting needle, while therapies involving supporting substrates avoid such stresses. The use of cells on scaffolds for transplantation to the retina thus has advantages over injection of cells as spheres or single-cell suspensions, where large numbers cells do not survive the transplantation procedure.

RPCs have been transplanted onto PLGA scaffolds oriented using uniaxial freezing, surviving at least 14 days after transplantation into the retina of the rat, even in the absence of immunosuppression

---

Neuroprostheses

An electric charge delivered to the blind eye produces a sensation of light.
LeRoy (1755)

Neuroprosthetic devices of the retina or the ear have been widely investigated with relatively very little overlap with tissue engineering strategies. Using electrodes to directly stimulate neurons, a range of sensory information is available to transmit to nervous tissue. The current systems are based upon open-loop stimulation, i.e. the devices are not modulated by feedback sensed by the system. Neuroprostheses, however, are still in the developmental stages.

There are two types of retinal prostheses, i.e. subretinal implants and epineural implants, and are implanted as shown in the figure (Cohen, 2007; Zrenner, 2002). Subretinal photodiodes are placed under the retina and contain light-sensitive photodiodes with hundreds of microelectrodes that stimulate the remaining neurons – testing is still in preclinical stages. Epineural implants have no light-sensitive elements and an image, captured by a camera, is transmitted to an implant on top of the retina and the RGC axons are then stimulated directly. The camera can be external to the eye or replaces the natural lens.

The cochlear implant, however, has been the 'success story' of neuroprosthetic devices developed over the last 30 years. As the technology has evolved, the profoundly deaf have been able to hear their first sounds with such devices, then some speech discrimination in a small percent of the patients and now open-set (auditory only) speech discrimination for the majority of current implant patients.

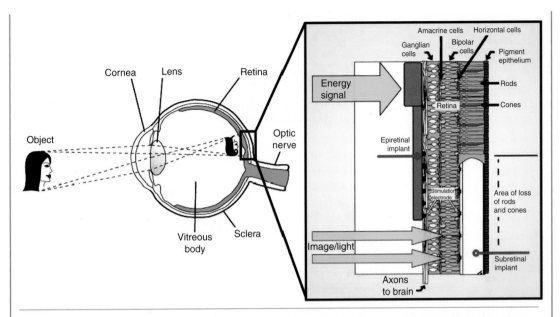

The two types of retinal implants – epiretinal and subretinal – are shown. With an epiretinal implant, a substrate directly stimulates the ganglion cells, using electrical signals from a camera, typically external to the eye. Subretinal implants are designed to replace the rods and cones of the retina, and consist of an array of microphotoelectrodes, each connected to a stimulation electrode. Incident light from an object is projected normally on the retina, and the subretinal implant resolves the light and stimulates the remaining neurons. Reprinted from Zrenner, E. (2002). Science, *295: 1022–1025, with permission from* Science Magazine.

The quality of hearing may improve further when combined with tissue engineering strategies to better integrate the auditory nerve with the electrode.

Other promising neuroprosthetic systems include the cortical implant BrainGate™ from Cyberkintics Neurotechnology Systems (Donoghue *et al.*, 2004). These and similar devices design to detect motor signals in the cerebral cortex and covert this signal into a useful output for paralyzed humans (Hochberg *et al.*, 2006). This area of the brain is activated when the movement action is imagined by paralyzed patient and the simple control of a computer has been demonstrated with implants in the cerebral cortex. The two research areas of neuroprosthetic devices and tissue engineering promise to acquire greater overlap as benefits from each approach are adopted in the respective fields.

(Lavik *et al.*, 2005). Other potential cell transplantation substrates included collagen, PLGA, poly(L-lactic acid), poly(hydroxypropyl methacrylamide), poly(hydroxybutyrate-*co*-valerate) and lens capsule material. Substrates of lens capsule material, microprinted with hexagonal regions of poly(vinyl alcohol), result in the formation of layers of RPE cells with morphology similar to that of natural tissue (Lee *et al.*, 2002). Substrates for subretinal implantation require a relatively thin thickness (Bruch's membrane has an average thickness of 5 μm), but require a certain amount of mechanical integrity to withstand the placement and implantation procedures. An ideal and effective cell transplantation substrate has not yet been found.

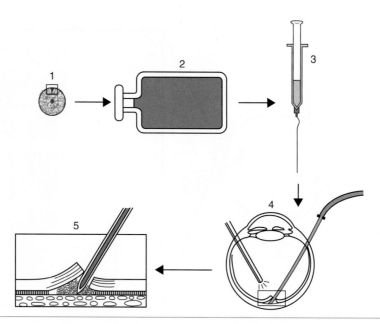

Figure 20.17 The procedure of IPE injection. After an iridectomy specimen (1), the cells are isolated and cultured (2), transferred into a delivery system (3), and injected into the subretinal space (4/5). In addition to the injection of such cell suspensions, tissue engineering approaches deliver cell monolayers on polymeric substrates. The image is reprinted from Thumann (2001) *Surv Ophthalmol*, 45: 345–354, with permission from Elsevier.

As an alternative to transplantation the use of viral vectors, particularly adeno-associated viral vectors (AAV), to introduce neuroprotective genes into photoreceptors and RPE has been the subject of intensive investigation in recent years (Auricchio and Rolling, 2005; Dinculescu *et al.*, 2005; Harvey *et al.*, 2006). In a variety of animal models of retinal disease this approach results in maintenance of retinal structure and visual function. Clinical trials are expected to be initiated in the near future. The use of AAV to induce ectopic expression of photosensitive pigments in the inner retina of animals deficient in photoreceptors has also been reported (Bi *et al.*, 2006).

20.6 CNS: brain

20.6.1 Trauma and stroke

Reconstruction of grey and white matter defects in the brain is especially difficult after large lesions or in the chronic situation, where an injury has occurred some time previously. In addition to neuronal death there is demyelination, retraction and/or aberrant sprouting of injured axons, glial/meningeal scarring, and often progressive tissue cavitation. In these circumstances, there is a need for cell replacement and some form of neural tissue engineering to develop scaffolds that facilitate reconstruction and restore continuity across the traumatized region.

Numerous repair strategies have been explored, including the use of fetal cell transplants, xenografts, genetically modified cells and the implantation of matrices that encourage endogenous tissue regeneration. Matrices tested include collagen, Matrigel™, and synthetic polymers such as nitrocellulose and porous hydrogels made from a variety of copolymers. The use of supporting scaffolds may allow the formation of a biohybrid tissue that acts as the structural and functional analogue of the original tissue (Woerly *et al.*, 1999). For stroke, in particular, there has been considerable recent interest in the possibility of using neural precursor cells or stem cells in novel cell therapies, perhaps in association with polymer scaffolds (Lindvall and Kokaia, 2004; Savitz *et al.*, 2004; Goldman, 2005). Assuming immunological problems can be overcome (Barker and Widner, 2004), such stem cells could be derived from various sources, including brain itself, bone marrow or umbilical cord, and applied directly or perhaps via intravenous infusion. Genetic modification of cells prior to transplantation may prove even more effective (Wei *et al.*, 2005).

20.6.2 Neurodegenerative diseases

Neurodegenerative diseases are often associated with substantial neuronal loss – the pattern of loss varying depending on the type of disease process. Supply of appropriate neurotrophic factors via grafted cells or from slow-release polymers may protect compromised neurons; alternatively, cell replacement strategies may be required.

Some of these neurodegenerative disorders, such as Huntington's disease and spinocerebellar ataxia, are inheritable and associated with single-gene mutations (see also section 20.5). Other motor neuron disorders, and diseases such as Alzheimer's disease and Parkinson's disease, have more complex etiology, with only a percentage directly associated with dysfunction of a specific gene. It is likely that many of these degenerative conditions are affected by environmental factors as well as some form of genetic susceptibility. Alzheimer's disease and Parkinson's disease, in particular, are diseases associated with the elderly, and are becoming increasingly common as life expectancy increases and the population ages.

Early diagnosis and prevention of further deterioration would be the optimal ways of treating these neurodegenerative disorders; however, it is often the case that when symptoms become apparent there has already been substantial neuronal death. Local release of neurotrophic factors, from polymers or from grafts of genetically modified cells encapsulated within polymer shells (Tan and Aebischer, 1996), may prevent further cell death. Direct gene therapy is also under intensive investigation, using viral vectors to introduce growth-promoting genes into compromised neurons, or using the vectors to introduce interference RNA into neurons to silence disease-causing genes (Xia *et al.*, 2004).

Other tissue engineering strategies involve attempts to replace the lost neurons with new cells that ideally become incorporated into host circuitries. Parkinson's disease is a multisystem disease – a major feature being the catastrophic loss of neurons in the midbrain that secrete dopamine. In numerous clinical trials, hundreds of Parkinson's disease patients have received intracerebral grafts of dopamine-rich fetal human brain tissue, many of these recipients experiencing some alleviation of their symptoms. Potentially alternative sources of tissue include xenografts and stem cells or neural progenitor cells modified to secrete dopamine (Goldman and Sim, 2005). Such approaches, perhaps combined with neuroprotective gene therapy, are potentially also of use in the treatment of motor neuron diseases and Huntington's disease (Melone *et al.*, 2005).

20.7 Animal models

The sciatic nerve of the rat is the most common model for induced regeneration in the PNS. Inconsistent parameters such as gap length (2–15 mm), implant times (1–150 weeks) and location of histological analysis (distal stump, center of gap, proximal stump) have resulted in the problematic comparison of various therapeutic strategies. A range of assessments for recovery of this model is reviewed by Nichols *et al.* (2005). Yannas and Hill (2004) have proposed a method for normalizing the animal model data from previous experiments, based on the assumption that therapies will have a sudden and dramatic reduction in regeneration success around a critical length gap, L_C.

The selection of an animal model is an important issue. Tissue engineering strategies for SCI studies have focused on the mammalian model, while invertebrates and nonmammalian vertebrates are often used for *in vivo* studies aimed at understanding fundamental biochemical and molecular processes associated with cell survival and regeneration. In SCI studies, the choice of model should largely be guided by our understanding of the neuropathology of the human injury, which at present is still limited (Bunge *et al.*, 1993, Kakulas, 1999). The pathology of SCI in humans appears to be largely similar to that in rodents, although important differences have been discerned, including reduced glial scarring, inflammation and demyelination, increased Schwann cell infiltration, and the protracted Wallerian degeneration in humans. Furthermore, there are critical differences in the organization and functional importance of different ascending (sensory) and descending (motor) pathways in human (see Figure 20.14) versus rodent brain and spinal cord – differences not always sufficiently taken into account by experimenters. Nevertheless, rodents offer an important and useful model in which to test a variety of SCI repair strategies.

For rats and mice, devices are available that reliably and consistently contuse the spinal cord, thereby mimicking the most common type of human SCI. Rats and mice are popular models for cell transplantation and tissue engineering strategies due to the ease of care and their availability; however, larger species are being investigated, including pigs and dogs. To assess the efficacy of specific approaches to repair the human spinal cord, it is generally considered that the development and use of primate models is necessary. Apart from ethical issues, the limitation with these models is that the maintenance of the injured primate is extremely labor-intensive and costly.

Cats have also been used to model SCI. An important difference over other models is that cats can be trained easily, which is due to sympathetic pathways that are unique to the feline model – excellent locomotor results such as walking or standing can be readily trained with this particular animal model (Fouad and Pearson, 2004). A disadvantage of the cat as a model for SCI is that locomotion in cats after injury is strongly influenced by the central pattern generator, which appears to be a crucial difference with humans.

Relevant anatomical and directional terms associated with nervous tissue	
Rostral or Anterior	Head or front end
Caudal or Posterior	Tail or hind end
Proximal	The injured part of the tissue closer to the neuron cell body
Distal	The injured part of the tissue away from the neuron cell body
Lateral	Away from the midline
Medial	Toward the midline
Dorsal	Back/top side
Ventral	Belly/bottom sided

20.8 Future approaches

The main issue for PNS regeneration remains the reconnection of severed nerves that are separated from each other above a critical gap length. Fibers, cell transplants and scaffolds will be investigated due to their success in bridging the larger injury model within nerve guides. *Ex vivo* genetic modification of grafted cells may also provide additional growth support for regenerating axons. Scaffolds formed via uniaxial freezing have morphological similarities to structures within which the bands of Büngner form distal to the PNI, and their full potential for large gap injuries is still to be determined. Ultimately the advances in other fabrication processes such as stereolithography or direct writing (see Chapter 14) may provide the highly resolved scaffold needed for 'fascicle to fascicle' guided regeneration.

The production of a successful regenerative bridging construct in the CNS requires knowledge of the dynamic temporal changes that occur after an injury, an understanding of the complex molecular and cellular environment in each injured area, and strategies to overcome this inhibitory environment. Combining cell transplantation in the contused spinal cord with other repair-promoting interventions such as pharmacotherapy and gene therapy allows substantially more 'tools' for neurological recovery. Considering the complex cellular and molecular interactions present within the spinal cord, tissue engineering approaches have predominantly addressed neuroregenerative aspects. Current *in vivo* tissue engineering approaches typically involve implantation into a fresh lesion; however, there are exceptions to this, where a secondary injury is induced 1–2 weeks post-lesion. While an *in vivo* construct may have properties for regeneration, it usually does not have any deliberate neuroprotective elements. There are many questions to ask regarding the excellent integration of many different materials into the spinal cord. The production of pro- or anti-inflammatory molecules by the invading macrophages may affect the healing of the spinal cord after filling a lesion with a scaffold or matrix. It is possible that the implantation of matrices/scaffolds/cells provides neuroprotective effects; however, the extent of these effects and their mechanisms of action are largely unknown. Surgery can also be performed on transgenic animals (particularly mice), providing insights on the role of certain genes in regeneration with tissue-engineered constructs. A major challenge will be to develop tissue engineering protocols that are suitable for use in chronic as well as acute injury situations. Further understanding of the effect of scaffold implantation will continue to provide tissue engineers with new concepts for bridging CNS defects and may one day provide an 'off-the-shelf' solution to spinal cord repair.

The retina is a tissue with surgical accessibility via the vitreous cavity; however, this can only be achieved practically with small puncture wounds of the eye. Strategies for delivering polymeric sheets for cell transplantation to the retina will have important ramifications in the delivery of tissue engineering strategies to the eye. Improved prosthetic devices that better activate central visual pathways are also likely to be developed (Cohen, 2007).

Minimally invasive methods will typically be preferred for tissue engineering strategies for the brain. *In situ* gelling materials that promote survival/differentiation of transplanted cells and the delivery of nanoparticles for targeted therapeutic delivery will increase in interest over the next decade. The types of cell available for transplantation are also likely to be greater in the future, including perhaps activation of the patient's own endogenous neural precursor cells. Approaches for the treatment of brain trauma and stroke are likely to involve combinations of therapies, and may be distinct from therapies developed to treat degenerative diseases.

Cochlear implants, and other neuroprosthetic devices, should also benefit from advances in tissue engineering strategies. Improved neuroprosthesis integration and reduced retraction (or greater regeneration) of the stimulated axons may assist with discrete signaling. The inner ear is a difficult-to-access organ, thus the application of *in situ* gelling matrices and nanoparticles is an important future development.

20.9 Summary

- A critical injury gap occurs in peripheral injury, above which the number of successful repairs rapidly declines.
- The Bands of Büngner are Schwann cell columns that form within natural laminin (oriented) scaffolds that are crucial for the regeneration of axons in the PNS.
- Neurotrophic factors play an important, but complicated, role in regenerative therapies.
- The limited regeneration in the central nervous system (CNS) is primarily due to the different cellular and molecular environment and less to the internal properties of the corresponding neurons.
- A contusion injury models the most frequent type of human SCI.
- A range of cell and scaffolds implanted into the injured spinal cord reduces astrocytic scarring, promotes tissue sparing, and facilitates axonal regeneration and myelination.
- As in other body systems, cell transplantation and tissue engineering in the CNS and PNS is influenced by inflammatory and immune-mediated events. Age and gender differences may also need to be taken into account.

- Tissue engineering therapies for treating traumatic injury (acute/chronic) versus degenerative diseases differ greatly (i.e. what to do and when to do it?).
- Genetic modification of cells prior to transplantation into the injured brain and spinal cord can increase neuronal survival and enhance the regeneration response.
- Neural tissue engineering strategies commonly adopt multifactorial strategies both in terms of diversity of therapies (i.e. scaffold *and* cells *and* drugs) and timing of interventions (parallel or sequential).
- The cochlear implant is a commercially and medically successful neuroprosthesis. Retinal and brain interfacing prostheses are in their developmental infancy, as are clinical treatments with neural tissue engineering strategies.

References

Aramant, R.B. and Seiler, M.J. (2005). *Prog Retin Eye Res*, 23: 474–494.

Aumailley, M. and Gayraud, B. (1998). *J Mol Med*, 76: 253–265.

Auricchio, A. and Rolling, F. (2005). *Curr Gene Ther*, 5: 339–348.

Bähr, M. (2000). *Trends Neurosci*, 23: 483–490.

Bamber, N.I., Li, H., Lu, X., *et al.* (2001). *Eur J Neurosci*, 13: 257–268.

Barker, R.A. and Widner, H. (2004). *NeuroRx*, 1: 472–481.

Berry, M., Gonzales, A.M., Clarke, W., *et al.* (2001). *Mol Cell Neurosci*, 17: 706–716.

Bertrand, J., Winton, M.J., Rodriguez-Hernandez, N., *et al.* (2005). *J Neurosci*, 25: 1113–1121.

Bharali, D.J., Klejbor, I., Stachowiak, E.K., *et al.* (2005). *Proc Natl Acad Sci USA*, 102: 11539–11544.

Bi, A., Cui, J., Ma, Y.P. *et al.* (2006). *Neuron*, 50: 23–33.

Blits, B., Carlstedt, T.P., Ruitenberg, M.J., *et al.* (2004). *Exp Neurol*, 189: 303–316.

Blits, B., Dijkhuizen, P.A., Carlstedt, T.P., *et al.* (1999). *Exp Neurol*, 160: 256–267.

Blits, B., Oudega, M., Boer, G.J., *et al.* (2003). *Neuroscience*, 118: 271–281.

Borkenhagen, M., Clémence, J.-F., Sigrist, H., *et al.* (1998). *J Biomed Mater Res*, 40: 392–400.

Bradbury, E.J., Moon, L.D., Popat, R.J., *et al.* (2002). *Nature*, 416: 636–640.

Bray, G.M., Villegas-Perez, M.P., Vidal-Sanz, M., *et al.* (1987). . *J Exp Biol*, 132: 5–19.

Bunge, M.B. and Pearse, D.D. (2003). *J Rehab Res Dev*, 40(Suppl): 55–62.

Bunge, R.P., Puckett, W.R., Becerra, J.L., *et al.* (1993) *Adv Neurol*, 59: 75–89.

Cao, L., Liu, L., Chen, Z.Y., *et al.* (2004). *Brain*, 127: 535–549.

Ceballos, D., Navarro, X., Dubey, N., *et al.* (1999). *Exp Neurol*, 158: 290–300.

Chamberlain, L.J., Yannas, I.V., Hsu, H.-P., *et al.* (1998). *Exp Neurol*, 154: 315–329.

Cheng, H., Cao, Y. and Olsen, L. (1996). *Science*, 273: 510–513.

Cohen, E.D. (2007). *J Neural Eng*, 4: R14–R31.

Crossin, K.L. and Krushel, L.A. (2000). *Dev Dyn*, 218: 260–279.

Cui, Q., Pollett, M.A., Symons, N.A., *et al.* (2003a). *J Neurotrauma*, 20: 17–31.

Cui, Q., Yip, H.K., Zhao, R.C.H., *et al.* (2003b). *Mol Cell Neurosci*, 22: 49–61.

Cui, Q., Hodgetts, S.I., Hu, Y. *et al.* (2007). *Neuroscience*, 146: 986–999.

Das, A.M., Zhao, X. and Ahmad, I. (2005). *Semin Ophthalmol*, 20: 3–10.

David, S. and Aguayo, A.J. (1981). *Science*, 214: 931–933.

Dijkhuizen, P.A., Pasterkamp, R.J., Hermens, W.T., *et al.* (1998). *J Neurotrauma*, 15: 387–397.

Dimou, L., Schnell, L., Montani, L. *et al.* (2006). *J Neurosci*, 26: 5591–5603.

Dinculescu, A., Glushakova, L., Min, S.H., *et al.* (2005). . *Hum Gene Ther*, 16: 649–663.

Doherty, P., Williams, E. and Walsh, F.S. (1995). *Neuron*, 14: 57–66.

Donoghue, J.P., Nurmikko, A.V., Friehs, G., *et al.* (2004). *J Clin Neurophys*, 57(Suppl): 588–602.

Dubey, N., Letourneau, P.C. and Tranquillo, R.T. (2001). *Biomaterials*, 22: 1065–1075.

Ellis-Behnke, R.G., Liang Y.X., You S,W. *et al.* (2006). *Proc Natl Acad Sci*, 103: 5054–5059.

Evans, G.R.D. (2001). *Anat Rec*, 263: 396–404.

Feron, F., Perry, C., Cochrane, J., *et al.* (2005). *Brain*, 128: 2951–2960.

Fine, E.G., Decostered, I., Papaliözos, M., *et al.* (2002). *Eur J Neurosci*, 15: 589–601.

Flynn, L., Dalton, P.D. and Shoichet, M.S. (2003). *Biomaterials*, 24: 4265–4272.

Fouad, K. and Pearson, K. (2004). *Prog Neurobiol*, 73: 107–126.

Gautier, S.E., Oudega, M., Fragoso, M., *et al.* (1998). *J Biomed Mater Res*, 42: 642–654.

Giannetti, S., Lauretti, L., Fernandez, E., *et al.* (2001). *Neurol Res*, 23: 405.

Goldman, S. (2005). *Nat Biotechnolnol*, 23: 862–871.

Goldman, S.A. and Sim, F. (2005). *Novartis Found Symp*, 265: 66–80.

Grill, R., Murai, K., Blesch, A., *et al.* (1997). *J Neurosci*, 17: 5560–5572.

Grimpe, B. and Silver, J. (2004). *J Neurosci*, 24: 1393–1397.

Guo, J., Su, H., Zeng, Y. *et al.* (2007) *Nanomedicine*, 3: 311–321.

Hadlock, T., Sundback, C. and Hunter, D. (2000). *Tissue Eng*, 6: 119–127.

Hadlock, T.A., Sundback, C.A., Hunter, D.A., *et al.* (2001). *Microsurgery*, 21: 96–101.

Harley, B.A., Spilker, M.H., Wu, J.W., *et al.* (2004). *Cells Tissues Organs*, 176: 153–165.

Harvey, A.R. (1999). In *Degeneration, Regeneration in the Nervous System* (Saunders, N.R. and Dziegielewska, M.M., eds). Amsterdam: Harwood Academic, pp. 191–203.

Harvey, A.R., Hu, Y., Leaver, S.G., *et al.* (2006). *Prog Retin Eye Res*, 25: 449–489.

Heath, C.A. and Rutkowski, G.E. (1998). *Trend Biotechnol*, 16: 163–168.

Hermens, W.T. and Verhaagen, J. (1998). *Prog Neurobiol*, 55: 399–432.

Hill, C.E., Moon, L.D., Wood, P.M., *et al.* (2006). *Glia*, 53: 338–343.

Holmes, T.C., de Lacalle, S., Su, X., *et al.* (2000). *Proc Natl Acad Sci USA*, 97: 6728–6733.

Hu, Y., Leaver, S.G., Plant, G.W., *et al.* (2005). *Mol Ther*, 11: 906–915.

Hu, Y., Arulpragasam, A., Plant, G.W., *et al.* (2007). *Exp Neurol*, 207: 314–328.

Huber, A.B., Ehrengruber, M.U., Schwab, M.E., *et al.* (2000). *Eur J Neurosci*, 12: 3437–3442.

Hurtado, A., Moon, L., Maquet, V., *et al.* (2006). *Biomaterials*, 27: 430–442.

Imaizumi, T., Lankford, K.L. and Kocsis, J.D. (2000). *Brain Res*, 854: 70–78.

Jakobsson, J. and Lundberg, C. (2006). *Mol Ther*, 13: 484–493.

Jain, A., Kim, Y.-T., McKeon, R.J., *et al.* (2006). *Biomaterials*, 27: 497–504.

Kakulas, B.A. (1999). *J Spinal Cord Med*, 22: 119–124.

Kamada, T., Koda, M., Dezawa, M., *et al.* (2005). *J Neuropathol Exp Neurol*, 64: 37–45.

Karimi-Abdolrezaee, S., Eftekharpour, E., Wang, J., *et al.* (2006). *J Neurosci*, 26: 3377–3389.

Keirstead, H.S., Nistor, G., Bernal, G., *et al.* (2005). *J Neurosci*, 25: 4694–4705.

Kigerl, K.A., McGaughy, V.M. and Popovich, P.G. (2006). *J Comp Neurol*, 494: 578–594.

King, V.R., Phillips, J.B., Hunt-Grubbe, H., *et al.* (2006). *Biomaterials*, 27: 485–496.

Kobayashi, N.R., Fan, D.P., Giehl, K.M., *et al.* (1997). *J Neurosci*, 17: 9583–9595.

Koeppen, A.H. (2004) . *J Neurol Sci*, 220: 115–117.

Labrador, R.O., Butý, M. and Navarro, X. (1998). *Exp Neurol*, 149: 243–252.

Lavik, E.B., Klassen, H. and Warfvinge, K. (2005). *Biomaterials*, 26: 3187–3196.

Lee, C.J., Huie, P., Leng, T., *et al.* (2002). *Arch Ophthalmol*, 120: 1714–1718.

Lee, A.C., Yu, V.M., Lowe, J.B., *et al.* (2003). *Exp Neurol*, 184: 295–303.

Lemmon, V., Farr, K.L. and Lagenaur, C. (1989). *Neuron*, 2: 1597–1603.

Leon, S., Yin, Y., Nguyen, J., *et al.* (2000). *J Neurosci*, 20: 4615–4626.

Lewin, S.L., Utley, D.S., Cheng, E.T., *et al.* (1997). *Laryngoscope*, 107: 992–999.

Li, Y., Sauve, Y., Li, D., *et al.* (2003). *J Neurosci*, 23: 7783–7788.

Lietz, M., Dreesman, L. and Hoss, M. (2006). *Biomaterials*, 27: 1236–1425.

Lindvall, O. and Kokaia, Z. (2004). *Stroke*, 35(Suppl 1): 2691–2694.

Liu, S., Qu, Y., Stewart, T.J., *et al.* (2000). *Proc Natl Acad Sci USA*, 97: 6126–6131.

Loh, N.K., Woerly, S., Bunt, S.M., *et al.* (2001). *Exp Neurol*, 170: 72–84.

Lu, P., Jones, L.L. and Tuszynski, M.H. (2005). *Exp Neurol*, 191: 344–360.

Lu, P., Yang, H., Jones, L.L., *et al.* (2004). *J Neurosci*, 24: 6402–6409.

Lund, R.D., Ono, S.J., Keegan, D.J., *et al.* (2003). *J Leukoc Biol*, 74: 151–160.

Lundborg, G. (2000). *J Hand Surg*, 25A: 391–414.

Lundborg, G. (2004). *Handchir Mikrochir Past Chir*, 36: 1–7.

Mackinnon, S.E., Dellon, A.L., Hudson, A.R., *et al.* (1985). *Plast Reconstr Surg*, 75: 833–841.

MacLaren, R.E., Pearson, R.A., MacNeil, A., *et al.* (2006). *Nature*, 444: 203–207.

Martin, K.R. and Quigley, H.A. (2004). *Eye*, 18: 1049–1055.

McDonald, J.W., Becker, D., Holekamp, T.F., *et al.* (2004). *J Neurotrauma*, 21: 383–393.

Meek, M.F. and Coert, J.H. (2002). *J Reconstr Microsurg*, 18: 97–109.

Meiners, S., Nur-e-Kamal, M.S. and Mercado, M.L. (2001). *J Neurosci*, 21: 7215–7225.

Melone, M.A., Jori, F.P. and Peluso, G. (2005). *Curr Drug Targets*, 6: 43–56.

Midha, R., Munroe, C., Dalton, P.D., *et al.* (2003). *J Neurosurg*, 99: 555–565.

Mitsui, T., Fischer, I., Shumsky, J.S., *et al.* (2005). *Exp Neurol*, 194: 410–431.

Moellers, S., Noth, J. and Brook, G. (2003). *Glia* (Suppl 2): 1–84. (P202).

Moon, L.D., Asher, R.A., Rhodes, K.E., *et al.* (2001). *Nat Neurosci*, 4: 465–466.

Nandoe Tewarie, R.D.S., Hurtado, A., Levi, A.D.O., Grotenhuis, A., *et al.* (2006). *Cell Transpl*, 15: 563–577.

Nichols, C.M., Myckatyn, T.M., Rickman, S.R., *et al.* (2005). *Behav Brain Res*, 163: 143–158.

Novikov, L.N., Novikova, L.N., Mosahebi, A., *et al.* (2002). *Biomaterials*, 23: 3369–3376.

Oudega, M. (2007) *Acta Physiol*, 189: 181–189.

Oudega, M. and Hagg, T. (1996). *Exp Neurol*, 140: 218–229.

Oudega, M., Moon, L.D. and de Almeida Leme, R.J. (2005). *Braz J Med Biol Res*, 38: 825–835.

Patist, C.M., Borgerhoff Mulder, M.B., Gautier, S.E., *et al.* (2004). *Biomaterials*, 25: 1569–1582.

Pearse, D.D., Pereira, F.C., Marcillo, A.E., *et al.* (2004). *Nat Med*, 10: 610–616.

Picard-Riera, N., Nait-Oumesmar, B. and Baron-Van Evercooren, A. (2004). *J Neurosci Res*, 76: 223–231.

Piotrowicz, A. and Shoichet, M.S. (2006). *Biomaterials*, 27: 2370–2379.

Plant, G.W., Christensen, C.L., Oudega, M., *et al.* (2003). *J Neurotrauma*, 20: 1–16.

Plant, G.W., Ramon-Cueto, A. and Bunge, M.B. (2001). In *Axonal Regeneration in the Central Nervous System* (Ingoglia, N. and Murray, M., eds). New York: Marcel Dekker, pp. 529–561.

Plant, G.W., Woerly, S. and Harvey, A.R. (1997). *Exp Neurol*, 145: 287–299.

Raisman, G. (1997). *Rev Neurol (Paris)*, 153: 521–525.

Ramon-Cueto, A. (2000). *Prog Brain Res*, 128: 265–272.

Ramón-Cueto, A., Plant, G.W., Avila, J., *et al.* (1998). *J Neurosci*, 18: 3803–3815.

Ruitenberg, M.J., Levison, D.B., Lee, S.V., *et al.* (2005). *Brain*, 128: 839–853.

Ruitenberg, M.J., Plant, G.W., Christensen, C.L., *et al.* (2002). *Gene Ther*, 9: 135–146.

Ruitenberg, M.J., Plant, G.W., Hamers, F.P., *et al.* (2003). *J Neurosci*, 23: 7045–7058.

Sahenk, Z., Seharaseyon, J., Mendell, J.R., *et al.* (1994). *Brain Res*, 655: 246–250.

Savitz, S.I., Dinsmore, J.H., Wechsler, L.R., *et al.* (2004). *NeuroR*, 1: 406–414.

Schafer, M., Fruttiger, M., Montag, D., *et al.* (1996). *Neuron*, 16: 1107–1113.

Schense, J.C., Bloch, J., Aebischer, P., *et al.* (2000). *Nat Biotechnol*, 18: 415–419.

Scherer, S.S. and Salzer, J.L. (2001). In *Glial Cell Development, 2nd edn* (Jessen, K.R. and Richardson, W.D., eds). Oxford: Oxford University Press, pp. 299–330.

Schnell, L. and Schwab, M.E. (1990). *Nature*, 343: 269–272.

Shen, Y.J., DeBellard, M.E., Salzer, J.L., *et al.* (1998). *Mol Cell Neurosci*, 12: 79–91.

Silva, G.A., Czeisler, C., Niece, K.L., *et al.* (2004). *Science*, 303: 1352–1355.

Silver, J. and Miller, J.H. (2004). *Nat Rev Neurosci*, 5: 146–156.

Smeal, S.M., Rabbitt, R., Biran, R., *et al.* (2005). *Ann Biomed Eng*, 33: 376–382.

Spencer, T. and Filbin, M.T. (2004). *J Anat*, 204: 49–55.

Spilker, M.H., Yannas, I.V., Kostyk, S.K., *et al.* (2001). *Restor Neurol Neurosci*, 18: 23–38.

Stokols, S. and Tuszynski, M. (2004). *Biomaterials*, 25: 5839–5846.

Stokols, S. and Tuszynski, M. (2006). *Biomaterials*, 27: 443–451.

Takami, T., Oudega, M., Bates, M.L., *et al.* (2002). *J Neurosci*, 22: 6670–6681.

Tan, S.A. and Aebischer, P. (1996). *Ciba Foundation Symp*, 196: 211–236.

Tashiro, K-I., Sephel, G.C., Weeks, B., *et al.* (1989). *J Biol Chem*, 264: 16174–16182.

Taylor, S.J., McDonald, J.W. III and Sakiyama-Elbert, S.E. (2004). *J Control Release*, 98: 281–294.

Teng, Y.D., Lavik, E.B., Qu, X., *et al.* (2002). *Proc Natl Acad Sci USA*, 99: 3024–3029.

Thumann, G. (2001). *Surv Ophthalmol*, 45: 345–354.

Torigoe, K. and Lundborg, G. (1998). *Exp Neurol*, 150: 254–262.

Tsai, E., Dalton, P.D., Shoichet, M.S., *et al.* (2004). *J Neurotrauma*, 21: 789–804.

Tsai, E., Dalton, P.D., Shoichet, M.S., *et al.* (2006). *Biomaterials*, 27: 519–533.

Tuszynski, M.H. (2002). *Lancet Neurol*, 1: 51–57.

Verdu, E., Labrador, R., Rodriguez, F., *et al.* (2002). *Restor Neurol Neurosci*, 20: 169–179.

Villegas-Perez, M.P., Vidal-Sanz, M., Rasminsky, M., *et al.* (1993). *J Neurobiol*, 24: 23–36.

Wei, L., Cui, L., Snider, B.J., *et al.* (2005). *Neurobiol Dis*, 19: 183–193.

Weidner, N., Blesch, A., Grill, R.J., *et al.* (1999). *J Comp Neurol*, 413: 495–506.

Woerly, S., Doan, V.D., Sosa, N., *et al.* (2004). *J Neurosci Res*, 75: 262–272.

Woerly, S., Petrov, T., Syková, E., *et al.* (1999). *Tissue Eng*, 5: 467–488.

Xia, H., Mao, Q., Eliason, S.L., *et al.* (2004). *Nat Med*, 10: 816–820.

Xu, X.M., Chen, A., Guénard, V., *et al.* (1997). *J Neurocytol*, 26: 1–16.

Xu, X.M., Zhang, S.-X., Li, H., *et al.* (1999). *Eur J Neurosci*, 11: 1723–1740.

Yannas, I.V. and Hill, B.J. (2004). *Biomaterials*, 25: 1593–1600.

Yin, Y., Cui, Q., Li, Y., *et al.* (2003). *J Neurosci*, 23: 2284–2293.

Yip, H.K. and So, K.-F. (2000). *Prog Ret Eye Res*, 19: 559–575.

Yoshii, S. and Oka, M. (2001). *J Biomed Mater Res*, 56: 400–405.

Yoshii, S., Oka, M., Shima, M., *et al.* (2004). *J Biomed Mater Res*, 70A: 569–575.

Yu, T.T. and Shoichet, M.S. (2005). *Biomaterials*, 26: 1507–1514.

Zhang, M. and Yannas, I.V. (2005). *Adv Biochem Eng Biotechnol*, 94: 67–89.

Zrenner, E. (2002). *Science*, 295: 1022–1025.

Chapter 21
Tissue engineering of organ systems

Steve Hodges, Peter Frey and Anthony Atala

Chapter objectives:

- To understand the basic anatomy and physiology of the organs involved in tissue engineering
- To understand organ failure and the respective diseases causing it
- To understand the importance of cell-cell and cell-matrix interaction in tissue engineered organs
- To be aware of the importance of adequate mechanical properties of the matrix in organ tissue engineering

- To understand the importance of using the human body as 'bioreactor' to allow final tissue differentiaton
- To know the limiting factors for successful *in vivo* implantation of *in vitro* engineered organs
- To be aware of the potential problems linked to implantation of tissue engineered organs

"Tissue engineering is an interdisciplinary field that applies the principles of engineering and the life sciences toward the development of biological substitutes that restore, maintain or improve tissue function."

Langer and Vacanti (1993)

"Tissue engineering is the basic science and development of biological substitutes for implantation into the body or the fostering of tissue remodeling for the purpose of replacing, repairing, regenerating, reconstructing, or enhancing function."

Galetti *et al.* (1995)

21.1 Introduction

Although engineering of two-dimensional (2D) and very small three-dimensional (3D) tissue structures has been achieved, construction of true 3D functional tissue structures, acting as organ systems, is still a big challenge. This even more so if the organ consists of a multitude of cell types each having a particular function and each depending on each other. Such engineered constructs depend on very complex cell carrier matrices, often in combination with morphogens, apt to allow adherence and growth of the respective cell types and, even more important, matrices promoting vascularization and allowing adequate nutritional support – a key issue in 3D constructs.

21.2 Urogenital tissue engineering

Congenital malformation extending from the severest forms of organ agenesis, dysplasia and hypoplasia to minor anatomical, structural and functional deficiencies may need entire or partial organ replacement. Furthermore the aging and the acutely or gradually diseased organ might require replacement. Individuals may suffer from cancer, trauma, infection, inflammation and iatrogenic injuries that may lead to genitourinary organ damage or loss and necessitate eventual reconstruction. The urogenital system comprises the urinary tract and the internal and external genital apparatus.

The urinary tract is subdivided into the upper tract (comprising the kidney and ureter) and the lower tract (comprising the bladder and urethra). In the female the ovaries, the uterus and the vagina belong to the internal genital apparatus. The external apparatus comprises the introitus, in the female the minor and major labia, and in the male the testicles and the penis and, in particular, the corporal structures (Figure 21.1).

Human organ donation is the classical way to replace diseased, missing or malformed organs. Three major problems are linked to human organ donation: tissue compatibility, tissue rejection and donor shortage.

Whenever there is a lack of native urologic tissue, reconstruction may be performed with nonurologic tissues such as intestinal or gastric segments, mucosa from several body sites and skin, with homologous tissues (e.g. cadaver fascia, cadaver or donor kidney), with heterologous tissues (e.g. bovine collagen) or with artificial substances [e.g. silicone, polyurethane or polytetrafluoroethylene (PTFE)].

The most commonly used tissues for urinary tract organ replacement or repair arise from the intestinal tract. Vascularized small or large bowel and gastric tissue patches are applied. These approaches are, however, often linked to complications, such as metabolic disturbances and even cancer, and rarely replace the entire function of the original tissues because of their inherently different functional parameters.

In most cases, the replacement of lost or diseased tissues with functionally equivalent tissues would improve the outcome for these patients. This goal may be attainable with the use of tissue engineering techniques.

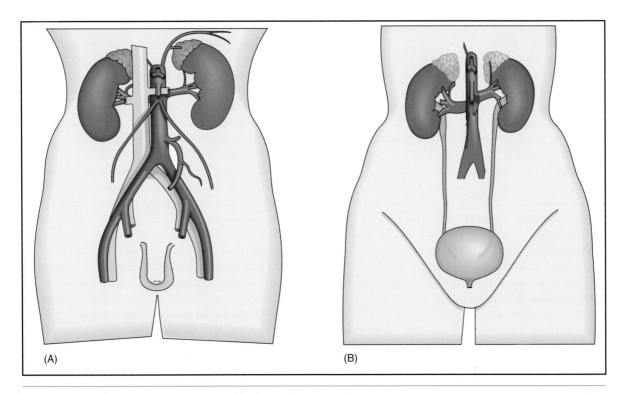

Figure 21.1 Schematic representation of the human urogenital system.

Tissue engineering follows the principles of cell biology, materials science and engineering toward the development of biologic substitutes that can partly or entirely restore and maintain normal function. Tissue engineering may involve matrices alone or in combination with growth factors, wherein the body's natural ability to regenerate is used to induce, orient or direct new tissue growth, or it may use matrices with seeded cells previously cultured *in vitro*.

When cells are used for tissue engineering, donor tissue is dissociated into individual cells, which are then implanted directly into the host or first expanded in culture, attached to a support matrix and reimplanted after expansion.

In tissue engineering, and in particular in tissue engineering of genitourinary tissues, biomaterials should act as an artificial extracellular matrix (ECM), and should offer biologic and mechanical functions of native ECM. The biomaterials should also allow controlled delivery of bioactive factors (e.g. cell adhesion peptides, growth factors) and should offer a three-dimensional space for the development of new tissues with appropriate structure and function (Kim and Mooney, 1998). Direct implantation of cells without biomaterial matrices has been tried (Brittberg *et al.*, 1994; Ponder *et al.*, 1991); however, it proved difficult to control the localization of the transplanted cells.

The optimal biomaterial should be biocompatible, promote cellular interaction and tissue development, and possess proper mechanical and physical properties. Three types of biomaterials can be used: naturally derived matrices (e.g. collagen and alginate), acellular tissue matrices [e.g. acellular bladder submucosa and small intestinal submucosa (SIS)] and synthetic polymers [e.g. poly(glycolic acid) (PGA), poly(lactic acid) (PLA) and poly(lactic-*co*-glycolic acid) (PLGA)].

Naturally derived materials and acellular tissue matrices have the advantage of being biologically recognized, but cannot be produced in a reproducible, controlled way. Synthetic polymers, however, can be produced reproducibly on a large scale with controlled properties of strength, degradation rate and microstructure.

According to Folkman and Hochberg, engineered tissues exceeding a volume of $3\,mm^3$ cannot survive adequately because of impaired nutrition and gas exchange (Folkman and Hochberg, 1973). Vascularization of the regenerating cells is mandatory to engineer large complex tissues.

This goal could be achieved by incorporation of angiogenic growth factors into the engineered tissue, by seeding endothelial cells in combination with the other cell types and by prevascularization prior to cell seeding. There is no doubt that the problem of tissue construct vascularization has to be solved before large entire solid organs can be regenerated.

The finally created cell–matrix construct can be of heterologous, allogeneic or ideally autologous origin and may allow lost tissue function to be at least partly restored (Amiel and Atala, 1999; Atala, 1995, 1997, 1998, 1999; Atala *et al.*, 1992, 1993a,b, 1994; Cilento *et al.*, 1994; Fauza *et al.*, 1998a,b; Kershen and Atala, 1999; Machluf and Atala, 1998; Oberpenning *et al.*, 1999; Park *et al.*, 1999; Yoo and Atala, 1997; Yoo *et al.*, 1998).

21.2.1 Bladder

Intestinal and, less frequently, gastric segments are normally used as tissues for bladder replacement or repair. However, intestinal tissues are designed to absorb, and when they come in contact with the urinary tract they also absorb toxic substance normally eliminated with the urine and therefore multiple complications may ensue, such as infection, metabolic disturbances, increased mucus production, stone formation and even malignancies (Kaefer *et al.*,

1997, 1998; McDougal *et al.*, 1992). The use of gastric tissue may induce metabolic disturbances or even perforation. Owing to the problems encountered with the use of gastrointestinal segments, numerous investigators have attempted alternative reconstructive procedures for bladder replacement or repair such as the use of seromuscular grafts, matrices for tissue regeneration and tissue engineering with cell transplantation.

Seromuscular grafts and de-epithelialized bowel segments, either alone or over a native urothelium, have been successfully attempted (Blandy *et al.*, 1961, 1964; Cheng *et al.*, 1994; Dewan, 1998; Harada *et al.*, 1965; Oesch, 1988; Salle *et al.*, 1990). Keeping the native urothelium intact avoids the complications associated with the use of bowel mucosa exposed to the urinary tract (Blandy *et al.*, 1961; Harada *et al.*, 1965). An example of this strategy is the combination of the techniques of auto-augmentation with those of enterocystoplasty. Auto-augmentation is performed by dissecting the seromuscular layer of the native bladder mucosa and letting the latter bulge, due to its elasticity, to then cover the diverticulum with a demucosalized intestinal or gastric segment (Figure 21.2) (Dewan, 1998).

Allogeneic acellular bladder matrices have served as scaffolds for host cellular bladder wall components. The matrices are prepared by mechanically and chemically removing all cellular components from the donor bladder tissue (Piechota *et al.*, 1998; Probst *et al.*, 1997; Sutherland *et al.*, 1996; Yoo *et al.*, 1998). The extracellular matrices serve as vehicles for partial bladder regeneration and relevant graft versus host reaction is not evident. Porcine SIS, a biodegradable, acellular, xenogeneic predominantly collagen-based tissue-matrix graft, was first used in the early 1980s as a matrix for tissue replacement in the vascular field. It has been shown to promote regeneration of a variety of host tissues, including blood vessels and ligaments (Badylak *et al.*, 1989). Animal experiments have shown that the noncell-seeded SIS matrix used for bladder augmentation is able to regenerate a bladder wall *in vivo* (Kropp *et al.*, 1996, 2004).

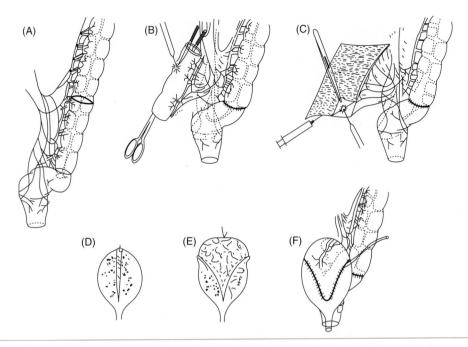

Figure 21.2 Steps involved in autoaugmentation colocystoplasty. (A) A segment of the sygmoid colon is mobilized on its vascular pedicle. (B) The colon is opened along the anti-mesanteric border. (C) The mucosa is removed from the colonic segment by diathermy dissection. (D) Dissection of the muscles from the bladder mucosa is commenced along the anterior wall. (E) The bladder layers are separated until a wide mouth diverticulum is created. (F) The denuded colonic surface is laid over the submucosa layer of the bladder and sutured in place.

In several studies using various materials as non-seeded grafts for cystoplasty, the urothelial layer was able to regenerate normally, but the muscle layer, although present, was not adequately developed (Kropp *et al.*, 2004; Probst *et al.*, 1997; Sutherland *et al.*, 1996; Yoo *et al.*, 1998). The grafts contracted to 60–70% of their original size (Portis *et al.*, 2000) with little increase in bladder capacity or compliance (Landman *et al.*, 2004). Studies involving acellular matrices that may provide the necessary environment to promote cell migration, adherence and growth, as well as differentiation have been conducted (Danielsson *et al.*, 2006a,b; Hubschmid *et al.*, 2005) (Figure 21.3). With continued research, these matrices may have a clinical role in bladder replacement in the future. Recently, bladder regeneration has been shown to be more reliable when the SIS was derived from the distal ileum (Kropp *et al.*, 2004).

Cell-seeded allogeneic acellular bladder matrices have been used for bladder augmentation in dogs. A group of experimental dogs underwent a trigone-sparing cystectomy and were randomly assigned to one of three groups. One group underwent closure of the trigone without a reconstructive procedure, another underwent reconstruction with a non-seeded bladder-shaped biodegradable scaffold, and the last underwent reconstruction using a bladder-shaped biodegradable scaffold that delivered seeded autologous urothelial cells and smooth muscle cells (Oberpenning *et al.*, 1999).

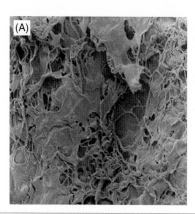
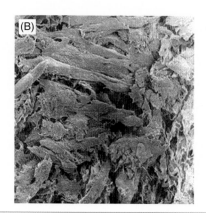

Figure 21.3 Scanning electron microscopy of native equine collagen matrix (TissuFleece®) (A) and with human smooth muscle cells cultured for 10 days (B). Original magnification × 1000.

Classical Experiment

After several reports in the past about successful culture of animal and human urothelial and bladder smooth muscle cells it was only in the 1990s that Southgate and her group (Southgate *et al.*, 1994) characterized in detail cultured human urothelial cells and were able to induce urothelial stratification *in vitro*. They summarized their research as follows:

The purpose of the work was to establish urothelium as an *in vitro* model for the study of proliferation, stratification, and differentiation in 'complex' epithelia. Normal human urothelial cells were cultured in a serum-free medium. The effects of epidermal growth factor (EGF), cholera toxin (CT), extracellular calcium and 13-*cis*-retinoic acid on cell growth, morphology, phenotype and cytodifferentiation were studied using phase-contrast microscopy and indirect immuno-fluorescence. Stratification-related changes were additionally analyzed by transmission electron microscopy. Under optimized conditions, long-term cultures were successful in 44 (74.5%) out of 59 specimens. Bacterial infection was the most common cause of failure (nine cases). Primary urothelial cells required an initial plating density of $\geq 10^4$ cells/cm^2 for survival; passaged cells survived

much lower plating densities ($\geq 2.5 \times 10^2$ cells/cm^2). CT significantly improved cell attachment, but neither CT nor EGF were essential for growth. By contrast, cells failed to proliferate without bovine pituitary extract. In media containing bovine pituitary extract, CT and EGF, cultures had a mean population doubling time of 14.7 ± 1.8 hours, maintained a nonstratified phenotype and expressed the cytokeratin (CK) profile of basal/intermediate urothelium: CK7, CK8, CK17, CK18 and CK19, with variable expression of CK13. CK20 was not expressed *in vitro*. CK14 and CK16 were also expressed, suggestive of squamous metaplasia in culture, which could be inhibited with 13-*cis*-retinoic acid. Increasing extracellular calcium from 0.09 to 0.9–4.0 mM slowed cell proliferation, induced stratification and desmosome formation, and increased expression of E-cadherin. High calcium, EGF, CT and retinoic acid did not induce markers of late/terminal urothelial cytodifferentiation. We describe a simplified technique for the isolation and long-term culture of human urothelial cells. Urothelial cells *in vitro* are capable of rapid proliferation and can be induced to form integrated stratifying cell layers in high calcium medium. Stratification-related changes are not necessarily accompanied by urothelial cell maturation and differentiation.

In the same time Baskin and his collaborators (Baskin *et al.*, 1993) described normal human bladder smooth muscle cells in culture and mentioned detailed methods of their identification and characterization.

This report documents the growth and culture characteristics of human and fetal bovine bladder smooth muscle cells *in vitro*. Bladder smooth muscle cell strains have been identified by their spindle shaped morphology, noncontact inhibited growth characteristics and the expression of smooth muscle cell specific α-actin. Extracellular matrix protein biosynthesis by these cells *in vitro*

has been characterized by metabolic labeling of proteins with (^{14}C) radiolabeled proline and analysis by [sodium dodecyl sulfate] gel electrophoresis. These studies demonstrate that bladder smooth muscle cells synthesize predominantly types I and III collagen, and fibronectin. In addition type III collagen exists in both a partially processed (pN α 1(III)) form and processed form. Complementary immunohistochemical studies show localization of type I, III and IV collagens, and fibronectin to bladder smooth muscle cell extracellular matrix. We conclude that both fetal bovine and human smooth muscle bladder cells are capable of secreting the classic components of the surrounding connective tissue.

The cystectomy-only and nonseeded controls maintained average capacities of 22 and 46% of preoperative values, respectively. An average bladder capacity of 95% of the original precystectomy volume was achieved in the cell-seeded tissue-engineered bladder replacements. The subtotal cystectomy reservoirs that were not reconstructed and the polymer-only reconstructed bladders showed a marked decrease in bladder compliance of 10 and 42% of the total compliance, respectively. The compliance of the cell-seeded tissue-engineered bladders showed almost no difference from preoperative values that were measured when the native bladder was present. Histologically, the nonseeded scaffold bladders presented a pattern of normal urothelial cells, however, with a thickened fibrotic submucosa and a thin layer of smooth muscle fibers. The retrieved tissue-engineered bladders showed a normal cellular organization, consisting of a trilayer of urothelium, submucosa and smooth muscle (Oberpenning *et al.*, 1999). Applying this technology, Atala *et al.* (2006) performed cystoplasty in seven patients with myelomeningocele with high-pressure or poorly compliant bladders. Urothelial and bladder smooth muscle cells, harvested by biopsy, were grown in culture and seeded on a biodegradable bladder-shaped scaffold

made of collagen or a composite of collagen and PGA. The autologous engineered bladder constructs were implanted either with or without an omental wrap. The mean bladder leak point pressure decreased, and the volume and compliance increased, and was greatest in the composite engineered bladders associated with an omental wrap. Bladder biopsies showed an adequate structural architecture and phenotype of the bladder wall.

21.2.2 Urethra

Different approaches have been suggested for the regeneration of urethral tissue. Woven PGA (Dexon®) meshes without cells have been used to reconstruct the urethra in dogs (Bazeed *et al.*, 1983; Olsen *et al.*, 1992). Furthermore, PGA has been used as a cell carrier to engineer tubular urothelium *in vivo*. SIS without cells and a homologous free graft of acellular urethral matrix were used as grafts for urethroplasty in rabbit models (Kropp *et al.*, 1998; Sievert *et al.*, 2000).

Bladder-derived acellular collagen matrix has proven to be a suitable graft for repair of urethral defects, and the created neourethras showed normal

State of the Art Experiment

The chosen state of the art experiment in the field of pediatric urology should be considered as an example representing the broad field of organ tissue engineering.

Apart form the numerous publications addressing the techniques of culture of human urothelial and bladder smooth muscle cells alone or in combination with biocompatible natural or synthetic matrices, two publications were predominantly responsible for building the bridge between pure basic research and clinical application in urology.

The first, by Frank Oberpenning et al., 'De novo reconstitution of a functional mammalian urinary bladder by tissue engineering', was published in Nature Biotechnology in 1999 (Oberpenning et al., 1999) and the second, by Antony Atala et al., 'Tissue-engineered autologous bladders for patients needing cystoplasty', was published in the Lancet in 2006 (Atala et al., 2006). The significance of their research is best summarized in the respective abstracts.

Human organ replacement is limited by a donor shortage, problems with tissue compatibility and rejection. Creation of an organ with autologous tissue would be advantageous. In this study, transplantable urinary bladder neo-organs were reproducibly created in vitro from urothelial and smooth muscle cells grown in culture from canine native bladder biopsies and seeded onto preformed bladder-shaped polymers. The native bladders were subsequently excised from canine donors and replaced with the tissue-engineered neo-organs. In functional evaluations for up to 11 months, the bladder neo-organs demonstrated a normal capacity to retain urine, normal elastic properties and histological architecture. This study demonstrates, for the first time, that successful reconstitution of an autonomous hollow organ is possible using tissue-engineering methods. From Oberpenning et al. (1999).

Patients with end-stage bladder disease can be treated with cystoplasty using gastrointestinal segments. The presence of such segments in the urinary tract has been associated with many complications. We explored an alternative approach using autologous engineered bladder tissues for reconstruction. Seven patients with myelomeningocele, aged 4–19 years, with high-pressure or poorly compliant bladders, were identified as candidates for cystoplasty. A bladder biopsy was obtained from each patient. Urothelial and muscle cells were grown in culture, and seeded on a biodegradable bladder-shaped scaffold made of collagen, or a composite of collagen and polyglycolic acid. About 7 weeks after the biopsy, the autologous engineered bladder constructs were used

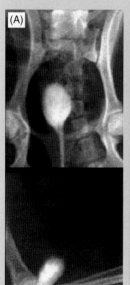 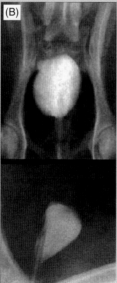 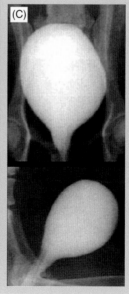

Radiographic cystograms 11 months after subtotal cystectomy. (A) Subtotal cystectomy without reconstruction (group A), (B) polymer-only implant (group B) and (C) tissue-engineered neo-organ (group C).

for reconstruction and implanted either with or without an omental wrap. Serial urodynamics, cystograms, ultrasounds, bladder biopsies and serum analyses were done. Follow-up range was 22–61 months with a mean of 46 months. Postoperatively, the mean bladder leak point pressure decrease at capacity, and the volume and compliance increase was greatest in the composite engineered bladders with an omental wrap 56%, 1.58-fold and 2.79-fold, respectively. Bowel function returned promptly after surgery. No metabolic consequences were noted, urinary calculi did not form, mucus production was normal and renal function was preserved. The engineered bladder biopsies showed an adequate structural architecture and phenotype. Engineered bladder tissues, created with autologous cells seeded on collagen–polyglycolic acid scaffolds, and wrapped in omentum after implantation, can be used in patients who need cystoplasty. From Atala *et al.* (2006).

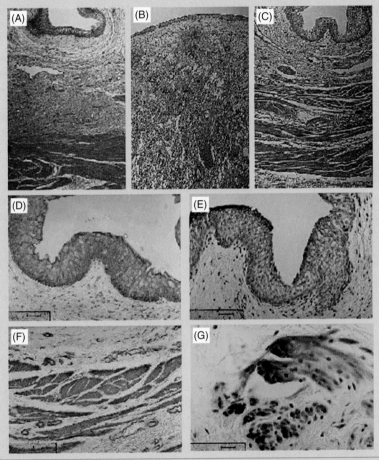

Histological and immunochemical analysis of implants 6 months after surgery. Hematoxylin & eosin histologic results (magnification × 140) are shown for (A) normal canine bladder (group A), (B) the bladder dome of the cell-free polymer reconstructed bladder (group B) and (C) the tissue-engineered neo-organ (group C). Immunocytochemical staining of tissue-engineered neo-organ revealed (D) positive staining of the epithelial cell layers with pancytokeratin AE1/AE3 antibodies (original magnification × 210); (E) urothelial differentiation-related membrane proteins, which constitute the apical plaques of the asymmetric unit membrane in normal urothelium (original magnification × 210); (F) anti-α smooth muscle actin staining of phenotypically normal smooth muscle (original magnification × 210); and (G) positive staining with S-100 antibodies binding to neural tissue and nerve sheaths (original magnification × 320).

urothelial lining and organized muscle bundles in the rabbit model (Chen *et al.*, 1999). The above results were confirmed clinically in a series of patients with a history of failed hypospadias repair wherein the urethral defects were repaired with human bladder acellular collagen matrices (Atala *et al.*, 1999).

The described techniques, using nonseeded acellular matrices, were applied experimentally and clinically in a successful manner for onlay urethral repairs. However, when tubularized urethral repairs were attempted experimentally, adequate functional urethral tissue regeneration was not achieved and complications, predominantly graft contracture and stricture formation, were noticed (Le Roux, 2005). On the contrary, seeded tubularized collagen matrices have performed better in animal studies. In a rabbit model, urethroplasty was performed with tubularized collagen matrices either seeded with cells or without cells. The matrices seeded with autologous cells formed new tissue, which was histologically comparable to native urethra. However, the tubularized collagen matrices without cells led to limited tissue development, fibrosis and stricture formation. These findings were also confirmed clinically (De Filippio *et al.*, 2002b).

21.2.3 Ureter

Ureteral nonseeded matrices have been used as scaffolds for growth of ureteral tissue in rats. The acellular matrices promoted the regeneration of the ureteral wall components (Dahms *et al.*, 1997). Ureteral replacement with polytetrafluoroethylene (Teflon®) grafts in dogs resulted in poor functional results (Baltaci *et al.*, 1998). Nonseeded ureteral collagen acellular matrices were tubularized and used to replace segments of canine ureters. The nonseeded acellular matrix tube was not able to successfully replace the resected segment of the ureter (Osman *et al.*, 2004).

Cell-seeded biodegradable polymer scaffolds have proved to be more successful in the reconstruction of ureteral tissues. In one study, urothelial and smooth muscle cells isolated from bladders and expanded *in vitro* were seeded onto tubular PGA scaffolds, and thereafter implanted subcutaneously into athymic mice. The urothelial cells proliferated to form a multilayered luminal lining, while the smooth muscle cells organized into multilayered structures surrounding the urothelial cells and pronounced angiogenesis was observed. The degradation of the polymer scaffolds resulted in formation of natural urothelial tissues. This approach was also used to replace ureters in dogs by transplantation of smooth muscle and urothelial cells on tubular polymeric scaffolds (Yoo *et al.*, 1995).

21.2.4 Kidney

The kidney is surely the most challenging organ in the urinary system to be reconstructed by tissue engineering methods because of its complex structure and function (Figure 21.4). To develop a bioartificial kidney some researchers are pursuing the replacement of isolated kidney function parameters with the use of extracorporal cellular units, while others are aiming at the replacement of total renal function by tissue-engineered bioartificial structures.

The commonly used dialysis to replace absent renal function is associated with a relatively high morbidity and mortality. For this reason investigators seek alternative solutions involving *ex vivo* cellular systems, this in the attempt to assess the viability and physiologic functionality of a cell-seeded device to replace the filtration, transport, metabolic and endocrinologic functions of the kidney. In acutely uremic dogs, a combination of synthetic hemofiltration and a device containing porcine renal tubules in an extracorporeal perfusion circuit was investigated. It could be shown that potassium and blood urea nitrogen were controlled during treatment with this device. Fractional reabsorption of sodium and water was possible, and active transport of potassium, bicarbonate and glucose, and a gradual ability to excrete

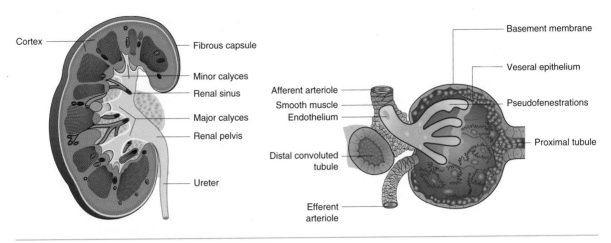

Figure 21.4 The complex renal structure.

ammonia were observed. These results showed the technologic feasibility of an extracorporeal assist device that is reinforced by the use of proximal tubular cells (Humes *et al.*, 1999).

Applying similar techniques, a tissue-engineered bioartificial kidney consisting of a conventional hemofiltration cartridge in connection with a device containing human renal proximal tubule cells was developed and used in patients suffering from acute renal failure. The initial clinical experience suggests that renal tubule cell therapy may provide a dynamic and individualized treatment program as assessed by acute physiologic and biochemical indices (Humes *et al.*, 2003).

Improvement of renal function can also be achieved by augmentation of renal tissue with kidney cell expansion *in vitro* and subsequent autologous transplantation. The feasibility of achieving renal cell growth, expansion and *in vivo* reconstitution of renal tissue with the use of tissue engineering techniques has been explored.

Recently, an attempt was made to harness the reconstitution of renal epithelial cells for the generation of functional nephron units. Renal cells were harvested and expanded in culture, to then be seeded onto a tubular device constructed from a polycarbonate membrane, connected with a silastic catheter that drained into a micro-reservoir. Histologic examination of the device, implanted in athymic mice, revealed extensive vascularization with the formation of glomeruli and highly organized tubule-like structures over time. Immunohistochemical staining confirmed the renal phenotype of the cells. Yellow urine-like fluid was collected from inside the implant, and the latter was comparable to dilute urine with regard to its creatinine and uric acid concentrations (Yoo *et al.*, 1996). Further studies have been performed showing the formation of renal structures in cows using nuclear transfer techniques (Lanza *et al.*, 2002). This technology will be further developed aiming at the expansion of this system to larger, 3D functional structures.

21.2.5 Uterus and vagina

Congenital malformations such as cloacal exstrophy and intersex disorders may not present sufficient uterine tissue for future reproduction. The possibility of engineering functional uterine tissue using autologous cells was investigated (Wang *et al.*, 2003). For this reason autologous rabbit uterine smooth muscle and epithelial cells were harvested, and then cultured and expanded *in vitro*. Thereafter, the cells were seeded onto uterine-shaped biodegradable polymer scaffolds, which were consequently used for subtotal uterine tissue replacement. Six months after implantation, histological, immunohistochemical and Western blot analyses confirmed the presence of normal uterine tissues. Biomechanical analyses and organ bath studies showed that the functional characteristics of these tissues were similar to those of normal uterine tissue.

Several pathologic conditions, including congenital malformations and malignancy, can adversely affect normal vaginal development or anatomy. Vaginal reconstruction is challenging due to the lack of native tissue. Vaginal epithelial and smooth muscle cells of female rabbits were harvested and expanded in culture. These cells were seeded onto biodegradable polymer scaffolds and the cell-seeded constructs were then implanted into nude mice for up to 6 weeks. Immunohistochemical, histological and Western blot analyses confirmed the presence of vaginal tissue phenotypes. Electrical field stimulation studies in the tissue-engineered constructs showed similar functional properties to those of normal vaginal tissue. When these constructs were used for autologous total vaginal replacement, functional vaginal structures were noted in the tissue-engineered specimens; however, implanting the noncell-seeded structures resulted in stenotic structures (De Filippio *et al.*, 2003).

21.2.6 Testicle

Androgen substitution is necessary in patients with testicular dysfunction to allow normal somatic development. Testicular dysfunction is treated with periodic intramuscular injections of chemically modified testosterone and, more recently, transcutaneously with skin patch applications. However, long-term nonpulsatile testosterone therapy can cause multiple problems, including erythropoiesis and bone density changes.

For these reasons a system was designed wherein Leydig cells were microencapsulated for controlled testosterone release. Purified Leydig cells were isolated and encapsulated in an alginate–poly-L-lysine solution. The encapsulated Leydig cells were injected into castrated animals and serum testosterone was measured regularly. The animals were able to maintain their testosterone levels during long-term observation (Machluf *et al.*, 1998). These studies suggest that microencapsulated Leydig cells may be feasible in future to replace or supplement testosterone in patients where anorchia or acquired testicular failure is present.

21.2.7 Penis

One of the major components of the phallus is corporal smooth muscle. The creation of autologous functional and structural corporal tissue *de novo* would be beneficial. In order to look at the functional parameters of the engineered corpora, acellular corporal collagen matrices were obtained from donor rabbit penis, and autologous corpus cavernosal smooth muscle and endothelial cells were harvested, expanded and seeded on the matrices. The entire rabbit corpora was removed and replaced with the engineered scaffolds. The experimental corporal bodies demonstrated intact structural integrity by cavernosography and showed similar pressure by cavernosometry when compared to the normal controls. The control rabbits without cells failed to show normal erectile function throughout the study period. Mating activity in the animals with the engineered corpora appeared normal by 1 month after implantation. The presence of sperm was confirmed during mating and was present in all the rabbits with the engineered corpora. The female rabbits mated with the animals implanted with engineered corpora, and also conceived and delivered healthy pups. Animals implanted with the matrix alone were unable to demonstrate normal mating activity and failed to ejaculate into the vagina (Chen *et al.*, 2005a).

Silicone is used as a biomaterial for penile prostheses; however, biocompatibility remains a concern (Nukui *et al.*, 1997; Thomalla *et al.*, 1987). The appliance of a natural prosthesis composed of autologous cells may be advantageous. The feasibility of using autologous engineered cartilage rods *in situ* was investigated (Yoo *et al.*, 1999). Chondrocytes were harvested from rabbit ears, and grown and expanded in culture. The cells were then seeded onto biodegradable poly-L-lactic acid-coated PGA polymer rods and implanted into the corporal spaces of the rabbit penis. At retrieval the presence of well-formed, milky-white cartilage structures within the corpora was noted at 1 month. The polymeric structures were fully degraded by 2 months. No evidence of erosion or infection in any of the implantation sites was observed. Further studies were performed to assess the long-term functionality of the cartilage penile rods *in vivo*. The animals could copulate and impregnate their female partners.

21.2.8 Engineered bulking agents in urology

Both vesicoureteral reflux and urinary incontinence are frequent conditions affecting the genitourinary system for which injectable bulking agents can be used for endoscopic treatment. The ideal substance for the treatment of reflux and incontinence should be easily injectable, nonantigenic, nonmigratory, volume preserving and safe for use in humans. In clinical practice hyaluronic acid–dextranomers or collagen are commonly used (Frey *et al.*, 1995; Läckgren *et al.*, 2001). Animal studies have shown that chondrocytes can be easily harvested and cultured together with alginate *in vitro*. The suspension can then be injected endoscopically and the elastic cartilage subsequently formed is able to act as bulking agent to correct vesicoureteral reflux without any evidence of obstruction (Atala, 1994). The

first human application of cell-based bulking agents occurred with the injection of chondrocytes for the correction of vesicoureteral reflux in children and for urinary incontinence in adults (Bent *et al.*, 2001; Diamond and Caldamone *et al.*, 1999).

The use of autologous smooth muscle cells was also explored for treatment of both urinary incontinence and vesicoureteral reflux (Cilento and Atala, 1995). The use of cultured myoblasts for the injection treatment of stress urinary incontinence has also been investigated (Chancellor *et al.*, 2000; Yokoyama *et al.*, 1999). The use of injectable muscle precursor cells has also been investigated as a potential treatment for urinary incontinence due to irreversible urethral sphincter injury or maldevelopment (Yiou *et al.*, 2003). A clinical trial involving the use muscle-derived stem cells (MDSC) to treat stress urinary incontinence has also been performed. Biopsies of skeletal muscle were obtained, and autologous myoblasts and fibroblasts were cultured. Under ultrasound guidance, myoblasts were injected into the rhabdosphincter and fibroblasts together with collagen were injected into the submucosal space. One year following injection, the thickness and function of the rhabdosphincter had significantly increased, and all patients were continent (Strasser *et al.*, 2004).

Injectable muscle-based gene therapy and tissue engineering were combined to improve detrusor function in a bladder injury model, and may potentially be a novel treatment option for urinary incontinence (Huard *et al.*, 2002).

21.3 Liver tissue engineering

The liver is a very complex organ that has the ability to regrow and regenerate (Figure 21.5). However, liver failure can still occur due to many different causes, including acquired liver disease, inborn errors of metabolism or drug toxicity. Currently, when a liver fails, the lost function can only be replaced fully by liver transplantation. The options include transplantation of the whole liver or parts of the liver

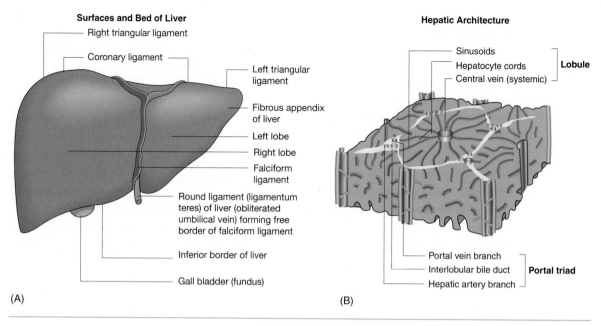

Surfaces and Bed of Liver
- Right triangular ligament
- Coronary ligament
- Left triangular ligament
- Fibrous appendix of liver
- Left lobe
- Right lobe
- Falciform ligament
- Round ligament (ligamentum teres) of liver (obliterated umbilical vein) forming free border of falciform ligament
- Inferior border of liver
- Gall bladder (fundus)

Hepatic Architecture
- Sinusoids
- Hepatocyte cords
- Central vein (systemic)
- **Lobule**
- Portal vein branch
- Interlobular bile duct
- Hepatic artery branch
- **Portal triad**

(A)   (B)

Figure 21.5 Liver structure.

from cadaveric donors, portions of the liver from living donors, or xenotransplantation, typically porcine. Unfortunately, there are far too few livers available for the need worldwide, with approximately 5 times as many recipients as donors (Pruett, 2002), and xenotransplantation is limited by immune and infectious complications. Even when human donors are available, liver transplantation requires continuous use of immunosuppressive medication to prevent rejection, and these agents have well known and severe side-effects. Compounding the problem is the high mortality of fulminate liver failure. Owing to the lack of suitable liver donors for all patients needing them, the mortality rate of liver failure, the high cost of treatment of patients awaiting liver transplants and the shortcomings of liver transplantation as a cure, scientists have searched for alternative means of liver function replacement (Park and Lee, 2005; van de Kerkhove *et al.*, 2004).

There have been numerous attempts at artificial replacement of liver function, with uniformly poor results. These devices attempt to simply detoxify the blood through hemodialysis, plasmapharesis and other methods by using adsorptive resins to bind toxins or, in some cases, by albumin dialysis. Purely artificial systems have not been successful because they simply filter the blood, and do not perform the many essential metabolic and synthetic functions of the liver (Millis and Losanoff, 2005).

Other attempts at liver replacement have focused on recreating the essential metabolic and synthetic functions of the liver. The primary functioning cell of the liver is the hepatocyte. The other cells composing the liver, including the hepatocyte precursor cells, stellate cells, Kuppfer cells, epithelial cells, sinusoidal epithelial cells, biliary epithelial cells and fibroblasts, have supporting and structural roles, but do not contribute directly to hepatic metabolism. The primary functions of the liver include metabolic activity, i.e. urea production, glycogen storage and metabolic waste removal. The key to liver replacement is to replicate these functions; since these primary functions are performed by hepatocytes, the method of liver substitution must harness hepatocyte function in a novel manner (Najimi and Sokal, 2005; Selden and Hodgson, 2004).

The simplest method of liver failure treatment using hepatocytes is cell transplantation. This method of hepatic replacement has been studied for over 25 years. The first liver cell transplantation was described in 1979, with the successful transplant of viable rat hepatocytes into the spleen in an animal study (Mito *et al.*, 1979). Shortly thereafter, additional studies demonstrated that hepatocytes transplanted into the spleen could survive and proliferate long term. It was also noted that during hepatocyte implantation, some cells would flow to the liver and implant there from the portal circulation. Other ectopic sites were investigated as sites for hepatocyte implantation, such as lungs, fat pads, subcutaneous areas and peritoneum, but were not as successful. These studies taught us valuable lessons about the importance of the microenvironment to cell survival, as the cells required conditions resembling the liver, including a matrix to grow on and a venous blood supply similar physiologically to the hepatic sinusoid to grow and function normally. They also demonstrated the role of nonhepatocyte cells in hepatocyte survival, as coculture with nonparenchymal cells was often required to ensure cell viability (Fox and Roy-Chowdhury, 2004; Selden and Hodgson, 2004).

The problems with hepatocyte transplantation include the inefficiency of cell engraftment and the delay in replacement of hepatic function by the implanted cells (Bilir *et al.*, 2000). Methods to improve the efficiency of cell engraftment have been developed and they all include damage to the liver with prevention of native cell recovery, allowing for the growth of the transplanted hepatocytes. Unfortunately, clinical experience shows that many patients treated with hepatocytes for acute liver

failure die of sepsis. This may be due to immuno-suppression during transplantation or perhaps the disruption of hepatic sinusoids, which allows for transplanted cell implantation, also allows for bacterial ingrowth (Selden and Hodgson, 2004).

In the setting of hepatic disease, transplanted cells should grow well because these cells will have a selective advantage over diseased cells. This theory has experimental evidence in animals and this approach of cell transplantation has been most successful clinically for metabolic disease. For example, glycogen storage disease, urea cycle deficiency, factor VII deficiency and Refsum disease have all been successfully treated with this strategy (Najimi and Sokal, 2005; Selden and Hodgson, 2004).

The number of cells required for therapy varies depending on the disease process, with less needed to replace an enzyme deficiency and more needed for total liver failure. Evidence shows that one may be able to restore hepatic function in patients with enzyme deficiencies by replacing only 1–5% of liver mass (Fox *et al.*, 1998; Muraca *et al.*, 2002; Selden *et al.*, 1995). This is likely due to the redundancy of the liver or enzyme induction. Treating acute or chronic liver failure is another issue, as transplanting more than 10% of the functioning liver mass is not possible, but one often needs more than 10% function to replace full liver function (Selden and Hodgson, 2004).

There have been several trials of treatment of fulminate liver failure with hepatocyte transplantation. Most of these studies have had equivocal results, with some patients showing signs of clinical improvement and surviving to transplantation, while others died shortly after cell infusion. In the setting of fulminate liver failure, there is no time for cells to engraft, repopulate and expand, so that a large number of cells must be infused initially to replace the rapidly deteriorating function. It has been hypothesized that 10^{10} cells are needed for transplantation to be successful and this is not possible with simple liver infusion. Studies are investigating the use of the peritoneal cavity as a site of cell transplant, although no successful human trials have been reported. The peritoneal cavity can accept large numbers of cells, however, and may be an ideal site for large volume cell transplantation. This method has shown some success in animal models via transplantation of cocultured hepatocytes and nonparenchymal cells (Baccarani *et al.*, 2005; Selden and Hodgson, 2004).

For cells to be useful in transplantation they must have a high proliferative capacity without loss of function and they must be economically feasible. There are several groups of possible cell sources for liver replacement. These include primary hepatocytes, immortalized cell lines, stem cells and transdifferentiated cells (Fox and Roy-Chowdhury, 2004; Selden and Hodgson, 2004).

Primary hepatocytes are the most desirable and commonly used cells. Unfortunately, hepatocyte function is closely tied to the orientation, location and surrounding environment of the cells. These environmental characteristics are lost in culture, so these cells rapidly and irreversibly lose their function. Therefore, novel methods of cell culture, or alternative sources of cells, must be developed to obtain the necessary hepatocytes (Selden and Hodgson, 2004).

In terms of novel cell culture techniques, most studies of long-term hepatocyte culture have focused on the ECM and cell–cell interactions. Some recent experimental work has focused on the control of substrate properties for modulating cell behavior. In addition, cocultivation of hepatocytes with other cells, such as endothelial cells, has been demonstrated to be crucial for hepatocyte viability and liver-specific functions (Mitaka, 1998; Selden and Hodgson, 2004).

Human hepatocytes would be the best choice for cell transplantation, but just as liver donors are limited, so are human hepatocytes. Fetal or neonatal hepatocytes may also provide a source of cells. They have finite proliferation capacity, but may be immortalized (Selden and Hodgson, 2004). Alternatives to human cells include primary porcine hepatocytes. These cells are plentiful and readily available, and have similar functional characteristics to the human cell (Barshes *et al.*, 2005). This strategy is not without its limitations, however, as there is the possibility of infectious complications such as xenozoonosis (e.g. porcine endogenous retroviral infection). Furthermore, rejection is a major problem in xenograft liver cell transplantation (Fox and Roy-Chowdhury, 2004).

Cell encapsulation is one strategy to prevent rejection. It does not prevent the development of antibodies to secreted proteins which cross the encapsulation barrier, but does prevent access of host cells to the transplanted cells. Systemic immunosuppressants are also useful if needed, but cotransplanting cells with Sertoli cells can also provide immunosuppression locally. Sertoli cells are central to the immune privilege demonstrated by the testes. Therefore, coculture of cells with Sertoli cells may provide a level of protection from the host immune response. A useful strategy would be encapsulation combined with coculture of these cells with Sertoli cells (Selden and Hodgson, 2004). Another option is genetically engineering less immunogenic cells. Scientists are breeding transgenic pathogen-free, low-immunogenicity pigs for the sole purpose of hepatocyte harvesting for use in humans (Barshes *et al.*, 2005; Selden and Hodgson, 2004).

Another option is the use of tumor-derived cell lines and immortalized lines. Immortalized hepatoblast carcinoma cell lines, e.g. HepG2/C3A or Sv40, can be cultured and expanded to large quantities, but do not have all the necessary functions of mature hepatocytes. Deficient functions, such as ammonia metabolism, may be restored by transfection with specific genes. However, since they are cancerous lines, they may not be used in humans for liver replacement. One can also attempt immortalization of hepatocytes in culture. This may be performed by genetic manipulation and an example is the NKNT-3 cell line, which may be used extracorporally but not inside the body (Baccarani *et al.*, 2005; Barshes *et al.*, 2005; Selden and Hodgson, 2004).

Another source of hepatocytes would be stem cells. Stem cells have an unlimited replication potential and can differentiate into various different cell types. Potential sources of stem cells for hepatocyte generation include embryonic stem (ES) cells, adult liver progenitor cells and transdifferentiated nonhepatic cells. ES cells offer great hope, but have been directed down a hepatocyte lineage only in animal models. Of course, there are ethical and legal concerns with ES cell use as well. The adult hepatic progenitor cells, or oval cells, may develop into hepatic cells or biliary epithelial cells, and they often provide for hepatocyte regeneration following liver injury. Unfortunately, these cells are difficult to isolate and culture, although this is being investigated. Whilst they may be cultured *in vitro*, studies show no benefit in cell expansion over mature hepatocytes. Certain nonoval hepatic progenitor cells have also been identified, with the promise of increased self-renewal capacity. Bone marrow-derived or hematopoietic stem cells have the potential to develop into hepatocytes, but the growth factors and extracellular milieu required for this development are not clearly understood and are areas of fertile research. Also, transdifferentiation of adult nonhepatic cells may also provide a novel cell source for liver replacement therapy (Baccarani *et al.*, 2005).

Another strategy of liver replacement in the setting of liver failure is bioartificial liver (BAL) replacement. As an evolution of the artificial liver devices, BAL constructs were developed to harness the function of hepatocytes in an extracorporeal system. BAL devices have been under development for over 40 years and usually consist of isolated hepatocytes with a membrane through which the plasma is circulated. There are four basic designs of BALs: hollow fiber, flat plate, perfused scaffolds and encapsulation chambers. Since 1990, several of these devices have been clinically tested and many are currently under testing for clinical trials (Allen *et al.*, 2002; Park and Lee, 2005) (Table 21.1).

Primary porcine hepatocytes are the most common cell types used in the BAL, although other cell sources are used as well. All BALs need to filter large amounts of blood efficiently and need to maintain the hepatocyte phenotype to be successful. A typical device will need to house approximately 15 billion hepatocytes to support a patient (Allen *et al.*, 2002; Park and Lee, 2005).

Examples of BAL devices include several currently undergoing clinical trials, such as the extracorporeal liver assist device (ELAD), which uses the human hepatocyte line C3A derived from a hepatoma cell line. Unfortunately it did not improve survival or blood toxin levels in a controlled study. The liver support system (LSS) uses a filtering mechanism combined with human hepatocytes harvested from discarded liver donors and porcine hepatocytes. In clinical trials, the LSS was used alone to treat patients with human hepatocytes and was also combined with an albumin dialysis device to form a modular extracorporeal liver support (MELS) device. Both these systems effectively maintained several transplant candidates and improved specific clinical parameters. The BLSS system uses porcine hepatocytes and has been shown to decrease blood ammonia levels in clinical trials, but failed to demonstrate clinical improvement. The radial flow bioreactor uses porcine hepatocytes in combination with a porous matrix system to filter plasma. Clinical trials have demonstrated an improvement in neurologic status clinically, with decreased ammonia and bilirubin levels. The academic medical center BAL used a combination of a polyester matrix for cell attachment and hollow fibers, with excellent improvements in measured parameters (Park and Lee, 2005).

The HepatAssist® device uses porcine hepatocytes and with the least number of hepatocytes of any of the devices it is able to treat patients in the shortest amount of time. It does not use direct interaction of the patient's plasma with the hepatocytes, but rather uses only diffusion and filtration. The HepatAssist® device, in a prospective, randomized, multicenter controlled clinical trial, demonstrated a statistically significant survival advantage compared to controls. Unfortunately, there was also an incidence of porcine endogenous retrovirus transmission following treatment (Allen *et al.*, 2002; Park and Lee, 2005).

From all the clinical and animal data from the liver assist device studies, certain trends have been noted. First, there is a tendency to create devices with more

Table 21.1 Overview of BAL systems

BAL system		Characteristics of six BAL systems						Clinical results		
		Bioreactor configuration	Shape of cells	Hepatocyte source	Immunological barrier	Perfusion (plasma separation rate, ml/min)	Reactor flow rate (ml/min)	Neurological improvement	Ammonia removal	Bilirubin elimination
Hollow fiber systems	ELAD (Amphioxas Cell Technologies)		Large aggregates	Human cell line (C3A) (400g)	Yes (70–120 kD)	Blood (NA)	200	Probably	−8% (increased)	−20% (increased)
	HepatAassist (Circe Biomedical)		Microcarrier attached irregular aggregates	Crypreserved porcine (5–7×10⁹)	None	Plasma (50)	400	Yes	18%	18%
	LSS(MELS) (Charite, Humbolt Univ., Germany)		Tissue-like organoids	Porcine (600g)	None (300kD)	Plasma (31)	100–200	Yes	Not reported	Not reported
	BLSS (Excorp Medical, Inc.)		Collagen gel entrapped	Porcine (70–120g)	Yes (100kD)	Bloos (NA)	100–250	No	33%	6%
Porous matrix systems	RFB-BAL (Univ. of Ferrara, Italy)		Aggregates	Porcine (200g)	None	Plasma (22)	200–300	Yes	33%	11%
	AMC-BAL (Univ. of Amsterdam, Netherlands)		Small aggregates	Porcine (1.2×10¹⁰)	None	Plasma (40–50)	150	Yes	44%	31%

cell mass to improve performance, with most using 10–20 billion hepatocytes. Also, immunologic barriers do not seem to be necessary. Finally, hepatocytes must be cultured as an organoid to improve cell–cell interactions and promote growth. Unfortunately, there are theoretical concerns about immunogenicity, xenozoonosis and tumorigenicity with these devices. Also, clinical trials have shown little significant clinical improvements (Allen *et al.*, 2002; Park and Lee, 2005).

Research has demonstrated that hepatocyte transplantation using polymeric matrices can be used as an alternative therapy for metabolic liver disease. Scaffolds are a crucial component of large-scale hepatocyte transplantation, as they are needed to provide a support framework and mode of delivery for the cells to sustain and guide cell growth, and to provide signals or growth factors needed for growth, differentiation and interaction (Baccarani *et al.*, 2005; Kulig and Vacanti, 2004; Selden and Hodgson, 2004).

Once a source of cells is obtained, the cells must be delivered to the host by some means to harness their function. Often a scaffold is used as the support structure for the cells and the composition of the scaffold can vary as much as the cell source. The scaffold must provide a biocompatible and biodegradable support for the cells. It must also be porous for nutrient and gas exchange, and provide growth factors for proper cell development (Baccarani *et al.*, 2005; Kulig and Vacanti, 2004; Selden and Hodgson, 2004).

The main problem for large organ engineering is angiogenesis and the development of 3D matrices. Regardless of composition, scaffolds may be modified to optimize cell growth and function. For example, scaffolds can be engineered with specific growth factors, may contains prevascularization or pores for angiogenesis and may use bioactive polymers, with the incorporation of proteins to improve cell adherence to the scaffold (Kulig and Vacanti, 2004).

The ultimate goal of tissue engineering for liver replacement involves the creation of a complete, implantable bioengineered liver. Many strategies have been explored as a means to achieve this goal, such as the injection of hepatocytes into vascular beds or the use of implantable seeded scaffolds. One of the main limitations of bioengineering large organs such as the liver is vascularization. Individual cells must be within close proximity to blood vessels for oxygenation and waste removal. To solve this problem, Vacanti's group has seeded cells around preformed vascular beds. Other options include seeding cells on porous scaffolds to allow for angiogenesis (Kulig and Vacanti, 2004; Millis and Losanoff, 2005).

Scaffolds must have an adequate surface area to hold the large number of cells needed for transplantation. In addition, they must allow for adequate nutrient and gas exchange, and for waste removal. They must also allow space for cellular reorganization and neovascularization. Studies have used a biodegradable polymer tube with a highly microporous 3D structure created from sheets of nonwoven PGA fiber mesh, covered with a 5% PLGA chloroform solution. Other methods of scaffold creation involve 3D printing, which can create complex scaffolds with interconnected vascular channels (Kulig and Vacanti, 2004).

Prior to transplantation, researchers must have a fully functional bioengineered liver, as it must immediately perform all hepatic functions. Thus, a method needed to be devised to create a functioning bioengineered liver *in vitro* and then to implant it *in vivo*. Vacanti's group used a prevascularized sponge device that was fabricated and seeded with adequate cells *in vitro*, and then implanted *in vivo*. Another strategy is to use an implantable scaffold with direct access to the portal blood. An example is a recent study that used SIS grafted between the portal vein and inferior vena cava as a venous conduit, which could be seeded with hepatocytes. A final strategy is microelectromechanical systems (MEMS), using silicon microfabrication technology to form large sheets of liver tissue, which can then be folded into 3D structures (Kulig and Vacanti, 2004).

21.4 Lung tissue engineering

As in other fields currently being investigated by tissue engineers, organ transplantation is the preferred

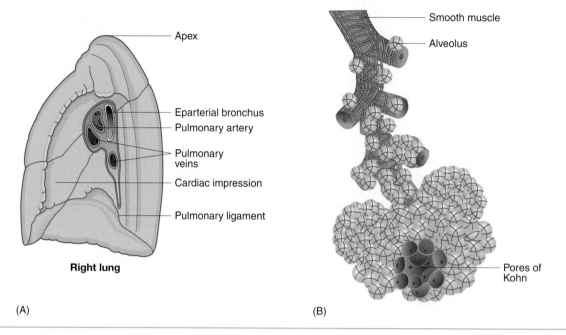

Figure 21.6 Anatomy of the lung.

treatment and standard of care for organ failure for the lung (Figure 21.6). The demand for human organs for transplantation is large and growing, and the supply of donor lungs (as with most transplanted organs) does not meet the demand. Therefore, scientists are fervently working to develop new technologies for the replacement or repair of damaged tissues and organs. Some of the technologies include stem cell biology, therapeutic cloning, tissue and organ bioengineering, and xenotransplantation (Ogle *et al.*, 2004).

The simplest form of tissue engineering to replace organ function involves cell-based therapies. This concept involves seeding newly engineered healthy cells at a site in an organ or tissue where damaged cells are inhibiting appropriate function. The key component of any tissue engineering model is the source of cells. Lung tissue engineering is no exception and research into sources of pulmonary cell

replacement has been an area of much exciting research recently. As in other areas of regenerative medicine, ES cells appear to be excellent sources due to their potential to develop into many different cell types and their limitless replicative capacity (Rippon *et al.*, 2004).

Studies have demonstrated that ES cells can be induced to differentiate *in vitro* into pulmonary epithelial cells, type II pneumocytes, using a specific medium designed for the maintenance of lung epithelial cells in culture (Rippon, 2004). The generation of differentiated airway epithelial tissue had not been reported, however, until Coraux *et al.* also demonstrated that type II alveolar epithelial cells could be derived from ES cells in a murine model (Coraux *et al.*, 2005). Coraux *et al.* went on to show that murine ES cells were able to differentiate into nonciliated secretory Clara cells and that type I collagen induced this development. In addition, when these cells were cultured at an air–liquid interface, ES cells were able to form fully differentiated airway epithelium. The cells were composed of basal, ciliated, intermediate and Clara cells, such as found in the tracheobronchial airway epithelium (Coraux *et al.*, 2005).

Another method of inducing the development of ES cells along the pulmonary epithelial lineage is the use of coculture techniques. Coculture of ES cells from a murine model with embryonic mesenchyme from the distal lung promoted the development of pneumocytes. The technique involved differentiating the stem cells first into embryoid bodies (EB) and then culturing these EBs for 1–2 weeks with embryonic pulmonary mesenchyme. The EBs demonstrated the development of channels lined with pulmonary epithelial cells including type II pneumocytes. Even more exciting was the observation in some animal studies that ES cells delivered systemically to animals with pulmonary damage seeded at the site of the diseased tissue and developed pulmonary phenotypes. These results encourage the search for an ideal use of ES cells as some form of cell therapy for pulmonary disease (Van Vranken *et al.*, 2005).

Other sources for pulmonary epithelial replacement have been investigated as well. For example, fetal rat lung cells have been studied as a source of cells for pulmonary tissue engineering. Lung cells from 16–19 day old Sprague-Dawley rat fetuses were harvested and seeded on collagen–glycosaminoglycan (GAG) scaffolds. The cells formed and maintained alveolar-like structures, and were composed mostly of type II pneumocytes. These cells demonstrated contractile characteristics that may have influenced of the development of the alveolar structures within the scaffold (Chen *et al.*, 2005b).

Cell-based tissue engineering could also be based on native and intrinsic tissue-specific stem cells of adult tissue. These cells are maintained in organs in specific microenvironments to allow for the replacement of damaged tissue (Stripp and Reynolds, 2005). The fully developed lung contains approximately 40 different cell types. These specific cell types vary along the pathway from proximal to distal airway.

These regions of the lung can be divided into the tracheobronchial, bronchiolar and alveolar zones of the lung, and the airway epithelium of each zone is unique. Most recent work, such as that of Coraux *et al.*, has described differentiating mouse ES cells into tracheobronchial epithelium – only one of the components of airway epithelium (Coraux *et al.*, 2005; Stripp and Reynolds, 2005).

In the hope of defining methods of developing all components of pulmonary epithelium, researchers have looked to the native stem cell niches of the lung. In order to take full advantage of these cell sources, our understanding of their physiology and pathophysiology must be expanded. For example, scientists need to define what causes these sites of regeneration to lose the capacity to replace damaged cells, leading to end-stage disease. Studies into similar issue have been performed with germline stem cells in *Drosophila*, but need to be expanded. Understanding the limiting factors of growth may enable us to understand the specific factors supporting their subsistence and expansion (Stripp and Reynolds, 2005).

An important area of future study would also be the examination of the ability of ES cells to produce distal airway cells (Ali *et al.*, 2002; Kubo *et al.*, 2004). Essential in the study of different components of the tracheobronchial tree is the ability to specifically identify the epithelial cells from different locations with unique identifying markers, such as specific secretory proteins (Stripp and Reynolds, 2005).

The next phase of lung tissue engineering involves adapting these technologies to cellular therapy for lung disease. Unfortunately, there are many crucial points that must be clarified prior to the establishment of such therapy. For example, the bioengineered tracheobronchial epithelia derived from ES cells may have been pushed so far in the direction of differentiation that they have exhausted the essential specific stem cell populations necessary for the newly engineered tissue to maintain itself in a healthy state. Thus, a method must be established to keep some cells in a niche maintaining tissue-specific stem cell properties for future cell replacement. In the human body, specific environments in the lung, near the opening of the submucosal glands and intercartilaginous regions, maintain tracheobronchial stem cells. Perhaps, the study of these sites and the growth factors/cellular signals typical of these environments will assist in modulating airway epithelial engineering. Second, how will these newly generated airway epithelial cells be seeded at the exact site of need in the pulmonary tree – an area not easily accessible to the surgeon? Also, the use of allograft tissue mandates a form of immunosuppression to prevent the rejection of the restorative tissue. Ideally, novel methods will be developed without the significant morbidity typically associated with systemic immunosuppression (Stripp and Reynolds, 2005).

The next step in lung tissue engineering involves moving past cellular therapy to actual bioengineered lung constructs. In recent work on models of airway disease, biopsies of lung were used to isolate epithelial and fibroblastic cells to produce 3D bronchial equivalents in culture. These constructs, while not used for lung replacement, but rather models of disease processes, provide a unique insight into the challenges of lung tissue engineering. Some important points noted from this work included that the cultured cells, without a specific ECM, could not organize themselves in a fashion similar to the native tissue. A 3D matrix was developed on which to seed the cells and which promoted their proper orientation. Also, the characteristics of the matrices, other than the overall shape, were crucial to proper cell development. For example, the matrices needed to be extremely thin to support mucociliary cell differentiation. As noted above, mesenchyme–epithelial interaction may be needed to regulate development, possibly due to paracrine influences. Also, retinoic acid was found to prevent squamous metaplasia in the cultured cells and, as seen previously, an air/liquid interface was required for human bronchial epithelial cells to develop a mucociliary phenotype. Finally, the stretch and strain of the culture medium, e.g. the stretch on collagen gels seeded with airway epithelial cells, was found to affect the ciliary beat frequency (Paquette *et al.*, 2003, 2004).

The final goal of all of this research is the ability to regenerate portions of the lung for replacement of damaged tissue. Clinically, the experience of this scale of bioengineering is limited. However, there have been some small trials investigating these technologies. In the surgery of the thorax, the replacement of airway segments following extirpative surgery is a vexing problem, as current reconstructive techniques are inadequate. A bioengineered airway would be a useful tool in cases such as tracheal replacement. A group from Hannover, Germany has used tissue engineering to create a bioartificial patch for tracheal replacement. In one reported case, a bioengineered airway patch was created for a patient who developed a dehiscence of his tracheobronchial anastamosis following a right completion carinal pneumonectomy. The patch was created by harvesting autologous muscle cells and fibroblasts from the patient, culturing and expanding these cells, and plating them on a collagen matrix derived from decellularized porcine jejunum. The patch was placed over an airway defect, satisfactorily closed the opening and was immediately airtight. The graft demonstrated neovascularization and excellent coverage with functional and viable ciliated respiratory epithelium. The cellularized patch worked well in this instance, although it is possible that an acellular patch could be used, as there was ingrowth of the normal respiratory epithelium over the patch (Figure 21.7) (Macchiarini *et al.*, 2004).

21.5 Gut tissue engineering

The gut or bowel is a crucial organ system in the human body. Never is this more apparent than when inadequate functioning intestinal tissue is present due to diseases such as intestinal ischemia and inflammatory bowel disease. The lack of adequate length of healthy intestinal tissue leads to the condition of 'short bowel syndrome', where nutrients are not adequately absorbed to support normal growth and development. The lack of bowel not only prevents the absorption of necessary nutrients, but also of fluids and electrolytes, leading to the development

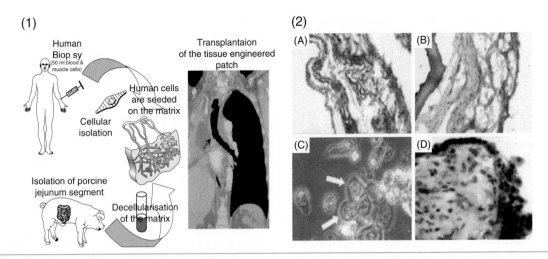

Figure 21.7 Engineering process (1) and histologic composition (2) of the bioartificial patch. Muscle cells and fibroblasts are isolated from a biopsy specimen obtained from the patient. These cells are seeded on a biologic matrix representing a collagen network generated from a decellularized porcine jejunal segment. During the incubation period, the cells start to remodel the xenogenic matrix and replace it with autologous connective tissue. Within 4 weeks, this autologous bioartificial implant can be clinically used. The computed tomographic scan of the chest 6 weeks after graft implantation shows the site (arrow) where the tissue-engineered patch was transplanted. (A) Acellular porcine collagen matrix. (Original magnification × 200; hematoxylin & eosin staining.) Notice the absence of cellular components and the red-stained collagen fibers. (B) Reseeded patch before implantation. Anti-vimentin immunohistology depicting fibroblast and smooth muscle cell seeding. (Original magnification ×200.) (C) Brush cytology from the patch surface showing respiratory epithelium (12th week). Arrows indicate epithelial cilia. (D) Immunhistochemical staining for respiratory epithelium (12th week). (Original magnification 200×; Villin antibody.) The asterisk marks the tracheal lumen.

of many varied and serious metabolic abnormalities (Chen and Badylak, 2001; Chen and Beierle, 2004).

The current best therapy for replacing diseased or absent intestine is organ transplantation. This involves an extensive operative procedure with its own morbidity and mortality, along with the well-described side-effects of chronic systemic immuno-supresion. There are also the concerns of inadequate numbers of organ donors and rejection of the transplanted tissue. Surgeons have attempted numerous creative surgical procedures to deal with inadequate bowel length, all designed to slow bowel transit time and thus allow for more absorption of nutrients, fluid, etc., but they have had little success. In 1967, Stanley Dudrick described the use of hyperalimentation for the nutritional support of patients without

the use of normal healthy intestine (Chen and Badylak, 2001). Although useful for select patients, this technology was not able to replace intestinal function entirely, as long-term hyperalimentation was found to be fraught with numerous serious complications. An excellent alternative to all these failed therapies would be to regenerate intestine using tissue engineering techniques that have been successfully applied to other organs (Chen and Badylak, 2001; Chen and Beierle, 2004).

Early attempts at tissue engineering bowel segments involved patch repairs of intestinal defects. As far back as 1963, Kobold and Thal used jejunal serosa to repair duodenal defects in dog models (Kobold and Thal, 1963). These patches were covered by duodenal epithelium in 2 months. In 1973, Binnington *et al.* repaired jejunal defects in animals with the serosa of the descending colon (Binnington *et al.*, 1973). The new mucosal covering was noted to grow in from the edges, covering the serosa. Expanding their work, they used a model of 'short bowel syndrome' in pigs to show that the jejunal patch was able to induce increased weight gain (Chen and Badylak, 2001). These 'patch' experiments were repeated using different models and strategies, such as colonic serosa or abdominal wall pedicle flap patches in rabbits, with similar results (Bragg and Thompson, 1986; Chen and Beierle, 2004; Lillemoe *et al.*, 1982). The studies concluded that these patches were all able to support the growth of functional neomucosa.

Based on the concept of mucosal ingrowth from the edges, groups looked into the ability of varying patch sizes to be covered with neomucosa. As expected, they found that multiple smaller patches were covered more quickly and completely than a single large patch. Also, tubular patches were only covered at the edges, with the centers of the tubes the most distant from normal mucosa remaining uncovered despite long observation periods and appropriate stenting (Bragg and Thompson, 1986).

All these studies share the need for intestinal mucosa to grow and spread across the patch surface. Therefore, the regulation of intestinal growth is an area of research crucial to the advancement of bowel tissue engineering. Numerous factors have been studied and proven to have an effect on intestinal mucosal growth. These growth factors and cytokines appear to have significant effects and must be harnessed in creative ways to maximize the ability to replace the damaged absorptive surface of diseased intestine (Thompson *et al.*, 1987).

Following the study of naturally occurring matrices for intestinal patching, scientists began to examine the possibility of engineering biomaterials for use in these circumstances. They searched for biocompatible, nontoxic and biodegradable substances that could support the growth of tissue and be absorbed following the establishment of the new tissue without significant side effects (Chen and Beierle, 2004). They discovered that biomaterials play a critical role

in the engineering of new functional tissue. They provide a supporting scaffold for tissue growth and they organize the cells into a 3D structure. Scientists have studied various biomaterials that can be separated into three distinct groups: natural (i.e. collagen), acellular (i.e. SIS) and synthetic (i.e. PGA, PLA and PLGA) (Kim and Vacanti, 1999).

Early work focused on synthetic scaffolds. In 1979, Harmon *et al.* used Dacron to repair ileal defects in rabbits (Harmon *et al.*, 1979). The matrix allowed for mucosal ingrowth and was extruded into the bowel after a short time. Watson *et al.* examined the use of a PTFE prosthetic in a dog model (Watson *et al.*, 1980). They used a tubular model, however, and as seen above there was minimal coverage of the lumen. These studies provided the first clues that nonabsorbable materials were inappropriate as scaffolds for tissue engineering. Thompson *et al.* were the first to report the use of an absorbable biomaterial (PGA) as a scaffold for ingrowth of neomucosa (Thompson *et al.*, 1986).

Generally, the ideal biomaterial should be biocompatible and biodegradable. Acellular and natural scaffolds are just that, and for those reasons have gained in popularity. An example of a widely used acellular biomaterial is SIS. SIS is primarily ECM material, following the removal of the mucosa, external muscle layers and other cells. It has been used in numerous tissue engineering applications. The SIS materials consist of naturally occurring ECM that have been shown to be rich in components that support angiogenesis, (i.e. fibronectin, GAGs, collagens types I, III, IV and V, basic fibroblast growth factor, and vascular endothelial cell growth factor). SIS has also been shown to be nonimmunogenic (Demirbilek *et al.*, 2003).

SIS has been extensively investigated as a scaffold for creating neointestine, due to its ease of use and ready availability. Matsumoto *et al.* first described the use of SIS for vein replacement in dogs in 1966 (Matsumoto *et al.*, 1966). The material is now processed in a standardized fashion and mass produced, making it readily available for tissue engineering studies (Chen and Badylak, 2001).

SIS has been used to cover defects in the esophagus and small bowel in dogs. In these studies the SIS patch allowed the mucosa to grow over the scaffold and the histology of the patch was similar to the native intestine (Chen and Badylak, 2001). In another recent study, tubular unseeded SIS grafts from Sprague-Dawley rat donors were used to create an ileostomy. These grafts initially were covered by inflammatory cells, but by 3 months were completely covered by mucosa. Also, the outsides of the tubular grafts were covered with bundles of smooth muscle-like cells and fibrovascular tissue. A similar study performed in the stomach showed similar results. In this study, rats underwent the removal of a full-thickness 1-cm segment of the stomach, with a patch repair using a double-layer patch of porcine-derived SIS. On examination 3 weeks postoperatively, the site was covered by granulation tissue and fibrosis, with some growth of normal gastric mucosa at the edges of the graft (de la Fuente, 2003). These studies demonstrated that SIS can develop structural features of the normal intestine and may provide an off-the-shelf option for replacement patches for intestinal tissue (Figure 21.8) (Wang *et al.*, 2005).

Other studies using various other novel scaffold materials, such as fibrin glue and PGA as a framework for tissue engineering neointestine, have had promising results as well. Hori *et al.* used a collagen sponge enzymatically processed from porcine skin as a scaffold for regeneration of stomach and small bowel with good success in a beagle model (Hori *et al.*, 2001). Isch *et al.* have also performed esophagoplasty by patching defects in the cervical esophagus in a dog with AlloDerm® (human decellularized skin) (Isch *et al.*, 2001).

The basic strategy and ultimate goal of engineering any organ is to culture and expand cells, implant these cells onto a biodegradable scaffold, and implant the newly engineered construct into the body to replace deficient organ functions. The cells provide the metabolic function, while the scaffold provides a support framework until the cells can produce their own, at which time the original scaffold dissolves (Rocha and Whang, 2004).

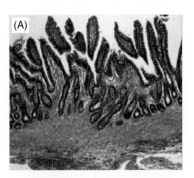

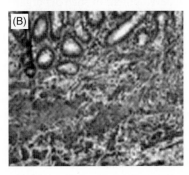

Figure 21.8 Histologic photomicrographs of SIS-regenerated rat small bowel tissue at 12 weeks. (A) well-organized mucosal epithelial layer with smooth muscle cells regeneration. (B) Immunohistochemically stained intestinal tissue suggesting bundles of well-formed smooth muscle fibers with less regularity (original magnification ×250).

The key components for this process include a cell source, and an appropriate biocompatible and biodegradable scaffold. Early work in intestinal tissue engineering by Vacanti used just such a strategy (Vacanti, 2003). Cells were obtained from an intestinal organ-oid unit, a scaffold of sheets of collagen-coated, microporous, nonwoven PGA was formed into a tubular shape, and these tubes once seeded with cells were implanted into the omenta of animals. Vacanti's group harvested cells from intestinal organoids. Tait *et al.* described these as an excellent source of cells for tissue engineering because of their unique composition, which includes progenitor cells, epithelium and mesenchymal stroma (Tait *et al.*, 1994). The implanted devices demonstrated angiogenesis from the omental vessels. On histological examination the constructs are similar to normal small bowel and have been anastomosed to the native bowel in animal studies without causing feeding problems (Chen and Beierle, 2004; Rocha and Whang, 2004).

The construct, once examined, shares many characteristics of normal intestine, with an epithelium, submucosa and muscularis layers. A major limitation to the development of such 3D organ structures involves the need for the delivery of nutrients and removal of waste products from these large structures, where simple diffusion is inadequate. The angiogenesis provided by the omental blood supply and the porous nature of the scaffold design allowed for the *in vivo* development of the above-mentioned intestinal construct. However, *in vitro* growth of such organs would be very useful for the study of the physiology and pathophysiology of the organs. Bioreactors are being developed to allow for the growth of these structures in the laboratory setting (Rocha and Whang, 2004).

The source of cells for gut tissue engineering is as crucial as it is to any field of regenerative medicine technology. It must be kept in mind, however, that there is always the option of modifying native cells in culture using gene therapy to ensure a normal phenotype. For example, as seen in early models of lung tissue engineering, patient-derived intestinal cells may be used to develop models for studying human intestinal physiology and pathophysiology. Just as the cell can be tailored to fit the appropriate phenotype, the scaffold may be 'engineered' to assist in the development of the neo-intestine. For example, the scaffold may be designed to secrete specific growth factors or hormones, etc., required for the rapid development of normal intestine (Rocha and Whang, 2004).

The concept of a bioengineered model intestine raises the possibility of the *in vitro* growth of intestinal tissue. These models provide not only a useful construct for the examination of the disease processes of the bowel, but also a possible 'off-the-shelf' source of intestinal tissues (Rocha and Whang, 2004).

21.6 Pancreas tissue engineering

Diabetes mellitus is a common disease, affecting 4% of the world, causing significant morbidity and mortality through its effects on vascular/microvascular disease. Diabetes is due to systemic hyperglycemia, from inadequate insulin production by the pancreas (Figure 21.9) and/or resistance to insulin (Kahn, 2004). There are two main types (a third has been described) of diabetes, type I and II, and they are caused by islet β cell dysfunction, with an absolute or relative decrease in insulin production. Type I diabetes results from the autoimmune destruction of β islet cells. Type II diabetes is characterized by an inability of the β cell to rapidly release sufficient insulin in response to oral nutrient stimulation and to impaired glucose tolerance. Also, β cells may not be able to convert pro-insulin to insulin (Kahn, 2004).

The most effective therapy for diabetes, since the discovery of insulin by Banting and Best in 1921, has been insulin replacement therapy (Bliss *et al.*, 1982). Although the ability to measure and replace insulin is a marvel of modern medicine, this therapy does not match insulin levels appropriately to the blood glucose concentration. Scientists therefore continue to search for new therapies to improve the treatment of diabetes mellitus, such as β cell replacement, which seeks to replace the deficient endogenous insulin production. The ultimate goal is to develop a permanent endogenous replacement for the lack of insulin (Kahn, 2004).

One of the first attempts at establishing a permanent endogenous insulin source was the transplantation of cadaveric islet cells, which in studies has been shown to restore normoglycemia. There are very few suitable donors, however, and this therapy requires chronic systemic immunosuppression. Other options also have shortcomings. For example, human adult islet β cells cannot be expanded beyond 10–15 divisions, and xenografts and pancreatic tumor lines have safety concerns. Thus, scientists have sought to use a regenerative medicine approach and develop insulin-producing structures *in vitro* utilizing different strategies. These strategies include engineering β cell-like

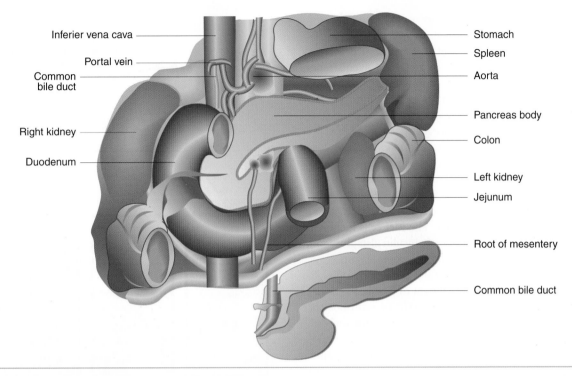

Inferier vena cava

Portal vein

Common
bile duct

Right kidney

Duodenum

Stomach

Spleen

Aorta

Pancreas body

Colon

Left kidney

Jejunum

Root of mesentery

Common bile duct

Figure 21.9 The pancreas anatomy and histology.

cells from somatic cells that are more plentiful and more easily expanded than β cells using genetic engineering or the development of β cells from progenitor or stem cells (Kahn, 2004).

Designing new β cells involves two approaches: gene therapy and cell replacement strategies. Engineered β cells need to be able to modulate basal insulin release for normal events such as exercise or infection. The β cells must also be able to regulate insulin release according to nutrient ingestion. Also, the normal islet cells are innervated, so that engineered β cells need to respond to neural input. Over life there are changing demands on insulin release that are the result of changes in insulin sensitivity, like puberty, pregnancy and old age. The engineered β cells need to be able to increase their replicative capacity according to need and involute when no longer needed. Regular exercise increases insulin sensitivity, as does weight loss.

Thus, in cases such as weight gain, β cells must be able to sustain insulin output. Acute modulation of insulin levels is also required during periods of stress such as infection, surgery and serious illness. This is due the modulating effects of the sympathetic nervous system on insulin resistance, so engineered β cells must be able to respond to these changes as well. Cells must also be able to produce insulin and appropriately process proinsulin to insulin to function properly. There has to also be a way to recreate the influence of gastrointestinal derived factors in the modulation of insulin release (Kahn, 2004).

A major concern for scientists in β cell engineering is recurrence of disease. Type I diabetes has an immunological cause, so it may recur with tissue-engineered cells. This immune response needs to be treated and prevented or the antigenic presentation of the engineered cells changed. The causes of type

II diabetes are not as clear, so recurrence could be a possibility here as well (Kahn, 2004).

Another concern is that engineered β cells may require immunosuppression, and these agents (specifically glucocorticoids) cause insulin resistance, suppress insulin release and prevent insulin processing. They also have other more severe risks, such as malignancy. Cells should be designed to release the hormone insulin in a pulsatile fashion, because continuous release can lead to insulin resistance. Also, pro-insulin may a have a cardiovascular risk, so proper processing of the hormones is essential (Kahn, 2004).

The pancreas is composed of acinar cells, ducts and islets of endocrine cells. The islet cells include β cells among others, and secrete insulin also glucagon, somatostatin and pancreatic polypeptide. Pancreatic cells develop from endodermal progenitor cells. To orchestrate *in vitro* development of β cells, the development of normal pancreatic cells *in vivo* must be well understood. To be used as therapy for diabetes mellitus, the cells developed must have certain physiologic functions, such as insulin gene expression, appropriate post-translational modification of insulin, glucose-sensitive insulin production and, ideally, an unlimited supply (Kahn, 2004).

These characteristics are typical of β cells, but that does not mean one needs β cells to replace them, just cells that acts like them. For example, modification of genes in cells that have similar origins to pancreatic cells may produce β cell-like qualities, such as the transfer of the *Pdx-1* gene into mouse liver cells, which can activate genes expressing insulin 1 and 2 and the pro-hormone convertase, but is not sensitive to glucose in the environment. It has also been demonstrated that ductal cells can be modified to produce insulin if modified with gene transfection (Kahn, 2004).

An alternative approach is to start with cells that have β cell characteristics and to modify these as needed to obtain the desired function. For example, Cheung *et al.* modified intestinal K cells, which normally secrete peptides in response to insulin secretion, using a transgenic model to express insulin in response to glucose sensation and these cells were able to restore

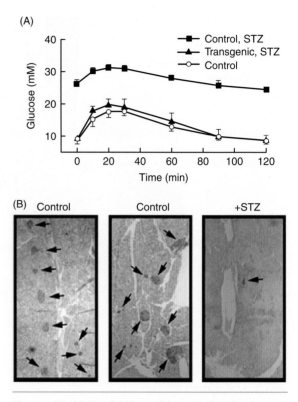

Figure 21.10 Production of human insulin from K cells protects transgenic mice from diabetes induced by destruction of pancreatic cells. (A) Oral glucose tolerance tests. Mice were given intraperitoneal injection of streptozotocin (STZ, 200 mg/kg) or an equal volume of saline. On the fifth day after treatment, after overnight food deprivation, glucose (1.5 g/kg body weight) was administered orally by feeding tube at 0 min. Results are means (±SEM) from at least three animals in each group. (B) Immunohistochemical staining for mouse insulin in pancreatic sections from control mice and an STZ-treated transgenic mouse. Arrows indicate mouse islets.

normoglycemia to streptozotocin-induced diabetic mice (Figure 21.10) (Cheung *et al.*, 2000). Another example is the modification of persistent hyperinsulinemic hypoglycemia of infancy-derived cells, a disease due to a defect in the β cell K_{atp} channel. These cells were transfected with genes for the K_{atp} channel and PDX-1, to produce insulin in a glucose-dependent manner. A unique strategy is to use constitutively insulin-secreting cells hybridized with glucose-responsive material, which possesses glucose-dependent changes in its permeability to insulin (Cheng *et al.*, 2004).

What about a stem cell approach to insulin replacement? The goal of stem cell therapy, as with all studies of multipotent cells in regenerative medicine, is the directed differentiation of these cells and the introduction of specific characteristics to these cells via genetic manipulation. There are several different choices of cells with which to work in this field of study, such as ES cells, adult stem cells, etc. However, unlike the lung, it is unclear if there is a pool of adult stem cells in the pancreas and whether the duct cells can act as islet progenitors (Ball and Barber, 2003; Kahn, 2004).

However, a study by Seaberg *et al.* (2004) described the isolation of pancreas-derived multipotent precursor cells. They isolated these cells by dissociating and plating pancreatic ductal and islet tissue. These tissues formed floating colonies typical of neurospheres. These colonies expressed markers such as PDX-1 and Nestin, characteristic of both pancreatic and neural precursors. It was discovered that these cells could develop into neurons, astrocytes and oligodendrocytes. These colonies could also produce insulin-producing β-like pancreatic cells. They were found to be glucose responsive, with increased insulin release in response to insulin. These findings are very encouraging, but require considerable research to prove clinical utility.

Another option is to transdifferentiate the adult stem cells from other organs into β cells. For example, the pancreas, liver and upper gastrointestinal tract all have similar embryological origins in the anterior endoderm and may derive from similar progenitor cells. In the absence of the *Ptf-1a* gene, endoderm-derived stem cells destined to be pancreas cells can develop into gastrointestinal cells. Also, pancreatic exocrine cells can transdifferentiate into hepatocytes under appropriate conditions. Several studies have proven this principle, by developing β-like cells from intestinal stem cells or hepatic oval cells, using the right developmental milieu (Lardon *et al.*, 2004; Song *et al.*, 2004).

Multipotent adult progenitor cells, such as bone marrow-derived stem cells, can develop into hematopoietic and stromal cells. They have also been shown to develop into different phenotypes such as neurons, cardiomyocytes, and hepatocytes. These cells have limited major histocompatibility complex expression, so may be immunoprivileged, and would be useful in transplantation for that very reason. Unfortunately, these properties are limited to stromal cells and much work must be done to determine they can effectively be modified for other uses (Ball and Barber, 2003).

Derived from the inner cell mass of developing blastocyst, ES cells have an unlimited potential for growth, stable karyotype and are pluripotent. These cells may be a potential supply of unlimited β cells, but it is no easy task to direct a pluripotent stem cells down a specific lineage, especially that of a β cell. Several early experiments have explored these avenues with some promise of good results in the future (Ball and Barber, 2003).

Injured pancreatic tissue undergoes repair via replacement of exocrine acini by duct-like structures. This is likely due to transdifferentiation. The exact process of pancreatic tissue repair involves dedifferentiation of epithelial cells followed by proliferation and then redifferentiation. Studying this process may provide insight into means of developing new technologies for β cell expansion. Also, many studies have demonstrated that ductal cells are involved in β cell neogenesis. In addition, recent studies have demonstrated evidence of adult acinar cells spontaneously transdifferentiating into insulin-expressing cells *in vitro* (Lardon *et al.*, 2004; Song *et al.*, 2004).

Other studies have investigated the use of a reversible immortalization strategy for expanding human islets cells with retention of function (Narushima *et al.*, 2005). These studies established a reversibly immortalized islet cell clone (NAKT-15) by transfection of human cells with a retroviral vector containing simian virus 40 large T-antigen (SV40T) and human telomerase reverse transcriptase (hTERT) cDNAs. These transfections were designed to allow the deletion of SV40T and TERT by Cre recombinase.

Transplantation of these transfected cells into strepto-zotocin-induced diabetic mice resulted in improved blood glucose control for over 30 weeks (Hohmeier and Newgard, 2004; Narushima *et al.*, 2005).

Other sources of insulin-producing cells have been investigated as well, including fetal human islets and porcine islets. Xenotransplantation, especially the use of pig islets, has decreased in recent years due to concerns about infectious complications and the possible immune response to the xenografts (Hohmeier and Newgard, 2004, 2005).

Scientists have developed many technologies to fight the immunologic rejection of transplanted insulin-secreting cells. For example, encapsulated cells have shown great promise as methods of immunoprotection. Mukherjee *et al.* investigated insulin-secreting mouse insulinoma βTC3 cells encapsulated in calcium alginate/poly-ʟ-lysine/alginate (APA) beads for immunoprotection. These cells may effectively produce insulin when placed in the body, and can be grown, produced and stored till needed by cryopreservation or vitrification (Mukherjee *et al.*, 2005). Another strategy to prevent the immunogenic reaction involved with transplanted insulin-producing cells is the use of the bioartificial pancreas. This device can provide blood glucose regulation without immunosuppressive medication. The strategy of microencapsulation of insulin-secreting cells can also be used, placing these cells in a matrix that provides support and immunoprotection. Choice of cells used varies, but mammalian islets and transformed β cell lines are the most frequently used, with encapsulation commonly by APA beads (Papas *et al.*, 1999a,b).

21.7 Future developments

Today's strategy for tissue engineering predominantly depends upon autologous cells from the diseased organ of the patient. In the case of extensive end-stage organ failure, a tissue biopsy may not yield enough normal cells for culture and ultimate transplantation. Primary autologous human cells cannot always be expanded from a particular organ, as it is

the case with the pancreas. In these cases, pluripotent human ES cells can become a source of cells from which the desired tissue can be derived.

Embryonic stem cells exhibit two remarkable properties: the ability to proliferate in an undifferentiated, but pluripotent state and the ability to differentiate into many specialized cell types (Brivanlou *et al.*, 2003).

Adult stem cells are ethically more acceptable and have the advantage of not transdifferentiating spontaneously into a malignant phenotype. Adult stem cells are difficult to expand into large quantities and their use for clinical therapy is therefore limited.

Fetal stem cells, harvested from amniotic fluid and placentas, may represent a novel source of stem cells. These cells express markers consistent with human ES cells, such as OCT4 and SSEA-4. They are multipotent and are able to differentiate into all three germ layers, and they do not form teratomas nor do they spontaneously transdifferentiate into a malignant phenotype. A further advantage is that the cells have a high replicative potential and they could be stored for future possible use in the patient, without the risks of rejection and ethical concerns (Siddiqui and Atala, 2004).

Furthermore, growth factors and other signaling molecules alone or in combination with matrices may be used for tissue guidance and regeneration, replacing cell-based tissue engineering. Once the growth factors to be delivered have been identified, there are several important criteria to take into consideration. First, the delivery must be controlled, and allow sustained and localized delivery of small amounts of growth factors to the tissue repair site, as systemic administration can cause toxicity in other organs. Further, the release profile must be controlled in ensuring frequent exposure to these factors for a relatively long time to obtain the desired effect. Growth factors typically have short half-lives once they are introduced into the body and are eliminated rapidly. Finally, the process of incorporating the factors must not involve harsh solvents or high temperatures that might denature or deactivate the proteins. To meet these criteria the growth factors could be entrapped within or covalently linked to a polymer matrix and released by matrix degradation.

As prenatal diagnosis of fetal abnormalities has become extremely accurate, theoretically, a small tissue biopsy could be obtained under ultrasound guidance. These tissues could then be processed and the various cell types could be cultured *in vitro*. Applying tissue engineering techniques, *in vitro* reconstituted structures could then be available for reconstruction at the time of birth.

21.8 Summary

- The engineering of simple hollow organs (e.g. the bladder) has been achieved, but the construction of more complex hollow organs (e.g. blood vessels) and functional solid organs (e.g. kidney, liver) is the main challenge facing the future of tissue engineering.
- Engineered bladder tissues, created with autologous cells seeded on collagen–PGA scaffolds and wrapped in omentum after implantation, are currently undergoing clinical trials in patients requiring bladder augmentation.
- Today's strategy for tissue engineering predominantly depends upon autologous cells from the diseased organ of the patient. In case of extensive end-stage organ failure, a tissue biopsy may not yield enough normal cells for culture and ultimate transplantation. Primary autologous human cells cannot always be expanded from a particular organ, as it is the case with the pancreas. In these cases, pluripotent human ES cells can become a source of cells from which the desired tissue can be derived.
- Nonseeded acellular matrices have functioned well clinically for onlay urethral repairs. However, when tubularized urethral repairs are performed, seeded tubularized collagen matrices are required.
- Similarly, nonseeded acellular matrix tubes are not able to successfully replace the ureteral segments of the ureter, while cell-seeded biodegradable polymer scaffolds have proven to be successful in reconstruction of ureteral tissues.

- The kidney is surely the most challenging organ in the urinary system to be reconstructed by tissue engineering methods because of its complex structure and function. To develop a bioartificial kidney some researchers are pursuing the replacement of isolated kidney function parameters with the use of extracorporal cellular units, while others are aiming at the replacement of total renal function by tissue-engineered bioartificial structures.
- As in renal replacement strategies, the complexity of liver tissue engineering has focused the most promising liver replacement therapies on BAL replacement, harnessing liver cell function in combination with an external dialyzing filter to temporarily replace hepatic function.
- Cellularized, and possibly acellular, segments of collagen scaffold material offer a possible source of tissue for patch repairs of the tracheobronchial tree.
- Studies into gut tissue engineering have demonstrated that SIS, AlloDerm® and other unseeded scaffolds can develop structural features of the normal intestine when implanted in the bowel, and may provide an off-the-shelf option for replacement patches for intestinal tissue. Additional studies have shown that cells obtained from intestines, seeded on a scaffold and formed into a tubular shape, developed tissues similar to normal small bowel and have been successfully anastomosed to the native bowel in animal studies.
- The most effective therapy for diabetes, since the discovery of insulin by Banting and Best in 1921, has been insulin replacement therapy, but this therapy does not match insulin levels appropriately to the blood glucose concentration. The ultimate goal, therefore, is to develop a permanent endogenous replacement for the lack of insulin, and many tissue engineering strategies have been directed toward this goal.

References

Ali, N.N., *et al.* (2002). Derivation of type II alveolar epithelial cells from murine embryonic stem cells. *Tissue Eng*, 84: 541.

Allen, J.W. and Bhatia, S.N. (2002). Improving the next generation of bioartificial liver devices. *Semin Cell Dev Biol*, 136: 447.

Amiel, G.E. and Atala, A. (1999). Current and future modalities for functional renal replacement. *Urol Clin North Am.*, 26: 235.

Atala, A. (1994). Use of non-autologous substances in VUR and incontinence treatment. *Dial Pediatr Urol*, 17: 11.

Atala, A. (1995). Commentary on the replacement of urologic associated mucosa. *J Urol*, 156: 338.

Atala, A. (1997). Tissue engineering in the genitourinary system. In (Atala, A. and Mooney, D., eds). *Tissue Engineering* Boston, MA: Birkhauser, p. 149.

Atala, A. (1998). Autologous cell transplantation for urologic reconstruction. *J Urol*, 159: 2.

Atala, A. (1999). Future perspectives in reconstructive surgery using tissue engineering. *Urol Clin North Am*, 26: 157. ix.

Atala, A., et al. (1992). Formation of urothelial structures *in vivo* from dissociated cells attached to biodegradable polymer scaffolds *in vitro*. *J Urol*, 148: 658.

Atala, A., et al. (1993a). Injectable alginate seeded with chondrocytes as a potential treatment for vesicoureteral reflux. *J Urol*, 150: 745.

Atala, A., et al. (1993b). Implantation *in vivo* and retrieval of artificial structures consisting of rabbit and human urothelium and human bladder muscle. *J Urol*, 150: 608.

Atala, A., et al. (1994). Endoscopic treatment of vesicoureteral reflux with chondrocyte–alginate suspension. *J Urol*, 152: 641.

Atala, A., et al. (1999). A novel inert collagen matrix for hypospadias repair. *J Urol*, 162: 1148.

Atala, A., et al. (2006). Tissue-engineered autologous bladders for patients needing cystoplasty. *Lancet*, 367: 1241.

Baccarani, U., et al. (2005). Human hepatocyte transplantation for acute liver failure: state of the art and analysis of cell sources. *Transplant Proc*, 376: 2702.

Badylak, S.F., et al. (1989). Small intestinal submucosa as a large diameter vascular graft in the dog. *J Surg Res*, 47: 74.

Ball, S.G. and Barber, T.M. (2003). Molecular development of the pancreatic beta cell: implications for cell replacement therapy. *Trends Endocrinol Metab*, 148: 349.

Baltaci, S., et al. (1998). Failure of ureteral replacement with Gore-Tex tube grafts. *Urology*, 51: 400.

Barshes, N.R., et al. (2005). Support for the acutely failing liver: a comprehensive review of historic and contemporary strategies. *J Am Coll Surg*, 2013: 458.

Bilir, B.M., et al. (2000). Hepatocyte transplantation in acute liver failure. *Liver Transpl*, 61: 32.

Baskin, L.S., et al. (1993). Bladder smooth muscle cells in culture: I. Identification and characterization. *J Urol*, 149: 190.

Bazeed, M.A., et al. (1983). New treatment for urethral strictures. *Urology*, 21: 53.

Bent, A., et al. (2001). Treatment of intrinsic sphincter deficiency using autologous ear chondrocytes as a bulking agent. *Neurourol Urodynam*, 20: 157.

Binnington, H.B., et al. (1973). A technique to increase jejunal mucosa surface area. *J Pediatr Surg*, 85: 765.

Blandy, J.P. (1961). Neal pouch with transitional epithelium and anal sphincter as a continent urinary reservoir. *J Urol*, 86: 749.

Blandy, J.P. (1964). The feasibility of preparing an ideal substitute for the urinary bladder. *Ann R Coll Surg*, 35: 287.

Bliss, M. (1982). Banting's, Best's, and Collip's accounts of the discovery of insulin. *Bull Hist Med*, 564: 554.

Bragg, L.E. and Thompson, J.S. (1986). The influence of serosal patch size on the growth of small intestinal neomucosa. *J Surg Res*, 405: 426.

Brittberg, M., et al. (1994). Treatment of deep cartilage defects in the knee with autologous chondrocyte transplantation. *N Engl J Med*, 331: 889.

Brivanlou, A.H., et al. (2003). Stem cells. Setting standards for human embryonic stem cells comment). *Science*, 300: 913.

Chancellor, M.B., et al. (2000). Preliminary results of myoblast injection into the urethra and bladder wall: a possible method for the treatment of stress urinary incontinence and impaired detrusor contractility. *Neurourol Urodynam*, 19: 279.

Chen, F., et al. (1999). Acellular collagen matrix as a possible 'off the shelf' biomaterial for urethral repair. *Urology*, 54: 407.

Chen, K. L., et al. (2005a). Total penile corpora cavernosa replacement using tissue engineering techniques. Abstract presented at the *Regenerate International Conference*, June 3–6, Atlanta, GA.

Chen, M.K. and Badylak, S.F. (2001). Small bowel tissue engineering using small intestinal submucosa as a scaffold. *J Surg Res*, 992: 352.

Chen, M.K. and Beierle, E.A. (2004). Animal models for intestinal tissue engineering. *Biomaterials*, 259: 1675.

Chen, P., et al. (2005b). Formation of lung alveolar-like structures in collagen–glycosaminoglycan scaffolds *in vitro*. *Tissue Eng*, 119–10: 1436.

Cheng, E., *et al.* (1994). Reversed seromuscular flaps in the urinary tract in dogs. *J Urol*, 152: 2252.

Cheng, S.Y., *et al.* (2004). Hybrid pancreatic tissue substitute consisting of recombinant insulin-secreting cells and glucose-responsive material. *Biotechnol Bioeng*, 877: 863.

Cheung, A.T., *et al.* (2000). Glucose-dependent insulin release from genetically engineered K cells. *Science*, 290: 1959.

Cilento, B.G. and Atala, A. (1995). Treatment of reflux and incontinence with autologous chondrocytes and bladder muscle cells. *Dial Pediatr Urol*, 18: 11.

Cilento, B.G., *et al.* (1994). Phenotypic and cytogenetic characterization of human bladder urothelia expanded *in vitro. J Urol*, 152: 655.

Coraux, C., *et al.* (2005). Embryonic stem cells generate airway epithelial tissue. *Am J Respir Cell Mol Biol*, 322: 87.

Dahms, S.E., *et al.* (1997). Free ureteral replacement in rats: regeneration of ureteral wall components in the acellular matrix graft. *Urology*, 50: 818.

Danielsson, C., *et al.* (2006a). Modified collagen fleece a scaffold for transplantation of human bladder smooth muscle cells. *Biomaterials*, 27: 1054.

Danielsson, C., *et al.* (2006b). Polyesterurethane foam scaffold for smooth muscle cell tissue engineering. *Biomaterials*, 27: 1410.

De Filippio, R.E., *et al.* (2002b). Total penile urethra replacement with autologous cell-seeded collagen matrices. *J Urol*, 167: 152.

De Filippio, R.E., *et al.* (2003). Engineering of vaginal tissue *in vivo. Tissue Eng*, 9: 301.

de la Fuente, S.G., *et al.* (2003). Evaluation of porcine-derived small intestine submucosa as a biodegradable graft for gastrointestinal healing. *J Gastrointest Surg*, 71: 96.

Demirbilek, S., *et al.* (2003). Using porcine small intestinal submucosa in intestinal regeneration. *Pediatr Surg Int*, 198: 588.

Dewan, P.A. (1998). Autoaugmentation demucosalized enterocystoplasty. *World J Urol*, 16: 255.

Diamond, D.A. and Caldamone, A.A. (1999). Endoscopic correction of vesicoureteral reflux in children using autologous chondrocytes: preliminary results. *J Urol*, 162: 1185.

Fauza, D.O., *et al.* (1998a). Videofetoscopically assisted fetal tissue engineering: skin replacement. *J Pediatr Surg*, 33: 377.

Fauza, D.O., *et al.* (1998b). Videofetoscopically assisted fetal tissue engineering: bladder augmentation. *J Pediatr Surg*, 33: 7.

Folkman, J. and Hochberg, M. (1973). Self-regulation of growth in three dimensions. *J Exp Med*, 138: 745.

Fox, I.J. and Roy-Chowdhury, J. (2004). Hepatocyte transplantation. *J Hepatol*, 406: 878.

Fox, I.J., *et al.* (1998). Treatment of the Crigler–Najjar syndrome type I with hepatocyte transplantation. *N Engl J Med*, 338: 1422.

Frey, P., *et al.* (1995). Endoscopic subureteral collagen injection for the treatment of vesicoureteral reflux in infants and children. *J Urol*, 154: 804.

Galetti, P.M., *et al.* (1995). Tissue engineering: from basic science to products: a preface. *Tissue Eng*, 1: 147.

Harada, N., *et al.* (1965). New surgical treatment of bladder tumors: mucosal denudation of the bladder. *Br J Urol*, 37: 545.

Harmon, J.W., *et al.* (1979). Fate of Dacron prostheses in the small bowel of rabbits. *Surg Forum*, 30: 365.

Hohmeier, H.E. and Newgard, C.B. (2004). Cell lines derived from pancreatic islets. *Mol Cell Endocrinol*, 2281/2: 121.

Hohmeier, H.E. and Newgard, C.B. (2005). Islets for all? *Nat Biotechnol*, 2310: 1231.

Hori, Y., *et al.* (2001). Tissue engineering of the small intestine by acellular collagen sponge scaffold grafting. *Int J Artif Organs*, 241: 50.

Huard, J., *et al.* (2002). Muscle-derived cell-mediated *ex vivo* gene therapy for urological dysfunction. *Gene Ther*, 9: 1617.

Hubschmid, U., *et al.* (2005). In-vitro growth of human urinary tract smooth muscle cells on laminin and collagen type I coated membranes under static and dynamic conditions. *Tissue Eng.*, 11: 161.

Humes, H.D., *et al.* (1999). Replacement of renal function in uremic animals with a tissue engineered kidney. *Nat Biotechnol*, 17: 451.

Humes, H.D., *et al.* (2003). Renal cell therapy is associated with dynamic and individualized responses in patients with acute renal failure. *Blood Purif*, 21: 64.

Isch, J.A., *et al.* (2001). Patch esophagoplasty using AlloDerm as a tissue scaffold. *J Pediatr Surg*, 362: 266.

Kaefer, M., *et al.* (1998). Reservoir calculi: a comparison of reservoirs constructed from stomach and other enteric segments. *J Urol*, 160: 2187.

Kaefer, M., *et al.* (1997). Continent urinary diversion: the Children's Hospital experience. *J Urol*, 157: 1394.

Kahn, S.E. (2004). Engineering a new beta-cell: a critical venture requiring special attention to constantly changing physiological needs. *Semin Cell Dev Biol*, 153: 359.

Kershen, R.T. and Atala, A. (1999). Advances in injectable therapies for the treatment of incontinence and vesicoureteral reflux. *Urol Clin North Am*, 26: 81.

Kim, B.S. and Mooney, D.J. (1998). Development of biocompatible synthetic extracellular matrices for tissue engineering. *Trends Biotechnol*, 16: 224.

Kim, S.S. and Vacanti, J.P. (1999). The current status of tissue engineering as potential therapy. *Semin Pediatr Surg*, 83: 119.

Kobold, E.E. and Thal, A.P. (1963). A simple method for the management of experimental wounds of the duodenum. *Surg Gynecol Obstet*, 116: 340.

Kropp, B.P., *et al.* (1996). Small intestinal submucosa: urodynamic and histopathologic evaluation in long term canine bladder augmentations. *J Urol*, 155: 2098.

Kropp, B.P., *et al.* (1998). Rabbit urethral regeneration using small intestinal submucosa onlay grafts. *Urology*, 52: 138.

Kropp, B.P., *et al.* (2004). Reliable and reproducible bladder regeneration using unseeded distal small intestinal submucosa. *J Urol*, 172: 1710.

Kubo, A., *et al.* (2004). Development of definitive endoderm from embryonic stem cells in culture. *Development*, 1317: 1651.

Kulig, K.M. and Vacanti, J.P. (2004). Hepatic tissue engineering. *Transpl Immunol*, 123–4: 303.

Lailas, N.G., *et al.* (1996). Progressive ureteral dilation for subsequent ureterocystoplasty. *J Urol*, 156: 1151.

Landman, J., *et al.* (2004). Laparoscopic mid sagittal hemicystectomy and bladder reconstruction with small intestinal submucosa and reimplantation of ureter into small intestinal submucosa: 1-year followup. *J Urol*, 171: 2450.

Langer, R. and Vacanti, J.P. (1993). Tissue engineering. *Science*, 260: 920.

Lanza, R.P., *et al.* (2002). Generation of histocompatible tissues using nuclear transplantation. *Nat Biotechnol*, 20: 689.

Lardon, J., *et al.* (2004). Exocrine cell transdifferentiation in dexamethasone-treated rat pancreas. *Virchows Arch*, 4441: 61.

Läckgren, G., *et al.* (2001). Long-term follow-up of children treated with dextranomer copolymer for vesicoureteral reflux. *J Urol*, 166: 1887.

Le Roux, P.J. (2005). Endoscopic urethroplasty with unseeded small intestinal submucosa collagen matrix grafts: a pilot study. *J Urol*, 173: 140.

Lillemoe, K.D., *et al.* (1982). Use of vascularized abdominal wall pedicle flaps to grow small bowel neomucosa. *Surgery*, 913: 293.

Macchiarini, P., *et al.* (2004). First human transplantation of a bioengineered airway tissue. *J Thorac Cardiovasc Surg*, 1284: 638.

Machluf, M. and Atala, A. (1998). Emerging concepts for tissue and organ transplantation. *Graft*, 1: 31.

Machluf, M., *et al.* (1998). Microencapsulation of Leydig cells: A new system for the therapeutic delivery of testosterone. *Pediatrics*, 102S: 32.

Matsumoto, T., *et al.* (1966). Replacement of large veins with free inverted segments of small bowel: autografts of submucosal membrane in dogs and clinical use. *Ann Surg*, 1645: 845.

McDougal, W.S. (1992). Metabolic complications of urinary intestinal diversion. *J Urol*, 147: 1199.

Millis, J.M. and Losanoff, J.E. (2005). Technology insight: liver support systems. *Nat Clin Pract Gastroenterol Hepatol*, 29: 398–405. quiz 434.

Mitaka, T. (1998). The current status of primary hepatocyte culture. *Int J Exp Pathol*, 796: 393.

Mito, M., *et al.* (1979). Studies on ectopic liver utilizing hepatocyte transplantation into the rat spleen. *Transplant Proc*, 111: 585.

Mukherjee, N., *et al.* (2005). Effects of cryopreservation on cell viability and insulin secretion in a model tissue-engineered pancreatic substitute (TEPS). *Cell Transplant*, 147: 449.

Muraca, M., *et al.* (2002). Intraportal hepatocyte transplantation in the pig: hemodynamic and histopathological study. *Transplantation*, 736: 890.

Najimi, M. and Sokal, E. (2005). Liver cell transplantation. *Minerva Pediatr*, 575: 243.

Narushima, M., *et al.* (2005). A human beta-cell line for transplantation therapy to control type 1 diabetes. *Nat Biotechnol*, 2310: 1274.

Nukui, F., *et al.* (1997). Complications and reimplantation of penile implants. *Int J Urol*, 4: 52.

Oberpenning, F.O., *et al.* (1999). De novo reconstitution of a functional urinary bladder by tissue engineering. *Nat Biotechnol*, 17: 2.

Oesch, I. (1988). Neourothelium in bladder augmentation: an experimental study in rats. *Eur Urol*, 14: 328.

Ogle, B., *et al.* (2004). Fusion of approaches to the treatment of organ failure. *Am J Transplant*, 4(Suppl 6): 74.

Olsen, L., *et al.* (1992). Urethral reconstruction with a new synthetic absorbable device. *Scand J Urol Nephrol*, 26: 323.

Osman, Y., *et al.* (2004). Canine ureteral replacement with long acellular matrix tube: is it clinically applicable? *J Urol*, 172: 1151.

Papas, K.K., *et al.* (1999a). Development of a bioartificial pancreas: I. *long-term propagation and basal and induced secretion from entrapped betaTC3 cell cultures. Biotechnol Bioeng*, 664: 219.

Papas, K.K., *et al.* (1999b). Development of a bioartificial pancreas: II. *Effects of oxygen on long-term entrapped betaTC3 cell cultures. Biotechnol Bioeng*, 664: 231.

Paquette, J.S., *et al.* (2003). Production of tissue-engineered three-dimensional human bronchial models. *In vitro Cell Dev Biol Anim*, 395–6: 213.

Paquette, J.S., *et al.* (2004). Tissue-engineered human asthmatic bronchial equivalents. *Eur Cell Mater*, 7: 1.

Park, H.J., *et al.* (1999). Reconstitution of human corporal smooth muscle and endothelial cells *in vivo. J Urol*, 162: 1106.

Park, J.K. and Lee, D.H. (2005). Bioartificial liver systems: current status and future perspective. *J Biosci Bioeng*, 994: 311.

Piechota, H.J., *et al.* (1998). *In vitro* functional properties of the rat bladder regenerated by the bladder acellular matrix graft. *J Urol*, 159: 1717.

Ponder, K.P., *et al.* (1991). Mouse hepatocytes migrate to liver parenchyma and function indefinitely after intrasplenic transplantation. *Proc Natl Acad Sci USA*, 88: 1217.

Portis, A.J., *et al.* (2000). Laparoscopic augmentation cystoplasty with different biodegradable grafts in an animal model. *J Urol*, 164: 1405.

Probst, M., *et al.* (1997). Reproduction of functional smooth muscle tissue and partial bladder replacement. *Br J Urol*, 79: 505.

Pruett, T.L. (2002). The allocation of livers for transplantation: a problem of Titanic consideration. *Hepatology*, 354: 960.

Rippon, H.J., *et al.* (2004). Initial observations on the effect of medium composition on the differentiation of murine embryonic stem cells to alveolar type II cells. *Cloning Stem Cells*, 62: 49.

Rocha, F.G. and Whang, E.E. (2004). Intestinal tissue engineering: from regenerative medicine to model systems. *J Surg Res*, 1202: 320.

Salle, J., *et al.* (1990). Seromuscular enterocystoplasty in dogs. *J Urol*, 144: 454.

Satar, N., *et al.* (1999). Progressive bladder dilation for subsequent augmentation cystoplasty. *J Urol*, 162: 829.

Seaberg, R.M., *et al.* (2004). Clonal identification of multipotent precursors from adult mouse pancreas that generate neural and pancreatic lineages. *Nat Biotechnol*, 229: 1115.

Selden, C. and Hodgson, H. (2004). Cellular therapies for liver replacement. *Transpl Immunol*, 123–4: 273.

Selden, C., *et al.* (1995). Histidinemia in mice: a metabolic defect treated using a novel approach to hepatocellular transplantation. *Hepatology*, 215: 1405.

Siddiqui, M.M. and Atala, A. (2004). Amniotic fluid-derived pluripotential cells. In Lanza RA (ed.). *Handbook of Stem Cells* Philadelphia, PA: Academic Press, vol. 2.

Sievert, K.D., *et al.* (2000). Homologous acellular matrix graft for urethral reconstruction in the rabbit: Histological and functional evaluation. *J Urol*, 163: 1958.

Song, K.H., *et al.* (2004). *In vitro* transdifferentiation of adult pancreatic acinar cells into insulin-expressing cells. *Biochem Biophys Res Commun*, 3164: 1094.

Southgate, J., *et al.* (1994). Normal human urothelial cells *in vitro*: proliferation and induction of stratification. *Lab Invest*, 71: 583.

Strasser, H., *et al.* (2004). Stem cell therapy for urinary stress incontinence. *Exp Gerontol*, 399: 1259.

Stripp, B. and Reynolds, S.D. (2005). Bioengineered lung epithelium: implications for basic and applied studies in lung tissue regeneration. *Am J Respir Cell Mol Biol*, 322: 85.

Sutherland, R.S., *et al.* (1996). Regeneration of bladder urothelium, smooth muscle, blood vessels, and nerves into an acellular tissue matrix. *J Urol*, 156: 571.

Tait, I.S., *et al.* (1994). Progressive morphogenesis *in vivo* after transplantation of cultured small bowel epithelium. *Cell Transplant*, 31: 33.

Thomalla, J.V., *et al.* (1987). Infectious complications of penile prosthetic implants. *J Urol*, 138: 65.

Thompson, J.S., *et al.* (1984). Comparison of techniques for growing small bowel neomucosa. *J Surg Res.*, 364: 401.

Thompson, J.S., *et al.* (1986). Growth of intestinal neomucosa on prosthetic materials. *J Surg Res*, 415: 484.

Thompson, J.S., *et al.* (1987). Stimulation of neomucosal growth by systemic urogastrone. *J Surg Res*, 424: 402.

Vacanti, J.P. (2003). Tissue and organ engineering: can we build intestine and vital organs? *J Gastrointest Surg*, 77: 831.

van de Kerkhove, M.P., *et al.* (2004). Large animal models of fulminant hepatic failure in artificial and bioartificial liver support research. *Biomaterials*, 259: 1613.

Van Vranken, B.E., *et al.* (2005). Coculture of embryonic stem cells with pulmonary mesenchyme: a microenvironment that promotes differentiation of pulmonary epithelium. *Tissue Eng*, 117–8: 1177.

Wang, T., et al. (2003). Creation of an engineered uterus for surgical reconstruction presented at the Proceedings of the American Academy of Pediatrics Section on *Urology*, New Orleans, LA.

Wang, Z.Q., *et al.* (2005). Morphologic evaluation of regenerated small bowel by small intestinal submucosa. *J Pediatr Surg*, 4012: 1898.

Watson, L.C., *et al.* (1980). Small bowel neomucosa. *J Surg Res*, 283: 280.

Yiou, R., *et al.* (2003). Restoration of functional motor units in a rat model of sphincter injury by muscle precursor cell autografts. *Transplantation*, 76: 1053.

Yokoyama, T., *et al.* (1999). Primary myoblasts injection into the urethra and bladder as a potential treatment of stress urinary incontinence and impaired detrusor contractility: long term survival without significant cytotoxicity. *J Urol*, 161: 307.

Yoo, J.J. and Atala, A. (1997). A novel gene delivery system using urothelial tissue engineered neo-organs. *J Urol*, 158: 1066.

Yoo, J.J., *et al.* (1995). Ureteral replacement using biodegradable polymer scaffolds seeded with urothelial and smooth muscle cells. *J Urol*, 153(Suppl): 375A.

Yoo, J.J., *et al.* (1996). Creation of functional kidney structures with excretion of urine-like fluid *in vivo*. *Pediatrics*, 98(Suppl): 605.

Yoo, J.J., *et al.* (1998). Bladder augmentation using allogenic bladder submucosa seeded with cells. *Urology*, 51: 221.

Yoo, J.J., *et al.* (1999). Autologous engineered cartilage rods for penile reconstruction. *J Urol*, 162: 1119.

Chapter 22
Ethical issues in tissue engineering

Stellan Welin

With illustrations by Karin Welin

Chapter objectives:

- To understand the distinction between values and facts
- To acquire a basic knowledge of consequentialism and deontology
- To distinguish between ethics and law
- To understand the main views of the human embryo and fetus
- To acquire basic knowledge of the Helsinki principles for research

  on donated human biological material
- To understand the ethical issues involved in commercialization of biological material
- To understand how new technological developments may create ethical problems but also may solve ethical problems

"Science can discuss the causes of desires, and the means for realizing them, but it cannot contain any genuinely ethical sentences ... "

Russell (1998, p. 25)

"The man who shuns and fears everything and stands up to nothing becomes a coward; the man who is afraid of nothing at all, but marches up to every danger, becomes foolhardy ... Thus temperance and courage are destroyed by excess and deficiency and preserved by the means."

Aristotle (1955, p. 94)

22.1 Introduction

Tissue engineering will hopefully be able to make many wishes come true. There may be biocompatible implants for repair of the body; there may be tissue and organs grown from human cells; there may be new possibilities of connecting the body and its nervous system to artificial limbs. Is not all of this good? Who can be against it?

There are many controversial areas, however. One is related to the use of human cells; in particular, the use of embryos to produce human embryonic stem cells. There are also questions about the commercialization of human biological material – after all, there are biotech firms who want to make a profit – and there are issues of donation and control of human biological material.

Completely different issues deal with the idea of 'naturalness' and the traditional medico-ethical concept that the right thing is to cure and help, not to improve and enhance human beings. Where is the limit when we stop being human? Who is to judge?

Questions like these cannot be answered by science alone. They deal with values and normative principles. Before turning to a discussion of the issues above I will give a short overview of ethics, the 'theory' of good and evil, right and wrong. In this chapter, I arrive at certain, sometimes controversial (but hopefully convincing), positions on various issues. Suffice it to say that other writers have arrived at other conclusions. In the end everyone has to think for themselves.

Box 1 Some ethics theories

There is a wide variety of theories of ethics with both normative statements and values. On the foundation of ethics one distinguishes between *deontological* and *consequentialist* theories. There are also ethical theories that regard ethics as being based on *agreements* under certain conditions.

Deontological theories are often characterized by 'right' precedes 'good'. One can do the right action (the duty) without looking at the consequences (the 'good'). Some deontological theories seek the foundation for ethics by rationally deriving norms for action. An example is given by the various versions of Kantian ethics. By pure reasoning every rational being can (in principle) find out what their duty is in various circumstances.

An example of a consequentialist theory is *utilitarianism*: 'Always act in such a way that your action produces as much good in the world as possible'. This is a version of 'the Greatest Happiness Principle' originally pronounced by the British philosopher Jeremy Bentham. There are no absolute duties; you have always to consider the consequences before acting.

Furthermore, there are various concepts of the 'good'. Some say the good is pleasure (hedonists), others think fulfillment of preferences is good, still others would pronounce self-fulfillment, etc. A well-know representative today of a kind of preference-utilitarianism is Peter Singer.

Basing ethics on *agreements* can be followed in two ways. One is hypothetical agreement in a hypothetical situation. A well-known example is John Rawls' *Theory of Justice* as a result of rational and prudent

k.w-06

agents agreeing on basic principles for society in a situation where they are aware that they have interests, but do not know the details nor basic facts about themselves. A prudent person, not sure whether they are a man or woman, will not endorse principles that discriminate against either gender.

Another idea is to base ethics on actual agreements reached in a free exchange of ideas without threats or power involved. This is basically the idea of the discursive ethics, developed by Jurgen Habermas. It is based upon agreements between actual persons, but in a very special situation.

Some prefer to view the agreement as something that has evolved as mutually beneficial rational agreements between persons pursuing their own interests. One example of this view is held by David Gauthier, a Canadian-American philosopher.

It is noteworthy that all the views presented in this fact box so far put special emphasis on consciousness (consequentialists often on all kind of consciousness; i.e. sentient beings), deontologists and agreement proponents often on human consciousness. The nonconscious world is a matter of ethics only in so far as it has affects on sentient beings and/or humans. There have been various attempts to develop ethics encompassing either all living organisms or maybe everything. Arne Naess and 'deep ecology' is an example of the first. Some examples of environmental ethics may exemplify the 'ethics of everything'.

Which view of all the ones presented here might be the correct one? There is no agreement, nor will there ever be. More and more work in bioethics, applied ethics and political philosophy might be characterized as developing a *political conception of ethics*, producing principles and values acceptable from a wide variety of more basic ethical theories.

22.2 Morality, ethics and values

With the help of language we can state facts, propose hypotheses, approve or disapprove, give promises, etc. Sometimes we describe the world. 'Today it will rain' expresses what is often called a *proposition*, which is either true or false. Some propositions are more complicated, e.g. a statement describing the whole human genome, but nevertheless are either true or false. Sometimes we just don't know of course. *A common idea is that science (including biomedicine) and social science is concerned with finding true propositions about the world and reliable predictions.* This is, however, not the whole story. Science and social science also try to give explanations, and to make the natural and social world intelligible.

Statements about values (what is good or bad, right or wrong) and norms (what one ought to do) are different. It is a common idea that there is a gap between statements of facts and statements of values and norms. The Scottish enlightenment philosopher David Hume stated that one can never derive or infer value or moral conclusions from statements of pure facts. There is an unbridgeable gap between 'ought' and 'is'. For example, from the fact that Sweden is a monarchy it does not follow that having a hereditary king as head of state is morally superior to having an elected president. (Of course, the opposite does not follow either.) From the (contested) fact that males are biologically determined for certain aggressive behavior, it does not follow that this behavior is morally right.

However, in many situations we use statements and concepts that both have a *factual* part (describing the world) and an *evaluative* or *normative* part (saying what is good or bad, making recommendations). For example, *democracy* is such a concept, *intelligence* is another. To make things even worse, very often people agree just on the evaluative part that democracy or intelligence is good, but disagree strongly about the factual parts. Unfortunately, this is the case with many of the key concepts in social science. In natural science, the concepts are usually less value-laden. In medicine, the concept of health obviously has this dual character as well as the even more contested notion of 'being normal' in a psychiatric or psychological sense. It is presumably a good thing to be normal, but what is really normal?

Science cannot make recommendations based *only* on facts. There must always be a normative or value component added, according to the view of the division of the realm of facts and values. Medicine can never – based solely on medical facts – show that a long life is better than a short one. At most, it may give recommendations on how to live and behave, if you *wish to have a long life.*

We use language not just for describing the world, but also for inducing behavior in others. 'Shut the door' is such an imperative, obviously neither true nor false. If we think that the other person should not just shut the door once, but every time, we may express this by saying 'always shut the door' or 'you ought to shut the door'. This is a rather trivial example, but most of the *moral norms* regulating (hopefully) our behavior can be expressed in similar ways. 'Always tell the truth' and 'one ought to keep one's promises' are typical examples of the everyday morality many people subscribe to, or at least believe they subscribe to, and follow.

There are many norms as explained above. Some are *conventions* like 'always drive on the right side of the road'. We have agreed to that norm; it is very dangerous if people differ locally on the issue on what side to drive. Such a convention can be changed (like in Sweden in 1967) and it can be different (like in the UK). A norm like 'always tell the truth' is different. It is not an agreement in the sense a convention is.

The simplest way to characterize a *moral norm for a person P* is that it is an *imperative or 'ought' statement* ('always tell the truth' or 'one ought to tell the truth') that:

(a) P tries to follow in his/her life – even if no-one can observe P.
(b) P is ashamed if he/she breaks the norm *even if no one finds out.*

If someone follows a particular norm ('always tells the truth', for instance) for fear of punishment, then it is not a moral norm *for that particular person* in the sense just stated. A moral norm is followed from 'inner pressure', not because of threats from others. The legal laws are other kinds of norms. The say, like the moral norms, what you should and should not do. However, in contradistinction to moral norms, legal norms are associated with sanctions like 'you will be punished if you do not follow the law'. In happy circumstances, the law incorporates the most important moral norms together with other prescriptions relating to conventions. Even if it is not a moral norm that we should drive on the right side of the road, the law nevertheless prescribes that we should all drive on the same side and will punish transgressors.

The reason for having laws and a legal system can be stated very easily: not everybody will follow the (most important) moral norms without the threat of punishment. In a society with perfectly moral citizens, one would not need a legal system. If all scientists were highly moral, one would not need legislation on matters related to research and technological development.

Not every moral norm finds its way into the legal system. You are not punished by the law if you lie to your partner or are unfaithful even if the behavior is morally blameworthy. That would be very intrusive and, apart from that, we do not agree on all moral norms. In fact, in modern western societies there has been a diminishing regulation of what we may call personal behavior. This is obvious in the area of sexual behavior, but also in matters of what you may say and think.

Different persons in the same society may adhere to different norms and have different values. Usually the difference is even bigger if we look to different cultures and other times. For example, the equal rights of men and women are not very old. One hundred years ago a woman could not decide herself if she wanted to marry a man (we assume here that the man wanted to marry the woman). The father of the bride to be – if he was alive – had that particular right. The woman must consent, but that was only a necessary, not a sufficient condition for marriage. The development towards an equal standing for women to men is considered a moral progress.

The different moral and value systems of individuals, social groups and culture can be investigated empirically. This is done by social science methods of interviews, observations of behavior and questionnaires. Over a period of 100 years there have been substantial shifts in the value system of the western world. Very few today would regard one human race as superior to another, few believe in different rights and behavioral norms for men and women, and animals have entered the frame as something to be included in moral considerations.

To find out about the existing moral systems and values in a society is very interesting. However, it is not an answer to the *normative* questions:

What should I do?
What values should I have?

Or phrased alternatively: *what moral should I have*? This is a question dealt with in the philosophical inquiries on ethics. Unfortunately, there is no consensus on this important question. Philosophers disagree among themselves on the ultimate foundation of ethics as well on the content. However, it has been possible to reach some consensus in some areas, even if that consensus has not reached the seminars of philosophers. For example, the Universal Declaration of Human Rights (http://www.un.org/Overview/rights.html) and the Helsinki Declaration on Biomedical Research on Humans (http://www.wma.net/e/policy/b3.htm) can be seen as kinds of principles that can be defended and understood from many different, more fundamental philosophical starting points. An influential attempt to present such a common framework for the biomedical area was the attempt to develop the 'Belmont principles', where respect for persons (autonomy), the principle of beneficence (do good, do not harm) and justice, play a central role, into a coherent whole (Beauchamp and Childress, 2001).

Following the Helsinki Declaration and various scandals in biomedical research there are now ethics committees in place who review biomedical research before it begins. There is also extensive regulation of pharmaceutical drugs and growing regulation on medical technologies and materials used for implantation in the human body. Much of this regulation can be viewed as an attempt to translate the ethical principles and values in the Helsinki Declaration into legal regulation. There are similar regulations on animal experimentation.

There is a growing consensus that the area of moral concern covers not just humans, but all *sentient beings*, i.e. beings that experience and feel pain and pleasure. The philosopher Peter Singer (1993) has famously written that we should show 'equal respect for interests' regardless of who has the interest [1]. It does not matter if the interest belongs to a white or black person, to a man or a woman – nor if it is an animal or a human being. Not showing equal respect in this way amounts, according to Singer, to *racism, sexism* or *speciesism*, respectively. Singer advocates a certain kind of equality between all sentient beings.

However, there are still distinctions to be made. A human being normally has a larger number of and/or more complex interests than, say, a mouse. If a human being is killed, more is lost than if a mouse dies. However, enough mice may outweigh a human being. This position is controversial and quite contrary to many ethical systems stressing the *dignity of humans*, where humans are seen as special and there is a clear dividing line between nonhuman animals and humans. Such a special position can be argued for from a religious perspective (God created man special) or from some property (like reason) which distinguish humans from animals.

22.3 Moral problems relating to the source of material for tissue engineering

In tissue engineering one usually needs to start from some biological material. It may be cells from an animal or a human being that are removed and, for example, placed on some matrix to grow into a structure. Most tissue-engineered structures are meant to eventually find their way into humans to be used as a treatment or replacement. This means that it is preferable to use human biological material as a starting point.

The use of nonhuman biological material will pose the same technical and medical problems as xenotransplantation, i.e. the idea of transplanting organs from animals into humans. One particular problem facing the use of xenogenic biological material is the risk of infection from retrovirus in the source material. In the transplantation area there is a serious shortage of human organs, which may be an argument to try xenotransplantation in a controlled form anyhow – when and if the source animal is suitable and the technology for treating rejections is more developed. No such shortage exists for human cells. From a moral point of view (for the good of the patient to have the material implanted) human material should be used to avoid a variety of technical problems.

There are *general moral problems* relating to the use of human material. They have to do with *donation, informed consent, ownership* and *control* of the resulting material. There are also *special moral problems* dealing with cells derived from human embryos, like embryonic stem cells, or from aborted human fetuses. The general problems also apply in the case of material from human embryos and aborted fetuses.

22.3.1 Biological material derived from human embryos

In 1998, the first human embryonic stem cell line was derived from an embryo (Thomson *et al.*, 1998). Since then there have been many more such human embryonic stem cell lines derived. In principle, such a human embryonic stem cell line may stay in the same condition forever. Other types of stem cells and cells can be derived from the embryonic stem cells by putting them in a suitable environment. This could be a very good source of human cells. There is still more to be learned about the development of the embryonic stem cell into an adult cell, but in due time we can expect such cell lines to be very good sources.

If you are interested in tissue engineering an extracorporeal human liver, you may of course start directly with human liver cells to be used in your structure. However, this will mean an invasive procedure to extract liver cells or you have to rely on donation of a cadaveric liver. It would be more convenient to derive liver cells from an embryonic stem cell line. Furthermore, the risk to the donor of the liver cells can be completely avoided. However, there are differing views of the early human embryo and some of these views rule out human embryonic stem cell derivations.

22.3.1.1. Conceptualism

To establish an embryonic stem cell line 'surplus' *in vitro* fertilization (IVF) embryos are usually used, after they have been cultured for some days. At the blastocyst stage (around 5 days), the inner cell mass containing the embryonic stem cells is extracted and transferred to another medium for further culturing. In this process the embryo is destroyed. One argument is that by destroying the embryo in this way it is morally wrong; the embryo has the same moral status as a newborn child, because a full human life begins at conception (known as *conceptualism*). The Catholic Church holds a version of conceptualism, as does the American President George Bush [2].

That 'surplus' embryos are used is no good defense, from the point of view of conceptualism. First, it would be possible to fertilize just one egg (or two) for implantation, then there would be no surplus embryos. Second, even if there were 'surplus children' it would be wrong to kill them, extract some cells and use the cells for therapy. In the same way, according to adherents of conceptualism, it is wrong to destroy surplus embryos.

Some conceptualists use an argument from potentiality to defend the full moral status of the newly formed embryo. An embryo, if left on its own in a suitable environment, i.e. a uterus, will develop by itself into a child unless interfered with. The embryo has an *active potentiality* to become a child. One may then argue 'backwards'. The newborn child is a full human being from a moral point of view. The fetus at an advanced stage, say just before birth, is very similar to the newborn child. And so it goes on. There is no specific break point when the fetus or embryo becomes human. To put it simply, the development from conception to a newborn child is the development *of* a human being – not *to* a human being. This position is relatively new, and has been strengthened by the modern theory of genetics and DNA [3]. The DNA of the grown-up is produced at conception and it starts working as a whole shortly afterwards.

The most obvious problem of this argument for active potentiality is, however, that the IVF embryo does not have active potentiality. It needs help to be implanted; if we just let 'nature proceed' it would never develop into a child on its own.

22.3.1.2 Gradualism

Another conception of the embryo is related to the sentient being approach. The simple idea is that the early embryo – as just a single cell – cannot feel or experience anything. It is not yet a sentient being and we do not owe the same kind of moral respect towards it as we do towards sentient beings. Somewhere along its development it becomes sentient. The belief is that this sentience is tightly coupled to the development of the brain. When is the fetus sentient enough to count as a human? It seems highly unlikely that there is a sudden jump just at birth. It occurs either before birth or after. Many people believe that it occurs before birth, while others do not [4]. Unfortunately, while it may eventually be possible to measure the degree of sentience (there are of course great problems in characterizing such a measure) this will not decide the morally important question: *what degree of sentience is enough to count as human sentience?*

Gradualism does not solve all problems. First, it is not clear from gradualism what moral respect we owe to the embryo. Is it different from what we owe to other human cells? Most people would think so. There is something special about an embryo. It has to do with the possibility to develop into a child. Do not all cells have this possibility? Cloning of animals actually proceeds from somatic cells. All moral differences must be based on *some* difference.

There is, however, a difference. The embryo is capable of this development on its own in a way that the unfertilized egg or the somatic cell is not. A somatic cell – even if its cell nucleus after fusion with a denucleated unfertilized egg may develop into a child – is analogous to the sperm or unfertilized egg.

Here, I am aware that this discussion is quite reminiscent of the Catholic one. However, the difference from conceptualism is that gradualism denies that the embryo has the same value as the born child. Gradualism holds, as does conceptualism, that the moral value of the embryo is higher than that of other cells, including a sperm and egg pair not yet fused. However, conceptualism and gradualism differ on the moral status of the newly conceived embryo.

But how high is the moral value of the embryo? Is it so high as to rule out destroying the embryo in order to derive human embryonic stem cells? While some people may not agree with this, there may be an amount of justification, e.g. a possible good outcome, which may outweigh the harm to the embryo. The good outcome that most easily comes to mind is that human embryonic stem cells may, in the future, be used for treating human illnesses. There are other uses that may outweigh the harm, e.g. substituting research human embryonic stem cells for research on sentient animals.

22.3.1.3 Using cells from aborted fetuses
In the gradualistic view it is also easier to accommodate abortion from a moral point of view than from the conceptualistic view. However, all arguments on abortion are based on the basic fact that the fetus is inside the body of the woman. These arguments do not apply to IVF embryos in Petri dishes, as they are not inside the body of a woman (they can be protected and safeguarded without infringement to the autonomy of the woman and her right to bodily self-determination). It is a consistent – although perhaps rare – position to hold that embryos in a Petri dish should be more protected than fetuses in a woman.

Suppose you subscribe to the gradualist theory. You accept (early) abortion because you believe in a woman's right to bodily self-determination and in the stage of pregnancy when the fetus is not sentient (in a human way) the right of bodily self-determination takes precedence. On the other hand, you believe that the value of a human embryo is too high to warrant its use for derivation of stem cells. Attempts have been made to defend its use anyhow, if it is a surplus embryo in an IVF context [5]. If not used for

implantation or stored for later infertility treatment, it will be discarded, wasted. Surely it is better to use it for something beneficial? This is often the reason why it is regarded as ethically less controversial to use donated embryos from IVF treatment than to use specifically produced embryos, whether from therapeutic cloning or from more conventional means.

The argument for surplus embryos is, however, not very convincing. Its proponents overlook the fact that there should be no surplus embryos at all – if you take the value of the embryo seriously. It is possible to do IVF in such a way that all of the eggs that are fertilized will be implanted. It may not be as effective, but we are used to ethical restrictions on treatment in other cases, e.g. transplantation of organs.

In the gradualist view most will regard a more developed fetus as more 'valuable' than the early embryo. Still, it is possible to make an argument that it is better to use fetal material than using embryonic material. The surplus argument can be applied. In a country where abortion is legal up to a certain gestational time, women decide for themselves if they want an abortion. After abortion, the aborted fetus is 'surplus' in different sense to the IVF embryo in the Petri dish – it is not the physicians or the scientist about to derive stem cells who produced the surplus material. In the case of the aborted fetus one may argue that it is better to use it than to simply waste it. The morally relevant distinction is that the physician and scientist in the IVF case bring about the surplus character, while this is not the case in abortion. Of course, as in the IVF case, the use of biological material should depend on the informed consent of the woman (in the IVF case consent is usually sought from the couple).

22.3.1.4 The risk of instrumentalization
Some critics of the use of embryonic and fetal material have warned about the risk of *instrumentalization* of human life: human beings may no longer be seen as goals in themselves, but as means to produce something else. In particular, human embryos may be seen in this way. Such critics claim that this will underline our moral community. Such a view has recently been forcefully argued by the German

philosopher and social scientist, Jurgen Habermas. His arguments apply to the harvesting and use of human embryos, even if they are designed primarily against reproductive cloning and genetic modification of the child to be. Such a cloned or modified person would not, according to Habermas (2003), be 'the author of his or her own life' and this will undermine any possibility to participate in the moral community.

Box 2 Science that shook the world

The first 'test-tube baby (1978)

Louise Brown was born in July 1978. She was the first child to be born from IVF treatment. Her mother had blocked Fallopian tubes, which means that the eggs cannot pass down from the ovaries and be fertilized. Dr Patrick Steptoe and Dr Robert Evans, in the UK, had been working on methods for IVF for many years. Some years earlier they had managed to fertilize eggs outside the female body, but they had not succeeded in having a pregnancy sustained for more than a few weeks. They had made about 80 such attempts.

In the unsuccessful attempts, the physicians had cultured the fertilized eggs until they had evolved into 64 cells. This time they implanted a younger fertilized egg after 2.5 days. The pregnancy continued normally, but some days before the expected birth the mother's blood pressure increased and Louise Brown were delivered by a Caesarean section.

This first instance has been followed by many more and IVF is now an established standard treatment. The first apprehended risk of health problems for the IVF babies can now be disregarded; the remaining problems seem to be due to a higher prevalence of twin births in older mothers. However, it

is worth noting that there were no regulated clinical trials – this development took place completely as clinical treatment.

Remaining ethical issues have to do with what happens to the fertilized eggs which are not implanted. They can be thrown away, frozen for later use, or donated for research, including derivation of human embryonic stem cells. In many countries IVF is heavily commercialized.

Baby Fae: receiving a baboon heart (1984)

In October 1984, a baby girl was born, known in the literature as Baby Fae. She had a lethal underdeveloped left side of the heart. Most babies with this problem die shortly after birth. Dr Leonard Bailey at Loma Linda University Center, US, decided to try to transplant a baboon heart in Baby Fae. Although heart transplants were successful in older people, there had been no successful infant-to-infant heart transplant (although this did happen 1 year later).

Baby Fae lived for a few days with the baboon heart. She died from massive organ failure when her blood started to clump together. The Baby Fae transplantation had been controversial from the start. First, the common scientific opinion was that the possibility of success was very low. Second, it had been questioned if and how sufficient the information was given to the mother of the baby girl. Third, a healthy baboon had been sacrificed even though the possibility of success was extremely low.

Xenotransplantation (cross-species transplantation) still remains an active field even if it has not yet hit the wider clinic community. Some cell transplantation protocols have been approved and executed, but no whole-organ transplantation is in sight. The field is still struggling to overcome rejection. It has also been discovered that there is a risk that the endogenous retrovirus in the pigs (the preferred 'donor') may infect humans. To avoid the worst rejections the pigs must be transgenic. It is fair to say that transgenic pigs at present are not good enough to avoid rejections. (Rejection is studied in baboons with transplanted transgenic pig organs.) The most hotly discussed ethical issues are related to animal rights and to the risk of spreading a disease by transplanting pig organs. Some are worried that xenotransplants will also affect our 'human-ness'.

The cloning of Dolly (1996)

Few scientific reports have affected the public more than the news of the cloning of a sheep. Making a genetic copy of an adult mammalian was believed to be impossible, but nevertheless Ian Wilmut and his colleagues at Roslin Institute, UK, managed to do it. They took a mammary cell from a Finn Dorset sheep and transferred its nucleus to an enucleated egg from a Scottish Blackface sheep. A tiny pulse of electricity made it start to develop. The egg was transferred to a surrogate mother, the egg donor, and Dolly was born after a normal pregnancy. Dolly is a genetic copy of the Finn Dorset sheep where the cell nucleus came from. (Her mitochondrial DNA came from the donor of the egg, however.)

Wilmut and coworkers performed 276 attempts before they succeeded. Since Dolly, many other mammalian species have been cloned. The birth of Dolly immediately started a big public discussion on cloning. In the US, the President asked the National Bioethics Advisory Board to prepare a report on cloning. They claimed that cloning should not be ruled out

from basic ethical principles, but should not be undertaken at this present time. The risk to the child is too high. Various health problems have been seen in the other cloned mammalians and Dolly developed arthritis at a surprisingly young age. She is now dead.

Those who favor human cloning tend to see it as a prolongation of IVF. It started with 'conventional IVF', as in the case of Louise Brown. The next step was the technology of injecting a single sperm into the egg in the case when the male has very little or nonmobile sperms. Cloning – according to this line of thought – would just be a case of IVF with no sperm at all. (The idea is to use an adult cell from the man and inject its nucleus into an enucleated egg from the woman, although other combinations do exist.) Most critics focus on health risks for the child and on the 'unnaturalness' of the procedure. It is less frequently discussed that the cloning technology may give a woman the opportunity of having children completely on her own. A woman has both cells and eggs.

22.3.2 Producing human embryonic stem cells and avoiding destroying embryos

In general, all sources that do not involve the destruction of embryos avoid the special problems associated with the human embryonic stem cells discussed above. Can it be possible to produce human embryonic stem cells in such a way as to avoid these problems? In April 2006, a commentary in *Nature Biotechnology* discussed the possibility to create pluripotent stem cells without first passing the stage of a human embryo (Snyder *et al.*, 2006). Some possible strategies are outlined as worthy of further research.

Most of the strategies involve inducing a change in either the male or female gametes precluding the formed oozyte to proceed further than, for example, the blastocyst stage. The oozyte, one could argue, is then not a real embryo as it cannot become a child. Perhaps we could call it a pseudo embryo. Less promising seems the idea of a post-fertilization change in the embryo, thus hindering its development beyond a certain point. That would amount to the destruction of an embryo, and would not be acceptable to those who opposed to such practices. The same can be said of the already well-known technique of replacing the nucleus of the oozyte with a somatic cell nucleus (therapeutic cloning). That would also result in an embryo which, in principle, could develop into a child. It would then be a case of reproductive cloning.

A better way would be – if possible – to modify the somatic cell nucleus in such a way that the pseudo embryo cannot develop beyond a certain stage.

22.3.2.1 *The importance of words*

The term 'pseudo embryo' was introduced above. The underlying idea in the commentary discussed above can be restated as: 'It may be morally wrong to destroy embryos. However, destroying pseudo embryos in order to harvest human embryonic stem cells is morally acceptable'. Some would probably say the same of an oozyte created from somatic cell nucleus transfer (therapeutic cloning). This is also a pseudo embryo. However, in this last case there is a possibility that this oozyte might develop into a child (reproductive cloning). Maintaining that this oozyte is not a real embryo amounts to the astonishing idea that while most humans emanate from embryos, some may (in the future) not.

In discussions on ethics, words and concepts play crucial roles. Perhaps we could agree to call an oozyte a pseudo embryo if it is not capable of developing beyond a certain early point. That would not be unethical or an attempt to fool the critics. However, it could be maintained that to call an oozyte produced by therapeutic cloning a pseudo embryo would be misleading and therefore unethical. Such a 'pseudo embryo' can give rise to a child exactly like a 'real' embryo.

There is a very interesting earlier example of such verbal exchanges that demanded a debate. In 1984 both Houses of the British Parliament rejected a proposal in the Warnock report to allow research on human embryos for the first 14 days of their existence. The members of parliament were very opposed to research on human embryos. However, after the research community came up with the clever idea that the oozyte in its early stages was not an embryo at all, but a 'pre-embryo', the pendulum swung. The British parliament approved research on pre-embryos and the MPs could continue to maintain their opposition to embryo research (Mulkay, 1994). This might pave the way to MPs approving research on human pseudo embryos while still rejecting research on human embryos.

22.3.2.2 Other sources of human cells

The uses of hematopoietic stem cells and other types of human stem cells are discussed elsewhere in the book. It seems at present that there is a problem of expanding such sources of cells up to large numbers. One may hope that a better understanding of the role of the niches of the various adult stem cells will allow for better culturing and expansion. The interesting point from an ethical point of view is that none of these sources entails the destruction of embryos. Such sources are thus acceptable to those who oppose the use of human embryonic stem cells. A clear illustration is the talk that Pope John Paul II gave to the International Transplantation Society Congress in Rome in 2000. In that talk, Pope John Paul II declared that donation of organs after death is an act of Christian love and endorsed the brain-related death criteria. However, while being somewhat flexible on organ donation at the end of life, Pope John Paul II condemned, in strong words, the destruction of human embryos to produces human embryonic stem cells. Instead, he urged scientists to use adult stem cells.

There have been reports of plasticity of stem cells, i.e. an adult stem cell of a certain type can be changed into another type from a different lineage. This is still somewhat controversial, but if it turns out to be possible, it will make the use of adult stem cells easier. As stated above, few will object to such use of human cells from an ethical point of view.

The only objection might be is that this transformation would be 'unnatural'. This is similar to the objection sometimes voiced against genetic modification – it, too, is unnatural. However, it is probably fair to say that most opponents of the present-day use of genetic modification object to it because of the perceived risks to health (food) or the environment. It is hard to see how that could be applied to the alteration of stem cells in the laboratory.

It is now time to turn to general ethical problems associated with using human cells for tissue engineering.

22.3.3 Donation and informed consent

Whether the biological material to be used comes from human embryos, aborted fetuses or adult persons, it originally belongs to someone other than the researchers in the tissue engineering area. They may of course use material from themselves, but usually it is material from someone else.

Even if, in most legal systems, you are not strictly the 'owner' of your body or its part, you have the ultimate authority over its use. This is the simple principle of *autonomy* enshrined in most medico-legal regulations and a very important point in the Helsinki Declaration – the principles guiding biomedical research on humans. You have to give consent to any invasive procedure involving your body and to all medical treatment in general, otherwise the procedure is illegal. There are some exceptions related to cases of emergency. If arrive unconscious at an emergency ward, the physicians do not of course have to wait until you wake up. They are allowed to act on *presumed consent*. For minors the parents will have a greater say.

When biological material from aborted fetuses is donated it is usually the woman who has to consent. This can be interpreted in two ways. The most straightforward interpretation is to view donation from an aborted fetus as similar to the donation of biological material that has been removed from the woman. The fetus is regarded as part of her body; the abortion is similar to an amputation. Consent has to be sought if consent is needed for the use of material from an amputated leg. In Sweden, for example, discarded biological material (that would otherwise be wasted) could formerly be used without consent (no one is asked), but is now included in the category where permission from the former 'owner' must be sought.

The second interpretation may be that a fetus is not simply a part of a woman's body but a separate individual. The material from an aborted fetus could be viewed as being similar to a donation from a dead child. Then it is clear that consent is needed.

In the case of embryo donation for stem cell research, in most cases it is the couple, wherever possible, who together have to consent to donation. The case of embryo donation for research, or derivation of stem cells, is best seen as being different from the interpretations of donation from an aborted fetus. Firstly, at the time of donation the IVF embryos are alive and, hence, the donation cannot be compared to donation from a deceased. Secondly, the IVF embryo is not a part of the woman's body so cannot be seen as an amputated limp. Thirdly, there are many alternative uses for an IVF embryo: it can be implanted in a uterus, frozen for later use in infertility treatment, donated to another couple (at present not legally allowed in many countries) or simply destroyed and discarded.

Usually more eggs are fertilized than will be implanted, so the couple is (or ought to be) aware that not all of the embryos can be used for a first implantation. No extra burden is based on them when they are asked about the various possibilities. There is a problem, however, regarding who is asking for donation. If it is someone from the IVF clinic the couple may feel 'forced' to consent.

In practice, one may say that embryo-donation is handled in a similar way to the donation of biological material from an adult. The difference is, of course, that there are two donors in the embryo case

and only one in the adult case. The embryo in the IVF case is, in a sense, viewed as an extension of the body of the two donors, who have contributed egg and sperm, respectively.

22.3.4 Ownership, control and commercialization of human biological material

As stated above, even if it is your own body, you do not strictly own it according to most legal systems. You are not allowed to sell the whole body or part of it. In some countries commerce is allowed with sperm and human eggs; buying and selling of human organs for transplantation is forbidden in nearly all countries.

22.3.4.1 The transplantation paradigm

There are many views to support this non-ownership with regard to the human body. One common argument is that it is *degrading* to do commerce with body parts, it is contrary to *human dignity*, etc. However, everyone agrees that an individual should decide on what will happen to his or her body and body parts. In transplantation it is the will of the deceased (before death) which is the most important (although in some cases the relatives might have a greater say). In all cases of living donations (whether of organs, tissue or cells), it is the individual who decides, unless they are, for example, a minor, or unconscious, etc.

There are, however, severe limitations in the area of autonomous choice in the donation of organ transplantation. Suppose someone, while still living, signs up for organ transplantation after death. There are many things that we do not let this person decide. The donor-to-be is not allowed to specify the race, political opinion, gender, etc. of the future recipient. Some may questions why; would it be unethical to donate organs only to persons with, for example, similar political views? Would it be morally worse not to donate at all?

In organ donation from a dead person, the only decision for the donor is whether to donate or not; the rest is up to the hospital (or in the case of cells

Box 3 Science out of control

A recurrent theme in many novels about science is that science becomes out of control – humans create something that cannot be controlled. Here are some examples.

Mary Shelley: *Frankenstein* (1831)

Victor Frankenstein, a scientist, creates a living creature in his laboratory from cadaveric human material from the newly buried dead. The new creature has many human traits. He is alone and demands that Frankenstein creates a woman for him. This is done, but then Frankenstein becomes afraid – he has created a new race, capable of procreating and multiplying, and perhaps even destroying humankind itself. He kills the female creature. The original monster promises revenge and threatens to kill Victor Frankenstein's future bride. This is done on the eve of the wedding. Frankenstein vows to kill the monster and the case takes him to the North Pole.

This is a remarkable – and still readable – book, written by the young Mary Shelley to entertain her poet-husband Percy Shelley, the poet Lord Byron and their friends. The metaphore of 'Frankenstein's monster' is still very alive, e.g. in the term 'Frankenfood'.

Aldous Huxley: *Brave New World* (1932)

The novel was inspired partly by a book by the biologist and friend – J. B. S. Haldane's *Daedalus or Science and the Future*. Huxley's novel is set 600 years after Ford – or Freud as he was mysteriously called when he pronounced on psychology.

In Huxley's tale we are taken to the London central hatchery, where human embryos are 'grown' on an industrial scale. Each fertilized egg gives rise to up to 96 identical individuals. The embryos are treated differently to create different people. Some are heavily radiated (the latest thing in the 1930s!) to lower intelligence. Society consists of different castes of humans and each has learned from earliest childhood to love their position. The world functions without friction. Nobody has to be unhappy – there is a drug 'Soma' which keeps them happy.

However, there is a reservation in the wilderness where people live 'natural lives'. The hero of the book, a member of the highest caste who is groomed for world leadership, is permitted to go into the wilderness. There he meets people who live differently to him, and have strange feelings and aspirations. He finds a woman at the reservation who was originally from the civilized world. She and her son are taken back to London but neither can adjust. The best part of the novel is when 'his Fordship' himself, one of the world governors, discusses the rational behind the smooth scientific running of the world.

Huxley's dystopia does not feature individual crazy scientists – as in Shelley's *Frankenstein*. Instead he paints a picture of a – at least in principle – benevolent society based completely on science. In that world, the individual is nothing and everything is for the collective happiness of everyone. It is a *utilitarian nightmare*.

Margaret Atwood: *Oryx and Crake* (2003)

This is a novel that starts after a huge catastrophe in an indeterminate future. As the story unfolds we learn that it may not be such a distant future. The cities of today are still there; however, some are abandoned after the rise of the oceans. It is a world of colossal social breakdown, where the rich live in specially guarded places and the areas outside are worse than the worst criminal downtown areas

of big American cities. There is Crake, a brilliant young boy who turns to biosciences, the young girl Oryx from a poor village somewhere in Asia, and the only survivor from the catastrophe, known from the start as Snowman.

The story is told backwards, but the following summary is the 'forward' version of it:

Crake advances quickly through the scientific world and is soon head of the 'Paradise project' of one of the big Pharmas. Here he is working on two things. One is a 'super pill' that cures most everything including sexual dysfunctioning. (It also makes people sterile, but that is not advertised.) The other project is the creation by gene technology of new – and better – animals and humans. In particular, humans are engineered to avoid all destructive passions, to be out of grasp of religion, etc. All these are things that Crake and his colleagues think are very dangerous and destructive. Oryx is a brought in to teach the new humans, and Jim – Snowman – is an old friend of Crake and responsible for security in 'Paradise'.

The super pills cause an intended (by Crake) pandemic that wipes out most of humanity. In the end only Snowman and the new humans are left together with the new animals. It all starts over again. Maybe it is in the end rather similar to what happened to our 'old' humanity. A scary book about what greed and hubris can do if combined with enough scientific talent and money.

for tissue engineering, the scientists) to decide. In the *transplantation paradigm* there is no influence or control of the future of your former cells (you may choose which organs to donate and if your organs may be used for other medical purposes if not suitable for transplantation).

22.3.4.2 The research paradigm

If you donate cells or tissues for use in tissue engineering, what kind of influence or control are you entitled to regarding the future fate of your (former) cells? It seems that there are two conflicting ways of thinking at present. One has been discussed above. The other may be called the *research paradigm*.

The idea propounded in the Helsinki Declaration is that donation should be done freely after full informed consent. It is of course far from clear what 'full and adequate information' is. The usual understanding is that it should include information about what kind of research is to be done, what the aim of the research is, and so on. Hence, the first difference from the transplantation paradigm is that you can choose the kind of research you donate to; in the transplantation paradigm the only choice is to donate or not to donate.

Sometimes there is a second difference as well. The donor of material may have a say in the future of the donated material. The cells or the tissue may not be used for research other than the one covered by the donation consent. The donor may revoke their consent and claim the material 'back' [6]. This idea has actually been incorporated in the Swedish Biobank Act, although it is somewhat vaguely phrased. The donor may at any time request that his or her donated material be destroyed. The head of the Biobank may then decide whether to do this or to make the material non-identifiable (how this can be done if the DNA is still there is indeed a mystery).

To summarize then, in the case of the transplantation paradigm, at organ (or tissue or cell) donation for transplantation, the control over the material is transferred to the recipient, and the donor has no more say over his or her former cells. The research paradigm and some of the ideas underlying it in the

Helsinki Declaration rest on the idea of the *extended body*. The materials donated for research (e.g. cells and tissues) are still part of the donor when they are cultured and experimented upon in the laboratory (it is part of your autonomy and bodily self-determination to have a continued say in what happen to these (former) cells and/or tissues). Many might think that this analogy is misleading and false, and it can also complicate life for the scientists who work with the donated material. While there should be regulations in place, they should not give the donor a veto in the use of the material.

22.3.4.3 Commercialization of cells and tissue

Many companies have been set up for commercial exploitation of tissue engineering based on donated human cells and tissue. Can this tissue-engineered product be sold for profit? Should the original donors be compensated in financial terms? When, if ever, are the donors' interests in the products engineered from 'their' cells to be considered void?

In both the transplantation and the research paradigm the altruistic and non-financial character of the donation is often stressed. Donors should not have a part of the future profit or any financial compensation. Whether rightly or wrongly, this is often deemed unethical. A better argument for non-compensation may be mustered from the *authorship criteria for scientific articles*. Usually there are many contributions made by various actors in the production of a scientific article. Some contributions are considered so small that it is enough to get a 'thank you' in the article. Some are considered so important that the person should be on the authorship list. Basically, an author should have made a *significant intellectual contribution* in the production of the article [7]. We can translate this to the donation process: *only those who make a significant contribution to the production of the resulting tissue engineering products have a right to be compensated*. The contribution of cellular material is not significant enough.

One school of thought is that the transplantation paradigm is preferable to the research paradigm. Hence, the donor gives away his or her material and then *the control is transferred* to the recipient (a research group, a university, a company). There is no ownership transferred if we do not believe in the ownership of human biological material. In Europe such ownership is not granted. Obviously, if a company does not own the human biological material it cannot be sold [8]. Indeed, just receiving the donated material for free and then selling it would be unfair to the donors.

There are, however, other ethically acceptable ways that a profit may be made. One is to put in work intellectual effort, knowledge and other resources to develop a new product. It is a common idea, going back at least to the British philosopher John Locke, that ownership can be based on the labor put into the process of extracting something from nature. It is thus fair that researchers and companies can make a profit from products *produced* from donated cells. Making profits from services provided is of course also acceptable.

There may be a problem if the donation was to a public institution, say a university, and the cells (or the products) are then transferred to a company and a profit is made. This is a problem relating to the public/private divide and there is no general answer. In Sweden, for example, the government and parliament, who basically pay for research at the universities, allows scientists to create companies and commercialize their products, where the companies get the profits. The basic idea is that this arrangement is good for society. In other countries, there may be different rules.

22.3.4.4 The morality of intellectual property rights in human biological material

One may believe that there are no *special* moral problems related to intellectual property rights in biological material in general and human biological material (removed from the body) in particular. (As long as we legally recognize ownership over animals, we cannot get upset if there are patents in animals.) Intellectual property rights, like patents, are smaller bundles of rights and entitlement, and do not give the holder the same right as the owner. Patents are furthermore limited in time. Patents in human biological material permanently removed from the body pose no *special* problems [9].

There are, however, *general* problems relating to intellectual property rights. The first is related to who should be the owner of the patent – is it the researcher, or the company or university that the researcher is working in? Is it morally appropriate that it is part of the employment contract that an employee signs away his or her intellectual property rights to the company? The rules for being an inventor/researcher are similar to *the authorship criteria for scientific articles*. The intellectual property rights themselves are, however, more analogous to copyrights. A copyright can be sold and bought; authorships hopefully are never sold and bought. It seems hard to find a unique position based on ethical arguments on the question where the ownership of the patent should be placed.

Another general problem is what you are entitled to do given the patent. A patent is a temporary monopoly – only the patent-holder has the right to exploit the invention commercially under the period when the patent is valid. A monopoly means that there is only one provider of the product or service. Can any price be charged? It is possible to develop an argument for an upper limit of that price. The argument runs as follows: First, there is a kind of consequentialist reason for the patent system. Without the patent system, certain useful products will not be developed [10]. Second, the monopoly price should only be enough to encourage investment and research resulting in the new product. Excess pricing is bad from the point of view of the public.

There should, therefore, be a limit to the price for the product and service provided. This is obviously more important in the health sector than it is in the beauty sector. The problem is similar to the pricing of pharmaceutical drugs.

22.4 Further moral considerations

New technologies introduce new risks and new responsibilities. For example, before the advent of fast and reliable transportation and the storage of food, it was not the responsibility of Europeans or North Americans to relieve starving people in Africa; indeed, it was not possible to do this. Today it is possible, and hence it is our responsibility to do what we can.

New technologies may thus enable us to help where it was previously impossible. It may also relieve us of some direct personal responsibility, and it may help us to avoid morally bad acts. Below I will give two different examples where successful tissue engineering may produce a moral gain.

22.4.1 Organs and tissues for transplantation

One objective of tissue engineering is to be able to produce human organs originating from human cells. The ability to produce organs and tissues will drastically change our lives. One obvious area is the possibility to remove the scarcity of organs for transplantation. Replacements can be made in time, without having to wait for a medical crisis; it would do away with the ugly issues of distributing scarce organs to waiting patients in desperate situations; and it would do away with the ethical problems of organ donation. Some people prefer to be buried with all their organs still in their body and this can then be realized without hurting someone else. (At present, if a potential organ suitable for transplantation is buried along with the person, at least one potential recipient of that organ is harmed.)

Such a regime of producing organs and tissues for transplantation would be a good thing. It will save lives and improve quality of living. It would also do away with an even more serious moral problem that we seldom discuss – that of organ donation. Some organs can be donated while we are still alive (e.g. a kidney). As we have two kidneys, the risk of removing one is not very high, while a life without any functioning kidneys at all is very difficult – and usually shorter. The dilemma, then, is should I join a donor scheme? Should you? However, if there were other sources available for kidneys we would not have to face this moral dilemma. Technological development produces ethical problems, but may also do away with others.

22.4.2 *In vitro* meat

If it will be possible to grow human organs, obviously animal organs may be grown as well. Why kill a pig in order to eat some pig muscles when the muscles can be grown by tissue engineering? Killing sentient beings, i.e. beings that can experience pain and suffering, have wishes, etc. can be seen as a morally serious issue (see Singer, 1993).

Some might say that it will be cheaper to feed the whole pig and then slaughter it than to tissue engineer the various edible parts. A simple guess is that when this technology develops it should prove cheaper to supply these nutrients and energy (which, by the way, do not move around), than to slaughter a complete pig (with a lot of unedible parts). Furthermore, even if tissue engineered meat turned out to be more expensive, you might argue that it can be worth paying extra for ethical reasons.

22.5 Some questions for the future

- Tissue engineering and generative medicine will prolong life, how will our societies cope with that? Will the Earth be overpopulated, will we be forced to have restrictions (permission) on conceiving a new child? What world will it be if there are many simultaneous generations living?
- Can we cope with a situation where the west will get healthier and live longer and the poor of the world will be denied that?
- Should those who pursue tissue engineering and generative medicine take responsibility for the future?

Notes

1. Unfortunately there are many such scandals. The Tuskagee syphilis study ending in 1972, where poor

black men were studied, not informed and not treated, and the Human Radiation Experiments, where the effect of radiation were tested on unaware subjects, are among the most well-known from the post-war era (e.g. see Faden *et al.*, 1996).

2. President Bush allowed American researchers to use federal funds when they worked with human embryonic stem cell lines established before August 2001, the time of his decision, but not use federal funds to develop new ones. In Spring of 2005, American researchers in the private sector were allowed to develop new human embryonic stem cell lines. To make thing more complex (at least for non-Americans) the Federal Drug Administration is prepared to review all kinds of stem cells – whether fulfilling Bush's criteria of acceptability or not – on an equal basis.

3. In the middle ages the Catholic Church, following Aristotle, regarded the quickening of the fetus as the significant point, not the conception.

4. There have been arguments that it is only some time after birth that the newborn child is sentient enough to count as a *human person*. Killing a child before that time would not be the same evil as killing an older person (see Kuhse and Singer, 1985).

5. Actually, in most of the early regulations on research on human embryonic stem cells this 'surplus' character of the extra IVF embryos was regarded as ethically more acceptable than, for example, creating an embryo for research. This was the case in the Swedish guidelines in 2001. This is presently changing to allow for therapeutic cloning.

6. This was the situation for Swedish human embryonic stem cell lines before the Biobank Act. Permission was granted by the Institutional Review Board for culturing for a limited time and then new permission had to be sought from the donors, who could refuse.

7. The rules given by the International Committee of Medical Journals Editors can be found at http://www.icmje.org/#author. Another important point is that all authors should read and approve the final version.

8. An in-depth treatment of the commercialization issue from an American point of view is given in Resnic (2002).

9. Patents in IVF embryos, donated organs, etc., are problematic, however.

10. A stronger consequentialist argument is perhaps that abolishing the patent system today may give rise to adverse effects. It is hard to know what would have happened if the patent system had never been introduced.

References

Aristotle (1955). *The Ethics of Aristotle: The Nicomachean Ethics.* London: Penguin Classics.

Beauchamp, T. and Childress, J. (2001). *Principles of Biomedical Ethics.* (5th edn.). New York: Oxford University Press.

Faden, R.R., Lederer, S.E. and Moreno, J.D. (1996). US medical researchers, the Nuremberg Doctors Trial, and the Nuremberg Code. A review of findings of the Advisory Committee on Human Radiation Experiments. *Journal of the American Medical Association*, 276: 1667–1671.

Habermas, J. (2003). *The Future of Human Nature.* Cambridge: Polity Press.

Kuhse, H. and Singer, P. (1985). *Should the Baby Live?* Oxford: Oxford University Press.

Mulkay, M. (1994). The Triumph of the pre-embryo: interpretation of the human embryo in parliamentary debate over embryo research. *Social Studies of Science*, 24: 611–639.

Resnic, D.B. (2002). The commercialization of human stem cells: ethical and policy issues. *Health Care Analysis*, 10: 127–154.

Russell, B. (1988). Science and ethics. In *Ethical Theory 1: The Question of Objectivity* (Rachel, J., ed.). Oxford: Oxford University Press.

Singer, P. (1993). *Practical Ethics.* Cambridge: Cambridge University Press.

Snyder, E.Y., Hinman, L.M. and Kalichman, M.W. (2006). Can science resolve the ethical impasse in stem cell research? *Nature Biotechnology*, 24: 397–400.

Thomson, J.A., Itskovitz-Eldor, J., Shapiro, S., *et al.* (1998). Embryonic stem cell lines derived from human blastocysts. *Science*, 282: 1145–1147.

Multiple Choice Questions

Chapter 1 Stem cells

1. Stem cells are an interesting cell source for tissue engineering because:
   A. They are abundant in the human body
   B. They display plasticity and can thus form any cell type of the human body
   C. They can make identical copies of differentiated cells thereby creating an unlimited supply of cells
   D. They can make identical copies of themselves and differentiate

2. Which of the following statements is true?
   A. Embryonic stem cells are totipotent, meaning they are able to form all tissues including germ cells
   B. Pluripotent embryonic stem cells do not exist in the human body
   C. Mesenchymal stem cells are pluripotent; they cannot form extra-embryonic tissue
   D. Multipotency means the ability to form multiple tissue from all three germ layers

3. Differentiation of stem cells:
   A. Occurs only when growth factors are added to the medium
   B. Does not necessarily result in the desired cell type
   C. Can easily be reversed
   D. Does not occur in the human body, because of the cell's capacity to self-renew

4. Which of the following statements is true?
   A. Two major epigenetic mechanisms are DNA modification and histone methylation
   B. Epigenetics can be defined as a stable change in DNA sequence resulting in altered gene expression

   C. More condensed chromatin is less accessible for gene transcription
   D. Cell fate cannot be reset by epigenetic reprogramming

5. Human embryonic stem cells:
   A. Can be isolated by flushing the ovary duct
   B. Can be cultured undifferentiated in the presence of LIF
   C. Can proliferate for years while remaining undifferentiated
   D. Are pluripotent, as proven by the formation of a chimera when injected into a mouse

6. Mouse embryonic stem cells:
   A. Express Oct4 in their undifferentiated state
   B. Have to be cultured on feeder layers to remain undifferentiated
   C. Differentiate when high levels of Nanog are expressed
   D. Are an interesting cell source for clinical application

7. Which of the following statements is true?
   A. Feeder-free growth of human ES cells is the last hurdle to be taken before human ES cells can be applied in the clinic
   B. Serum-replacement medium will not result in the uptake of animal products by human ES cells
   C. Teratomas cannot be formed, when ES cells are differentiated and the desired cell type is present in the population
   D. None of the above

8. Adult stem cells:
   A. Have tissue-dependent mechanisms for self-renewal
   B. Can only be found in their respective mature tissue

C. Are present in high numbers in the body
D. Can be purified into a homogeneous cell population by using CD markers

9. The stem cell niche:
   A. Has only been recently discovered
   B. Is a sheltering environment which only contains non dividing stem cells
   C. Is simple in structure and can easily be simulated *in vitro* to keep stem cells undifferentiated
   D. Positions the stem cells to receive signals to control growth and inhibit differentiation

10. Which of the following statements are true?
    A. Embryonic stem cells have telomerase activity and therefore will not have chromosomal instability
    B. Mesenchymal stem cells in the body go into replicative senescence
    C. Both
    D. None

Chapter 2 Morphogenesis, generation of tissue in the embryo

1. During embryonic development extracardiac cells contribute to heart. Which of the following cell populations have a contribution of neural crest cells?
   A. Myocardium
   B. Endocardium
   C. Endocardial cushions

2. Stabilization of newly formed vessels is regulated by:
   A. Angiopoetin-Tie2 signaling pathway
   B. Hox genes
   C. Gata4/5/6 and nkx2.5 interactions

3. After myocardial infarction the scar tissue will be formed by cells from:
   A. Epicardial derived cells
   B. Cardiomyocytes
   C. Endocardial cushion cells

4. During development the secondary heart field provides additional cardiac tissue to:
   A. The arterial pole of the heart
   B. The venous pole
   C. Both arterial and venous pole

5. What is gastrulation of mammalian embryos?
   A. The formation of the hypoblast and the epiblast
   B. The formation of the definitive germ layers
   C. The growth of the embryo before implantation

6. Neural crest cells are derived from which tissue type?
   A. Blood cells
   B. Surface ectoderm
   C. Neural plate border

7. Which cells and structures develop from the mesoderm?
   A. Dermal hair papillae, hair, blood vessels
   B. Dermal hair papillae, blood vessels, subcutaneous fat
   C. Sebaceous gland, apocrine grand, fibroblasts, blood vessels

8. Which cells develop from the ectoderm?
   A. Merkel cells
   B. Melanocytes
   C. Langerhans Cells

9. Which statement about peripheral nerve development is true?
   A. Schwann cells are derived from the neural crest and have a transient role in peripheral nerve development
   B. Schwann cell differentiate from the mesoderm under the influence of Nrg1-signaling
   C. Schwann cells are derived from the neural crest and provide trophic signals to the peripheral nerve

10. Which statement about bone formation is true?
    A. The resorption of a cartilagenous matrix by osteoclasts is a crucial step in intramembranous ossification
    B. Early steps in chondrogenesis are controlled by the sox family of transcription factors, while hypertrophic differentiation is controlled by the osteoblast specific transcription factor RunX2
    C. Osteoblasts and osteoclasts are derived from a common bipotential precursor. Lineage

biforcation is controlled by wnt/β-catenin signaling

11. Which statement is NOT true?
    A. Chondrocytes regulate the formation of cartilage degrading osteoclasts by the secretion of the cytokine RANKL
    B. The activity of the transcription factors sox9 and RunX2 is regulated by β-catenin and controls lineage commitment of the osteochondroprogenitor cell
    C. The flat bones in the skull are formed by intramembranous ossification, while the appendicular skeleton is formed by endochondral ossification

Chapter 3 Tissue homeostasis

1. The salamander is able to regenerate limbs by a process of:
   A. Apoptosis
   B. Homeostasis
   C. Epimorphosis
   D. Echinosis
2. Which of the following tissue has a high regeneration capacity:
   A. Skin
   B. Bone
   C. Cartilage
   D. Brain
3. In cartilage a certain cell population is involved in the regulation of joint development and repair. The cell population is located in:
   A. The middle zone
   B. The deep zone
   C. The surface zone
   D. The calcified zone
4. The multipotent stem cells of the hair give rise to:
   A. Skin
   B. Sebaceous gland
   C. Hair follicle
   D. All of them

5. Which of the following statements are true:
   A. The heart is not a regenerative organ
   B. No evidence for brain regeneration existed before 1998
   C. The regenerative potential of the human heart was demonstrated by BrdU labeling
   D. All above
   E. None above
6. Well-defined stem cell niches have been identified in the:
   A. Eye
   B. Brain
   C. Intestine
   D. Bone
7. Which of the following statements is true:
   A. Stem cell niches is a specialized microenvironment
   B. Different stem cell niches have different regulatory system
   C. Insulin like growth factor-I is a central controlling growth factor in stem cell niches
   D. None of the above
8. Transient amplifying cells move by means of:
   A. The extracellular matrix
   B. Cell division
   C. Apoptosis
   D. Gene expression
9. Tissue regeneration has consequence for Tissue Engineering because:
   A. Constructs could interfere with cell migration
   B. Stem cell function could be hampered by TE implants
   C. Resorption time of a TE construct could be to slow in relation to regenerative capacity of the treated tissue
   D. All of the above
10. Skin transplantation using cultured cells for burn patients:
    A. Was introduced clinically 2001
    B. Is dependent on large skin biopsies
    C. Is cultured on a feeder layer of stem cells
    D. Is cultured on a feeder layer of fibroblasts

3. Hyaluronan molecules are degraded by hyaluronidase, which hydrolyzes their β(1→4) linkages. Hyaluronan is composed of 250-25,000 units of:
   A. β(1→4)-linked *N*-acetyl-D-glucosamine and *N*-acetyl-D-galactosamine residues
   B. β(1→4)-linked D-glucuronic acid and β(1→3) *N*-acetyl-D-glucosamine residues
   C. β(1→4)-linked D-galactose and *N*-acetyl-D-glucosamine-6-sulphate residues

4. HA interacts with cells through:
   A. Binding to specific cell-surface receptors
   B. Binding to specific cell-surface receptors and sustained transmembrane interactions with its synthetases
   C. Binding to specific intracellular receptors in the cytoplasm

5. What are the main factors that affect the protein properties?
   A. Chain entanglement and the sum of intermolecular forces
   B. Cross-linking of the chain
   C. Amino acid sequence, configuration and structural conformation

6. Soy protein is formed by:
   A. Conglycinin (7S) and glycinin (11S) subunits
   B. Glycosaminoglycans
   C. Amino acids

7. The degradation of cellulose is especially difficult because:
   A. The macromolecule does not have charged groups
   B. Of its special fibrilar structure
   C. Of the especial conformation of the glycosidic bonds

8. Cellulose:
   A. May be just obtained from vegetable sources
   B. Are mainly obtained from vegetable sources, but some bacteria and fungi can also synthesized it
   C. May be also produced by animals

9. Cellulose is soluble:
   A. In water at high temperature
   B. In many organic solvents
   C. Only in a few solvents

10. Cellulose and its derivatives have been proposed for several biomedical applications, including (several answers can be correct):
    A. Scaffolds for tissue engineering
    B. Injectable and self-setting hydrogels
    C. Hard materials for bone replacement
    D. Regeneration of blood vessels
    E. Drug delivery systems

11. Alginate and other polysaccharides have several properties that make them interesting candidates for Tissue Engineering applications:
    A. Bioinert, abundant sources and with a low cost
    B. Biocompatible, abundant sources and extensively used in other industries
    C. Biocompatible and abundant, but with a high cost

12. Crosslinking polysaccharides allows us to:
    A. Decrease the degradation time of the material
    B. Increase the degradation time of the material
    C. Increase the flexibility of the material
    D. None of the above

13. Alginate can be crosslinked with divalent
    A. Cations
    B. Anions
    C. Both of the above

14. After crosslinking, alginate forms a stiff structure known as:
    A. Alginate box
    B. Egg drop box
    C. Egg box

15. Dextran microspheres have been commercially available to be used as:
    A. Separation matrices
    B. Cell microcarriers
    C. Both of the above
    D. None of the above

16. Alginate and dextran are, respectively, polysaccharides:
    A. From plant origin
    B. From microbial origin
    C. From plant and microbial origin
    D. From algae and microbial origin
    E. None of the above

17. Polyhydroxyalkanoates (PHAs) are natural occurring polyester found that:
    A. Are found as inclusions in the cytoplasm of several bacteria
    B. Are used as carbon and energy reserve materials
    C. The two previous answers are correct

18. The introduction of other units in the PHB – has a significant effect on mechanical behavior of the polyester:
    A. PHB homopolymer is a ductile material, while the the HB-co-HV copolymer is a stiff polymer that exhibits increasing stiffness upon the increase in 3-hydroxyvalerate content
    B. PHB homopolymer is a brittle material, while the HB-co-HV copolymer is elastic
    C. PHB homopolymer is a brittle material, while the the HB-co-HV copolymer is a ductile polymer that exhibits increasing ductility upon the increase in 3-hydroxyvalerate content

19. Several studies have investigated PHAs to a diverse number of applications. The biomedical potential of PHAs arises from:
    A. The wide range of mechanical properties, the biodegradable behavior and overall biocompatible character
    B. The ability in controlling biodegradation and cell response through the crystallinity of the polymer or blend
    C. None of the above

20. Several studies have assessed the effect of different types of PHAs on cell response. Which of the following statements can be inferred from the chapter?
    A. Cell response does vary with the type of PHA (homopolymer/copolymer/blend), but appears to independent from the crystallinity of the polymer
    B. Variations in cell response are mostly attributable to differences in the morphology and degradation behavior between different PHAs

    C. Several factors affect cell response, including the crystallinity of the polymer

21. Polysacharides can be classified according with their origin, composition and sequence of sugar units in polysaccharide chains, presence of ionizing groups. According with this classification, carrageenans can be described as:
    A. Microbial, homopolysacharide, linear, anionic
    B. Algal, heteropolysacharide, linear, anionic
    C. Algal, heteropolysacharide, branched, anionic

22. What modifications/changes in the chemical structure of collagen can improve the stability of collagen scaffolds?
    A. Crosslinking
    B. Hydroslysis

Chapter 7 Degradable polymers for tissue engineering

1. Degradable polymers for tissue engineering should:
    A. Be biocompatible, have tailorable degradation rate and result in non toxic degradation products
    B. Consist of monomers that are part of the Krebs metabolic cycle
    C. Start degrading after tissue regeneration is complete

2. Which statement about the synthesis of degradable polymers is *not* true?
    A. Copolymerization can be used tune the material properties.
    B. Block copolymers are faster degradable compared to random copolymers
    C. Step growth polymerization requires monomers containing two or more functional groups

3. Labile bonds in degradable polymers such as ester, anhydride and carbonate bonds can be broken by
    A. Hydrolysis
    B. Oxidation reactions
    C. Both A and B are true

4. The most important factors that influence the rate of hydrolysis of the degradable bonds are a type of:
   A. Labile bond, dissolved oxygen concentration and solubility of oligomers
   B. Labile bond, solubility of oligomers and water uptake
   C. Labile bond, water uptake and pH
5. Which factor is *not* determining the erosion process?
   A. The solubility of the oligomers in the surrounding medium
   B. The transport of the oligomers in the lymphatic system
   C. The hydrolysis rate constant of the labile bonds
6. When a degradable polymer undergoes bulk erosion, the:
   A. Molecular weight decreases immediately in time, whereas mass loss starts at a later stage
   B. Molecular weight and mass decrease at the same time
   C. Mass reduces immediately, whereas molecular weight is not affected initially
7. Surface erosion of the degradation polymer results in:
   A. Parallel mass loss and molecular weight decrease
   B. Immediate mass loss, whereas the molecular weight not affected initially
   C. Immediate decrease in molecular weight, whereas mass loss starts at a later stage
8. What makes the in-vivo degradation generally faster than the in-vitro degradation?
   A. The tissue response increases chain scission by due to the presence of reactive species
   B. Mass loss is increased due to higher solubility of the oligomers and by active transport by phagocytes
   C. Both A and B are true
9. What specific characteristic makes PLA, PLGA and their copolymers suitable as degradable polymer for tissue engineering?
   A. Their degradation products are present in the metabolic path way in the human body

B. They contain crystal domains that the slowly degrade
C. Their chain scission by hydrolysis can be auto-catalyzed by the carboxylic end groups
10. What is the challenge regarding the application of degradable polymers in tissue engineering?
    A. To correlate the in-vitro and in-vivo degradation
    B. To balance the mechanical properties and degradation rate
    C. To create sufficiently strong materials

Chapter 8 Degradation of bioceramics

1. Calcium phosphate formation occurs from extracellular fluids. What is the physico-chemical reason behind this ability?
   A. They are supersaturated and they contain bone forming cells
   B. They are supersaturated with regard to calcium and phosphate ions
   C. They are supersaturated with regard to proteins
2. The staining of the osteoclast-specific gene Tartrate Resistant Acid Phosphatase allows:
   A. The identification of osteoclasts?
   B. the phagocytic action of the osteoclasts?
   C. The resorptive activity of the osteoclasts?
3. Degradable bioceramics cannot be used in load bearing-applications because:
   A. In humid conditions, they have a low compressive strength compared to bone
   B. They are fatigue-resistant in humid conditions but they are brittle
   C. They are brittle and they are not fatigue-resistant under humid and dry conditions
4. Under normal physiological conditions, phase transformations occur in vitro and in vivo at the surface of the bioceramics. What is the mineral phase?
   A. Carbonated apatite for both for bioactive glasses and calcium phosphate ceramics
   B. Calcium carbonate for bioactive glasses and hydroxyapatite for calcium phosphate ceramic
   C. Calcium carbonate for bioactive glasses and carbonated apatite for calcium phosphate ceramics

5. Some porous calcium phosphate ceramics are found osteoinductive in large animals (goats, baboons), their degradation behavior is:
   A. Stimulated by osteoclastic activity
   B. Mainly mechanical because of the constant motion in the muscles
   C. Proposed to play a role in osteoinduction mechanism

6. *In vivo*, cell- and solution-mediated degradation mechanism occur:
   A. Simultaneously
   B. One after the other
   C. It depends on the bioceramic

7. Among these propositions, which ones are an advantage in bone tissue engineering in the degradation of bioceramics
   A. They are not stocked in the body, but excreted in a natural manner
   B. A too fast degradation rate cannot support bone formation
   C. A too slow degradation rate does not stimulate bone formation

8. Regarding calcium phosphate ceramics, we can tune the degradation kinetics *in vivo* by. One of the propositions below is not true. Which one?
   A. Changing the porosity of the ceramics
   B. Creating an amorphous phase
   C. Adding zinc in order to stimulate osteoclastic activity
   D. Adding fluoride in hydroxyapatite structure
   E. Mixing hydroxyapatite with tricalcium phosphate

9. The bone-bonding capacity of bioglasses is:
   A. Directly proportional to the CaO/SiO_2 ratio
   B. Directly correlated to the degradation kinetic
   C. For an approximate ratio of 50% CaO:50% SiO_2

10. Tuning in vivo degradation mechanisms of bioceramics results in: (several answers can be correct)
    A. Tuning bone formation
    B. Controlling osteoclastic resorption
    C. Controlling dissolution kinetics

Chapter 9 Biocompatibility

1. The mechanisms of biomaterial – host reactions involve:
   A. Different material degradation processes to those found outside the body
   B. Different physiological processes to those found in therapies that do not involve biomaterials
   C. No new material or physiological processes

2. Elastic constants of a biomaterial may influence biocompatibility because:
   A. They influence the strength of the material
   B. They control strain distribution between material and tissue
   C. They influence degradation mechanisms over time

3. Chronic inflammation in soft tissues involves the activation of:
   A. Macrophages
   B. Complement
   C. The clotting cascade

4. For a biodegradable tissue engineering scaffold, the optimal degradation period is:
   A. One week
   B. Three months
   C. Dependent on the precise circumstances

5. The inflammatory response to a biodegradable scaffold is mostly influenced by:
   A. The precise chemical nature of the material
   B. The physical nature of the particulate degradation products
   C. The porosity of the scaffold

6. Polyurethanes are a family of polymers that:
   A. Are biostable
   B. Are biodegradable
   C. May be biostable or biodegradable depending on chemical structure

7. The most important bacterially produced polymers used as scaffolds are:
   A. Polyhydroxyalkanoates
   B. Polyolefins
   C. Aromatic polyesters

8. Carbon nanotube reinforced biodegradable composites:
   A. Are already in use in commercial tissue engineering scaffolds
   B. May have significant difficulties in tissue engineering because of the unknown fate of nanoparticles
   C. Have considerable potential in cardiovascular tissue engineering.

9. An optimal biocompatible biomaterial may be described as:
   A. A material with a precisely defined surface energy
   B. A material with a precisely defined porosity
   C. Non-existent, since there is no such thing

10. Surface modified polymeric biomaterials have most potential in:
    A. Long term implantable devices
    B. Tissue engineering scaffolds
    C. Drug delivery systems

Chapter 10 Cell source

1. What are the two defining features of all post-natal stem cells?
   A. They are all able to differentiate into cell types outside of the lineage from which they developed
   B. They are able to self-renew and the progeny of a single cell are able to reform all cell types in the tissue from which they were derived
   C. They can easily be expanded in culture

2. Which type of clones derived from epidermis contain the epidermal stem cell?
   A. Paraclones
   B. Meroclones
   C. Holoclones

3. In nerve tissue, what type of cell may not only contribute to the neural stem cell niche, but also serve as a stem cell itself?
   A. Ependymal cells
   B. A-type neuroblasts
   C. Certain types of glial cells

4. How can the skeletal stem cell be prospectively isolated from post-natal bone marrow?
   A. Currently, there is no single marker, or set of markers that can purify skeletal stem cells
   B. FACS cell sorting using the Stro-1 antibody
   C. Plating a single cell suspension at clonal density

5. How can multipotency of a stem cell population be determined?
   A. By in vitro assay under various differentiation conditions
   B. By determination of markers for specific cell types
   C. By in vivo transplantation of the progeny of a single cell along with an appropriate scaffold, if needed

6. 'Mesenchymal' stem cells derived from various connective tissues are identical
   A. Yes
   B. No
   C. A precise comparative study has yet to be performed, but it appears that they are related but not identical

7. Bone marrow stromal cells can modulate the immune response:
   A. By direct cell contract with immune cells
   B. By secretion of cytokines and growth factors that suppress the immune response.
   C. Both A and B

8. Xenogenic transplants of embryonic porcine tissue is feasible due to the similar structure of organs and tissues of human an pig tissue
   A. True
   B. False

9. Stem cell self-renewal is:
   A. Deterministic (asymmetric division with one daughter cell remaining a stem cell, the other daughter cell becoming more committed)
   B. Stochastic (division is sometimes symmetric, with both daughters remaining stem cells, or both daughters becoming more committed)
   C. It is not yet known in mammalian systems, but stem cell self-renewal may be a mixture of both deterministic and stochastic mechanisms

10. What was the first, and to date, the only routine use of post-natal stem cells in regenerative medicine?
    A. Use of neural stem cells for the treatment of Alzheimer's disease
    B. Bone marrow transplantation (which contains the hematopoietic stem cell) for treatment of blood disorders
    C. Use of molecularly engineered muscle-derived stem cells for the treatment of muscular dystrophy

Chapter 11 Cell culture: harvest, selection, expansion and differentiation

1. What is meant by the term 'autologous' transplantation?
   A. The body is automatically responding to the implant
   B. Cells are harvested form a patient with the purpose of implanting the cells back to the same patient
   C. Cells are harvested from a close relative and used for transplantation back to a patient
2. What is the advantage of using cells in TE implants?
   A. The cells can function as continuous producers of a missing component in a tissue
   B. The cells can take an active part in restoring the function in the inherent cells of the surrounding tissue
   C. Both a and b
3. The bone marrow is a common cell source for many TE applications. What is the most common way for harvest of bone marrow?
   A. By flush-out
   B. By a core biopsy
   C. By an aspirate from the iliac crest
4. When a skin biopsy is harvested as a startup material for primary culture, the cells has to be released from their extra cellular matrix (ECM). What method is the fastest and crudest way to remove the ECM surrounding the cells?

A. By mechanical cutting and vortexing
B. By placing the cells over night in buffers containing chelating agents
C. By incubation in combinations of enzymes such as collagenases and proteases

5. From population of cells, contaminated with fibroblasts, which method would be of preference (from a cost effective and aseptic processing view) for obtaining a purified population of cells for direct implantation to a patient in a clinical TE application?
   A. Fluorescence activated cell sorting (FACS)
   B. Positive selection using magnetic activated cell sorting
   C. Negative selection using magnetic activated cell sorting
6. Which separation method requires least addition of specific biochemicals?
   A. Preplating
   B. Gradient centrifugation
   C. Antibody driven separation
7. Which of the cells prefers to be expanded on plastic surfaces?
   D. Embryonic stem cells
   E. Chondrocytes
   C. Hematopoietic cells
8. What is the purpose of coating surfaces with ECM components during the expansion of cells?
   A. Increase the attachment of cells
   B. Regulate the grade of dedifferentiation
   C. Both a and b
9. Choose the right ending of the following sentence, during dedifferentiation, cells increase the proliferation rate and
   A. Lose their original phenotype
   B. Turns into another phenotype (i.e. fat into bone)
   C. Starts to synthesize high amounts of extracellular matrix components
10. What would be a typical cell culture condition for differentiation?
    A. High cell density and low serum concentration
    B. High cell density and high serum concentration
    C. Low cell density and high serum concentration

Chapter 12 Cell nutrition

1. Nutrient gradients occur in tissue-engineered constructs of clinically relevant sizes. What parameters influence the steepness of these gradients?
   A. The diffusion coefficient of the nutrient inside the construct
   B. The cellularity of the construct
   C. The consumption rate of the cells within the construct
   D. All of the above

2. The most commonly used buffer for cell culture media is bicarbonate. What is the main prerequisite for this buffer to function properly?
   A. The presence of CO_2 in the gas phase
   B. A pH of 7.4
   C. The medium must be co-buffered with HEPES

3. What are the 3 different stages of cell culture, once the appropriate cells are isolated?
   A. (1) initiation of differentiation, (2) expansion of the cells, (3) maintenance of differentiation
   B. (1) initiation of differentiation, (2) maintenance of differentiation, (3) expansion of the cells
   C. (1) expansion of the cells, (2) initiation of differentiation, (3) maintenance of differentiation

4. What is the difference between *convection* and *diffusion*?
   A. Convection is mass transfer due to the motion of the fluid, whereas diffusion is the movement of component molecules in a mixture in the direction of a nutrient gradient
   B. Diffusion is mass transfer due to the motion of the fluid, whereas confection is the movement of component molecules in a mixture in the direction of a nutrient gradient
   C. Convection is mass transfer due to the motion of the fluid and the movement of component molecules in a mixture in the direction of a nutrient gradient, whereas diffusion only occurs from fluid motion

5. Why should the OTR be larger, or at least be equal to the OUR:
   A. This prevents the waste of nutrients
   B. This allows both convection and diffusion at the same time

   C. This ensures that sufficient oxygen is supplied to all cells

6. How can bioreactor culture improves the nutrient supply by:
   A. Decreasing the diffusion distance
   B. Decreasing the convection
   C. Decreasing the mass transfer

7. Transport of oxygen to the cells within tissue engineered constructs can be enhanced by:
   A. Flow of culture medium around the constructs, which reduces the boundary layer in the liquid phase
   B. Perfusion through constructs, which decreases the diffusional distances for oxygen transport,
   C. Both of the above

8. The effects of adding oxygen carriers to culture medium are similar to those that hemoglobin has in oxygen transport in our blood. These effects are (check all that applies):
   A. Concentration of oxygen in culture medium
   B. The rate at which any oxygen used by the cells is replenished
   C. The total amount of oxygen available to the cells in culture medium

Chapter 13 Cryobiology

1. Preservation of native tissue and tissue engineered constructs involving cells is feasible by the following methods:
   A. Freezing, vitrification and freeze drying
   B. Freezing, vitrification and hypothermia
   C. Freezing and vitrification

2. Cryoprotectants can de divided into penetrating agents and non-penetrating agents. Which group is more likely to be less toxic?
   A. Both groups are equally non-toxic at subzero temperatures
   B. Non-penetrating agents
   C. Penetrating agents

3. Vitrification avoids ice crystal formation during:
   A. Cooling, storage and warming
   B. Cooling
   C. Storage and warming

4. Vitrification properties of solutions used in cryopreservation are commonly evaluated by differential scanning calorimetry (DSC) and cryo-microscopy. What can be detected from the DSC thermograms on warming?
   A. Homogeneous nucleation
   B. Melting, devitrification, vitrification
   C. All of the above
5. In order to create conditions for vitrification, certain knowledge concerning the physical properties of cryoprotective solutions candidates is required. What should be correct definition of devitrification?
   A. Devitrifiation means melting of vitrified samples
   B. Devitrification is formation of ice from super-cooled liquid.
6. What requirements should be taken into consideration during the design of cryopreservation protocols? (several answers can be correct)
   A. Cryoprotectants (component selection)
   B. Cooling/warming rates
   C. Temperature of equilibration and dilution procedures
   D. Duration of equilibration and dilution procedures
   E. Humidity
   F. Size of constructs
7. There are several methods for eliminating the risk of cross-contamination during cryopreservation. What strategy should be applied for small specimens?
   A. Vapour storage
   B. 'STRAW IN STRAW' system
   C. All of the above
8. How can the efficiency of a cryopreservation method be established for a encapsulated cells in an *in vitro* experiment?
   A. By determining survival of cells and integrity of capsule
   B. By determining viability and function of cells in a long term culture
   C. All of the above
9. Successful vitrification protocols were reported for:
   A. Native heart valves
   B. Tissue engineered and native blood vessels

C. Tissue engineered organs
D. All the above
10. Vitrification is preferable method for cryopreservation in tissue engineering research and for transplantation purposes because: (several answers can be correct)
   A. There is no cell damage caused by ice crystal formation
   B. The method is non-damaging for the integrity of the construct
   C. The method is cost effective
   D. The method is less time consuming than traditional freezing methods

Chapter 14 Scaffold design and fabrication

1. Which of the following criteria would be considered the **least** critical for design of a scaffold for bone tissue engineering?
   A. Pore size
   B. Mechanical properties
   C. Biodegradation properties and products
   D. Interconnecting pore size
   E. Pore morphology
2. Certain scaffold processing techniques result in scaffolds with randomly organized pore architectures. Which of the following processing techniques allows direct control over pore shape and orientation, resulting in an organized and highly reproducible pore architecture?
   A. Electrospinning
   B. Phase separation
   C. Fused deposition modeling
3. To produce fine fibres 1–10 μm in diameter, 'direct-writing' techniques rely on the accurate formulation of which of the following?
   A. Hydrogels
   B. Polyelectrolyte and colloidal inks
   C. Collagen
4. When adopting extrusion-based fabrication methods (e.g. FDM, 3D Plotting), where successive layers of fibres are deposited in the x–y plane, which of the following statements about the pore architecture is TRUE?

A. It is not possible to control pore architecture in the x, y and z-direction

B. Pore architecture will always be identical in the x, y and z-direction

C. Pore architecture can only be controlled by changing the fiber diameter

D. Pore architecture in the z-direction can be controlled by extruding one identical layer on top of another

5. In all tissues, cells typically reside and interact within a surrounding 3D extracellular matrix (ECM) made up of various types of natural polymers. Having in mind to mimic the size of the collagen type I structure in the ECM, which of the following processing techniques would you use to generate an artificial ECM?

A. Electrospinning

B. 3D Plotting

C. Porogen leaching

D. Selective laser sintering

6. What is the biological reasoning behind designing scaffolds for tissue engineering with mechanical properties similar to those of the host tissue?

A. To protect cells within the construct from excessive strain/loading

B. To provide cells within the construct a mechanical stimulus similar to physiological levels

C. To allow time for ingress of host tissue/vasculature and tissue integration prior to excessive biodegradation and reduction in scaffold properties

D. Aid implant handling, fixation and shape retention during surgery

E. All of the above

7. What are the main advantages of organ printing compared to other methods described in this chapter?

A. Cells are embedded directly into the biomaterial during scaffold fabrication

B. Control of different scaffold materials and cell types in 3D

C. Constructs built by organ printing can be directly connected to the host vasculature

D. Organ printing is simple and easy to apply in any tissue engineering lab

E. All of the above

8. From a biological point of view scaffolds should serve various functions?

A. Acting as an immobilization site for transplanted cells

B. Forming a protective space to prevent unwanted tissue growth into the wound bed and allow healing via regeneration of differentiated tissue

C. Directing migration or growth of cells via surface properties of the scaffold

D. Directing migration or growth of cells via release of soluble molecules such as growth factors, hormones and/or cytokines

E. All of the above

9. Which of the following can be included in material to be electrospun?

A. Cells

B. Carbon nanotubes

C. DNA

D. Antibiotics

E. All of the above

10. To make a cell invasive scaffold from 2-hydroxyethyl methacrylate via phase separation, in which order should the following typically occur?

A. Phase separation/Initiation/Gelation

B. Initiation/Phase separation/Gelation

C. Initiation/Gelation/Phase separation

Chapter 15 Controlled release strategies in tissue engineering

1. What is meant by the term 'therapeutic window'?

A. The time over which a drug is released

B. The concentration range between and effective drug concentration and a toxic drug concentration

C. The spatial region around a depot in which drug is present at an effective concentration

2. What is meant by the term 'solid solution'?

A. A drug distributed in a material depot as solid particles

B. A drug distributed in a material homogeneously mixed with the material

C. A drug that is soluble in water, distributed in a material before release

3. By what mechanism would a material like a collagen sponge degrade *in vivo*?
   A. By bulk enzymatic hydrolysis
   B. By surface-mediated hydrolysis
   C. By cell-mediated enzymatic hydrolysis

4. By what mechanism would a material like poly(ε-caprolactone) degrade *in vivo*?
   A. By bulk hydrolysis
   B. By bulk enzymatic hydrolysis
   C. By surface-mediated hydrolysis

5. In order to obtain a material that degrades by surface erosion, what physicochemical characteristics should the material have?
   A. The material should be hydrophobic
   B. The material should be hydrophilic
   C. The material should be soluble in water

6. When a material degrades by enzymatic action, what determines whether erosion is surface-mediated or bulk?
   A. The source of the enzyme, namely present in body fluids or cell-associated
   B. The diffusion coefficient of the enzyme in the material
   C. Both A and B

7. What kinds of molecules are most morphogenetic compounds?
   A. Peptides and proteins
   B. DNA
   C. Both A and B

8. What is meant by protein denaturation?
   A. When a protein partially unfolds and loses its activity
   B. When a protein is too dilute to be active
   C. When a protein adsorbs to a material surface

9. Some classes of morphogenetic molecules must be immobilized to be bioactive. What class(es) are these?
   A. Growth factors
   B. Adhesion factors
   C. Both A and B

10. Would release of a protein morphogen like BMP-2 from a proteinaceous material like a fibrin gel be expected to be linear?
    A. Yes
    B. No
    C. It depends on cellular activity within the gel

Chapter 16 Bioreactors for tissue engineering

1. Static culture systems may be unsuitable for thick TE constructs because:
   A. Cells will not attach
   B. Preferential flow paths might develop
   C. Diffusion limitations result in nutrient and oxygen deprivation in central regions

2. Materials found to be biocompatabile with a given cell type should be:
   A. Assumed biocompatible with other cell types
   B. Retested for each cell type of interest

3. Mechanical stimulation during cultivation of TE constructs has been shown to: (check all that applies)
   A. Enhance cell/tissue organisation
   B. Promote cell differentiation
   C. Stimulate ECM production D. Enhance tissue specific function

4. In a perfusion chamber of 6 mm diameter, fluid flows at a superficial velocity of 1 mm/sec. What is the flow rate in the system?
   A. Around 28 µl/sec
   B. Around 35 µl/sec
   C. Cannot be determined

5. Which is a possible contribution of computational fluid dynamic modelling to the development of a bioreactor system?
   A. To predict local patterns of shear stress
   B. To predict local profiles of oxygen consumption
   C. To predict local efficiency of glucose utilization

6. The market of a company producing tissue culture bioreactors includes: (check all that applies)
   A. Fundamental cell biology research groups
   B. Pharmaceutical companies

C. Biomanufacturing companies

D. Manufacturers of drug delivery vehicles

7. Rotating bioreactor vessels have hydrodynamically active environment with multiple effects on the growth of engineered cartilage. These effects are (check all that applies):

A. Enhanced mass transport at construct surfaces

B. Enhanced transport of oxygen throughout the constructs

C. Physical stimulation of the cells

D. Enhanced control of medium pH

8. Direct perfusion of culture medium through channeled scaffolds increases the viability and improves function of cardiac tissue constructs because of (check all that applies):

A. Enhanced transport coefficients of oxygen to the construct surfaces

B. Smaller distances for diffusion of oxygen to the cells inside constructs

C. Higher porosity of the scaffold due to the channeling

9. The concentration of viable cells in channeled, perfused cardiac constructs will increase if:

A. The diameter and spacing of channels are decreased

B. Medium flow rate is decreased

C. Scaffold thickness is increased

10. How can tissue engineering benefit from bioreactors (check all that applies)?

A. Cells can be seeded more homogeneously

B. Scale-up is facilitated

C. Supply of nutrients can be improved

D. Sterile culture is no longer required

Chapter 17 Tissue Engineering for Skin Transplantation

1. Which of the following statements is not true

A. Epithelial and dermal replacements take the form of biological dressings

B. Stem cell technology is the future of skin wound repair

C. Tissue engineered skin is available as an epidermal component only

D. None

2. Which of the following statements is true?

A. Allogenic keratinocytes are incorporated permanently into a healed wound

B. The stratum basale is the only proliferative layer of keratinocytes

C. Porcine collagen has the same biochemical structure as human

D. Stem cells have short cell cycle times

3. The following is not a recognized method of Keratinocyte transfer *in vivo*:

A. Stratified sheets

B. Adsorbed onto a V.A.C. closure device

C. Preconfluent, inverted on a carrier membrane

D. Aerosol spray delivery

4. Which of the following statements is false?

A. The first successful culture of human keratinocytes relied upon co-culture with myocytes

B. Apligraf® is constructed from male cells only

C. Integra® is formed of shark and bovine tissue

D. Hemidesmosomes are a feature of the basement membrane

5. Hyaluronic Acid

E. Is stabilized by thermal induction

F. Contains an amino acid sequence that is highly conserved across species

G. Contains binding domains for Fibroblasts

H. Is presented as a solid piece which must be shaped to the wound

6. Keratins

F. Are usually expressed in pairs

G. Are not found in glabrous skin

H. Are coded for on the X chromosome

I. Are localized to the epibasal region of the keratinocyte

7. Keratinocyte culture

A. Is optimal at 33.8 degrees Celcius

B. Requires Porcine Serum after the first passage

C. Culture solution includes penicillin-based antibiotics

D. None of the above

8. Stem cells are an interesting cell source for tissue engineering because
   A. They are abundant in the human body
   B. They display plasticity and can thus form any cell type of the human body
   C. They can make identical copies of differentiated cells thereby creating an unlimited supply of cells
   D. They can make identical copies of themselves and differentiate.

9. Which of the following statements is true?
   A. Embryonic stem cells are totipotent, meaning they are able to form all tissues including germ cells
   B. Pluripotent embryonic stem cells do not exist in the human body
   C. Mesenchymal stem cells are pluripotent; they cannot form extra-embryonic tissue
   D. Multipotency means the ability to form multiple tissue from all three germ layers

10. Differentiation of stem cells
   A. Occurs only when growth factors are added to the medium
   B. Does not necessarily result in the desired cell type
   C. Can easily be reversed
   D. Does not occur in the human body, because of the cell's capacity to self-renew

11. Which of the following statements is true?
   A. Feeder-free growth of human ES cells is the last hurdle to be taken before human ES cells can be applied in the clinic
   B. Serum-replacement medium will not result in the uptake of animal products by human ES cells
   C. Teratomas cannot be formed, when ES cells are differentiated and the desired cell type is present in the population
   D. None of the above

12. Adult stem cells
   A. Have tissue-dependent mechanisms for self-renewal
   B. Can only be found in their respective mature tissue

C. Are present in high numbers in the body
   D. Can be purified into a homogeneous cell population by using CD markers

13. The stem cell niche
   A. Has only been recently discovered
   B. Is a sheltering environment which only contains non dividing stem cells
   C. Is simple in structure and can easily be simulated *in vitro* to keep stem cells undifferentiated
   D. Positions the stem cells to receive signals to control growth and inhibit differentiation

14. Skin is an important tissue engineering target because
   A. It may be used for reconstructive surgery of burns victims
   B. It is one cellular level thick
   C. It is unable to differentiate
   D. Its cell are unable to make identical copies of themselves.

15. Which of the following statements is true?
   A. Skin does not provide a barrier to the outside environment
   B. The keratinized layer of skin provides the barrier to the outside environment
   C. Skin is a modified squamous tissue
   D. The epidermis makes up the entirety of skin

16. Which are the following statements relating to Xenogenic cells is true?
   A. Cells derived from a another species
   B. Occurs only when growth factors are added to the medium
   C. Occur in the human body
   D. Are not viable for tissue engineering

17. Which of the following statements relating to epithelial cell culture is true?
   A. Oxygen is important for the survival and growth of epithelial cells
   B. Epithelial cells require a growth medium in order to be cultured in a laboratory
   C. Epithelial cells grow out into colonies
   D. All of the above

18. Which of the following statements is more correct?

A. Epithelial growth media contain hormones
B. Epithelial growth media contain growth factor
C. Epithelial growth media contain specific metabolites
D. All of the above

19. Normal skin
A. Has a dermis and epidermis
B. Has an epidermis which contains only undifferentiated epithelial cells
C. Does not have a basement membrane
D. Contains chondrocytes

20. Which of the following statements is true?
A. Caderveric skin cannot be used as a wound dressing
B. Glycerol preserved caderveric skin may be used to provide a de-epidermalised dermis
C. The antigenic component of caderveric skin is unable to cause 'rejection'
D. None of the above

21. Integra
E. Has a silicone layer
A. Cannot be used for wound healing
B. Occurs as a natural component of the skin
C. Is derived from porcine tendon collagen

22. One of the following is not a dermal replacement:
A. Trancyte
B. Integra
C. Hyalograft 3D
D. Laserskin

Chapter 18 Tissue engineering of cartilage

1. In which tissue(s) can hyaline cartilage can be found:
A. Epiglottis
B. Ear
C. Articular cartilage
D. Rib cartilage
E. None
F. C and D

2. Which statement is correct? Fetal cartilage is characterized by:
A. Repair capacity
B. Low cell density

C. Collagen type II
D. Chondroitin-4-sulfate

3. What is a correct statement for collagen?
A. Collagens only forms fibrils
B. The main type of collagen in cartilage is type V
C. Collagen type II can be found in two splicing variants
D. The collagen turnover in cartilage is less than 10 years

4. The cell density in normal cartilage is:
A. 70000 to 240000 cells/mm^3
B. 70 to 2400 cells/dm^3
C. 7000 to 24000 cells/mm^3
D. 24000 cells/mm^2

5. Which statement is wrong? Fetal cartilage has an ability to repair isolated lesions due to:
A. A higher amount of hyaluronic acid
B. Higher cell concentration
C. Lower matrix content
D. All above
E. None above

6. What is not true about osteoarthritis?
A. Untreated cartilage defects could lead to osteoarthritis
B. Osteoarthritis only affects people over 60 years of age
C. Overweight is a risk factor for osteoarthritis
D. Over one million joint replacements are done each year due to osteoarthritis

7. Which of the following statements is true regarding ACT?
A. Cultured articular chondrocytes have been in clinical use since 2004
B. Cells are cultured for 8 weeks before implantation
C. Cells are cultured in the presence of autologous serum
D. Articular chondrocytes has no capacity of redifferentiation

8. The most suitable specimens for cartilage tissue engineering is:
A. Cells isolated from the superficial zone of cartilage

B. Cells isolated from the middle zone of cartilage

C. Cells isolated from deep zone of cartilage

D. Full thickness biopsy needed

9. What is/are correct statements regarding space grown cartilage
   A. Are inferior constructs grown in 1 g gravity
   B. Yielded cartilage constructs
   C. Produced less GAG
   D. All of the above

10. Which of the following scaffold materials are absorbable:
    A. Dacron
    B. Teflon
    C. Polyester
    D. Dexon

Chapter 19 Tissue engineering of bone

1. What is osteoinduction?
   A. The generation of bone by a certain tissue/cell by itself, i.e. in the absence of host cells e.g. in diffusion chambers
   B. The ability of a material/graft to allow ingrowth of vessels and osteoprogenitor cells from the recipient bed, or spreading of bone over the surface proceeded by ordered migration of differentiating osteogenic cells
   C. The capacity of undifferentiated inducible osteoprogenitor cells (IOPCs), that are not yet committed to the osteogenic lineage, to form osteoprogenitor cells
   D. The production of organic bone matrix and its subsequent mineralization or calcification

2. What is true about living bone tissue?
   A. Compact bone is the immature unorganized type of bone
   B. Haversian bone is active in growth, calcium homeostasis and haematopoiesis, and its supportive function is mainly in locations with a predominantly compression type of loading
   C. Lamellar bone is the remodeled form of woven bone
   D. Woven bone contains Haversian channels

3. What is not a valid reason to use alternatives autologous bone as a graft?
   A. The amount of autologous bone in patients is limited
   B. The use of autografts can prolong surgery time
   C. The use of autograft bone can be painful
   D. Many cheap and easy-to-use alternatives are available that are better than autograft bone

4. An ideal scaffold for the use in bone tissue engineering:
   A. Should have more mechanical strength than the host bone and be porous
   B. Should have similar mechanical strength as the host bone and be biocompatible
   C. Should have similar mechanical strength as the host bone and not dissolve
   D. Should have less mechanical strength as the host bone

5. What is true for BMPs?
   A. First described by Friendenstein and later named by Urist, over 20 different BMPs are known to stimulate bone formation
   B. The same concentration and exposure time of BMP-2 and BMP-7 have been shown to induce similar successful results in different species
   C. The recombinant BMPs are cheaper than the bone matrix – derived BMPs and equally as potent for osteoinduction
   D. BMPs have been used in human spine studies and have shown higher rates of spine fusion in several studies as compared to the use of autograft bone

6. The highest concentration of osteoprogenitor cells are most likely obtained from bone marrow aspirations:
   A. After flushing the femurs (thigh bones) of a fresh young rat cadaver
   B. Insertion of a needle in the pelvic bone of a young rabbit to aspirate 5 cc at once
   C. Insertion of a big needle in the pelvic bone of a 40 year old man to aspirate 5 cc at once
   D. Insertion of a big needle in the pelvic bone of a 40 year old man to aspirate 10 cc at once

7. When cells are grown 'to confluency' in a culture dish, one is advised to:
   A. Wait 3 more days until calcification of the cell layer is visible and then use trypsine to release the cells
   B. Use a cell scraper the same day to release the cell layer and resuspend the cells with warm medium with a syringe and small needle to obtain a homogenous cell suspension
   C. Wash the cells with warm PBS and use typsine to release the cells the same day
   D. Wash the cells with warm PBS and use DMSO to release the cells the same day

8. What kind of study is needed to become informed about the influence of dexamethasone on the growth rate of goat periosteum cells?
   A. Efficacy studies
   B. *In vitro* studies
   C. Proof of the concept studies
   D. Feasibility studies

9. When bone formation is observed inside the pores of a cell-seeded scaffold that was located 4 weeks inside a diffusion chamber in a rat, we can conclude that this bone was derived from:
   A. The seeded cells (osteogenesis)
   B. The host osteoprogenitor cells delivered by newly formed blood vessels in the scaffold (osteoinduction)
   C. The host bone growing into the scaffolds (osteoconduction)
   D. None of the above

10. Potential problems anticipated in the use of bone tissue engineering in humans are:
    A. The upscaling from cubic mm-range (in mice) to the clinical cubic cm-range (1000 fold)
    B. The inferior quality of the human bone marrow aspirates as compared to e.g. rats and goats
    C. The relatively high age at which the treatment will be performed
    D. All of the above

Chapter 20 Tissue engineering of the nervous system

1. A neuron that transmits an impulse away from the CNS is called a:
   A. Motor neuron
   B. Sensory neruon
   C. Afferent neuron

2. Approximately how many neuron cell bodies are in the forearm of a rat?
   A. 20,000
   B. 100,000,000
   C. none

3. What is the critical gap defect calculated for a silicone nerve guide in the sciatic nerve of a mouse?
   A. 9.7 ± 1.8 mm
   B. 5.4 ± 1.0 mm
   C. 14.5 ± 2.1 mm

4. BDNF will bind with the p75 low affinity nerve growth factor receptor and which other membrane receptor on neurons?
   A. TrkA
   B. TrkB
   C. TrkC

5. What is the most common form of spinal cord injury in humans?
   A. Hemisection
   B. Transection
   C. Contusion

6. After implanting a fibronectin scaffold in the spinal cord as a hemisection, what is one of the first events to occur?
   A. Axon penetration
   B. Astrocyte migration
   C. Inflammation

7. Which of the planes in the spinal cord is NEVER symmetrical?
   A. Coronal
   B. Saggital
   C. Longitudinal

8. The axons in an optic nerve are predominantly from

A. Motor neurons
B. Efferent neurons
C. Sensory neurons

9. Which of the following CNS cells are myelinating glia?
   A. Oligodendrocyte
   B. Astrocyte
   C. Microglia

10. Which of the following peptide sequences are found in laminin?
    A. VFDNFVLK
    B. YIRGS
    C. IKVAV

Chapter 21 Tissue engineering of organ systems

1. An engineered cell-matrix construct is ideally of:
   A. Allogeneic origin
   B. Heterologous origin
   C. Autologous origin

2. Which of the following constructs achieved so far the best results in ureteral reconstruction:
   A. Collagen tubes alone
   B. Polytetrafluoroethylene tubes
   C. Cell-seeded polymer tubes

3. Anorchia or acquired testicular failure can be treated by:
   A. Iimplantation of tissue engineered testicle
   B. Implantation of encapsulated Leydig cells
   C. Implantation of encapsulated spermatozoids

4. For the treatment of urinary incontinence the following bulking agent is commonly used in clinical practice:
   A. Hyaluronic acid-dextranomer
   B. Chondrocytes
   C. Myoblasts

5. In an effort to replace lost liver function hepatocytes have been successfully implanted into:
   A. Lungs
   B. Peritoneum
   C. Spleen

6. The fully developed lung contains approximately:
   A. Eight different cell types
   B. Forty different cell types
   C. Ninety different cell types

7. In gut tissue engineering which of the following scaffold was only used with limited success:
   A. Natural, processed collagen scaffold
   B. Acellular tissue (ECM) constructs
   C. Synthetic matrices

8. In pancreatic tissue engineering the following correct statement has to be considered:
   A. In beta cell engineering the recurrence of the disease is a major concern
   B. Type II diabetes has an immunological cause
   C. For glucose sensitive insulin production beta cells are mandatory

9. Spontaneous malignant transformation is most unlikely in:
   A. Fetal stem cells
   B. Embryonic stem cells
   C. Adult stem cells

10. In future the optimal approach to tissue engineering could be:
    A. The development of stem cells loaded matrices
    B. The development of growth factor releasing matrices
    C. The development of in vitro differentiated tissue construct